MW00845660

# Reliability and Failure of
# ELECTRONIC MATERIALS AND DEVICES

# Reliability and Failure of ELECTRONIC MATERIALS AND DEVICES

## Second Edition

**MILTON OHRING**
Department of Materials Science and Engineering
Stevens Institute of Technology
Hoboken, New Jersey

With

**LUCIAN KASPRZAK**
IBM Retired
Bear, Delaware

AMSTERDAM • BOSTON • HEIDELBERG • LONDON
NEW YORK • OXFORD • PARIS • SAN DIEGO
SAN FRANCISCO • SINGAPORE • SYDNEY • TOKYO
Academic Press is an imprint of Elsevier

Academic Press is an imprint of Elsevier
32 Jamestown Road, London NW1 7BY, UK
525 B Street, Suite 1800, San Diego, CA 92101-4495, USA
225 Wyman Street, Waltham, MA 02451, USA
The Boulevard, Langford Lane, Kidlington, Oxford OX5 1GB, UK



ISBN: 978-0-12-088574-9

**Library of Congress Cataloging-in-Publication Data**
Ohring, Milton, 1936-
Reliability and failure of electronic materials and devices / Milton Ohring, Lucian Kasprzak. – Second edition.
pages cm
Includes index.
ISBN 978-0-12-088574-9 (hardback)
1. Electronic apparatus and appliances–Reliability. 2. System failures (Engineering) I. Kasprzak, Lucian. II. Title.
TK7870.23.O37 2014
621.381–dc21
2014031265

**British Library Cataloguing in Publication Data**
A catalogue record for this book is available from the British Library

For information on all Academic Press publications
visit our web site at http://store.elsevier.com/

# DEDICATION

He whose works exceed his wisdom, his wisdom will endure.
Ethics of the Fathers

In honor of my father, Max,
... a very reliable dad.

- *Milton Ohring*

# CONTENTS

# PREFACE TO THE SECOND EDITION

The first edition is a classic in form, content, and execution. The form of each chapter takes the reader on a trip through a basic phenomena and its application to failure in electronics (from simple physics to the complicated math often involved). The content of the book covers all the relevant disciplines necessary to do a thorough job of discovery of the true cause of failure. The execution includes relevant examples from the macroscopic to the microscopic to the atomic, where necessary. The book is truly a masterpiece—due in large part to Milt's knowledge and experience in physics, chemistry, electronics, materials, and most of all his unique ability to frame difficult problems in the appropriate mathematics.

As such then, only changes were made where necessary, to keep the book current and useful to the reader (a researcher struggling to determine the true cause of failure so that it can be remedied and never happen again). By the way that is what has made this industry so successful—failure analysis to true root cause.

**L. Kasprzak**

# PREFACE TO THE FIRST EDITION

Reliability is an important attribute of all engineering endeavors that conjures up notions of dependability, trustworthiness, and confidence. It spells out a prescription for competitiveness in manufacturing, because the lack of reliability means service failures that often result in inconvenience, personal injury, and financial loss. More pointedly, our survival in a high-tech future increasingly hinges on products based on microelectronics and optoelectronics where reliability and related manufacturing concerns of defects, quality, and yield are of critical interest. Despite their unquestioned importance, these subjects are largely ignored in engineering curricula. Industry compensates for this educational neglect through on-the-job training. In the process, scientists and engineers of all backgrounds are recast into reliability "specialists" with many different attitudes toward reliability problems, and varied skills in solving them. The reliability practitioner must additionally be a detective, a statistician, and a judge capable of distinguishing between often-conflicting data and strongly held opinions. This book attempts to systematically weave together those strands that compose the fabric of the training and practice of reliability engineers.

The multifaceted issues surrounding the reliability and failure of electronic materials and devices lie at the confluence of a large number of disciplinary streams that include materials science, physics, electrical engineering, chemistry, and mechanical engineering, as well as probability and statistics. I have tried to integrate the derivative subject matter of these disciplines in a coherent order and in the right proportions to usefully serve the following audiences:

- advanced undergraduate and first year graduate engineering and science students who are being introduced to the field,
- reliability professionals and technicians who may find it a useful reference and guide to the literature on the subject, and
- technical personnel undergoing a career change.

While the emphasis of the book is on silicon microelectronics technology, reliability issues in compound semiconductor and electro-optical devices, optical fibers, and associated components are also addressed. The book starts with an introductory chapter that briefly defines the subject of semiconductor reliability, its concerns, and historical evolution. Chapter 2 introduces electronic materials and devices, the way they are processed, how they are expected to behave, and the way they sometimes malfunction. The important subjects of intrinsic and manufacturing defects, contamination, and product yield are the focus of Chapter 3.

Without Chapter 4 on the mathematics of reliability it is doubtful that the book title could include the word reliability. Historically, reliability has been inextricably intertwined with statistics and probability theory. Even today a large segment of the reliability literature bears a strong hereditary relationship to these mathematical and philosophical antecedents. Nevertheless, "The failure of devices occurs due to natural laws of change, not to the finger of fate landing at random on one of group of devices and commanding *fail*" (R.G. Stewart, *IEEE Transactions on Reliability,* R-15, No. 3, 95 (1966)). In a less charitable vein, R.A. Evans has pointed out that probability and statistical inference help us "quantify our ignorance" of failure mechanisms. Uncovering truth should be the objective instead. That is why the primary focus of the book and most of its contents deal with the physics of failure as refracted through the lenses of physical and materials science. With this understanding, our treatment of reliability mathematics is largely limited to the elementary statistical handling of failure data and the simple implications that flow from such analyses. Nevertheless, reliability mathematics permeates the book since failure data are normally presented in these terms.

The midsection of the book spanning Chapters 5 through 10 is devoted to a treatment of the specific ways materials and devices degrade and fail both on the chip and packaging levels. Failure mechanisms discussed and modeled include those due to mass and electron transport, environmental and corrosion degradation, mechanical stress, and optical as well as nuclear radiation damage. Most failures occurring within interconnections, dielectrics and insulation, contacts, semiconductor junctions, solders, and packaging materials can be sorted into one of these categories. Grouping according to operative failure mechanism, rather than specific device, material, or circuit element, underscores the fundamental generic approach taken.

Important practical concerns regarding characterizing electronic materials and devices in the laboratory, in order to expose defects and elucidate failure mechanisms, is the subject of Chapter 11. Finally, the last chapter speculates about the future through a discussion of device-shrinkage trends and limits and their reliability implications.

Due to continual and rapid advances in semiconductor technology, the shelf life of any particular product is very short, thus raising the question of how to convey information that may be quickly outdated. Because concepts are more powerful than facts, I have tried to stress fundamentals and a physical approach that may have applicability to new generations of devices. Within this approach the dilemma arose whether to emphasize breadth or depth of subject matter. Breadth is a sensible direction for an audience having varied academic backgrounds desirous of a comprehensive but qualitative treatment; on the other hand, depth is necessary to enable practitioners to confront specific challenges within the evolving electronics industries. As a compromise I have attempted to present a balanced treatment incorporating both attributes and sincerely hope neither audience will be disappointed by the outcome. Nevertheless, space limitations often preclude development of a given subject from its most elementary foundations.

I assume readers of this book are familiar with introductory aspects of electronic materials and possess a cultural knowledge of such subjects as modern physics, thermodynamics, mass transport, solid mechanics, and statistics. If not, the tutorial treatment of subject matter hopefully will ease your understanding of these subjects. Questions and problems have been included at the end of each chapter in an attempt to create a true textbook.

If this book contributes to popularizing this neglected subject and facilitates the assimilation of willing as well as unconvinced converts into the field of reliability, it will have succeeded in its purpose.

**M. Ohring**

# ACKNOWLEDGMENTS

The idea for this book germinated in the classroom of Bell Laboratories, both at AT&T and Lucent Technologies, where I taught and organized several courses on reliability and failure of electronics over the past decade. Distilling the essence of the vast scattered information on this subject, first for course purposes and then into the text and figures that have emerged between these covers, would not have been possible without the resources of Bell Labs and the generous assistance of some very special people. Of these, two of my dear friends, Lucian (Lou) Kasprzak and Frank Nash, must be acknowledged first. Lou, an early observer of transistor hot-electron effects at IBM, is a member of the board of directors of the IEEE International Reliability Physics Symposia, while Frank, an expert in laser reliability at Bell Labs, wrote the incisive book, *Estimating Device Reliability: Assessment of Credibility,* Kluwer Academic publishers, (1993). Both stimulated my thinking on the direction of the book and helped me to acquire the research literature that was indispensable to its writing. Importantly, they dispelled some of my naïveté about reliability and helped me to fashion a credible philosophy of the subject, something acquired only after lengthy grappling with the issues.

Students enrolled in courses for which books are simultaneously being written often leave their imprint on them. This is true of several of the students in the Stevens Institute–Bell Laboratories on-premises graduate program. Therefore, I want to thank both Gary Steiner and Jeff Murdock for critiquing various versions of the text. Gary additionally reviewed the exercises and enhanced the quality of several figures; Jeff assured me that the text was "Murdock-proof." Other students who tangibly contributed to the book are Ken Jola, Jim Bitetto, Ron Ernst, and Jim Reinhart.

Ephraim Suhir, Walter Brown, Matt Marcus, Alan English, King Tai, Sho Nakahara, George Chu, Reggie Farrow, Clyde Bethea, and Dave Barr, all of Bell Laboratories and Bob Rosenberg of IBM deserve thanks for their encouragement, helpful discussions, and publications. Despite their efforts I am solely responsible for any residual errors and holes in my understanding of the subject.

I am also grateful to the ever-helpful Nick Ciampa of Bell Laboratories for his technical assistance during the preparation of this book and generosity over the years. For help at important times I wish to acknowledge

Pat Downes and Krisda Siangchaew. Lastly, thanks are owed to Zvi Ruder for supporting this project at Academic Press and to Linda Hamilton and Julie Champagne for successfully guiding it through the process of publication.

My dear wife, Ahrona, has once again survived the birth of a book in the family, this time during a joyous period when our three grandchildren, Jake, Max, and Geffen, were born. She and they were unfailing sources of inspiration.

**M. Ohring**

# CHAPTER 1

# An Overview of Electronic Devices and Their Reliability

## 1.1 ELECTRONIC PRODUCTS

### 1.1.1 Historical Perspective

Never in human existence have scientific and technological advances transformed our lives more profoundly, and in so short a time, as during what may be broadly termed the Age of Electricity and Electronics.[1] From the telegraph in 1837 (which was in a sense digital, although clearly electromechanical) to the telephone and teletype, television and the personal computer, the cell phone and the digital camera, and the World Wide Web (WWW), the progress has been truly breathtaking. All these technologies have been focused on communicating information at ever increasing speeds. In contrast to the millennia-long metal ages of antiquity, this age is only little more than a century old. Instead of showing signs of abatement, there is every evidence that its pace of progress is accelerating. In both a practical and theoretical sense, a case can be made for dating the origin of this age to the eighth decade of the nineteenth century [1]. The legacy of tinkering with voltaic cells, electromagnets, and heating elements culminated in the inventions of the telephone in 1876 by Alexander Graham Bell, and the incandescent light bulb 3 years later by Thomas Alva Edison. Despite the fact that James Clerk Maxwell published his monumental work Treatise on Electricity and Magnetism in 1873, the inventors probably did not know of its existence. With little in the way of "science" to guide them, innovation came from wonderfully creative and persistent individuals who incrementally improved devices to the point of useful and reliable function. This was the case with the telephone and incandescent lamp, perhaps the two products that had the greatest influence in launching the widespread use of electricity. After darkness was illuminated and

[1] *Electricity* means those advances capitalizing on electromagnetics and electromechanics, e.g., generators and motors. In contrast, *Electronics* relates to the broad range of devices, e.g., vacuum tubes and transistors, which function by controlling the flow of electrical charges in a vacuum, gas, solid, liquid, or plasma.

*Reliability and Failure of Electronic Materials and Devices*
ISBN 978-0-12-088574-9
http://dx.doi.org/10.1016/B978-0-12-088574-9.00001-X

communication over distance demonstrated, the pressing need for electric generators and systems to distribute electricity was apparent. Once this infrastructure was in place, other inventions and products capitalizing on electromagnetic-mechanical phenomena quickly followed. Today, texting from a cell phone has replaced the telegraph for the ultimate person-to-person real-time digital conversation. Literally, the telegraph of 1837 has become texting in 2007. Both use letters to interact with someone on the other end (of the wire, so to speak). The rate is about the same, possibly a letter or so a second, when you consider composition for texting, which is real time versus predefined on a form for the telegraph. Both the telegraph (1837) and texting (2007) have roughly the same data entry rate of about two letters a second.

Irrespective of the particular invention, however, materials played a critical role. At first, conducting metals and insulating nonmetals were the only materials required. Although a reasonable number of metals and insulators were potentially available, few were produced in quantity or had the requisite properties. The incandescent lamp is a case in point [2,3]. In the 40 years prior to 1879, some 20 inventors tried assorted filaments (e.g., carbon, platinum, iridium) in various atmospheres (e.g., vacuum, air, nitrogen, hydrocarbon). Frustrating trials with carbon spirals and filaments composed of carbonized fiber, tar, lampblack, paper, fish line, cotton, and assorted woods paved the way to Edison's crowning achievement. His patent revealed that the filament that worked was carbonized cardboard bent in the shape of a horseshoe. Despite the fact that an industry based on incandescent lamps grew rapidly, the filaments were brittle and hard to handle. The glass envelopes darkened rapidly with time, and the bulbs were short lived. Salvation occurred around 1910 when the Coolidge process [4] for making fine tungsten filament wire was developed. Well beyond a century after the original Edison patent, filaments continue to be improved and lamp life extended and today we are increasingly using light emitting diodes (LEDs) as the next generation of efficient illumination.

With the ability to generate electromagnetic waves around the turn of the century, the era of vacuum electronics was born. The invention of vacuum tubes enabled electric waves to be generated, transmitted, detected, and amplified, making wireless communication possible. In particular, the three-electrode vacuum tube invented by Lee de Forest in 1906 became the foundation of electronics for the first half of the twentieth century [5]. Throughout the second half of the twentieth century, electronics has been transformed both by the transistor, which was invented in 1947, and

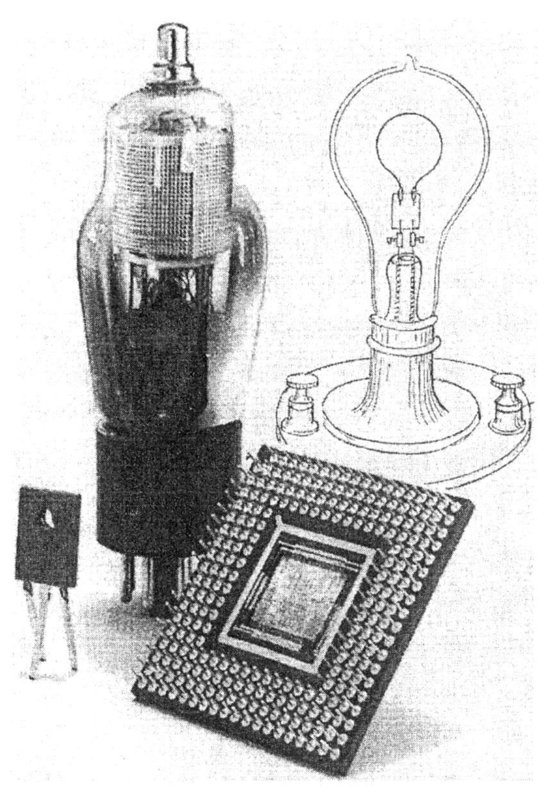

**Figure 1.1** Edison's horseshoe filament lamp sketched by patent draftsman Samuel D. Mott serves as the backdrop to the vacuum tube, discrete transistor, and integrated circuit. *Courtesy of FSI International, Inc.*

integrated circuits (ICs), which appeared a decade later. The juxtaposition of these milestone devices in Figure 1.1 demonstrates how far we have come in so short a time.

A pattern can be discerned in the development of not only electrical devices and equipment, but also all types of products. First, the genius of invention envisions a practical use for a particular physical phenomenon. The design and analysis of the components and devices are then executed, and finally, the materials and manufacturing processes are selected. Usage invariably exposes defects in design and manufacturing, causing failure or the wearing out of the product. Subsequent iterations of design or materials processing improve the reliability or probability of operating the product for a given time period under specified conditions without failure. In a sense, new reliability issues replace old ones, but incremental progress is made. Ultimately, new technologies replace obsolete ones, and the above sequence of events repeats once again.

Well into the Age of Electricity and Electronics, we still use vastly improved versions of some of those early inventions. As you change your next light bulb, however, you will realize that reliability concerns are still an issue. But solid-state electronic products also fail in service, and often with far greater consequences than a burned-out light bulb. While other books are concerned with the theory of phenomena and the practice of designing useful electrical and electronic products based on them, this book focuses on the largely unheralded activities that insure they possess adequate reliability during use.

### 1.1.2 Solid-State Devices

The scientific flowering of solid-state device electronics has been due to the synergism between the quantum theory of matter and classical electromagnetic theory. As a result, our scientific understanding of the behavior of electronic, magnetic, and optical materials has dramatically increased. In ways that continue undiminished to the present day, exploitation of these solid-state devices, particularly semiconductors, has revolutionized virtually every activity of mankind, e.g., manufacturing, communications, the practice of medicine, transportation, and entertainment [6].

Before focusing on electronics, it is worth noting parallel developments in the field of photonics. The latter has been defined as "the technology for generating, amplifying, detecting, guiding, modulating, or modifying by nonlinear effects, optical radiation, and applying it from energy generation to communication and information processing" [7]. Building on the theoretical base laid by Maxwell, Planck, and Einstein, optoelectronics (or electro-optics) applications include lasers, fiber optics, integrated optics, acousto-optics, nonlinear optics, and optical data storage. Destined to merit an historic age of its own, photonics is projected to dominate applications, as fundamental physical limits to electronic function are reached for example compact disks (CDs) and digital video disks (DVDs).

As already noted, two major revolutions in electronics occurred during the time from the late 1940s to the 1970s. In the first, transistors replaced vacuum tubes. These new solid-state devices, which consumed tens of milliwatts and operated at a few volts, eliminated the need for several watts of filament heater power and hundreds of volts on the tube anode. The second advance, the invention of the IC in 1958, ushered in multidevice chips to replace discrete solid-state diodes and transistors and electronic circuits. Early solid-state devices were not appreciably more reliable than their vacuum tube predecessors, but power and weight savings were impressive. While discrete devices were advantageously deployed in a wide variety of traditional electronics applications, information processing required far greater densities of transistors. This need was satisfied by ICs. Is it possible that the productivity gains fueling our gross domestic product are really the flowering of ICs and computers on the WWW?

## 1.1.3 Integrated Circuits

### *1.1.3.1 Applications*

Starting in the early 1960s, the IC market was based primarily on bipolar transistors. Since the mid-1970s, however, ICs composed of metal-oxide-silicon (MOS) field effect transistors prevailed because they possessed the advantages of device miniaturization, high yield, and low power dissipation. (Both bipolar and MOS transistors will be discussed in Chapter 2.) Then in successive waves, IC generations based on MOS transistors arose, flowered, and were superseded by small-scale integration, medium-scale integration, large-scale integration, very-large-scale integration, ultra-large-scale integration (ULSI), and extreme large-scale integration. The latter, consisting of more than $10^{10}$ devices per chip, provides an entire system on a chip. In the early 1970s, sales of ICs reached $\$1 \times 10^9$; today, digital MOS ICs dominate a market worth $10^{12}$ dollars a year; likewise, the broader market for electronic equipment exceeds a $10^{14}$ dollars.

ICs are pervasive in computers, consumer goods, and communications, where the total demand for each is roughly the same. Worldwide IC revenues are currently about 30% in these three sectors, with an additional 9% accounted for by industry and 2% by the military [7]. The role of the computer industry and defense electronics as driving forces for IC production has been well documented. Less appreciated is the generally equal and at times even more powerful stimulus provided by consumer electronics in fostering the IC revolution.

Consumer products are those purchased by the individual or a family unit. Cell phones, digital cameras, iPods, wireless everything, webcams, Global Positioning System devices, high-definition televisions (HD TVs), home theater systems, DVD recorders, wristwatches, computers, LCDs, cordless phones, smoke detectors, and automobiles are examples that contain ICs. The automobile is an interesting case in point. There are presently more than 25 microprocessor IC chips in every car, subject to a very hostile environment, to control engine ignition timing, gas–air mixtures, idle speed, and emissions, as well as chassis suspension, antiskid brakes, and four-wheel steering. Furthermore, recent trends have made vehicles intelligent through special application-specific ICs that provide computerized navigation and guidance assistance on highways, self parking, and electronic map displays. In the near future, cars will be remotely controlled by satellite from home to your destination. Advances in one sector of applications ripple through others. Thus, video phones and conferencing, multimedia computers, three-dimensional movies, and digital HD TV have permeated the consumer marketplace.

The *Bell Labs News* issue of February 7, 1997, marking the 50-year anniversary of the invention of the transistor, roughly estimated that "there are some 200 million billion transistors in the world today—about 40 million for every woman, man, and child." In the United States alone, more than 200 ICs are used per household. The transistor count is rapidly swelling as the standard of living rises in China and India. With some $5 \times 10^9$ humans on earth, there is much room for continued expansion of IC consumption in Third World countries. What is remarkable is how inexpensive the fruits of this technology have become. In the 1950s, the cost of a transistor dropped from $45 to $2. Today's transistors cost less than a hundred-thousandth of a cent each!

#### 1.1.3.2 Trends

Meeting the needs of burgeoning markets for digital electronic products has necessitated a corresponding growth in the number of transistors and complexity of the ICs that house them. To gain a perspective of IC chip progress made in the 37 years from their inception to 1994, the time span analyzed, consider the following measures of advance [8]:

1. Minimum feature size ($F$) has decreased by a factor of 50.
2. Die area ($D^2$) has increased by a factor of approximately 170.
3. Packing efficiency ($PE$), defined as the number of transistors per minimum feature area, has multiplied by over 100.

The composite number of transistors per chip $N$ is the product of these three terms, or $N = F^{-2} \times D^2 \times PE$. Substitution indicates a staggering increase by a factor of close to $5 \times 10^7$, a number that is manifest in Moore's law. In the mid-1970s, Gordon Moore, a founder of Intel, observed an annual doubling of the transistors per IC chip, which held true for about 15 years until about 1990, and a reduction in the rate of increase to 1.5 per year since then. By the end of the twentieth century, a staggering one billion devices per chip was realized, and by 2020, the count will be close to a trillion. Moore probably overstated the initial progress made, which is, nevertheless, extraordinarily impressive. In view of the fact that the price of chips has not changed appreciably, while their reliability has considerably improved over this time, these phenomenal advances are unprecedented in technological history.

Both microprocessor and memory chips are the indispensable drivers of computer technology. Microprocessor ICs perform the numerous logic functions in circuits, while memory chips essentially store the information generated. Trends showing rapidly increasing transistor numbers in IC chips are presented in Figure 1.2(a) for processor and memory chips. Kurzweil has expanded Moore's law as seen in Figure 1.2(b), where progress is plotted as a cost of calculation for the entire twentieth century. When coupled with

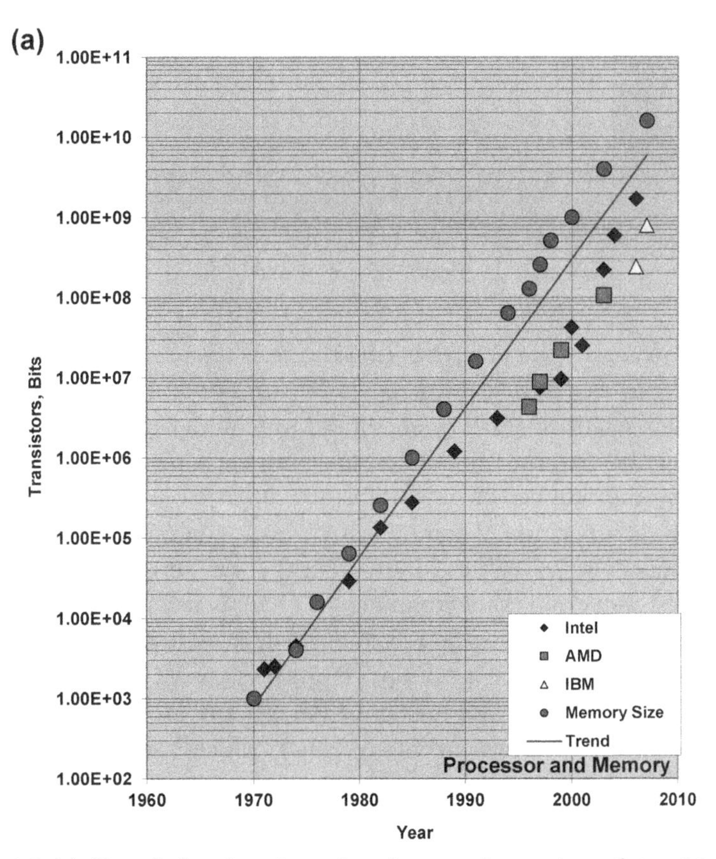


**Figure 1.2** (a) Plot of the time-dependent increase in number of transistors on microprocessor and memory integrated circuit chips. (b) Kurzweil expansion of Moore's law—Fifth Paradigm, see Moore's Law in Wikipedia on the Internet.

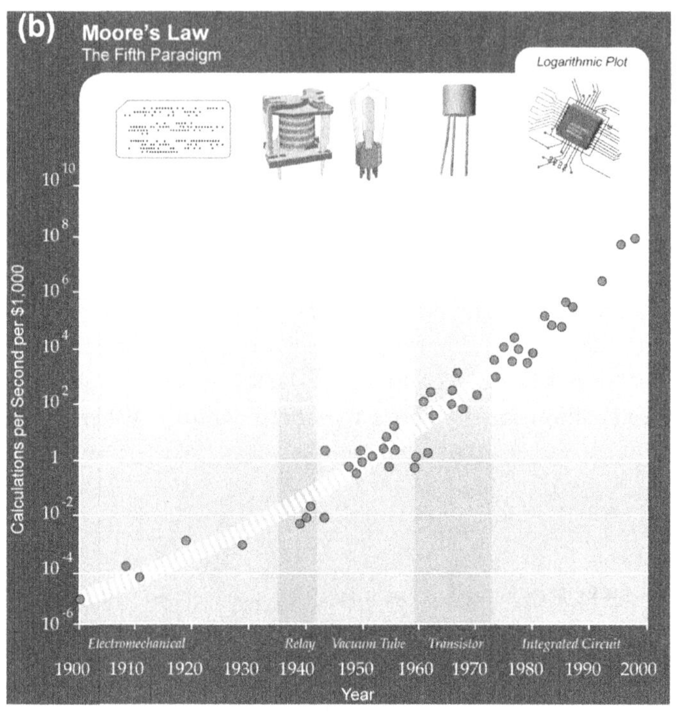


**Figure 1.2** (b)—Cont'd.

the far slower growth in chip size over the years (Figure 1.3), one can begin to appreciate the huge growth in device density and the magnitude of feature (linewidth, Figure 1.4(a)) shrinkage required to make it possible. Increasing the transistor count improves microprocessor performance as gauged by the number of instructions executed per second. Measured in frequency of operation, Figure 1.4(b) reveals a hundredfold increase in performance every 10 years. However, the key driving parameter, as mentioned above and as shown in Figure 1.4(a), is actually the reduction in linewidth used to fabricate the ICs. We will discuss this parameter in greater detail in Chapter 3, when we deal with processing yield and defects. Also of importance is the increase in hard drive storage capacity achieved in the past quarter century, as shown in Figure 1.4(c), which has improved by five orders of magnitude. Relying on the same lithography technology, magnetic read/write heads have taken advantage of the reduction in linewidth to create ever smaller heads that fly ever closer to the magnetic disk surface.

It is instructive to quantitatively sketch the time evolution of high-tech consumer electronics in terms of product size. In the past few decades, for

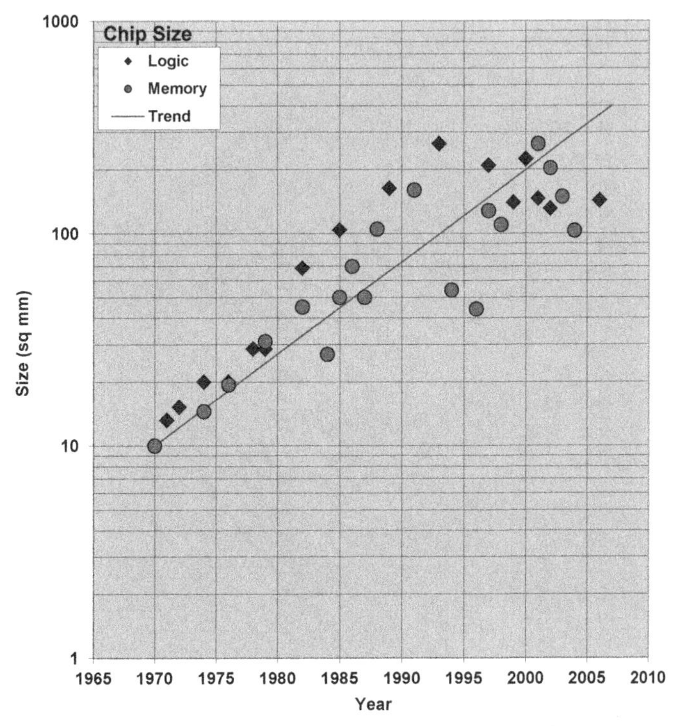


**Figure 1.3** Plot of the time-dependent increase in chip area of microprocessor and memory circuit chips.

example, the overall volume of mobile communications and personal computers has been reduced by an order of magnitude or more, as shown in Figure 1.5(a) and 1.5(b); corresponding weight savings have also been realized. The latest digital cell phones are less than 100 g, the smallest laptop computers are just about 2 pounds, and digital cameras, at about 4 ounces, have gone from 1 to 10 Mega-pixels in less than a decade (Figure 1.5(c)). It has been estimated that if a digital cell phone contained vacuum tubes instead of transistors, it would be larger than the Washington Monument! This volume reduction is possible because of the increasing power of ICs.

The source of these advances has been our ability to shrink the dimensions of critical device features. As a dramatic example of this, consider the width of aluminum interconnections in the high-device-density IC chips shown in Figure 1.6. The indicated dimensions have shrunk over the years from several microns to 0.35 μm presently. By 2010, conductor linewidths of 0.07 μm are projected. More will be said about shrinkage of

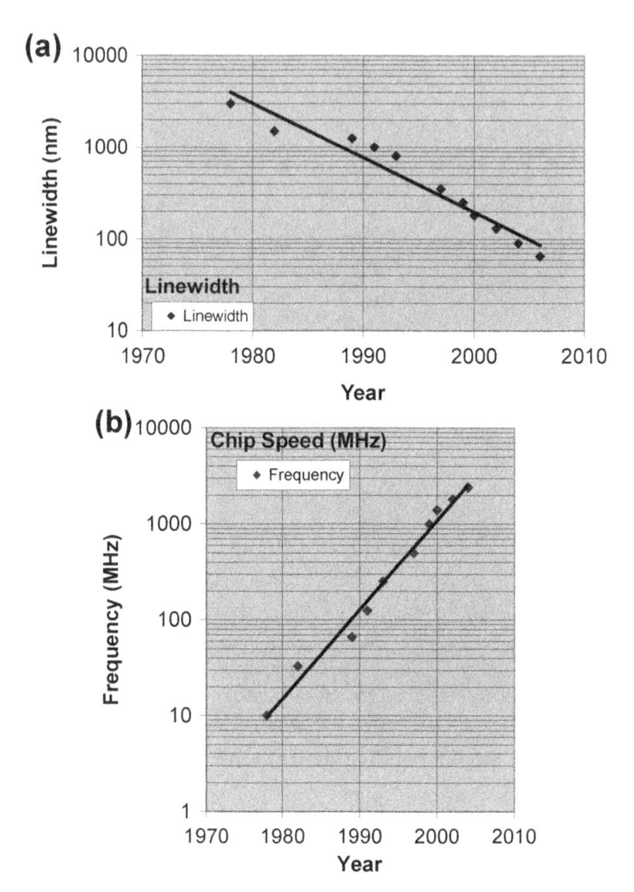


**Figure 1.4** (a) Plot of the time-dependent decrease in lithography linewidth in nanometers. (b) Plot of the time-dependent increase in microprocessor chip speed measured in megahertz. (c) Plot of the increase in hard drive storage capacity.

other device features throughout the book, and trends will be summarized in Chapter 12.

Each increase in packing density introduces a variety of new quality and reliability issues associated with specific devices, and more importantly, with the rapidly increasing number of contacts, interconnections, and solder joints. Reduced dimensions have made circuits increasingly sensitive to more subtle defects, which had not previously threatened chip reliability. Confounding physical intuition, IC chip reliability as measured by the number of circuit failures, has, in fact, dramatically increased over the last two decades. As revealed by Figure 1.7, the number of failures in FITs (1 failure in $10^9$ h of service = 1 FIT) has been reduced by a factor of 10 in

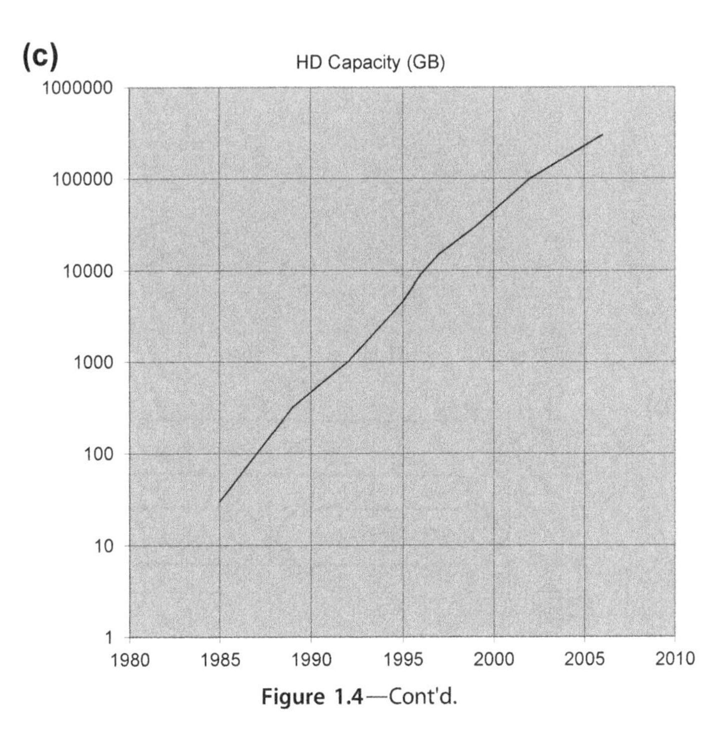


**Figure 1.4**—Cont'd.

this same period of time. Shown are the long-term failure rates as well as those that occur shortly after use commences, i.e., infant mortality (see Section 1.3.8). These remarkable reliability trends will probably continue in response to both customer needs and stiff competition, which have aggressively forced manufacturing to reduce the number of defects.

## 1.1.4 Yield of Electronic Products

Before we can derive the benefits of electronic products, they must first be manufactured in satisfactory numbers. Ideally, one might naively hope that after carefully processing a wafer containing a large number of ICs, all would function properly when tested. In reality, however, the number of good, or functional, circuits may range from close to zero for a new technology to almost 100% for a mature technology where the "bugs" have been worked out. The very important term yield refers to the fraction of successful products that survive final screening and meet specifications relative to the total number that started out to be processed or manufactured. Yield is essentially the bottom-line concern of manufacturing technology. In general, yields are never 100% because of some combination

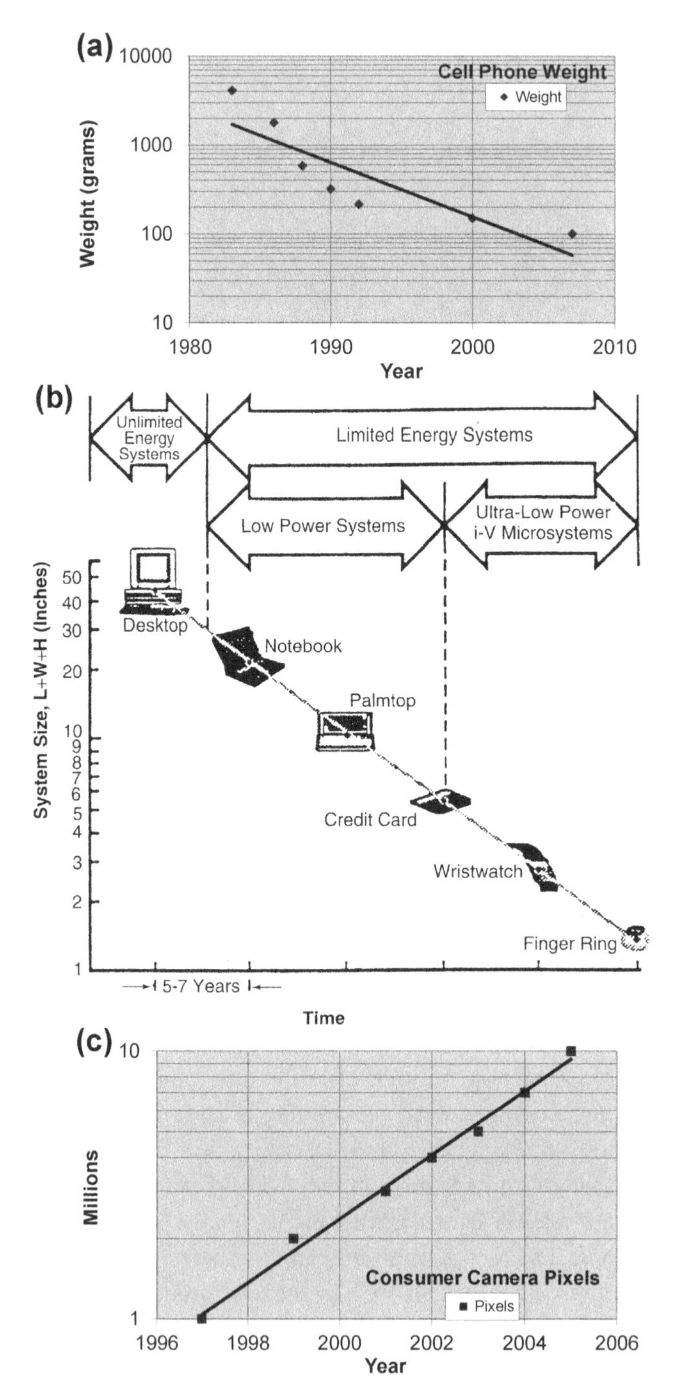


**Figure 1.5** (a) Plot of the time-dependent decrease in portable/cell phone weight in grams. (b) Reduction in the dimensions of personal computing systems with time. *(From Ref. [9].)* (c) Plot of the time-dependent increase in pixels for consumer digital cameras.

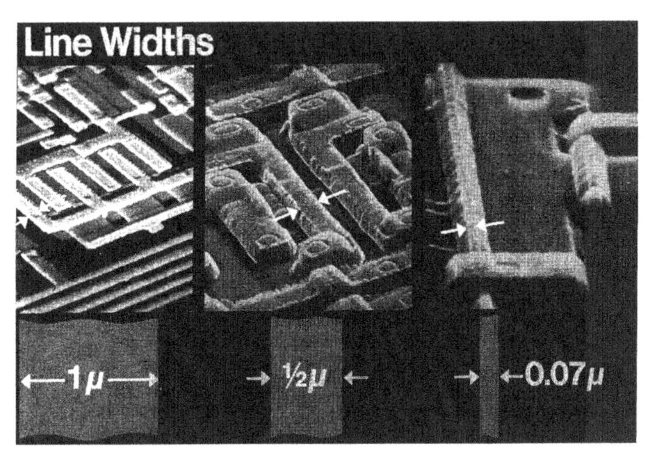


**Figure 1.6** Shrinkage in the aluminum interconnection linewidth in high device density IC chips. *Courtesy of P. Chaudhari, IBM Corp.*

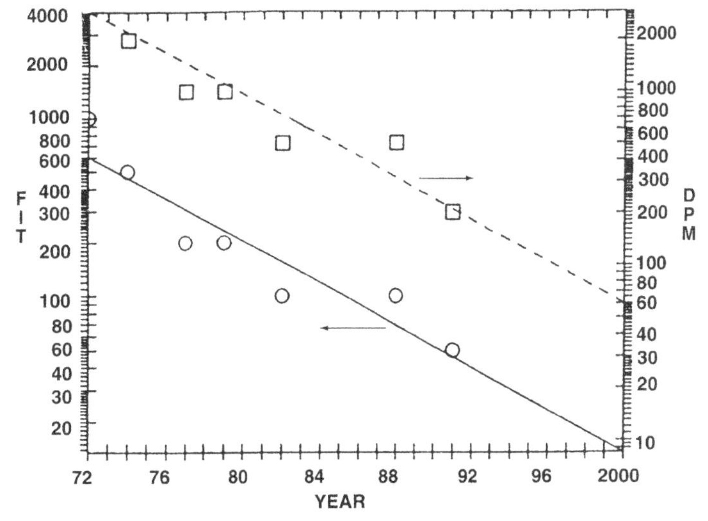


**Figure 1.7** Long- and early-term failure-rate goals at Intel. The units for long-term failures are FITs, while those for short-term failures are DPMs. (DPM = defects per million). *From Ref. [10].*

of problems associated with circuit design, processing, and random defects. The first may simply be due to errors in circuit design, while the others usually result from lack of process control and contamination. Specific examples of each will be deferred until Chapter 3.

An excellent way to appreciate the broader factors that influence yield is through study of Figure 1.8. The first important conclusion that can be

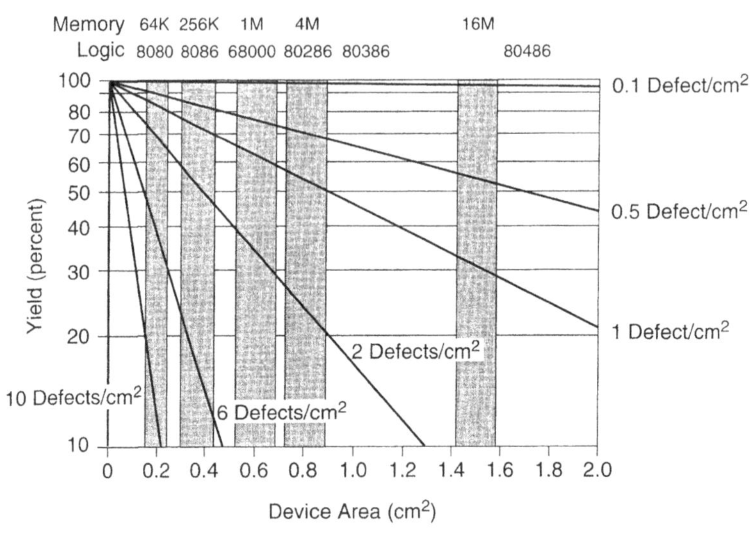


**Figure 1.8** Yield and defect trends in memory and logic chips as a function of device area. *From Intel/Future Horizons as reproduced in Ref. [11].*

drawn is that yield drops as the chip area increases. Second, for chips of a given area, the highest yields occur for those that have the fewest defects. Here we note that every new generation of memory and logic chips invariably means an increase in device area. In DRAM memory chips, for example, increasing the capacity to 64 megabit (64 Mb) implies a die size of roughly 2.6 $cm^2$. This means that many fewer defects per unit area can be tolerated if the same yield level is to be achieved in successive generations of product. Simultaneously, the defect that signifies outright die rejection if present, the so-called killer defect, becomes smaller and smaller in size because device features have shrunk in size. What aggravates the problem considerably is that process complexity increases as the device count rises. One-megabit DRAMs require over 300 process steps, whereas more than 500 steps are necessary for 64-Mb DRAM production. The chance of incorporating defects is thus proportionately raised.

What do defects and yield have to do with reliability? Although yield and reliability have quite different meanings, there is a link between them. As we have just seen, yield depends on minimizing the presence of defects. Reliability, an issue that concerns the performance of "yielded" products in service, also depends on defects. The fact that yielded products have passed the last screen does not mean they do not contain potential or latent defects. Too small to be troublesome initially, the latter may prove fatal under subsequent use involving a stressing regimen that enlarges their size or influence. Their mutual association with defects implies the logic of connection between yield and reliability.

The subject of yield is addressed again in Chapter 3, where defects, their sources, and how they are statistically modeled are extensively treated. In addition, both theoretical and experimental associations between yield and reliability are discussed.

## 1.2 RELIABILITY, OTHER "...ILITIES," AND DEFINITIONS

### 1.2.1 Reliability

Earlier we defined *reliability* as the *probability* of operating a product for a given *time* period under specified *conditions* without *failure*. Clearly, reliability reflects the physical performance of products over time and is taken as a measure of their future dependability and trustworthiness. Each of the above words in italics is significant and deserves further commentary.

Under *probability*, we include the sampling of products for testing, the statistical analysis of failure test results, and the use of distribution functions to describe populations that fail. Even though it is not feasible to predict the lifetime of an individual electronic component, an average lifetime can be estimated based on treating large populations of components probabilistically. Interestingly, projections of future behavior can often be made without specific reference to the underlying physical mechanism of degradation.

The *time* dependence of reliability is implied by the definition, and therefore the time variable appears in all of the failure distribution functions that will be subsequently defined. Acceptable times for reliable operation are often quoted in decades for communications equipment; for guided missiles, however, only minutes are of interest. The time dependence of degradation of some measurable quantity such as voltage, current, strength, or light emission is often the cardinal concern, and examples of how such changes are modeled are presented throughout the book.

Specification of the product testing or operating conditions is essential in predicting reliability. For example, elevating the temperature is the universal way to accelerate failure. It therefore makes a big difference whether the test temperature is 25 °C, −25 °C, or 125 °C. Similarly, other specifications might be the level of voltage, humidity, etc., during testing or use.

Lastly, the question of what is meant by failure must be addressed. Does it mean actual fracture or breakage of a component? Or does it mean some degree of performance degradation? If so, by how much? Does it mean that a specification is not being met, and if so, whose specification? Despite these uncertainties it is generally accepted that product failure occurs when the required function is not being performed. A key quantitative measure of reliability is the failure rate. This is the rate at which a device or component can be expected to fail under known use conditions. More will be said about failure rates in Chapter 4, but clearly such information is important to both manufacturers and purchasers of electronic products. With it, engineers can often predict the reliability that can be expected from systems during operation. Failure rates in semiconductor devices are generally determined by device design, the number of in-line process control inspections used, their levels of rejection, and the extent of postprocess screening. Very often, the performance of an IC depends critically on the material's properties achieved during a particular process step. This requires statistical process control of the process to carefully chosen limits that will guarantee the material's properties, such that the device performs properly. Therefore, extensive characterization of how the material's properties depend on process variables is required. An example of this sensitivity is found in the preparation of an extremely thin (angstroms thick) indium tin oxide (ITO) layer used to control the barrier height to carrier injection on a surface or interface used for electrical contact. The barrier height at the surface of the ITO is critically dependent on the details of the preparation process of the ITO.

### 1.2.2 A Brief History of Reliability

Now that various aspects of reliability have been defined, it is instructive to briefly trace their history with regard to the electronics industry. In few applications are the implications of unreliability of electronic (as well as mechanical) equipment as critical as they are for military operations. While World War I reflected the importance of chemistry, World War II and all subsequent wars clearly capitalized on advances in electronics to control the machinery of war-fare. However, early military electronics were far from reliable [12,13]. For example, it was found that 50% of all stored airborne electronics became unserviceable prior to use. The Air Force reported a 20-h maximum period of failure-free operation on bomber electronics. Similarly, the Army was plagued by high truck and power-plant mortalities, while the Navy did not have a dependable torpedo until 1943. Horror stories persisted (and still do) about the large fraction of military electronics that failed to operate successfully when required. Until the early 1950s, the problem of unreliability was met by increasing spare-parts inventories. However, this created the new problem of logistics and the big business associated with the growth of military logistics commands. With this as a background, the US

Defense Department and electronics industry jointly established an Advisory Group on Reliability of Electronic Equipment in 1952 to seek ways of meliorating the problem. The realization that massive duplication was too expensive drove the military to quantify reliability goals. A several order of magnitude enhancement in product reliability was required to reduce the logistics monster and increase the operational time of equipment. The next 20 years witnessed the introduction of solid-state electronics and much research on all aspects of the subject of reliability. A cumulative reference list of published books dealing with microelectronics and reliability in the years up to 1970 can be found in the journal of the same name [14].

### 1.2.3 Military Handbook 217

In this section, we shall see how reliability issues have been addressed in a quantitative way by the Defense Department. To meet reliability standards for electronic components and system hardware being developed for the US defense agencies, contractors were obliged, until June 1994, to conform to guidelines in the Military Handbook 217 (MIL-HDBK-217) entitled "Reliability Prediction of Electronic Equipment" [15]. In it there is information on how to predict failure rates of specific devices and components including discrete semiconductor devices, ICs, vacuum tubes, lasers, resistors, capacitors, switches, rotating devices, connectors, lamps, and so on. Failure rate formulas are given for each product in terms of the involved variables, or factors, that influence reliability, e.g., temperature, voltage, environment, number of connector pins, and year of manufacture, so that lifetimes may be predicted.

Updated periodically, it has achieved the status of a bible in the field; other nations and many individual companies have differing, but essentially similar, guidelines. However, the practice of using MIL-HDBK-217 and other handbooks has been the subject of intense debate over the years. Supporters claim that it enables feasibility evaluation, comparisons of competing designs, and identification of potential reliability problems. Detractors, and they are probably in the majority and certainly more vociferous, maintain that MIL-HDBK-217 reliability predictions do not compare well with field experience and generally serve to stifle good engineering judgment. Encyclopedic in its coverage of practical reliability concerns, examples of its use coupled with a discussion of its validity will be deferred to Chapter 4. MIL-HDBK-217 is no longer a conformance document but rather a guidance document for contractors to use as they see fit. However, contractors still have to prove they meet reliability requirements. Hence, great attention is still paid to the methods and techniques taught in MIL-HDBK-217, which is still available.

### 1.2.4 Long-term Nonoperating Reliability

We can all appreciate that electrical products might suffer reliability problems as a result of service. But what are we to make of products that are stored in a nonoperating condition for long periods of time, and then fail when turned on. Equipment that is not in direct use is either in a state of dormancy or storage. Dormancy is defined as the state in which equipment is connected and in an operational mode but not operating. According to this definition, dormancy rates for domestic appliances, professional and industrial equipment, and transportation vehicles may be well over 90%. On the other hand, equipment that is totally inactivated is said to be in storage; such products may have to be unpacked, set up, and connected to power supplies to become operational. The consequences of nonfunctioning electrical equipment can be severe in the case of warning systems that fail to protect against fire, nuclear radiation, burglary, etc. when called into service unexpectedly. Military success on the battlefield can hang in the balance because of such failure to function.

The above are all examples of nonoperating reliability, the subject of an entire book by Pecht and Pecht [16]. These authors contend that what are usually understood to be relatively benign dormant and storage environments for electronics may, in fact, be quite stressful. Frequent handling, relocating and transport of equipment, and kitting (packing products in kits), as well as unkitting are sources of impact loading, vibration, and generally harmful stresses. An entirely quiescent environment, involving no handling at all, has even proven deleterious to some products. The materials within the latter are often not in chemical or mechanical equilibrium after processing, and residual stresses may be present. Like a compressed spring that naturally tends to uncoil, the material approaches equilibrium through atomic diffusion or mechanical relaxation processes. Battery corrosion due to current leakage is a prime example of this kind of degradation. Less common examples include failure of presoldered components and open circuiting of ICs, both while being stored on the shelf prior to final assembly into electronic products. The former is due to the lack of solderability, presumably because of oxidation, during aging on the shelf. In the latter case, the vexing phenomenon of stress voiding results in slitlike crack formation in the aluminum grains of interconnections, causing ICs to fail. This effect, attributed to internal stress relief, is aggravated when current is passed through the chip. Stress voiding will be discussed more fully in Chapter 5.

With few exceptions, the environments electronic products are exposed to in operational, dormant, and storage modes are not very different. The primary difference is accelerated damage due to the influence of applied electric fields, the currents that flow, and the temperature rise produced. Several authors have proposed a correlation between the operating and nonoperating reliabilities of electronics. A multiplicative factor "*K*" defining the ratio of these respective failure rates has been suggested. Values of *K* ranging anywhere from 10 to 100 have been reported. Such numbers appear to be little more than guesswork and have limited value. A more intelligent approach would be to extrapolate accelerated test results to conditions prevailing during dormancy and storage.

### 1.2.5 Availability, Maintainability, and Survivability

Important terms related to reliability but having different meanings are availability, maintainability, and survivability. The first two refer to cases where there is the possibility of repairing a failed component. Availability is a measure of the degree to which repaired components will operate when called upon to perform. Maintainability refers to the maintenance process associated with retaining or restoring a component or system to a specified operating condition. In situations where repair of failures is not an option (e.g., in a missile), an appropriate measure of reliability is survivability. Here, performance under stated conditions for a specified time period without failure is required. Normally these concepts are applied to the reliability of systems.

## 1.3 FAILURE PHYSICS

### 1.3.1 Failure Modes and Mechanisms, Reliable and Failed States

We now turn our attention to the subject of actual physical failure (or failure physics) of electronic devices and components, a concern that will occupy much of the book. At the outset, there is a distinction between failure mode and failure mechanism that should be appreciated, although both terms have been used interchangeably in the literature. A failure mode is the recognizable electrical symptom by which failure is observed. Thus, a short or open circuit, an electrically open device input, and increased current are all modes of electrical failure. Each mode could, in principle, be caused by one or more different failure mechanisms, however. The latter are the specific microscopic physical, chemical, metallurgical, environmental phenomena or processes that cause device degradation or malfunction. For example, open circuiting of an interconnect could occur because of corrosion or because of too much current flow. High-electric-field dielectric breakdown in insulators is another example of a failure mechanism.

Components function reliably as long as each of their response parameters, i.e., resistance, voltage, current gain, and capacitance, has values that remain within specified design limits. Each of the response parameters in turn depends on other variables, e.g., temperature, humidity, and semiconductor doping level, that define a multidimensional space consisting of reliable states and failed states. Failure consists of a transition from reliable to failed states [17]. Irrespective of the specific mechanism, failure virtually always begins through a time-dependent movement of atoms, ions, or electronic charge from benign sites in the device or component to harmful sites. If these atoms or electrons accumulate in sufficient numbers at harmful sites, damage ensues. The challenge is to understand the nature of the driving forces that compel matter or charge to move and to predict how long it will take to create a critical amount of damage. In a nutshell, this is what the modeling of failure mechanisms ideally attempts to do. A substantial portion of the latter part of the book is devoted to developing this thought process in assorted applications. Interestingly, there are relatively few categories of driving forces that are operative. Important examples include concentration gradients to spur atomic diffusion or foster chemical reactions, electric fields to propel charge, and applied or residual stress to promote bulk plastic-deformation effects. And, as we shall see, elevated temperatures invariably hasten each of these processes. In fact, the single most universally influential variable that accelerates damage and reduces reliability is temperature. Other variables such as humidity, stress, and radiation also accelerate damage in certain devices.

Additionally, the time to a given failure is related to processing and latent defects. If the harmful site already has a minor defect, then the failure happens more quickly because less damage is necessary to cause a failure to occur. Failure rate predictions can be flawed by such occurrences if they are unknown, for example, because test units were not examined before stress or when a root cause is not determined for each failure occurrence.

### 1.3.2 Conditions for Change

Much can be learned about the way materials degrade and fail by studying chemical reactions. As a result of reaction, the original functional materials are replaced by new, unintended ones that impair function. Thermodynamics teaches that reactions naturally proceed when energy is reduced. Thus, in the transition from reliable to failed behavior, the energies of reliable states exceed those of the failed states; furthermore, the process must proceed at an appreciable rate. Similarly, in chemical systems, two conditions are necessary in order for a reaction to readily proceed at constant temperature and pressure:

1. First, there is a thermodynamic requirement that the energy of the system be minimized. For this purpose, the Gibbs free energy ($G$) is universally employed as a measure of the chemical energy associated with atoms or molecules. In order for a reaction to occur among chemical species, the change in free energy ($\Delta G$) must be minimized. Therefore, $\Delta G = G_f - G_i$ must be negative ($\Delta G < 0$), where $G_f$ is the free energy of the final (e.g., failed) states or products and $G_i$ the free energy of the initial (e.g., reliable) states or reactants. When $\Delta G$ attains the most negative value possible, all of the substances present are in a state of thermodynamic equilibrium. For chemical reactions, thermodynamic data are often tabulated as $\Delta G$, or in terms of its constituent factors, $\Delta H$ (the enthalpy change) and $\Delta S$ (the entropy change). These contributions are related by

$$\Delta G = \Delta H - T\Delta S. \tag{1.1}$$

---

## Example 1.1

Consider the problem of selecting a contact material for a device made of the high-temperature superconductor $YBa_2Cu_3O_7$. In addition to the usual attributes of low contact resistance, adhesion, matching of thermal expansion, etc., the contact must be chemically stable. Silver and aluminum are potential contact candidates. Which metal contact is likely to be more stable?

**Answer** We first make the assumption that $\Delta G \approx \Delta H$ in Eqn (1.1) when $T\Delta S$ is small. This is generally true in the present case. Therefore, the criterion for possible chemical reaction is that $\Delta H < 0$. The reactions between Ag or Al and common oxides of YBCO superconductors are given below. Using published enthalpies at 298 K for the involved chemical species, the following $\Delta H_{298}$ values are calculated for potential reactions between the contact metals and assorted superconductor oxides. In doing so, we recall from elementary thermochemistry that the net enthalpy change is the sum of the enthalpies of the products minus the sum of enthalpies for the reactants, i.e., $\Delta H_{298}$ (reaction) $= \Sigma\ \Delta H_{298}$ (products) $- \Sigma\ \Delta H_{298}$ (reactants).

---

| Reaction | Enthalpy |
|---|---|
| $\underset{(-40\ \text{kcal/mol})}{Cu_2O} + \underset{2(0)}{2Ag} = \underset{(-7.3\ \text{kcal/mol})}{Ag_2O} + \underset{2(0)}{2Cu}$ | $\Delta H_{298} = +32.7\ \text{kcal}$ |
| $\underset{(-450\ \text{kcal/mol})}{Y_2O_3} + \underset{8(0)}{8Ag} = \underset{3(-7.3\ \text{kcal/mol})}{3Ag_2O} + \underset{2(-23.87\ \text{kcal/mol})}{2AgY}$ | $\Delta H_{298} = +380\ \text{kcal}$ |
| $\underset{(-133\ \text{kcal/mol})}{BaO} + \underset{3(0)}{3Ag} = \underset{(-7.3\ \text{kcal/mol})}{Ag_2O} + \underset{(-20.05\ \text{kcal/mol})}{AgBa}$ | $\Delta H_{298} = +106\ \text{kcal}$ |
| $\underset{(-40\ \text{kcal/mol})}{Cu_2O} + \underset{\frac{5}{3}(0)}{\frac{5}{3}Al} = \underset{(-16.1\ \text{kcal/mol})}{AlCu_2} + \underset{\frac{1}{3}(-400\ \text{kcal/mol})}{\frac{1}{3}Al_2O_3}$ | $\Delta H_{298} = -109\ \text{kcal}$ |
| $\underset{(-450\ \text{kcal/mol})}{Y_2O_3} + \underset{2(0)}{2Al} = \underset{(-400\ \text{kcal/mol})}{Al_2O_3} + \underset{2(0)}{2Y}$ | $\Delta H_{298} = +50\ \text{kcal}$ |
| $\underset{(-133\ \text{kcal/mol})}{BaO} + \underset{\frac{2}{3}(0)}{\frac{2}{3}Al} = \underset{\frac{1}{3}(-400\ \text{kcal/mol})}{\frac{1}{3}Al_2O_3} + \underset{(0)}{Ba}$ | $\Delta H_{298} = -0.3\ \text{kcal}$ |

The positive enthalpy values for Ag mean that this metal is less reactive than Al and therefore likely to be a more stable contact.

2. The second requirement is a practical one that involves kinetics. Once the reaction is thermodynamically possible, it must proceed at an appreciable rate. Otherwise, there will be little actual chemical change and degradation. The rate at which chemical reactions occur is proportional to the Maxwell–Boltzmann factor (also known as the Boltzmann or Arrhenius factor) according to

$$\text{Rate} = \text{A} \exp - \left(\frac{\Delta G^*}{RT}\right); \text{A} = \text{constant} \tag{1.2}$$

where $\Delta G^\star$ is the activation free energy (in the usual units of J/mol or cal/mol), $T$ is the absolute temperature, and $R$ is the gas constant (=8.314 J/mol-K or 1.987 cal/mol-K). Because of its ubiquitous presence whenever degradation is thermally accelerated, Eqn (1.2) is easily one of the most important equations in this book and one worth remembering.

Note that $\Delta G^\star$ does not equal $\Delta G$, and in fact has a very different interpretation, as shown in Figure 1.9(a). The energy barrier that atoms or molecules must surmount to squeeze by neighbors, or push them apart, so that reaction can occur is essentially $\Delta G^\star$; its magnitude hinges on the particular path between initial (unstable) and final (stable) states. If a catalyst is present, $\Delta G^\star$ may be effectively reduced. In contrast, $\Delta G$ is the fixed energy difference between the final and initial states, irrespective of path.

### 1.3.3 Atom Movements and Driving Forces

The language of diffusion in solids is similar to that used in chemistry when describing chemical reactions. We speak of atoms surmounting energy barriers established along the reaction coordinate. In the periodic lattice of Figure 1.9(b), a model is presented for diffusing atoms positioned between neighboring planes within a crystalline matrix. For atoms to move from one site to another, two conditions must be fulfilled; a site must first be available for atoms to move into (e.g., a vacancy) and second, the atom must possess sufficient energy to push atoms aside in executing the required jump. Thus, rates of diffusion and chemical reactions are mathematically expressed by the same Boltzmann-type formula (Eqn (1.2)).

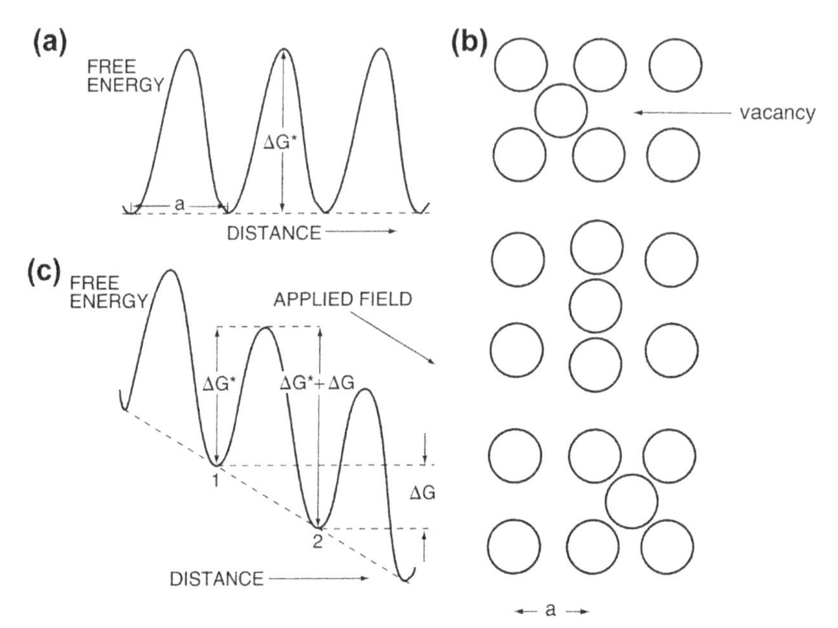


**Figure 1.9** Potential energy of atoms in a periodic lattice: (a) in the absence of a force, (b) model of a diffusing atom jumping between neighboring plane sites into a vacancy, (c) in the presence of a force (field). The energy barrier for reaction is $\Delta G^*$. *From Ref. [18].*

Specifically, the diffusion constant $D$, in units of square centimeters per second, can be written as

$$D = D_0 \exp - \left(\frac{E_D}{RT}\right), \tag{1.3}$$

where $D_0$ is a constant. For simplicity throughout the remainder of the book we will generally replace $\Delta G^\star$ by the relevant activation energy barrier, in this case $E_D$. Both $D_0$ and $E_D$ are dependent on the nature of the diffusant and matrix and the specific path taken by the atoms. Values of $D$ as a function of $T$ for atomic diffusion in metals, semiconductors, and oxides are given in various sources [18].

Now we consider the transport of atoms in systems where generalized driving forces exist. If a driving force or bias did not exist in the case of diffusion, atoms would migrate with equal probability in opposite directions and there would be no net displacement. When driving forces (or fields) exist over atomic dimensions $a$, the periodic array of lattice energy barriers assumes a tilt, or bias, as shown in Figure 1.9(c). Now, atoms in site

2 have a lower energy by an amount $\Delta G$ than those in site 1, a distance $a$ away. In addition, the energy barrier to atomic motion from 1 to 2 is lower than that from 2 to 1. In fact, the slope of $G$ with distance, or the free energy gradient, is caused by and is physically equal to the applied driving force ($F$). This force effectively acts over distance $a/2$, so that $F = 2\Delta Ga$ with units of force per mole (or force/atom). Since atoms vibrate in place with frequency $\nu$ ($\approx 10^{13}\ \mathrm{s}^{-1}$) awaiting a successful jump, the rate at which atoms move to the right is given by

$$r_{12} = \nu \exp\left[-\frac{(\Delta G^{*} - \Delta G)}{RT}\right] (\mathrm{s}^{-1}), \tag{1.4a}$$

while in the latter case,

$$r_{21} = \nu \exp\left[-\frac{(\Delta G^{*} + \Delta G)}{RT}\right] (\mathrm{s}^{-1}). \tag{1.4b}$$

The net rate, or difference of individual rates, is the quantity of significance, and it is given by

$$r_{\mathrm{net}} = r_{12} - r_{21} = \nu \exp\left[-\frac{\Delta G^{*}}{RT}\right]\left[2 \sinh\frac{\Delta G}{RT}\right]. \tag{1.5}$$

It is instructive to consider the case where $\Delta G^{\star} > \Delta G$. If additionally, $\Delta G$ is small compared to $RT$, $\sinh \Delta G/RT \approx \Delta G/RT$. After substituting $\Delta G = Fa/2$ and noting that the atomic velocity $v = a\, r_{\mathrm{net}}$, Eqn (1.5) becomes

$$v = \frac{DF}{RT}, \tag{1.6}$$

where $D$ is associated with $a^2 \nu \exp\left[-\Delta G^{\star}/RT\right]$ (or $a^2 \nu \exp\left[-E_D/RT\right]$.

This important formula is known as the Nernst–Einstein equation and is valid in describing both chemical and physical changes induced through diffusional motion of atoms. It is to the dynamics of microscopic chemical systems what Newton's law relating force to acceleration is to macroscopic mechanical systems. Later we will model degradation phenomena like electromigration, mechanical creep, and compound growth using this equation. When the driving force is large, however, the higher powers of $\sinh \Delta G/RT$ cannot be neglected. Atomic velocities or rates of reaction may then vary as the hyperbolic sine or some power $\beta$ of the driving force, i.e., $F^{\beta}$. Alternatively, it is common to absorb $\Delta G$, or the driving forces, into $\Delta G^\star$, so that in effect a reduced activation energy is operative. As a result, hybrid expressions of the general form

$$v = KF^{\beta} \exp\left[-\frac{(\Delta G^* - aF)}{RT}\right] \quad (1.7)$$

($K$ = constant) are sometimes used in modeling high field damage or phenomena (e.g., dielectric breakdown, high electric-field conduction in insulators). The mean time to failure (MTTF) is usually assumed to be proportional to the reciprocal of the degradation reaction velocity, i.e., MTTF $\approx v^{-1}$.

It is common to have more than one operative driving force acting in concert with temperature to speed reactions. Note that temperature is not a true driving force, even though terms like "temperature stressing" are frequently used; true driving forces like temperature gradients establish a system bias. Electric fields, electric currents, mechanical stress, and humidity gradients are examples of forces that have served to hasten failure of components and devices. Empirical formulas combining products of forces and modified Boltzmann factors (e.g., $\ldots F_i^{\beta i} F_j^{\beta j} \ldots, \ldots \exp[-(\Delta G^* - aF_i - bF_j \ldots)/RT)$ might, for example, be used to account for dielectric breakdown at a given electric field ($F_i$) and humidity level ($F_j$).

Whether small or large driving forces are operative is dependent on the physical process in question. Small energy changes, corresponding to bond shifting, or breaking and reforming in a single material, are involved in diffusion, grain growth, and dislocation motion. This contrasts with chemical reactions, for example, where high-energy primary bonds break in reactants and reform in products.

### 1.3.4 Failure Times and the Acceleration Factor

At the outset, it is important to distinguish among the various times that are associated with failure. Imagine a number ($n$) of identical units that fail in service after successively longer times $t_1$, $t_2$, $t_3$, $t_4$, $t_5$, ... $t_n$. The MTTF is simply defined as

$$\text{MTTF} = \frac{t_1 + t_2 + t_3 + t_4 + t_5 + \ldots + t_n}{n}. \quad (1.8)$$

This time, however, differs from the median time to failure ($t_{50}$), which is defined to occur when 50% have failed; thus half of the failures happen prior to $t_{50}$ and the remaining half after $t_{50}$. Lastly, there is the mean time between failures, or MTBF, defined as

$$\text{MTBF} = \frac{(t_2 - t_1) + (t_3 - t_2) + \ldots + (t_n - t_{n-1})}{n} = (t_n - t_1)/n \qquad (1.9)$$

In this book, we shall usually use MTTF to describe or model physical failure times. On the other hand, $t_{50}$ will mostly arise in the statistical or mathematical analysis of failure distributions.

In the reliability literature, there is a widely used term known as the acceleration factor (AF), which is based on explicit formulas for failure rate, failure time, or the Nernst–Einstein equation in its various versions and modifications. AF is defined as the ratio of a degradation rate at an elevated temperature $T_2$ relative to that at a lower base temperature $T_1$, or conversely, as the ratio of times to failure at $T_1$ and $T_2$. Through application of Eqn (1.2), the AF is easily calculated in the temperature range between $T_1$ and $T_2$:

$$\text{AF} = \frac{\text{MTTF}(T_1)}{\text{MTTF}(T_2)} = \frac{\exp[\Delta G^*/RT_1]}{\exp[\Delta G^*/RT_2]} = \exp\left[\left(\frac{\Delta G^*}{R}\right)\left(\frac{1}{T_1} - \frac{1}{T_2}\right)\right]. \qquad (1.10)$$

Thus, if the failure time is known at $T_2$, assuming one knows $\Delta G^{\star}$, then it can also be calculated at $T_1$. Additionally, these Arrhenius expressions can be used to determine the activation energy if AF is known experimentally.

When one driving force having magnitudes $F_1$ and $F_2$ is operative at the two temperatures, we may tacitly assume that degradation reaction rates proceed with the velocity given by Eqn (1.6). With such an assumption, the AF is given by

$$\text{AF} = \frac{\text{MTTF}(T_1)}{\text{MTTF}(T_2)} = \frac{(T_1/F_1)\exp\left[\frac{E_D}{RT_1}\right]}{(T_2/F_2)\exp\left[\frac{E_D}{RT_2}\right]} = \frac{T_1 F_2}{T_2 F_1}\exp\left[\frac{E_D}{R}\left(\frac{1}{T_1} - \frac{1}{T_2}\right)\right]. \qquad (1.11)$$

We shall return to AFs in Section 4.5.5.

A final note about temperature is necessary in the context of rates. In Figure 1.9(b), as temperature increases, the atoms vibrate at higher frequency and move farther apart. This dual effect causes two sources of reaction rate increase. First, the higher frequency of vibrational energy (kT) means the atoms bump into the spaces between the atoms more frequently with higher energy. Second, the spaces between the atoms are larger on average because the center-to-center spacing increases as temperature rises. Thus, temperature accelerates the rate at which the bad effect will take place. In a sense, temperature has a twofold effect as described above. First, there is more energy driving the reaction (diffusion) rate, and second, the reaction (diffusion) rate can take place more easily because the center-to-center dilation results in larger spaces for the atoms to move through.

### 1.3.5 Load-Strength Interference

An instructive way to close this section on reliability physics is to consider the concept of load-strength interference [19]. Every product has a collection of individual strengths ($S_i$) or stress levels of resistance to corresponding loads ($L_i$) applied in service. Note that in the context of this discussion, the loads are more correctly, applied stresses. Thus, a ceramic has an ultimate tensile strength beyond which it fractures, a critical electric field level that causes breakdown, a specific resistance to thermal-shock loading, and so on. As long as the mechanical and electrical service loads challenge, but do not exceed, the respective strengths, the ceramic is safe. But, should any one of its strengths be exceeded by the corresponding load, the ceramic undergoes catastrophic failure by that particular mechanism.

These ideas are illustrated in Figure 1.10(a), where the particular set of load and strength variables have discrete values. In this case, the strength exceeds the load, and there is no failure. It is more realistic to assume that the strength of materials is statistically distributed, as shown in Figure 1.10(b) and indicated on the vertical axis. For example, a given ceramic has a probability of failing over a wide distribution of tensile strength levels, while in metals that distribution is extremely narrow. Similarly, the applied loads are often not constant but distributed over a range of values. Since the magnitude of the statistically lowest strength level is significantly higher than the largest applied load by a large margin, failure is again impossible. With time, however, the strength of materials may wane due to degradation (e.g., thermal softening due to creep, corrosion), but the applied stress distribution still remains unchanged. Conditions for marginal behavior followed by failure may then arise because service load levels overlap or interfere with the strengths of the weaker members. This subject will be addressed again in Section 12.5.2.

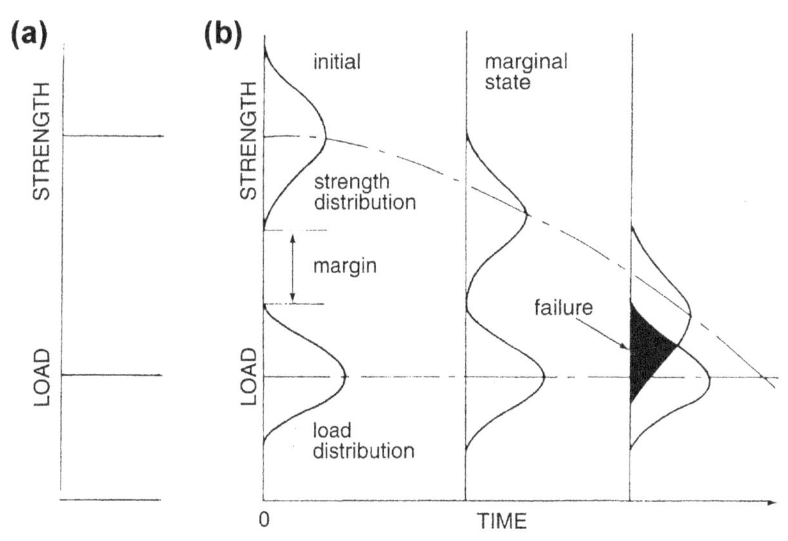


**Figure 1.10** (a) Discrete set of load and strength values. There is no failure because the strength exceeds the load. (b) Both strengths and loads are statistically distributed. With time at constant load, decreasing strength causes increasing overlap, or interference, and eventual failure.

### 1.3.6 Semiconductor Device Degradation and Failure

The mechanisms for semiconductor failure can be grouped into three main categories [20]:

1. Chip level or intrinsic failures
2. Packaging or extrinsic failures
3. Electrical stress induced (in-circuit) failures

Intrinsic failure mechanisms originate from the silicon chip or die and the processing performed in the front end[2] of manufacturing line (FEOL-front end of line). Potential defects and sources for failure exist in each material used and in every process step that alters composition and device features. Defect sources can reside in the silicon wafer substrate itself; in the dielectrics like $SiO_2$ and other insulating films (silicon nitride, phosphorous, and boron-containing oxide glasses); in the metallizations and passivation used in the back-end processing, which include aluminum-base alloy or

[2] Front end means silicon wafer processing before metallization, i.e. silicon, dielectric and contact holes. Back end means wafer processing starting with first metallization and continuing with passivation of as many metal levels of metallization, as are needed for wiring to the chip terminals, which are wire bonded to a lead frame or flip chip mounted to a package surface.

copper film interconnections between devices and components; and in the semiconductor contacts composed of assorted metal alloys and compounds (Al, Al–Si, various metal silicides, Ti–W, TiN, etc.). Design errors plus the combination of lithography (patterning, tolerances, registration, and overlay) and processing defects, the ubiquitous presence of contamination, and the intrinsic limitations of material properties and thickness all sow the seeds for eventual chip-level failures. Temperature, voltage, current, stressing during use, combined with environmental influences such as humidity and radiation, then accelerate the transition from reliable to failed states. Mass-transport-induced failure mechanisms on the chip include interdiffusion that results in unwanted compounds, precipitates or void formation at contacts, electromigration or a current-induced migration of atoms in interconnections, and corrosion at unprotected contacts. Examples of deleterious charge transport effects include $Na^+$ contamination, dielectric breakdown, penetration of $SiO_2$ by so-called hot electrons that originate in the underlying silicon, and "soft errors" in logic or memory caused by ionizing radiation. Chapters 5, 6, 7, and 10 address such failures in greater detail.

*Extrinsic* failures are identified with the interconnections formed in the back-end (BEOL-back end of line) processing and packaging of chips. Potential failures now arise in the steps associated with ensuring that IC contact pads are electrically connected to external power sources and signal processing circuits. Problems arise in attaching dies to metal frames and wires to contact pads, encapsulating chips into packages, and soldering them to printed circuit boards. Specific failure mechanisms include die fracture, open bond joints, moisture-induced swelling and cracking of plastic packages, corrosion, contamination by impurities, formation of intermetallic compounds and voids at bonds, and fatigue, as well as creep of solder connections. In addition to mass-transport driving forces, thermal mismatch stresses between components play an important role in effecting solder degradation and failure. Chapters 8 and 9 are devoted to addressing failure in packages and interconnections, and particularly the role played by mechanical stresses.

*Electrical stress* induced failures are event dependent and largely caused by overstressing and excessive static charge generated during handling in service. Electrostatic discharge (ESD) and electrical overstress (EOS) are the operative mechanisms that often destroy sensitive electronic components. In addition to mishandling, damage is frequently caused by poor initial design of circuits or equipment. The subject of ESD is treated in Chapter 6.

### 1.3.7 Failure Frequency

Now that an overview of failure in ICs and associated packages has been presented, we may well ask how frequently failure occurs. This is a complex question that depends not only on the more-or-less objective classification of device or circuit type, level of technology, and manufacturer, but also on the subjective assessment of what constitutes failure and the often problematical way it is reported. In this regard, we consider both the frequency of failure modes and failure mechanisms. Both display wide quantitative variations, and that is why only a qualitative discussion of them is merited.

The basic failure modes have been classified as short circuits, open circuits, degraded performance, and functional [21]. By functional failures we mean situations where the output signal is different from the one expected. Thus, diodes and transistors, for example, often fail through short circuiting, as do ceramic, foil, and dry tantalum capacitors. Optoelectronic devices, piezoelectric crystals, and resistors, on the other hand, frequently open circuit. Coils, ICs, and relays are examples that largely exhibit functional failures. Virtually all devices and components suffer more or less performance degradation prior to activation of the particular failure mode.

Many pie charts, histograms, and tables dealing with the frequency of failure mechanism occurrence permeate the literature. They will not be reproduced here because of their very transitory value and narrow utility. In some, intrinsic failures associated with metallizations and dielectrics dominate, while in others, extrinsic failures associated with packaging defects and solder joints are in the majority. Electrical stress induced damage always seems to be well represented.

In addition, it has also been found that the responsibility for failure can be allocated between the manufacturer and the user. ESD and EOS failures caused by mishandling and those due to environmental factors (humidity) are examples of user responsibility. For some of the reasons noted above, the proportionate share of fault between manufacturer and user varies widely. However, a summary [21] of such data shows roughly equal fault.

The literature of reliability physics generally devotes coverage of failure mechanisms in proportion to their frequency of occurrence or their perceived importance; this is also reflected in the attention given to them in this book. While failure frequencies of the past and present are no indication of future performance, they may provide some perspective of what to expect. In this sense, Chapter 1 through 11 are prerequisite to Chapter 12, which attempts to peer into the reliability implications of future technology trends.

### 1.3.8 The Bathtub Curve and Failure

This overview ends with an introduction to the most widely recognizable graphic representation of reliability—the bathtub curve—followed by some thoughts on failure. Originally designed to display the rate at which humans die with age, the bathtub curve has been adopted to describe the life of assorted products. It is reproduced in Figure 1.11 and represents the failure rate as a function of time. Although a more rigorous definition of failure rate will be given in Chapter 4, the number of failures per unit time will suffice here. Components and devices usually live through one or more stages of this curve. During early life the failure rate is high, since the weak products containing gross or macroscopic manufacturing defects do not survive. This short-time portion of the curve is therefore known as the early failure, or more commonly as the infant mortality, region. Eliminating this region of the curve is effected by the widely employed practice of screening out obviously defective components, as well as weak ones with a high potential for failure. In such screening processes, products must survive the ordeal of burn-in (e.g., stressing at high temperature, application of EOS, temperature cycling) prior to shipment.

In contrast to the declining failure rate of infant mortality, the next region is characterized by a roughly constant failure rate. Since random failures occur that are not intrinsic to devices but rather to external circumstances (e.g., lightning, floods) during the indicated time span, this region normally corresponds to the working life of the components. Lastly,

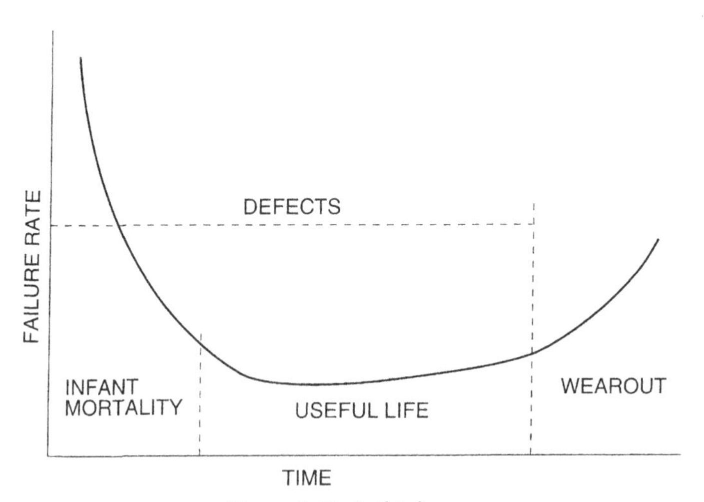


**Figure 1.11** Bathtub curve.

the failure rate increases in the last stages of life due to wear out. More subtle, microscopic defects grow over time, curtailing life in these latter stages. The reader will well appreciate that many mechanical components and biological systems seem to track similar life histories.

Approaches to achieving high reliability differ significantly in the infant mortality and wear-out stages. In the former, the sources, or root causes, of defects must be exposed, followed by action to eliminate them. If defects cannot be eliminated at the source, then screening by burn-in is conducted. Wear-out failure is combated by developing robust technologies with improved circuit and device designs that employ damage-resistant materials. Nowadays, IC chips rarely wear out; rather, new technology usually forces them into retirement. In Section 4.5.9, the bathtub curve will be revisited to capitalize on the statistical functions introduced there.

Perhaps the three most important words raised about failure are when, how, and why; in one form or other, the remainder of the book is largely devoted to addressing them. The connections between these terms are neatly depicted in the triangular representation of Figure 1.12 [22]. At the triangle's

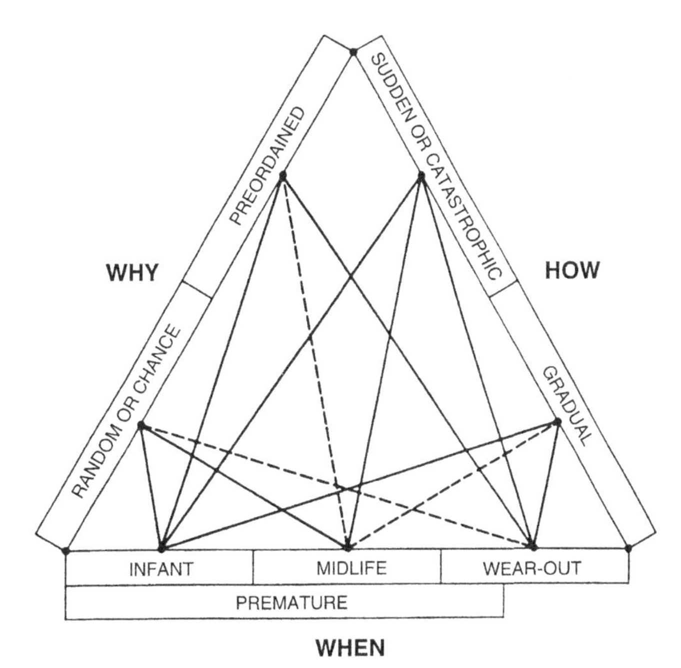


**Figure 1.12** Depiction of connections between the when, why, and how of failures. Solid lines denote a connection, and dashed lines indicate the absence of a connection. *From Ref. [22].*

base we essentially have the divisions of the bathtub curve we have just considered. Corresponding failures can be gradual, the way soles on shoes wear, or sudden and catastrophic, such as when light-bulb filaments burn out. Failures can be random, say due to ESD, or preordained, where some internal mechanism causes the product to fail deterministically at a set rate from the moment of birth. Preordained failures are thus time dependent. These descriptors may be directly related, as indicated by the solid lines, or show the absence of a connection in the case of a dashed line. Thus, wear-out failures may be gradual or sudden, but not random. Infant mortality failures display all characteristics, and this makes it difficult to deal with them.

The individual chapters of the book will also deal with two more questions, namely, what happened and where did it happen. The answers are equally important and sort of corollaries to the how and why. Both sets of answers taken together define the root cause of material failure. These additional questions characterize the specific mechanism of degradation of a particular material, which ultimately caused the observed failure at some specific location in the IC, package, or printed circuit board (PCB). Our endeavor is to elucidate the known mechanisms and the kinetics of their behavior with an eye toward eliminating their occurrence in future product designs and processes.

## 1.4 SUMMARY AND PERSPECTIVE

As its title suggests, this chapter has provided a brief overview of the reliability of electronic materials and devices. These important concerns are intimately tied to the performance of present electronic products as well as to the creation of improved successors. Repeated cycles of manufacture, field use, failure, and product redesign have characterized the Age of Electricity and Electronics since its inception, and will probably be coterminous with its demise, if that ever occurs. Issues of defects, product yield, reliability, and degradation are important considerations in these cycles. In this chapter, a broad footprint of these and related topics treated in the book has been broadly sketched.

Appreciation of the causal connections between the types of electronic products and the way they fail requires an understanding of the nature of the materials they are composed of. Recommended texts on materials science and engineering [18,23–25] will assist the reader to acquire this background. Next, the behavior of specific devices (Chapter 2) and how they were processed with particular reference made to sources of defects and contamination (Chapter 3) serve to establish the nature of the product prior to service. Product yield measures the effectiveness of eliminating defects during processing. While readers may be knowledgeable about aspects of electronic materials and the devices, processing and the issues surrounding defects and yield are likely to be less familiar.

Now the stage is set for Chapter 4, which is devoted to a quantitative treatment of reliability. Provided that sufficient failure data are available from field experience or testing, there are several statistical approaches for dealing with them. The purpose is to obtain failure rates, the MTTF, and to predict future failure behavior. No book on reliability would be complete without treating these concepts. But, by the same token, no book on reliability would be balanced if it did not present some of the current criticisms and controversies surrounding long-held beliefs on failure and the lifetime predictions that stem from them.

Much of the remainder of the book is concerned with failure physics on the chip as well as packaging levels. Here, well-established methods of scientific investigation are employed to uncover the mechanisms of failure and how they depend on the involved driving forces. Obtaining expressions for failure or degradation times in terms of the operative stresses is a particular objective of these failure modeling studies. Irrespective of the specific mechanism, failure virtually always begins through a time-dependent movement of atoms, electrons, or charge from benign sites in the device or component to harmful sites, or failed states. If these atoms or electrons accumulate in sufficient number at harmful sites, damage results. As noted earlier, the challenge is to understand the nature of the driving forces that compel matter or charge to move and to predict how long it will take to create a critical amount of damage.

## EXERCISES

1. "On October 21, 1879, Thomas Edison's famous (tungsten filament) lamp attained an incredible life of 40 h. The entire electrical industry has as its foundation this single invention ... the incandescent lamp." J.W. Guyon, GEC (1979).
   a. Support the validity of the second sentence.
   b. How many different materials can you identify in a common light bulb that pose reliability problems?

2. Provide examples of products combining
   a. electrical and mechanical technologies.
   b. electrical and electronic technologies.
   c. electronic and optical technologies.
3. The law of personal computing systems scaling suggests that the logarithm of the sum of the length ($L$) plus width ($W$) plus height ($H$) varies linearly with time. If $L + W + H$ is halved every 6 years, sketch the variation for the following computers: desktop, notebook, palmtop, credit card, wristwatch, and finger ring. When can finger ring computers be expected?
4. The number of transistors used per person in the United States has grown from about 5 to $4 \times 10^7$ during the years 1955–1997. Sketch this growth in usage on a graph using appropriate axes. If the trend continues, estimate the number of transistors each person will be using in 2010.
5. The growth in the number of devices per chip ($N$) appears to obey a law of the form $dN/dt \approx kN$ (k = constant), i.e., growth proportional to how many devices are instantaneously present. Chemical reactions and radioactive decay also suggest similar trends. Discuss physical reasons for the exponential rise or exponential decay.
6. Write an equation that approximates the decrease in size of mobile phones. In what year can we expect to see the Dick Tracy comics become reality by incorporating telephones into wristwatches? Assume that the watch dimensions are $3.5 \times 2 \times 1$ cm.
7. The implications of unreliability in the medical profession are at least as severe as they are in aerospace technology. Comment on possible differences and similarities in the reliability issues faced by each.
8. By considering Figures 1.2 and 1.3, calculate the *density* (number per unit area) of devices as a function of time. Is the rate of growth in device density higher in microprocessor or memory chips?
9. Comment on the thermodynamic stability of silicon, aluminum, and silicon dioxide that are often either in contact with or located within a few nanometers of each other on IC chips.
10. Contrast the shape of the bathtub curve for human death rates with age in 1900 with that at the end of the twentieth century. What are the reasons for the difference?
11. The energy barrier that must be surmounted to convert "reliable" states to "failed" states located a distance $a$ apart is $E_{r-f}$. Furthermore, the energies of the reliable and failed states are $E_r$ and $E_f$, respectively.

**a.** Represent these states and the energy that separates them graphically.
**b.** Write an expression for the rate of production of failed states as a function of temperature.
**c.** Suppose that a driving force acts to produce failed states. Write an expression for the net rate of failed-state production. Represent the action of the driving force graphically.

**12.** For a particular failure mechanism, two expressions for the MTTF have been suggested, namely,

$$\text{MTTF} = AF^{-n}\exp\left[\frac{E}{RT}\right] \text{ and } \text{MTTF} = B\exp\left[\frac{E-\alpha F}{RT}\right], \text{ where } F \text{ is the}$$

driving force. If $A$, $B$, $E$, $n$, and $\alpha$ are positive constants independent of temperature and force,
**a.** Write expressions for the AF in each case when the temperature is raised from $T_1$ to $T_2$ at constant $F$.
**b.** Write expressions for the AF in each case when the force is increased from $F_1$ to $F_2$ at constant $T$.
**c.** Which of the two formulas for MTTF gives a higher value of AF for temperature acceleration?
**d.** Which formula gives a higher value of AF for field acceleration?

**13.** It has been claimed that an AF of two governs chemical reaction rates when the temperature is raised by 10 °C. What activation energy is required for this to happen at room temperature?

**14.** Contrast the applicability of Newton's laws and the Nernst–Einstein equation to the motion of atoms in solids.

**15.** Suppose the strength ($S$) of a material has a Gaussian-like spread about mean value $S_0$ such that the distribution of $S$ values is given by $\exp -\left[\frac{10(S-S_0)^2}{S_0^2}\right]$. The applied load ($L$), or stress distribution, is similarly given by $\exp -\left[\frac{8(L-L_0)^2}{L_0^2}\right]$, where $L_0$ is the mean load. Failure will occur when the strength and load overlap such that a common value of 0.04 is reached in each distribution.
**a.** What are the values of $L$ and $S$ that would cause failure?
**b.** Calculate the ratio $L_0/S_0$ at failure.

**16.** Initially, the extent of strength–load overlap in a material occurs at a distribution value of 0.001. During service the applied load remains constant, but the mean strength of the material declines exponentially as $S_0(t) = S_0 \exp[-kt]$, where $k$ is a constant. If the load and strength distributions are described by Gaussians as in Exercise 1–15, and failure occurs when the value for each is 0.04, calculate the failure time in terms of $L_0$, $S_0$, and $k$.

## REFERENCES


[1] J.A. Fleming, Fifty Years of Electricity—The Memories of an Electrical Engineer, The Wireless Press, London, 1921.
[2] R. Friedel, P. Israel, B.S. Fein, Edison's Electric Light, Rutgers University Press, New Brunswick, 1986.
[3] M. Josephson, Edison, McGraw Hill, New York, 1959.
[4] C.L. Briant, B.P. Bewlay, MRS Bull. XX (8) (1995) 67.
[5] S. Okamura (Ed.), History of Vacuum Tubes, IOS Press, Amsterdam, 1994.
[6] A.J. DeMaria, IEEE Circuit. Devic. 7 (3) (1991) 36.
[7] E.A. Sack, Proc. IEEE 82 (1994) 465.
[8] J.D. Meindl, Proc. IEEE 83 (1995) 619.
[9] S. Malhi, P. Chatterjee, Circuit. Devic. 10 (3) (1994) 13.
[10] D.L. Crook, in: 28th Annual Proceedings of the IEEE Reliability Physics Symposium, 1990, p. 2.
[11] M. Penn, Microelectr. J. 23 (1992) 255.
[12] R.M. Alexander, Tutorial Notes, IEEE International Reliability Physics Symposium 4.1, 1984.
[13] J. Vaccaro, Proc. IEEE 61 (1974) 169.
[14] G.W.A. Dummer, Microelectronics and Reliability 9 (1970) 359.
[15] Military Handbook (MIL-HDBK-217F) Reliability Prediction of Electronic Equipment, Department of Defense, 1991.
[16] M. Pecht, J. Pecht, Long-term Non-operating Reliability of Electronic Products, CRC Press, Boca Raton, 1995.
[17] A.V. Ferris-Prabhu, Tutorial Notes, IEEE International Reliability Physics Symposium 3.1, 1984.
[18] M. Ohring, Engineering Materials Science, Academic Press, San Diego, 1995.
[19] P.D.T. O'Connor, Practical Reliability Engineering, second ed., Wiley, Chichester, 1985.
[20] E.A. Amerasekera, D.S. Campbell, Failure Mechanisms in Semiconductor Devices, Wiley, Chichester, 1987.
[21] F. Jensen, Electronic Component Reliability, Fundamentals, Modeling, Evaluation and Assurance, Wiley, Chichester, 1995.
[22] F.R. Nash, W.B. Joyce, R.L. Hartman, E.I. Gordon, R.W. Dixon, AT&T Tech. J. 64 (3) (1985) 671.
[23] C. Newey, G. Weaver, Materials in Action Series—Materials Principles and Practice, Butterworths, London, 1990.
[24] M.F. Ashby, D.R.H. Jones, Engineering Materials 1—An Introduction to Their Properties and Applications, Pergamon Press, Oxford, 1980.
[25] C.R.M. Grovenor, Microelectronic Materials, Adam Hilger, Bristol, 1989.

CHAPTER 2

# Electronic Devices: How They Operate and Are Fabricated

## 2.1 INTRODUCTION

Books on reliability ideally focus on issues related to component and device failure within the context of how they are intended to function. Because it is often not necessary to know about the latter to address the former, these two bodies of knowledge and expertise often remain separate concerns. This is especially so because most failures involve materials other than the semiconductors, e.g., metallizations, insulation, and packaging materials. Nevertheless, it is fair to say that malfunction, failure, and reliability can be more intelligently confronted through some knowledge of device behavior and fabrication methods employed. For example, awareness of device current–voltage characteristics is certainly helpful in diagnosing failure modes in some instances. Similarly, an appreciation of the processing steps involved in device manufacture may suggest ways to avoid future failures. In this spirit, the chapter is primarily devoted to elevating the consciousness and sensitivity of readers to electronic devices and their uses, characteristics, and instabilities. Many aspects of the degradation and reliability of specific devices will be introduced here in a qualitative way. We shall return to these same issues in later chapters, where degradation and the physics of failure will be analytically modeled at a deeper level.

A broad tutorial survey of electronic materials initiates the chapter, which then goes on to treat the operation of diodes, transistors, and optical devices. For the most part, the silicon and compound semiconductor devices introduced are those that have been used most broadly in practice. Instead of mathematical rigor in describing device behavior, equations will be used sparingly and introduced without derivation. Qualitative and intuitive meanings will be emphasized. In doing so it is hoped that the language and terminology used to describe devices will make references to them more understandable later in the book as well as in the reliability literature. Following this, several of the processing steps involved in device manufacture are introduced. Defects and processing problems, the probable

*Reliability and Failure of Electronic Materials and Devices*
ISBN 978-0-12-088574-9
http://dx.doi.org/10.1016/B978-0-12-088574-9.00002-1

sources of a number of reliability concerns, are discussed in the next chapter.

Electrical engineers will probably choose to skip much of this chapter because of its admittedly shallow coverage; however, the latter part dealing with processing may cover new ground. Conversely, the nomenclature and qualitative descriptions of electronic materials and devices may prove instructive to those trained in other disciplines. In either case, several excellent books have been written on the physics of solid state electronic devices [1–5] and on processing technology [6–10] that all readers can benefit from.

## 2.2 ELECTRONIC MATERIALS

### 2.2.1 Introduction

At the outset it is important to realize that a very large number of materials have been employed at one time or another in experimental and commercial electronic devices and equipment. They form the sum and substance of active devices and passive components, the packages that enclose them, the substrates and boards on which they are mounted, and the solder and wires that join and connect them. All commercial electronic materials and devices have typically survived a rigorous selection ordeal whose hurdles include:

1. meeting stringent electrical specifications and requirements,
2. developing reproducible processing and achieving high yields in manufacturing,
3. extensive testing to assure reliable operation.

We are concerned with the properties of these surviving materials and devices and their fates during subsequent use. The survivors, however, are typically but a fraction of the total number of potential candidates that were initially considered. One only has to recall Edison's experimentation with lamp filament materials to appreciate what is involved.

A list of the materials employed in commercial electronic devices is entered in Tables 2.1 and 2.2. Although incomplete, the list includes most of the materials addressed in this book. These materials can be conveniently classified into three categories, namely, *conductors*, *semiconductors*, and *insulators*. Each category is distinguished both by the magnitude of the electrical resistivity as indicated in Figure 2.1 and by the sign of the temperature coefficient of resistivity. By conductors we mean metals with resistivities ranging from $10^{-6}$ to perhaps $10^{-4}$ $\Omega$-cm. Semiconductor

**Table 2.1** Physical properties of electronic materials

| **Metals** | **Melting point (°C)** | **Electrical resistivity (Ω-cm)** | **Temp. coeff. resistivity (α) (°C$^{-1}$)** | **Density (g/cm$^3$)** |
|---|---|---|---|---|
| Copper | 1083 | $1.7 \times 10^{-6}$ | 0.0039 | 8.93 |
| Silver | 962 | $1.6 \times 10^{-6}$ | 0.0041 | 10.5 |
| Gold | 1063 | $2.2 \times 10^{-6}$ | 0.0034 | 19.3 |
| Aluminum | 660 | $2.7 \times 10^{-6}$ | 0.0039 | 2.7 |
| Tungsten | 3415 | $5.5 \times 10^{-6}$ | 0.0045 | 19.3 |
| Platinum | 1772 | $10.6 \times 10^{-6}$ | 0.0039 | 21.5 |
| Palladium | 1552 | $10.8 \times 10^{-6}$ | 0.0038 | 12.0 |
| Nickel | 1455 | $6.8 \times 10^{-6}$ | 0.0069 | 8.9 |
| Chromium | 1875 | $13 \times 10^{-6}$ | 0.003 | 7.2 |
| Invar (36Ni-64Fe) | 1425 | $80 \times 10^{-6}$ | 0.0014 | 8.0 |
| Alloy 42 (43Ni-58Fe) | 1425 | $68 \times 10^{-6}$ | 0.0014 | 8.1 |
| Kovar | 1450 | $50 \times 10^{-6}$ | | 8.3 |
| 63Sn-37 Pb | 183 | $15 \times 10^{-6}$ | | 8.42 |
| 95 Pb-5Sn | 310 | $19 \times 10^{-6}$ | | 11 |
| Cu-W (20%Cu) | 1083 | $2.5 \times 10^{-6}$ | | |
| $CoSi_2$ | 550[a] | $15 \times 10^{-6}$ | | |
| $TiSi_2$ (C49) | 650[a] | $60 \times 10^{-6}$ | | |
| $TiSi_2$ (C54) | 870[a] | $15 \times 10^{-6}$ | | |
| $WSi_2$ | 650[a] | $100 \times 10^{-6}$ | | |
| PtSi | 300[a] | $35 \times 10^{-6}$ | | |

| **Ceramics/glasses** | **Melting point (°C)** | **Electrical resistivity (Ω-cm)** | **Dielectric constant** | **Thermal conductivity (W/m-K)** | **Density (g/cm$^3$)** |
|---|---|---|---|---|---|
| Silicon dioxide | 1980 | $>10 \times 10^{15}$ | 3.8 | 7 | 2.2 |
| Aluminum oxide | 2323 | $>10 \times 10^{14}$ | 9.9 | 37 | 3.9 |
| Silicon nitride | 2173 | $>10 \times 10^{14}$ | 7.0 | 30 | 3.2 |

*Continued*

**Table 2.1** Physical properties of electronic materials—Cont'd

| Ceramics/glasses | Melting point (°C) | Electrical resistivity (Ω-cm) | Dielectric constant | Thermal conductivity (W/m-K) | Density (g/cm$^3$) |
|---|---|---|---|---|---|
| Silicon carbide | 3110 | $>10 \times 10^{14}$ | 42 | 270 | 3.2 |
| Aluminum nitride | | $>10 \times 10^{11}$ | 8.8 | 140—230 | 3.3 |
| Beryllium oxide | 2570 | $>10 \times 10^{14}$ | 6.8 | 240 | 2.9 |
| Boron nitride | | $>10 \times 10^{10}$ | 6.5 | 600 | 1.9 |
| Borosilicate glass | 1075 | $>10 \times 10^{10}$ | 4.1 | 5 | |
| ***Plastics*** | | | | | |
| Epoxy-Kevlar | 200[a] | $>10 \times 10^{14}$ | 3.6 | 0.2 | |
| Polyimide | 200[a] | $>10 \times 10^{14}$ | 4.0 | 0.35 | |
| Epoxy FR-4 | 175[a] | $>10 \times 10^{14}$ | 4.7 | 0.2 | 1.9 |
| Polyimide | 300—400[a] | $>10 \times 10^{15}$ | 3.0—3.5 | 0.2 | 1.5 |
| Teflon | 400[a] | $>10 \times 10^{14}$ | 2.2 | 0.1 | 2.2 |
| Silicone (high-purity) | 150[a] | $>10 \times 10^{14}$ | 2.7—3.0 | 0.16 | |
| Parylene N (vapor deposited) | | $>10 \times 10^{14}$ | 2.65 | 0.12S | |

Note: Additional properties are listed in Tables 8–2.
[a]Processing temperature.
Data collected from various sources.

**Table 2.2** Properties of elemental and compound semiconductors

| Material | Lattice parameter (nm) | Melting point (K) | Energy gap (eV at 25 °C) | Electron mobility ($cm^2$/V-s) | Hole mobility ($cm^2$/V-s) |
|---|---|---|---|---|---|
| Diamond | 0.3560 | ~4300 | 5.4 | 1800 | 1400 |
| Si | 0.5431 | 1685 | 1.12 I | 1450 | 450 |
| Ge | 0.5657 | 1231 | 0.68 I | 3600 | 1900 |
| ZnS | 0.5409 | 3200 | 3.54 D | 120 | 5 |
| ZnSe | 0.5669 | 1790 | 2.58 D | 530 | 28 |
| CdS | 0.5832 | 1750 | 2.42 D | 340 | 50 |
| AlAs | 0.5661 | 1870 | 2.16 I | 1200 | 420 |
| GaP | 0.5451 | 1750 | 2.26 I | 110 | 75 |
| GaAs | 0.5653 | 1510 | 1.43 D | 8500 | 400 |
| InP | 0.5869 | 1338 | 1.27 D | 4600 | 150 |
| InAs | 0.6068 | 1215 | 0.36 D | 30,000 | 460 |

I refers to indirect band gap.
D refers to direct band gap.
From Refs [1, 11].

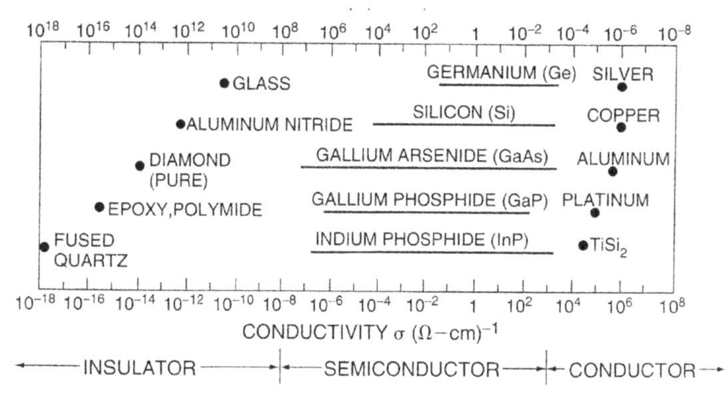


**Figure 2.1** Electrical resistivities of electronic materials.

resistivities depend very strongly on doping level and span a typical range from $10^{-3}$ to $10^{6}$ Ω-cm. However, extremely heavily doped Si has metallic resistivities, while semi-insulating GaAs has a resistivity of ~$10^{8}$ Ω-cm. Insulator resistivities are the highest of all and can reach values in excess of $10^{15}$ Ω-cm. In electronic applications, insulators are represented by oxides, e.g., $SiO_2$ in IC chips, $Al_2O_3$ substrates on which devices are mounted, and polymers, e.g., IC encapsulants and printed circuit boards.

Physically, metals can be distinguished from all other classes of materials because they become *poorer* conductors as the temperature is raised, implying a positive temperature coefficient of resistivity. Conversely, semiconductors and insulators become *better* conductors as they become hotter. Even here there are rare exceptions to the rule over small temperature excursions.

## 2.2.2 Semiconductors

### *2.2.2.1 Structure*

Semiconductors and the devices based on them are a primary focus of the book. Some, but certainly by no means all, of the reasons for device failure and reliability problems can be attributed to malfunction of semiconductors. The special attributes of semiconductors derive from both their physical and electronic structures. In Figure 2.2 the crystal structures of both Si and GaAs, the two most important semiconductors used in electronics, are depicted. Silicon has a diamond-cubic structure that is derived from the face-centered cubic lattice but with a two-atom motif per lattice point; this means that there are eight atoms per unit cell. On the other hand, in GaAs the Ga and As atoms alternate as shown within the basic diamond cubic lattice.

In cubic structures there are three important crystallographic planes and directions. The (100) planes are the cube faces and the [100] directions are the cube edges. Planes passing through diagonally opposite cube edges have (110) indices, while (111) planes pass through three diagonally opposite cube corners. The [110] directions are those between the cube center to the midpoint of each of the 12 cube edges; similarly, [111] directions radiate from the cube center to each of the eight cube vertices. Almost every semiconductor electronic device requires either bulk or thin film semiconductor single crystals of high crystallographic perfection, except amorphous thin film transistors, used in LCDs (liquid crystal displays).

Critical to many of the structural and electronic properties of elemental Si is the tetrahedral atomic grouping, in which each atom is surrounded by four identical atoms. The bonds between atoms, represented by lines, are covalent in nature, reflecting the fact that a pair of outer electrons, one from each of the neighboring atoms, comprises the bond. Covalent bonds are typically strong and resistant to deformation. This accounts for silicon's appreciable hardness, relatively high Young's modulus, lack of ductility, and high melting point. Although covalent bonds are mechanically strong, the bonding, or *valence,* electrons within them can be easily detached at relatively low temperatures to become more energetic *conduction* electrons. Holes, with an effective positive charge, are then left behind in the broken valence bonds.

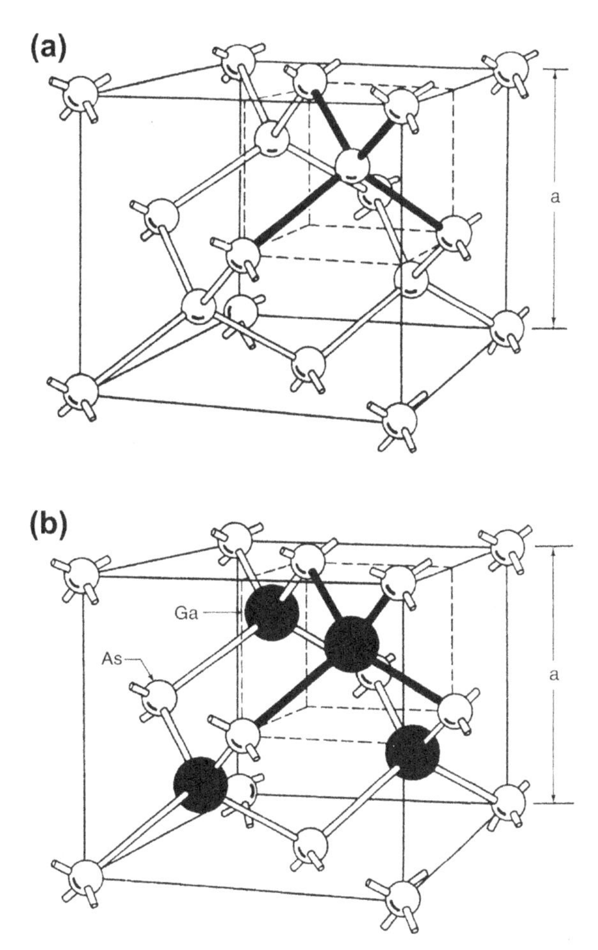


**Figure 2.2** (a) Diamond cubic crystal structure of silicon. (b) Model of the zinc blende crystal structure of GaAs. Both Ga and As populate face-centered cubic lattices that interpenetrate one another.

Both holes and conduction electrons can readily move under the influence of electric fields. The semiconductor property known as mobility ($\mu$) is defined as the ratio of the carrier velocity to the applied electric field and has units of ($cm^2$/V-s). Mobility values for carriers in semiconductors are listed in Table 2.2. Undesirable reductions in carrier mobility occur in the presence of impurities and lattice defects. The combination of carrier concentrations and velocities influences and limits all electrical conduction phenomena.

### 2.2.2.2 *Bonds and Bands*

#### 2.2.2.2.1 Bonds

Several representations of electronic structure have evolved over the years as aids in interpreting the behavior of semiconductor materials and devices. The simplest of these, depicted in Figure 2.3, is a physical-bonding model where the tetrahedra are flattened to two dimensions, and each bond between atoms now represents a couple of shared valence electrons. Introduction of n-type dopants (P, As, and Sb) and p-type dopants (B, In, and Al) into the Si lattice creates extra donor electrons and extra acceptor holes, respectively. Through doping, the number of electron and hole charge carriers can be dramatically enlarged in a controllable way. In this model, conduction electrons hop around in the open spaces between bonds; holes, on the other hand, are confined to, and migrate through, the

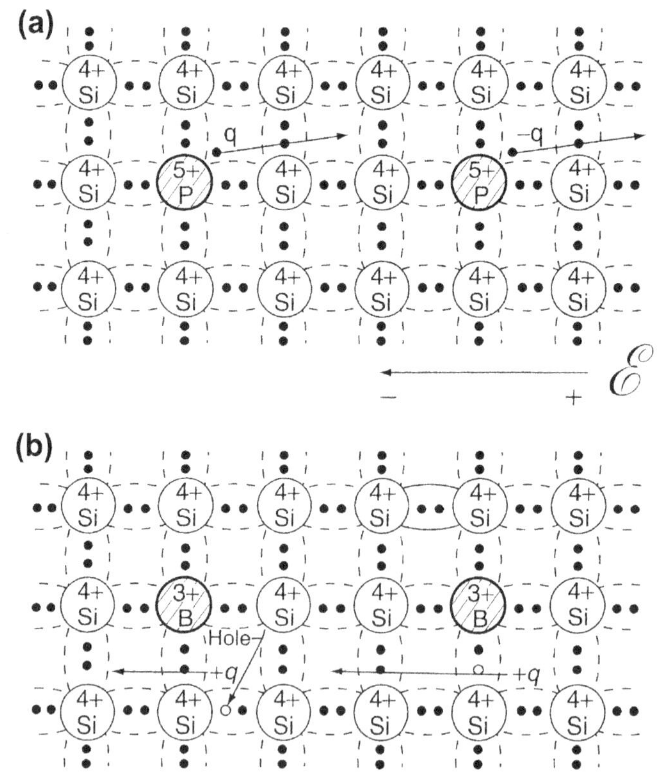


**Figure 2.3** (a) Representation of a two dimensional silicon lattice containing donor ($P^{+5}$) atoms. (b) Two-dimensional silicon lattice containing acceptor ($B^{+3}$) atoms. Each bond consists of two valence electrons. Direction of carrier motion in electric field E is shown.

bond network. An applied electric field causes these charge carriers to move in opposite directions. Keep in mind the semiconductor does not possess an excess of charge, but rather is effectively neutral and macroscopically uncharged.

### 2.2.2.2.2 Bands

The electronic structure of undoped as well as doped semiconductors is popularly represented in terms of the energy band diagrams of Figure 2.4. When isolated atoms of Si are brought together to form the solid, individual electron energy levels collectively broaden to create bands containing an astounding $10^{22}$ or so distinct levels (one for each electron) per cubic centimeter. They are packed so tightly that the energy difference between the highest and lowest levels is only a few electron volts. Thus all of the electrons comprising valence bonds form the *valence band,* and all of the conduction electrons enter the *conduction band.* The energies of electron levels within states are noted on the ordinate, while the horizontal axis can be loosely identified with some macroscopic lattice distance having much larger than atomic dimensions. A key feature of this diagram is the *energy gap* ($E_g$) that separates valence and conduction bands. In Si, $E_g = 1.12$ eV, while in GaAs, $E_g = 1.43$ eV at 300 K. Electron energy levels and states within the energy gap are forbidden in pure, structurally perfect Si. Any perturbation (e.g., dopant atom, impurity located either on lattice or interstitial

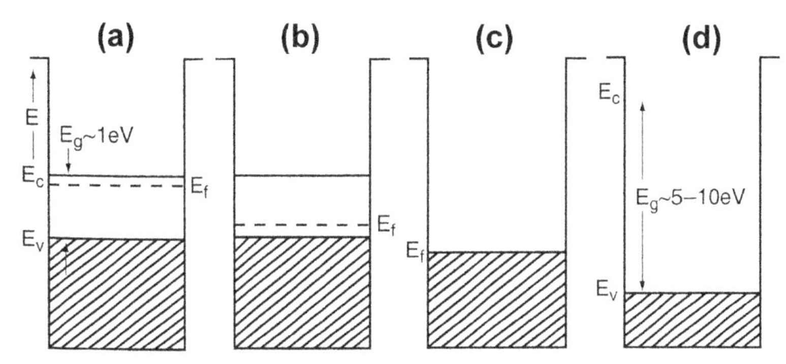


**Figure 2.4** Electron energy band diagrams for semiconductors, metals, and insulators, (a) n-type semiconductor. (b) p-type semiconductor. Largely filled valence band lies below $E_V$ and largely empty conduction band-lies above $E_C$. The Fermi level lies closer to $E_C$ in the n-type semiconductor and closer to $E_V$ in the p-type semiconductor. (c) Band diagram of a metal. The uppermost occupied electron state in the distribution corresponds to the Fermi level, above which there are always accessible empty electron states. There is no energy gap in metals. (d) Band diagram for insulators. Electron energy gaps range from 5 to 10 eV.

sites, vacancy or dislocation defects, precipitates) that disturbs this ideal order is reflected in the creation of new quantum states with characteristic energy levels. Such states usually fall somewhere in the gap.

On this basis it is easy to see that filled n-type dopant donor states (Figure 2.4(a)) lie close to the bottom of the conduction band, to which they readily donate electrons. Similarly, empty p-type dopant acceptor states (Figure 2.4(b)) lie just above the top of the valence band, where they accept electrons, leaving holes behind in the process. In essence, this is how electrical conduction occurs in semiconductors. An energy level of interest that reflects the probability of carrier population of states is known as the Fermi ($E_F$) level. It is defined to be the energy level whose probability of occupation is 0.5. In n- and p-type semiconductors, $E_F$ lies closer to either the conduction ($E_C$) or valence ($E_V$) band edges, respectively, but usually within the gap.

Band diagrams are helpful in distinguishing the behavior of semiconductors from those of metals and insulators. In metals (Figure 2.4(c)) the Fermi level caps the electron distribution. Below it all levels are occupied by electrons, and above it they are unoccupied. It takes very little energy to access these closely spaced empty levels if the electrons are stimulated electrically, thermally, and optically; this accounts for the excellent electrical and thermal conductivity as well as high reflectivity of metals. Band diagrams for insulators (Figure 2.4(d)) resemble those for semiconductors, except for the larger electron energy gaps ranging from 5 to 10 eV.

#### 2.2.2.2.3 Energy–Momentum Representation

Another widely used representation of semiconductor bands is shown in Figure 2.5(a) and b. Here the energy ($E$) of electron states is plotted as a function of the magnitude and direction of the electron momentum, or equivalently, versus wave vector ($k$). The resultant parabola-like $E$ vs $k$ curve is important because it provides a useful way to distinguish between Si and Ge on the one hand, and GaAs and similar compound semiconductors on the other [1]. While the former are known as *indirect band-gap* semiconductors, the latter are *direct band-gap* semiconductors. Only direct-gap semiconductors are useful in optical devices like light-emitting diodes (LED) or lasers. Photons of the band-gap energy are emitted when electrons descend from the bottom of the conduction band to recombine with holes at the top of the valence band. The transition between the two involved states must conserve electron momentum. Similarly, when a photon whose energy is equal to or greater than the gap

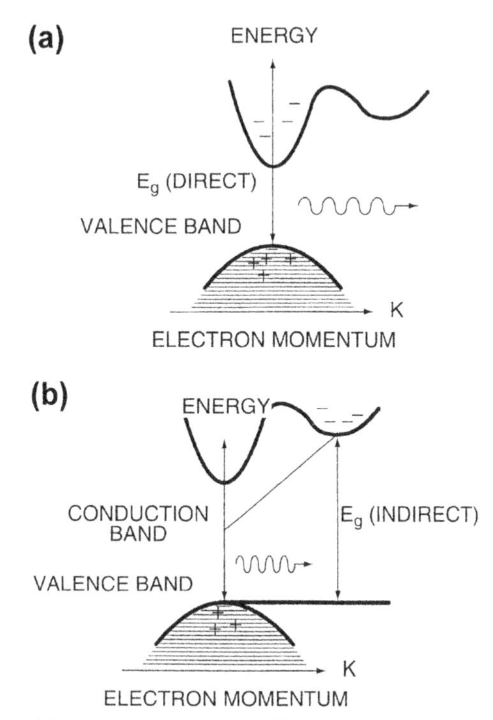


**Figure 2.5** Energy (*E*) versus wave vector (*k*) representation of band structure. (a) direct band-gap semiconductor. (b) indirect band-gap semiconductor.

energy is absorbed, momentum conservation governs the electron ascent between the corresponding valence and conduction band states. In direct band-gap materials like GaAs and InP these electron descents and ascents are vertical, conserving momentum; but in Si and other indirect semiconductors the transitions are "slanted" and involve momentum changes that effectively excite lattice vibrations which heat the solid. These wasteful processes dramatically reduce the probability that photons will either be emitted or absorbed during respective electron–hole recombination or generation.

### *2.2.2.3 Compound Semiconductors*

The important III–V compound semiconductors are derived from atomic combinations arising from columns III A (e.g., Al, Ga, In) and V A (e.g., N, P, As) of the periodic table. Other II–VI compound semiconductors are composed of elements from columns II B (Zn, Cd, Hg) and VI A (S, Se, Te). A summary of the electrical properties of elemental and compound semiconductors is reproduced in Table 2.2, where lattice parameters, energy gaps, and mobilities are indicated for the more widely used materials. Ternary (3 atom) and quaternary (4 atom) semiconductors are also widely used in electro-optical applications. For example, AlGaAs ternaries are derived from AlAs and GaAs mixtures. Similarly, InGaAs and GaInAsP devices used in optical communications can be understood as combinations of binary compounds, e.g., GaAs and InAs, and GaAs and InP. In many instances the resultant ternary energy gap and lattice parameter values are simply weighted averages derived from the constituent binaries. Thus by alloying GaAs and GaP, families of light-emitting diodes with compositions $GaAs_{1-x}P_x$ become available, provided that the ternary semiconductor alloys are still direct-gap materials. For $x = 0.4$, interpolation yields an energy gap of 1.9 eV, which from Eqn (2.1) below corresponds to the red portion of the spectrum.

When films of $Al_xGa_{1-x}As$ deposited on GaAs substrates, or InGaAs deposited on InP, are involved, it is essential that the film be epitaxial. This means that film and substrate must be "lattice matched" or have as nearly identical lattice parameters as possible. Otherwise, defects that severely limit device function will tend to be present at the film–substrate interface. Typically, mismatches in lattice parameter of no more than 0.1% can be tolerated.

#### *2.2.2.4 Electrical and Optical Function*

Semiconductors have several unique properties that enable them to exhibit unusual electrical and optical effects. Like metallic conductors they possess sufficient numbers of high-mobility charge carriers to pass large currents. Unlike conductors, but like insulators, they are capable of supporting large electric fields; this happens at p–n junctions under reverse bias where charge carriers are depleted. Remarkably, in semiconductor devices either attribute can be emphasized in applications requiring rectification (diode) and amplification (transistor) of current, or storage of charge (metal oxide semiconductor capacitor). The application of the requisite bias voltages to the device serves to control the magnitude of current flow.

Optical activity is possible in semiconductors because they possess an important pair of electron levels which span the energy gap. Many states of very nearly the same energy are either full or empty at levels near the bottom of the conduction band and the top of the valence band; thus electron transitions can readily occur. The wavelength ($\lambda$) associated with the band-gap energy is derived from the well known Planck equation,

$E = hc/\lambda$ (i.e., $E_g = hc/\lambda$), where $h$ is Planck's constant and $c$ is the speed of light. A useful formula connecting $E_g$ and $\lambda$ is.

$$\lambda(\text{microns}) = 1.24/E_g(\text{eV}). \tag{2.1}$$

Three basic types of electro-optical phenomena can be simply understood by referring to Figure 2.6. Depicted are the valence and conduction bands of the involved semiconductor or active electro-optical medium, and the radiative transitions that occur between energy levels in them. Either photons or electronic signals can trigger mutual electro-optical responses at p–n junctions.

1. *Stimulated absorption* of incident photons (Figure 2.6(a)) excites electrons from the valence to the conduction band, generating mobile carriers or an electric signal. After photon detection (e.g., by a photodiode) electron circuitry is activated to amplify the signal, or deliver external power (e.g., a solar cell).
2. The reverse process of photon emission occurs during deexcitation of electrons from the conduction-to-valence band (Figure 2.6(b)). In light-emitting diodes (LED), for example, the emission is triggered by an electric current that passes through the junction causing electron–hole recombination and a corresponding *spontaneous emission* of monochromatic light. The light can now be viewed directly in a

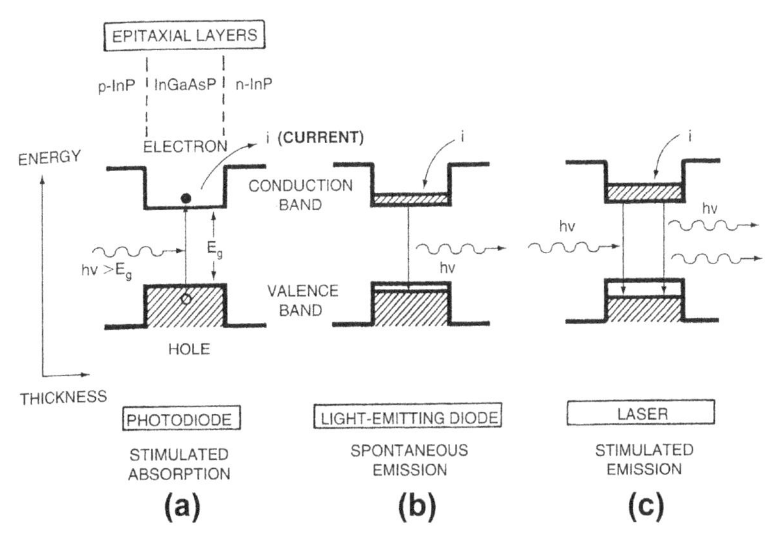


**Figure 2.6** Basic electron transitions between valence and conduction bands in semiconductors shown within the context of junction devices. (a) Stimulated absorption. (b) Spontaneous emission. (c) Stimulated emission.

display, or made to pass through lenses of an optical system, or enter an optical fiber.

3. The last case (Figure 2.6(c)) combines features of the previous two. Through absorption of light (pumping), electrons ascend from the valence band to an excited state, causing a normally unnatural, inverted population of electron levels. Incident photons that impinge on such excited atoms may now induce *stimulated emission* of photons from them. These photons have the same energy and phase as those that excited them in the first place. If instead of allowing the light to exit the medium it is made to reflect back and forth in phase, while impinging on other atoms in excited states, the light intensity is amplified. This is the basis of laser action.

Just how the electro-optical effects are capitalized upon in actual devices is addressed later in the chapter.

### 2.2.3 Conductors

Conductors are used in circuits to speed the transmission of electronic signals from where they are generated to where they are processed, e.g., detected, amplified, stored, or displayed, without distortion or any loss of information. In solid-state circuits this means that conductors must first make low-resistance contact to various n and p doped regions, and gate electrodes of devices, and then be connected to other active devices and passive components via a network of interconnections. The latter are found in great profusion on the IC chip itself as well as at the (packaging) interface between the chip and the outside world. Degradation of both contacts and interconnections may alter the fidelity of the intended signal by changing its amplitude, time constant, sign, and shape; frequently, open circuiting is the final manifestation of conductor degradation.

Properties required of conductors include high electrical conductivity (to minimize both Joule heating and RC time constants), chemical stability (absence of corrosion), ability to be easily processed into required shapes with desired dimensions, compatibility with other device processing and fabrication steps, and lastly, resistance to loss of dimensional and physical integrity due to stress or the passage of large current densities (electromigration). Because the conductivity requirement is often the critical one, conductors are usually limited to metals. These materials have face-centered and body-centered cubic crystal structures for the most part. Unlike semiconductors, in all practical applications metals are polycrystalline. Relevant properties of metals are given in Table 2.1.

Of the 80 or so metallic elements only a few (e.g., Al, Cu, Au, Ag) have seen extensive use in commercial electronic devices and systems. On IC chips, interconnections are basically limited to Al, Cu and Al alloy (Al–Cu, Al–Cu–Si) metallizations at any given level, and to W/TiN via combinations to connect levels. But, when contacts to semiconductors and external connectors are considered, a much richer assortment of metals, alloys (Ti–W, PdSi, Sn–Pb solder), and intermetallic compounds ($TiSi_2$) is utilized. Metal contacts and interconnections at several levels within ICs are schematically indicated in Figure 2.7 together with typical sites and types of degradation.

Metals are good conductors of electricity because they contain huge numbers of free or conduction electrons; i.e., Cu has 1 and Al has 3 per atom, so we are dealing with $\sim 10^{22}$ per $cm^3$. Their band structures are complex, but importantly, metals possess either overlapping valence and conduction bands or an unfilled conduction band. Further, the absence of energy gaps means that electrons never face any appreciable barrier in accessing the numerous, closely spaced, empty higher energy states under the stimulus of an applied electric field or incident photon. Two simple relationships govern the resistivity, $\rho$, of metals:

$$\rho_{\text{Total}} = \rho_T + \rho_i + \rho_d \tag{2.2}$$

and

$$\rho = \rho_0(1 + \alpha \Delta T). \tag{2.3}$$

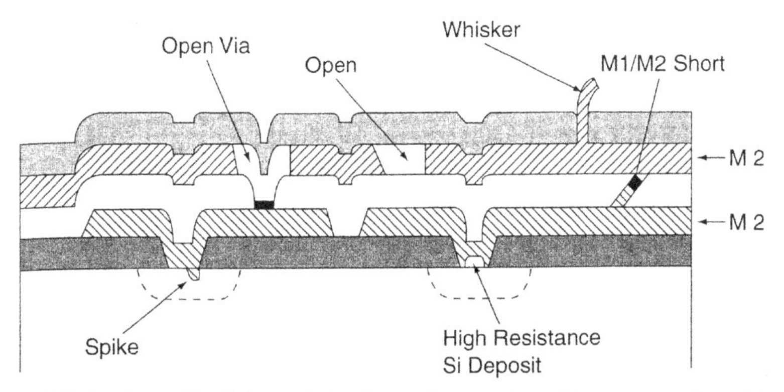


**Figure 2.7** A schematic of the variety of metal contacts and interconnections (shown in cross-hatch) in integrated circuits together with typical degradation phenomena and sites of occurrence.

The first is a statement of Matthiessen's rule (Eqn (2.2)), which reflects the fact that the total resistivity of a metal has three contributions:

1. a thermal one, $\rho_T$, which increases linearly with temperature in the vicinity of 25 °C. Since $\rho_T$ generally dominates the other two contributions in this temperature range, Eqn (2.3) describes the observed resistivity ($\rho$) of metals for all practical purposes. In this formula $\rho_0$ is the resistivity at 25 °C, $\alpha$ is the temperature coefficient of resistivity, and $\Delta T$ is the difference in operating temperature relative to 25 °C.
2. $\rho_i$, the resistivity due to atomic impurities (dissolved or in compound form).
3. $\rho_d$, the resistivity due to defects.

Alloying always increases the resistivity of the base metal. But compared to orders of magnitude change in semiconductor resistivity with minute doping levels, metals suffer relatively little conductivity loss with much larger alloying additions. Lattice defects such as vacancies, dislocations, and grain boundaries raise the resistivity of metals, but only slightly. When impurity or defect concentrations are small, their effect on the total resistivity can be measured only at cryogenic temperatures, where the thermal contribution ($\rho_T$) is very small by comparison, i.e., frozen out. During arcing or Joule heating the conductivity of metals may degrade due to oxidation or interdiffusion reactions that cause extensive compositional change. Interconnections suffering electromigration damage invariably show an increase in resistance.

## 2.2.4 Insulators

We have already noted that in the scheme of conductors, metals and insulators are at opposite extremes, with semiconductors occupying a broad middle ground. When the latter are very pure or lightly doped, they have high resistivities, and their dielectric character is emphasized. Interestingly, similarities between semiconductors and insulators exist. Energy gaps exist in both, and the response of conductivity to doping and temperature is similar. Insulators are used to isolate conducting regions at different electric potentials so that no current flows between them. They also serve vital dielectric functions that enable circuits to store charge in capacitors that behave as memory elements. High dielectric constant ferroelectric ceramics such as $BaTiO_3$ are important in this regard for potential inclusion in memory chips. Dielectric property change upon cyclic electrical stressing is a concern, however.

The insulating materials we are primarily concerned with are oxides (e.g., $SiO_2$, $Al_2O_3$) and polymers. Invariably these insulators have amorphous structures. Interestingly, solid insulators are not the only ones of interest in electrical equipment. Transformer oil is an example, and its breakdown is an important consideration in the reliability of very high voltage equipment. On the IC chip level, thin films and coatings of silicon dioxide, silicon nitride, and silica-based glasses are used in assorted insulating, dielectric, processing, and device passivation roles. These film materials are deposited at relatively low temperatures by both physical and chemical vapor deposition techniques. The most important insulator in this book is $SiO_2$ because it plays critical roles in two important diverse electronic technologies, for quite different reasons. Not only does $SiO_2$ serve as the thin film gate insulator in field effect transistors, but in fiber form it is the medium through which light propagates in optical communications systems. In these applications it has been the focus of dielectric as well as mechanical breakdown investigations that will be respectively discussed in Chapters 6 and 10.

Since insulators have very few electron and hole carriers, conduction by these carriers is limited, a fact due to the large value of $E_g$. Thin oxide and glass films are actually quite a bit more conductive than bulk insulators. For one thing, due to a lack of crystal perfection, there are fuzzy electron-distribution tails at both the top of the valence band and the bottom of the conduction band. These extend into the normally state-free gap and overlap to create a broad continuum of electron states. Secondly, the metal electrode contacts, due to the metal workfunction[1], influence the nature of electron injection into the insulator. Lastly, because dimensions are small, modest applied voltages all too often lead to large electric fields that, in turn, may foster nonlinear conduction behavior, charge trapping, and sometimes breakdown. Later, in Chapter 6, other conduction mechanisms will be introduced. Questions of how electrons interact with matrix atoms and other electrons in the insulator to generate damage will be addressed then.

## 2.3 DIODES

### 2.3.1 The p–n Junction

The fundamental building block of most semiconductor devices is the p–n junction, i.e., p and n doped regions joined at a junction. Diodes contain one such junction, transistors two, and there are devices that have three or

[1]Energy necessary to free an electron from the Fermi Level in the metal to vacuum.

more junctions. Understanding the operations of rectification, amplification, and switching implies some knowledge of junction behavior. Immediately after p and n doped regions come into contact or are shorted together, holes and electrons rapidly traverse the junction and, as shown in Figure 2.8(a), three things happen.

1. Carrier equilibrium results in equalization of the respective Fermi levels.
2. A built-in electric contact potential ($V_0$) or field (E) is established that discourages any further charge transport.
3. In effect, a space charge or depletion region (a kind of dielectric, carrier-free zone) develops.

The extent of the space charge region varies inversely with doping level. When the latter is high, the material is highly conductive and cannot

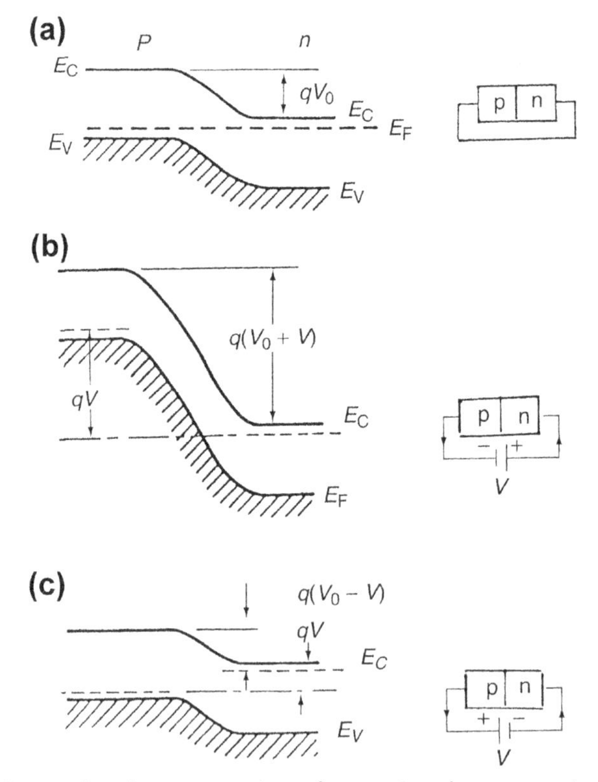


**Figure 2.8** Energy band representation of a semiconductor p–n junction under different conditions of bias. (a) Zero bias voltage—The Fermi levels are equilibrated on both sides of the junction, creating an internal electric field. (b) Reverse bias—The Fermi levels are displaced as carriers are depleted from the junction region. A small reverse current flows. (c) Forward bias—Large forward current flows as carriers recombine at the junction.

support high electric fields. Just the opposite is true at low doping levels. Thus, devices that function at low voltages are generally more highly doped than those exposed to high voltage levels.

Importantly for device behavior, the internal barrier to charge flow can be manipulated by applying an external bias voltage. When the diode is reverse biased by connecting n to the positive supply voltage ($V$) and p to the negative supply voltage (Figure 2.8(b)), all of the p-type Si electron levels, including the Fermi level, rigidly rise relative to the n-type Si levels by an amount $qV$. Electrons and holes are drawn away from the junction in the respective n and p regions. We speak of a *depletion* region free of carriers. This has the effect of widening the space charge region as well as raising the magnitude of the potential barrier and electric field across the junction. As a result, only a very small reverse current ($i_R$) flows. Alternatively, when the diode is forward biased by connecting n to the negative supply voltage and p to the positive supply voltage (Figure 2.8(c)), electron levels in the two semiconductors now rigidly shift in the other direction relative to each other. Electrons and holes now readily flow across the junction where the carriers recombine. This has the effect of reducing the space charge region as well as lowering the magnitude of the potential barrier and electric field across the junction. Importantly, electron–hole recombination means a large forward current ($i$). The p–n junction thus enables currents of vastly differing magnitude to flow by simply changing the bias voltage polarity. This is the behavior desired of rectifiers. In contrast, ohmic characteristics are linear and display the same current magnitudes, but of opposite sign, when the voltage polarity reverses.

The ideal junction characteristics, depicted in Figure 2.9, are succinctly governed by an equation that describes the current–voltage variation as.

$$i = i_R \left[ \exp\left( \frac{qV}{kT} \right) - 1 \right], \tag{2.4}$$

where $q$ is the electronic charge, and $kT$ has the usual meaning. Although DC operation has been described, junctions in devices generally function under ac transient conditions. This causes voltages, currents, and internal fields to be time dependent.

## 2.3.2 Contacts

When metal contacts are applied to the p and n semiconductor regions, low-resistance ohmic behavior is normally desired. The band diagram for

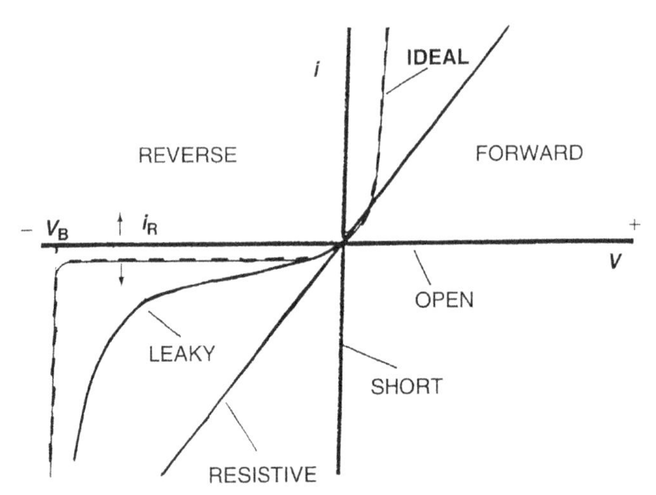


**Figure 2.9** Ideal and nonideal $i$–$V$ junction characteristics.

an idealized ohmic contact is shown in Figure 2.10(a). In practice, however, ohmic contacts are prepared by heavily doping the underlying semiconductor in order to minimize its resistivity. This considerably narrows the region of band bending, promoting bidirectional tunneling of charge across the metal–semiconductor interface (Figure 2.10(b)), and resultant linear $i$–$V$ behavior.

In addition, metal–semiconductor contacts can also display nonlinear or rectifying characteristics, resulting in unexpected changes in circuit behavior. Schottky-barrier or metal–semiconductor diodes are based on the rectifying properties of certain metal contacts to semiconductors. If a band diagram is drawn for a metal with a work function larger than that of the n-type semiconductor it contacts; the resultant structure of Figure 2.11 behaves very much like the p–n junction diode of Figure 2.8 upon biasing. There is, of course, no band bending in the metal. A barrier potential of height $\Phi_B$ now governs the asymmetric transport of charge across the junction. Despite the fact that the $i$–$V$ characteristics appear identical, Schottky-barrier and p–n diodes differ in one significant way. Under forward bias, majority electrons injected over the barrier are hotter than normal electrons in the metal. Their excess energy is very rapidly quenched, and such thermalized electrons cannot flow back into the semiconductor upon reverse bias because they face a large energy barrier in that direction. The short recovery time of $10^{-11}$ s makes Schottky diodes capable of very fast switching in microwave circuits.

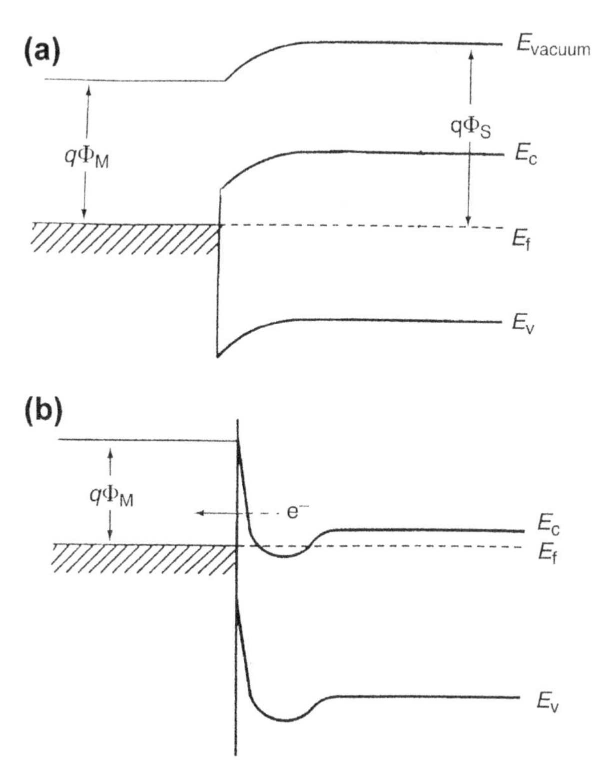


**Figure 2.10** (a) Band diagram representation for an idealized ohmic contact between a metal and semiconductor. (b) Band diagram for an ohmic contact illustrating tunneling. $\Phi_M$ and $\Phi_S$ are respective work functions.

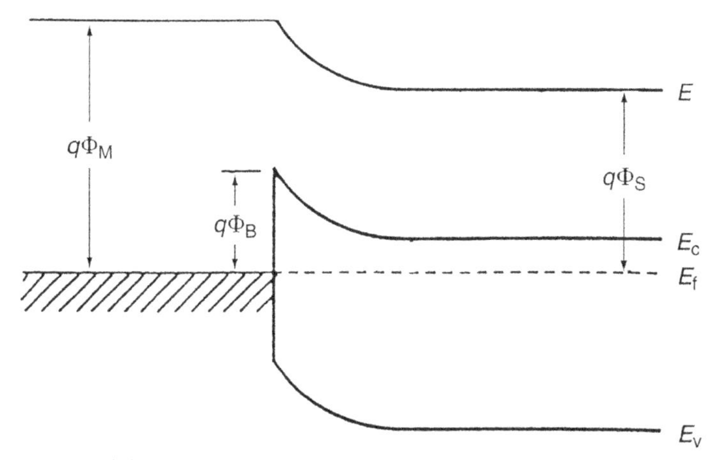


**Figure 2.11** Band diagram for idealized Schottky-barrier contact between a metal and semiconductor.

Applications normally require either ohmic or Schottky behavior; interestingly, certain compound-semiconductor field-effect transistors contain both ohmic and Schottky contacts (Section 2.7.4). Irrespective of whether they are ohmic or rectifying, solid-state reactions between Al contacts and silicon as well as assorted metals (e.g., Au, Ti) and GaAs have been troublesome sources of degradation and failure in devices. These reliability problems are exacerbated when high current densities flow through contacts.

## 2.3.3 Deviations from Ideal Junction Behavior

There are a number ways that actual junction behavior differs from the ideal.

### *2.3.3.1 Characteristics*

Rather than by Eqn (2.4), diode $i$–$V$ characteristics are more nearly described by $i = i_R\left[exp\left(\frac{qV}{nkT}\right) - 1\right]$, where $n$, the ideality factor, varies between 1 and 2. In addition, the contact potential due to work function differences between the involved semiconductors makes the $i$–$V$ forward characteristics appear to be more squarish. Thus, very little forward current flows until a critical voltage somewhat below $E_g$ is applied, after which the current rises very steeply.

Several effects serve to weaken rectifying behavior, resulting in the deviations from good $i$–$V$ characteristics displayed in Figure 2.9. Ohmic losses in diodes are manifested by passage of less current at high forward bias levels. Carrier generation outside of the junction region means higher reverse currents. These effects cause the diode to be more resistive and display linear $i$–$V$ behavior as a limiting case. In poorly made diodes the reverse current may sometimes have a surface component that leaks around the junction. Charged contaminants and semiconductor surface states foster electrical leakage paths. The proportions of bulk and surface currents depend on the electrical character of the junction at the point where the bulk meets the surface. Passivating Si diodes with $SiO_2$ layers minimizes such leakage. In good diodes the reverse current remains very small and then sharply increases at the point of breakdown, a subject discussed in the next section. Such characteristics are "hard," distinguishing them from the "softer" $i$–$V$ behavior exhibited by junctions possessing weak spots where localized breakdowns cause excessive leakage current.

Finally, there are the characteristics of the open diode that passes no current irrespective of voltage, and the shorted diode through which large currents flow because the device has a low resistive value and is not rectifying.

#### 2.3.3.2 Breakdown

Diode breakdown is much less alarming than it sounds and does not necessarily mean damage. When the applied *reverse* voltage is increased beyond certain limits (Figure 2.9) diodes undergo breakdown. Depending on doping level, electric fields in the junction ranging from 200 to 1000 kV/cm can develop in Si. Under such conditions minority carriers that produce the usual reverse or leakage current are accelerated to kinetic energies high enough for them to ionize matrix atoms upon collisional impact. In the case of a single incident electron, two electrons (the original one and the impacted valence electron) will now reside in the conduction band after collision. Each can impact-ionize lattice atoms repeatedly, and with sufficient electron multiplication, *avalanche breakdown* may occur. The reverse current rises to a value limited only by the external resistance of the circuit. In this state it may operate satisfactorily as a reverse-biased diode. However, if the reverse current is excessive, the junction can overheat, causing the diode to burn out. Zener diodes capitalize on operation in the breakdown state and are used as voltage references in rectifier and control circuits.

#### 2.3.3.3 Punch-throughs

Although the term sounds as ominous as breakdown, punch-through is not necessarily threatening to continued diode function. Punch-through occurs when lightly doped regions are also physically thin. Upon reverse biasing, the depletion region widens and may extend directly to the ohmic contact, thus shorting out the diode or transistor. Provided that the current flow is limited by the circuit resistance and no overheating occurs, the diode will recover its former properties when the reverse voltage is reduced. As we shall see later, punch-through also occurs in transistors.

## 2.4 BIPOLAR TRANSISTORS

### 2.4.1 Transistors in General

Transistors are the most important solid-state devices. Because they are also the most numerous, device reliability and failure issues tend to be overwhelmingly focused on them. Since transistors contain two p–n junctions, have three contacts or terminals, are complex geometrically, and frequently require the action of perpendicular electrical fields, their behavior is more difficult to understand than that of diodes. To complicate matters further there are two fundamentally different types; in the bipolar junction transistor (BJT) current is amplified, while in the field effect transistor (FET) a voltage controls a current. The former relies on the participation of both types of carriers, hence the descriptor "bipolar." In contrast, only one carrier plays a prominent role in FETs. Transistors perform the basic electrical functions of amplification and switching in both discrete and IC configurations involving low as well as large power applications.

With regard to function it is helpful to distinguish between *linear* and on–off *digital* logic circuits. Linear circuits typically operate with low input voltage levels of perhaps ~1 μV or so (e.g., audio equipment) and outputs having small deviations about a quiescent operating DC level. Digital circuits are constrained to operate at fixed voltage levels (e.g., 0 or 5 V) and are impervious to noise levels that would present difficulty in linear circuitry. The latter are much less tolerant of variations in electrical component characteristics than is the case for digital circuits. With change in operating temperature, linear circuits may drift in and out of specifications, whereas digital circuits will either function properly or abruptly fail. Both BJTs and FETs are used in linear and digital circuits; computer logic and memory circuits are primarily composed of FETs, however, and these therefore dominate the transistor market.

## 2.4.2 Bipolar Junction Transistors

### *2.4.2.1 Operation*

It is common to start discussions of BJTs with the p–n–p type shown in Figure 2.12 because the direction of majority carrier (hole) motion coincides with that of the current flow. There is also an n–p–n version of this device that is easier to fabricate. Because the electron mobility in Si ($\mu_e = 1450$ cm$^2$/V-s) is some three times higher than for holes, these faster n–p–n devices are preferred in most applications. By reversing the doping type, bias voltages, and current direction, the treatment applies to n–p–n devices as well. These three-terminal devices contain two junctions separating three semiconducting regions, i.e., the emitter, base, and collector. In the case under discussion these are doped p, n, and p, respectively, and are wired together in a common base configuration. When unbiased, the Fermi

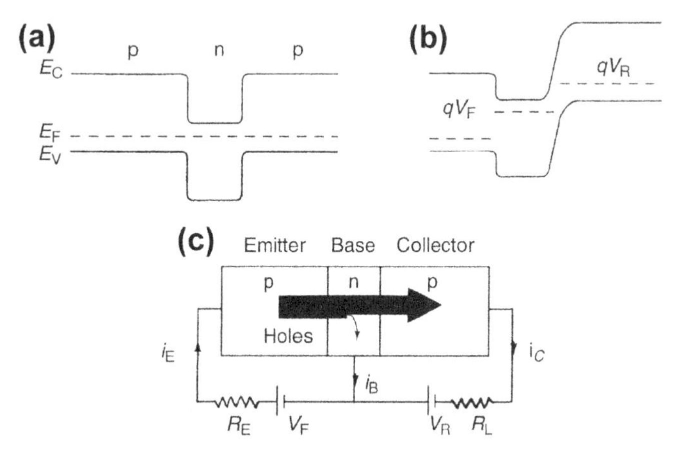


**Figure 2.12** (a) Band diagram representation of a p–n–p bipolar junction transistor in the common base configuration. Under zero bias, the Fermi levels line up across both junctions. (b) Band diagram representation of transistor under forward bias. (c) Transistor with bias voltages $V_F$ and $V_R$ applied.

levels line up across the band diagram representation of the device (Figure 2.12(a)).

During transistor operation the p–n junction on the left is forward biased, while the one on the right is reverse biased. The bands bend and move up or down as shown (Figure 2.12(b)). First the heavily doped emitter injects majority (hole) carriers into the lightly doped base region. There the holes become minority carriers and would normally recombine with majority electrons. However, because the base is lightly doped and made very thin (e.g., 1 μm thick), holes diffuse right through it before they have a chance to recombine. Those holes that traverse the base–collector interface are now efficiently swept to the collector contact; equivalently, holes float uphill on the band diagram, which is their way of reducing energy.

In well-designed transistors the collector current, $i_C$, is only slightly smaller than the emitter current, $i_E$; for example, $i_B$ is typically 1% of $i_E$ or $i_C$. Therefore, the common base current gain, defined as $\alpha = i_C/i_E$, is slightly less than unity. It is not difficult to calculate the AC power gain in the circuit shown in Figure 2.12(c). Due to the exponential character of Eqn (2.4), a small change in emitter–base voltage ($\Delta V_{EB}$) will cause a large change ($\Delta i_E$) in emitter current. The power supplied to the forward-biased emitter junction, whose resistance is $R_E$, is $R_E(\Delta i_E)^2$. Likewise, the power delivered to the load resistor $R_L$ is $R_L(\Delta i_C)^2$, where $\Delta i_C$ is the change in collector current. Because $\Delta i_E$ and $\Delta i_C$ are approximately equal, the power gain ($G$) is simply.

$$G = \frac{R_L(\Delta i_C)^2}{R_E(\Delta i_E)^2} \approx \frac{R_L}{R_E}. \tag{2.5}$$

Typically, for $R_E = 25\ \Omega$ and $R_L = 10\ \mathrm{k}\Omega$, a value of $G$ equal to 400 can be anticipated.

Transistor action essentially occurs because the emitter–base voltage in one part of the device controls the collector current elsewhere in the device. In this way small changes in the current at one terminal enable the current through the other two terminals to be controlled. Similar transistor action occurs in other wiring configurations, e.g., common emitter. Specific doping levels and base dimensions are required for optimal transistor performance.

#### 2.4.2.2 Planar Bipolar Transistor

Discrete, or single bipolar, transistors of the type just discussed were the first to reach the market. Later the concept of integrating many bipolar transistors on a single wafer meant a new processing philosophy. Rather than stacking bulk semiconductor regions "side to side" as in Figure 2.12, a "top to bottom" thin film technology was adopted. By sequentially doping through the wafer thickness, the n–p–n (or p–n–p) junction depths, and the base thickness in particular, can be very well controlled. The problem is how to make contact to the emitter, base, and collector regions. Obviously, contacts can only be made on the top surface of the wafer. Novel methods have been developed to direct the current flow, as shown in Figure 2.13. Electrons injected from the emitter cross the narrow base region vertically down into a highly conducting, heavily doped buried layer that serves as a horizontal extension of the collector. They flow laterally through the buried layer and then veer up to the collector contact at the top surface. Provision must also be made to isolate (insulate) the transistor from other nearby circuit elements and devices. The operating principle of the planar bipolar transistor is the same as that for the discrete bipolar transistor. High operational speeds and current capacities have earmarked these devices for digital logic as well as linear input/output circuits. State of the art bipolar transistors are using germanium in the base region of a device called, SiGe-based HBT (heterojunction bipolar transistor) [12].

The most complex digital system can be constructed from circuits that perform relatively few operations such as AND, OR, and INVERT.

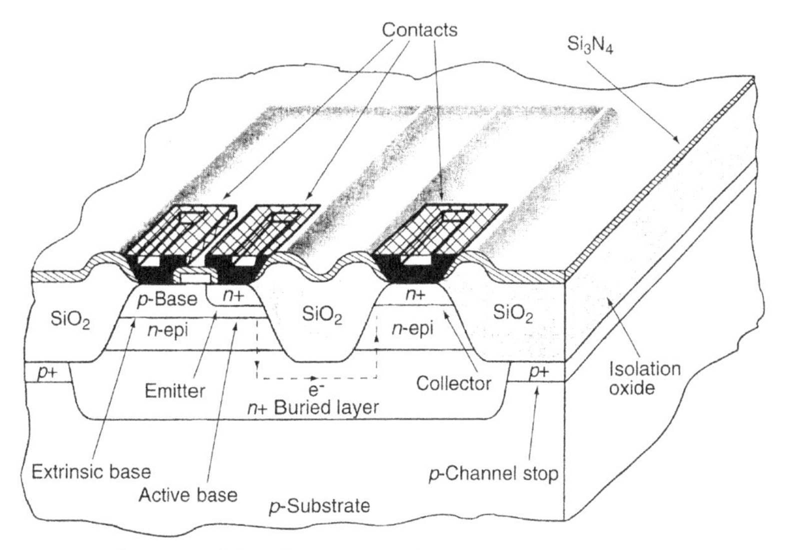


**Figure 2.13** Schematic of the planar n–p–n bipolar transistor structure utilizing oxide isolation. The current path from emitter to base to collector is indicated. *From Ref. [7].*

Individual circuits that perform these functions are known as logic gates. Over the years different logic gate circuits containing collections of bipolar transistors have been designed and incorporated into a wide variety of digital electronics. We just have time to mention some of the popular ones, namely emitter-coupled logic (ECL), transistor–transistor logic ($T^2L$), and integrated-injection logic ($I^2L$). The reader is referred to electrical engineering texts for descriptions of their operation and relative merits.

Logic functions can also be carried out more efficiently by field effect transistors, and it is to these devices that we now turn our attention.

## 2.5 FIELD EFFECT TRANSISTORS

### 2.5.1 Introduction

Like BJTs, field effect transistors (FETs) are three terminal devices; unlike them, a gate voltage rather than a current controls the current at the other two terminals. Field effect transistors are easily the most widely used of all solid-state devices. They currently dominate computer logic as well as memory integrated circuit chips and are projected to widen this dominance. At the outset, two distinct types of field effect transistor must be distinguished. The first and most important is the MOS type, where the metal (M) gate electrode is separated from the semiconductor (S) by a thin insulator film (i.e., $SiO_2$). Voltage applied to the gate controls the current flow between underlying semiconductor regions (source and drain) by either creating or eliminating a conducting channel. Junction field effect transistors (JFET) are the second type, and they function by applying a gate voltage to vary the depletion width of a reverse-biased p–n junction.

### 2.5.2 The MOS Capacitor

Knowing how the MOS capacitor structure responds to applied voltage is prerequisite to understanding MOS field effect transistor (MOSFET) behavior. This capacitor structure, which is the essential building block of silicon MOSFETs as well as charge-coupled devices, actually consists of two capacitors in series. The first has a fixed value of capacitance determined by the thin $SiO_2$ film sandwiched between planar metal (or other conductor, e.g., polysilicon or silicide) and semiconductor surface. Underlying this is the capacitance associated with Si, which varies depending on the extent of charge depletion in the semiconductor. Now consider the "ideal" MOS structure for p-type Si, shown in Figure 2.14(a). With no gate voltage or bias applied, Fermi level equilibration leaves all bands horizontal, or flat; this condition arises when the work functions of the metal and semiconductor are the same, and there is no oxide charge (which we assumed for this illustration). When a negative gate voltage ($-V_G$) is applied to the metal, the energy of all of its conduction electrons is raised uniformly by an amount $qV$ relative to Si. Holes are attracted to the $Si/SiO_2$ interface, making Si there appear to be even more p-type than originally. This causes band bending at the interface (Figure 2.14(b)), which graphically describes a state known as *accumulation*. The high interfacial hole density effectively serves to make the Si a positively conducting capacitor plate. For high $V_G$ values, the MOS capacitance ($C_o$), which depends inversely on $SiO_2$ thickness, is large and solely due to the oxide.

As the polarity of $V_G$ is reversed, the overall capacitance decreases. The reason is that holes are repelled from the $Si/SiO_2$ interface. In this *depletion* regime Si acts like a dielectric with a capacitance governed by the width of the depletion region (Figure 2.14(c)). The two series capacitors have a lower overall capacitance than that of either the semiconductor or oxide alone. Now the bands bend the other way; the slope of the oxide conduction band edge or electric field reverses sign, and near the interface the semiconductor acquires an n-like character. At still higher positive voltages, electrons are drawn to the interfacial region, where the semiconductor is locally inverted to n-type. Extreme band bending is characteristic of large

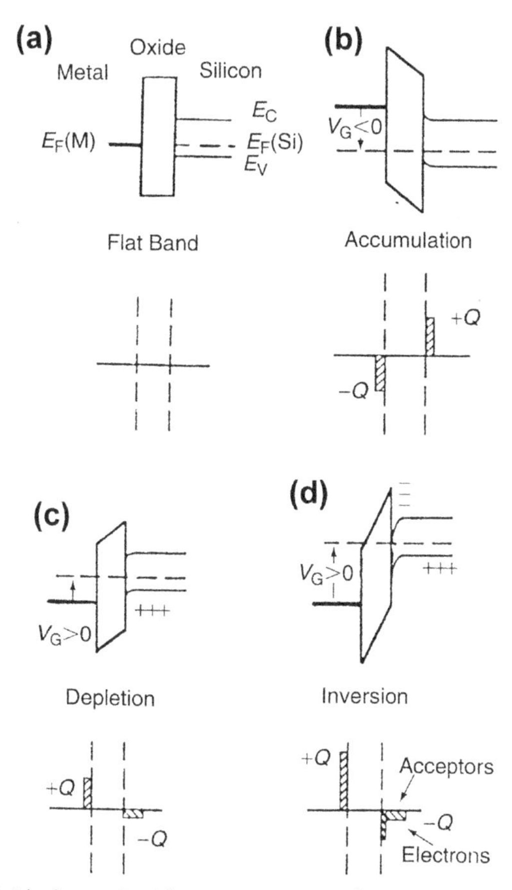


**Figure 2.14** (a) Ideal metal-oxide-*p*-type semiconductor (MOS) capacitor structure. Without applied gate voltage there is no oxide field, and a flat-band condition prevails. (b) Negative gate-voltage—accumulation of holes results in band bending at the interface. (c) Positive gate-voltage—expulsion of holes causes Si to act like a dielectric with a capacitance governed by the width of the depletion region. (d) Large positive gate voltage—inversion regime is created and Si interface is effectively n-type.

values of $V_G$. An *inversion* regime (Figure 2.14(d)) is created where the overall capacitance reaches a minimum. Effective charge distributions ($Q$) on the metal and in the semiconductor are depicted below each of the MOS bias states.

Ideal capacitance–voltage (C–V) characteristics derived from biasing the MOS structure are displayed in Figure 2.15; an entirely complementary behavior is exhibited by n-type Si but with opposite gate voltage polarities. The MOS capacitor not only plays an active role in device function, but has

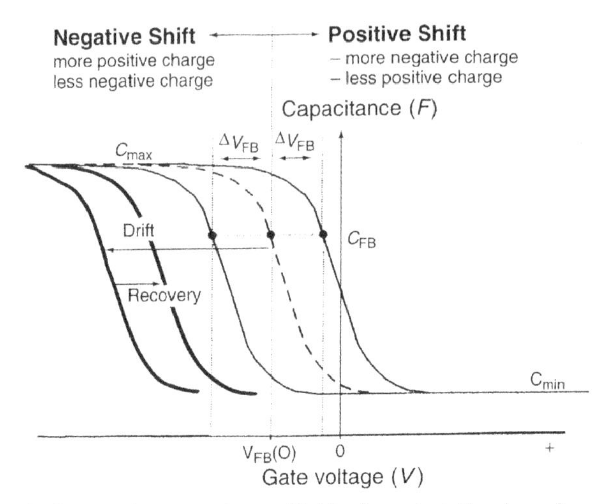


**Figure 2.15** Ideal capacitance–voltage (C–V) characteristics based on the biasing conditions in Figure 2.18. Flat-band voltage ($V_{FB}$) shifts from uncharged capacitor (dotted) due to oxide charge are shown. More positive charge (less negative charge) produces negative voltage shifts, and more negative charge (less positive charge) produces positive voltage shifts.

proven valuable in assessing bulk ionic contamination (e.g., $Na^+$) and interfacial electronic charge ($Q_i$) introduced during gate-oxide processing. When such charge exists it can produce large voltage shifts (i.e., $V = Q_i/C_o$) in the C–V characteristics (Figure 2.15). Charging of gate oxides during plasma processing or exposure to ionizing radiation produces similar voltage shifts. Charge drift and recovery in electric fields cause operational instabilities in devices.

## 2.5.3 MOS Field Effect Transistor

The n-type MOS (NMOS) field effect transistor is shown in Figure 2.16(a), where it is evident that n-doped source and drain regions flank a central MOS capacitor section composed of p-type silicon. Beneath the oxide there are essentially two p–n junctions connected back to back. When no gate voltage is applied, only a small reverse leakage current flows between the source and drain. Recalling the previous section, we note that application of a sufficiently large $+V_G$ causes inversion of the MOS structure. So-called "strong inversion" essentially occurs at a threshold gate-voltage level $V_T$ given by:

$$V_T = \Delta\Phi_{MS} - Q_i/C_o + Q_d/C_o + 2\phi_F. \qquad (2.6)$$

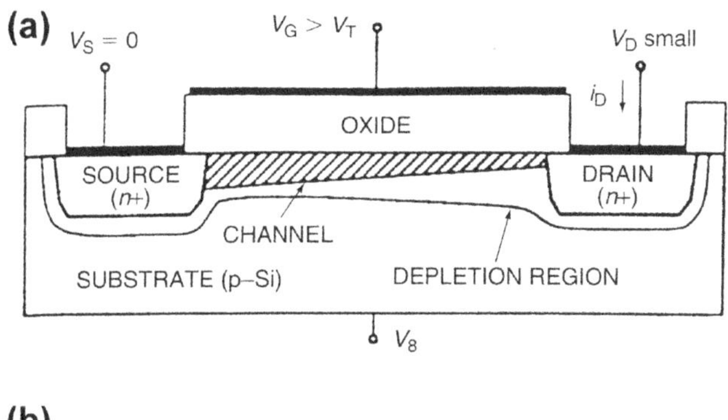


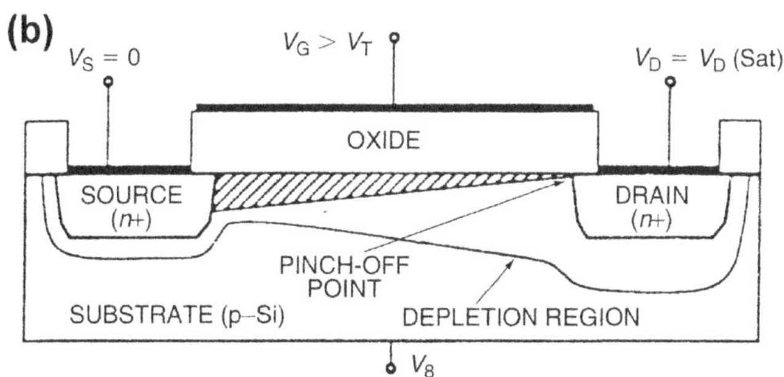


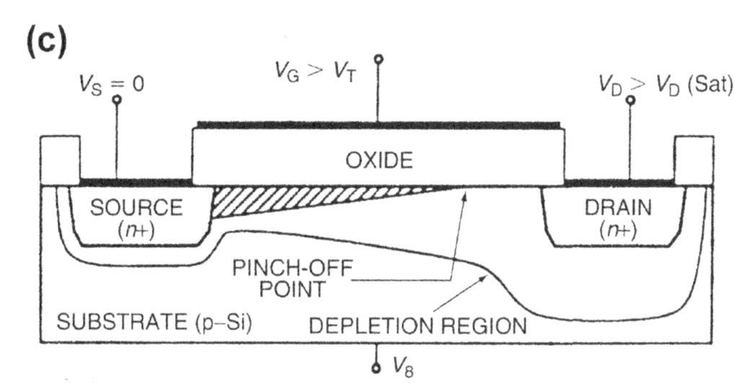


**Figure 2.16** Behavior of n-MOS field effect transistor. (a) Cross-sectional view showing operation in the linear region. (b) Operation at the edge of saturation (pinch-off). (c) Operation beyond saturation.

At this voltage a thin surface inversion layer or conducting n-channel forms joining the two heavily doped $n^+$ regions, and the transistor is turned on. This equation basically expresses that bands must first be flat, and the effective voltage contributions to $V_T$ by the first two terms ensure this. The work function difference between the gate electrode and semiconductor is $\Delta\Phi_{MS}$, while the second term was introduced above, to reflect bulk and interface charge $Q_i$. Next, the p-type Si must be depleted of hole charge,

$Q_d$ (term 3), and finally, strong inversion (term 4) must occur. The quantity $\phi_F$ is essentially the difference in magnitude between the midgap (intrinsic) level and the actual Fermi energy, $E_F$. Both $Q_d$ and $2\phi_F$ as well as $E_F$ depend on doping level.

As written, this formula refers to n-channel devices, but it also applies to p-channel devices as well if the signs of the last two terms on the right are changed from + to −. It should be noted that large contributions to $V_T$ are made by both $Q_i$ and oxide capacitance, $C_o$. Advantage is taken of this by controllably adjusting $V_T$ through ion implantation in order to effectively change $Q_i$, and by growing thick oxides to electrically isolate devices. By the same token, even slight time-dependent electron charging of the oxide in service increases $Q_i$ and hence alters $V_T$. In fact, the same factors that cause shifts in C–V characteristics induce threshold voltage shifts. These cause reliability problems because MOSFET action may be unintentionally triggered.

Once the underlying Si is inverted, the electron current flow from the source to drain through the surface n-channel can be controlled by varying the drain voltage ($V_D$). The remaining portions of Figure 2.16 reveal what happens with successively higher drain voltages. If $V_D$ is small, electron flow yields a drain current, $i_D = V_D/R$, where $R$ is the channel resistance. This linear region is indicated on the $i_D$–$V_D$ characteristics of Figure 2.17(a). As $V_D$ increases, it eventually reaches a point where the inversion layer width at the drain shrinks to zero, the pinch-off state. Beyond pinch-off, $i_D$ essentially saturates, but the channel width is effectively reduced. Electron injection into the drain depletion region is now similar to that of carrier injection from an emitter–base junction into the base–collector depletion region of a bipolar transistor. In addition to these n- and p-channel "enhancement mode" MOSFETS, there are "depletion mode" MOSFETS. These are fabricated containing the channel in the conducting state, and are thus switched on even when $V_G = 0$; during operation the conducting channel must be turned off by applying the proper polarity voltage.

In both n- and p-type MOSFETs degraded by hot-carrier induced oxide damage, the $i_D$–$V_D$ characteristics deviate from the prestressed state as shown in Figure 2.17(b) and c. In essence, drain currents decline in n-MOSFETs and rise in p-MOSFETs. Hot-carrier effects are addressed again in Section 6.4.

### 2.5.4 CMOS Devices

The combination of MOSFETs, one n-type and the other p-type, is known as a complementary MOS or CMOS device (Figure 2.18). By

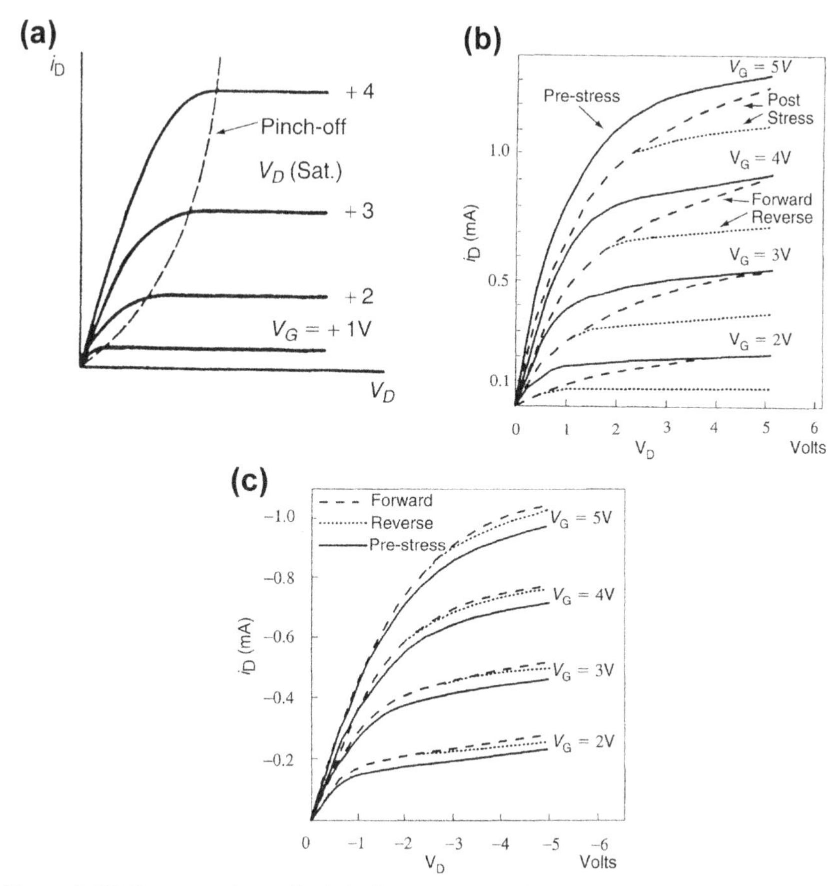


**Figure 2.17** Current-voltage ($i_D$–$V_D$) characteristics of the MOSFET. (a) Ideal behavior. (b) Forward and reverse n-MOSFET $i_D$–$V_D$ characteristics after hot-carrier oxide damage. (c) Forward and reverse p-MOSFET $i_D$–$V_D$ characteristics after hot-carrier oxide damage. *From Ref. [13].*

connecting gates in common as the input and drains in common as the output as shown, this device operates as an inverter in digital circuits. We first recall that $V_T$ is positive in the n-channel transistor and negative in the p-channel transistor. When the input voltage ($V_{in}$) is "high," the n-channel device is in the "on" state and the *p*-channel device is in the "off" state. Therefore, the channel resistance of the p-device is much higher than that of the n-device. This causes voltage $V$ to drop mostly across the p-channel, and the output voltage ($V_{out}$) is "low." If the input voltage is now "low," the n-device is off and most of $V_{out}$ stems from $V$ which represents the "high" state. Therefore, in either state a transistor is off, and there is no DC

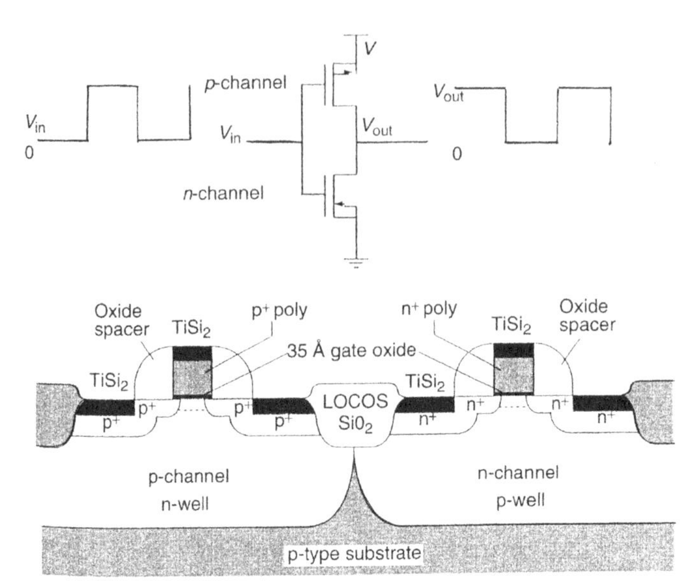


**Figure 2.18** *Top*: Circuit connection of transistors in a complementary MOS (CMOS) device. The indicated inverter function requires common gates as the input and common drains as the output. *Bottom*: CMOS structure.

path to ground other than leakage current. No drain current flows except during the very short switching period. The fact that little power is drawn is a primary reason why CMOS dominates the market for low-power logic and memory ICs where switching is the principal function.

## 2.5.5 MOSFET Instabilities and Malfunction

### *2.5.5.1 Oxide Charge*

The critical gate oxide is quite hospitable to both ionic and electronic charge. In addition to *mobile* alkali ion contaminants, there is both bulk oxide positive and negative *trapped* charge, as well as *interfacial* charge trapped at the Si–$SiO_2$ interface/boundary, and *fixed* charge some 3 nm away. Area densities for the latter two are $10^{10}$ to $10^{11}/cm^2$. Sources of the charge are thought to be structural defects related to bond breaking and metallic impurity contamination in the oxidation process. During device operation, energetic, or hot, electrons in the channel may also get scattered into the oxide. Time-dependent dielectric breakdown and shifts in $V_T$ are two important reliability consequences of such oxide charge buildup. This subject will be addressed more fully in Chapter 6.

### *2.5.5.2 Short Channel Effects*

A concern in VLSI and ULSI circuits is the existence of anomalous MOS transistors with shorter source–drain spacing than minimum design rules specify. The creation of a completely depleted area under the gate will then occur at lower gate voltages, and $V_T$ will be effectively reduced. Electric fields are also higher in the drain region, making such transistors much more susceptible to hot electron effects.

### *2.5.5.3 Parasitic Bipolar Transistors: Latchup*

Despite efforts to confine MOS transistor action to a designated region of the substrate, sometimes unintended parasitic bipolar transistor behavior intrudes into the same or a neighboring location. So long as no p–n junction is forward biased, the bipolar transistor is inoperative. But for high enough substrate–source voltages, a large emitter current is injected, resulting in large device currents.

A more serious parasitic phenomenon known as *latchup* occurs in CMOS devices. Here the closely spaced sources and drains of n and p channel MOSFETs give rise to two parasitic bipolar transistors, one n–p–n and one p–n–p, as shown in Figure 2.19. These transistors are cross connected so that the base–collector junctions are common. Under bias, the p–n–p collector delivers current to the n–p–n base, and the n–p–n collector

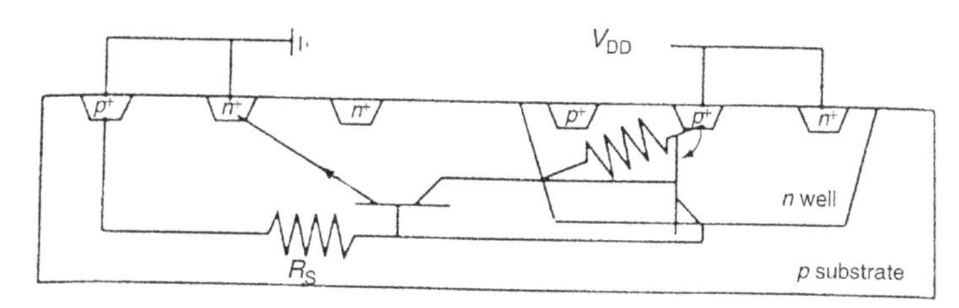


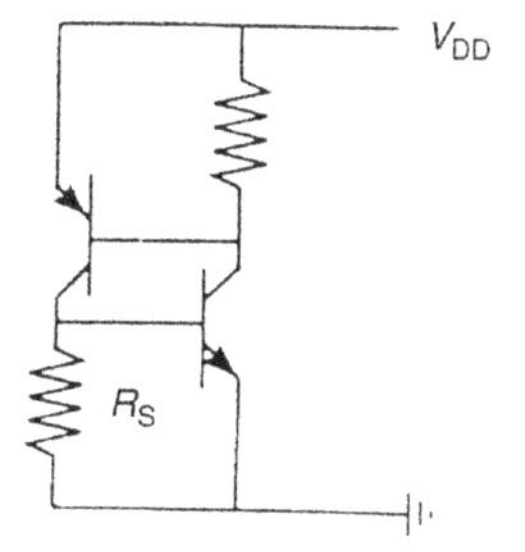


**Figure 2.19** Latchup arising in CMOS devices due the creation of closely spaced and coupled n–p–n and p–n–p parasitic bipolar transistors.

delivers power to the p–n–p base. If the gains of these transistors are high, both are driven into saturation, and the power supply voltages become connected through a low resistance path. In essence, a device known as a thyristor, which contains p–n–p–n doped regions, is effectively created. Carriers generated in unwanted regions by leakage currents, substrate current, or even photon irradiation can trigger latchup. When this happens, short-circuit currents rapidly overheat the IC chip, destroying it. Better design of transistor isolation is effective in combating latchup.

### 2.5.6 Bipolar versus CMOS

Now that the two major transistor technologies have been introduced, it is pedagogically worthwhile to compare the characteristics and attributes of each type of device [14].

1. *Size*. Using the same lithography tools, the footprint of bipolar transistors is typically 20–40% larger than the corresponding MOS transistor. A key advantage of CMOS technology is the smaller device area, which translates to a higher packing density.
2. *Power consumption*. As indicated above, standby power consumption of CMOS transistors is much less than that for bipolar transistors. This advantage, plus high packing densities makes CMOS the obvious choice for ULSI and ELSI applications.
3. *Complexity of fabrication process*. CMOS devices, with typically 20% fewer processing steps required, are easier to fabricate than bipolar transistors. Part of the simplicity and smaller size of MOS devices stems from the less stringent isolation required. To isolate or insulate devices from one another on the same chip, more complex and deeper dielectric structures (trenches) are required for bipolar than for MOS devices.
4. *Current carrying capacity*. An important advantage of the bipolar transistor is its decidedly superior gain because of its high transconductance. Defined as the derivative of the output current to input voltage, the transconductance ($G_m$) of a transistor is a measure of its quality as an amplifying device. In bipolar transistors, $G_m = qi_c/(kT)$, where $q$ is the electronic charge, $i_c$ is the collector current, and $kT$ has the usual meaning. For MOS transistors $G_m = (2\mu m\ i_{DS}\ C_{ox}\ Z/L)^{1/2}$, where $\mu$ is the electron mobility, $i_{DS}$ is the drain–source current, $C_{ox}$ is the gate–oxide capacitance, and $L$ and $Z$ are the channel length and width, respectively. Substitution of typical values reveals that to match the bipolar transconductance, the dimensions of MOS transistors must be a few orders of magnitude larger.

5. *Sensitivity and speed.* The exponential dependence of the current on the base–emitter voltage gives the bipolar transistor superior signal detection capabilities relative to MOS devices. Bipolar devices are also faster than MOS transistors and exhibit higher frequency response and gain. This makes them well suited for high performance analog applications.
6. *Market share.* Currently, NMOS and CMOS technologies account for about 80% and bipolar 15% of the IC market share. Compound semiconductors and BiCMOS make up the remainder.

For flexible circuit design, BICMOS technology combines the high density of CMOS with the speed of bipolar transistors. This technology has been used in the Intel Pentium microprocessor, a chip containing 3.3 million transistors. In general, BICMOS circuits employ CMOS transistors to perform the logic functions and bipolar transistors to drive the output loads.

### 2.5.7 Junction Field Effect Transistor

Like the MOSFET, the JFET, depicted in Figure 2.20(a), also consists of a gate–source–drain combination. Rather than an oxide/semiconductor structure, the active region beneath the gate consists of three vertically stacked semiconductor layers. In operation the upper and lower gates (the top and bottom heavily doped p-layers) are connected together, while the lightly doped n-channel in between has the drain voltage ($V_D$) applied across it. Both p–n junctions are slightly reverse biased initially, so that depletion layers extend into the channel. As $+V_D$ is applied, a channel current ($i_D$) flows whose magnitude is simply given by Ohm's law, or $V_D$ divided by the channel resistance. At larger values of $V_D$, the voltage drop (i.e., bias) that progressively rises from source to drain increases both $i_D$ and the depletion width in concert. Finally, the two depletion regions merge at the drain, and the channel pinches. At that point the current saturates just as it does in MOSFETs and the $i_D$–$V_D$ characteristics resemble those displayed in Figure 2.17(a).

Similar devices known as MESFETs are widely used in high-frequency circuits. Fabricated in III–V compound semiconductors, they are discussed in Section 2.7.3.

## 2.6 MEMORIES

### 2.6.1 Types of Memories

Together with logic chips, memory chips are critical for computer circuit applications. Memories are circuits that enable the storage of digital

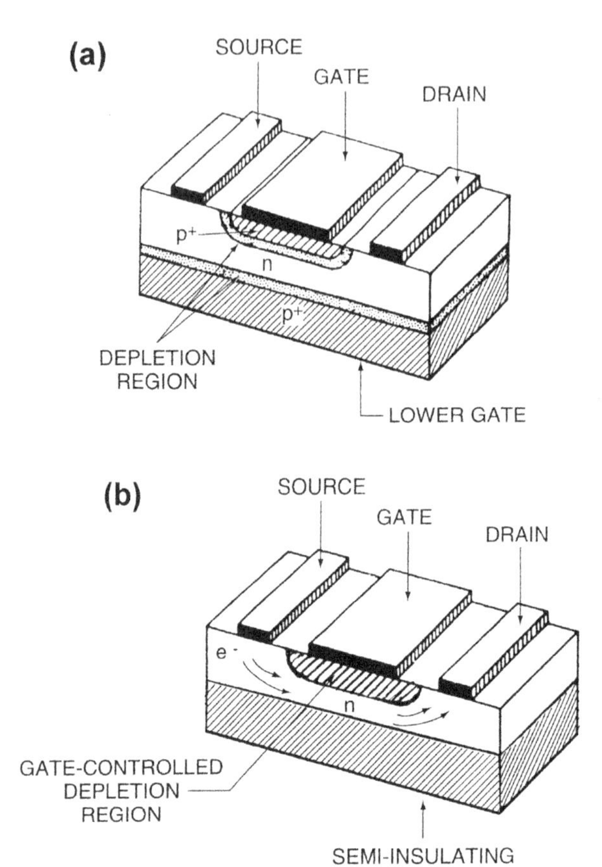


**Figure 2.20** (a) View of a junction field-effect transistor with a detail of the central channel region. (b) Schematic of the MESFET. Current flow through the channel is controlled by the gate voltage, which varies the depletion layer width. *From Ref. [1], Copyright by Bell Telephone Laboratories, Inc.*

information, or data, in terms of bits (binary digits). They now account for close to 40% of the integrated circuit market and can stand alone on a single IC chip or be embedded within a larger IC. A road map through the broad variety of memories used, and the categories they belong to, is shown in Figure 2.21. Memories that lose their data when the supply voltage is removed are said to be *volatile*, while those that retain data under such conditions are *nonvolatile*. Another subdivision reveals that there are three kinds of memory—serial memory, random access memory (RAM), and read only memory (ROM). Volatile memories include serial, static (SRAM), and dynamic (DRAM) types. Three-quarters of the total memory

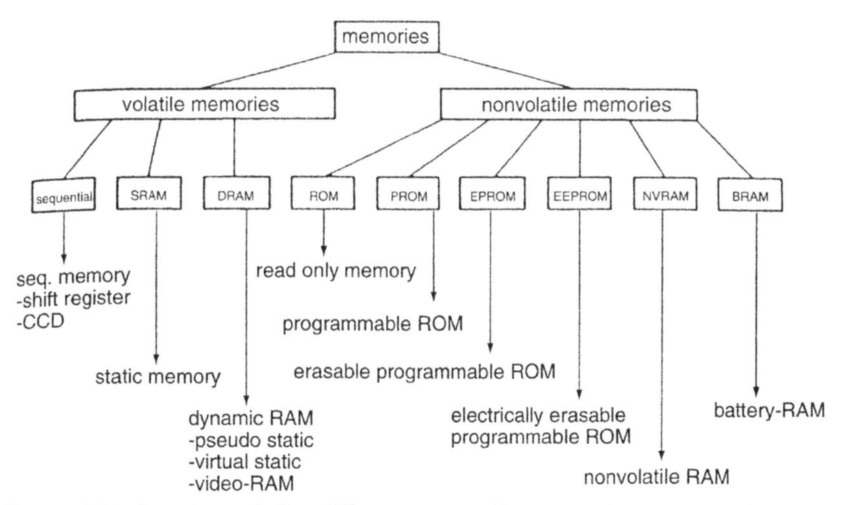


**Figure 2.21** Overview of the different types of semiconductor memories. *From Ref. [15].*

market, including both volatile and nonvolatile, is dominated by DRAMs. In serial memories, data enters and leaves in the same sequence, so the "first in" is the "first out." Charge coupled devices (CCD) described later are used in such memories for video applications.

Random access memories, which come in SRAM and DRAM varieties, are arranged in a matrix array and require both read–write data and address inputs and data outputs. Data can either be written or read directly in a random way, unlike memory locations on magnetic tape, which must be lined up with the read–write head. Storage of data occurs within memory cells. In SRAMs these cells contain inverter circuits, and combinations of them are known as flip-flops, which are capable of assuming two states. While SRAMs are the fastest of all MOS memories, this is accomplished at the expense of requiring four to six transistors per cell. In contrast, DRAM cells contain an integrated transistor–capacitor unit, enabling greater packing densities at lower cost. The operation of a DRAM will be described in the next section. Since charge is stored in a capacitor within DRAM cells, gradual leakage necessitates refreshing approximately every 2–16 ms. In contrast, static RAMs ensure that voltage levels are maintained as long as the power supply is on.

Read only memories are, in fact, random access memories that are permanently written during the fabrication process. Therefore, the information is lasting (nonvolatile) and can be read but not altered. There are also several kinds of ROMs that can be programmed by the user.

1. PROMs are ROMs that can only be programmed once. They usually consist of bipolar transistors, and the act of programming causes fuse links to be electrically blown, permanently fixing the memory circuit.
2. EPROMs are similar to PROMs but are erasable by exposure to ultraviolet (UV) radiation. Their operation will be discussed in the next section.
3. EEPROMs are PROMs that are electrically erasable and can be reprogrammed repeatedly.

## 2.6.2 Memories Are Made of This!

In this section the structure and operation of typical serial, RAM, and ROM memory devices will be briefly described.

1. *Charge Coupled Device* (*CCD*). The CCD of Figure 2.22 consists of closely spaced MOS capacitors arrayed on a continuous $SiO_2$ layer that covers the underlying p-type semiconductor. If sufficiently large

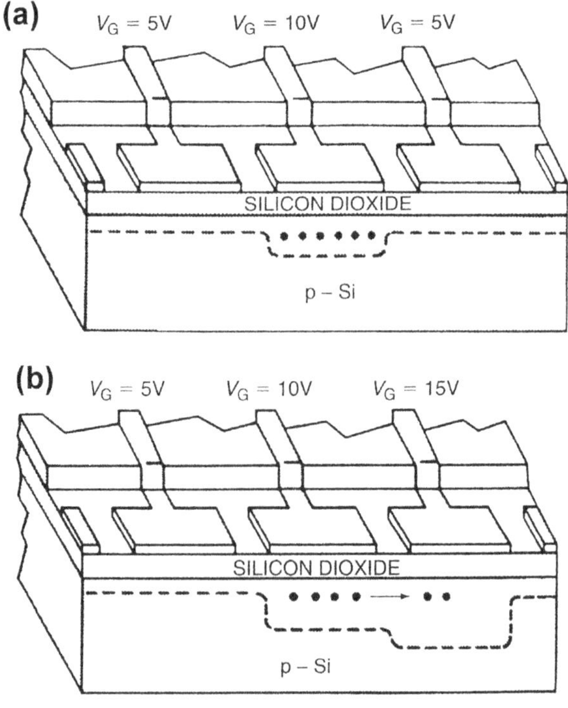


**Figure 2.22** Schematic of a charge-coupled device. Charge originally stored on center capacitor is transferred to neighboring cell by increasing bias voltage. *From Ref. [1], Copyright by Bell Telephone Laboratories, Inc.*

positive voltage pulses are applied to all of the electrodes, each capacitor will be depleted. In order to transfer charge to one of the cells, a higher bias is applied there to create greater depletion. In effect, a potential well develops there that captures injected electrons. By pulsing the neighboring cell to the right more positively, electrons can be made to transfer between cells. This lively transfer of charge can proceed in both directions. CCDs have found use in imaging devices that ultimately produce electrical video signals.

2. *DRAM.* The drive to minimize the number of devices per memory cell has led to the widespread use of the NMOS transistor/capacitor combination. A cell cross-section and the way it is wired into the circuit are shown in Figure 2.23. In 256 Mb DRAMs the cell size is $<0.7\ \mu m^2$, while in 1 Gb DRAMs, areas less than $0.3\ \mu m^2$ are involved. But, for 4 Gb DRAMs the cell size is only $<0.1\ \mu m^2$! Shrinkage of DRAM cells mean that thicknesses of gate and capacitor oxide dielectrics must scale accordingly [16]. The drain of the MOSFET serves as the conductive path between the inversion layers under the capacitor and the transfer gates. Electrodes of the storage capacitor are the drain/conducting polysilicon fill and the substrate ground; they sandwich the $SiO_2$ dielectric in between. High packing density requirements have motivated clever approaches to simultaneously reduce chip surface area while increasing charge storage capacity. By fabricating deep trench capacitors, the capacitance is markedly increased. Trench technology poses processing challenges in filling deep holes without the formation of defective cells. Further, as the size of the storage capacitor has continued to decrease, the need has arisen for dielectric materials with higher dielectric constant (Hi k), so that more charge can be stored in a smaller space. Materials such as hafnium silicate, oxidized nitride (ON), oxidized nitride/oxide (ONO), tantalum oxide ($Ta_2O_5$), barium strontium titanate ($Ba_xSr_{1-x}$-$TiO_3$), $HfO_2$, TaO, and $TiO_2$ have begun to replace $SiO_2$, because of their higher dielectric constant. We shall return to the issue of defects in these important devices in Section 3.2.5. Important main modes of failure include gate shorts, and the loss of stored charge due to leakage or recombination.

3. *EPROM.* These read only devices are basically MOS transistors that contain an additional polysilicon gate layer embedded within the oxide beneath the gate electrode. Because the embedded gate is unconnected it is electrically floating. Through application of a high source–drain voltage pulse, an avalanche of electrons is initiated near the

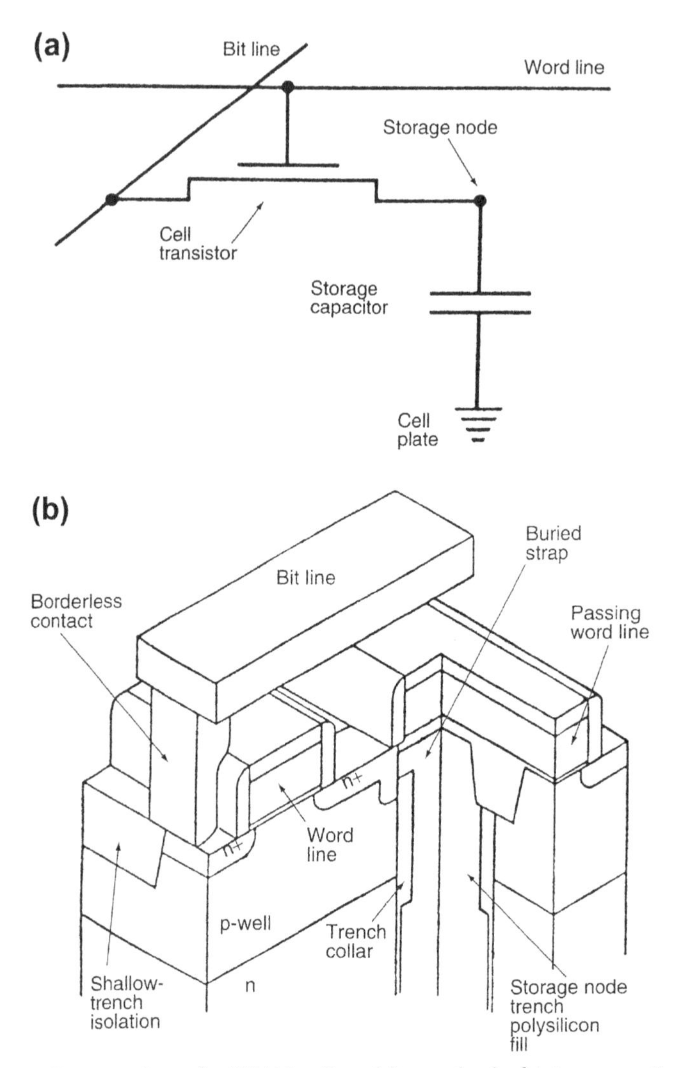


**Figure 2.23** Cross-section of a DRAM cell and its method of interconnection in the circuit. To write, the bit line is driven to a high or low logic level with the cell transistor on. Then the transistor is shut off leaving the capacitor charged high or low. (Charge leakage requires refreshing.) To read, the bit line is left floating as the transistor is turned on; the small change in bit-line potential is sensed and amplified to a full logic level. *From Ref. [16].*

pinched-off region. Some electrons acquire enough energy this way to enter the oxide and charge the floating gate. It is for this reason that these MOSFETs are known as floating avalanche, or FAMOS, devices. In packaging these devices a window is installed over the chip to enable

UV light to shine through and erase the memory. By locating the floating gate only a few nanometers from the silicon substrate, tunneling currents can charge the floating gate and also erase information; this is done in EEPROM devices.

### 2.6.3 Reliability Problems in Memories

Memory reliability problems can be divided into three categories [17], namely (1) processing or fabrication related, (2) operation related, and (3) environment related. Most concerns are common to SRAMs and DRAMs and to a lesser extent to EPROMs.

Processing difficulties include control of the conductor line width and semiconductor doping levels, defects due to contamination, stress, and process-induced radiation. Yields are adversely impacted, and poor performance and degradation often show up in the infant mortality population. Adding redundant columns or rows of devices that can be programmed to substitute for failing memory cells is one of the strategies adopted to combat failed devices.

Operational reliability problems stem from the nonuniform scaling of device features to achieve high packing densities. Reduction in transistor gate lengths without corresponding changes in operating voltages has resulted in higher drain electric fields. These promote time-dependent changes in dielectric properties and threshold voltage shifts.

Environmental damage is largely related to packaging issues. On the chip level, nuclear radiation in the form of gamma rays and alpha particles, which may originate in the packaging materials (e.g., Pb–Sn solder), impinge on memory cells and ionize atoms, causing local soft errors. Chapter 7 addresses such reliability issues.

## 2.7 GAAS DEVICES

### 2.7.1 Why Compound Semiconductor Devices?

Hailed as the electronic materials of the future, advances in silicon technology have perpetually consigned compound semiconductors to a second-place status. Nevertheless, there are electronic functions that are performed inefficiently or not at all by Si devices; this is the reason for our interest in III–V materials, specifically GaAs, InP, and ternary and quaternary alloys based on these materials. A comparison with Si indicates several important reasons for their use in devices.

1. The speed (frequency) of devices made from compound semiconductors exceeds that for Si devices simply because electron mobilities are

higher. For example, Table 2.2 reveals that $\mu_e$ (GaAs) is some six times that for $\mu_e$(Si). This shortens electron transit times and enhances high-frequency operation. Very high speed microwave devices and integrated circuits require these attributes of GaAs devices.

2. The larger band gap in GaAs relative to Si means that it can withstand higher working temperatures, an attractive property in small geometry power devices used in microwave applications. Low power devices that operate at room temperature have correspondingly low thermal generation of carriers and leakage currents, hence low noise.
3. Direct band-gap transitions in compound semiconductors make optical emission and absorption processes more efficient than in Si (Section 2.2.2.4). This has enabled the production of efficient lasers, light emitting diodes, and photodetectors.
4. Optical devices can often be made to operate at specific wavelengths within a broad spectral domain. In Si, only the wavelength corresponding to the band-gap energy is easily accessible. By alloying compound semiconductors, ternary and quaternary alloys emerge with energy gaps controllably tuned to a desired operating wavelength. Optical communications technology capitalizes on this to create high-performance light sources and detectors that operate at wavelengths where minimum optical losses occur.
5. A desirable attribute of GaAs is its ability to be produced in semi-insulating form with resistivities of $\sim 10^8$ Ω-cm. Such insulating single crystals make excellent substrates for GaAs microwave devices and integrated circuits.

Advantages seldom come without drawbacks. In GaAs we are dealing with a brittle, expensive material that is not easy to process and fabricate into devices. On the other hand, Si is mechanically robust and cheap, and significantly easier to process than GaAs. Silicon can also be protected and passivated by an easily grown, high-quality $SiO_2$ film, something that cannot be done in GaAs.

The subsequent discussion reflects the division of compound semiconductor devices according to microwave and opto-electronic applications. General tutorial reviews of the operation, reliability, and failure of compound semiconductor devices can be found in references [18–20].

## 2.7.2 Microwave Applications

Microwave frequencies range between $10^9$ Hz (1 GHz) to 1000 GHz with respective wavelengths of 30 to 0.03 cm. Within this spectral domain are a number of communication systems applications that are important in both the military and civilian sectors. Military applications include ground and airborne radar, electronic warfare including guided weapons, and satellite communications; civilian applications include microwave radar (for police, small boats, intruder alarms, and door openers), direct broadcast satellites (12 GHz), and mobile (1–3 GHz band) as well as cellular (~1 GHz) communications. To make these systems function, devices that generate, detect, and amplify microwave signals are required. Since the limit for silicon devices is ~1.5 GHz, GaAs devices are preferred above this frequency. Microwave circuits are sometimes difficult to design because conductors such as stray solder act like antennas at these frequencies.

### 2.7.3 The GaAs MESFET

By replacing the upper p–n junction of a JFET (Figure 2.20(a)) by a reverse-biased Schottky barrier, the MESFET emerges. The first part of the acronym for this device, which is deposited on a semi-insulating GaAs substrate, stands for (Schottky) metal (contact) epitaxial (compound) semiconductor. In the MESFET, schematically depicted in Figure 2.20(b), source–gate–drain electrodes combine to control current flow through a lightly n-doped single crystal (epitaxial) film of GaAs that serves as the channel. Sandwiching the channel are the metal contacts above and the semi-insulating GaAs substrate below. Device behavior and characteristics parallel those of the Si JFET. But GaAs MESFETS differ from their lower-frequency Si cousins because the channels are thinner (~0.1 μm) and gate lengths shorter (<1 μm). For channels of length $L_c$, the device cutoff frequency is given by $v_s/(2\pi L_c)$, where $v_s$ is the electron saturation velocity ($\sim 2 \times 10^7$ cm/s in GaAs). Thus for a 1 μm channel length, ~35 GHZ operation is possible.

In addition to the low-power, small-signal MESFETs used in receiver front ends and digital integrated circuits, there are power MESFET devices that operate at higher voltages. These devices contain an interdigitated, or comb-like, array of contacts that encompasses a larger area with a deep active channel. The contrast in geometries and operating conditions is reflected in different failure mechanisms for low-signal relative to high-power devices.

In contrast to the Schottky gate contact, source and drain contacts in these transistors must be ohmic (e.g., Au–Ge). Accelerated high-temperature testing of these devices has revealed failures due to the degradation of the metal employed in these contacts. "Gate sinking," a common failure mechanism involving metal interdiffusion into the GaAs, is discussed further in Chapter 5. Decrease in drain current at zero gate bias, reduction in pinch-off voltage, and degradation of channel properties (e.g., increase in drain to source resistance) are some of the reliability problems associated with GaAs transistors. Improvements in gate metallizations over the past two decades have extended device lifetimes from thousands of hours employing Al to millions of hours with TiPtAu.

### 2.7.4 High Electron Mobility Transistor

These quantum devices operate much like MOSFETs and exhibit gains and speeds that are among the highest for any transistors [21,22]. A common figure of merit for FETs is the transconductance, and it can be increased by reducing the channel resistance through heavy doping. But this strategy degrades electron mobility because of carrier scattering by the ionized impurities. In high electron mobility transistosr (HEMTs), the dual desiderata of high carrier concentrations and mobilities are simultaneously achieved. This is basically accomplished at the interface between n-doped AlGaAs and undoped GaAs, as indicated in Figure 2.24(a). Electrons in the conduction band of the doped AlGaAs (larger energy band gap) spill over into the GaAs (smaller energy band gap), where they are now trapped in a quantum-mechanical well that is free of ionized donor scatterers. As a result of band bending at the junction, the electrons in the well are essentially confined to a thin sheet. For this reason we speak of a two-dimensional electron gas (2-DEG). An electric field applied normal to the page will result in enhanced carrier transport in the well channel. The effect of this so-called modulation doping is magnified at low temperatures because thermal scattering by vibrating lattice atoms is minimized. Incorporated within the structure of the GaAs-base HEMT shown in Figure 2.24(b) is such a quantum well. Today, HEMTs also known as MODFETs have been fabricated with gate-lengths of approximately 0.25 μm in InP-base semiconductors. These devices demonstrably outperform GaAs MESFETs with respect to gain and low noise. Short gate-length GaAs and InGaAs MODFETs have frequency responses above 100 GHz with switching delays of $\sim$5 ps [19].

Sinking of Schottky gates and degradation of ohmic source and drain contacts are major failure mechanisms in both MESFETs and HEMTs. Electromigration, surface oxidation of GaAs, burnout, diffusion of dopants from channels, and intermetallic phase formation in Al–Au metallizations [18] are additional degradation mechanisms.

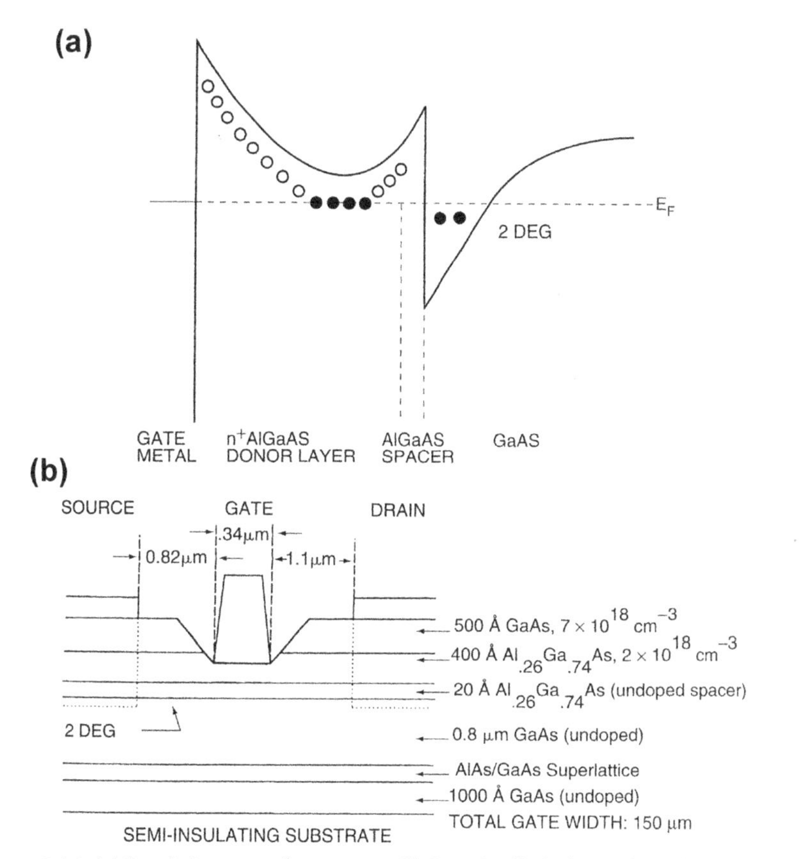


**Figure 2.24** (a) Band diagram of an n-type $Al_xGa_{1-x}As$–GaAs heterojunction structure. Free electrons originating in the $Al_xGa_{1-x}As$ spill over into the quantum well on the GaAs side, creating a highly mobile two-dimensional electron gas or carrier density. An electric field is applied normal to the paper. (b) Cross-section of a HEMT structure with dimensions to scale. The gate metal is TiPtAu, while the source and drain contacts are AuGeNi. *From Ref. [23].*

## 2.7.5 GaAs Integrated Circuits

Many of the same reasons for the integration of discreet Si devices have driven the development of both analog and digital memory and logic GaAs ICs. Except for microwave applications, these products have lagged behind their Si counterparts. In addition, there have been imperatives to fabricate monolithic microwave (and millimeter wave) integrated circuits, or MMICs. But microwave circuits are not simply large collections of FETs; in fact, some MMICs do not contain FETs at all! In some MMICs the number of passive circuit elements (resistors, inductors, diodes, transmission lines, interdigitated and overlay capacitors) may outnumber the FETs by a factor of 10 to 1 [24]. Recently, an extensive evaluation of MMICs composed of 0.5 μm GaAs MESFETs (with Au/Pd/Ti metallizations) tested at temperatures of 200–225 °C revealed that the main failure mechanisms were diffusion related, e.g., gate sinking and ohmic contact degradation. Activation energies for failures ranged between 1.5 and 2.2 eV for the most part [25].

## 2.8 ELECTRO-OPTICAL DEVICES

### 2.8.1 Introduction

Our main concern with compound semiconductors in this book is related to the failure and reliability of electro-optical devices. The three basic phenomena capitalized upon in these devices-absorption, spontaneous emission, and stimulated emission—were already introduced in Section 2.2.2.4 and Figure 2.6. Since so much is at stake in ensuring the reliability of optical communication systems, much of our concern here is with the individual active semiconductor devices employed. Today, long-distance under sea (submarine) systems employ infrared wavelengths of 1.3 and 1.55 μm, where fiber transmission losses are minimal. This means that the lasers and LEDs that launch light at one end of the fiber and the photodiodes that detect it at the other end must operate at the same wavelength. To tune devices this way, semiconductors must be tailored to have the necessary energy gaps as well as a good lattice match to the substrate. Guiding the design of the band gap is Eqn (2.1). For example, by alloying Ga, As, In, and P, ternary $In_{0.47}$ $Ga_{0.53}$ As and quaternary InGaAsP devices lattice-matched to InP can be fabricated to serve the needs of light sources and detectors in optical communication systems.

### 2.8.2 Solar Cells

A good example of devices capitalizing on absorption processes are solar cells, in which solar energy is absorbed and directly converted into electricity. These devices have a p–n junction close enough to the surface so that incident photons can easily penetrate to it. Once the junction is illuminated, carriers that were immobile in the space charge region of the prior unilluminated, or dark, junction, now move. Electrons lose energy by occupying lower conduction band levels in the n-type material; similarly, holes lose energy by floating upward to fill valence band levels in *p*-type material. Some of these carriers will promptly recombine. However, enough of them separate further to produce the equivalent of positive and negative battery electrodes, or a photovoltaic effect. The illuminated junction then acts like a photon-activated electrical pump, generating a current in the diode that can power an external circuit.

Reliability of terrestrial solar cells is adversely affected by environmentally generated surface deposits that reduce incident light and enhance corrosion of contact metals. Amorphous, hydrogenated silicon (a-Si:H), extensively employed in thin-film transistors for LCDs, has also been used to fabricate solar cells. In such cells, low operating temperatures and exposure to high light intensities degrade the conversion efficiency in ways that are not entirely understood.

### 2.8.3 PIN and Avalanche Photodiodes

Both of these diodes are normally employed as photon (absorption) detectors in optical systems and operate under reverse bias. Compound semiconductor PIN diodes are available in planar and mesa configurations [26], which are schematically depicted in Figures 2.25(a) and 2.25(b). These devices rely on InP and InGaAs film layers to enable 1.3–1.5 mm wavelength operation in high reliability, long haul, and high bit-rate optical

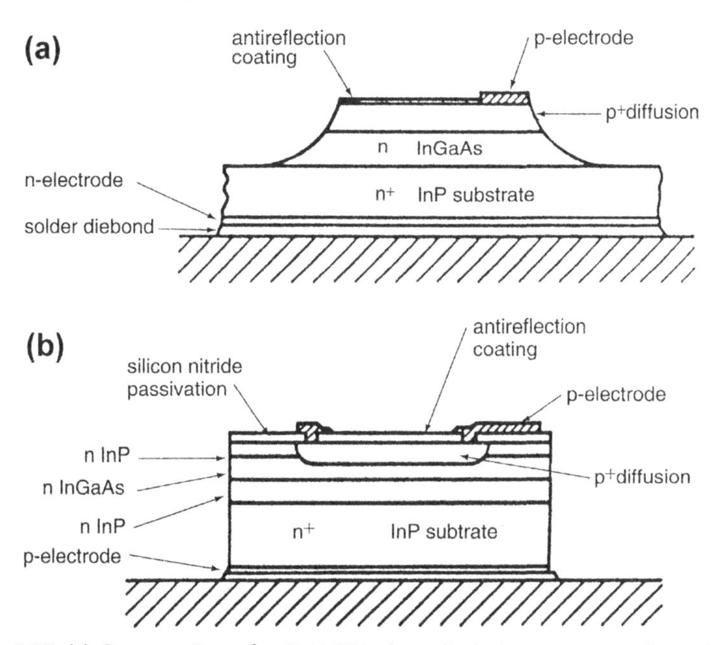


**Figure 2.25** (a) Cross-section of a III–V PIN photodiode in a mesa configuration. (b) Cross-section of a planar III–V PIN photodiode structure.

communication systems. The main objectives in designing PINs are to reduce capacitance to achieve high bandwidth, eliminate noise, and minimize reverse-bias dark current, which limits receiver sensitivity. Easy-to-fabricate mesa structures enable the first objective to be met, but such devices exhibit dark current instabilities. Planar PIN structures have a passivation layer that protects the junction, and a p-doped InP cap that minimizes leakage current. As a result, planar PINs are inherently more reliable than mesa devices.

Avalanche photodiodes (APD) are similar to planar III–V semiconductor PINs but operate at much higher voltages to obtain multiplication of the photogenerated carriers. Light is absorbed in the InGaAs layer, and holes are swept to an InP junction where the avalanche multiplication of carriers initiates. The separation of absorption and multiplication regions reduces the junction leakage current. Reliability problems arise from high electric fields, which induce surface charge motion [27]. Other failure mechanisms that lead to increased dark currents stem from contact metal penetration of the junction region and localized breakdown of the junction edge. Although lifetimes of APDs are shorter than for PIN devices, the high failure activation energies of 1 eV result in reliable operation under normal conditions.

## 2.8.4 Light Emitting Diodes

When these diodes are forward biased, they spontaneously emit light as conduction-band electrons recombine with valence-band holes. This phenomenon of injection electroluminescence has been exploited in LEDs to produce the familiar colored numbers and letters in displays. LEDs are also a viable alternative to lasers in some short to medium distance optical communications systems, where their lower light output is not a disadvantage. Such an application is illustrated in Figure 2.26, where the LED, composed of InGaAsP lattice-matched to the InP substrate, launches 1.3 μm infrared light down the optical fiber.

Degradation in these devices stems from both the semiconductor materials and the contact metals [28,29]. Typical p-side metallizations employed are Au–Zn or Au/Pt/Ti. Devices using the latter trimetal contacts generally exhibit less loss in light output during aging at temperatures of 120–200 °C. For the Au–Zn LEDs the activation energy for failure has been determined to be in the range 0.75–0.85 eV. At high light power, degradation is primarily caused by dark spot and dark line defect formation, a mode that also afflicts lasers. Precipitation of host atoms and the familiar

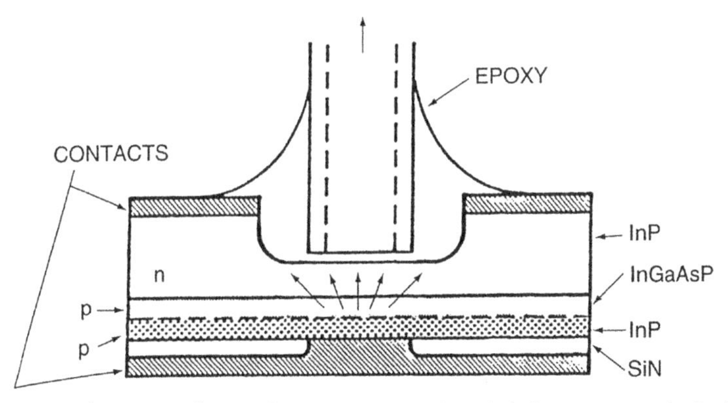


**Figure 2.26** Schematic of a surface-emitting InGaAsP light-emitting diode lattice-matched to an InP substrate. The 1.3 μm infrared light emitted enters an optical fiber.

migration (sinking) of contact metal into the semiconductor have been identified as degradation mechanisms. Nevertheless, edge-emitting diodes (ELED) are considered to be highly reliable devices, especially at low power levels, where lifetimes of $\sim 10^7$ h at 25 °C have been projected.

Organic light-emitting diodes (OLED), composed of a two layer (conductive and emissive) structure of organic semiconductors, emit visible light in response to an applied voltage. The color of the light depends on the choice of organic semiconductors. Typical applications include small displays and general lighting. When used in displays there is no need for a fluorescent backlight, needed in active matrix LCD displays. The build up of layers in the structure are the anode (often Al), conductive layer, emissive layer and cathode. In Figure 2.27 electrons flow from the cathode to the anode. Holes have a higher mobility in organic semiconductors and hence some holes, from the conductive layer, diffuse into the emissive layer. Recombination of electrons with holes in the emissive layer, near the boundary to the conductive layer, produces light. The cathode, on top of the emissive layer, is typically ITO (indium tin oxide), which is transparent

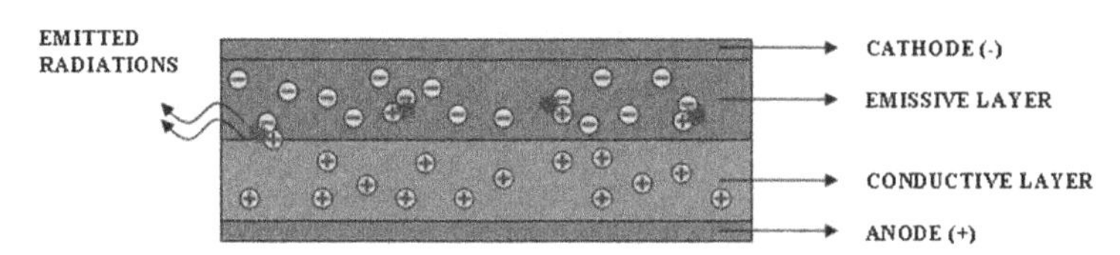


**Figure 2.27** OLED structure.

and allows the emitted light to escape. The OLED emits under forward bias. In the reverse direction no light is emitted. The emission of light at low voltage is an advantage because it requires low power and leads to long battery life. Arial light intensity, although lower than LEDs, is bright and greater than LCDs. OLEDs are typically used in small (eventually) flexible displays for mobile phones and portable MP3 players, car radios and in high resolution head-mount displays, which are used much less frequently than 24/7 computer displays.

Emission in the blue was plagued by limited life until a phosphorescent organic semiconductor was substituted. Lifetimes of 20,000 h (well beyond tungsten filaments) can now be achieved. The organic semiconductors used are also degraded by moisture, hence the need for hermetic seals to insure a long lived display. An example from Sony is shown in Figure 2.28.

## 2.8.5 Semiconductor Lasers

Due to the many critical laser applications there has been far more published information on the reliability of lasers than of other opto-electronic devices. While light is emitted from LED junctions in all directions, intense, directed beams exit lasers. In order for spontaneous emission of photons and laser action to occur, however, it is first necessary to have a population inversion of carriers. This is accomplished in a semiconductor diode structure where both p and n regions are so heavily doped that the Fermi levels lie inside the valence and conduction bands, respectively. Under conditions of high forward bias, large electron and hole concentrations exist

**Figure 2.28** Sony's flexible OLED display.

within respective bands that now overlap in the vicinity of the junction. A carrier population inversion is created, and stimulated emission occurs. Parallel facets are cleaved and polished, forming the Fabry–Perot cavity in which the photons multiply, while the other faces are roughened to discourage lasing in these directions.

Emitted laser light has a wavelength corresponding to the band-gap energy (Eqn (2.1)) and exhibits the so-called $L$–$i$ ($L =$ optical power, $i =$ current) output characteristics schematically indicated in Figure 2.29. Up to a critical threshold current density ($j_{th}$), the light output is very low, but beyond $j_{th}$ it rises rapidly. Higher operating temperatures raise $j_{th}$ and lead to lower light outputs. Laser diodes displaying the kinked and nonlinear characteristics shown in Figure 2.29 are considered to be faulty [30]. Peaked kinks are often, but not always, indicative of internal defects and damage, while internal reflections frequently, but not always, cause the second type of kink characteristics. Kinks are also associated with higher-order modes of laser emission and changes in $j_{th}$.

The simplest of all laser structures, the *homojunction* configuration, contains the same semiconductor material, e.g., GaAs, on both sides of the junction. Good only as a pedagogical tool, such lasers require enormous threshold current densities ($\sim 50$ kA/cm$^2$ at 300 K and $\sim 5$ kA/cm$^2$ at 77 K) to initiate lasing. The severe reliability problems implied by Joule heating thus rule out continuous homojunction laser operation at room temperature. Furthermore, carriers can wander from the active region and recombine, resulting in a broadened emission. Output light stems from a considerably narrower active region in *double-heterojunction* lasers. First, the

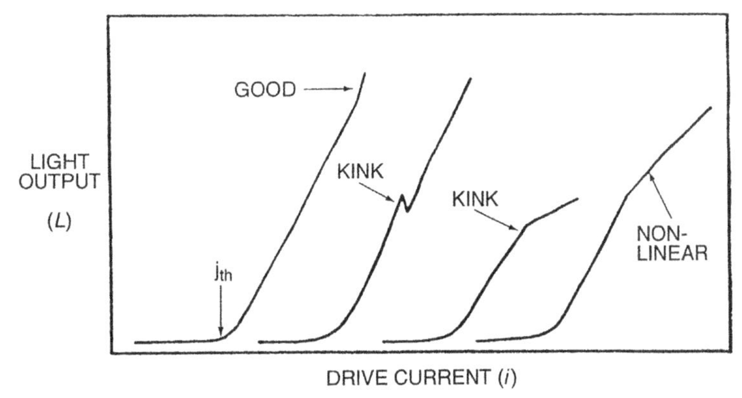


**Figure 2.29** Assorted laser light output current characteristics depicted schematically. *From Ref. [30].*

carriers are confined by the two heterojunction GaAs and $Al_xGa_{1-x}$ As barriers. Second, light is confined to the active region by virtue of the higher index of refraction of GaAs relative to AlGaAs. Both effects lower the lasing threshold an order of magnitude, while narrow stripe contacts (20 μm wide) reduce $j_{th}$ further, to the point where reliable continuous operation can occur at room temperature. The evolution of these GaAs-based lasers, designed to operate in the range 0.8–0.9 μm, is depicted in Figure 2.30.

Popular laser geometries for optical communication systems are shown in Figure 2.31 and employ InGaAsP semiconductor materials [30]. Known as BH or buried-heterojunction lasers, the light generated in the buried active regions is strongly guided in these devices. Other features include low threshold currents and high quantum efficiencies. Instead of using only the reflective facets for controlling light output, an extended diffraction grating is incorporated in so-called distributed feedback (DFB) lasers for this

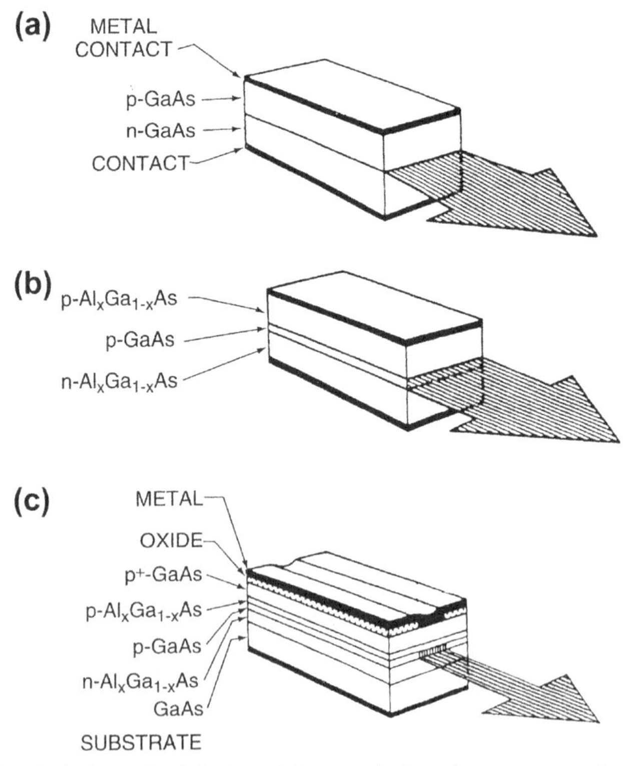


**Figure 2.30** Evolution of GaAs-based lasers, designed to operate in the range 0.8–0.9 μm. *From Ref. [1], Copyright by Bell Telephone Laboratories, Inc.*

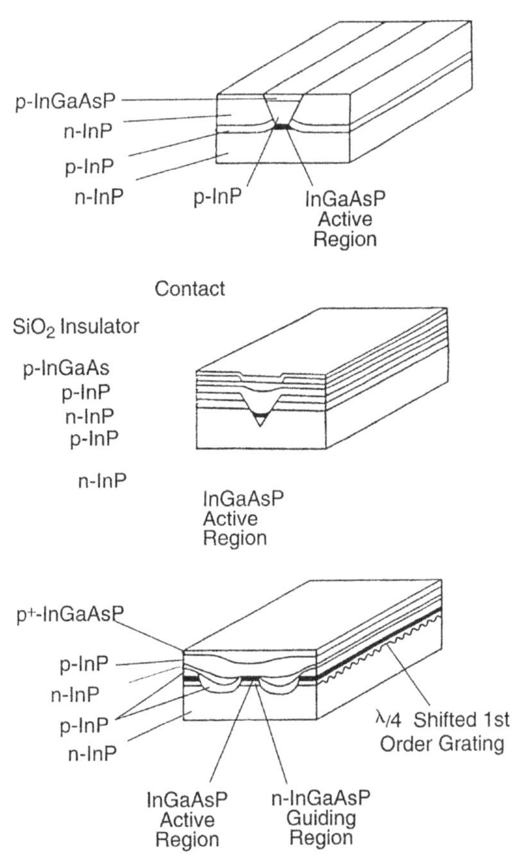


**Figure 2.31** Popular laser geometries for optical communication systems. *Top:* Etched mesa buried heterostructure (EMBH). *Middle:* Channeled substrate buried heterostructure (CSBH). *Bottom:* Double channel planar buried heterostructure distributed-feedback. *From Ref. [30].*

purpose. In this way a single laser wavelength from the multipeaked laser output can be selected and amplified.

A number of failure modes commonly observed in lasers include damage to facets, dark line and dark spot defect formation, aging effects, electrostatic discharge, and current-confinement problems. The latter are manifested as a drop in laser efficiency because current leaks or is shunted away from the active region. Poor design, processing, and power surges are often the cause of current confinement problems, to which some lasers structures are more prone than others. Interestingly, InGaAsP lasers have a higher immunity to laser degradation effects than AlGaAs/GaAs-type lasers.

Detailed mechanisms for the failure of lasers and other electro-optical devices and components will be presented in Chapter 10.

## 2.9 PROCESSING—THE CHIP LEVEL

### 2.9.1 Introduction

It is much easier to draw schematics of devices and IC circuit diagrams on paper or with the aid of computers than to realize them in physical terms. The production of devices requires different electrical materials to be shaped or patterned and creatively merged to form the active devices and passive components that must then be interconnected, insulated, and finally packaged. These issues and demands are at the core of a very broad interdisciplinary manufacturing technology that designs, fabricates, and packages solid-state devices and circuits.

Much of the glamour associated with semiconductor manufacturing occurs on the Si chip level, and that is what we shall address here. To appreciate what is involved, consider the state-of-the-art CMOS technology employed in the fabrication of advanced microprocessor chips. At the base of the stunning image of Figure 2.32, beneath the skeleton of the exposed interconnects, contacts, and first aluminum wire level, are the CMOS transistors. These MOSFETs have effective channel lengths of 0.25 μm and gate-oxide thicknesses of 7.7 nm [31].

Issues related to the processing and fabrication of both Si and compound semiconductor devices have been very widely treated in the literature and

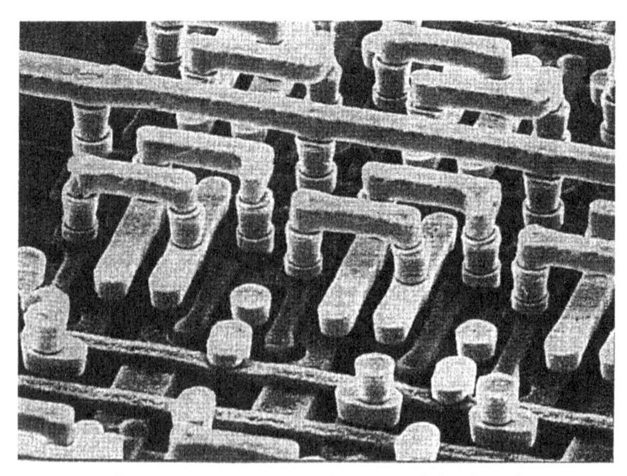

**Figure 2.32** Scanning electron microscope image of the local interconnect and first-level aluminum wires in an SRAM cell. *From Ref. [31].*

form the subject of many excellent books [6–10]. In this section some of the more important individual process steps in the fabrication of silicon IC chips are briefly introduced. The intent is to convey a sense of the physical and chemical principles involved, and how potential defects may be introduced into the product. Processes discussed are number-keyed to the schematic of the MOSFET shown in Figure 2.33. In the following chapter specific processing defects will be addressed in greater detail.

## 2.9.2 Silicon Crystal Growth [1]

When defects or reliability problems surface in any manufactured product, our first reaction is to suspect the quality of the starting materials. Wafers are the starting materials, but they are sliced from large single crystals. Therefore, growing crystals is the first, and in many ways a critical step, in ensuring product quality. Bulk single crystals of silicon are grown by melting a Si charge in a $SiO_2$ crucible, inserting a small cooled single-crystal seed into it, and pulling out the large solidified crystal. This is the Czochralski technique for growing "CZ" crystals. Successive solidification occurs amid complex heat and mass transfer processes between and among the growing single crystal, molten Si, and surrounding inert gas ambient. Even more complex is the simultaneous convective fluid motion that disperses melt dopants and impurities. This fluid flow is dependent on melt temperature and relative speed of rotation between crystal and melt as the former is being pulled.

Simultaneously, compositional change occurs in both the melt and the crystal. To see why, look at almost any binary Si-dopant (P, B, As) or impurity phase diagram. Note that the liquidus and solidus phase

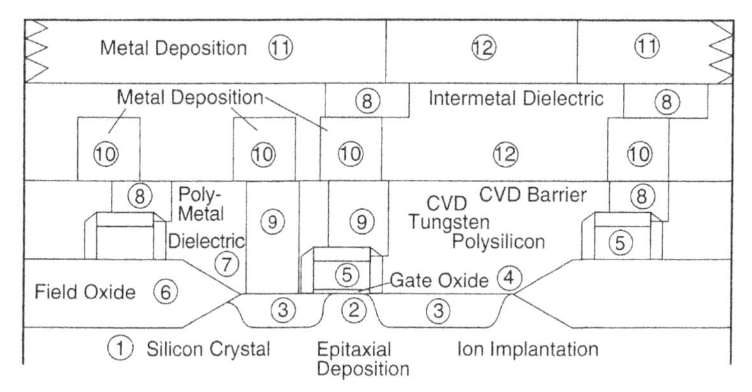


**Figure 2.33** Schematic of the MOSFET indicating key processing steps.

boundary lines drop for small solute additions. Equilibrium at the crystal–melt growth interface implies that the solid is always purer than the liquid. Since the effect is cumulative during directional solidification, axial gradients in composition are predicted [32]. Grown crystals not only exhibit varying concentration profiles of intentionally added dopants, but contain unwanted elements like oxygen. But it is the transition metal impurities that are viewed with particular alarm; they cause the most severe problems, and great efforts are made to eliminate them or neutralize their influence. In addition, defect swirl patterns are not infrequently observed on planes sliced perpendicular to the crystal axis. Furthermore, surface striations that are several microns wide and alternately dopant rich and dopant poor sometimes decorate the length of bulk Si and compound semiconductor single crystals. Both features apparently stem from temperature fluctuations at the growth interface and complex convective flow patterns that cause a periodic partitioning of solute. From the standpoint of the MOSFET, variations in dopant level across the plane of a wafer must be minimized to ensure uniform device processing and chip performance.

Following growth, the crystals are shaped round by grinding and then sawed and polished to produce wafers. Presently, many integrated circuit production lines still process 15–20 cm diameter Si wafers, but large 30 cm diameter wafers have become the world standard with a projection of 45 cm in the future.

### 2.9.3 Epitaxy [2]

This important term refers to the ordered crystallographic registry of a single crystal film deposit on an underlying crystalline substrate template. If film and substrate are the same (Si on Si) we speak of *homoepitaxy*. In electro-optical devices they are different (InGaAsP on InP), and *heteroepitaxy* is the appropriate term. The yield of early thin-film bipolar transistors fabricated within homoepitaxially deposited Si was dramatically improved relative to those produced within the original, or uncoated, Si wafers. Chemical vapor deposition (CVD) or some variant of it is used to deposit epitaxial semiconductor films on wafer substrates. The process typically entails reaction of a chlorinated silane, e.g., $SiCl_4$, and $H_2$ at temperatures of $\sim 1100\ ^\circ C$ and a pressure of $\sim 1$ atm. Epitaxial processing in CMOS technology helps assure device isolation and minimizes junction leakage currents.

Epitaxial deposition for bipolar devices often involves ion implanted or diffused areas (for isolation (*p*-type) and subcollectors (n type)) in the wafer prior to the epitaxial deposition. These buried layers can cause auto-doping control problems.

Critical to insuring single crystal formation is a combination of a low film nucleation rate and a high film growth rate. If many nuclei form on the substrate, a worthless polycrystalline film deposits. A low gas supersaturation and deposition rate will create few nuclei. On the other hand, high substrate temperatures enhance diffusion rates and facilitate atomic incorporation into lattice sites. This promotes the early lateral extension of the single crystal film, and growth perfection as it thickens. Epitaxial Si films are purer and frequently more defect-free than the substrate wafer; furthermore, they can be doped independently of it. Nevertheless, they do contain the assorted point, line, and surface defects noted later in Figure 3.15 that occasionally cause device malfunction.

## 2.9.4 Ion Implantation [3]

For high-performance devices with shallow junction depths, ion implantation, rather than diffusion, is the preferred method of doping. Unlike diffusion, where the surface dopant concentration is higher than in the crystal interior, the reverse is true in ion implantation. The subsurface wafer penetration of dopant ions is due to energies of ~50–200 keV acquired when they are accelerated in the high electric field of the implanter. Possessing energies thousands of times larger than the binding energy of lattice atoms, the dopant ions wreak havoc as they are embedded into Si. At first a lightning-like ion current excites the atomic electrons and ionizes atoms, absorbing energy in the process. When the dopants lose enough energy, they slow sufficiently and then begin to violently collide with Si nuclei, knocking many off their lattice sites. Each dopant ion executes a different odyssey and comes to rest at a different distance from the wafer surface. Spatially, dopant concentrations are statistically distributed as a Gaussian function both normal to the surface and transverse to it. The depth of the peak and spread of the Gaussian depends on the dopant and the nature of the matrix it passes through, e.g., Si, $SiO_2$, SiN, photoresist. Ion beam damage to the silicon matrix creates disordered and even amorphous regions. These must be repaired by rapid thermal annealing at ~900 °C, where atoms can diffuse sufficiently to restore lattice crystallinity and full electrical activity to the dopants.

In modern IC chips some 15 to 20 or more ion implantation steps are carried out to dope large volumes (tubs), small regions (sources and drains), and to adjust the threshold voltage of field effect transistors. This latter operation, which controllably alters the interface charge ($Q_i$ in Eqn (2.6)) at the Si–$SiO_2$ boundary, is quite critical and requires precise ion doses and spatial distributions. Tailoring dopant profiles by ion implantation has additionally enabled minimization of hot-electron damage and latchup effects, as well as gettering of heavy metal (Fe) contaminants. In the latter two cases, high-energy implants are required to drive dopants deep beneath the surface layer of devices.

Ion implanters are one of the most complex pieces of microelectronics processing equipment; they are also potential and actual sources of contamination and defects. Ideally, a chemically pure, monoenergetic beam is desired. In a practical sense one may imagine defects arising from variations in ion beam energy, depth, lateral spatial distribution and ion dose, as well as from the presence of chemical contaminants. Examples are given in Chapter 3.

All of the processes to this point have been concerned with the underlying Si. Above the Si surface are assorted grown and deposited dielectric and conducting films that we now consider.

## 2.9.5 Grown Gate, Field, and Isolation Oxide [4]

Today's ULSI technology is due in large part to the excellent properties of thermally grown gate, field and isolation oxides. This critical $SiO_2$ film, used for gate oxide, the smallest fabricated dimension in MOS devices, has been the focus of much study and reliability testing. Currently, gate oxide is less than 10 nm thick, it is made by controllably oxidizing silicon wafers in dry $O_2$. The simple reaction, $Si + O_2 \rightarrow SiO_2$, at temperatures between 900 and 1100 °C, produces a dense amorphous film. Thick oxide growth has traditionally been modeled by linear kinetics at short time (i.e., $d_o \approx t$) and by parabolic kinetics at long time, i.e., $d_o \approx t^{1/2}$, where $d_o$ is the oxide thickness and $t$ is the time. These formulas may not hold for the ultra thin gate oxides now grown.

Growth in the presence of $N_2O$ is a recent advance that has improved the quality of gate oxides. Nitrided oxides are apparently more resistant to (charge) trap generation and to plasma-induced charging, give rise to smaller electric-field-induced leakage currents, and are generally longer lived than conventional oxides [33,34]. Oxide charging during operation, and the associated reliability problems of breakdown and hot electron degradation are topics discussed in Chapter 6.

## 2.9.6 Polysilicon and Metal Gates [5,35,36]

The gate oxide is contacted by a polysilicon electrode. Long ago, polysilicon replaced the much more unreliable aluminum electrodes that caused early dielectric breakdown in underlying oxides. Polysilicon films are deposited in low-pressure CVD reactors through the reduction of silane by hydrogen. Below ~590 °C amorphous films form, while above ~610 °C polycrystalline films deposit. This narrow temperature range defines a useful processing window. At higher temperatures, gas phase reactions result in rough, loosely adhering deposits and silane depletion that causes poor film uniformity. Impracticably low deposition rates occur at lower temperatures.

More recently, polysilicon has started to be replaced by metal gates such as TiN, WN and TaN which are compatible with the new hi-k dielectrics [35]. Because the threshold voltage depends on choice of metal (work function again), different metals are needed for the n-channel and *p*-channel devices used in CMOS structure [36].

## 2.9.7 Deposited and Etched Dielectric Films [6,7,16]

Since all devices and interconnections support voltages and carry current while sitting on a conducting Si wafer, strategically placed dielectrics are needed to isolate them electrically. Aside from the critical (grown) gate oxide, an assortment of other insulator films are *deposited*. They include $SiO_2$, silicon nitride, and phospho- and boro-silicate glasses produced by CVD methods, often with the assistance of a plasma discharge. The dielectric films must then be etched to expose windows to semiconductor contacts or to confine subsequently deposited metal films and vias (see next section). After microelectronic chips are fabricated they are usually encapsulated by a silicon nitride film, an effective barrier that prevents moisture and alkali ion ($Na^+$) penetration. All of the above dielectrics are amorphous in structure and of the order of 1 μm thick. Typical reactions with precursor gases (on the left) and product films (on the right) include:

for $SiO_2$, $Si\,(OC_2H_5)_4$ (TEOS) $\rightarrow SiO_2 +$ byproducts at 700 °C

for silicon nitride, $8NH_3 + 3Si_2Cl_2H_2 \rightarrow 2Si_3N_4 + 6HCl + 12H_2$ at ~ 750 °C

$SiH_4 + NH_3 \rightarrow SiNH + 3H_2$ at 300 °C

for phosphosilicate glass, $SiH_4 + 4PH_3 + 6O_2 \rightarrow SiO_2 - 2P_2O_5 + 8H_2$ at 450 °C

In general, dielectric film deposition steps are carried out at successively lower temperatures so as not to alter the prior processed and doped materials.

The composition, structure, and properties of such CVD films vary widely with deposition conditions. Silicon nitride is an example. At very high temperatures dense, crystalline $Si_3N_4$ forms, but at lower temperatures considerable hydrogen is incorporated, particularly during plasma assisted deposition. Differences in dielectric strength, internal stress, and resistance to permeation of water and ionic impurities occur, but generally not to the point of compromising reliability. High phosphorous or arsenic glass compositions also corrode aluminum lines when moisture is present. More significant are the problems associated with surface coverage by dielectric films and maintaining a planar terrain for subsequent metal depositions. Incomplete filling of deep trenches and regions between metal lines can lead to structural voids and occasional short circuiting.

Maintaining a flat planar topography is extremely important if subsequent lithography patterning errors and defects in deposited structures are to be avoided. The relatively new technique of chemical-mechanical polishing (CMP) has enabled the production of planar interfaces between successive metal and intermetal dielectric levels.

### 2.9.8 Metallization [8–10,14]

There is a need to contact and interconnect all of the doped semiconductor electrodes, i.e., source, drain, and gate, to other components and devices on the chip. Assorted metal and alloy conductors are employed for contact structures including Al and Al–Si, PtSi, Ti–W, $Pd_2Si$, $TiSi_2$. Interconnections, on the other hand, have historically been composed of Al thin-film alloys with dilute Cu or Cu–Si additions. Copper has begun to replace Al, because it has higher conductivity and therefore less IR drop, point to point. In high-density ICs, efficient packing of devices on the chip real estate requires placing interconnect stripes at successive metal levels [37] commonly denoted by M1, M2, M3, etc., as shown in Figure 2.34. High performance chips have from eight to 10 levels of metallization with metal studs filling vias between layers to maintain a planar surface. The highest level is composed of bus-bars that carry current to the chip contact pads. Vias in the form of studs or plugs are like staircases that connect neighboring floors, or levels, of interconnects. Tungsten plugs or vias are often deposited by CVD methods to vertically link adjacent interconnect levels. As shown in Figure 2.33, these vias are surrounded by TiN barrier films that prevent mobile ions from reaching sensitive junctions and reactive gases from

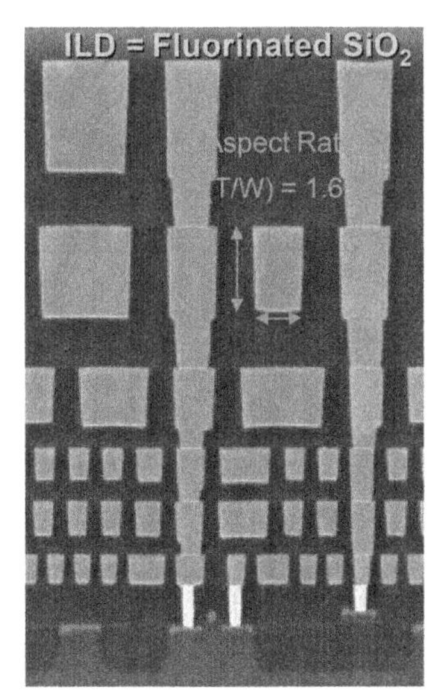


**Figure 2.34** Scanning electron microscope image of cross-section with six levels of metallization. *From Ref. [38].*

etching exposed oxides. The above metals possess the best collective basket of desirable properties, including (1) high electrical conductivity; (2) low contact resistance to Si; (3) little tendency to chemically react with or interdiffuse in Si; (4) resistance to corrosion and environmental degradation; (5) electromigration resistance; (6) ease of deposition; (7) compatibility with other materials and steps in the fabrication process.

The physical vapor deposition (PVD) techniques of *evaporation* and *sputtering* are the means employed to deposit metal films. Sequential atom-by-atom deposition results in film nucleation and subsequent growth. Unlike epitaxial deposition, both methods yield polycrystalline films typically having submicron grain sizes, due in part to substrates that are often amorphous (e.g., $SiO_2$).

Evaporation is the simpler of these metallization methods and involves heating of the source metal until it evaporates at appreciable rates. The process is carried out in high vacuum while wafer substrates, carefully positioned relative to the evaporation source, are rotated to yield uniform metal coverage. In sputtering, the metal or alloy to be deposited is fashioned into a cathode plate electrode and generally placed parallel to the wafers (anode) within a chamber containing inert sputtering gas. At low pressures less than ~5 mTorr, and potential differences of a few kilovolts DC, a glow discharge is sustained between electrodes. Within the discharge, energetic positive gas ions bombard the target, dislodging atoms that then deposit on the wafers as a film. In magnetron sputtering, magnets placed behind the cathode confine electrons to hop around the target and create a richer plasma that increases film deposition rates.

The two metallization processes compete in many applications, and not only in semiconductor technology, but other industries as well. Issues that must be considered in selecting the best one for a particular application include (1) adhesion to the substrate; (2) film stress; (3) stoichiometry of alloys films; (4) uniform coverage of the topography; and (5) electrical damage to substrate oxides. Sputtering is preferred in most semiconductor applications because stoichiometry of alloys is maintained and step coverage is usually superior. However, evaporated Al films appear to be more resistant to electromigration damage.

Many reliability problems are associated with contact and interconnection metallizations. They include changes in contact resistance due to interdiffusion reactions between Al and Si, electromigration-induced degradation of contacts and interconnections, and general corrosion damage. Contacts between Si and Al, two indispensible elements in microelectronics, are particularly troublesome. This results from mutual atomic interdiffusion (Section 5.4.2) and is remedied through alloying (e.g., Al–1%Si) or the use of diffusion barriers (e.g., metal silicides, TiW). These must be imposed between the silicon and Al interconnections to prevent reaction. Other problems stem from stress. When surrounded on all sides by oxide, as interconnects are in the various metallization levels, tensile stresses develop in the metal. The reason, as we shall see in the next chapter, is due to the larger thermal expansion coefficient of Al relative to $SiO_2$. These thermal stresses are sometimes sufficiently large to generate voids that open interconnects.

Since 2000 dual damascene copper metallization [39] has been introduced notably by IBM (Figure 2.35). In this process copper is plated into trenches formed in the low k dielectric passivation separating metal layers. After inlaying the copper in the preformed trenches the excess copper is removed using chemical mechanical planarization (CMP). The higher conductivity of the copper is a necessity as the line widths become narrower and the number of metal levels increases.

### 2.9.9 Plasma Etching

Plasma processes are widely employed in the fabrication of semiconductor devices. Examples include plasma-enhanced CVD, photoresist

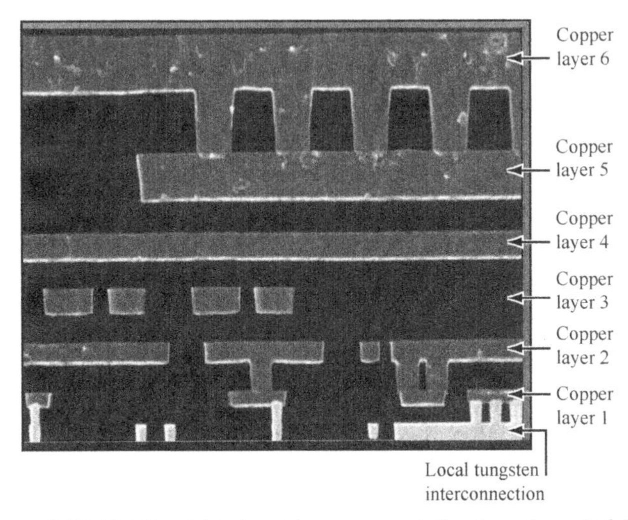


**Figure 2.35** Multilevel (six levels) copper metallization. *From Ref. [39].*

removal, and the etching of polysilicon, dielectric, and metal thin films. Within the plasma a very reactive environment consisting of high energy electrons, ions, and precursor gas free radicals is created by applied voltages of a few kV. This combination of energetic reactants speeds up otherwise sluggish vapor deposition or material removal (etching) reactions, and effectively makes them proceed at lower temperatures. In our MOSFET, the need to etch contacts and vias in $SiO_2$ is met by radio-frequency (RF) plasmas employing process gases containing fluorine, carbon, and oxygen. Minimum dimensions, high aspect ratio features, and selectivity of the etch are necessary attributes of the etching process. The last attribute requires that neighboring polysilicon, metal, or silicides be immune to etching attack.

Damage to materials during plasma etching includes the buildup of charge and dielectric wearout due to bond breaking and charge trapping.

## 2.9.10 Lithography

Perhaps the most critical process in microelectronics is lithography, a technology that embodies a collection of critical steps. Lithography involves the use of exposure tools, masks, photoresist materials, and etching processes to enable such blanket area processing steps as film deposition, etching, diffusion, and ion implantation to effectively occur only in preselected areas of a wafer surface. While there are obvious parallels to photography, instead of producing flat images on paper, a three-dimensional relief topography is created in materials like Si, $SiO_2$, silicon nitride, aluminum, and photoresist. After a mask (reticle) is generated with the desired geometric features, lithographic patterning involves spinning a photoresist film on the wafer surface. As the name implies, photoresists are sensitive to photons and when developed are resistant to chemical attack by liquid or gaseous etchants. Some are also sensitive to electrons and X-rays. In general, they consist of specially compounded polymers and display two broad types of behavior. When exposed to light, the cross-linked polymer chains of *positive resists* undergo scission, or cutting, rendering them more soluble in a developer, while unexposed areas resist chemical attack. Under the same stimulus, *negative resists* molecules polymerize, while unexposed regions are soluble and dissolve in the developer.

If, for example, it is desired to dope silicon, a blanket film of $SiO_2$ is first deposited and then covered with a photoresist layer. After exposure and development, the unprotected $SiO_2$ film is removed by etching, and the photoresist is stripped away. This leaves the bare Si surface geometrically patterned by the remaining $SiO_2$ film mask. Now the exposed Si regions may be ion implanted. Similar steps must be repeated at many different depths, or levels, on the wafer as actual integrated circuits unfold. Alignment, tolerance and overlay of successive levels, e.g., contact metal over a contact hole, is absolutely critical so that correct registry is maintained simultaneously over millions of device features on large wafers. The magnitude of the problem can be appreciated when it is understood that 10 to 20 lithography levels are required in advanced IC chips.

Another method of using photoresist coupled with an underlay film is known as the lift off process, primarily used for metallization. A dual layer of underlay and photoresist is patterned and the underlay is overetched, causing the photoresist to overhang the underlay. After metal deposition the undesired metal pattern on top of the photoresist is lifted off by dissolving the underlay film.

Poor product yields have often been associated with lithography defects. These arise predominantly from mask pattern defects that are discussed in Section 3.4.2.4. Metallization patterns defined after etching may then, for example, contain either missing or added metal, which results in open or shorted conductors, respectively. Mask defects can be more readily tolerated by the expedient of employing 5× or 10× reduction stepper tools that project the patterns onto photoresists. The 1× systems do not offer this opportunity and therefore require more defect-free masks.

## 2.10 MICROELECTROMECHANICAL SYSTEMS

### 2.10.1 Microelectromechanical Systems [40–42]

With the advent of very sophisticated and very precise techniques of semiconductor fabrication it became possible to construct miniature electromechanical systems of silicon or polymers and connect them, at the same time, to ICs for both control and sensing their mechanical movement or response to external stimuli. The birth of Microelectromechanical Systems provided the ability to construct tiny systems of millimeter size, micron size or even nanometer sizes, bridging to Nanotechnology, another emerging field. The applications for these tiny machines are almost ubiquitous and yet very little known. Injet (piezoelectric) print heads, optical data communication switches, projector displays, accelerometers in airbags, air pressure sensors for automobile tires, fingerprint sensors, gyroscopes, all contain microelectromechanical systems (MEMS) of one kind or another specifically designed for the unique application employed. Figure 2.36 shows the tiny mirrors used in digital light processing chips known as DMD. These chips are used in high definition TV as well as the projectors used in meeting rooms all around the globe. The failure of these systems is quite dependent on their application, however hermiticity of the package, fatigue, wear, delamination and environmental induced failure has been observed.

## EXERCISES

1. What are some differences between materials used for electrically and mechanically functional applications?
2. Metals are the best conductors of electricity and heat, reflect light most efficiently, and are the most malleable of materials. Explain these attributes of metals in terms of their electronic and atomic structure.
3. Copper wiring in a telephone switching station was exposed to a fire. After the fire the electrical resistivity of the wiring was measured. Some wiring was observed to have a lower resistivity, while other wiring exhibited an increase in resistivity.
   a. Describe a scenario that would result in a resistivity decrease.
   b. Describe a scenario that would result in a resistivity increase.
4. A good dielectric is a good insulator, but a good insulator is not necessarily a good dielectric. Explain why.
5. a. Rectifiers conduct large currents in one direction. Metals in contact also conduct large currents, but in both directions. Why can't rectifiers be fabricated using metals?

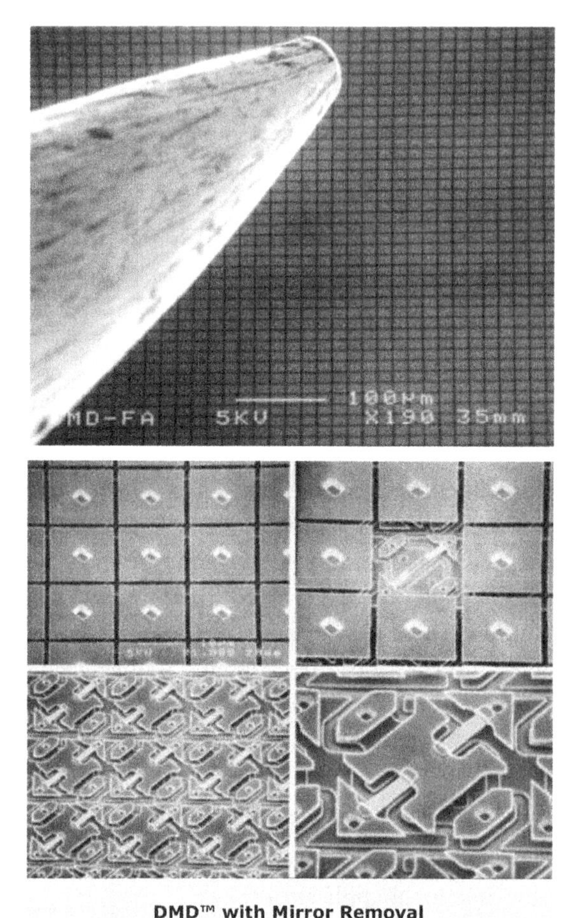


**Figure 2.36** Digital light processing technology. *Courtesy: TI Texas Instruments.*

   b. An electric field is built in at a semiconductor p–n junction. Insulators can support applied electric fields. Why can't useful rectifiers be fabricated using insulators?
   c. What properties make semiconductors indispensable for junction devices.

6. Electronic materials that will probably play important roles in the future are GaN for light sources, $YBa_2Cu_3O_7$ for interconnections, and titanates for capacitors. Approximately where on the scale of conductivities in Figure 2.1 do these materials belong?
7. The p-type regions of two discrete diodes are connected in common. Why does the resultant n–p–n structure not behave like a transistor?

8. Distinguish between (a) bonds and bands, (b) thermal and impurity scattering, (c) spontaneous and stimulated emission, (d) Fermi energy and energy gap, (e) direct and indirect band-gap semiconductors, (f) electronic and opto-electronic devices, (g) metal–semiconductor and p–n junction diodes, (h) ohmic and Schottky contacts, (i) MESFETs and HEMTs, (j) bipolar and MOS transistors.
9. High local electric fields promote avalanche breakdown effects in both semiconductors and dielectrics. Such breakdown is usually harmless in semiconductors but can be destructive in dielectrics. Why? Electron-avalanche effects do not occur in metals. Why?
10. High-power semiconductor diodes are usually fabricated containing a heat sink. To understand why, sketch the effect of high operating temperatures on diode current–voltage characteristics.
11. When ohmic and Schottky contacts suffer electrical degradation, some alteration of the metal–semiconductor band diagram is expected. Indicate possible band-diagram changes in the vicinity of a contact and explain why they occur.
12. Boeing aerospace conducted reliability studies on some 25 million semiconductor devices and found that bipolar power devices failed significantly less often than NMOS and CMOS integrated circuits. What are possible reasons for this observation?
13. In a real, or nonideal, MOS capacitor structure, the work functions of the metal and semiconductor differ. Sketch the band structures of an unbiased metal–$SiO_2$–p-type Si capacitor if $\phi_M > \phi_{Si}$.
    a. Illustrate depletion, accumulation and inversion for this structure.
    b. Sketch the band structures for depletion, accumulation, and inversion in an ideal metal–SiO2–n-type Si capacitor structure.
14. One strategy to minimize the probability of latchup in CMOS devices is to fabricate them within a silicon film grown epitaxially on an insulator (SOI). Illustrate how SOI structures suppress latchup.
15. Distinguish between the functions of (a) microprocessor and memory chips, (b) DRAMs and SRAMs, (c) charge-coupled devices and DRAMs, (d) gate capacitor and trench capacitor.
16. When zero bias is applied in the HEMT structure, the conduction band edge in the GaAs lies below the Fermi level, implying a large density of the 2-D electron gas.
    a. What do you expect to happen when the gate is biased negatively?
    b. Draw band diagrams for both zero and negative bias.

**17.** When compared to silicon device technology, compound-semiconductor laser production involves may fewer steps and generally more art. If this is so, what are the reliability implications of such products?

**18.** Some polymers degrade upon exposure to radiation due to bond breaking. Covalent bond energies for some important atomic pairings are:

| Bond pair | Energy (kJ/mol) | Bond pair | Energy (kJ/mol) |
|---|---|---|---|
| C−C | 340 | C−H | 420 |
| C−C | 620 | C−Cl | 330 |

What portion of the electromagnetic spectrum would be expected to cause radiation-damage effects in each bond type?

**19.** Damage to opto-electronic devices and materials often stems from recombination of electron–hole pairs without photon emission (nonradiative recombination) and from self-absorption of the involved radiation. Explain how each phenomenon could cause degradation.

## REFERENCES


[1] S.M. Sze, Semiconductor Devices—physics and Technology, Wiley, New York, 1985.
[2] B.G. Streetman, Solid State Electronic Devices, fourth ed., Prentice Hall, Engelwood Cliffs, NJ, 1995.
[3] J.M. Mayer, S.S. Lau, Electronic Materials Science—for Integrated Circuits in Si and GaAs, Macmillan, New York, 1990.
[4] A. Bar-Lev, Semiconductors and Electronic Devices, third ed., Prentice Hall, Engelwood Cliffs, NJ, 1993.
[5] T.M. Fredericksen, Intuitive IC Electronics, second ed., McGraw Hill, New York, 1989.
[6] S.M. Sze (Ed.), VLSI Technology, second ed., McGraw Hill, New York, 1988.
[7] S. Middleman, A.K. Hochberg, Process Engineering Analysis in Semiconductor Device Fabrication, McGraw Hill, New York, 1993.
[8] W.S. Ruska, Microelectronic Processing, McGraw Hill, New York, 1987.
[9] G.E. Anner, Planar Processing Primer, Van Nostrand Reinhold, New York, 1990.
[10] R.C. Jaeger, Introduction to Microelectronic Fabrication, vol. V, Addison-Wesley, Menlo Park, CA, 1988.
[11] B.R. Pamplin, in: R.C. Weast (Ed.), Handbook of Chemistry and Physics, CRC Press, Boca Raton, FL, 1980.
[12] B. Myerson, IBM J. Res. Develop. 44 (3) (2000) 391.
[13] A. Schwerin, W. Hansch, W. Weber, IEEE Trans. on Electron Devices, ED-34 (1987) 2493.
[14] A.A. Iranmanesh, B. Bastani, IEEE Circuits Devices 8 (3) (1992) 14.
[15] H.J.M. Veendrick, MOS ICs-from Basics to ASICs, VCH Publishers, Weinheim, 1992.
[16] E. Adler, et al., IBM J. Res. Devel. 39 (1/2) (1995) 167.

[17] H.E. Mass, G. Groeseneken, H. Labon, J. Witters, Microelectronics J. 20 (1989) 9.
[18] W.T. Anderson, Tutorial Notes, IEEE International Reliability Physics Symposium, 7.1, 1992.
[19] W.J. Roesch, Tutorial Notes, IEEE International Reliability Physics Symposium, 1.1, 1988.
[20] A. Christou, Tutorial Notes, IEEE International Reliability Physics Symposium, 3.1, 1986.
[21] H. Morkoc, IEEE Circuits and Devices 7 (1991) 15.
[22] F. Magistrali, C. Tedesco, E. Zanoni, C. Canali, in: A. Christou (Ed.), Reliability of Gallium Arsenide MMICs, Wiley, Chichester, 1992.
[23] W.T. Anderson, K.A. Christianson, C. Moglestue, Quality and Reliability Eng. Int. 9 (1993) 367.
[24] K. Wilson, GEC J. Res. 4 (1986) 126.
[25] A. Bensoussan, P. Coval, W.J. Roesch, T. Rubalcava. In: 32nd Annual Proceedings of the IEEE Reliability Physics Symposium, 1994, p. 434.
[26] C.P. Skrimshire, J.R. Farr, D.F. Sloan, M.J. Robertson, P.A. Putland, J.C.D. Stokoe, R.R. Sutherland, IEE Proc. Pt J 137 (1) (1990) 74.
[27] S.P. Sim, in: A. Christou, B.A. Unger (Eds.), Semiconductor Device Reliability, Kluwer, Amsterdam, 1990.
[28] M. Fukuda, IEEE J. Lightwave Technol. 6 (1988) 1488.
[29] M. Fukuda, Reliability and Degradation of Semiconductor Lasers and LEDs, Artech, Boston, 1991.
[30] J. Spencer, R. Koelbl, Tutorial Notes, IEEE International Reliability Physics Symposium, 1.1, 1989.
[31] R.W. Mann, L.A. Clevinger, P.D. Agnello, F.R. White, IBM J. Res. Develop. 39 (1995) 403.
[32] W. Zulehner, in: G. Harbeke, M.J. Schulz (Eds.), Semiconductor Silicon, P.2, Springer Verlag, Berlin, 1989.
[33] A.B. Joshi, R. Mann, L. Chung, M. Bhat, H.T. Cho, B.W. Min, D.L. Kwong. In: 33rd Annual Proceedings of the IEEE Reliability Physics Symposium, 1995, p. 156.
[34] C. Felsch, E. Rosenbaum. In: 33rd Annual Proceedings of the IEEE Reliability Physics Symposium, 1995, p. 142.
[35] O. Joubert, A. Legouil, R. Ramos, M. Helot, O. Luere, E. Richard, G. Cunge, T. Chevolleau, E. Pargon, L. Vallier, T. Morel, S. Barnola, T. Lill, J. Holland, A. Patterson, AVS 22 (2006).
[36] Y.-C. Yeo, Q. Lu, P. Ranade, H. Takeuchi, K.J. Yang, I. Polischchuk, T.-J. King, C. Hu, S.C. Song, H.F. Luan, D.-M. Kwong, IEEE Electron Device Lett. 22 (5) (2001) 227.
[37] J.G. Ryan, R.M. Geffken, N.R. Poulin, J.R. Paraszczak, IBM J. Res. Develop. 39 (1995) 371.
[38] Intel Tech. J. 6 (2), 2002, p.10.
[39] T.N. Theis, IBM J. Res. Develop. 44 (3) (2000) 379.
[40] D. Sparks, S. Massoud-Ansari, N. Najafi, Reliability, testing and characterization of MEMS/MOEMS III, in: D. Tanner, R. Ramesham (Eds.), Proceeding SPIE 5343 (2004) 70.
[41] A. Margomenos, D. Peroulis, K.J. Herrick, and L.P.B. Katehi. European Microwave Conference, 2001, p. 1.
[42] W.R. Ashurst, C. Carraro, R. Maboudian, IEEE Trans. Device Mater. Reliability 3 (4) (2003) 173.

# CHAPTER 3

# Defects, Contaminants, and Yield

## 3.1 SCOPE

There are defects and *defects*, and there are *killer defects*!! From the relatively innocuous to those that immediately disqualify an integrated circuit (IC) chip, this chapter is largely concerned with the nature of defects, their generation during processing, strategies used to minimize their influence, and how they adversely influence product yield. In Chapter 2 we described how a failure could be hastened by a damaged area. We will now negotiate the mirky waters between an IC that fails at final test and one that passes but is doomed to early failure. The harmful effects of crystallographic, structural, compositional, or stoichiometric defects strongly depend on the nature of the material they inhabit. For example, metals are usually far more tolerant of point defects, dislocations, and grain boundaries (GB) than are semiconductors. Semiconductors must be free of dislocations and GBs, whereas metallizations contain these in relatively large densities with little ill effect. And, since the matrices of most ceramics and dielectrics used in microelectronics are already amorphous or fine grained, these same crystallographic defects usually affect properties only marginally, if at all.

At the next level there are *defects* that represent gross deviations in the structural and chemical integrity of the material over macroscopic dimensions. They include voids, precipitates, missing or misformed structural features because of lithographic errors, deformed or even cracked structures due to stress, and particulate contaminants. Such defects may or may not cause devices or ICs to immediately fail inspection, and might or might not be a source of reliability problems.

*Killer defects*, however, immediately cause rejection of the product. It does not take a very large defect to become a killer defect if it is strategically located, such as at a semiconductor junction or on a lithography mask. The most troubling aspect is that as circuit features continue to get ever smaller, killer defect sizes shrink in concert. As a rough guide, the "1/3 rule" [1] suggests that the killer defect size scales as one-third the minimum feature size.

The general problem of defects can be put into perspective by considering DRAM chips, the primary driver of microelectronics technology. On the

*Reliability and Failure of Electronic Materials and Devices*
ISBN 978-0-12-088574-9
http://dx.doi.org/10.1016/B978-0-12-088574-9.00003-3

basis of the 1/3 rule, the killer defect size is estimated to be ~0.06 μm for 256-Mb DRAMs (2000), ~0.05 μm for 1-Gb DRAMs (2003), and only ~0.03 μm for 4-Gb DRAMs (2007). About 700 distinct processing steps are now required to produce advanced DRAM chips. Among these are over 25 lithographic steps, which must be carried out in a manner such that production yields of chips passing final inspection are high. At every step in the process there are obstacles, or shall we say opportunities, that have the potential to cause defective products.

Concern over processing defects is perhaps the chief technical preoccupation of manufacturing industries. Survival in the marketplace often hinges on identifying defects and developing and implementing strategies to eliminate them. Defects are intrinsic to processing, and a process devoid of defects does not yet exist.

In what must be regarded as one of the more important chapters in the book, aspects of IC manufacturing are explored again with particular emphasis placed on processing defects and their consequences. Defects not only influence production costs directly but also reliability concerns more indirectly; those that go undetected in qualified products at final test are like seeds that can sprout under certain circumstances to yield a bitter reliability harvest.

This chapter starts by introducing crystallographic defects and their role in Si and GaAs. Subsequent topics continue the treatment of chip-level processing introduced at the end of Chapter 2, paying particular attention to cataloging processing defects. Specific examples are incorporated explaining the role of stress and highlighting the deleterious role of ubiquitous contaminants. Lastly, the integrated statistical effect of defects on process and product yield is modeled.

While it is impossible to carry out manufacturing processes without incorporating defects, the latter are for the most part removed by inspection and final test, or by screening. And those that pass into the released (sold) product are often "cosmetic" and more benign in service than the impression sometimes conveyed by this chapter. With this caveat we start with a discussion of defects in crystalline solids and semiconductors.

## 3.2 DEFECTS IN CRYSTALLINE SOLIDS AND SEMICONDUCTORS

### 3.2.1 General Considerations

All crystalline materials contain *intrinsic* (lattice) defects such as vacancies, dislocations, and GBs. These defects are defined with examples in Table 3.1. For the uninitiated it appears tempting to attribute failures and reliability

**Table 3.1** Imperfections in condensed matter

| Classification | Definition | Examples |
|---|---|---|
| Vibrating atom | Temporary, small displacement from ideal position | Zero point, thermal |
| Electronic charge | Charge carrier excited from ground-state bonding configuration | Electron (−), hole (+), exciton |
| Chemical impurity | Foreign atom of differing size, valence, electronegativity, and/or structure relative to host atoms/structure | Substitutional, interstitial |
| Point lattice defect | Missing host atom, extra host atom, atom occupying wrong lattice site | Lattice vacancy, self-interstitial, antisite defect (compounds) |
| | Nonbridging bond | (Noncrystalline materials) |
| One-dimensional defect | Row of atoms at the edge of extra half-plane of atoms | Dislocation (edge and screw) |
| Two-dimensional defect | Boundary separating an error in stacking sequence of atomic planes, boundary between two crystals of different relative orientation | Stacking fault, GB |
| Three-dimensional defect | Macroscopic region of different density, chemical content, coordination, etc., from host | Void, free volume for glasses, disorder, defect cluster, precipitate |

From Ref. [2].

problems in electronic materials to the role of such defects, and at one level such an assumption has validity. In practice, however, the evidence leading from microscopic defects to macroscopic failures is usually very tenuous, and few of the latter can unambiguously be attributed to the former. Complicating matters further are the *processing* defects, introduced during various steps of the manufacturing sequence. These are grosser in scale than lattice defects and strongly influence yield and malfunction of devices. Returning to lattice defects, the rationale for studying them is to intelligently assess their potential and actual roles as root causes of failures, even though it may be difficult to draw causal relationships.

### 3.2.2 Point Defects

The vacancy or absence of an atom at a lattice site is the most basic defect. Vacancies occur in all classes of solids. Not surprisingly, because the movement of atoms from one location to another (diffusion) requires vacancies, and because diffusion is easier at higher temperature, we find that the fraction or concentration of vacant lattice sites ($C_v$) is given by the familiar temperature-dependent Boltzmann expression

$$C_v = \exp - \frac{E_f}{kT'} \tag{3.1}$$

where $E_f$ is the energy required to form a vacancy, i.e., remove an atom from the bulk and place it on the surface. This equation, predicted from thermodynamic considerations, has been experimentally verified in all classes of solids. As temperature increases so does the atomic vibrational frequency, atom spacing, and concentration of vacancies. One of the most damaging effects vacancies exhibit is the formation of voids, usually at an interface between differing materials. Such damage is generally attributed to the condensation of very large numbers of vacancies.

The chief difference between point defects in metals and semiconductors is that the latter are generally electrically active. Charge activity arises from the four severed dangling covalent bonds surrounding the vacancy in the diamond cubic lattice of Figure 3.1. In contrast, the very mobile free electrons rapidly smooth any charge singularities associated with vacancies in metals.

An initially uncharged vacancy $V^o$ in a Si lattice can act as an acceptor by acquiring an electron so that it becomes $V^-$. Similarly, it can acquire a positive charge by combining with a hole and become $V^+$. One can also envision neutral vacancies acquiring more than a single net charge, so that

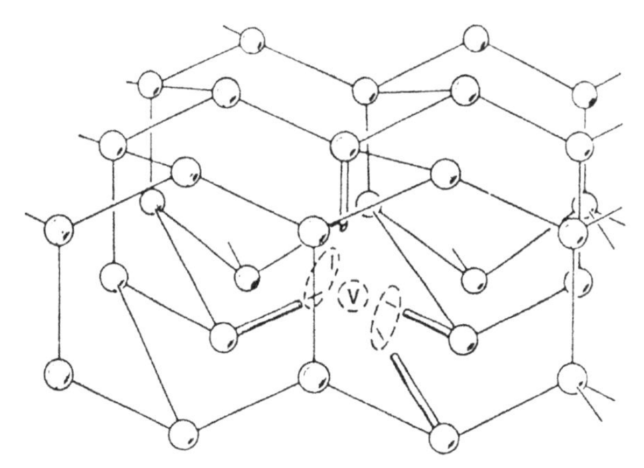


**Figure 3.1** Schematic of diamond cubic lattice and broken bonds around a vacancy (V).

$V^{2-}$, $V^{3-}$, $V^{2+}$, $V^{3+}$, etc., species can exist. Concentrations of such vacancies depend on carrier concentrations because vacancy–charge interactions can be viewed from the standpoint of chemical equilibrium, e.g., $[V^{-}]/(n[V^{o}]) = K$. Here we note that $n$ and $p$ are electron and hole concentrations, and that $K$ is the equilibrium constant.

There are other defects associated with impurity atoms (Au, $Fe_i$, Mo, Ni) or impurity–impurity pairs ($Fe_i$–B) that are electrically active in Si. The subscript i denotes an atom occupying an interstitial site. Similar impurity atom defects have been identified in GaAs (e.g., Fe, Cu) and InP (e.g., Zn). Just as with traditional dopants, these defects have energy levels within the forbidden gap and can act as donors or acceptors. A difference is the much deeper (ionization) energies in the gap relative to standard dopants. To study such low defect concentrations (i.e., $10^{10}$–$10^{14}/cm^3$) use has been made of deep-level transient spectroscopy (DLTS). This technique relies on excitation of charge emission from defects located within the capacitance region of a p–n junction and detection of the resulting current decay to equilibrium [3]. Significantly, DLTS has been installed in the manufacturing environment to monitor trace contamination during deposition of epitaxial films.

Other semiconductor defects include vacancy–interstitial pairs or Frenkel defects and antisite defects, which occur in II–VI and III–V compounds. These are characterized, for example, by a Ga atom on an As site within the As sublattice, and similarly misplaced As atoms on the Ga sublattice.

### 3.2.2.1 Effect on Diffusion

Point defects play a role in elevated temperature doping, sintering of metal–semiconductor contacts, postannealing of ion implants, and epitaxial deposition. All involve diffusion of atoms, and such processes are complicated in semiconductors. During doping, for example, concentration-dependent diffusion effects arise that can be attributed to pair interactions between charged as well as uncharged vacancies and dopants or impurities. To illustrate the complexity involved, an expression for the extrinsic (impurity) diffusion coefficient ($D_{ex}$) is [4]

$$D_{ex} = D_i^o + D_i^+ (p/n_i) + D_i^- (n/n_i) + D_i^{2-} (n/n_i)^2 + D_i^{3-} (n/n_i)^3. \quad (3.2)$$

Thus $D_{ex}$ depends on the doping concentration levels that yield $n$ or $p$, as well as the carrier concentration ($n_i$) in dopant-free or intrinsic Si. In order of appearance in Eqn (3.2), the various $D_i$ terms represent the intrinsic diffusivity of impurity interaction with neutral point defects, a singly-charged donor point defect, a singly-charged acceptor point defect, a doubly-charged acceptor point defect, and a triply-charged acceptor point defect. Since $n$ and $p$ directly reflect the doping concentration ($C$), this complex expression suggests that $D_{ex}$ depends nonlinearly on $C$ such that $D_{ex} \sim C^{\alpha}$, with $\alpha = 1,2,3\ldots$. Therefore, doping profiles are far more complex than those identified with a linear diffusion theory, where $D$ is independent of concentration (see Section 5.2).

Examples of anomalous dopant diffusion effects and profiles are shown in Figure 3.2(a) and (b). It is evident that the concentration profiles of arsenic, boron, and phosphorous deviate from the well-known complementary error function solution that characterizes linear diffusion (Eqn (5.3)) from a constant surface source. The case of the heavily doped phosphorus profile in a diffused emitter (Figure 3.2(b)) is particularly interesting. Near the surface, phosphorus atoms associate with vacancies. But dissociation of this $(PV)^-$ complex via the reaction $(PV)^- \rightarrow P^+ + V^- + e^-$ releases vacancies in the profile tail. The enhanced diffusion induces the observed kink in the profile that causes the so-called emitter-push effect in n–p–n narrow-base, diffused-bipolar transistors. Typically, the boron-diffused base just under the emitter is ~0.2 μm deeper than outside the emitter region (Figure 3.3(a)) because vacancies enhance boron diffusion as well.

Dopant diffusion profiles are also altered at junctions or near surfaces and interfaces particularly when oxygen is present. Consider a junction near the surface of a Si wafer patterned by a silicon nitride mask to expose windows for oxidation. For dopants like B, P, and As, diffusion is enhanced,

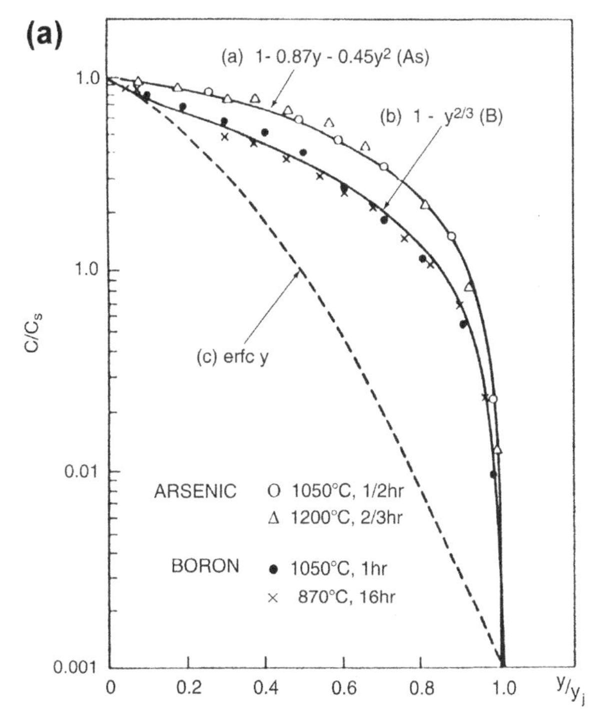


**Figure 3.2** (a) Normalized diffusion profiles for arsenic and boron in silicon. The error function (erfc) solution is compared with actual profiles corresponding to concentration-dependent $D$ values. Note that $C_s$ is the surface concentration and $y/y_j$ is related to $x/x_j$ where $x_j$ is the junction depth. (b) Phosphorous diffusion profiles in silicon. Deviation from the erfc profile (a) occurs at successively higher surface concentrations (b, c, d), where ultimately kinks develop. *(From Ref. [5].)*

and the junction moves deeper into the wafer under the window than under the nitride, which is an effective barrier to oxygen penetration (Figure 3.3(b)). Under the same conditions, however, the diffusion of Sb is retarded (Figure 3.3(c)), resulting in a shallower junction. In another effect, shown in Figure 3.3(d), lateral diffusion of dopant enhances the junction depth ($x_j$) under nitride films during oxidation. The smaller $w$ is, the larger is $x_j$. An explanation of these effects assumes that Si (self) interstitials are injected into Si during oxidation in order to simultaneously accommodate the volume expansion stemming from oxide formation and reduce the interfacial stress that is generated. Semiconductor point defects include interstitials, and like vacancies, they may occur in nonequilibrium concentrations. Dopants (B and P) that prefer to use interstitials to facilitate

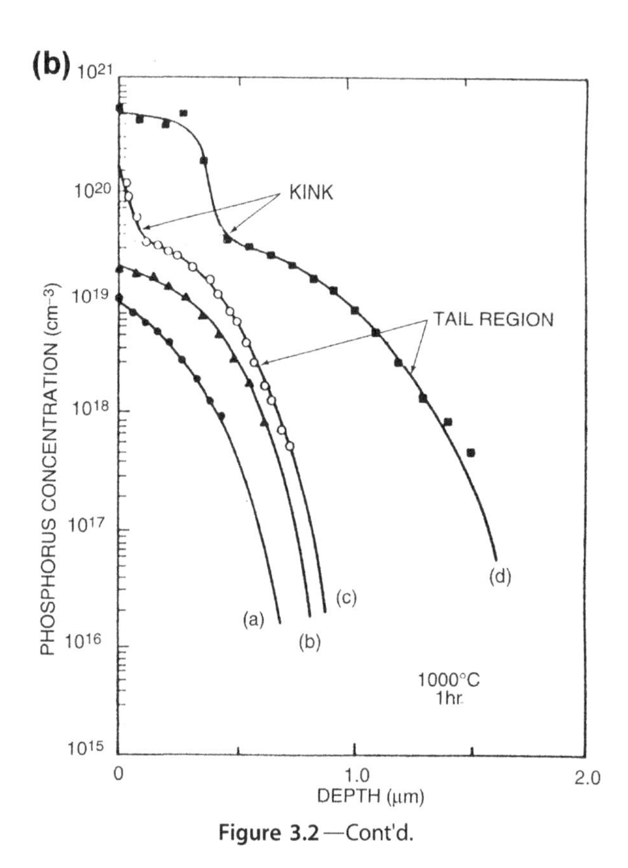


**Figure 3.2**—Cont'd.

mass transport show oxidation-enhanced diffusion, while the opposite is true for Sb, which diffuses via a vacancy mechanism.

Calling these enhancements and retardations of dopant concentration processing defects is perhaps too strong an attribution. Even though they alter the intended device behavior, their effects can be designed around in most cases.

### *3.2.2.2 Oxygen in Crystalline Silicon*

It was the fate of silicon, aluminum, and oxygen, the most abundant elements on earth, to be brought together in the minutest of amounts to make ICs possible. While these elements play unambiguously beneficial roles in semiconductor, conductor, and insulator functions, the presence of oxygen within silicon is like a "two-edged sword" [6] as far as device behavior is concerned. Oxygen enters Czochralski process (CZ)-grown crystals because the melt continually erodes and dissolves the $SiO_2$ crucible

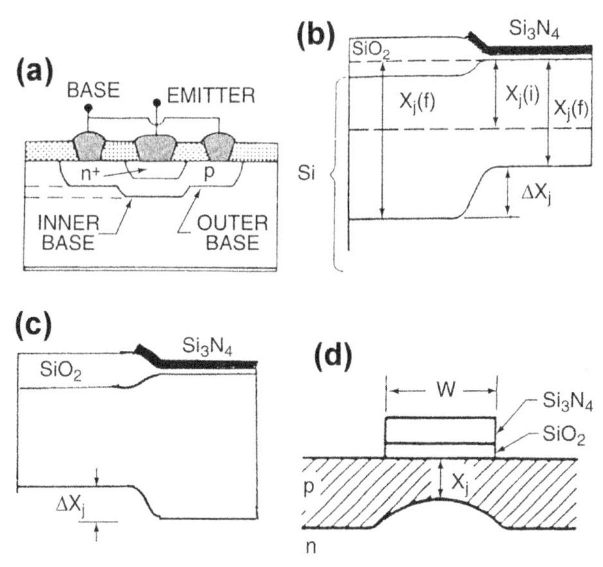


**Figure 3.3** Collection of anomalous diffusion effects and profiles, where $x_j$ (i) and $x_j$ (f) are the initial and final junction positions, respectively. (a) Emitter-push effect. (b) Dopant diffusion enhancement under the window relative to the nitride. (c) Dopant diffusion retardation under the window relative to the nitride. (d) Lateral diffusion enhancement.

wall during crystal pulling. A snapshot of the way defects enter the pulled crystal is schematically conveyed in Figure 3.4. Point defects present in solid solution at the highest temperatures promote the nucleation and growth of dislocation defects and oxide precipitates as the crystal cools. Within crystalline Si, oxygen reveals its many-faceted behavior depending on temperature. When dissolved, oxygen occupies an interstitial site and is electrically inactive. Upon heating silicon to various temperature ranges, the changes that occur are summarized in Table 3.2. When oxygen-induced defects are retained in wafers at room temperature, they can be a source of considerable concern.

#### *3.2.2.3 Gettering*

Treatments that remove process-induced contaminants from the active regions of the Si wafer, where devices are being fabricated, are known as gettering [2]. Metal contaminants are the particular targets of gettering processes because they, in combination with precipitates at dislocations and stacking faults, cause excessive leakage currents and low junction breakdown voltages. There are two types of gettering.

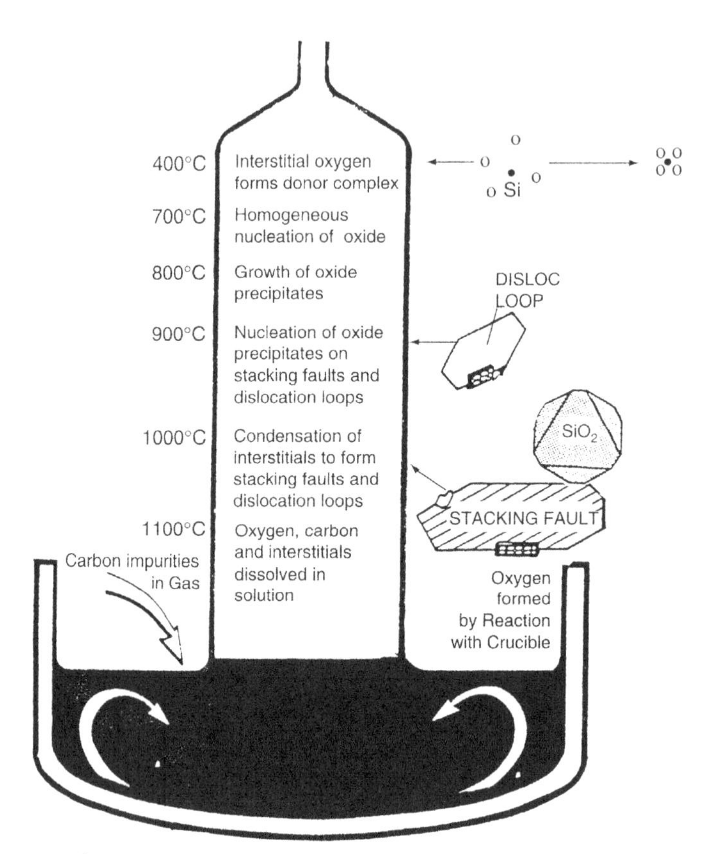


**Figure 3.4** Defect interactions and precipitation reactions in CZ-grown silicon single crystals as a function of temperature.

**Table 3.2** Oxygen effects in CZ-grown silicon

| Temperature range | State of oxygen |
|---|---|
| 400–500 °C | Electrically active, oxygen-related donors formed due to small oxygen complexes |
| 600–700 °C | First visible evidence of oxygen precipitation, rodlike coesite ($SiO_2$) structures and 2-nm amorphous precipitates identified |
| 800–1000 °C | Platelike amorphous structures in the size range 10–500 nm |
| 1000 °C | Platelike precipitates accompanied by punched-out prismatic dislocation loops and ESF with associated precipitate colonies |
| 1200 °C | Large polyhedral precipitates identified as being amorphous, plus large stacking faults |

From Ref. [2], p. 223.

#### 3.2.2.3.1 Extrinsic (or External)

In this case gettering is achieved by deliberately introducing trapping sites in the backside of the Si wafer. Through surface damage produced by mechanical abrasion, ion implantation, or laser beam heating and cooling, dislocations are generated. At high temperatures (~1000 °C) these defects attract harmful metals such as Cu, Fe, Ni, and Cr, which diffuse to them from the active device regions. The kinetics of gettering depends on the rates of impurity release from the associated state, as well as diffusion to and capture by the getter site.

#### 3.2.2.3.2 Intrinsic (or internal)

Intrinsic gettering is employed to remove both harmful oxygen and metallic elements through heat treatment alone. First, wafers should ideally contain an oxygen concentration of about $8.5 \times 10^{17}$ $cm^{-3}$. After an elevated temperature solution heat treatment the wafers are cooled, and oxygen precipitates from solid solution in the form of $SiO_2$ particles. These plastically compress the lattice, locally generating dislocations and stacking faults that tie up impurities in the manner described above for extrinsic gettering. In bipolar and complementary metal oxide–semiconductor (CMOS) technologies, which require higher process temperatures and a larger number of processing steps compared to conventional NMOS technology, attention must be paid to the problem of wafer warpage during thermal treatments [7].

The trick in either case is to locate the precipitates far from the devices. This can be accomplished in a three-step, high–low–high temperature annealing sequence [8]. First the wafer surface is deoxidized at ~1100 °C. Outdiffusion and evaporation of oxygen reduces its level to the solubility limit, creating a so-called denuded zone. Annealing times are chosen to extend the denuded zone beyond the position of the deepest p–n junctions to be fabricated. Next, the wafer temperature is reduced to ~700 °C, which forces oxygen to be expelled from supersaturated Si to nucleate precipitates. Times ranging from a few hours to a few days are required to do this optimally. Lastly, a high-temperature anneal in the range 1000–1200 °C is required for precipitate growth, dislocation formation, and diffusion of atoms to traps.

Often this complex gettering treatment is performed in conjunction with extrinsic gettering.

### 3.2.3 Dislocations

These well-known defects extend along a *line* of atoms in a crystalline matrix. Dislocations exist in all classes of solids but are most easily visualized

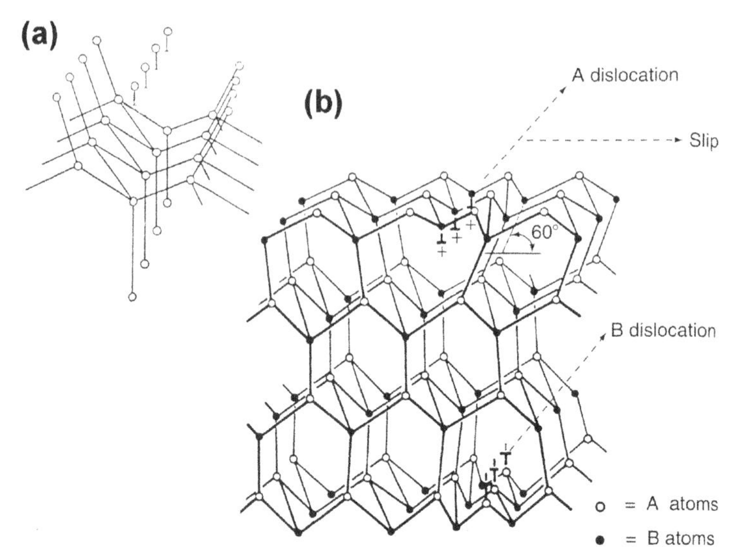


**Figure 3.5** (a) Edge dislocation in the diamond lattice. (b) Dislocations in AB compound semiconductor lattice. *(From Ref. [9].)*

in cubic lattices. There are two fundamental types of dislocations—the edge and screw. Forming an edge dislocation by inserting an extra row of atoms into the diamond cubic lattice of silicon leaves a line of dangling bonds in its wake as shown in Figure 3.5(a). These bonds are generally viewed as acceptors that capture electrons to lower their overall energy. But now a hole can be trapped, and in the recombination that occurs, the captured electron is annihilated. The original high electron energy state is restored, and the process can repeat. This is the reason that dislocations are viewed as carrier generation–recombination centers. Cores of such dislocations are thus negatively charged and surrounded by a cylinder of positive space charge. This would confer inverted *p*-type behavior near the dislocation in an otherwise n-type matrix.

Dislocations in compound semiconductor lattices, depicted in Figure 3.5(b), are more complex than in Si because two atoms (A and B) are now involved. There are now different core charges depending on whether A or B surfaces are exposed. Dangling A bonds, for example, might produce a positive core charge. Dislocations in compound semiconductors have been known to reduce the lifetime of laser diodes, affect the threshold voltage of field-effect transistors, and reduce electron–hole recombination processes that lead to radiation of light.

### 3.2.3.1 *Electrical Effects of Dislocations*

To better understand the electrical effects associated with dislocations, let us estimate the effective size of the dislocation depletion radius ($r_D$) due to n-type doping to a level ($N_D$). Following Zolper and Barnett [10], Poisson's equation in cylindrical coordinates, used to describe the electrical potential ($V$) that exists around the dislocation pipe of Figure 3.6(a) at any radius $r$, has the form

$$\nabla^2 V = \frac{1}{r}\frac{\mathrm{d}}{\mathrm{d}r}\left(r\frac{\mathrm{d}V}{\mathrm{d}r}\right) = \frac{-\rho}{\varepsilon\varepsilon_o} = \frac{-eN_D}{\varepsilon\varepsilon_o}. \tag{3.3}$$

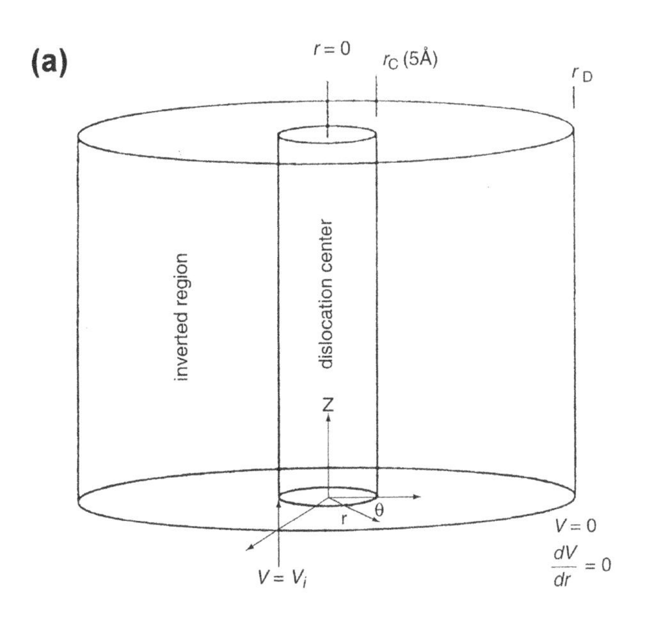


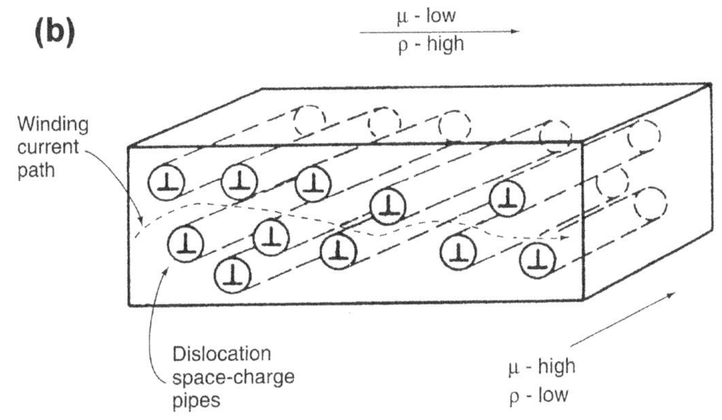


**Figure 3.6** (a) Charged dislocation pipe in cylindrical coordinates. *(From Ref. [10].)* (b) Current flow around charged dislocations in semiconductors. The electron mobility is $\mu$ and the resistivity is $\rho$. *(From Ref. [9].)*

Here $\rho$ is the charge density, e is the electronic charge, and $\varepsilon$ and $\varepsilon_o$ are the dielectric constant and permittivity of free space, respectively. As boundary conditions, it is assumed that $V(r_D) = 0$, and $dV(r_D)/dr = 0$. Upon double integration and substitution of these boundary conditions,

$$V(r) = \frac{-eN_D}{\varepsilon\varepsilon_o}\left\{\left(\frac{r_D^2}{2} - ln\, r_D\right) - \left(\frac{r^2}{2} - ln\, r\right)\right\} \tag{3.4}$$

Finally, at the dislocation core radius ($r_c$), typically 0.5 nm, a potential $V_i$ exists that causes inversion. Assuming $V_i$ at $r_c$ (i.e., $V(r_c) = V_i$) we can then solve for $r_D$ at any doping level by trial and error methods. Calculation shows that if $V_i = 1$ V, and $N_D = 2 \times 10^{17}$ cm$^{-3}$, then $r_D = 44$ nm. Physically, the dislocation core traps electrons and in the process is surrounded by a positively charged space charge region.

The implications of charged dislocations on current flow are depicted in Figure 3.6(b). Tortuous current paths result in increased carrier scattering and reduced mobility. Anisotropy develops such that carrier mobility and conductivity normal to dislocation lines are both reduced relative to their behavior parallel to them. Dislocations have also long been identified as charge recombination centers that degrade the minority carrier diffusion length and lifetime. Table 3.3 indicates the extent to which dislocations degrade the efficiency of silicon ribbon solar cells. Perhaps more significant than the effect of dislocations on carrier motion is their ability to interact with assorted local impurities and point defects. When metallic impurities segregate at dislocations, problems due to excessive leakage currents and shorting of junctions arise. We have already noted such interactions in the practice of gettering metallic impurities by intentionally introducing dislocations.

**Table 3.3** Correlation between dislocation density and solar cell characteristics

| Dislocation density (cm$^{-2}$) | Minority carrier diffusion length (μm) | Cell efficiency % |
|---|---|---|
| Region | | |
| At twins bulk | | |
| $2 \times 10^8 \sim 10^6$ | 12 | 9.5 |
| $3 \times 10^8 \sim 10^6$ | 19 | 10.0 |
| None observed | 135 | 14.3 |
| None observed | 156 | 14.9 |

After Ref. [11].

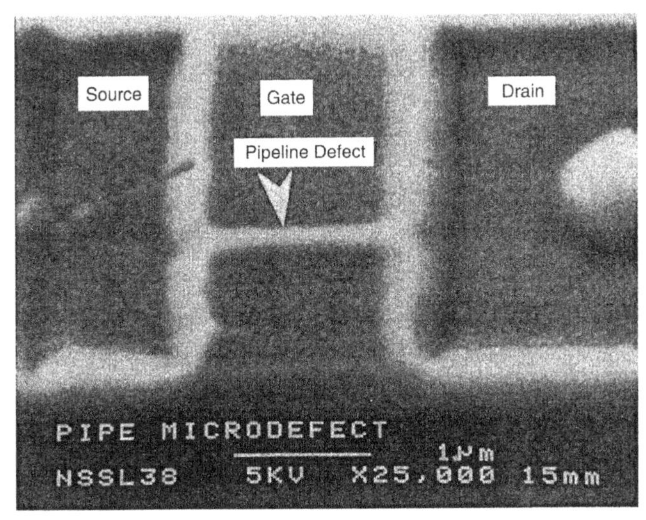


**Figure 3.7** Scanning electron microscope image of source–drain pipeline defect in a field-effect transistor. *From Ref. [12].*

Dislocations present in the thin base region of epitaxial bipolar transistors are particularly troublesome defects. Such dislocation "pipes" in effect are conducting paths between the emitter and collector that electrically short the base. Pipeline defects have also been observed [12] in CMOS devices, e.g., EEPROMs, where they serve to short sources to drain regions (Figure 3.7). Dopant implant damage coupled with compressive stress from the surrounding oxide appears to be the source of this yield-limiting defect.

### *3.2.3.2 Stacking Faults*

So far we have focused on bond-breaking defects that include dislocations, surfaces, or vacancies. On the other hand, stacking faults correspond to errors in perfect stacking sequences and do not generally involve breaking nearest neighbor bonds. Therefore, charge distributions around stacking fault defect states are only mildly perturbed from those of a perfect lattice. If we consider the (111) close-packed plane in face-centered cubic structures, atoms appear to be arrayed as racked billiard balls on a pool table. Let us call this the A plane. As new atoms or balls are stacked above this layer they generate a parallel (111) plane designated as B, displaced slightly from this first layer depending upon which depressions or interstices they nest in. A third stacked row of atoms now has a choice of lying directly above the A layer or occupying a set of interstices that lie above neither the A nor the B layer; such a new layer is the C plane. In perfect FCC structures, atomic layers are ordered in sequential ... ABCABCABC ... fashion. But if the planar ordering is disrupted slightly and the sequence is ... ABCBABCABC ... (by adding an extra B row) or is ... ABCACABC ... (by withdrawing a B row), then stacking faults are created.

Due to the presence of two atoms in the primitive unit cell of silicon, the sequence of {111} planes is now ...AA′BB′CC′AA′BB′CC′..., etc. Intrinsic stacking faults (ISF), extrinsic stacking faults (ESF), and twin stacking faults (TSF) may be viewed as simple variations from this ideal stacking sequence. For example, in the ISF a pair of atomic planes (such as AA′) is removed from the ideal stacking sequence, while the ESF is formed by inserting a pair of AA′ planes between the BB′ and CC′ layers and the TSF is produced by imposing reflection symmetry through a plane midway between the A and A′ layers, i.e., ... BB′CC′AA′C′CB′B. ...

Stacking faults are formed during oxidation. This process requires accumulation of oxygen at the wafer surface, diffusion through the growing oxide, and reaction at the Si interface, where, as noted earlier, an excess of Si interstitials and a deficiency of vacancies develop. Oxidation-induced stacking fault (OSF) defects nucleate at wafer surface damage sites, contamination, and oxide precipitates. They grow by absorbing Si interstitials or emitting vacancies, and their length increases with oxidation time and temperature. Small chlorine additions to the oxidizing ambience markedly decrease their density. Leakage currents in metal oxide–semiconductor devices are enhanced by many orders of magnitude in the presence of OSFs, especially if the associated dislocations are decorated by metallic precipitates [13]. The cross-section of an epitaxially grown Si layer (Figure 3.8(a)) shows a stacking fault defect possessing a stripelike character. Nucleating at impurities or oxides, stacking faults in Si fan out and often appear as triangles or squares when they intercept the wafer surface normally. OSFs have resulted in high reverse leakage currents at p–n junctions in charge-coupled devices, as the correlation of Figure 3.8(b) reveals.

### 3.2.4 Grain Boundaries

These surface or area defects are the interfaces that separate individual single crystal grains from one another. In the hierarchy of defects, GBs are the most structurally complex because they encompass dislocations, point defects, and displaced atoms, all in a complex admixture. There are several effects that GBs display in semiconductors. Aside from extending light absorption to slightly lower energies in some materials (e.g., CdS), GBs are always unambiguously deleterious to device behavior. Their structural

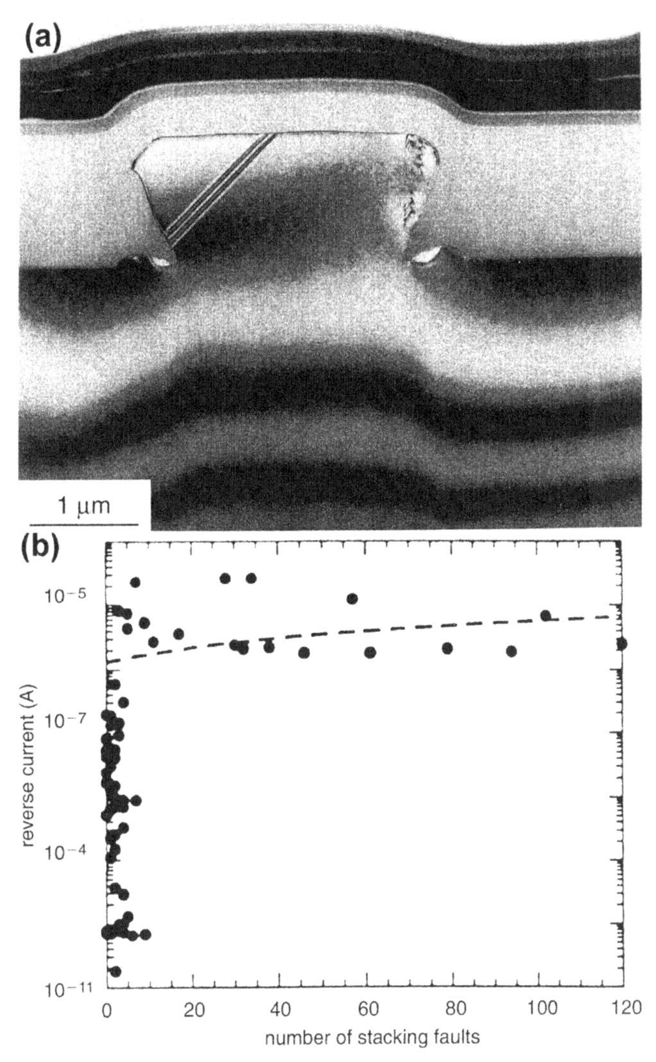


**Figure 3.8** (a) Stacking fault (parallel stripes) and dislocations in the epitaxial tub of a high-speed silicon transistor. The entire area is covered by a dielectric layer. Cross-section transmission electron micrograph. *(Courtesy of S. Nakahara, Lucent Technologies, Bell Laboratories Innovations.)* (b) Reverse leakage currents in charge-coupled devices as a function of the number of OSFs. *(From Ref.* [14].*)*

complexity means that the electrical behavior will likewise be complex. The simplest, or tilt, GBs are merely vertical arrays of edge dislocations. Thus a model similar to that proposed earlier for dislocations holds for GBs, as suggested by Figure 3.9. In this model [15] for n-type material a planar,

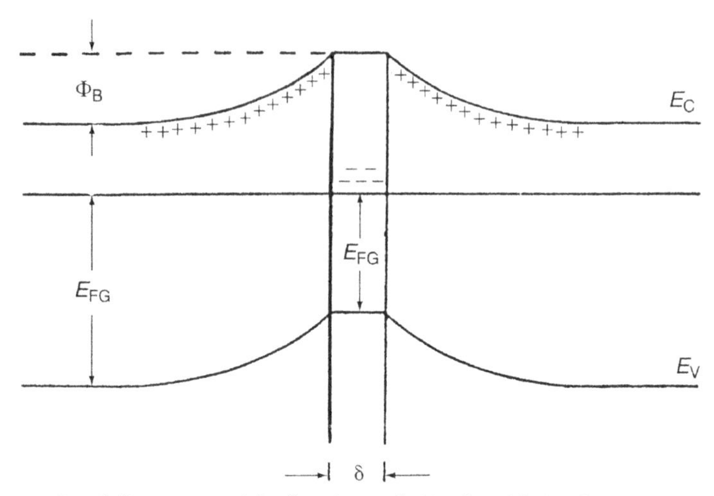


**Figure 3.9** Band diagram model of a charged GB of width δ. Charge is transferred from the grains to fill GB trapping states, establishing a Schottky barrier of height $\Phi_B$. The Fermi levels in the grains and GB are $E_{FG}$ and $E_{FB}$, respectively. *From Ref. [15].*

uniform array of trapped majority carriers (electrons) at the boundary creates positively charged depletion layers in the grains on either side. As the Fermi levels in the grains and boundary equilibrate, a Schottky barrier is established. Experiments involving transport properties in bicrystal and polycrystalline semiconductors have shown that GB defects raise the bulk resistivity, reduce carrier mobility, and enhance carrier recombination rates. These are reasons why the optical performance of polycrystalline light-emitting diodes and solar cells is decidedly inferior to that of similar crystalline devices.

GBs are the preferred location for chemical reactions (e.g., etching, corrosion) as well as the solid-state mass transport effects discussed in Chapter 5 (e.g., diffusion, electromigration). Because boundaries are relatively open structurally, atoms attached to them tend to be energetic. This enables atoms and vacancies to migrate far more rapidly along GBs than through the bulk lattice. Furthermore, GBs are often stronger than the bulk at low temperatures, and they help resist mechanical deformation under stress. But at elevated temperatures, they are weaker than the bulk, and when the material breaks, the fracture often propagates along a GB path. The dividing line is approximately half the melting temperature ($T_M$, in degrees K); above $T_M$ atomic motion and reactions are favored at GBs, weakening their integrity.

Great efforts are made to eliminate any hint of GBs in semiconductor materials. An exception is Si solar cells, where polycrystalline cells are often suitable for the application, and are cheaper to produce, and of course the polysilicon gates of MOSFET devices.

### 3.2.5 Dislocation Defects in DRAMs—A Case Study

The main technology driver in microelectronics is considered to be DRAMs because their fabrication demands the cutting edge in fine-line lithography and employs the largest wafer diameters. For these reasons DRAM technology is the conduit for innovations in materials, processes, and equipment into IC technology. The periodic architecture of memory cell arrays facilitates detection and localization of defective memory cells in fully automated electrical tests. Thus DRAMs provide ideal devices for studying the influence of defects on manufacturing yield.

A schematic of a 4-Mb DRAM cell is reproduced in Figure 3.10(a). Dislocations with distinctive shapes were detected near the capacitor trench using the transmission electron microscope. These defects, shown in Figure 3.10(b), occurred in a region that received a high dose of implanted boron when doping the transistor source and drain [16]. As suggested by Figure 3.10(a), stresses generated during processing were large enough to initiate plastic slip on (111) planes of Si. The role of stress is addressed again in Section 3.3.2.

Recalling that DRAMs must be refreshed periodically to compensate for capacitor charge leakage, we may ask whether dislocations shorten the charge retention time or the refresh delay time. If so, a direct connection between these defects and DRAM performance can be demonstrated. With this motivation, tests that correlated the refresh delay time (or time between refresh cycles) and number of dislocations detected were performed, and typical results are plotted in Figure 3.11. Good cells are represented by the steep, or intrinsic, portion of the data, which extrapolates to a greater than 200 ms charge retention time [17]. The inescapable conclusion is that cells with a retention time of less than this fail by dislocation leakage currents.

## 3.3 PROCESSING DEFECTS

### 3.3.1 Scope

In this section we revisit IC manufacturing with an emphasis on defects that arise during chip-level processing. Schnable [18] has conveniently summarized processing defects, the ways they are manifested, and some of their

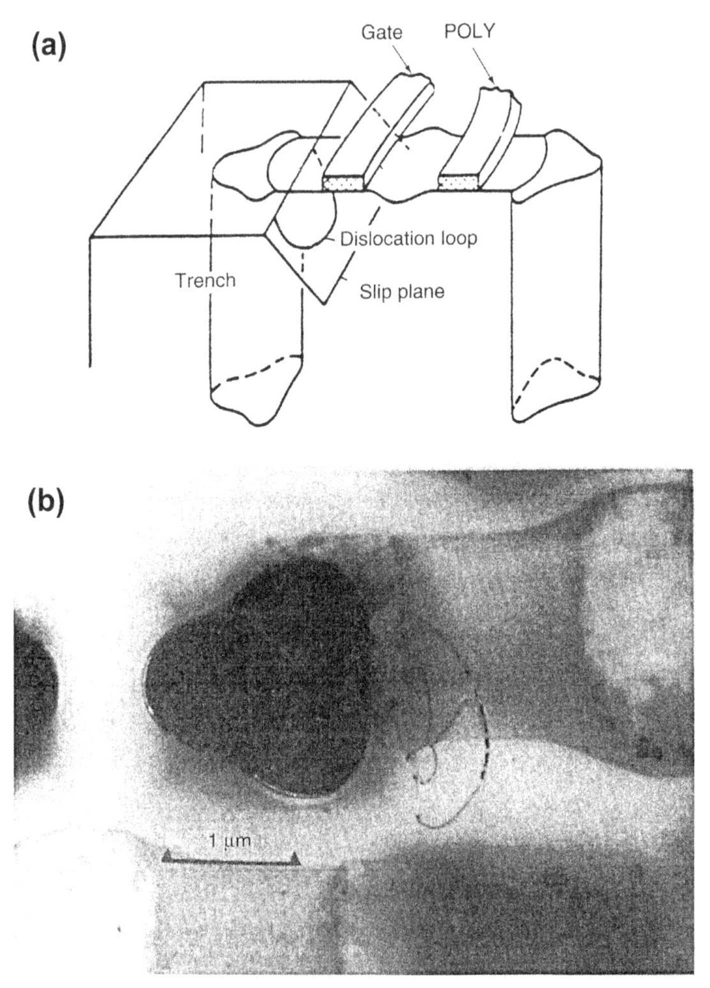


**Figure 3.10** (a) Schematic of DRAM cell. Two device cells with trench capacitors, transfer device pedestal outline, and (111) glide plane containing a dislocation are shown. (b) Plan-view transmission electron micrograph of 4-Mb DRAM cell with dislocations having distinctive shapes generated next to the trench capacitor (dark) and transfer device drain. *(From Ref.* [16].*)*

consequences in the concise compilation of Table 3.4. Defects of a crystallographic as well as structural, electronic, chemical, and morphological nature are included. High on the list of defects in virtually every processing step is contamination, a topic that is comprehensively treated in Section 3.4. We note that other defects have common origins such as stress and the lack of geometric coverage of IC features during film deposition and etching.

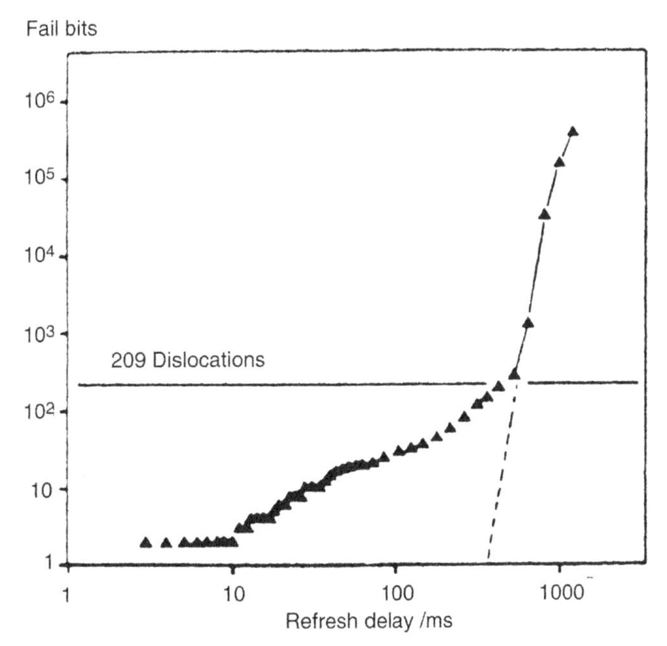


**Figure 3.11** Cumulative number of bits in a 4-Mb DRAM that fail a refresh test as a function of refresh delay time. *From Ref. [17].*

Therefore, these are topics that will be treated in some detail. Lastly, lithography defects deserve separate consideration and will be discussed again in connection with yield. Later, in Chapter 8, there is a similar compilation of defects arising from packaging processes.

## 3.3.2 Stress and Defects

Stress in IC chips is not viewed as a defect per se, because it is an intrinsic consequence of processing dissimilar materials in contact. Nevertheless, stress has an important influence in generating or exacerbating many Table 3.4 defects such as hillocks, stacking faults, slip, internal stress, microcracking, and corrosion. In process steps involving film growth (oxidation) or film deposition (chemical vapor deposition (CVD), metallization), internal stresses develop in the bilayer film–substrate structure.

Internal stresses arise in various manufacturing operations. When a homogeneous metal strip is reduced slightly by rolling, the surface fibers are extended more than the interior bulk. The latter resists the fiber extension and places the surface in compression, while the interior is stressed in tension. This residual stress distribution is locked into the metal but can be

**Table 3.4** Defects associated with microelectronics manufacturing processes at the chip level

1. **Epitaxy defects:** contamination; excessive lateral autodoping; excessive vertical autodoping; film too thick or too thin; haze; heavy metal contamination; hillocks; inadequate resistivity uniformity; inadequate thickness uniformity; low minority carrier lifetime; poor crystallographic quality; pattern shift; resistivity too high; roughness; scratches; stacking faults
2. **Oxidation defects:** alkali ion contamination; dopant penetration through oxide mask; excessive dopant pileup; fixed oxide charge incorrect; heavy metal contamination; inadequate radiation hardness; interface state density too high; localized oxide thinning; OSFs; oxide too thick or too thin; oxide-trapped charge too high; particulate contamination; slip in substrate; slow trapping instability; thickness nonuniform; white ring (Kool effect)
3. **Photoresist patterning defects:** alkali ion contamination; bridging; inadequate planarization; inadequate resist adhesion; inadequate resist etch resistance; inadequate resolution; inadequate step coverage; incorrect edge contour; incorrect linewidth; intrusions; misalignment; missing images; mouse bites; neckdowns; notching; opaque spots; overexposure; particulate contamination; pattern discontinuities; pattern distortion; pinholes; poor resist etch resistance; poor resist thickness uniformity; protrusions; resist too thick or too thin; rough edges; rounding; scratches; scumming; spacing violations; underexposure
4. **Diffusion:** alkali contamination; bandgap narrowing; boron penetration through oxide; crystallographic damage; diffusion pipes; dislocations; dopant precipitation; emitter push; heavy metal contamination; inadequate uniformity; laterally enhanced diffusion; liquid-phase penetration of oxide mask; popcorn noise; sheet resistivity too high; sheet resistivity too low; slip; junctions too deep or too shallow

5. **Ion implantation:** charge buildup; contaminants in implant beam; inadequate dose uniformity; incorrect depth; incorrect dopant species; incorrect dose; incorrect implant angle; metallic contamination; neutral species in implant beam; overheating; particulate contamination; penetration into underlying device; semiconductor lattice damage
6. **Thermal treatment:** alkali ion contamination; carbon contamination; heavy metal contamination; incorrect ambient; incorrect cooling or heating rate; incorrect temperature; incorrect time; particulate contamination; slip; thermal pitting
7. **Cleaning:** alkali ion contamination; excessive etching; heavy metal contamination; incomplete contaminant removal; particulate contamination; pitting; radiation damage; scratches; silicon crystallographic damage
8. **Wet chemical etching:** alkali ion contamination; corrosive residue; freckles (unetched precipitates); heavy metal contamination; insoluble residue; local cell etching; loss of resist adhesion; nonuniform etching; overetching; particulate contamination; photomask erosion; photovoltage effects; pitting; poor etch ratio; poor wetting; staining; undercutting; underetching; wrong etch rate
9. **Plasma etching:** charge buildup; corrosive residues; excessive etching of substrate; excessive loading effects; excessive resist erosion; excessive variation in induction time; freckles (unetched precipitates); grass; hydrogen donor effects in Si; inadequate etch ratios; incorrect edge angle; insufficient amount of etching; ionic contamination; low selectivity; metallic contamination; nonuniform etching; overetching; poor etching uniformity; radiation damage; radiation damage to $SiO_2$; residue in etched openings; residue on sidewalls; silicon crystal damage; stringers; undercutting; underetching
10. **CVD:** devitrification; etch rate too high; excessive compressive stress; excessive hydrogen content; excessive tensile stress; excessively hydroscopic; insufficient conformality; localized etch rate at step too high; nonuniform thickness; not stoichiometric; too thick or too thin; wrong composition
11. **Metallization:** black gold; contains voids; ductility too low; excessive native oxide in contacts; excessive oxide content in metal; excessive oxygen in initially deposited region; excessive roughness; high sheet resistivity; hillocks; in-process radiation damage; low resistivity ratio; microcracks; poor adhesion; poor step coverage; poor thickness uniformity; susceptible to corrosion; susceptible to electromigration; too thick or thin; wrong alloy composition

From Ref. [18].

released like a jack-in-the-box. Simply machining a thin surface layer from the rolled metal surface will upset the mechanical equilibrium and cause the remaining material to bow. Residual stresses also arise in castings, heat-treated glass, and machined and ground materials.

A more pertinent example of residual stress occurs in silicon wafers whose backsides are thinned by grinding prior to dicing [19]. Thinner wafers facilitate dicing, dissipate heat better, and make for thinner packages. Unfortunately, grinding induces stresses that cause wafers to warp and more easily break during the dicing process. Furthermore, warped dies are more difficult to electrically contact and mount in the package. All in all, yield can suffer appreciably due to warped wafers. This underscores the importance of understanding and controlling stress.

### 3.3.2.1 Internal Stress

We all know that stresses and strains will develop throughout a body when forces are applied to it. Under elastic loading conditions, the stresses and strains disappear upon unloading. But materials that are plastically deformed or heated may sometimes contain residual stresses even when the plastic load or heat is removed. Furthermore, what are we to make of processed wafers that are internally stressed, even though no mechanical forces were ever applied to them? Such situations arise all too frequently in thin films attached to substrates, and in these cases internal or residual stresses are said to exist. Internal stresses have important implications in influencing product yield as well as subsequent reliability. There are two main sources of internal stress, and each will be discussed in turn.

### 3.3.2.2 Extrinsic Residual Stress

These stresses, commonly called thermal stresses, originate from the thermal expansion or contraction of materials when they are mechanically constrained. To see how thermal stress can develop in a homogeneous body, consider a rod of length $L_o$ clamped at both ends. If its temperature is raised from $T_o$ to $T$, the rod would tend to expand in length by an amount equal to $\alpha\,(T - T_o)\,L_o$, where $\alpha$ is the coefficient of thermal expansion defined by $\alpha = \Delta L/L_o(T - T_o)$ $(K^{-1})$. But the rod is constrained and is, therefore, effectively shortened in compression. The compressive strain is simply $e = \Delta L_o/L_o = -\alpha(T - T_o)\;L_o/L_o$ or $-\alpha(T - T_o)$. By Hooke's law a stress $\sigma$ develops that is given by $Ee$, where $E$ is the modulus of elasticity. Therefore, combining terms (neglecting signs) yields

$$\sigma = E\alpha(T - T_o). \tag{3.5}$$

Processing of IC chips generally involves bi- and multilayers of different materials, and a number of failure and reliability problems discussed in this book can be attributed to internal or residual stresses in such structures. Even the outwardly homogeneous ground silicon wafer discussed above is an effective bilayer consisting of a thin stressed surface layer and the contiguous bulk with a stress of opposite sign. Internal stresses invariably develop in film/substrate combinations because of thermal mismatch between the film and substrate at the film formation and use temperatures. To see why, we assume that the film is stress free at the elevated deposition temperature $T$, and that $\alpha_f > \alpha_s$. As the film cools to room temperature $T_o$, it contracts more than the substrate, as shown in Figure 3.12(a). Regardless of the resultant stress distribution that prevails, the structure must be in mechanical equilibrium. This requires that both the net force ($F$) and bending moment ($M$) vanish on every film/substrate free body cross-section. Thus

$$F = \int \sigma \mathrm{d}A = 0 \tag{3.6}$$

and

$$M = \int \sigma y \mathrm{d}A = 0, \tag{3.7}$$

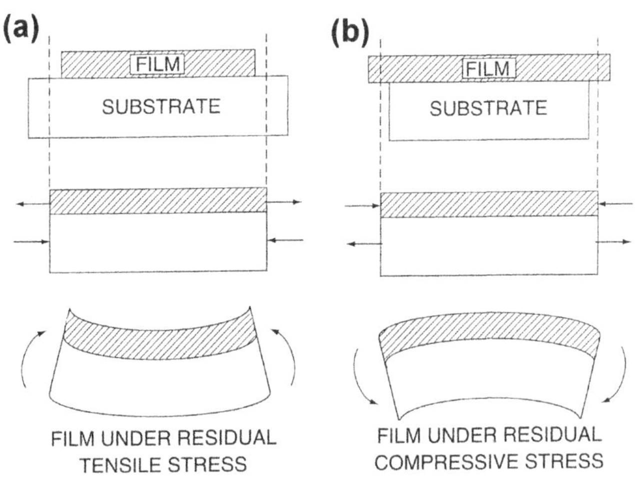


**Figure 3.12** (a) Development of residual tensile stress in a film when $\alpha_f > \alpha_s$. (b) Development of residual compressive stress in a film when $\alpha_f < \alpha_s$. *(From Ref.* [20].*)*

where $A$ is the sectional area and $y$ is the moment arm. Compatibility demands that both the film and the substrate have the same length. A compromise is struck, and the film stretches while the substrate contracts by an equal amount to keep the overall length identical in each.

However, the combination is still not in mechanical equilibrium because there is still an unbalanced end moment. To compensate, the film/substrate pair elastically bends as shown if not restrained from moving. Thus, films containing internal tensile stresses bend the substrate concave upward. In an entirely analogous fashion, compressive stresses develop in films that tend to thermally contract less than the substrate. Internal compressive film stresses, therefore, now bow the substrate convex downward (Figure 3.12(b)). These results are perfectly general regardless of the specific mechanisms that cause films to stretch or shrink relative to substrates. Sometimes, the tensile stresses are sufficiently large to cause film fracture. Similarly, excessively high compressive stresses can cause film wrinkling and local loss of adhesion to the substrate. Examples of both effects are shown in Figure 3.13.

The magnitude of the thermal stress developed in the film is given by [20]

$$\sigma = \frac{E}{1 - \nu_f}[\alpha_f - \alpha_s]\,(T - T_o), \tag{3.8}$$

where $\nu_f$ is the Poisson ratio of the film. Stresses of lower magnitude and opposite sign develop in the substrate. It should be noted that across a film/substrate couple the strains are continuous, but there is no such restriction on the stresses. Because the elastic moduli of film and substrate differ, there is a stress discontinuity at the interface. Since two or more materials bonded to each other are involved in many applications, Eqn (3.8) is an important formula that is frequently quoted in the reliability literature. The right-hand side of this equation appears as a factor in virtually every thermal stress formula developed in connection with packaging and thermal cycling of IC chips during processing and use (see Chapter 8).

#### *3.3.2.3 Intrinsic Residual Stress*

In addition to thermally induced stresses, there are intrinsic sources of stress that are operative at any given temperature. In films these stresses stem from growth processes, i.e., during epitaxy and structural evolution. Tensile stresses usually arise in metal and silicide films from densification

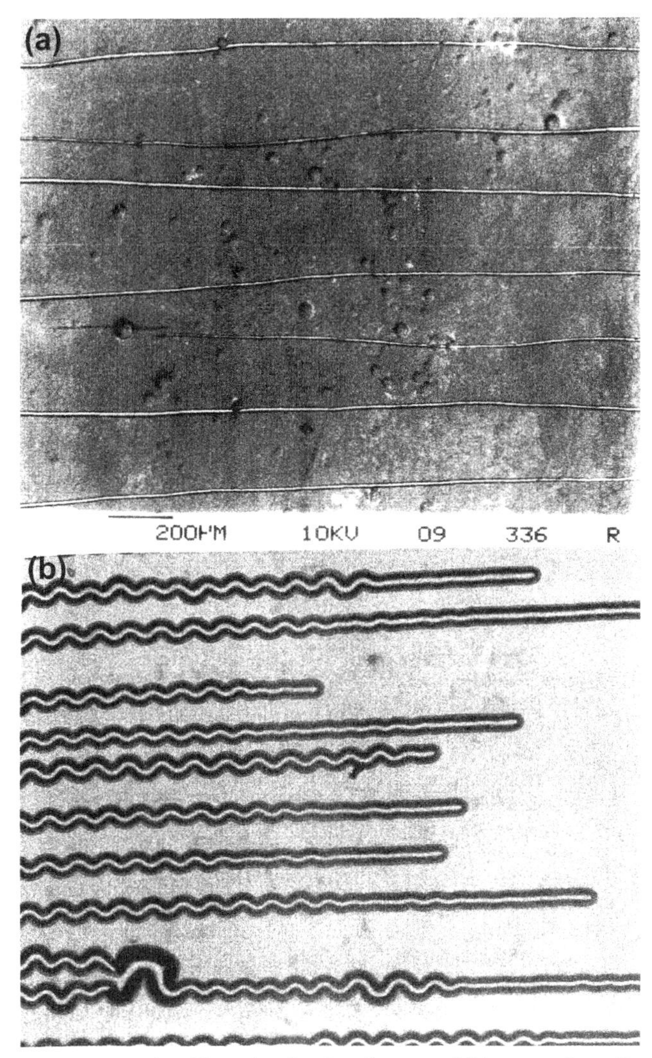


**Figure 3.13** Stresses in thin films that lead to fracture. (a) Image of tensile cracks in a chromium film electrodeposited on a nickel substrate that was subsequently pulled in tension. (b) Optical micrograph of crazes in a tungsten thin film sputtered on a silicon substrate. Isotropic, compressive residual stresses with magnitudes of ~1–3 GPa led to wormlike, wavy, and straight crack traces. Width of straight cracks is ~60 μm. *(Courtesy of I.C. Noyan, IBM Corp.)*

phenomena arising from coalescence of nuclei, recrystallization, and grain growth. In each case the film shrinks relative to the substrate. On the other hand, hydration and ion implantation lead to compressive stress because they expand the film relative to the substrate.

There are complex reasons for the development of intrinsic residual strains, but in epitaxial films they include differences in lattice constant or lattice misfit between film and substrate. In producing high-quality epitaxial films of compound semiconductors it is critical that the lattice constants of the film ($a_f$) and substrate ($a_s$) be as closely matched as possible. Lattice mismatch or misfit ($f$) is defined as

$$f = \frac{a_s - a_f}{a_f}, \tag{3.9}$$

which is the negative of the elastic strain, $e_E$, required to make the film register perfectly or epitaxially with the substrate. What happens during epitaxy is schematically indicated in Figure 3.14. When $f$ is zero there is perfect homoepitaxy and no strain or stress. For small values of $f$, straining of interfacial film–substrate atomic bonds is not sufficient to nucleate dislocations, and the interface is defect free. However, for larger misfits the only way to reduce accumulated bond strain energy is to generate undesirable misfit dislocations at the interface. For devices not to display degraded electrical behavior, misfit values less than 0.001 are preferred. Otherwise, misfit strain energy will accumulate internally as the film grows until microstructural rearrangements in the form of dislocations

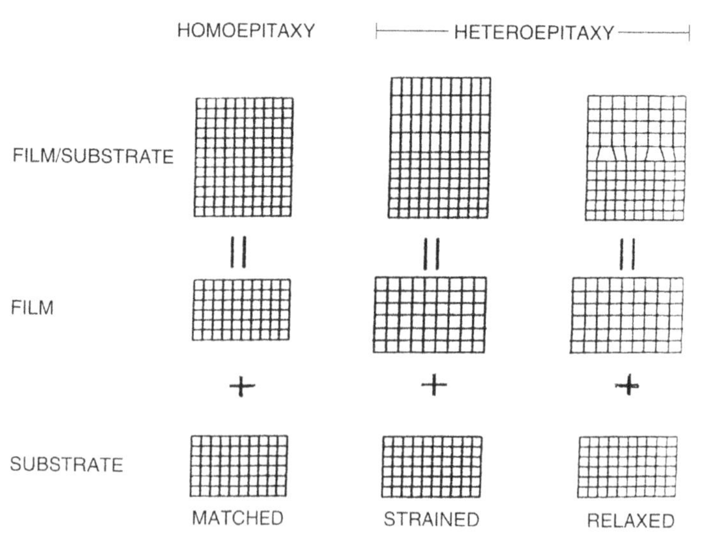


**Figure 3.14** Schematic illustration of lattice matched, strained, and relaxed heteroepitaxial structures.

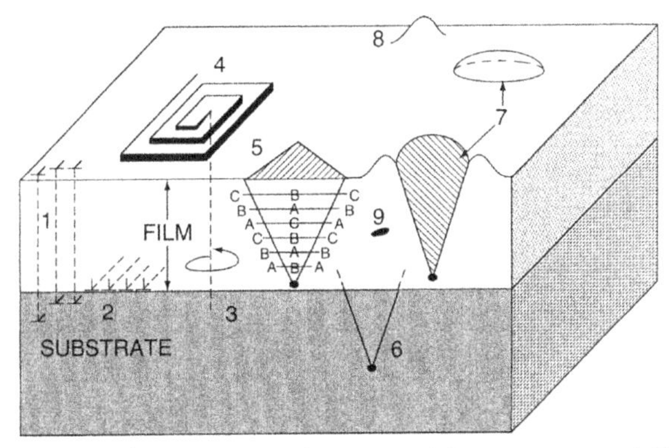


**Figure 3.15** Collection of crystallographic matrix defects in epitaxial films displayed schematically. 1, Threading edge dislocations; 2, interfacial misfit dislocations; 3, threading screw dislocations; 4, growth spiral; 5, stacking fault in film; 6, stacking fault in substrate; 7, oval defect; 8, hillock; 9, precipitate or void.

or stacking faults relieve this energy build up. These undesirable defects form at film thicknesses of approximately $b/(2f)$, where $b$ is the dislocation Burgers vector. This means that defects will typically appear after ~200 nm of film growth if $f = 0.001$.

A collection of defects that have been observed in epitaxial films is displayed schematically in Figure 3.15. All, except for the oval defects, probably need no additional comment. The troublesome oval defects are faceted growth hillocks that nucleate at the film–substrate interface, and they nest in the epitaxial layer. With a polycrystalline core bounded by four {111} stacking fault planes, they are present in densities as high as 1000/$cm^2$ with sizes exceeding 1 μm. Their presence in GaAs films has been attributed to surface contamination.

### *3.3.2.4 Manifestations of Internal Stress*

#### 3.3.2.4.1 Stress Magnitude

The magnitudes of the sum of the thermal and intrinsic internal or residual stresses in films commonly employed in silicon technology are entered in Table 3.5. These data are representative, and it should be realized that stress values are subject to considerable scatter depending on how the films were deposited. In dielectric films, both positive and negative stresses less than 1 GPa in magnitude can be expected. Silicides, on the other hand, are stressed in tension to a level of several gigapascals.

## Example 3.1

What is the thermal stress in thermally grown $SiO_2$ films on Si when the difference between growth and ambient temperatures is $T - T_o = 900K$?

**Answer** The extrinsic stress is given by $\sigma = E_{SiO_2}/(1 - \nu_{SiO_2})\ [\alpha_{SiO_2} - \alpha_{Si})]\ (T - T_o)$. Taking the values $E_{SiO_2} = 83$ GPa, $\nu_{SiO_2} = 0.167$, $\alpha_{SiO_2} = 0.5 \times 10^{-6}$ K$^{-1}$, and $\alpha_{Si} = 4 \times 10^{-6}$ K$^{-1}$, the stress is calculated to be $\sigma = -0.31$ GPa. Comparison with Table 3.5 shows that virtually all the stress is accounted for by differential thermal expansion which places $SiO_2$ into compression.

It should be noted that the tabulated stress levels are rather high. In fact, they typically range between $10^{-3}$ and $10^{-2}$ $E$, and this means that levels comparable to the yield stress for these materials are reached. Therefore, instances of stress-induced plastic flow and damage in device structures should not come as a surprise.

**Table 3.5** Residual stress values in films encountered in Si devices

| Film | Process | Conditions | Stress (GPa) |
|---|---|---|---|
| $SiO_2$ | Thermal | 900–1200 °C | −0.2 to −0.3 |
| $SiO_2$ | CVD 400 °C | 40 nm/min | +0.13 |
| $SiO_2$ | $SiH_4 + O_2$ | 400 nm/min | +0.38 |
| $SiO_2$ | CVD | 450 °C | +0.15 |
| | TEOS | 725 °C | +0.02 |
| $SiO_2$ | TEOS | 685 °C | +0.38 |
| | TEOS +25% B,P | | −0.02 |
| $SiO_2$ | Sputtered | | −0.15 |
| $Si_3N_4$ | CVD | 450–900 °C | +0.7 to +1.2 |
| $Si_3N_4$ | Plasma | 400 °C | −0.7 |
| | | 700 °C | +0.6 |
| $Si_3N_4$ | Plasma 13.56 MHz | 150 °C | −0.3 |
| | | 300 °C | +0.02 |
| $Si_3N_4$ | Plasma 50 kHz | 350 °C | −1.1 |
| Poly Si | LPCVD | 560–670 °C | −0.1 to −0.3 |
| $TiSi_2$ | PECVD | As-deposited | +0.4 |
| | | Annealed | +1.2 |
| $TiSi_2$ | Sputtered | | +2.3 |
| $CoSi_2$ | Sputtered | | +1.3 |
| $TaSi_2$ | Sputtered | 800 °C anneal | +3.0 |
| $TaSi_2$ | Sputtered | | +1.2 |
| W | Sputtered | 200–400 W power | +2 to −2 |
| W | Sputtered | 5–15 mTorr Ar pressure | −3 to +3 |
| Al | | | +0.5 to ~+1 |

B,P; PECVD; TEOS, tetraethyl orthosilicate.
Data are mostly from Ref. [21], R60.

#### 3.3.2.4.2 Trench Dislocations

Among the suggested sources for DRAM trench dislocations are mechanical deformation and microcrack formation caused by particulate impact during processing, high implanted dopant levels, and silicide precipitates. In studying device isolation trenches, which have certain similarities to DRAM trenches, Hu [21] has proposed the following three sources for stress.

1. Compressive stress in thermal oxides grown on nonplanar surfaces such as trench corners. Thermal oxide linings are preferred vis-à-vis other dielectric films because they have the lowest interface charge state and trap densities.
2. The mismatch in thermal expansion coefficients between the trench dielectric ($SiO_2$) and substrate Si.
3. Intrinsic stress in CVD deposited polysilicon and $SiO_2$.

#### 3.3.2.4.3 Hillock and Whisker Formation

In multilayer integrated devices as well as passive electronic components, the metal film conductors employed sometimes develop troublesome surface protuberances known as "hillocks" and "whiskers." The two probably originate from similar growth and relaxation mechanisms under the action of compressive stresses. They arise both during processing and stressing in service, and therefore adversely affect yield as well as reliability. Hillocks are much shorter, and may be viewed as stunted and more rounded whisker growths. Sometimes observed in thin film aluminum interconnects and on contact pads, they are detrimental because their penetration of passivating films can lead to electrical short circuits.

Whiskers are filamentary growths of pure metals, solid solution alloys, and compounds that have a high degree of crystalline perfection and uniformity. Most whiskers are typically $\sim 1\ \mu m$ thick, straight, parallel-sided single crystals with low index directions. Between straight segments, whiskers can bend sharply. Whiskers are threatening because they can grow to many times the film thickness (Figure 3.16(a)) and short relatively remote conductors (Figure 3.16(b)). In addition to vapor deposition processes that leave deposits in compression, electrodeposited tin, cadmium, and indium surfaces commonly exhibit whiskers. Therefore, bright Sn deposits on brass should be avoided, and Ni-rather than Cd-plated screw fasteners are preferred. A rule of thumb by Hitch [22] suggests that whiskers generally form at temperatures of 0.5 $T_M$, where $T_M$ is the melting point of the metal.

Hillocks and whiskers have been observed to sprout during electromigration, a subject treated at length in Chapter 5. The $SiO_2$ used to

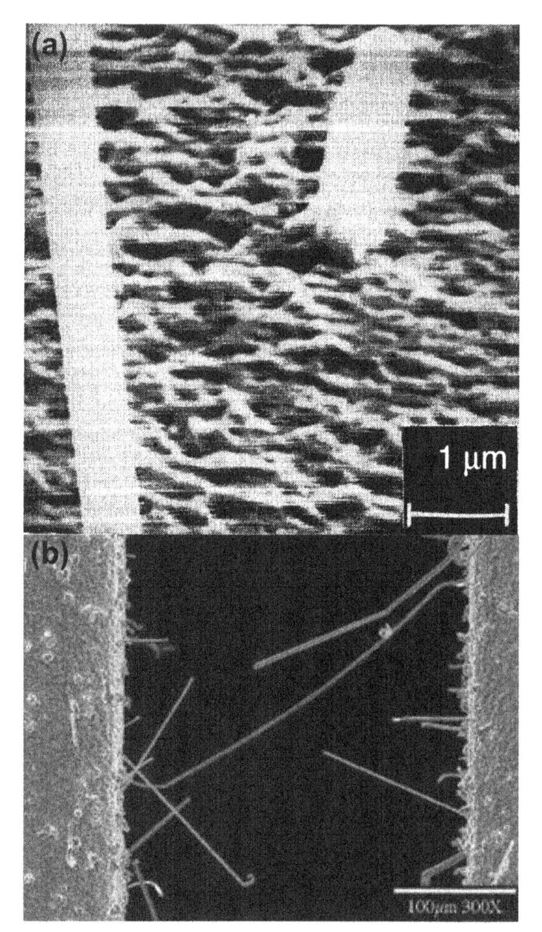


**Figure 3.16** (a) Scanning electron microscope image of whiskers in thin tin films. *(Courtesy of P.H. Sun.)* (b) Tin whisker shorting calacitors. *(Courtesy DfR Solutions.)*

passivate interconnections serves to conformally constrain the powered conductors. Compressive stresses induced in the metal by electrotransport can be relieved through extrusion of hillocks or whiskers that sometimes crack and pierce the confining, insulating dielectric overlayer [20]. Interestingly, compression reduction or creation of tensile stresses through current reversal during electromigration or thermal cycling sometimes causes the hillocks to shrink in size.

#### 3.3.2.4.4 Wafer Bow

The convex or concave bowing of film–substrate combinations as a result of internal stress has already been noted in Section 3.3.2. Bowing may result in lithography mask registration, tolerance, and overlay errors because processed feature sizes at the wafer center differ from those at the wafer circumference. However, the use of vacuum chucks tends to eliminate wafer warpage problems. The celebrated Stoney formula (16), relates the film stress to the bow radius of curvature $R$ and the thicknesses $d$ of both film (f) and substrate (s) as

$$\sigma_{\mathrm{f}} = \frac{E_{\mathrm{s}} d_{\mathrm{s}}^2}{6R(1 - \nu_{\mathrm{s}}) d_{\mathrm{f}}}. \tag{3.10}$$

Over the years a host of experimental techniques, including optical interference, X-ray diffraction, and beam deflection, have been employed to measure the stress in films and coatings [20]. A recent nondestructive method employs a laser beam that is scanned across the film/Si wafer combination. The radius of curvature, which may be hundreds to thousands of meters long, is determined by a (light) position-sensitive detector. In this way, for example, the simple radially symmetric dishlike bowing can be easily distinguished from the potato-chip-like twisting distortion sometimes seen. We are usually only interested in film stress because the relatively massive substrate is only stressed to low levels.

## 3.3.3 Step Coverage

### *3.3.3.1 Metallization Coverage*

The terrain of ICs consists of an array of stepped regions due to the many stripes, trenches, and openings in dielectrics. These must be coated with a uniform thickness of metal for reliable interconnect and contact performance. But as Figure 3.17 suggests, uniform coverage of a step is sometimes difficult to achieve [23]. For the case shown, wafers rotating in a planetary holder are metallized with aluminum evaporated from an electron beam source. A computer simulation of the deposited film, based on line-of-sight impingement and shadowing effects, reveals thinning near the oxide edge and secondary crack formation. The actual metallization topography compares well with prediction. It is not difficult to imagine that such a conductor might pass electrical test and inspection; however, a reliability problem looms if local heating at the constriction and metal transport due to electromigration are excessive.

Both the extent of thinning and crack depth are a function of the wafer surface orientation. In the case of a rectangular trench with steep walls such that the surface normal of the trench bottom is collinear with the Al source, metal will uniformly cover the bottom and thinly coat the

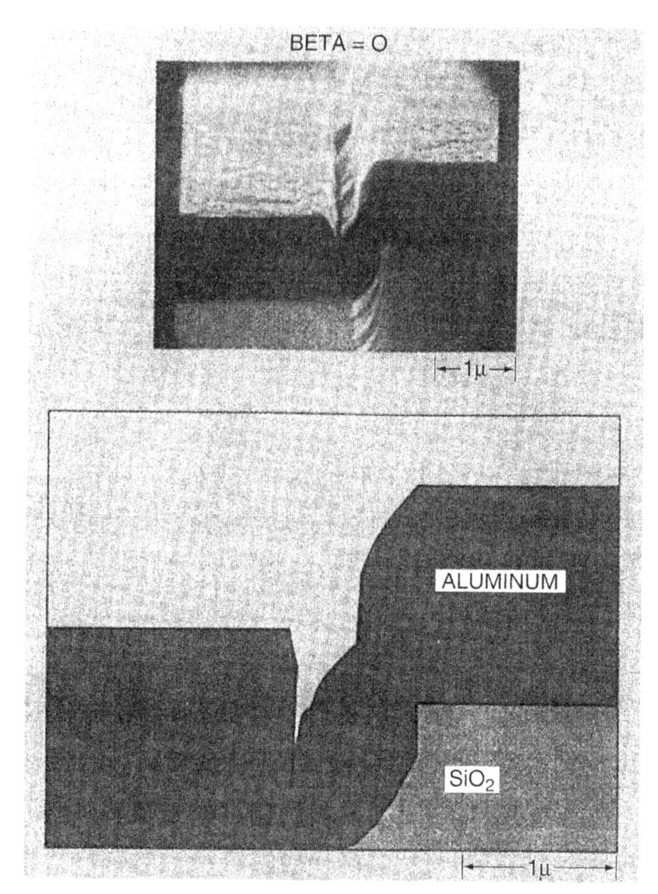


**Figure 3.17** Scanning electron micrograph and computer simulation of aluminum film coverage of a 1-μm step. Thinning and secondary crack formation are evident. *From Ref. [23].*

side walls. However, for deep trenches and large wafer tilts, asymmetric and discontinuous metal side wall coverage may occur with open circuiting a potential danger. One way to minimize or even eliminate metallization cracking is to raise the temperature of the substrate. This fosters increased surface diffusion of metal and filling of crevices, which leads to better overall coverage.

Sputtering employing large area cathodes generally results in metallizations with improved step coverage [24]. The reasons are a combination of effects involving ion bombardment–induced defect production and enhanced diffusion, plus resputtering and redeposition of metal atoms on the surface.

#### *3.3.3.2 Dielectric Film Coverage*

Lack of conformal coverage is not only limited to metals but also occurs in the deposition of dielectric films. Consider how the multilayer interconnect structure typical of Figure 2.32 is fabricated. At a given level, metal stripes are patterned on a flat $SiO_2$ substrate, and after etching, a gap is left with metal on either side. The gap and metal are now filled and covered with a plasma-enhanced CVD oxide grown from silane and nitrous oxide as shown in Figure 3.18. Filling is incomplete, however, because less oxide deposits on the vertical side walls than on horizontal surfaces. There is a considerable amount of cusping, giving it a "bread loaf" profile (Figure 3.18(a)). As deposition continues, the oxide pinches, trapping a void, or "keyhole," between the metal lines. This void might be opened during chemical–mechanical polishing (CMP) planarization (Figure 3.18(b)). If so, there is danger of creating a short when the void is filled during subsequent metal deposition (Figure 3.18(c)). Problems of dielectric film coverage and gap filling have occurred in several generations of DRAM chip fabrication and have become increasingly challenging as linewidths continue to shrink.

CMP and planarization of surfaces containing metals (vias) and dielectrics is critical in the realization of multilevel metallization structures on chips with device feature sizes less than 0.25 μm. Defects associated with the polishing process, e.g., differential polishing rates, embedded abrasives, residual surface particles, damage to oxide glasses, etc., are illustrated in Figure 3.19. The prohibitively small allowable (killer) defect sizes (less than 0.08 μm) and low defect densities ($\sim 0.15/cm^2$) that are projected mean that their elimination during processing, and their detection during electrical test and inspection, will present significant challenges.

## 3.4 CONTAMINATION

### 3.4.1 Introduction

High on the list of defects in virtually every processing step entered in Table 3.4 is contamination. By this we usually mean foreign atoms, molecules, and particles that are chemically and physically different from the desired materials they inhabit. Contaminants can be gaseous, liquid, or solid in nature and can contaminate gaseous, liquid, and solid matrices. For example, gaseous matrices are contaminated by other gases, liquid droplets, and solid aerosol particles. Liquid matrices are contaminated by dissolved gas bubbles, droplets, and solid colloid particles, and solid matrices by

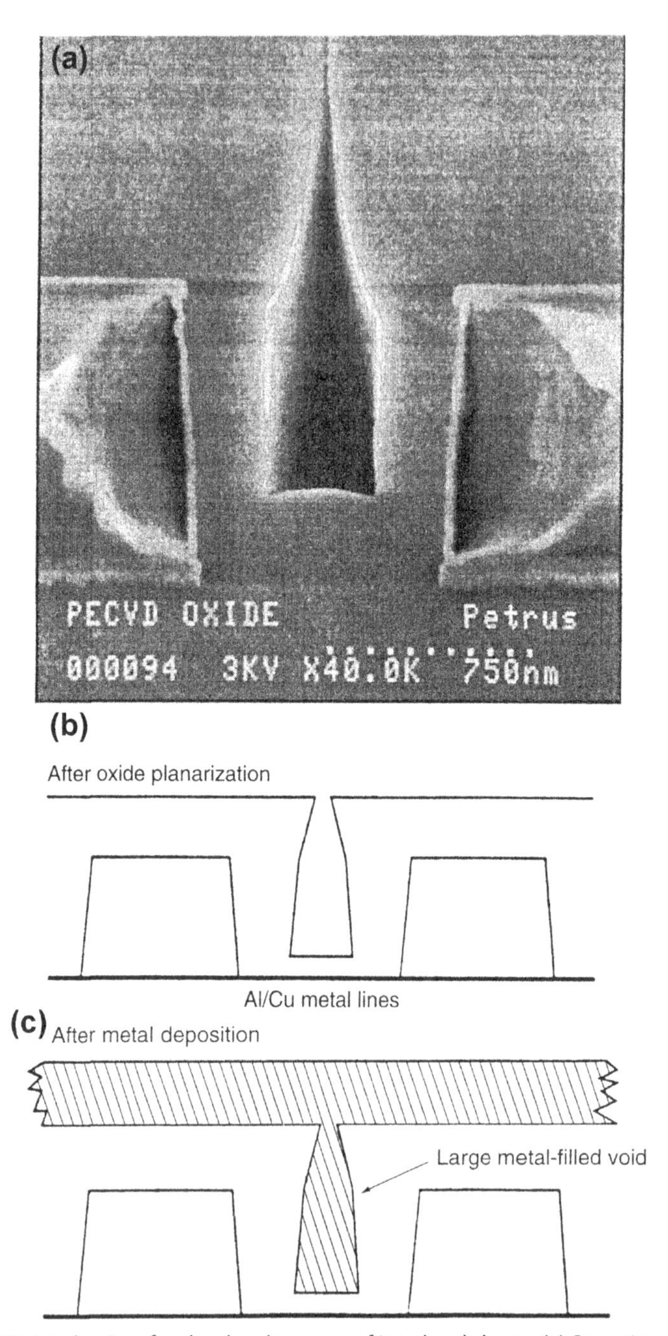


**Figure 3.18** Mechanism for the development of intralevel shorts. (a) Scanning electron micrograph of peaked void trapped within the CVD dielectric that is deposited between neighboring metal lines. (b) Void peak is removed during oxide planarization by CMP. (c) Subsequent deposition of metal raises the danger of creating a short between metallization levels. *(From Ref.* [25].*)*

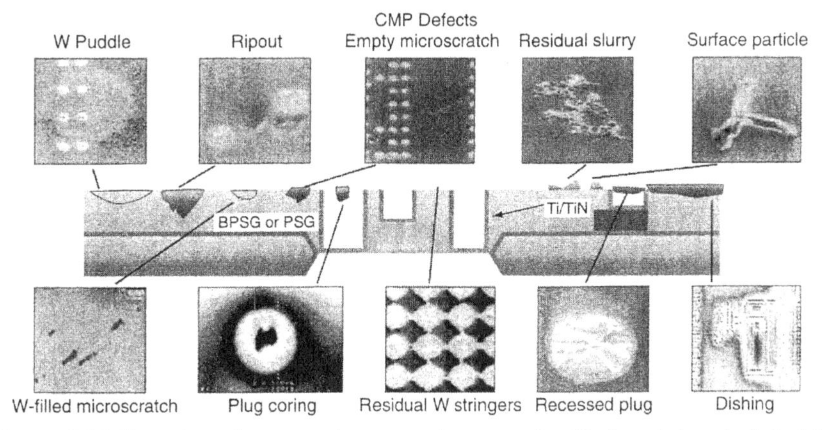


**Figure 3.19** Scanning electron microscope images of critical metal and dielectric defects induced by CMP. Defect sites on the wafer are noted. *From Ref. [26]. Courtesy of KLA-Tencor.*

adsorbed gaseous and liquid surface films as well as particulates. Furthermore, contaminants can be ionic or nonionic, organic, or inorganic.

Defects are manifested primarily in the form of particulate contaminants. For example, particulates have been estimated to be responsible for 75% or more of the loss in yield of volume-manufactured VLSI chips [27]. In view of this, it is not surprising that billions($10^9$) of dollars are spent annually by microelectronic chip manufacturers to combat contamination problems [28]. A significant proportion of current articles and advertising in trade magazines like *Solid State Technology* and *Semiconductor International* is devoted to contamination and defect detection or elimination issues. This reflects the magnitude and compelling nature of the problem. The sources of contamination in processing, the nature of particulate contaminants, and the steps and strategies adopted to eliminate or minimize their influence are all addressed in this section.

Contamination of electronic products stems from essentially four sources: people, clean rooms, equipment, and processes. The impact of these on chip yield will be our primary focus of attention. Exposure to the environment also contaminates electrical products and sometimes leads to reliability problems. Environmental contamination used to adversely affect chip yields, but its impact has declined over the years through the widespread use of clean rooms. Nevertheless, package processing, which is not always done in clean rooms, is more susceptible to atmospheric contamination. Gone are the days when workers contaminated wafers with the cosmetics they wore, the dandruff they shed, or the salt they transported on their shoes during snowy mornings. Over the years the contaminant size has roughly reflected the finest features being processed. But the smaller the particle size, the more numerous and diverse are their sources. Concurrently, circuit shrinkage has significantly reduced the size of contaminant particles that can be tolerated.

In the past, the distribution of contaminant sources was roughly evenly divided. Recent trends, however, show that people and clean rooms are less important contributors to contamination problems than equipment and processes. This has happened because of the improvement in clean room technology and the covering of operators in almost space suit like garb, before entry to the clean room. Nowadays, contamination sources are quite subtle. Examples include airborne molecules, process gases (containing $O_2$, $N_2$, $H_2O$, etc., at ~0.1–10 ppm levels), liquids (e.g., etching and cleaning solutions, even deionized water), as well as process chambers, tools, robots, and hardware (e.g., desorbed matter, wear particles). For example, in a variety of VLSI process tools involving vacuum hardware, baseline particle levels are constant to a first approximation [29]. However, this is punctuated by short bursts where particle counts may rise severalfold above background as a result of opening or closing load locks or transferring wafers. No tool, chemical, or rinsing solution is above suspicion. With this realization we now consider some examples of process-generated contaminants and their roles in fostering damage and yield loss. In contrast, environmental sources of contamination, e.g., corrodents, and their effect on long-term reliability will be treated in Chapter 7.

### 3.4.2 Process-Induced Contamination

Many processes have proven to be culprits in generating contaminants ranging in size from the atomic to the particulate level. We shall also broaden our definition to include charging of surfaces as a type of "contamination," in order to present the damage introduced during plasma etching.

It is instructive to view process contamination as a kind of "food chain," illustrated in Figure 3.20, that first involves generation and transport of particles, and then deposition on the wafer. Implementing control measures to eradicate either their generation or their transport means that particles will not deposit.

Several processes known to generate contamination are briefly treated in turn.

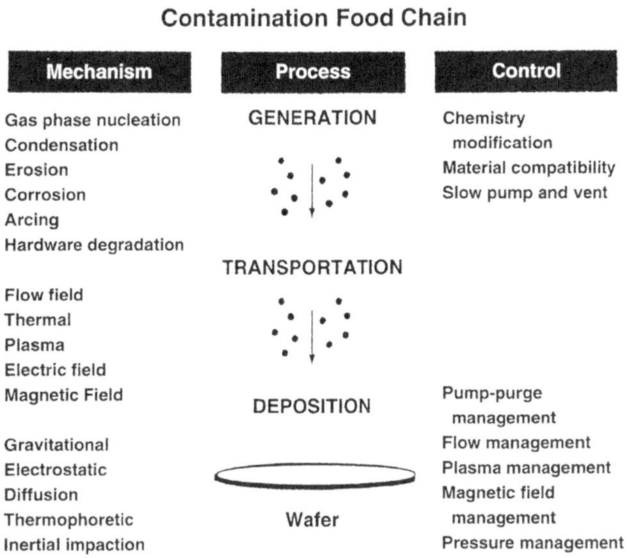

**Contamination Food Chain**

| Mechanism | Process | Control |
|---|---|---|
| Gas phase nucleation<br>Condensation<br>Erosion<br>Corrosion<br>Arcing<br>Hardware degradation | GENERATION | Chemistry modification<br>Material compatibility<br>Slow pump and vent |
| Flow field<br>Thermal<br>Plasma<br>Electric field<br>Magnetic Field | TRANSPORTATION | |
| Gravitational<br>Electrostatic<br>Diffusion<br>Thermophoretic<br>Inertial impaction | DEPOSITION<br>Wafer | Pump-purge management<br>Flow management<br>Plasma management<br>Magnetic field management<br>Pressure management |

**Figure 3.20** Contamination food chain. *From Ref. [30].*

### *3.4.2.1 Chemical Vapor Deposition*

It is believed that perhaps as much as 25% of all particle-induced yield loss is derived from low-pressure CVD (LPCVD) processing [31]. Among the films produced by LPCVD are silicon nitride, polysilicon, and $SiO_2$ from tetraethyl orthosilicate. Thin-film deposits naturally build up on the tubes of these hot-wall reactors during deposition. Film stress, along with the turbulence of loading wafers, causes shedding of these particulate deposits onto the wafer. Rubbing of quartzware in furnaces and oxidation and flaking of heaters are additional contamination sources.

Aside from impurities in the precursor gases, CVD processes can also generate particulates from the gas-phase environment. The basic gas-to-solid film reaction is designed to occur at wafer substrates. However, when the supersaturation of reactants is too high, nucleation of the solid product may occur homogeneously in the gas phase. In the form of "snow," the vapor-borne particulates settle on the substrate and are incorporated into the growing film. Generation of dislocations and stacking faults, regions of poor adhesion, and roughened film topographies are some of the adverse effects produced.

### *3.4.2.2 Ion Implantation*

Even the perceived "clean" process of ion implantation is not above contaminating wafers with particularly damaging elements such as Fe, Ni,

Cr, and Mo. Because of them undesirable MOS threshold voltage drift, defects, junction depth shifts, and reduced carrier lifetimes result [32]. Specific problems arise from the following sources:

1. Ion-beam-sputtered atoms from photoresist masks, accelerator apertures, vacuum hardware, wafer fixtures, etc., are incorporated in the beam and implant.
2. Chemically different ion species may have the same mass ($m$) to charge ($q$) ratio. The selection of a particular ion for implantation is done with magnets and depends on $m/q$. If, for example, $^{11}B^{19}F_2^+$ dopant and $^{98}Mo^{2+}$ ion contaminants are simultaneously present, both B and Mo would be implanted into the wafer because $m/q = 49$ for both species.
3. Due to charge exchange processes in the beam, ions can either acquire or lose charge; this results in deeper or shallower implant depths, respectively, because the energy of an ion in the beam depends on its valence.
4. Cross-contamination is common when implanters are called on to provide high fluences or fluxes for several different dopants. Thus, faster diffusing boron and phosphorous dopants that contaminate arsenic and antimony implants displace junction depths. For example, a 0.01% contaminant level can cause a junction depth shift of 5%.

### *3.4.2.3 Plasma Processing*

During various film deposition and etching processes, devices are often directly exposed to an radio frequency(rf) plasma discharge ambience of reactive ions, which are further accelerated by the wafer self-bias voltage. Enhanced and guided by the metallization or polysilicon patterns that act as "antennas," the result of these ion collisions is charging of the surface. Particularly large antenna surface areas may ultimately cause sufficient charge injection to degrade dielectric properties.

Damage to materials during plasma etching has been classified into two broad categories [33,34]:

1. Particle current-flow-induced damage. The buildup of charge produced by the plasma and the current driven by electric and magnetic fields are sources of this damage. Current flow through the dielectric produces wear-out damage due to bond breaking and charge trapping.
2. Plasma exposure damage. This category of damage is a side effect of particle or photon impingement on the surface. Residue layers, impurity permeation of the surface, and film adhesion damage due to photons or chemical attack are all examples of damage due to plasma exposure.

Particularly sensitive to plasma damage are gate oxides that have been shown to charge up, exhibit altered C–V curves (Section 2.5.2), and suffer earlier breakdowns as a result. Fortunately, plasma damage effects often occur during overetching and not during the main etching [35]. The subject of plasma damage to gate oxides is treated again in Section 6.3.8.

### 3.4.2.4 Lithography

Just one glance at the defective mask in Figure 3.21 reveals how patterned features on an IC chip can be severely damaged through incorporation of the indicated imperfections. Particulate contamination in the lithographic process is a chief contributor to such mask defects [36]. Lithography defects of differing severity have been classified and include killers, unreliable, hard, soft, and repairable [37]. Examples of these lethal defects are conductor opens and shorts through insulators between neighboring conductors. Unreliable defects do not cause immediate failures but may

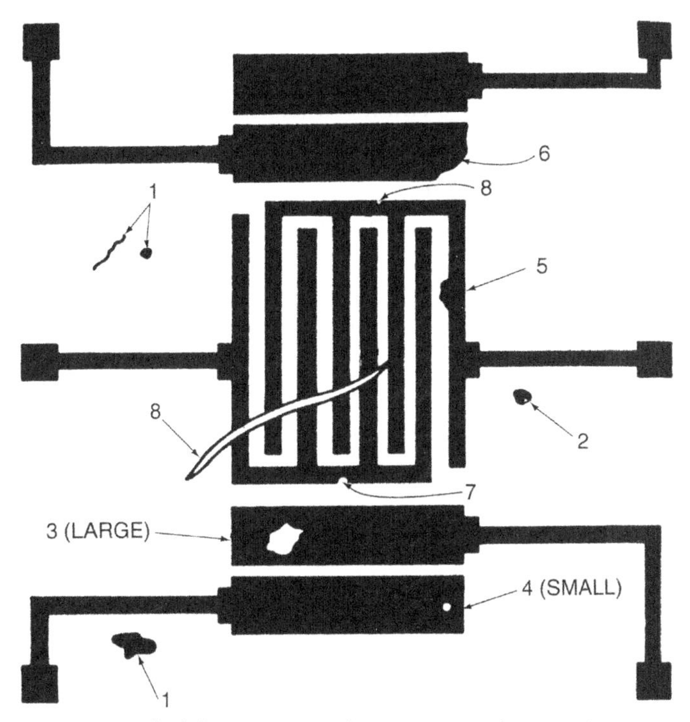


**Figure 3.21** Types of defects commonly encountered in resist processing. 1, Contamination; 2, opaque spot; 3, large hole; 4, pinhole; 5, excess material; 6, lack of adhesion; 7, mouse nips (intrusion); 8, scratch. *From Ref. [36].*

lead to long-term reliability problems. Pinholes and mouse bites that remove a small fraction of a patterned line are examples. Other unreliable defects include image density variations, leftover chemicals or photoresist, and out-of-focus images. Hard defects are composed of excess or missing mask metal film, e.g., chromium, while soft defects consist of particulate contaminants, clean room residues, and transparent material. Dust and watermarks that can be removed through cleaning the mask plate are examples of soft defect sources.

Lithography masks usually consist of patterned chromium films on glass plates, and these materials as well as photoresists are among potential sources of defects [38]. For example, the glass may contain bubbles, scratches, pits, and surface microcracks. Pits prevent Cr adhesion, resulting in pinholes, while scratches cause nonuniform etching at pattern edges. The chrome film defects include pinholes and voids, particulate inclusions, and invisible chemical anomalies. Chromium oxides, nitrides, and carbides are examples of the latter, and collectively they cause erratic local etching and undesired patterns.

Photoresists and the assorted chemical solutions used in lithography are also sources of contamination. They undergo extensive filtration to minimize the presence of particle and metal (e.g., Na, K) contamination.

### *3.4.2.5 Metal Contamination*

Since the early days of Si device technology metallic contaminants have limited yield and have been a reliability risk as well. In this regard, Kolbesen [16] has estimated that the iron present in a pinhead, if spread uniformly across the entire annual production of the world's leading wafer manufacturer, would contaminate the wafers with a lethal dose of $10^{13}$ Fe atoms per $cm^3$. The gate oxide is particularly vulnerable to Fe, which lowers the dielectric breakdown voltage. In reducing the oxide thickness from 20 nm to 10 nm, corresponding to a translation from 1 to 0.5 μm linewidth technology, the Fe concentration had to be lowered from $10^{13}$ Fe atoms per $cm^3$ to $10^{11}$ Fe atoms per $cm^3$, or a factor of 100 [39]. The sources of metal contamination in production lines have been traced to virtually all tools, including furnaces and epitaxial reactors, ion implanters, plasma and reactive ion etching equipment, wet cleaning and etching chemicals, and wafer handlers and equipment. A combination of harmful properties listed below causes metals to play a detrimental role in devices.

1. Transition metals such as Cu, Ni, Fe, and Cr are fast diffusers in Si. All have diffusivities exceeding $10^{-6}$ cm$^2$/s at temperatures of 1000 °C. In particular, Cu, which is currently used in metallizations, has a D value exceeding $10^{-6}$ cm$^2$/s even at 500 °C, and an activation energy of less than 1 eV. This compares with D values of $\sim 10^{-13}$ cm$^2$/s at 1000 °C, and activation energies of 3–4 eV for the common silicon dopants, i.e., B, P, and As.
2. Metals have low solubility in Si (in the part per million and below range) with a steep temperature dependence. Trace impurities may dissolve in wafers at elevated processing temperatures but be supersaturated when cooled to room temperature. Their rejection from solution leads to precipitation of silicides, e.g., $FeSi_2$ and $NiSi_2$, at lattice defect nucleation sites.
3. Many transition metals are effective minority carrier recombination and generation centers because they form levels within the Si band gap. In addition, these metals react with acceptor impurities.
4. Metal precipitates serve as nuclei for the generation of extended defects such as dislocations and OSFs in the Si substrate and gate oxide.
5. Metal decoration of existing defects makes them far more conducting. Dislocations and stacking faults so affected have caused excessive leakage currents in p–n junctions and retention time failures in DRAMs.

A chronic metal contaminant of interest in microelectronics is sodium. In the ionic form it inhabits the various oxides and glasses of the IC chip. Sodium is now largely eliminated from gate oxides, where it causes threshold voltage instabilities in MOSFET operation at surface densities above $1 \times 10^{10}$/cm$^2$. Presently, Na concentrations ranging from $10^{17}$ to $10^{19}$ atoms/cm$^3$ are found above the metallization levels, and are primarily introduced via photoresists [40].

### 3.4.3 Introduction to Particle Science

To develop strategies for reducing particulate contamination levels in microelectronics, it is important to understand the nature of particles, the way they are transported, and their mechanisms of adhesion to surfaces and wafers in particular [41]. The most important attribute of a particle is its size, because properties such as particle velocity, adhesion force, and ability to scatter light depend on it. There are vast differences in the behavior of large-scale objects compared to the submicron-sized particles that are our present concern. For example, the latter will adhere quite strongly to the ceiling of a chamber, but a 1-mm particle will immediately drop because gravitational forces far exceed the very weak adhesion forces. In microelectronic processing, drag, gravitational, and diffusional forces spread contamination by driving particles to wafer surfaces where various forces of adhesion (e.g., van der Waals (vdW), electrostatic, capillary) cause them to stick. Overcoming the influence of particle adhesion forces is a way to combat contamination. For these reasons, each of the forces will be briefly discussed in turn.

#### *3.4.3.1 Drag*

As particles move through a fluid (gas or liquid) they are opposed by frictional drag forces. For a small spherical particle of diameter $d$, moving slowly with velocity $v$, the drag force is given by Stokes' law, or

$$F_{\mathrm{d}} = 3\pi\eta d\ v/C_{\mathrm{c}}. \tag{3.11}$$

Here $\eta$ is the coefficient of viscosity, and $C_c$ is the Cunningham slip correction factor. For air at 1 atm and 20 °C, $C_c = 1 + 0.16/d$, with $d$ given in micrometer units. Thus, for particles smaller than 0.5 μm, $C_c$ becomes significant. Equation (3.11) is well satisfied when the Reynolds number for gas flow, defined by $d\,v\rho_f/\eta$ ($\rho_{ff}$ = fluid density), is less than 1; this is generally a good assumption for microelectronic processing.

#### *3.4.3.2 Gravity*

Of all the forces that act on aerosol particles, gravity is especially important because it acts all the time. The gravitational force is given by

$$F_{\mathrm{g}} = \frac{\pi d^3}{6}(\rho_p - \rho_f)g, \tag{3.12}$$

where $\rho_p$ is the particle density and $g$ is the gravitational constant. By equating $F_d$ and $F_g$, the terminal velocity with which particles fall by gravity is given by

$$v = \frac{d^2(\rho_p - \rho_f)C_c g}{18\eta}. \tag{3.13}$$

Gravity always enhances the rate of particle impingement on the surface of a wafer. The deposition flux, which is the product of the particle concentration and velocity, will be small if $v$ is small. Importantly, particle impingement rates fall off as the square of the particle size.

### 3.4.3.3 Diffusion

The constant buffeting and bombardment by fluid molecules causes small particles to execute Brownian diffusive motion. Such random motion becomes important when the particle is less than 1 μm in size. Diffusion distances are given by the familiar formula, $x^2 = 4Dt$, (see Eqn (5.6)), where $D$ is the diffusion coefficient and $t$ is the time. Theory shows that $D$ is given by $D = C_c\ kT/(3\pi\eta d)$, where $kT$ has the usual meaning. Clearly, smaller particles are expected to travel further. We are often concerned with the rate of particle impingement on wafers as a result of diffusion through the surrounding gas phase. Because small particles readily adhere to surfaces, their concentration ($C$) is lower there than in the bulk gas. This gradient ($dC/dx$) establishes a diffusive flux ($J$) of particles toward the surface given by $D\ dC/dx$. Assuming that the wafer is immersed in an ambience where the particle concentration is $C_o$, the rate of particle impingement (number per unit area per unit time) due to diffusion is given by

$$J = C_o\left(\frac{D}{\pi t}\right)^{1/2} \tag{3.14}$$

a result that can be obtained from Eqn (5.4) by differentiating. The number of particles $N(t)$ that accumulate in time is simply the integral of this, or

$$N(t) = C_o\int_o^t\left(\frac{D}{\pi z}\right)^{1/2}dz = 2C_o\left(\frac{Dt}{\pi}\right)^{1/2}, \tag{3.15}$$

where $C_o$ is the particle concentration far from the surface. A comparison of the particle densities that settle on a surface by diffusion relative to gravity is shown in Figure 3.22. Gravitational settling dominates for large particle diameters, whereas diffusional impingement controls when the particle size is small. It is important to note that Brownian motion distances are the same order of magnitude as the thin boundary layer thickness of gas flowing by surfaces. Thus diffusion enables particles to penetrate this stagnant layer, reach the surface, and collect there. The expected particle flux is also indicated in Figure 3.22. Due to the low concentration of large particles, there is a decline in impingement flux at these particle sizes.

### 3.4.3.4 Electrostatic Forces

Initially uncharged particles often acquire charge through interaction with ionic species and gas molecules that become charged by ionizing radiation.

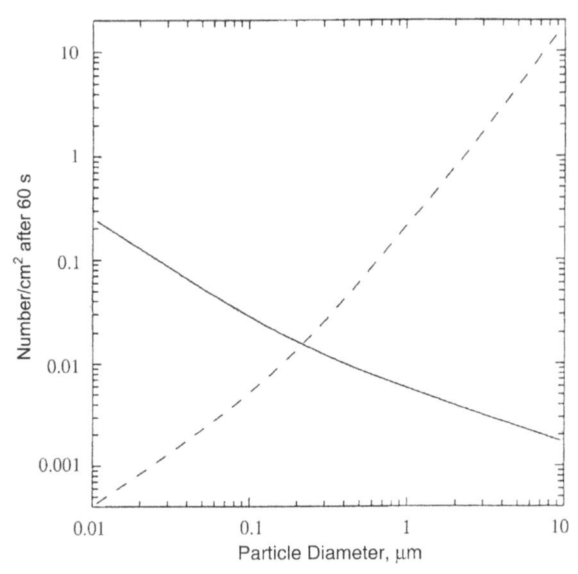


**Figure 3.22** Number of particles collected per square centimeter of horizontal surface by diffusion (solid line) and gravity (dashed line), as a function of particle size. Particle density = 1 g/cm$^3$, particle concentration = 1/cm$^3$, temperature = 20 °C. *From Ref. [41].*

The probability ($P_e$) that $n$ electronic charges (both positive and negative) are associated with particles is governed by their size ($d$) according to a Maxwell-Boltzmann-type expression

$$P = \frac{\text{const}}{(dkT)^{1/2}} \exp\left[\frac{-n^2 q^2}{dkT}\right]. \tag{3.16}$$

Thus, very small particles will be uncharged. For particles larger than 0.1 μm, the average $n$ increases with $d$ as $n = 2.37\ (d_{\mu m})^{1/2}$ [42]. Charged particles will interact with one another according to Coulomb's law.

### *3.4.3.5 Particle Adhesion*

Particles adhere to surfaces through vdW, electrostatic, and capillary forces as well as chemical (hydrophobic/hydrophilic) interactions. Soluble contaminants are effectively removed by chemical solutions that have little influence on insoluble particles. In what follows, we focus on the latter particles and the nature of the physical forces that bind them to and remove them from surfaces. For the case of spherical particles of diameter $d$, a

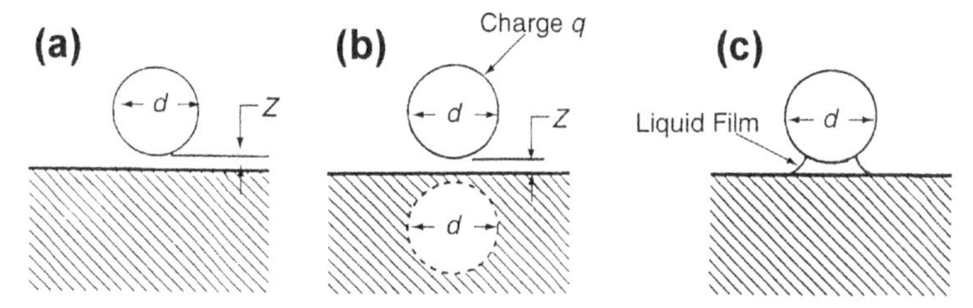


**Figure 3.23** Types of adhesion forces between particles and surfaces. (a) vdW; (b) electrostatic; (c) capillary.

distance $z$ away from a planar surface (Figure 3.23(a)), the vdW adhesion force is given by

$$F_{ad}(vdW) = \frac{A_H d}{12z^2}, \tag{3.17}$$

where $A_H$ is known as the Hamaker constant. The latter depends on the particle and surface compositions, their hardnesses, and the shape and topography of the interacting surfaces. For 1-μm glass microspheres on bare Si in air, $A_H = 14.5 \times 10^{-13}$ erg, while on $SiO_2$, $A_H = 8.5 \times 10^{-13}$ erg. This translates into vdW forces of $\approx 1$ mdyn; surface pressures of $\approx 10^8$ dyn/cm$^2$ generated are large enough to deform both particle and surface.

Electrostatic forces of adhesion between particles and surfaces (Figure 3.22(b)) have Coulombic and contact potential (CP) components. The former is given by

$$F_{ad}(C) = \frac{q^2}{4\pi\varepsilon\varepsilon_o[d + 2z]^2}, \tag{3.18}$$

and the latter by

$$F_{ad}(CP) = \frac{\pi\varepsilon_o d V_{cp}^2}{2z}, \tag{3.19}$$

where $\varepsilon$ is the dielectric constant of the intervening medium and $\varepsilon_o$ is the permittivity of free space. $V_{cp}$ is the CP between the particle and surface, a quantity that typically ranges between 0 and 0.5 V. The charge transfer equilibrates the Fermi energies between the two bodies, establishing a capacitor-like double layer at the interface. Generally for small particles, adhesion forces due to CP exceed those due to electrostatic attraction. For a 1-μm particle, with $z = 4 \times 10^{-8}$ cm and $V_{cp} = 0.5$ V, $F_{ad}$

(cp) ≈ 1 mdyn. This force is roughly comparable to that for vdW interactions.

In humid environments adhesion between particles and the surface is promoted by the intervening water film that develops between them because of capillary attraction (Figure 3.23(c)). A capillary force ($F_c$) of amount

$$F_c = 2\pi d\gamma \tag{3.20}$$

can be expected, where $\gamma$ is the surface tension of water (72 dyn/cm). For a 1-μm particle, $F_c = 4.5 \times 10^{-2}$ dyn, a force that is larger than others considered above.

### *3.4.3.6 Cleaning*

Particle removal from wafer surfaces means establishing external forces that exceed the forces of adhesion. These external forces can be divided into three categories [42]: vibratory or mechanical action, hydrodynamic or aerodynamic drag, and wet or dry chemical action. Specific wafer cleaning techniques include scrubbing, high-pressure gas or liquid jets, wet chemical methods, and ultrasonic, ultraviolet, or ozone cleans.

#### 3.4.3.6.1 Scrubbing

In this mechanical technique a brush rotates across the surface, removing inorganic particles and organic films. Like a car hydroplaning on a wet highway, the hydrophilic (water-attracting) brush bristles are separated from the wafer surface by the thin layer of scrubbing solution. Scrubbing effectively removes large particles (>1 μm) from hydrophobic (water repelling) wafer surfaces. But scrubbing is unsuitable for submicron particle removal, especially from patterned wafers, and may even cause damage.

#### 3.4.3.6.2 Pressurized Fluid Jets

This method of cleaning will remove small particles from patterned wafers. As the fluid moves past particles with velocity $v$, it drags them with a force given by Eqn (3.11). Higher velocities and viscosity coefficients are more effective in cleaning. As a consequence, the drag force is considerably higher with fluid than with gas jets. Troublesome static charge buildup on wafers and the possibility of mechanical damage are disadvantages.

#### 3.4.3.6.3 Ultrasonics

High-intensity ultrasound waves at frequencies above 20 kHz are employed in this technique. They produce mechanical pressure fluctuations that result in the formation of cavitation bubbles. During the pressure cycle, the bubbles alternately expand and collapse. At the collapsing bubble interface, sufficient energy is released locally at particles to overcome the force of adhesion and loosen them. While small particles are efficiently removed, it is difficult to control cavitation without causing possible wafer damage.

##### 3.4.3.6.4 Wet Chemical Cleaning

This widely used wafer decontamination method effectively removes metal ions and soluble impurities, but is not good for particle removal. Because assorted acids and bases are used, rinsing and drying steps are required. In addition, particulate contamination is often present in chemical solutions.

##### 3.4.3.6.5 Ultraviolet or Ozone

Dry processing and contactless removal of organic contamination are advantages of this cleaning technique. However, this method is not proven for particle removal.

### 3.4.4 Combating Contamination

#### *3.4.4.1 Particle Detection*

The first step in developing a plan for contamination control is to be able to confidently detect and analyze particle size and chemistry of samples. For this purpose instruments have been developed to measure particles in gases and liquids as well as on solid surfaces (wafers). Instrument capabilities include sensitivity (e.g., how small a particle can be detected), selectivity (e.g., ability to not mistakenly respond to something other than what is sought), rate of sampling and dynamic range, or the ratio of the highest to lowest level (e.g., particle size, concentration). Needless to say, high sensitivity, selectivity, rate of sampling, and dynamic range are essential.

In general, optical scattering techniques are employed for particle size and distribution measurements. Larger particles and those of higher refractive index than the surrounding media tend to scatter more light than smaller and lower refractive index particles (however, this is not always true). Rayleigh (dipole) scattering dominates when the particles are much smaller than the wavelength of light used, and yields a signal proportional to the square of the particle volume. Geometrical optics prevails when the particles are much larger than the wavelength, and scattering then occurs by reflection and refraction. Between these size range extremes there is Mie scattering. Particle sizes of less than 100 nm can be detected by available instruments, which typically employ computerized image analysis or holography techniques. This equipment and questions of what to sample for, where to sample, how many samples to take, how to interpret a count of zero, and how to analyze data are discussed by Cooper [28].

### 3.4.4.2 Clean Rooms

The use of clean rooms has traditionally been the chief strategy in combating contamination and particle-induced defects. Monitoring particulate contamination and controlling processing variables, through feedback and feedforward of defect information, help establish high yields for process steps carried out in the clean room. Air whose temperature and humidity is tightly controlled enters clean rooms through efficient particulate air filters and then is made to flow as laminarly as possible. This minimizes the sideways dispersion of contaminants generated in the room.

The *class* of a clean room is determined by how many particles of a given size are contained in a volume of its air. For example, a class 100 clean room contains 100 (or 3500) particles 0.5 μm or larger in size, per cubic foot (or cubic meter) of air volume. People in offices work in a class 100,000 clean room. Wafers in an office room would have 1000 times more particles impacting them than those placed in a class 100 clean room. Particle size distribution curves for various classes of clean rooms are depicted in Figure 3.24(a). Because the number of particles increases the smaller they are, more stringent clean room requirements are demanded as minimum IC features shrink. In general, class 10 clean rooms are sufficient for IC processing (oxidation, implantation), but lithographic steps require a more particle-free environment, e.g., a class 1 clean room. Evidence of the need for space-suit-type protection can be seen in the garb worn at Intel (Figure 3.24(b)). As the need develops, the measures taken will be even more elaborate. As an aside it is also important to note that according to Intel the class 1 clean room is 10,000 times cleaner than the cleanest operating room [43]. We also note that the water used in IC fabrication is the cleanest on the planet, because even the tiniest microbe will cause a massive defect in an IC.

### 3.4.4.3 Integrated Processing

A considerable advancement beyond clean rooms is integrated processing [44]. In this new manufacturing approach, sequences of two or more chemical or physical IC processing steps are carried out while wafers are maintained in the clean or controlled ambience of cluster tools. For processes carried out at low temperatures, e.g., plasma etching, evaporation,

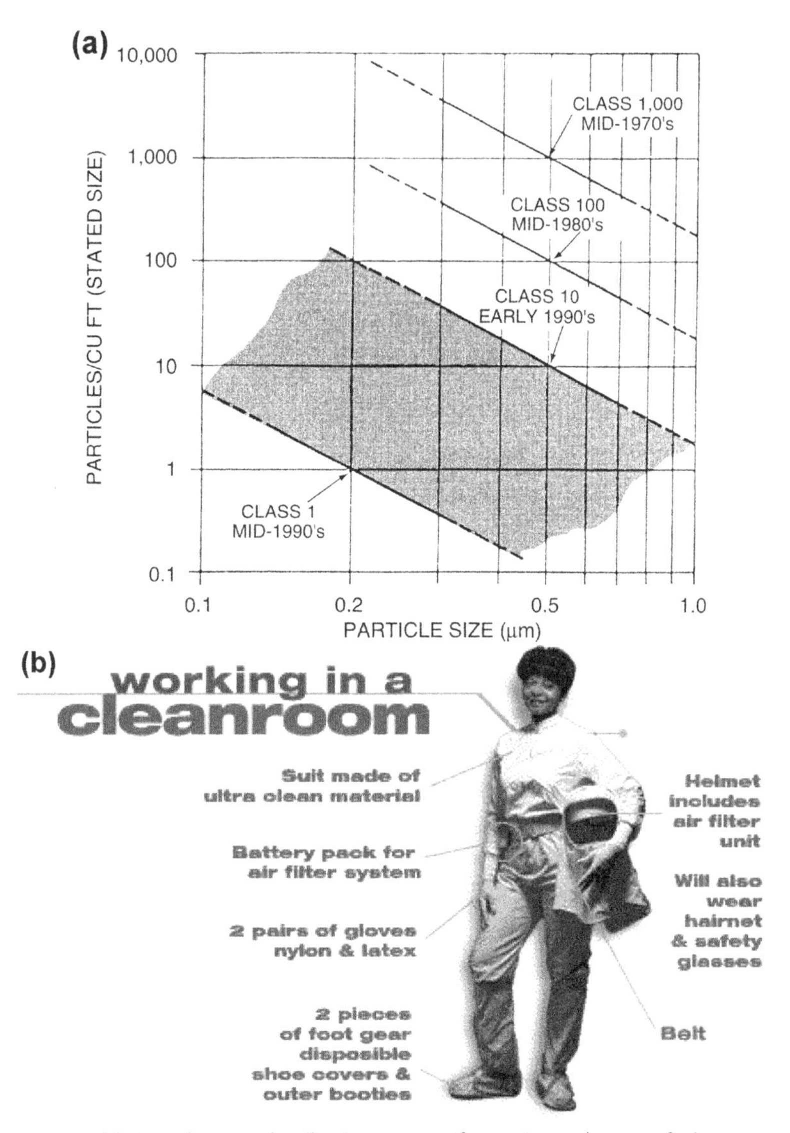


**Figure 3.24** (a) Particle size distribution curves for various classes of clean rooms. *(From Ref. [5].)* (b) Spacesuit-like garb worn in clean room. *(Courtesy of Intel.)*

sputtering, and LPCVD, integrated processing is structured around a vacuum system. At the center of the cluster tool is a central wafer handler capable of delivering the wafer to any one of a number of surrounding chambers or process modules (Figure 3.25). In addition, vacuum load locks permit exchange of wafers between any of the cluster tool modules and

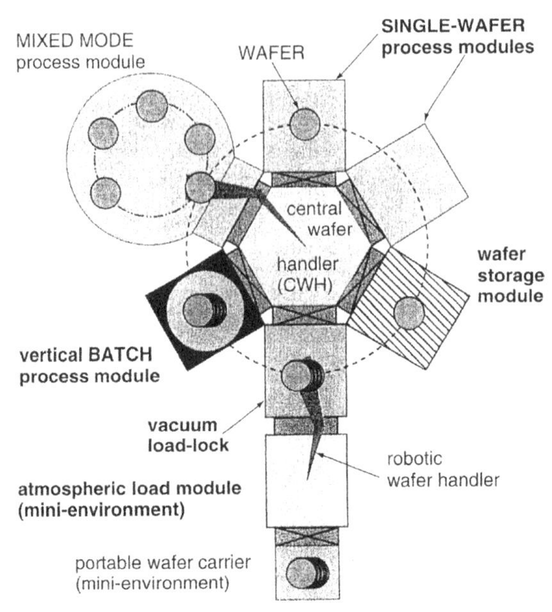


**Figure 3.25** Schematic of an integrated vacuum processing or cluster tool. *From Ref. [44].*

external clean rooms where lithography is practiced. Similarly, other controlled minienvironments, e.g., with vacuum, oxygen-free, inert, or controlled gas ambience, enable cleaning, CVD deposition of dielectric films, reactive ion etching, rapid thermal annealing, and oxidation to be carried out. The greatly enhanced cleanliness results in lower contamination levels of particulates and reactive impurities. In effect, processing in cluster tools occurs in a class 0.1 clean room.

## 3.5 YIELD

### 3.5.1 Definitions and Scope

Yield is generally defined as the ratio of the number of good products produced to the total number of units started in any manufacturing operation [45]. The denominator of this ratio is the sum of the good products and those that had the potential of being good but for some reason were rejected. It should be clear that yield affects the cost of semiconductor products and our ability to predict their price in advance. Yield and cost are the crucial issues that determine the economic success or failure of semiconductor processing and manufacturing industries. The expected inverse correlation between yield and processing defects is depicted in Figure 3.26. Despite wide swings in the monthly tracking of product yield, the trend is inexorably upward as learning occurs. It has been suggested that there are links between defects and reliability as well. Thus logic would dictate that yield and reliability are related despite the fact that the latter presumably refers solely to the performance of good, or yielded, products.

Product yields are generally less than 100% for the following reasons;

1. **Circuit design problems.** This may simply be due to errors in circuit design. The celebrated discovery in 1994 of omitted circuits within the 3 million transistor Pentium microprocessor chip is an example. Although the manufacturing yield of this chip was certainly satisfactory initially, the subsequent recall and substitution had the same economic impact as drastically reduced yield. Other problems arise from alteration of circuit parameters beyond acceptable limits due to processing difficulties, which are then not accommodated in the design. New device behavior or circuit characteristics, unforeseen by the designer, sometimes surface to the detriment of yield. Some circuits are relatively tolerant to variations in circuit parameters, while others are very sensitive. In such cases an interactive effort between the designer (who identifies sensitive parameters) and process engineer (who optimizes the value and range of these parameters) is required for redesign and tuning of one or more process steps. Circuit design problems are well beyond the scope of this book and will not be discussed further.

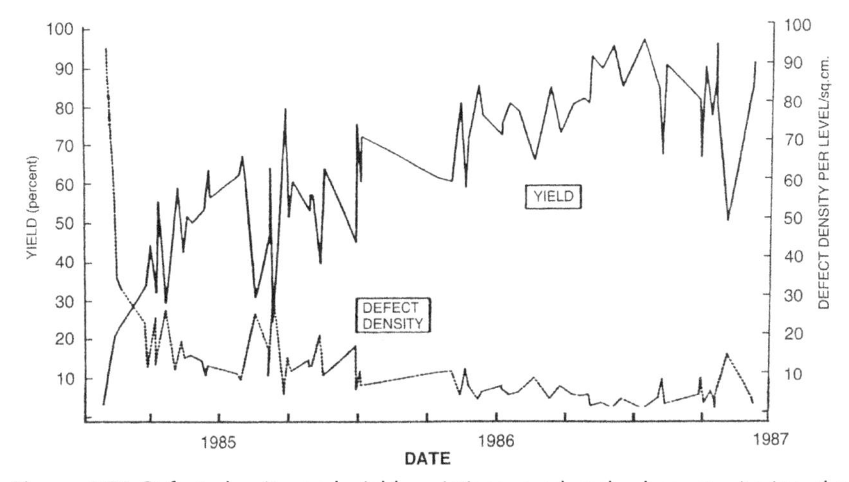


**Figure 3.26** Defect density and yield variations produced when monitoring the impact of yield enhancement actions in a 1.25-μm process. *From Ref. [46].*

2. **Processing defects.** Processing is frequently accompanied by inadvertent changes in the dimensions and composition of device features and components that adversely affect yield. Examples of such processing defects and their implications were discussed at length throughout this chapter.
3. **Random point defects.** Even though processing parameters are well within design specifications, the yield may still be below 100%. The reason is usually attributed to point defects or particles of macroscopic dimensions that settle on the water or obstruct the proper execution of a given processing step. Dust particles on a lithography mask is an example of the latter. These and other random defects can be killers. Often, however, a less-than-killer defect survives screening and becomes the nucleus for eventual failure in service.

For any series of IC manufacturing steps, the overall, or *cumulative,* yield ($Y$) is the product of the *wafer process yield* ($Y_w$), the *wafer test yield* ($Y_t$), and the *module test yield* ($Y_m$), such that

$$Y = (Y_w) \times (Y_t) \times (Y_m). \tag{3.21}$$

Rejects occur at each stage where yield is evaluated, i.e., during individual processing steps including epitaxy, oxidation, metallization, metal removal, electrical testing of scribed and diced fabricated chips, and final testing of packaged and assembled modules. Therefore, $Y_w$, $Y_t$, and $Y_m$ are each similarly defined by the above generic yield definition. Because of the many hurdles that must be surmounted during these rigorous evaluation procedures it is remarkable that the quality and reliability of semiconductor electronics is excellent. However, there is always room for improvement, and this section is dedicated to that end.

### 3.5.2 Statistical Basis for Yield

A functional definition of yield is the probability of producing chips having no fatal defects that would cause immediate rejection. Yield prediction, therefore, requires calculating the probability of finding a particular state (a chip with no defects) out of all possible states (chips with zero, one, two, or more defects) when the events or defects are statistically distributed over all states according to some distribution function [47].

It is instructive to start by introducing the statistics of the binomial distribution, which can be physically described by the tossing of coins that have two outcomes—heads or tails. This subject is discussed in virtually every text on statistics and probability. If the probability of getting a head in one trial toss is $p$ and the probability of getting a tail is $q$, then it is evident that $p + q = 1$ (in an unbiased coin $p = q = \frac{1}{2}$). The same holds for $N$ tosses, where the probability of obtaining *at least* one head or *at least* one tail is unity, or $(p + q)^N = 1$. Expansion yields

$$(p+q)^N = \sum_{k=1}^{N} \frac{N!}{k!(N-k)!} p^k (1-p)^{N-k}, \tag{3.22}$$

an equation known as the binomial distribution. Extending the result to $N$ coins each flipped once or one coin flipped $N$ times, the probability ($P$) that $n$ heads will be obtained in $N$ tosses is

$$P(n; N, p) = \frac{N!}{n!(N-n)!} p^n (1-p)^{N-n}. \tag{3.23}$$

In a simple example, consider three coins that are tossed once. What is the probability of getting exactly two heads? There are eight possible outcomes (hhh), (hht), (hth), (htt), (ttt), (tth), (tht), and (thh), and of these, three have two heads. Thus the probability we seek is $\frac{3}{8}$. This answer also results from direct substitution of $N = 3$, $n = 2$, $p = \frac{1}{2}$, in which case $P = 3!/(2!1!) \times (\frac{1}{2})^2 (\frac{1}{2})^1 = \frac{3}{8}$.

With no loss in generality let us consider Eqn (3.23) under conditions where $N$ is large, i.e., $N >> 1$, $N >> n$, and $p$ is small; but the product $N_p = \alpha$ is of moderate size. Next, the following quotient of factorials and polynomials is expanded,

$$\begin{aligned}\frac{N!}{(N-n)!} &= N(N-1)(N-2)\ldots(N-[n-1]) \\ &= N^n\left(1-\frac{1}{N}\right)\left(1-\frac{2}{N}\right)\cdots\left(1-\left[\frac{n-1}{N}\right]\right) \approx N^n.\end{aligned}$$

$$\begin{aligned}(1-p)^{N-n} &= \exp\{(N-n)\ln(1-p)\} \\ &= \exp\left\{-(N-n)\left(p+\frac{1}{2}p^2+\frac{1}{3}p^3+\ldots\right)\right\} \\ &= \exp\left\{-Np\left(1+\frac{1}{2}p+\frac{1}{3}p^2+\ldots\right)\right\} \\ &= \exp[-Np] \text{ or } \exp[-\alpha].\end{aligned}$$

Upon substitution in Eqn (3.23),

$$P\left(n; N, p\right) = \frac{n}{n!}\exp[\,-\,]. \tag{3.24}$$

This result is the Poisson approximation to the binomial distribution.

What does all of this have to do with yield? Suppose now that a number of events (defects) occur randomly in adjacent, nonoverlapping domains, and that these events are independent of events in other areas. The probability of a chip containing $n$ defects is then given by Eqn (3.24). Thus the probability of a chip containing zero defects, a quantity previously defined as the yield, is given by

$$P(0, Np) = P(0, \alpha) = Y = \exp[-\alpha]. \tag{3.25}$$

Furthermore, suppose the domain is the chip of area $A$, and $\alpha$ is the average number of events or fatal defects per chip. If the defect density or number of defects per unit area is $D$, then $\alpha = DA$, and

$$Y = \exp[-DA]. \tag{3.26}$$

This important equation expresses the probability of a chip having no defects provided that the defect density is constant. The strong exponential decrease in yield with increase in $A$ and $D$ should be appreciated. As a consequence of enlarging the die area from 1 to 1.5 $cm^2$ the yield will drop by a factor equal to exp [$-$ (1.5$-$1.0)$D$], assuming the same $D$; for example, the respective factors are 0.78 and 0.61 when $D = 0.5$ and 1 defect/$cm^2$. Imperatives to lower $D$ when increasing the wafer size are thus evident. The fundamental relationship expressed in Eqn (3.26) is the basis for yield modeling, a subject treated next.

### 3.5.3 Yield Modeling

Instead of being constant, assume now that the defect density varies according to the function $f(D)$. Then the probability that a chip will have $n$ defects is given by the Poisson equation, i.e.,

$$P(n, DA) = \int_0^{\infty} \frac{f(D)(DA)^n}{n!}\exp[-DA]\mathrm{d}D. \tag{3.27}$$

When $n = 0$ the yield is calculated as

$$Y = \int_0^{\infty} f(D)\exp[-DA]\mathrm{d}D. \tag{3.28}$$

This integral, proposed by Murphy [48], plays an important role in yield modeling. Depending on the form of the function $f(D)$, different yields are obtained when evaluating the integral. Among the many defect density functions suggested over the years, the following have been found useful:

1. Constant D. In the case where $D$ is a constant (i.e., $D_o$), we require that $f(D) = 0$ everywhere except for $D = D_o$. Mathematically, $f(D) = \delta(D - D_o)$, and $f$ behaves like a delta function ($\delta$). Substitution into Eqn (3.28) and integration yields our expected formula

$$Y = \exp[-D_oA]. \tag{3.29}$$

2. Triangular. This approximation to a Gaussian distribution has the form $f(D) = D/D_o^2$ for $D_o \geq D \geq 0$, and $f(D) = 1/D_o\ (2 - D/D_o)$ for $2D_o \geq D \geq D_o$. When $D \geq 2D_o$, $f(D) = 0$. Substitution into Eqn (3.28) and integration yields

$$Y = \frac{(1 - \exp[-D_oA])^2}{(D_oA)^2} \tag{3.30}$$

3. Exponential. In this case $f(D) = 1/D_o \exp[-D/D_o]$, and substitution into Eqn (3.28) yields

$$Y = (1 + D_oA)^{-1} \tag{3.31}$$

after integration.

Graphic representations of these as well as Gaussian and gamma distribution functions of $D$ are shown in Figure 3.27. The shape of these functions depends on the value of $\sigma$, the standard deviation normalized with respect to $D_o$. In the case of the gamma function, when $\sigma = 1$ we have the simple exponential function. Plots of normalized yield behaviors for the three-parameter gamma distribution as a function of the defect density–area product are displayed in Figure 3.28. For values of $\sigma$ different from zero, yields are always greater than that given by Eqn (3.29).

The reader should not be lulled into thinking that just because mathematical equations for yield exist, $f(D)$ can actually be predicted. In fact, quite the opposite is true. Presently, there is no way to determine $f(D)$ directly. Instead, it is inferred from actual production yields by curve fitting. Once obtained, $f(D)$ can be said *not* to be wrong if calculated yields derived from it match the production experience. Nevertheless, there is no reason to assume that $f(D)$ derived for product A will apply to product B, or even to product A in another time frame and environment.

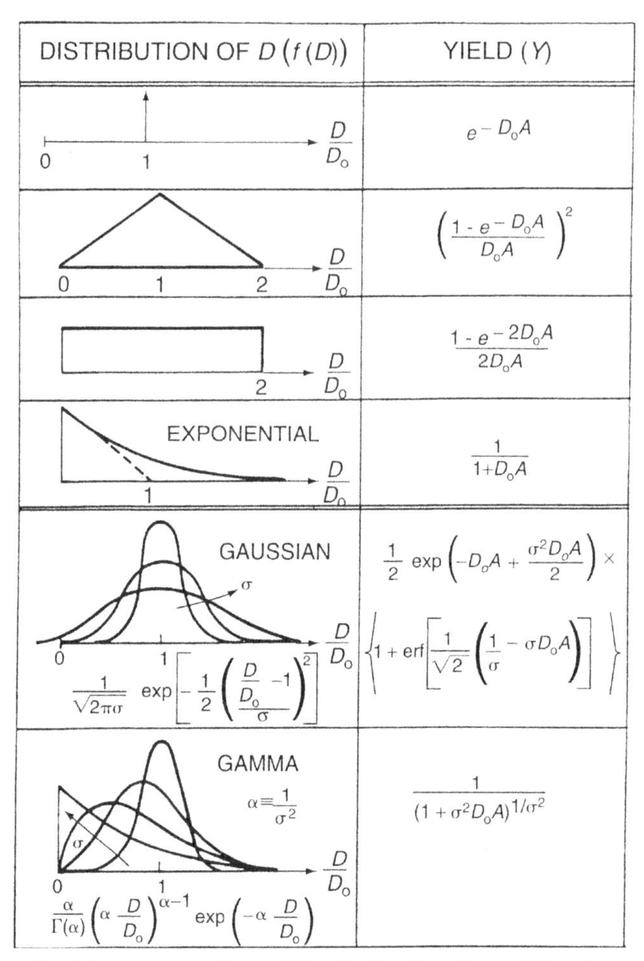


**Figure 3.27** Graphic representations of distribution functions of *D* and corresponding expressions for yields. *From Ref. [49].*

## 3.5.4 Comparison with Experience

Yields generally improve over time as learning occurs. For new products the initial yield may be close to zero. But, by establishing yield targets, monitoring actual performance, prioritizing yield detractors, implementing action, and then iterating improvement cycles of successively shorter duration, yields rise slowly at first and then more rapidly, until product yields level off at ~90% or more. Simultaneously, defect densities are observed to fall in concert.

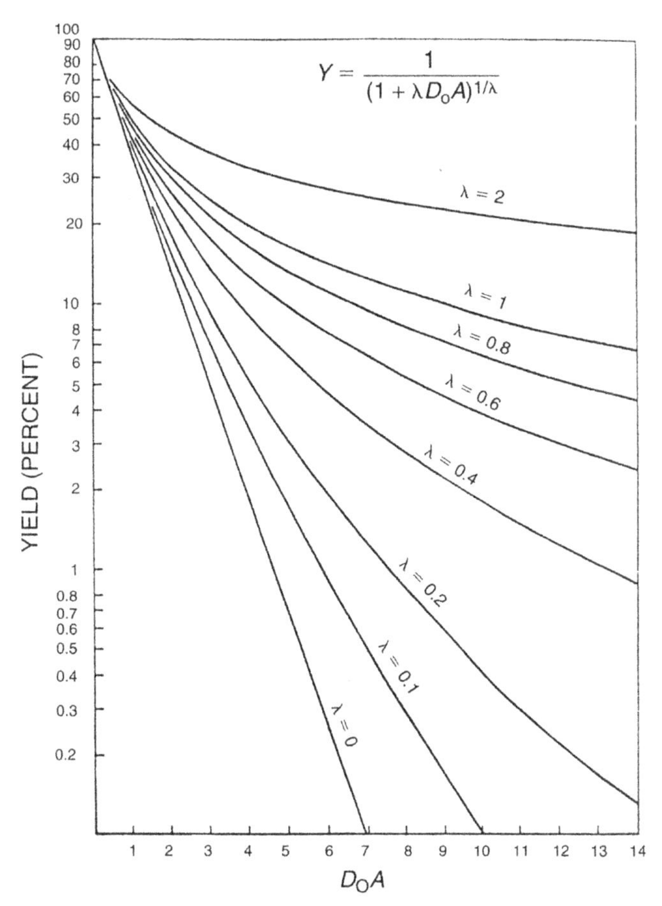


**Figure 3.28** Plots of three-parameter-yield model. $A$ = chip area, $D_0$ = mean defect density, and $\lambda$ = variance of $D$ distribution. (Note: $\lambda = \sigma^2$). *From Ref. [49].*

Among other findings concerning yield are the following:

1. Yield generally declines in dies located at the outer periphery of wafers, as shown in Figure 3.29.
2. As the number of process steps for a given product increases, yield, not surprisingly, is expected to drop. A graphic illustration of how severe the manufacturing yield drop is when the total number of process steps reaches several hundred is depicted in Figure 3.30.
3. Yield projection using Eqn (3.29) is pessimistic because defects preferentially congregate at certain locations on the wafer or *cluster,* rather than distribute randomly. This tends to leave unexpected defect-free areas that raise yield.

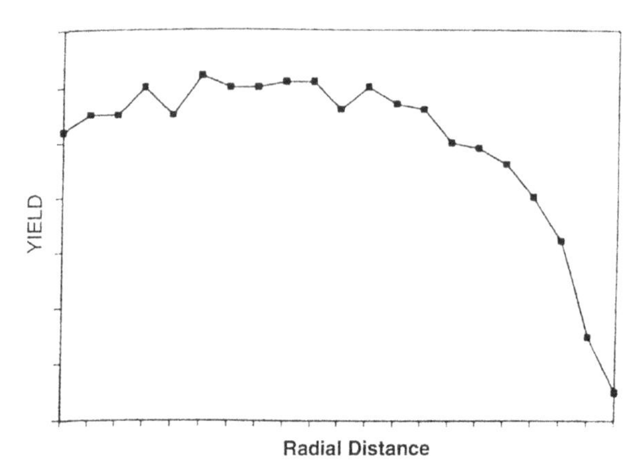


**Figure 3.29** Relative chip yield versus radial distance from center of silicon wafer. *From Ref. [45].*

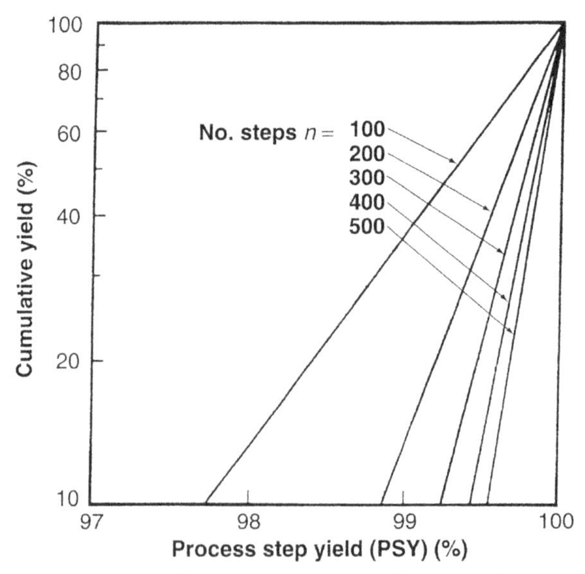


**Figure 3.30** Cumulative manufacturing yield dependence on the process step yield. Total manufacturing yield decreases rapidly with yield for each process step as the number of process steps gets large. *From Ref. [44].*

### 3.5.5 Yield and Reliability—What Is the Link?

There is a quantitative link between yield and reliability that we now explore. A clear distinction was drawn between these separate subjects in Chapter 1. Yield issues fall within the province of the manufacturer;

reliability, on the other hand, is of primary concern to the user and is measured from the time the product is put in service. Nevertheless, it is known that low-yielding product lots usually have lower reliability, as evidenced by higher failure rates during burn-in. The defects, which are the cause of yield loss, can also cause failure during burn in or early in product life(latent reliability defects), because these defects are so close to total failure that they can only survive for a short time, e.g. a 95% notch in a metal line fails due to electromigration during the first few minutes of service. Correlations between yield and reliability have been recognized in practice [50,51] and developed theoretically [51,52]. In the latter regard a quantitative correlation between "reliability defects" and yield defects developed by Huston and Clarke [53] is instructive, and will be sketched here. While we all have a feeling for yield defects, reliability defects are those that do not cause immediate rejection or failure, but change their characteristics under use conditions. These latent defects cause either "near opens" or "near shorts," and since they are neither completely electrically open nor shorted during the time of manufacture, they are difficult to detect, monitor, or control. Photolithography defects in the aluminum interconnects of Figure 3.31 are examples of near opens. Similarly, defects in gate oxides and passivating insulators are near shorts. Other examples of

**Figure 3.31** Photolithography defects in two parallel aluminum interconnects. *From Ref. [53].*

either near-open or near-short defects may exist at contacts, steps, vias, or in polysilicon and between metallization layers.

In the yield modeling literature the concept of critical area ($A_Y$) arises as a mathematical quantity that describes the probability that a defect will result in yield loss, e.g., due to a complete open or short. Therefore, we may write

$$Y = Y_o \exp[-DA_Y], \tag{3.32}$$

where $Y_o$ is the yield for nonrandom defects. In an analogous manner it is possible to define a reliability critical area ($A_R$), so that the reliability function ($R$) would be

$$R = R_o \exp[-DA_R], \tag{3.33}$$

where $R_o$ is a constant. If the two critical areas are simply related by $A_R/A_Y = K$ (a constant), and $Y_o$ and $R_o = 1$, then

$$R = Y^K. \tag{3.34}$$

An experimental test was made of this relation by visually counting both kinds of defects in a collection of processed wafers. In Figure 3.32(a) the number of reliability defects is plotted versus yield defects, while the log–log plot of Figure 3.32(b) yields a value of $K = 0.30$.

To determine $A_R$, it is necessary to first calculate $A_Y$. By definition, the average critical area is given by

$$A_Y = \int_0^\infty P_Y(x)D(x)\mathrm{d}x, \tag{3.35}$$

where, $P_Y(x)$ is the probability that a defect of size $x$ will cause a yield defect, and $D(x)$ is the defect density function. A similar expression holds for reliability defects. Now consider a pattern of parallel metal stripes ($w$ wide with a space $s$ between them) as shown in Figure 3.33. A yield defect, or complete open, occurs if the particle of dimension $x$ falls on the pattern and completely covers a metal line. On the other hand, a reliability defect, or near open, occurs if the particle reduces the cross-sectional area of the stripe below some arbitrary value. In the case where reliability defects are assumed, when the cross-sectional

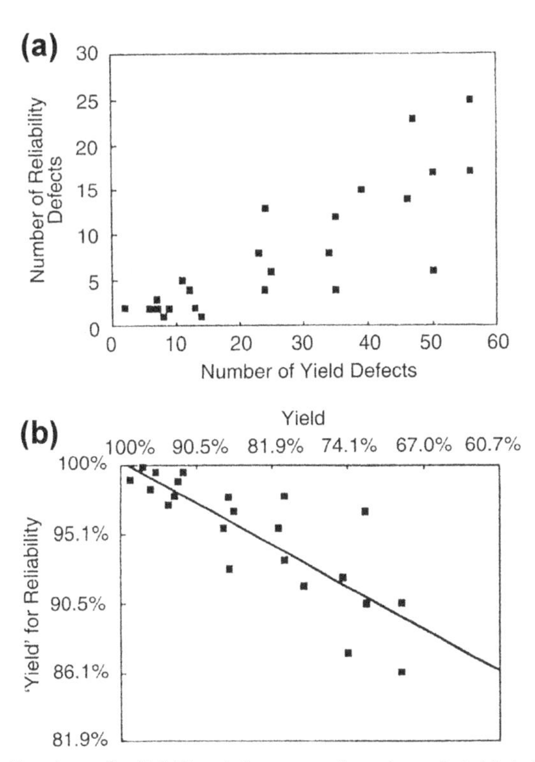


**Figure 3.32** (a) Number of reliability defects as a function of yield defects. Each point represents the number of observed defects for a single wafer. (b) Log–log plot of reliability yield versus yield. The slope yields a value of $K = 0.30$. *(From Ref. [53].)*

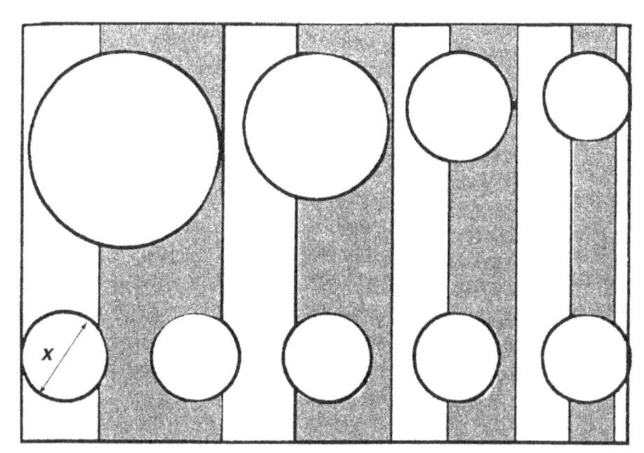


**Figure 3.33** A defect modeled across a metal pattern. The effect of small, medium, and large defects is shown. *From Ref. [53].*

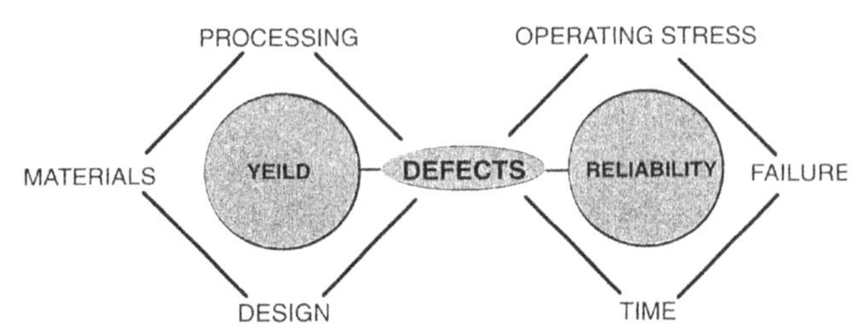


**Figure 3.34** A representation of the way yield and reliability concerns are interconnected through defects.

area is reduced below $w/4$, it is not too hard to show [53,54] the following results:

| Particle size (x) | $P_R(x)$ | $P_Y(x)$ |
|---|---|---|
| $0 \leq x \leq 3w/4$ | 0 | 0 |
| $3w/4 \leq x \leq w$ | $(x - w/2)/(w + s)$ | 0 |
| $w \leq x \leq w + s$ | $\frac{1}{2}w/(w+s)$ | $(x - w)/(w + s)$ |
| $w + s \leq x \leq 2w + s$ | $[w + \frac{1}{2}(s - x)]/(w + s)$ | $(x - w)/(w + s)$ |
| $2w + s \leq x$ | 0 | 1 |

Next, a form for $D(x)$ must be selected, and it is usually taken as $D(x) \sim 1/x^3$, the experimentally observed form. Lastly, the integrals are evaluated through piecewise integration in the different size ranges. In this way both $P_Y(x)$ and $P_R(x)$ can be evaluated and $K$ determined.

### 3.5.6 Conclusion

A graphic way to conclude the chapter is to consider the representation depicted in Figure 3.34 connecting yield and reliability through a common concern, namely, defects. Until now, our chief interest has been with factors that characterize and influence yield. Starting with the next chapter, interest shifts to the right side of this diagram as we consider those factors that relate to reliability.

## EXERCISES

1. Values of the energy to form a vacancy ($E_f$) of about 1 eV have been measured in the noble metals Cu, Ag, and Au. In Si, $E_f \approx 2.3$ eV.
   a. What is the equilibrium concentration of vacancies at 1000 K and at 300 K?

**b.** What is the vacancy concentration in Si at room temperature?

**c.** Comment on whether such concentrations pose reliability problems.

**2.** Oxygen concentrations are generally greater near the seed end of grown crystals than near the bottom. Why? What implication does this have for wafer yield?

**3.** Approximately what temperature is required to getter iron and copper atoms located 100 μm from extrinsically introduced dislocations if the treatment time is limited to 1 hour? The respective diffusion coefficients are $D(\mathrm{Fe}) = 10^{-3}$ exp $[-(0.68\ \mathrm{eV/kT})]$ $\mathrm{cm^2/s}$ and $D(\mathrm{Cu}) = 4.7 \times 10^{-3}$ exp $[-(0.43\ \mathrm{eV/kT})]$ $\mathrm{cm^2/s}$.

**4.** The diffusion coefficient of oxygen in Si is given by $D = 0.17$ exp $[-2.54\ \mathrm{eV/kT}]$ $\mathrm{cm^2/s}$. Suppose the O concentration in Si is initially $7 \times 10^{17}/\mathrm{cm^3}$. How long should deoxidation be carried out at 1100 °C to create a denuded zone containing an O concentration of $7 \times 10^{16}/\mathrm{cm^3}$ a distance 2 μm beneath the wafer surface? Assume that diffusion with an error-function-type solution governs deoxidation (see Chapter 5).

**5.** Differences in atomic radii (R) between silicon ($R_{Si}$) and dopant atoms ($R_D$) give rise to misfit stresses. In this case a measure of the misfit (f) has been suggested to be $f = 1 - (R_D/R_{Si})^3$. Values for $R_D$ and solubilities (S) at 1200 °C for antimony and boron are $R_{Sb} = 0.136$ nm and $S(\mathrm{Sb}) = 5.8 \times 10^{19}\ \mathrm{cm^{-3}}$ and $R_B = 0.098$ nm and $S(\mathrm{B}) = 5 \times 1020\ \mathrm{cm^{-3}}$, respectively. For silicon, $R_{Si} = 0.117$ nm, and the atomic density is $5 \times 10^{22}\ \mathrm{cm^{-3}}$.

**a.** Compare this expression for *f* with that of *f* in Eqn (3.9).

**b.** Estimate the residual stresses in Si containing maximum dissolved levels of antimony and boron.

**6.** An electric field acts normal to the GB of an n-type silicon bicrystal. Sketch the resulting band structure.

**7.** A film/substrate couple is residually stressed. Sketch the stress and strain variation across the interface. Specifically, are the stresses and strains continuous or discontinuous in going from film to substrate?

**8.** In the furnace processing of silicon wafers that are being cooled from elevated temperatures, radiation occurs more readily from the wafer periphery than from the center. As a result, a difference of 190 °C has been measured between these sites when the wafer is at 1100 °C.

**a.** What is the sign of the residual stresses expected at the edge relative to the center of the wafer?

**b.** Estimate the magnitude of the thermal stress. Will dislocations form if the critical yield strength of Si is $2 \times 10^6\ \mathrm{g/cm^2}$?

9. A critical thin film in an electronic device is prepared by a process in which two different thermal mechanisms compete. The first mechanism has a high activation energy and yields few product defects, while the second mechanism proceeds with low activation energy but generates many defects. Is it more desirable to operate the process at high or low temperatures? Why?
10. In a wafer fabrication facility that produces flash memory chips for portable computers, silicon nitride is deposited at high temperatures under low pressure, and chlorine as well as ammonia gas is liberated. Silicone rubber O-ring gaskets used in the door of the vacuum furnace reach a temperature of 218 °C.
    a. Comment on potential yield problems arising from the use of these gaskets.
    b. What properties of gaskets would you recommend for this application?
11. In making flat-panel display screens for laptop computers it is determined that defects measuring 0.5 μm in size can be tolerated if their density does not exceed 0.0005 per $cm^2$. A study made in the semiconductor industry revealed that a density of 5 defects per $m^2$ was achievable, for a defect size of 0.1 μm. The environment in the semiconductor study contained 10,000 particles measuring 0.1 μm, per cu ft. What class clean room would you recommend for the manufacture of flat-panel display screens?
12. A clean room contains a concentration of 100, 0.5-μm particles in a cubic foot of air
    a. What is the class of the clean room?
    b. Wafers are placed on a horizontal surface in this clean room. How long will it take before the surface (number) concentration of particles reaches $0.01/cm^2$?
    c. Suppose the same wafers are stored vertically. How long will it take for the same surface concentration of particles to be reached? Make reasonable assumptions for particle properties.
13. Suppose we wish to calculate the number of chips on a wafer
    a. For a 20-cm-diameter wafer with square chips that measure 8 mm on a side how many chips are there on the wafer?
    b. For the same wafer containing rectangular chips that measure 10.67 mm × 6 mm how many chips are there?
    c. A formula to calculate the number ($N$) of wafers on a chip is given by $N = \pi(R - A^{1/2})^2 A^{-1}$, where $R$ is the effective wafer radius and $A$ is the chip area. Repeat parts a and b.

**d.** Another formula that distinguishes between chips with the same area but different aspect ratio ($\alpha$) is $N = (\pi R2/\alpha E^2)\exp[\alpha E/R]$, where $E$ is the chip edge length. (The chip area is $\alpha E^2$.) Recalculate $N$ for parts a and b using this formula.

**e.** To see what accounts for the difference in these answers consult A.V. Ferris-Prabhu, IEEE Circuits and Devices, (January 1989), p. 37.

**14.** It is desired to double the area of a chip while keeping the processing yield constant. By what factor must the defect density be changed to accomplish this in the case of a. constant, b. triangular, and c. exponential defect distributions?

**15.** Consider the half-Gaussian defect density distribution given by$f(D) = (2D_o/\pi)\exp[-1/\pi(D/D_o)^2]$. Through the use of Eqn (3.28) show that the yield is given by $Y = \exp[\pi(D_oA)^2/4]\{1 - erf(\pi 1/2D_oA/2)\}$

**16.** In the chapter it was stated that

**a.** yield generally declines in dies located at the outer periphery of wafers, and

**b.** defects preferentially congregate or cluster at certain locations on the wafer rather than distribute randomly.

Provide possible explanations for these observations.

**17.** An IC process produces three defects per chip. Which of the yield models—constant, triangular, or exponential—is the most optimistic about yield?

**18.** Determine the number of fatal defects on a wafer containing laser chips if the yield of good chips is 80%. There are 2000 chips measuring 0.1 mm × 0.25 mm per wafer. Assume that the defects are randomly distributed.

**19.** The defect trend in DRAMs is as follows:

| Defects Number/cm$^2$ | Generation of DRAM |
|---|---|
| 6 | 16 Kb |
| 1 | 256 Kb |
| 0.6 | 1 Mb |
| 0.1 | 16 Mb |

If the trend continues, what defect density will be tolerated in the 1-Gb DRAMs? Estimate the number of defects on a 20-cm-diameter wafer containing 1-Gb DRAMs if the chip dimensions measure 1.5 cm × 1.5 cm.

**20.** A multichip module consists of six chips. These are listed with their yields in early production runs as well as later in mature products

| Chip type | Number | Early yield | Later yield |
|---|---|---|---|
| Microprocessor | 1 | 55 | 90 |
| ASIC | 2 | 50 | 85 |
| Memory | 3 | 60 | 99 |

What are the overall early and long-term yields for these modules?

## REFERENCES


[1] S.D. Wolter, M.B. Ferrara, K.A. Welch, Solid State Technol. (October 1995) 85.
[2] L.C. Kimerling, J.R. Patel, in: N. Einspruch (Ed.), VLSI Science, Microstructure Science, vol. 12, Academic Press, Boston, 1985.
[3] G.L. Miller, D.V. Lang, L.C. Kimerling, Ann. Rev. Mater. Sci. (1977).
[4] J.C.C. Tsai, in: S.M. Sze (Ed.), VLSI Technology, McGraw Hill, New York, 1988.
[5] S.M. Sze, Semiconductor Devices, Physics and Technology, J. Wiley and Sons, New York, 1985. Copyright by Bell Telephone Laboratories, Inc.
[6] H. Tsuya, Semiconduct. Semimet., 42, Academic Press, 1994, 619.
[7] B.O. Kolbesen, H.P. Strunk, in: N. Einspruch (Ed.), VLSI Science, Microstructure Science, vol. 12, Academic Press, Boston, 1985.
[8] J.W. Mayer, S.S. Lau, Electronic Materials Science: For Integrated Circuits in Si and GaAs, Macmillan, New York, 1990.
[9] H.F. Matare, Defect Electronics in Semiconductors, Wiley-Interscience, New York, 1971.
[10] J.C. Zolper, A.M. Barnett, IEEE Trans. Elec. Dev. 37 (1990) 478.
[11] S. Mahajan, Progr. Mater. Sci. 33 (1) (1989).
[12] S.-H. Soh, J. Lari, S. Hunt, T. Davies, M. Kuo, U. Kim, M. Cheung, E. Lucero, 33rd Annual Proceedings of the IEEE Reliability Physics Symposium, p. 244, 1995.
[13] G.J. Declerck, in: R.A. Levy (Ed.), Microelectronic Materials and Processing, Kluwer Academic Publ., Dordrecht, 1989.
[14] P. Schley, G. Kissinger, R. Barth, K.-E. Ehwald, Solid State Phenomena, 1993, pp. 32–33, 353.
[15] C.H. Seager, Ann. Rev. Mater. Sci. 15 (1985) 271.
[16] P.M. Fahey, S.R. Mader, S.R. Stiffler, R.L. Mohler, J.D. Mis, J.A. Slinkman, IBM J. Res. Develop. 36 (1992) 158.
[17] B.O. Kolbesen, in: S. Coffa, J.M. Poate (Eds.), Crucial Issues in Semiconductor Materials and Processing Technologies, Kluwer, 1992.
[18] G.L. Schnable, in: A.H. Landsberg (Ed.), Microelectronics Manufacturing Diagnostics Handbook, Van Nostrand Reinhold, New York, 1993.
[19] I. Blech, D. Dang, Solid State. Tech. 37 (8) (1994) 74.
[20] M. Ohring, The Materials Science of Thin Films, Academic Press, Boston, 1992.
[21] S.M. Hu, J. Appl. Phys. 70 (R53) (1991).
[22] T.T. Hitch, Circuit World 16 (2) (1990).
[23] I.A. Blech, D.B. Fraser, S.E. Hasko, J. Vac. Sci. Tech. 15 (1978) 13.
[24] W.D. Westwood, in: R. Levy (Ed.), Microelectronic Materials and Processes, Kluwer, Dordrecht, 1989.

[25] D.R. Cote, S.V. Nguyen, W.J. Cote, S.L. Pennington, A.K. Stamper, D.V. Podlesnik, IBM J. Res. Develop. 39 (1995) 437.
[26] Semiconductor International, November 1997, p. 55.
[27] T. Hattori, Solid State Tech. 33 (1990) 7. S-1.
[28] D.W. Cooper, in: A.H. Landsberg (Ed.), Microelectronics Manufacturing Diagnostics Handbook, Van Nostrand Reinhold, New York, 1993.
[29] P. Borden, IEEE Trans. Semiconduc. Manuf. 3 (4) (1990) 189.
[30] T. Francis, Semiconductor International, October 1993, p. 62.
[31] P. Burggraaf, Semiconduc. Int. 18 (5) (1995) 69.
[32] F.A. Stevie, R.G. Wilson, D.S. Simons, M.I. Current, P.C. Zalm, J. Vac. Sci. Technol. B 12 (4) (1994) 2263.
[33] S.J. Fonash, C.R. Viswanathan, Y.D. Chan, Solid State Tech. 37 (7) (1994) 99.
[34] H. Shin, C. Hu, IEEE Trans. Semiconduc. Manuf. 6 (1993) 96.
[35] Y. Uraoka, K. Eriguchi, T. Tamaki, K. Tsuji, IEEE Trans. Semiconduc. Manuf. 7 (1994) 293.
[36] L. F. Thompson and M. J. Bowden, in Introduction to microlithography, ed. L. F. Thompson, C. G. Wilson, and M. J. Bowden, American Chemical Society, Washington, DC (1983)
[37] S.N. Gupta, A.K. Bagchi, N.N. Kundu, Microelectron. J. 16 (1) (1985) 22.
[38] S. Wolf, R.N. Tauber, Silicon Processing for the VLSI Era, Lattice Press, Sunset Beach, CA, 1986.
[39] W.B. Henley, L. Jastrebski, N.F. Haddad, 31st Annual Proceedings of the IEEE Reliability Physics Symposium, p. 22, 1993.
[40] J. Chinn, Y.S. Ho, M. Chang, T. Turner, 32nd Annual Proceedings of the IEEE Reliability Physics Symposium, p. 249, 1994.
[41] D. Leith, in: R.P. Donovan, M. Dekkar (Eds.), Particle Control for Semiconductor Manufacturing, 1990. New York.
[42] V.B. Menon, in: R.P. Donovan, M. Dekkar (Eds.), Particle Control for Semiconductor Manufacturing, 1990. New York.
[43] See: Inside the Intel manufacturing process, what is a clean room?, Intel Web Site.
[44] G.W. Rublof, D.T. Bordonaro, IBM J. Res. Develop. 36 (1992) 233.
[45] P.J. Bonk, M.R. Gupta, R.A. Hamilton, A.V. Satya, in: A.W. Landzberg (Ed.), Microelectronics Manufacturing Diagnostics Handbook, Van Nostrand Reinhold, New York, 1993.
[46] J.M. Pimbley, M. Ghezzo, H.G. Parks, D.M. Brown (Eds.), VLSI Electronics Microstructure Science, vol. 19, Academic Press, 1989.
[47] A.V. Ferris Prabhu, Introduction to Semiconductor Device Modeling, Artech House, Boston, 1992.
[48] B.T. Murphy, Proc. IEEE 52 (1964) 1537.
[49] D. Koehler, W.E. Beadle, in: W.E. Beadle, J.C.C. Tsai, R.D. Plummer (Eds.), Quick Reference Manual for Silicon Integrated Circuit Technology, Published by John Wiley and Sons, 1985. Copyright Bell Laboratories.
[50] J.G. Prendergast, Proc. 31st Annual Proceedings of the IEEE Reliability Physics Symposium, p. 87, 1993.
[51] F. Kuper, J. van der Pol, E. Ooms, T. Johnson, R. Wijburg, W. Koster, D. Johnston, 34th Annual IEEE International Reliability Physics Symposium, p. 17, 1996.
[52] C.G. Shirley, A Defect Model of Reliability, Tutorial Notes, IEEE International Reliability Physics Symposium, p. 3.1, 1995.
[53] H.H. Huston, C.P. Clarke, 30th Annual Proceedings of the IEEE Reliability Physics Symposium, p. 268, 1992.
[54] C.H. Stapper, IBM J. Res. Develop. 27 (1983) 549.

CHAPTER 4

# The Mathematics of Failure and Reliability

## 4.1 INTRODUCTION

This chapter focuses on the mathematical methods employed in analyzing product quality and failure data. These data are obtained through measurement, testing, and service experience with the objective of obtaining the most probable values of parameters that characterize quality control, failure rates, and projected lifetimes of components and devices. As we shall see, statistical analysis is part of all quality and yield assessments associated with the processing and manufacture of devices. In particular, the use of reliability mathematics and analysis in predicting the fate of products during service is a recurring theme of this book.

We may well ask what are the sources for the data we analyze. Assorted information on the quality of a product or component is invariably gathered during stages of its manufacture through testing and screening. Then finally, prior to shipment, products undergo final inspection and qualification. The purpose is to weed out obvious rejects as well as weak products that would probably fail early during service. Testing to uncover these rejects may, at a minimum, involve nondestructive inspection and measurement of electrical characteristics. For more robust products, the weeding process requires that they survive the ordeal of exposure to conditions harsher than those they will experience in actual use. Burn-in is such an example. Here a simple pass–fail (or go–no go) decision is made as a result of response to stressing at elevated temperatures, voltages, currents, etc. Information gathered from such screening ultimately relates to either process or product yield, and is used in the yield models presented previously. Because of its pass–fail nature, the data are essentially *time independent* and are normally handled by simple Gaussian or normal distribution function statistics. Burn-in tests, however, do have a time dependence that is usually recorded.

In order to predict the reliability or fate of products during service, more extensive *time-dependent* destructive testing will now be required to allow values of failure rates to be determined. These are extracted from

*Reliability and Failure of Electronic Materials and Devices*
ISBN 978-0-12-088574-9
http://dx.doi.org/10.1016/B978-0-12-088574-9.00004-5

other mathematical functions, e.g., exponential, lognormal, and Weibull distributions that are used to fit failure data. Failure times, which depend on the testing conditions, e.g., temperature, voltage, etc., will vary from sample to sample and lot to lot. When the product is inexpensive (e.g., a light bulb) and there are many, one has the luxury of testing numerous samples to failure, and good statistical information will generally emerge. However, if the product is costly (e.g., a laser) an extensive failure testing program may not be feasible. Normally, reliability testing consumes a considerable portion of the vendor's research and development budget. When one considers the pharmaceutical industry, it is apparent that extensive reliability testing is not limited only to electronic products.

In addition to vendor-initiated product evaluation, there is the customer service experience to contend with. Depending on the nature of the product, field failure data may be more problematic, statistically, because the data are less abundant and subject to an often uncontrolled operating or testing environment. Nevertheless, this information is extremely valuable because it provides a powerful incentive to the vendor to conduct a failure analysis study that may uncover an overlooked product defect. By merging all of this vendor/customer-generated reliability information, intelligent predictions of product lifetimes are possible.

It must be emphasized, however, that the puzzling causes and attributes of physical degradation and failure mechanisms are generally not illuminated through statistical analysis of lifetime data. Neither are any clues revealed with regard to minimizing or eliminating failure. Rather, statistical analysis of the kind addressed in this chapter in a sense quantifies and institutionalizes our ignorance of the causes of failure. As pointed out by Pecht and Nash [1], reliability assessment based on root-cause analysis of failure mechanisms, damage sites, and dependence on operating stress has proven to be effective in the prevention, detection, and correction of failures associated with the design, manufacture, and operation of products.

Traditional reliability prediction methods based on statistical analysis and curve fitting of failure data have also addressed these same issues of design, manufacture, and operation of products, but not as effectively. Nevertheless, virtually all failure test data reported in the literature are analyzed by the mathematics of statistics and associated probability theory. As we shall see, however, there is controversy involved in doing this and there are questions related to the credibility of the conclusions.

This chapter's discussion of statistics and probability, failure distributions, reliability estimates, and confidence of predictions is treated more extensively in Refs [2–5]. They are recommended for their lucid presentations.

## 4.2 STATISTICS AND DEFINITIONS

### 4.2.1 Normal Distribution Function

A useful skill for an engineer concerned with product reliability is to be able to convert a mass (mess?) of failure data into a form amenable to analysis without having a FIT (an acronym of "failure unit"). As an example, consider a company that makes electrical fuses for trucks that are rated at 10 A. Fuses fail by melting when current-induced Joule heat cannot be effectively removed. Because fuses protect other circuitry from electrical overload failure, it is vital that they function within the rated value, say between 9 and 11 A. Circuit damage or less-than optimum performance may be the outcome of operating outside these current extremes. Fuses are produced in lots of tens of thousands, and it is not practical to test them all and reject those that fail. To determine if fuses operate within specifications without excessive variability, a random sample from the population of 10,000 fuses is chosen and tested by increasing the current until they fail or open the circuit. The failure currents measured for such a population of 100 fuses are entered in Table 4.1. It is clear that although values cluster at about 10 A, there is a spread that warrants further statistical analysis [2].

To proceed, it is convenient to first create a failure frequency distribution, as indicated in Table 4.2. For this purpose the range represented by minimum and maximum current values is arbitrarily subdivided into a number of cells (11 in this example) and the number of fuses destroyed in each current interval is noted. The results are best represented in the form

**Table 4.1** Fuse test currents

| | | | | | | | | | |
|---|---|---|---|---|---|---|---|---|---|
| 9.60 | 9.76 | 9.46 | 10.42 | 9.34 | 10.20 | 10.06 | 10.36 | 9.16 | 9.54 |
| 10.22 | 10.41 | 10.19 | 9.61 | 10.20 | 9.76 | 10.32 | 10.04 | 9.88 | 10.02 |
| 9.46 | 9.72 | 9.28 | 10.02 | 9.88 | 10.11 | 9.72 | 10.24 | 10.24 | 10.85 |
| 9.58 | 9.76 | 9.51 | 9.68 | 9.82 | 8.86 | 10.04 | 9.88 | 10.09 | 10.06 |
| 9.76 | 9.22 | 10.01 | 10.44 | 9.74 | 9.88 | 9.92 | 9.56 | 10.08 | 10.08 |
| 10.12 | 9.69 | 10.24 | 10.23 | 10.89 | 10.22 | 9.92 | 10.21 | 9.60 | 10.19 |
| 9.76 | 10.32 | 10.04 | 9.88 | 10.02 | 9.62 | 9.76 | 9.46 | 10.42 | 9.34 |
| 10.20 | 10.06 | 10.36 | 9.16 | 9.54 | 9.76 | 9.20 | 10.00 | 10.44 | 10.48 |
| 9.88 | 9.92 | 9.56 | 10.08 | 10.09 | 9.58 | 9.76 | 9.51 | 9.68 | 9.82 |
| 8.84 | 10.04 | 9.88 | 10.11 | 10.06 | 9.46 | 9.72 | 9.28 | 10.02 | 9.89 |

**Table 4.2** Frequency and cumulative distributions of fuse test currents

| Frequency distribution | | | Cumulative distribution | |
|---|---|---|---|---|
| Cell | Current limits | Number in cell | Upper current | Cumulative number |
| 1 | 8.70–8.90 | 2 | 8.90 | 2 |
| 2 | 8.90–9.10 | 0 | 9.10 | 2 |
| 3 | 9.10–9.30 | 7 | 9.30 | 9 |
| 4 | 9.30–9.50 | 6 | 9.50 | 15 |
| 5 | 9.50–9.70 | 15 | 9.70 | 30 |
| 6 | 9.70–9.90 | 22 | 9.90 | 52 |
| 7 | 9.90–10.10 | 22 | 10.10 | 74 |
| 8 | 10.10–10.30 | 14 | 10.30 | 88 |
| 9 | 10.30–10.50 | 10 | 10.50 | 98 |
| 10 | 10.50–10.70 | 0 | 10.70 | 98 |
| 11 | 10.70–10.90 | 2 | 10.90 | 100 |

of a histogram, shown in Figure 4.1, where the ordinate is the number (or percent) of failures. Since the histogram is approximately symmetric about 10 A with only a ±5% spread, the distribution of failures is deemed acceptable. It is common to plot failure data in another way. The number of failures that occur at less than or equal to a given value is noted, then a *cumulative distribution function* (CDF) can be defined. Such cumulative distribution numbers for the indicated fuse current values are also listed in Table 4.2 and plotted in Figure 4.2.

The continuous normal, or Gaussian, distribution function is ideally suited to mathematically treat these data. This widely used, two-parameter function was used previously in connection with yield (see Figure 3.26). Discovered in 1733 by De Moivre as the limiting form of the binomial distribution for discrete random variables, the normal distribution deals with the statistics of random events, numbers, and phenomena. Underlying the validity of this function is the central limit theorem, which states that when the sample size is very large, the distribution of means approximates a normal distribution. Thus a superposition of the independent random variables, each of which contributes a small amount to the total, always tends toward normality. As the number of contributing factors increases, the approximation becomes more exact. This condition certainly obtains in IC manufacturing with tens of photomask steps and hundreds of individual process steps, each of which can introduce defects.

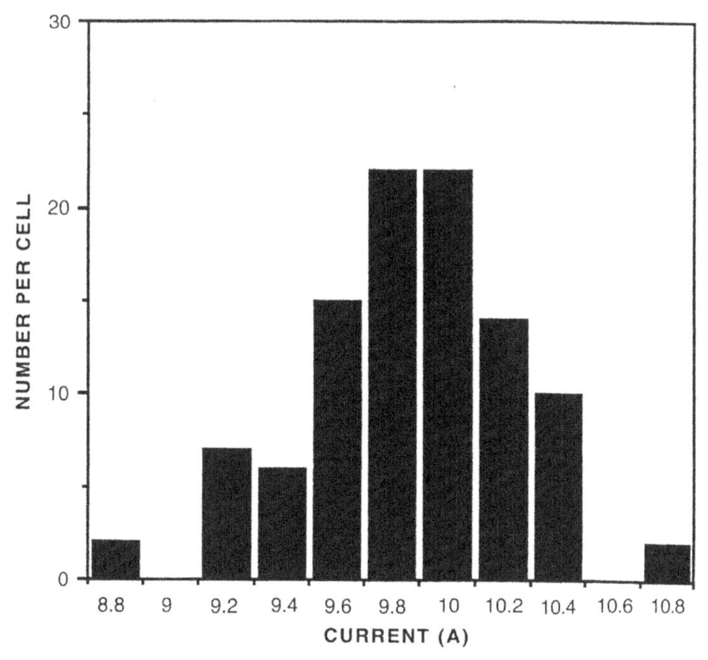


**Figure 4.1** Frequency-distribution function for fuse failure data plotted in the form of a histogram representing the number of failed fuses in discrete current intervals.

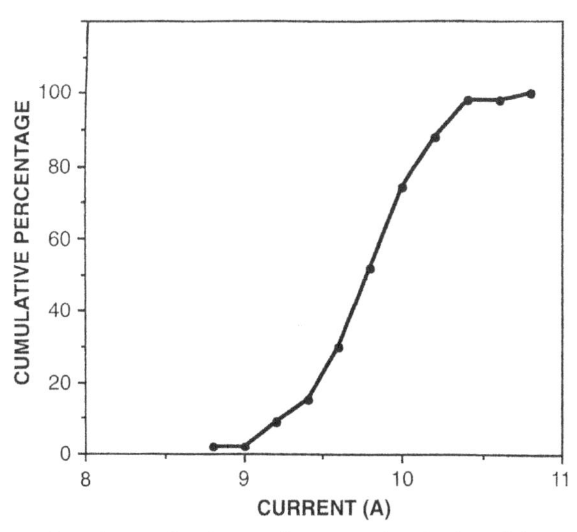


**Figure 4.2** Cumulative distribution function plot for fuse-failure currents.

The *probability distribution function* (PDF) that defines the normal distribution is $f(x)$ has the familiar mathematical form

$$f(x) = \frac{1}{\sigma(2\pi)^{1/2}} \exp\left[-\frac{(x-\mu)^2}{2\sigma^2}\right], \tag{4.1}$$

where the quantities $\sigma$ and $\mu$ are defined in the bell-shaped curve of unit area (Figure 4.3(a)). For a distribution of values it is evident that $\mu$ is the mean value and $\sigma$ is a measure of their spread about the mean known as the standard deviation. From statistics we know that $\mu$ is the sum of the individual $x_i$ values divided by the number of data points ($n$), i.e., $\mu = \Sigma x_i/n$, where the sum is over $i$. Similarly, $\sigma$ is defined as $\{\Sigma(\mu - x_i)^2/(n-1)\}^{1/2}$. Using these definitions, $\mu$ and $\sigma$ for the fuse current data can be readily evaluated.

It is customary to express $x$ in standard units, $z = (x - \mu)/\sigma$, because the simpler mathematical form, based symmetrically about the origin,

$$f(z) = \frac{1}{(2\pi)^{1/2}} \exp\left(\frac{-z^2}{2}\right), \tag{4.2}$$

results. This equation suggests that $z$ is normally distributed with $\mu = 0$ and $\sigma = 1$.

### 4.2.2 Statistical Process Control, $C_{pk}$, and "Six Sigma"

The application of statistics to IC fabrication can be for either discrete variables, such as defects, or continuous variables, such as threshold voltage of MOSFETs. A state-of-the-art IC fabrication process has approximately 25 mask levels and about 600 individual process steps. Each process step and masking level can introduce defects, all of which can potentially impact the yield and reliability of an IC. Defects at any process step will impact yield and reliability. For example, slight variations in oxide thickness, stored oxide charge, semiconductor doping in the channel, or even channel length will influence the threshold voltage of the 100 million CMOS FETs in a chip, as well as, the hundreds of chips in a wafer. Variations within a chip and from chip to chip will be manifest in final test yield at the wafer level as we will discuss next. So the question is how to handle the control of the process and the defects introduced by processing. First we need to understand and appreciate some of the features of Eqn (4.1), which is widely employed in industry to monitor the extent of process and production

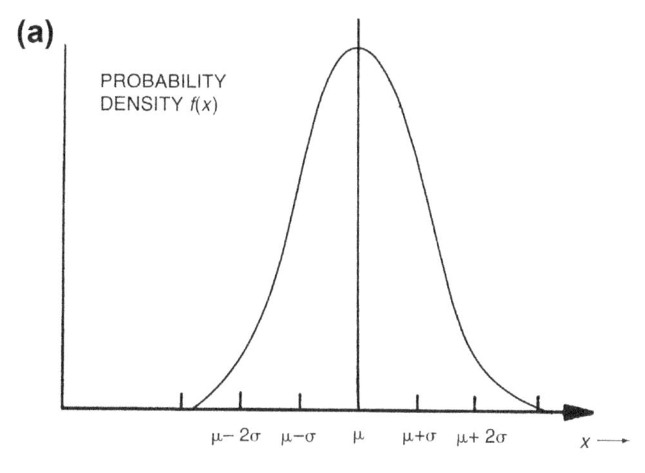


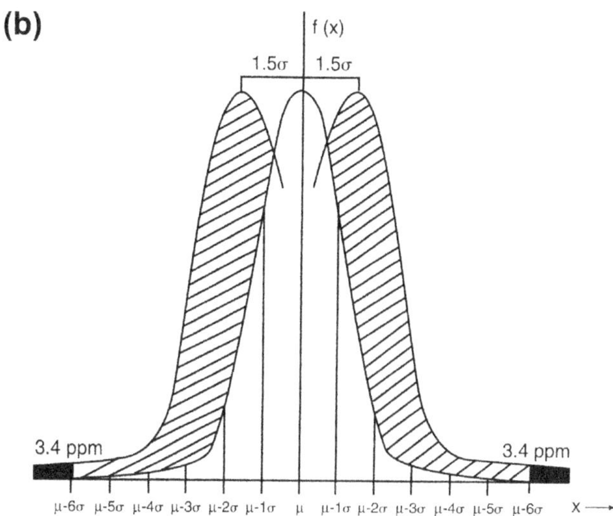


**Figure 4.3** (a) Plot of the normal probability distribution function $f(x) = 1/\sigma(2\pi)^{1/2}$ $\exp -[(x - \mu)^2/2\sigma^2]$ versus $x$. (b) Six Sigma quality statistics depicted in terms of positive and negative process shifts about the nominal target mean. The numbers of rejects that arise are shaded in at the extremes of the displaced distributions. *From Ref. [6].*

quality control. Sigma ($\sigma$) is a measure of the spread of a measured process variable, say $V_t$ (threshold voltage), a parameter used to infer yield. $V_t$ has limits beyond which a chip or entire wafer will be scrapped (yield = 0). So for critical parameters, such as $V_t$, oxide charge, oxide thickness, etc., these limits define an acceptable range of the parameter in question. Ideally, we would like the parameter mean centered between the upper limit and the

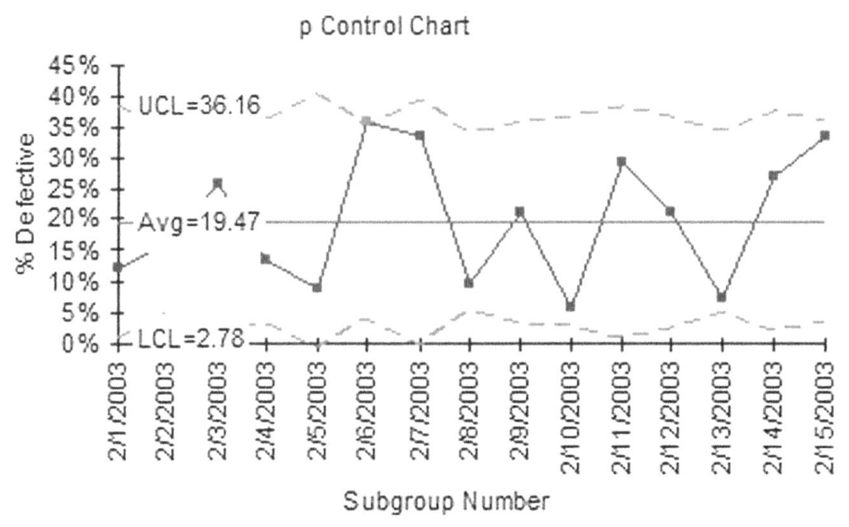


**Figure 4.4** Statistical process control chart.

lower limit that we will accept. We say the parameter mean should lie half way between the upper control limit (UCL) and the lower control limit (LCL) (see Figure 4.4). Then, if $\sigma$ is small, we expect the entire distribution (normal) for a single process step or mask level to lie within the UCL and the LCL for that parameter. Well, for a normal distribution ~68.26% lies between $+\sigma$ and $-\sigma$, and ~95% lies between $+2\sigma$ and $-2\sigma$. Approximately 99.7% lies between $+3\sigma$ and $-3\sigma$. If now the spread from $+3\sigma$ to $-3\sigma$ is small compared to the spread between UCL and LCL, then we have high certainty that the parameter is controlled well enough for our purposes. But wait again, we have hundreds of chips in a wafer and hundreds of millions of FETs on each chip. So if we only control to $+3\sigma$ and $-3\sigma$ we will possibly make scrap 0.3% of the time either on a chip or wafer basis, which on a chip basis implies we may never make a good chip. It became very important therefore to control parameters to much tighter levels in order to guarantee some yield. What we described is statistical process control (SPC), a method to ascertain that a process step is controlled well enough to yield greater than zero.

A more sophisticated method measures terms $C_p$ and $C_{pk}$ to determine the capability of a process. $C_p = (\text{UCL} - \text{LCL})/6\sigma$ and $C_{pk} = C_p(1 - k)$, where $k$ is the ratio of the mean drift to $(\text{UCL} - \text{LCL})/2$. $C_p$ measures short-term capability (see Figure 4.5) and $C_{pk}$ measures long-term process capability (see Figure 4.6). The interested reader is referred to an excellent

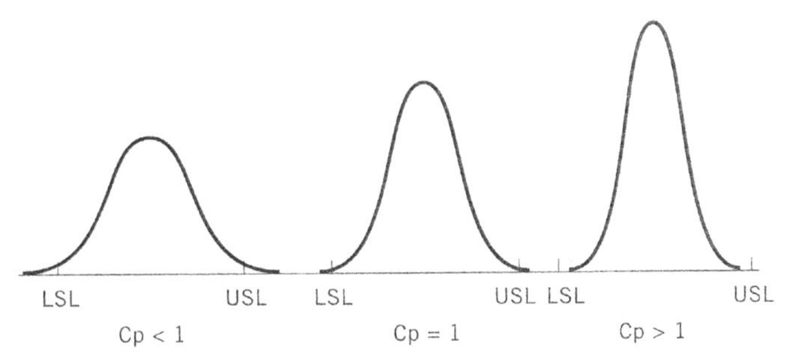


**Figure 4.5** Capability index $C_p$ for normal distribution from Ref. [7].

book by E.E. Lewis [7]. These parameters measure how much of the process window (UCL − LCL) is used by the process and how well the process is centered. Values of $C_p$ and $C_{pk}$ of 1 are marginal, whereas 2 or 3 are much better because less of the process window is covered by the spread of the parameter in question.

The next step in SPC is called "Six Sigma." For example, "Six Sigma in everything we do..." [6] was a Motorola Corporation rallying call to increase competitiveness by enhancing the quality of products. Whereas

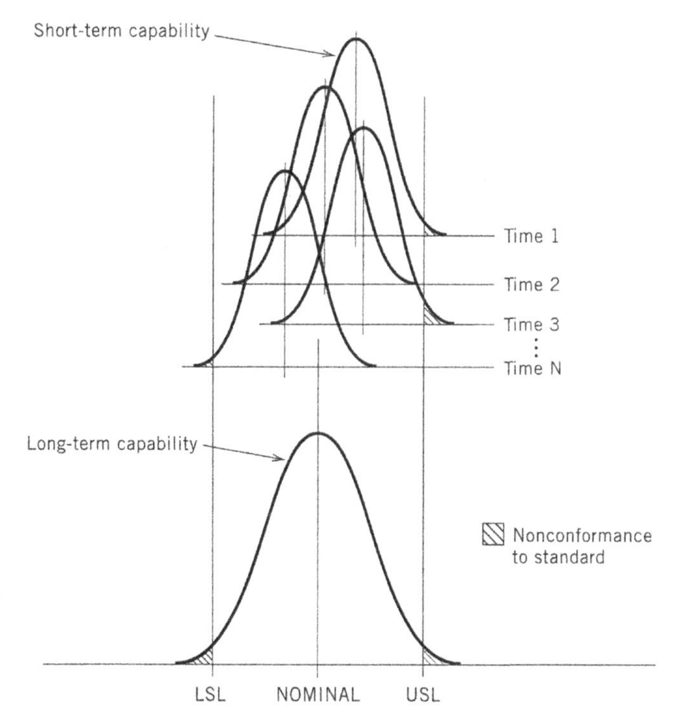


**Figure 4.6** Long term process capability from Ref. [7].

68.26% of the area under the normal distribution or product distribution curve falls within $1\sigma$ ($-\sigma$ to $+\sigma$), 99.7% within $3\sigma$, and 99.9999998% falls within $6\sigma$ ($-6\sigma$ to $+6\sigma$). Manufacturing processes, however, experience shifts in the mean as great as $\pm 1.5\sigma$ in practice because equipment, operators, and environmental conditions are never constant. Six Sigma quality is therefore defined as the number of rejects that occur when such shifts happen (Figure 4.3(b)). This, in effect, means tolerating no more than 3.4 defective parts per million produced. It amounts to $\sim 4.5\sigma$ of allowable variability. A strict Six Sigma would result in two defective parts in $1 \times 10^9$. IC chips in the near future will contain $1 \times 10^9$ FETs and most likely require the strict Six Sigma control of defects. An example of long-term capability is shown in Figure 4.7. Of course, achieving Six Sigma quality is easier said than done, but it remains a vital challenge for industry. Those inclined can calculate what seven sigma implies!

### 4.2.3 Accuracy and Precision

We have just seen that failure testing is subject to a Gaussian spread; the same is true of all property measurements and manufacturing-quality control monitoring. With respect to measurements the most often asked questions are, how *accurate* are they, and with what *precision* are they made The distinction between these two terms can be immediately grasped by viewing Figure 4.8. High shooting accuracy is achieved in both Figure 4.8(a) and (b) because the target hits are centered around the bull's-eye. Thus the mean value $\mu$ is the same in both cases and not far from the distribution center. But the precision of the shooter is better in a than in b because all the bullet holes cluster and have a small dispersion, or spread. High precision is thus characterized by a small $\sigma$; equivalently, we speak of a small random uncertainty. Poor accuracy is attained in Figures 4.8(c) and (d) because all holes are far off the bull's-eye; here we speak of a large systematic uncertainty. Again, shooter precision is better in c because all the bullet holes cluster more tightly. The case in d is worst of all because neither accuracy nor precision is achieved; i.e., both $\sigma$ and $\mu$ are large.

Is it more desirable to have high accuracy or high precision? Good shooting instructors know that it is better first to work on attaining smaller clusters or groupings. Thus the path to becoming a sharpshooter is to be precise first. Accuracy can then be gained by adjusting or calibrating the movable sights.

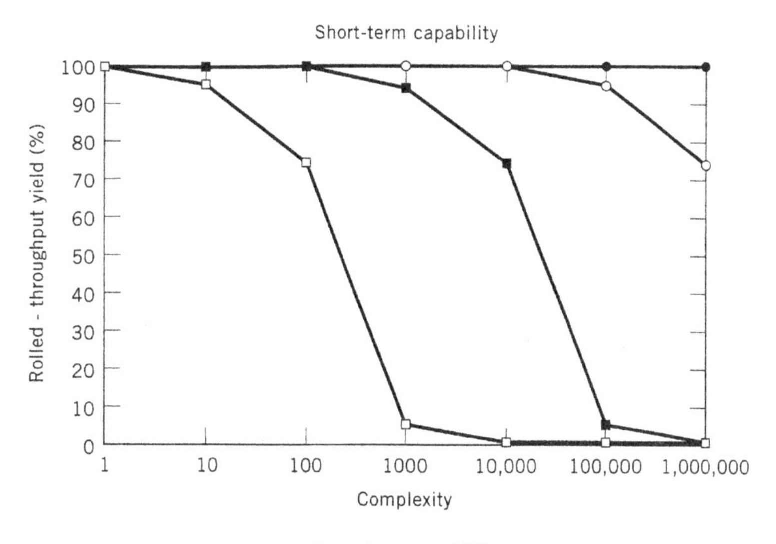


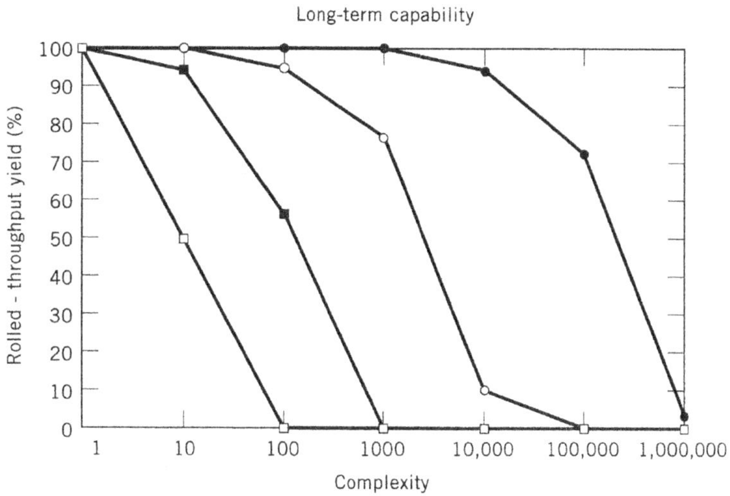


● Six sigma
○ Five sigma
■ Four sigma
□ Three sigma

Note: Long-term capability based on an equivalent, one-sided 1.5$\sigma$ mean shift

**Figure 4.7** Long term process capability from Ref. [7].

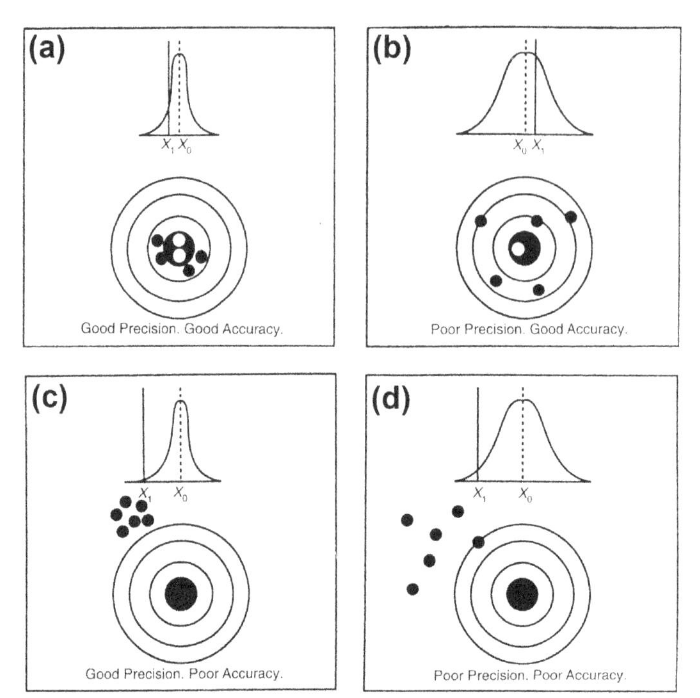


**Figure 4.8** Four target-shooting characteristics depicted schematically in both graphical and statistical terms. (a) Good precision–good accuracy; small $\sigma$ and small shift ($X_1$) from target center ($X_0$). (b) Poor precision–good accuracy; large $\sigma$ and small $X_1$. (c) Good precision–poor accuracy; small $\sigma$ and large $X_1$. (d) Poor precision–poor accuracy; large $\sigma$ and large $X_1$. *From Ref. [8].*

### 4.2.4 Failure Rates

The fuse current data treated above were essentially time independent. Reliability, however, implies time dependence of failure, with time reckoned from the instant the product is put into service. To include time ($t$) we first define the PDF to be $f(t)$. Therefore, the CDF, defined by $F(t)$, is given by the integral

$$F(t) = \int_{-\infty}^{t} f(y)\mathrm{d}y, \tag{4.3}$$

where the lower time limit is taken to be $-\infty$ instead of 0 for mathematical purposes, and $y$ is a dummy variable of integration. For a given population, $F(t)$ is the fraction that fails up to time $t$, and it is related to the PDF by

$f(t) = dF(t)/dt$. A reliability, or *survival*, function, $R(t)$, can then be defined to indicate the surviving fraction, i.e.,

$$R(t) = 1 - F(t). \tag{4.4}$$

The *failure rate* (also known as the *hazard rate*), $\lambda(t)$, is defined as the number of products (e.g., devices) that failed between $t$ and $t + \Delta t$, per time increment $\Delta t$, as a fraction of those that survived to time $t$. This complex definition of the failure rate is written as

$$\lambda(t) = \frac{F(t+\Delta t) - F(t)}{\Delta t(1 - F(t))} = \frac{f(t)}{1 - F(t)} = \frac{f(t)}{R(t)} \tag{4.5}$$

by employing $F(t)$ and its derivative. From this it is easy to show that $\lambda(t) = -d\ln R(t)/dt$.

A good way to appreciate how failure rates are calculated is through a couple of practical examples.

## Example 4.1

Suppose that a population of 525,477 solid-state devices is burned in at elevated temperature, and the number of failed devices are weeded out as a function of time is recorded in the following table. Determine the failure rates, $\lambda(t)$, at the indicated times.

### Answer

Results of the calculations are tabulated.

| Time (h) | Number failed, $f(t)$ | Cumulative failures, $F(t)$ | $\lambda(t)$ ($h^{-1}$) | Number of device hours | FITs |
|---|---|---|---|---|---|
| 1 | 6253 | 6253 | 0.0119 | 525,477 × 1 | $1.19 \times 10^7$ |
| 2 | 1034 | 7287 | 0.00199 | 519,224 × 2 | $9.96 \times 10^5$ |
| 3 | 617 | 7904 | 0.00119 | 518,190 × 3 | $3.97 \times 10^5$ |
| 4 | 419 | 8323 | 0.000810 | 517,573 × 4 | $2.02 \times 10^5$ |
| 5 | 502 | 8825 | 0.000971 | 517,154 × 5 | $1.94 \times 10^5$ |
| 6 | 401 | 9226 | 0.000777 | 516,652 × 6 | $1.29 \times 10^5$ |
| 7 | 297 | 9523 | 0.000577 | 516,252 × 7 | $8.22 \times 10^4$ |
| 8 | 214 | 9737 | 0.000415 | 515,954 × 8 | $5.18 \times 10^4$ |
| 9 | 206 | 9943 | 0.000400 | 515,740 × 9 | $4.44 \times 10^4$ |
| 10 | 193 | 10,136 | 0.000375 | 515,534 × 10 | $3.74 \times 10^4$ |

Sample calculation at 5 h using Eqn (4.5): $\lambda(t) = 502/525{,}477 \times (1 - 8323/525{,}477)^{-1} = 0.0009710/\text{h}$.

Because the failure rates are very low, a new numerically larger unit, the FIT, has been defined and is widely used in the microelectronics industry. The FIT is not an acronym, but a contraction of "failure unit"; it equals the number of failures in $10^9$ device-hours. For example, 1 FIT is one failure in $10^9$ device-hours, one failure in $10^7$

devices after 100 h of operation, or one failure in $10^6$ devices after 1000 h, etc. Employing this definition, failure rates in terms of FITs were generated, and the values are displayed in the last column of the above table. After 5 h, for example, $\lambda(t) = (502/517{,}154 \times 5) \times 10^9 = 1.94 \times 10^5$ FITs.

---

## 4.3 ALL ABOUT EXPONENTIAL, LOGNORMAL, AND WEIBULL DISTRIBUTIONS

### 4.3.1 Exponential Distribution Function

#### *4.3.1.1 Mathematics*

We have already introduced the normal distribution function, whose PDF was given by Eqn (4.1). It was used to assess product properties or quality, and no time dependence was attached to the variable of interest. There are three additional PDFs, the exponential, lognormal, and Weibull, that can also be used like the normal distribution function in evaluating quality control. But in this book they have a more vital role; they are called upon to analyze and characterize failure and reliability data in material systems with the objective of predicting future behavior. Therefore, time, not a property relating to quality, is now the variable of interest. Of these functions the most simple mathematically is the exponential distribution. The exponential PDF is defined by

$$f(t) = \lambda_o \exp(-\lambda_o t), \tag{4.6}$$

where $\lambda_o$ is a constant. Through simple integration (Eqn (4.2)) the CDF is evaluated to be

$$F(t) = \int_{-\infty}^{t} \lambda_o \exp(-\lambda_o y)\mathrm{d}y = 1 - \exp(-\lambda_o t). \tag{4.7}$$

Finally, the failure rate is given by

$$\lambda(t) = \frac{f(t)}{1 - F(t)} = \frac{\lambda_o \exp(-\lambda_o t)}{[1-(1- \exp - \lambda_o t)]} = \lambda_o, \tag{4.8}$$

and, importantly, is seen to be *constant*. Therefore, the mean time to failure (MTTF) is the reciprocal, or $1/\lambda_o$. Plots of the functions $f(t)$, $F(t)$, and $\lambda(t)$ are depicted in Figure 4.9 as a function of time (in $1/\lambda_o$ units).

## Example 4.2

Suppose the population of devices is described by the life distribution $F(t) = 1 - \exp(-0.0001t)$.

1. What is the probability that a new device will fail after 1000 h? After 5000 h? Between these times?
2. What is the failure rate at 1000 h? After 5000 h?

### Answer

1. By substitution, $F(t) = 1 - \exp[-0.0001(1000)] = 0.095$. Similarly, at 5000 h, $F(t) = 0.39$. The probability of device failure between 1000 and 5000 h is $F(5000) - F(1000) = 0.39 - 0.095 = 0.30$.
2. $f(t) = dF(t)/dt = 0.0001 \exp(-0.0001)$. The failure rate is $\lambda(t) = f(t)/[1 - F(t)] = 0.0001 \exp(-0.0001t)/\exp(-0.000t) = 0.0001$. For this distribution the failure rate is seen to be time-independent.

### 4.3.1.2 Constant Failure Rate Modeling

Due to ease in dealing with a constant failure rate, the exponential distribution function has proven popular as the traditional basis for reliability

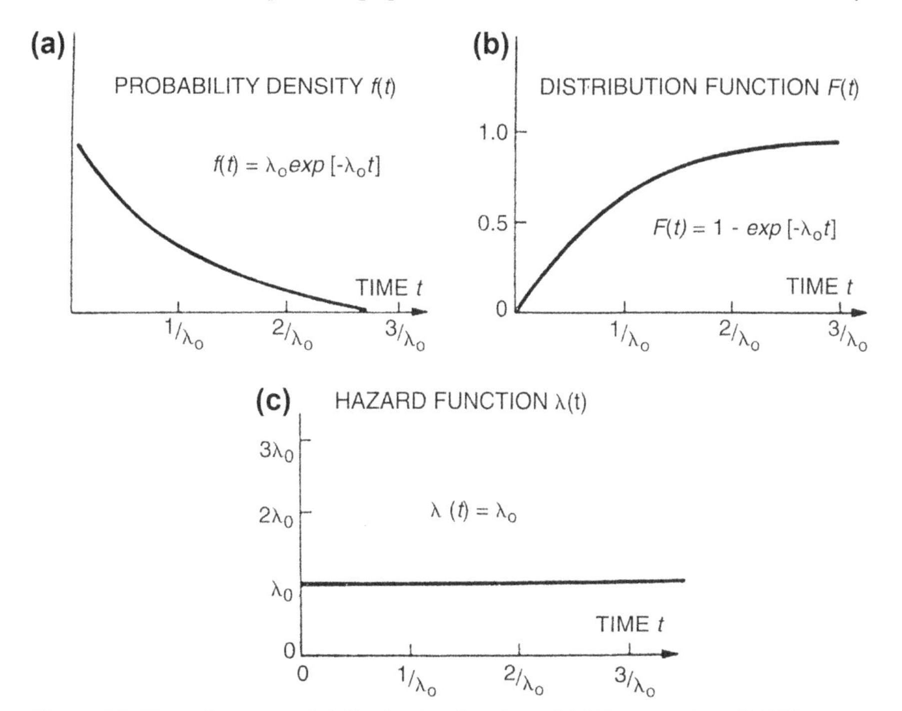


**Figure 4.9** Plots of exponential distribution functions. (a) $f(t)$ versus time. (b) $F(t)$ versus time. (c) $\lambda(t)$ versus time. Time is measured in units of $1/\lambda_o$.

modeling. For the reasons enumerated below, some of which are historical in nature, it is not difficult to see why the constant failure rate model has been so widely used [1].

1. The first generalized reliability models of the 1950s were based on electron vacuum tubes, and these exhibited constant failure rates.
2. Failure data acquired several decades ago were "tainted by equipment accidents, repair blunders, inadequate failure reporting, reporting of mixed age equipment, defective records of equipment operating times, mixed operational environmental conditions ..." [9]. The net effect was to produce what appeared to be a random constant failure rate.
3. Early generations of electronic devices contained many intrinsically high failure rate mechanisms. This was reflected in different infant mortality and wear-out failure rates in subpopulations, and contributed to the appearance of a constant failure rate for products in service.
4. Even in the absence of significant intrinsic failure mechanisms, early fragile devices responded to random environmental overstressing by failing at a roughly constant rate.
5. It often happens that equipment repeatedly overhauled or repaired contains a variety of components in a variable state of wear. Even though each of the components probably obeys time-dependent failure distributions, e.g., lognormal or Weibull, the admixture of varying projected lifetimes may conspire to yield a roughly time-independent rate of failure.
6. The simple addition of a decreasing infant mortality rate and an increasing wear-out failure rate results in a roughly constant failure over a limited time span.

Given these reasons it is not difficult to see why the U.S. Department of Defense and its associated agencies (e.g., Rome Air Development Center, Navy) and assorted military electronics contractors (e.g., RCA, Boeing Aircraft Company) adopted the exponential model as a basis for reliability prediction and assessment. In the Military Handbook (MIL-HDBK-217), cited in Chapter 1, failure rates for devices and components are generally given in the form

$$\lambda_p = \lambda_b \pi_T \pi_Q \pi_E \ldots = \lambda_b \Pi \pi_i, \tag{4.9}$$

where $\lambda_p$ is the part failure rate and $\lambda_b$ is the base failure rate usually expressed by a model relating the influence of electrical and temperature stresses on the part. The individual $\pi$ factors take into account the roles of temperature (T), quality (Q), and environment (E) as well as other variables that may influence failure rates (e.g., voltage-stress factor, forward-current factor).

A page from MIL-HDBK-217 is reproduced in Figure 4.10, enabling us to calculate failure rates for low-frequency, silicon FETs. The temperature factor is easily recognized to be the thermally activated Maxwell–Boltzmann factor,

MIL - HDBK-217F

6.4 TRANSISTORS, LOW FREQUENCY, Si FET

SPECIFICATION
MIL-S-19500

DESCRIPTION
N-Channel and P-Channel Si FET (Frequency ≤ 400 MHz)

$$\lambda_p = \lambda_b \pi_T \pi_A \pi_Q \pi_E \quad \text{Failures}/10^6 \text{ Hours}$$

Base Failure Rate - $\lambda_b$

| Transistor Type | $\lambda_b$ |
|---|---|
| MOSFET | .012 |
| JFET | .0045 |

Temperature Factor - $\pi_T$

| $T_J$ (°C) | $\pi_T$ | $T_J$ (°C) | $\pi_T$ |
|---|---|---|---|
| 25 | 1.0 | 105 | 3.9 |
| 30 | 1.1 | 110 | 4.2 |
| 35 | 1.2 | 115 | 4.5 |
| 40 | 1.4 | 120 | 4.8 |
| 45 | 1.5 | 125 | 5.1 |
| 50 | 1.6 | 130 | 5.4 |
| 55 | 1.8 | 135 | 5.7 |
| 60 | 2.0 | 140 | 6.0 |
| 65 | 2.1 | 145 | 6.4 |
| 70 | 2.3 | 150 | 6.7 |
| 75 | 2.5 | 155 | 7.1 |
| 80 | 2.7 | 160 | 7.5 |
| 85 | 3.0 | 165 | 7.9 |
| 90 | 3.2 | 170 | 8.3 |
| 95 | 3.4 | 175 | 8.7 |
| 100 | 3.7 | | |

$$\pi_T = \exp\left(-1925\left(\frac{1}{T_J + 273} - \frac{1}{298}\right)\right)$$

$T_J$ = Junction Temperature (°C)

Quality Factor - $\pi_Q$

| Quality | $\pi_Q$ |
|---|---|
| JANTXV | .70 |
| JANTX | 1.0 |
| JAN | 2.4 |
| Lower | 5.5 |
| Plastic | 8.0 |

Application Factor - $\pi_A$

| Application ($P_r$, Rated Output Power) | $\pi_A$ |
|---|---|
| Linear Amplification ($P_r$ < 2W) | 1.5 |
| Smell Signal Switching | .70 |
| Power FETs (Non - linear, $P_r \geq 2W$) | |
| $2 \leq P_r < 5W$ | 2.0 |
| $5 \leq P_r < 50W$ | 4.0 |
| $50 \leq P_r < 250W$ | 8.0 |
| $P_r \geq 250W$ | 10 |

Environment Factor - $\pi_E$

| Environment | $\pi_E$ |
|---|---|
| $G_B$ | 1.0 |
| $G_F$ | 6.0 |
| $G_M$ | 9.0 |
| $N_S$ | 9.0 |
| $N_U$ | 19 |
| $A_{IC}$ | 13 |
| $A_{IF}$ | 29 |
| $A_{UC}$ | 20 |
| $A_{UF}$ | 43 |
| $A_{RW}$ | 24 |
| $S_F$ | .50 |
| $M_F$ | 14 |
| $M_L$ | 32 |
| $C_L$ | 320 |

**Figure 4.10** Failure rate for low-frequency field-effect transistors. *Pages 6–8 reproduced from MIL-HDBK-217F.*

while the quality factor applies to the specific device model and the type of package. Environmental factors vary widely between the extremes of "ground benign" conditions ($G_B = 1$) and a cannon launch ($C_L = 450$). In addition, there is a fourth application factor $\pi_A$ that depends on the power level. For example, in the case of a plastic encapsulated small signal switching MOSFET operating at 30 °C, and used in space flight ($S_F$), a failure rate of $\lambda_p = 0.012 \times 1.1 \times 8.0 \times 0.50 \times 0.70 = 0.037$ per $10^6$ h, or 37 FITs is predicted. Be forewarned that the Handbook's precision greatly exceeds its accuracy by several orders of magnitude!

#### *4.3.1.3 Is Constant Failure Rate Modeling a Paradigm in Transition?*

This question has been discussed by McLinn [10], who suggested that it is time to create a new and better paradigm to replace the "defunct" exponential distribution that has served electronics for the past 45 years. Based on a 1991 critical review [1] of proposed reliability models for advanced microelectronic devices of high gate count (e.g., VLSI, VHSIC), it was recommended that the assumption of a constant failure rate be considered invalid. This was particularly true for device failures caused by electromigration and dielectric breakdown phenomena. In addition, years of experience has shown that failure-rate predictions using existing MIL-HDBK-217 specifications can differ by several orders of magnitude for the same device depending on the manufacturer. Far from exhibiting a constant failure rate, it is well known that infant-mortality failures of semiconductor products display a decreasing failure rate with time. In addition, recent generations of semiconductor products are significantly more reliable than their predecessors and exhibit low failure rates.

To address these facts, MIL-HDBK-217 has incorporated additional terms such as the learning factor $\pi_L$, which attempts to weigh the effect of the number of years ($Y$) the product has been in production. One such explicit form is $\pi_L = 0.01 \exp(5.35 - 0.35Y)$. Inclusion of this time factor effectively results in a decreasing rather than constant hazard rate. Alternative constant failure-rate-reliability prediction procedures by various companies (e.g., AT&T, Bellcore, British Telecom, France Telecom, Nippon Telegraph, Telephone) have been developed. Finally, total abandonment of MIL-HDBK-217 specifications by some US device manufacturers as well as by Japan and other Pacific Rim nations in favor of physical investigations of the root causes of failure has become increasingly popular. Primarily because it is well known that, on two different process lines in the same company, the same product with the same process may exhibit quite different process variability, which may change the root cause of failure, the failure rate, and its model. Therefore the origin of the product can sometimes influence its behavior.

In conclusion, two broad viewpoints of addressing reliability, MIL-HDBK-217 versus the physics of failure, are compared item by item in Table 4.3 to expose relative advantages and shortcomings.

## 4.3.2 Lognormal Distribution

Shortcomings in the exponential distribution function have prompted the use of alternative distribution functions to model reliability data. One of the most popular of these is the lognormal distribution function. Mathematically, the lognormal is not a separate distribution, because by taking natural logarithms of all data points, the transformed data can be analyzed as a normal distribution. The lognormal PDF is given by

$$f(t) = \frac{1}{t\sigma(2\pi)^{1/2}} \exp\left(\frac{-[(\ln(t) - \ln(\mu)]^2}{2\sigma^2}\right) \tag{4.10}$$

while the lognormal CDF has the form

$$F(t) = \Phi\left[\sigma^{-1} \ln\left(\frac{t}{\mu}\right)\right]. \tag{4.11}$$

Here $\Phi(z) = \frac{1}{2}[1 + \mathrm{Erf}(z/2^{1/2})]$ or $\frac{1}{2}[2 - \mathrm{Erfc}(z/2^{1/2})]$, since $\mathrm{Erfc}(z/2^{1/2}) = 1 - \mathrm{Erf}(z/2^{1/2})$.

Substitution in Eqn (4.8) shows that the hazard function is given by

$$\lambda(t) = \frac{2^{1/2} \exp\left[-1/2\sigma^2 \ln(t/\mu)^2\right]}{\pi^{1/2} t\sigma \mathrm{Erfc}\left[\ln(t/\mu)/\sigma(1/2)^{1/2}\right]}. \tag{4.12}$$

In these expressions the error function (Erf) and complementary error function (Erfc) are related to the integral of the Gaussian function and appear in connection with solutions to diffusion problems (see Eqn (5.4)). Both range from 0 to 1, while their arguments assume any values. Comparison with Eqn (4.1) reveals that the lognormal distribution is derived from the normal distribution by substituting $\ln t$ for $x$. Just as there are two parameters that define the normal distribution, two parameters also define the lognormal distribution. The first is $\mu$, which has the physical significance of representing the median time, or time when 50% of the

**Table 4.3** Comparison of reliability prediction methodologies

| Issue | MIL-HDBK-217 | Physics of failure |
|---|---|---|
| Relative cost of analysis | Low. Many computer programs available. Implementation costs less than 1% of hardware cost. | Very high. Cost to implement complex system may exceed that of hardware. |
| Device coverage | Covers 19 major electronic device categories. | Addresses microcircuits only. No models for other devices. |
| Device defect modeling | Assumes field failures are caused by random defects accelerated by operating stresses. | Failure mechanisms generally neglect impact of manufacturing defects. |
| Device design modeling | Assumes design is perfect (based on qualification testing). Assumes all failures are caused by manufacturing defects. | Assumes field failures are caused by poor device designs. Ignores handling and manufacturing defects. |
| Model development | Models based on a statistical analysis of actual field failures (statistical models). | Develops models based on principles of physics, chemistry, mechanics and materials (deterministic models). |
| DoD/ Industry | Used successfully on thousands of military development efforts over past 30 years. | No complete system analysis ever documented. |
| Coordination | All updates undergo DoD, tri-service and industry coordination | Developed models have never been coordinated. |
| Input data required | Minimal. Readily available to the system designer in terms of part type, stress, characteristics | Extremely extensive. Information difficult to obtain without testing. Variations in material property can cause wide swings in model predictions. |
| Output data | Predicts relative frequency of device failures in a population. Failure rates can be summed and scaled to yield system behavior. Alternate designs can be compared. | Predicts the time of occurrence of specific failure mode in a specific device. Reliability is based on weakest part. Comparisons only on weakest links. |

**Table 4.3** Comparison of reliability prediction methodologies—Cont'd

| Issue | MIL-HDBK-217 | Physics of failure |
|---|---|---|
| Arrhenius model | Used to empirically model temperature versus reliability. | Used to obtain temperature acceleration of corrosion, electromigration, solder creep, etc. |
| Appropriate application | System design. Parts count can be applied with very limited information. | Component design. |
| Device screening | Recognizes that device screening reduces latent defects and field failures. | Screening has no impact on component design. Assumes defect-free components. |
| Operating temperature | Direct impact on the frequency of failure but not as large as critics claim. | Promoters claim microcircuit device reliability is independent of temperature below 150 °C. |
| Temperature cycling | Accounted for in the environmental factor. | Models focus on temperature cycling as primary cause of failure. |

After Ref. [11].

distribution will fail, i.e., $\mu = \ln t_{50}$. Second, there is the shape parameter $\sigma$, which strongly influences the shape of $f(t)$ and $F(t)$, as shown in Figure 4.11(a) and (b). Note that $\sigma$ is *not* the standard deviation of a population of lifetimes as in the normal distribution. Also shown in Figure 4.11(c) is the variation in lognormal failure rates as a function of $\sigma$.

Later, in Section 4.3.5, it is shown how to extract $\sigma$ values from reliability plots. Large values of $\sigma$ ($\sigma > 2$) reflect initially high failure rates that decrease with time. When a semiconductor product is in early manufacture, and quality control is still a problem, typical values of $\sigma$ are as high as 3 or 4. Such values are indicative of processing and failures that are out of control. Later, when manufacturing has advanced along the learning curve and the product is a candidate for high-reliability applications, $\sigma \approx 1$ or less [12]. When $\sigma$ is close to 1, the failure rate is roughly constant, but it increases for values of less than ~0.5. Thus the lognormal distribution can represent the early failure, steady state, or wear-out time regimes in the life of devices. As an example, failure times of GaAs–AlGaAs lasers aged at 70 °C are plotted in lognormal fashion in Figure 4.12.

A useful and widely employed way to display the hazard rate for lognormal distributions, known as the Goldthwaite plot, is shown in Figure 4.13 [14]. Based on Eqn (4.12), this representation plots the failure

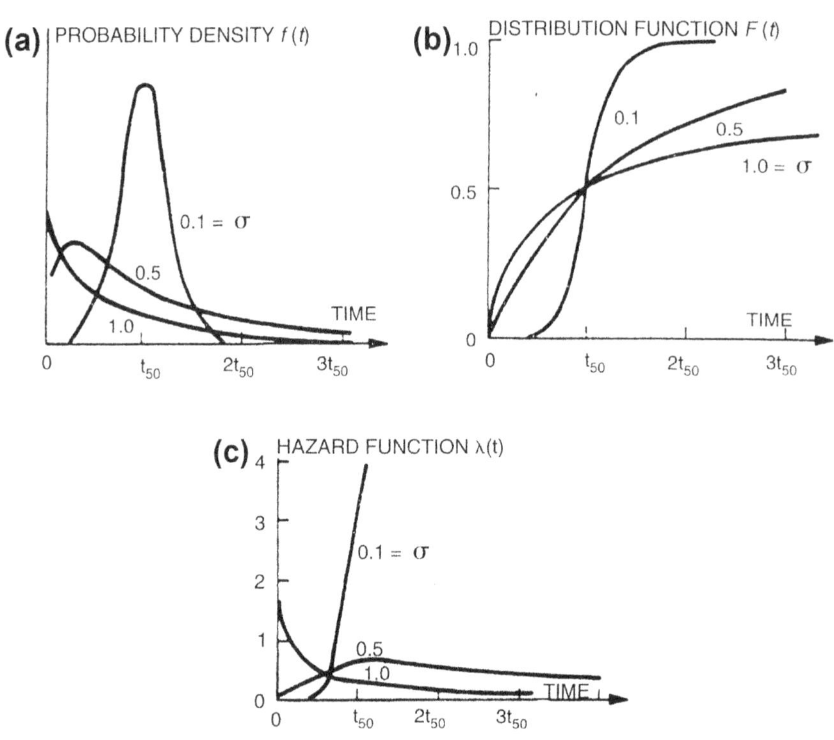


**Figure 4.11** Plots of lognormal distribution functions. (a) *f*(*t*) versus time. (b) *F*(*t*) versus time. (c) $\lambda(t)$ versus time. Time is measured in units of $t_{50}$. Variations in lognormal distributions are shown as a function of $\sigma$.

rate (in units of FITs $\times$ $t_{50}$) versus the normalized time $t'$ (a ratio of $t/t_{50}$) for different values of $\sigma$. It is apparent that an increase in $\sigma$ has the effect of moving the failure-rate peak to shorter times; that is why small $\sigma$ values are desired. Superimposed on the figure are plots of the locus of constant failure rate ($\lambda_o$) at a given time $t_o$. Thus the ordinate $\lambda\ t_{50}$ is taken as $\lambda_o\ t_{50}$, so that $\lambda\ t_{50} = (\lambda_o\ t_o)/(t_o/t_{50})$. This equation plots as a straight line with a slope of $-1$ in the log–log plot of Figure 4.13. The interested reader should determine what sample size and data could be used to measure $t_{50} = 10^8$ h and a life of 40 years. Often projections are far more precise than we can measure. Are they then accurate? What confidence is there in their use?

As an application [16] of this figure, consider a failure distribution with $\sigma = 2$. The failure rate is low initially, but it rises with time. When $t/t_{50} \approx 3 \times 10^{-3}$, the failure rate is 10 FITs, and $t$ is equal to 40 years. For this to happen $t_{50}$ must be equal to $1.2 \times 10^8$ h; if $t_{50}$ is less than this value, the failure rate will increase to more than 10 FITs in 40 years.

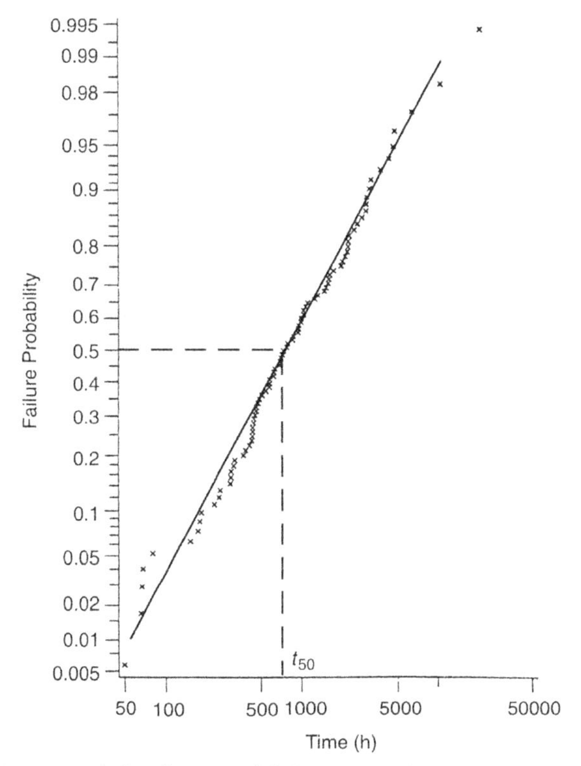


**Figure 4.12** Lognormal distribution of failure times for early vintage GaAs–AlGaAs lasers aged at 70 °C. *From Ref. [13].*

### 4.3.3 Weibull Distribution

Like the lognormal, the Weibull is a two-parameter distribution. By adjusting a time scale parameter $\alpha$ and a shape parameter $\beta$, a variety of functional behaviors can be generated to fit a wide range of experimental data. For example, the Weibull distribution applies to both decreasing failure rates, typical of early failures, and increasing failure rates that describe long-time wear out. The forms of the probability and CDFs are explicitly given by

$$f(t) = \frac{\beta t^{\beta-1} \exp\left[-(t/\alpha)^{\beta}\right]}{\alpha^{\beta}} (\beta, \alpha \text{ are constants} > 0) \tag{4.13}$$

and

$$F(t) = 1 - \exp\left[-(t/\alpha)^{\beta}\right]. \tag{4.14}$$

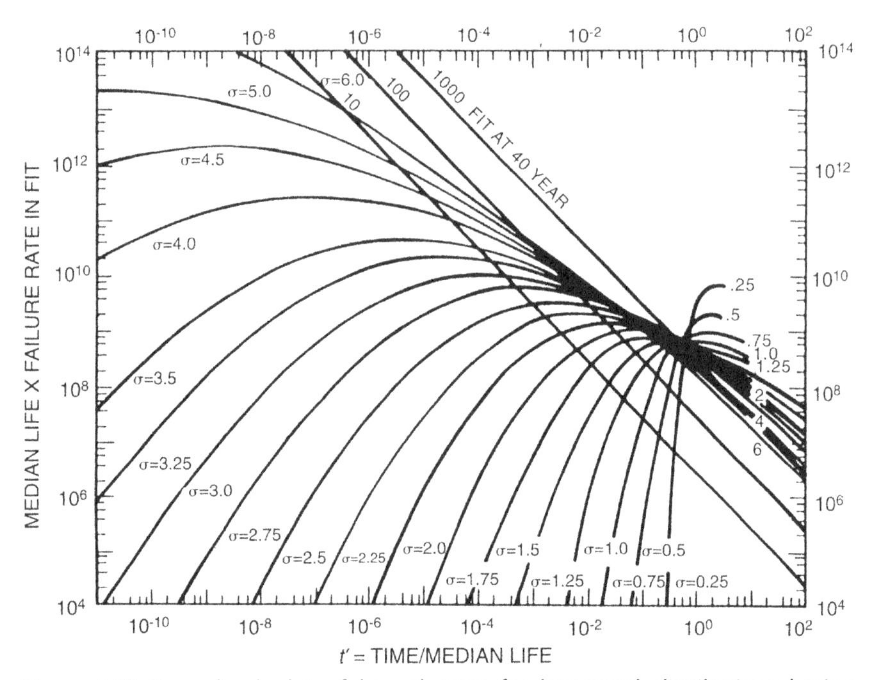


**Figure 4.13** Normalized plot of hazard rates for lognormal distributions having different values of $\sigma$. *After Ref. [15].*

These functions are plotted in Figure 4.14(a) and (b), together with the Weibull failure rate, (Figure 4.14(c)) which has the functional form

$$\lambda(t) = \frac{\beta t^{\beta-1}}{\alpha^{\beta}}. \tag{4.15}$$

There are a few cases of the Weibull distribution that are worth noting. If, for example, $\beta = 1$, we have the case of the exponential distribution. Explicit forms for the important reliability functions are then $f(t) = 1/\alpha \exp[-(t/\alpha)]$, $F(t) = 1 - \exp[-(t/\alpha)]$, and $\lambda(t) = 1/\alpha$.

When $\beta = 2$, the *Rayleigh* distribution, a special case of the Weibull distribution, emerges. The functions that characterize this distribution are given by $f(t) = (2t/\alpha)\exp[-(t/\alpha)]$, $F(t) = 1 - \exp[-(t/\alpha)^2]$, and $\lambda(t) = 2t/\alpha^2$. Rayleigh distributions are not normally used to describe the reliability of electronic products; however, this distribution function appears in an application to corrosion failure in Section 7.4.4. Significantly, the hazard rate increases linearly with time.

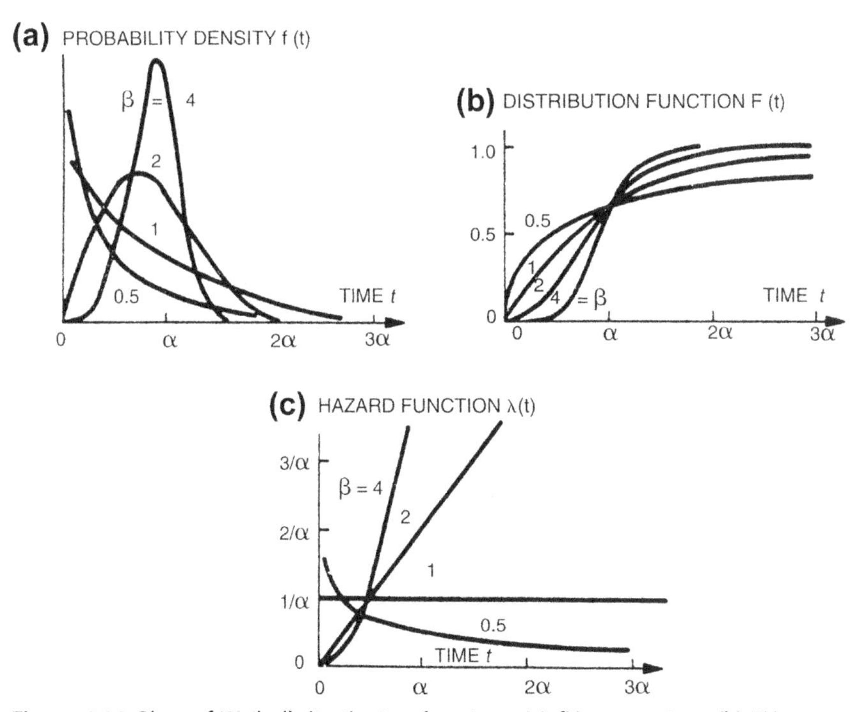


**Figure 4.14** Plots of Weibull distribution functions. (a) $f(t)$ versus time. (b) $F(t)$ versus time. (c) $\lambda(t)$ versus time. Time is measured in units of $\alpha$. Variations in Weibull distributions are shown as a function of $\beta$.

Time-dependent corrosion failures of bonding pads are plotted in Weibull form in Figure 4.15.

### 4.3.4 Lognormal versus Weibull

A comparative examination of Figures 4.11 and 4.14 reveals that the PDF, CDF, and failure rates of both of these important functions can be made to approximate one another by adjusting the respective parameters. If this is the case, then what are the distinctions among these distributions? Does it make a difference on which function is chosen to describe failure data? Which failure phenomena are best plotted in lognormal form? and Which are better described by a Weibull distribution? And finally, do these functions have any physical significance? This last question, which underlies the previous three, is addressed separately in Section 4.5.

Which distribution to use for a given set of data is an important question in reliability modeling. Both lognormal and Weibull distributions outwardly appear to fit most sets of data equally, as illustrated in Figure 4.16

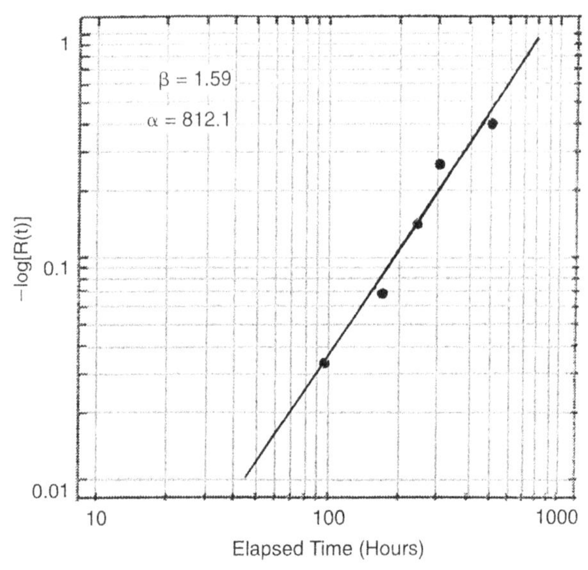


**Figure 4.15** Weibull plot of bonding-pad corrosion failures. Increasing failure rate is indicated by the value of $\beta = 1.59$. *From Ref. [17].*

for the case of light bulb wear-out failures [3]. Closer examination, however, reveals that the Weibull distribution provides a better fit to short time failures, while the lognormal plot is better at predicting longer lifetimes. However, in this application long-time projections rather than infant-mortality failures are usually of interest; thus a lognormal characterization of bulb life is preferred.

Lognormal distributions tend to apply when gradual degradation occurs over time because of diffusion effects, corrosion processes, and chemical reactions. Early p–n–p mesa transistors, bipolar and MOS transistors, light-emitting diodes (LEDs), and lasers are examples of devices whose failures have been modeled by lognormal statistics [18]. On the other hand, Weibull distributions appear to be applicable in cases where the weak link, or the first of many flaws, propagates to failure. Dielectric breakdown, capacitor failures, and fracture in ceramics are typically described by Weibull distributions.

Plotting failure data that fit lognormal and Weibull distributions are facilitated with the use of special lognormal or Weibull graph paper because of the desired linear plots result. When predictions must be projected beyond the range of recorded failure data, the choice of the distribution function becomes important. A more advanced knowledge of statistics,

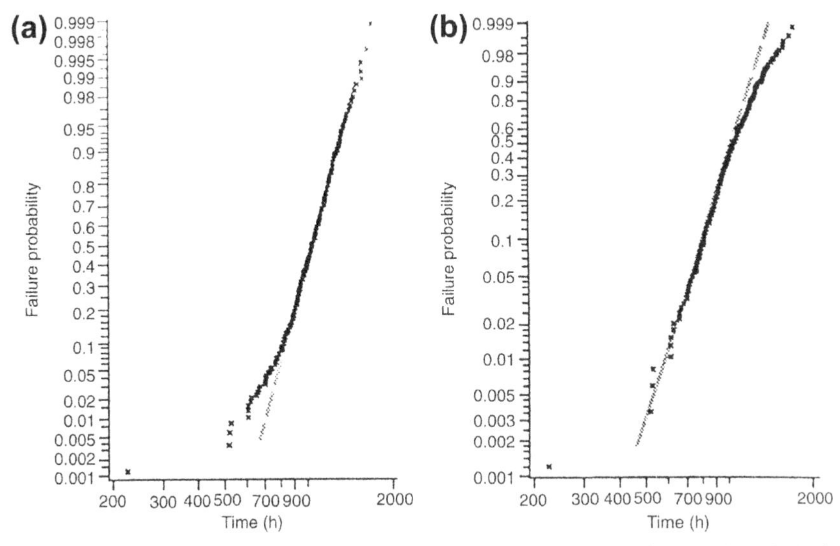


**Figure 4.16** Light-bulb failure times plotted in terms of lognormal (a) and Weibull (b) distribution functions. *From Ref. [13].*

probability theory, and failure physics may be required to make intelligent life projections in such cases.

If the data in Figure 4.16 were available for only a small population, say 10 test units, then the shape of the plot would be much in question and projections would be very problematic. This could be the case for a very expensive device made only in small quantities. Under such circumstances it would be very difficult to determine an actual fit to either lognormal or Weibull distributions. Further, if the device was required to exhibit high reliability and not be replaced, then the first occurrence of failure would be paramount and its prevention essential to success of the program. In such cases analysis of each failure would be necessary, to determine root cause, before risk and probability of failure could be projected.

## 4.3.5 Plotting Probability Functions as Straight Lines

Virtually every reliability study concludes with some sort of linear plot of the number of failures versus time from which reliability parameters are derived. Just what constitutes the failed population, or a quantity equivalent to or proportional to it, depends on the particular failure phenomenon in question. The rate of current increase is such a measure for laser failures in Figure 4.12. Since a good linear fit of failure time data is required, the question arises of how to make nonlinear reliability functions plot as straight lines. Equivalently, How are Weibull and lognormal plot axes created, so that lines emerge while plotting failure-statistics data versus time? Reliability analysis is then essentially reduced to selecting the right plotting paper or computer program that generates it.

The problem of linearizing the Weibull function is easy because no complex functions are involved. Starting with Eqn (4.14) we note that taking successive logs of both sides yields

$$\ln\left[1 - F(t)\right] = -(t/\alpha)^{\beta} \tag{4.16}$$

and

$$\ln\{-\ln\left[1 - F(t)\right]\} = \beta \ln(t) - \beta \ln\alpha. \tag{4.17}$$

The simple algebraic equation of a straight line is $y = mx + b$, where $y$ is the ordinate, $x$ is the abscissa, $m$ is equal to the slope, and $b$ represents the intercept on the ordinate axis. In comparison with Eqn (4.17), $y = \ln\{-\ln[1 - (t)]\}$, $x = \ln(t)$, $m = \beta$, and $b = -\beta \ln \alpha$. Thus Weibull paper is generated by plotting $\ln\{-\ln[1 - F(t)]\}$ or $\log\{-\log[1 - F(t)]\}$ usually expressed as cumulative percentage, versus either $\ln(t)$ or $\log(t)$, respectively. If the CDF or $F(t)$ data that ranges from slightly above zero to slightly below 1.0 obeys Weibull statistics as a function of time, then a straight line will be obtained when plotted on this paper. It is not a difficult matter to extract $\beta$ and $\alpha$ values using standard algebraic methods.

In the case of the lognormal function, Eqn (4.11) suggests that a straight line results if $\ln(t)$ is plotted versus $\Phi^{-1}F(t)$; i.e., $\ln(t) = \mu + \sigma\Phi^{-1}F(t)$. From this plot a slope of $\sigma$ and an intercept equal to $\mu$ are obtained. The slope can also be estimated from the literal definition

$$\sigma = \frac{\ln t_2 - \ln t_1}{\Phi^{-1}F(t_2) - \Phi^{-1}F(t_1)}. \tag{4.18}$$

This last expression simplifies if points on the graph corresponding to $F = 50\%$, $t_{50\%}$ and $F = 15.9\%$, $t_{15.9\%}$ are selected. Letting $\Phi^{-1} F(t_2) = \Phi(0)$, and therefore $F(t_2) = 0.5$; similarly $\Phi^{-1}F(t_1) = -1.00$ when $F(t_1) = \Phi(-1) = 0.159$. (Note $\mathrm{Erf}(-z) = -\mathrm{Erf}(z)$.) Substituting these values yields the simple result $\sigma = \ln[t_{0.5}/t_{0.159}]$. For example, in the case of Figure 4.12, $\sigma = 1.1$.

When failure data are abundant and there are many $F$ values to plot, little difficulty in generating straight lines is experienced. However, what if there are few failures, so that obtaining a good estimate of statistical parameters is problematical? Under these circumstances $F$ can be obtained by ranking the exact failure times. For example, if the sample size is $n$, then $1/n$ is an estimate of $F(t)$ for the first ordered failure time, $2/n$ for the second failure time, etc. Thus, for the $i$th failure, $F(t_i) = i/n$. There is, however, a recommended way to estimate a statistically more accurate CDF or $F$ value for the population [2], and that is by using Eqn (4.19); i.e.,

$$F(t_i) = \frac{i - 0.3}{n + 0.4}. \tag{4.19}$$

To see how to use this formula, consider the following illustrative problem:

## Example 4.3

A collection of 60 laser diodes were tested, and 7 failures occurred after 1000 h. The lifetimes obtained are tabulated below.

| Rank number ($i$) | Lifetime ($h$) | $F(t_i) = (i - 0.3)/(60 + 0.4)$ | |
|---|---|---|---|
| 1 | 181 | 0.012 | (1.2%) |
| 2 | 299 | 0.028 | (2.8%) |
| 3 | 389 | 0.045 | (4.5%) |
| 4 | 430 | 0.061 | (6.1%) |
| 5 | 535 | 0.078 | (7.8%) |
| 6 | 610 | 0.094 | (9.4%) |
| 7 | 805 | 0.111 | (11.1%) |

1. Plot the results in both Weibull and lognormal fashion.
2. What is the MTTF for each?
3. For the lognormal plot, what is the value of $\sigma$?
4. For the Weibull plot, what is $\alpha$?

### Answer

1. First, $F(t_i)$ is calculated for $n = 60$ and $i = 1, 2, 3, \ldots, 7$ and entered in the table. The $F$ values are plotted versus log time on Weibull and lognormal paper as shown in Figure 4.17(a) and (b).
2. At a value of $F = 0.5$, MTTF $= 2170$ h (Weibull) and 3600 h (lognormal).
3. It was shown above that $\sigma = \ln [t_{0.5}/t_{0.159}]$. Substituting, $\sigma = \ln [3600/980] = 1.30$.
4. From Eqn (4.16) we note that when $t = \alpha$, $\ln [1 - F(t)] = -1$. Thus, $1 - F(t) = 0.368$, and $F(t) = 0.632$. At this value, which is noted on the ordinate axis, $\alpha = t = 2990$ h.

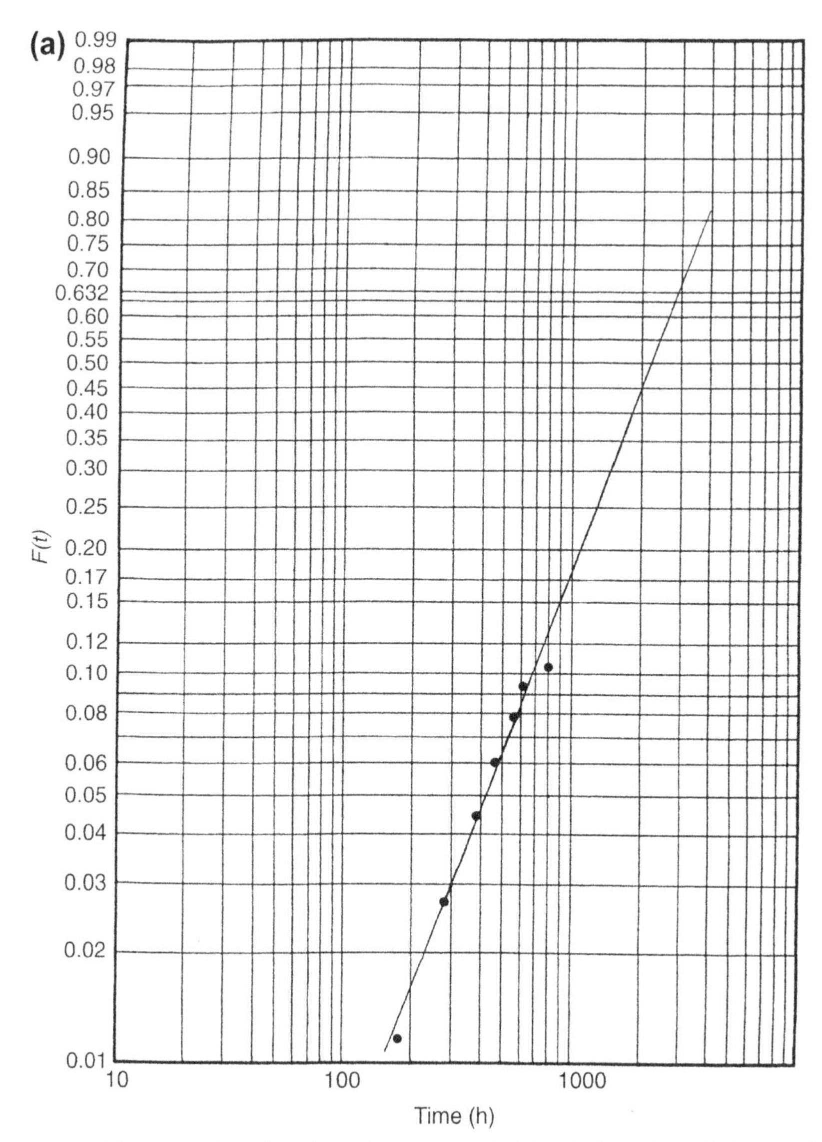


**Figure 4.17** Laser diode failure times from Example 4.3 (a) plotted on Weibull probability paper and (b) plotted on Lognormal probability paper.

This example clearly shows the differences and difficulty in projecting failure rates with limited data, the normal situation for a working reliability engineer. Using the models as plotted we would project 99% failure at 7500 h for the Weibull and 75,000 h for the lognormal, quite a

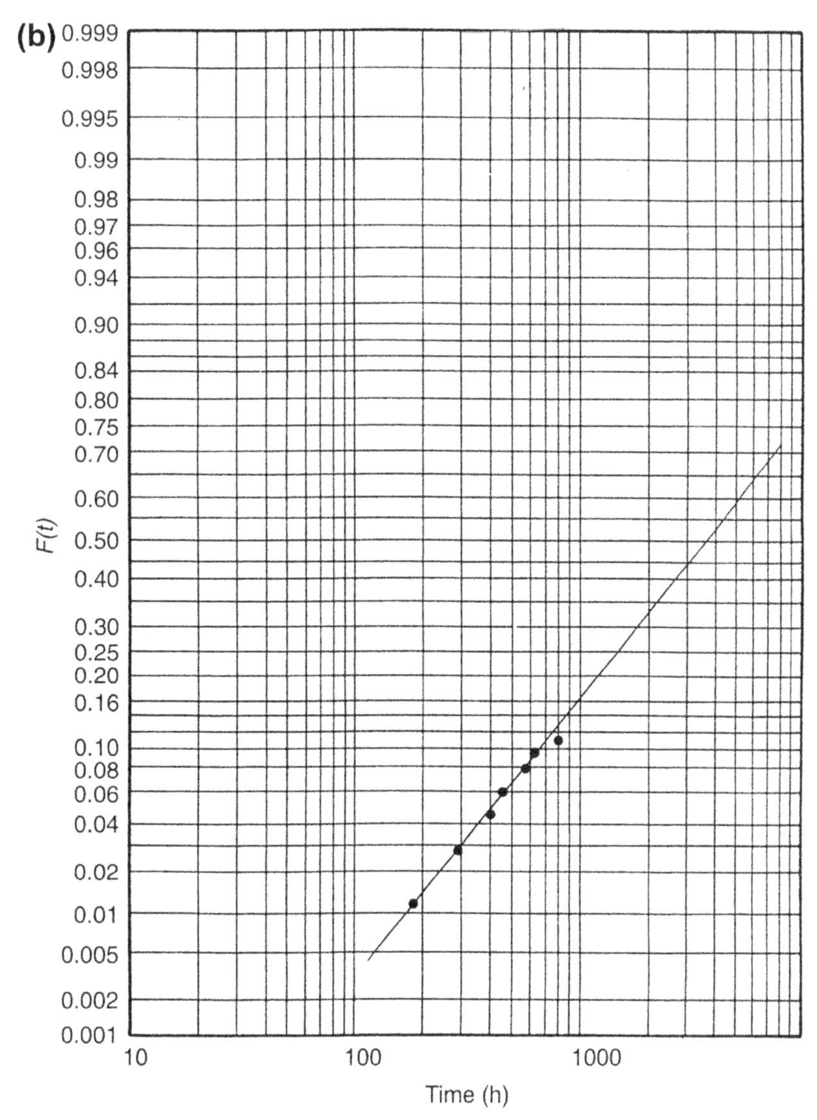


**Figure 4.17** —Continued

difference! And maybe more important, for a mission critical application, which demanded very high reliability, the Weibull would predict a much shorter time to first failure than the lognormal. The decision would be very difficult and possibly require redundancy to ensure the execution of the mission.

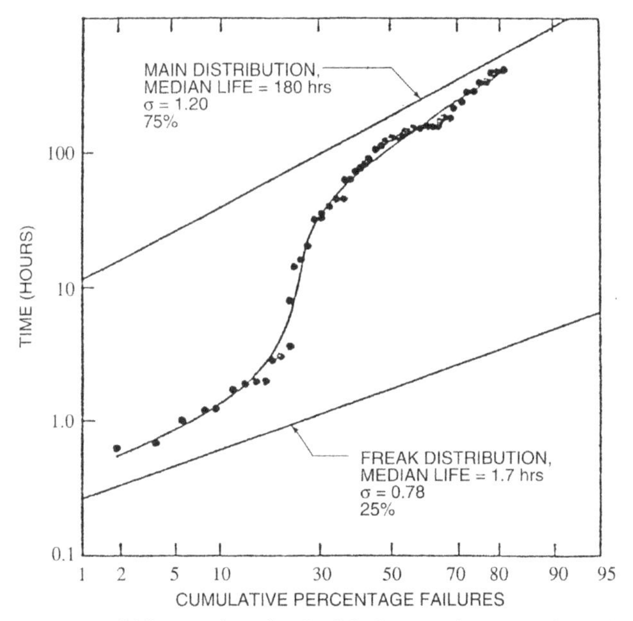


**Figure 4.18** Lognormal lifetime data for CMOS ICs tested at 200 °C consist of a main and a freak distribution. Roughly 75% of failures belong to the main distribution, which has a median life of 180 h and $\sigma = 1.20$. The remaining 25% constitute the freak distribution, with a median life of 1.7 h and $\sigma = 0.78$. *Data from AT&T Bell Laboratories.*

## 4.3.6 Freak Behavior

Although it inspires confidence to have failure data plot linearly on one of the probability papers, quite often this is not the case [19]. In Figure 4.18 such an S-shaped example of lognormal data is shown for CMOS integrated circuits, suggesting that two distinct failure mechanisms are operative. The observed plot was decomposed into a main distribution operative at long times as well as a "sport," or freak, population that exhibits untimely failures at short times. Each component behavior in this plot is linear, and a superposition of the two weighted according to the individual relative proportions yields the observed bimodal failure behavior. While the sport population had a median time to failure of ~1.7 h, the value for the main distribution was ~180 h.

Freak failures are addressed again later in the chapter.

# 4.4 SYSTEM RELIABILITY

## 4.4.1 Introduction

Thus far we have dealt only with the reliability and statistical analysis of the failure of an individual device (e.g., transistor), component (capacitor), or constituent region within these (insulation). However, in practice, we always use multidevice and multicomponent structures that are connected together and interact in complex ways within a system. How do we deal with reliability of the system as a whole? There is a large literature on the subject of system reliability [2,4,20]. We shall deal only with several relatively simple systems that build on our prior statistical experience with individual products. As has been the custom in this chapter, mathematical rigor will be sacrificed in an attempt to convey meanings and significance.

Two rules are needed to calculate the reliability of the simplest types of systems [2]:

1. The *multiplication rule* states that the probability that several independent events will all occur is the product of the individual event probabilities.
2. The *complement rule* states that the probability that an event will not occur is one minus the probability of the event occurring.

Let us now consider the system shown in Figure 4.19(a) composed of $n$ independent elements connected in *series*, or in a chain, such that system failure occurs when any one of the elements fails. If $R_i(t)$ is the reliability or

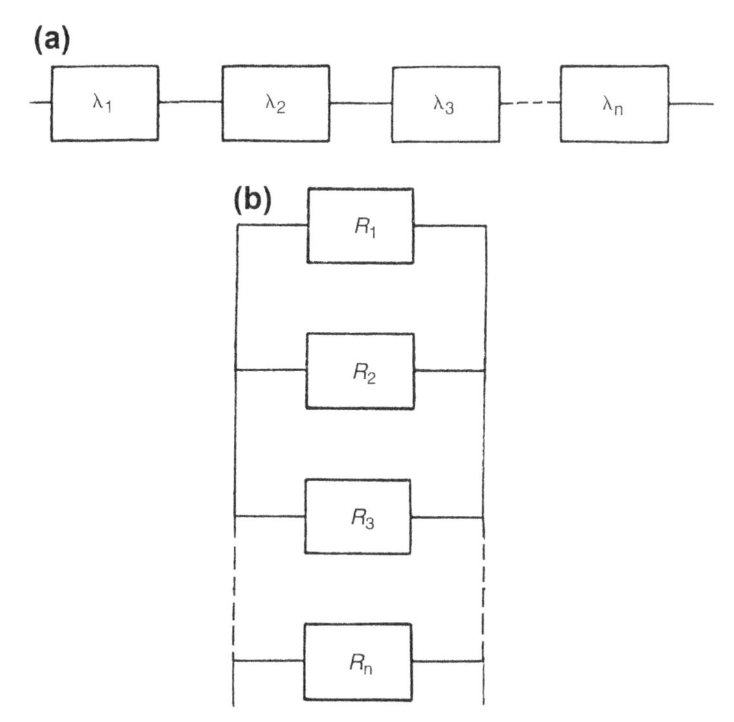


**Figure 4.19** (a) System composed of $n$ independent elements with failure rate $\lambda_i$ connected in series. Note: $n > i \geq 1$. (b) A parallel, or redundant, system of $n$ components each with reliability $R_i$.

survival function of a given element, the overall system reliability, $R_s(t)$, is given by the multiplication rule, or

$$R_s(t) = R_1(t) \times R_2(t) \times R_3(t) \times \ldots R_n(t). \tag{4.20}$$

In terms of the CDF formulation (Eqn (4.4)), the system unreliability is given by the complement rule, or

$$F_s(t) = 1 - \prod_{i=1}^{n}(1 - F_i(t)). \tag{4.21}$$

Since the hazard function $\lambda = -\mathrm{d}[\ln R(t)]/\mathrm{d}t$, the system failure rate, $\lambda_s$, is the sum of the individual failure rates,

$$\lambda_s = \lambda_1 + \lambda_2 + \lambda_3 + \ldots + \lambda_n. \tag{4.22}$$

This means that the reliability of a series system is always worse than that of the weakest component.

In contrast, the *parallel*, or redundant, system of $n$ components (Figure 4.19(b)) continues to operate until the last unit fails. The corresponding functional relationships for $R_s(t)$ and $F_s(t)$ in a parallel system are

$$R_s(t) = 1 - \prod_{i=1}^{n}[1 - R_i(t)] \tag{4.23}$$

and

$$F_s(t) = F_1(t) \times F_2(t) \times F_3(t) \times \ldots F_n(t). \tag{4.24}$$

It can be shown [2] that if a single component with a CDF of $F$ is replaced by $n$ components, the ratio of the system to component failure rates is lowered by a factor equal to

$$\frac{\lambda_i}{\lambda_s} = \frac{1 + F + F^2 + F^3 + \ldots F^{n-1}}{nF^{n-1}}. \tag{4.25}$$

The simple formulas given above illustrate that the reliability of individual units govern the life of the system in a bottom-up approach. In general, for a system of $n$ elements whose individual reliabilities are known but whose interconnection configurations are unknown, the lower and upper bounds for system reliability are given by the series and parallel models, respectively.

### Example 4.4

Consider a 50-pin module such that the probability of failure upon inserting it into a mating connector is $1.8 \times 10^{-3}\%$ per pin. What is the probability of failure of the module?

**Answer**

In this problem the module is the system (s), and the pins (p) are the elements. If we assume that a single pin failure causes the module to fail, then by Eqn (4.21), $F_s = 1 - (1 - F_p)^n$, where $n$ is the number of pins per module. Substituting, $F_p = 1.8 \times 10^{-5}$ and $n = 50$ yields $F_s = 1 - (1 - 1.8 \times 10^{-5})^{50} = 0.00090$.

Therefore, the failure probability is predicted to be 0.09%, or 9 module failures in 10,000 insertions.

## 4.4.2 Redundancy in a Two-Laser System

Transoceanic cables for optical communications systems typically contain a hundred or so repeaters stretched across the ocean floor. Their function is to boost, or regenerate, the laser light signal, which attenuates with distance of transmission. Because of the enormous expense of submarine cable repair, redundant lasers are used in repeater systems to achieve the high reliability that is so critical in this application. Redundancy, or the presence of spares that can be pressed into service should they be needed, is a common strategy used to achieve low system failure rates. In this section we consider the reliability of a regenerator that contains two lasers, one sparing the other, where each is assumed to fail according to lognormal statistics. Joyce and Anthony [12] considered two different models for the two-laser regenerator, and they are treated in turn.

### *4.4.2.1 Cold-Spared Assembly*

In the so-called "cold-spared" assembly the second laser is called into service and begins to degrade after the first laser fails. The duration of system survival depends on the first component lasting until turn-on of the second, and the probability that the second will operate to time $t$. A plot of the cold spared failure rate, $\lambda_s(c)$, is shown in Figure 4.20, and clearly resembles the Goldthwaite plot (Figure 4.13).

As an example, let us assume the lasers have a median lifetime of 100 years and $\sigma = 1$. After 10 years of system life, the abscissa of Figure 4.20 is 1/10, in which case $\lambda_s(c)$ (FITs) $\times$ $t_{50} = 3 \times 10^5$. Therefore,

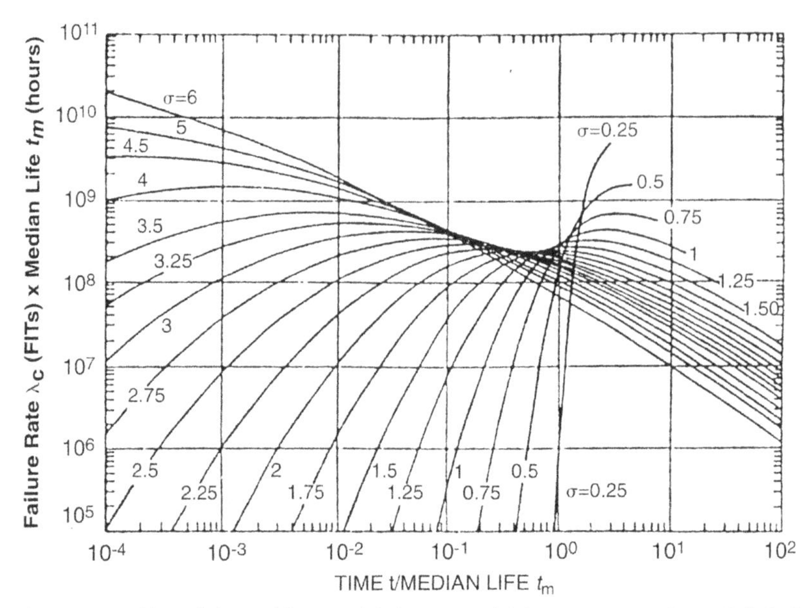


**Figure 4.20** Plot of the cold spared failure rate, $\lambda_s(c)$ versus time. *From Ref. [12].*

$\lambda_s$ (c) $= 3 \times 10^5/8.76 \times 10^5 = 0.34$ FITs. With the passage of time the advantages of redundancy diminish.

### *4.4.2.2 Hot-Spared (Parallel) Assembly*

This assembly consists of lasers configured in parallel. Thus, lasers degrade concurrently, and the assembly fails if and only if both independent components fail. For this simpler case, the appropriate lognormal distribution and hazard functions of Eqns (4.11) and (4.12) apply. The hot-spared failure rate (Eqn (4.25)) is simply $\lambda_s$ (h) $= (2F/1 + F)\ \lambda_i$; when plotted, it is similar in appearance to that for $\lambda_s$ (c).

What is more revealing is the ratio of the two-system failure rates, i.e., $\lambda_s$ (c)/$\lambda_s$(h), which is plotted in Figure 4.21. Cold sparing is predicted to have a considerable advantage over hot sparing, particularly at earlier times. In the example given above, $\lambda_s$ (c)/$\lambda_s$ (h) $= 7 \times 10^{-2}$ after 10 years. Despite these predictions, it may be physically difficult to realize the benefits of cold sparing in practice. For example, the regenerator could fail not because of wear out, which is what has been tacitly assumed all along. Rather, material creep may slowly misalign each laser with respect to its fiber and allow less light to enter it. Since this creep mechanism occurs whether or not the laser is turned on, the system is essentially hot spared.

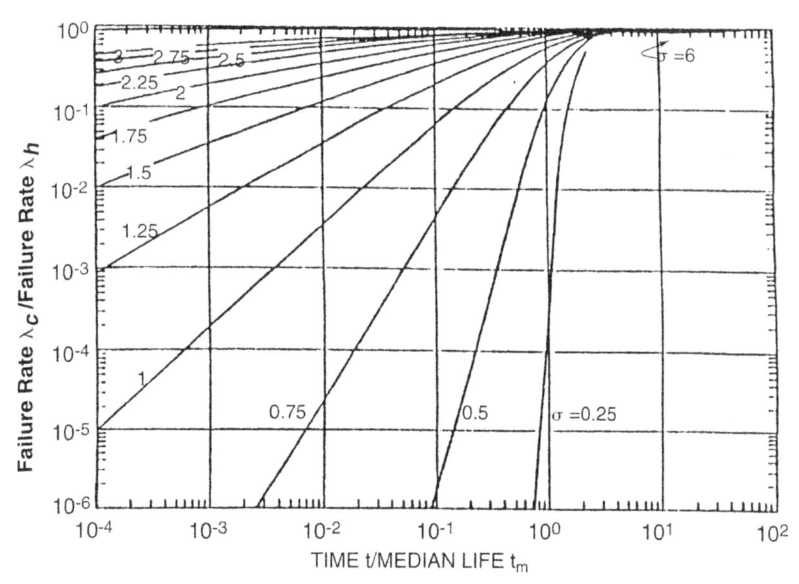


**Figure 4.21** Plot of the ratio of the cold-spared to hot-spared failure rates, i.e., $\lambda_s(c)/\lambda_s(h)$ versus time. *From Ref. [12].*

## 4.4.3 How MIL-HDBK-217 Treats System Reliability

Among the electrical products whose reliability MIL-HDBK-217 attempts to predict are microelectronic circuits including monolithic bipolar and MOS digital as well as linear gate logic arrays, microprocessors, MOS memories, GaAs MMICs, and assorted hybrid microcircuits. In these as well as other integrated circuits we are speaking of systems composed of thousands and even millions of identical transistor or gate units. The basic difference in the failure rate formulas for these systems, compared to individual devices (see Section 4.3.1.2), is that account is made of both the chip and package complexity. As a typical example, the predicted failure rate for an MOS gate array circuit can be calculated using the formula (p. 5–3 of MIL-HDBK-217)

$$\lambda_p = \left(C_1\pi_T + C_2\pi_E\right)\pi_Q\pi_L \text{ failures}/10^6\text{h}. \tag{4.26}$$

Among the terms in this equation is the learning factor $\pi_L$, discussed in Section 4.3.1.3. The constant $C_1$ depends on the number of gates, which in turn is approximated by the number of transistors divided by 4. As expected, actual $C_1$ values rise with gate count. Values of $C_2$ reflect the complexity and type of package and are given by an expression of the type,

$C_2 = k_1(Np)_2^k$. $N_p$ is the number of pins present, while $k_1$ and $k_2$ are constants that depend on type of package employed, i.e., hermetic, dual in-line, flatpack, cans, etc.

The shortcomings associated with MIL-HDBK-217 failure-rate modeling raised earlier persist when we are dealing with systems. Simply because of the added system complexity, one suspects that life predictions will be even more problematic than for individual devices.

## 4.5 ON THE PHYSICAL SIGNIFICANCE OF FAILURE DISTRIBUTION FUNCTIONS

### 4.5.1 Introduction

A question raised earlier inquired whether the widely used failure distributions possess any physical significance. If they do, is their mathematical form predictable? Thus far they appear to be a useful guide to plotting failure time data and predicting, through projection, what lies ahead. However, they do not seem to have any link to the microscopic mechanisms that cause failure or to be rooted in physical terms. In the following sections an attempt is made to provide a physical rationale for the form of failure distribution functions. The effects of acceleration factors (AFs), the role of temperature, and how they are incorporated into the reliability functions are topics treated in the last of these sections.

### 4.5.2 The Weakest Link

Phenomena like the fracture of silica optical fibers (Section 10.4) and dielectric breakdown of insulators (Section 6.3) are thought to originate at a fatal flaw that grows with time. In the same way that a weak link threatens the integrity of a chain, so too defects of critical size will cause failure in electronic components. A simple derivation of the weakest link theory assumes that for the involved component of size $L$ to survive under some condition of stressing, each of its differential elements $\mathrm{d}L$ must survive. In the case of the mechanical fracture of an optical fiber, $L$ would be the length of fiber subjected to loading. If $F$ is the probability of length $L$ breaking, then $1 - F$ is the probability of survival. Similarly, the survival probability of an infinitesimal link is $(1 - \mathrm{d}F)$ where $\mathrm{d}F$ is the probability that length $\mathrm{d}L$ breaks. Thus we may equate the survival probability of length $L$ to that of all of its constituent links, or

$$1 - F = (1 - \mathrm{d}F)^{L/\mathrm{d}L}. \tag{4.27}$$

Note the multiplicative nature of the survival probability, where the exponent is simply the number of differential links. By taking logs, this equation is simply recast into the form

$$F = 1 - \exp\left\{\frac{L}{dL}\ln\left(1 - dF\right)\right\}. \tag{4.28}$$

When d$F$ is small, expansion of ln (1 − d$F$) yields −d$F$, and therefore, $F = 1 - \exp\{(L/dL(-dF)\}$, or.

$$F = 1 - \exp\left\{\frac{-LdF}{dL}\right\}. \tag{4.29}$$

We expect that $dF/dL$ will physically depend on the applied stress $\sigma$. The dependence of $F$ on $L$ can be modeled in various ways, but it is common to assume a power law relationship, i.e., $dF/dL = 1/L_o\,(\sigma/\sigma_o)^m$, where $L_o$ and $m$ are constants. This leads to the expression

$$F = 1 - \exp\left\{-\frac{L}{L_o}\left(\frac{\sigma}{\sigma_o}\right)^m\right\}, \tag{4.30}$$

which has the Weibull form of Eqn (4.14) when $\sigma$ replaces time as the variable of interest. This equation is widely used in the reliability analysis of the strength of optical fibers and ceramic materials.

In the case of dielectric breakdown the operative stress is the electric field $E$, and electrode areas ($A$) rather than lengths are relevant. Therefore, the failure probability for dielectrics as a function of E is predicted to have the form

$$F = 1 - \exp\left\{\frac{-A}{A_o}\left(\frac{E}{E_o}\right)^n\right\}, \tag{4.31}$$

where $n$ is a constant. Again it has the Weibull form in situations where $E$ rather than time is the variable with a statistical spread. This equation will be referred again in Chapter 6.

### 4.5.3 The Weibull Distribution: Is There a Weak Link to the Avrami Equation?

There is an interesting, and apparently unreported, connection between Weibull distributions and the Avrami equation of metallurgical literature fame. To see why, let us consider an initially defect-free matrix of volume $V_o$ within a component that will eventually fail. With time, the seeds of damage, e.g., precipitates, voids, charged traps, continuously germinate throughout the matrix with a constant nucleation rate $\dot{N}$ (in units of number of nuclei/cm$^3$ − s). If a nucleus of damage (or defect) grows as a sphere from time $t = 0$, then its volume is given by $V = 4\pi r^3/3$. Assuming that the sphere of radius $r$ expands with constant growth velocity $G$, then at time $t$, $r = Gt$; therefore, the volume is $V = 4\pi(Gt)^3/3$. Physically, damage nucleates at different times; if the event occurs at time $\tau$, then the defect volume $V'$ is given by $V' = 4\pi G^3(t - \tau)^3/3$. The number of such damage defects formed per unit volume of undamaged matrix in time $d\tau$ is $\dot{N}d\tau$. If the damage entities do not overlap, the total damaged volume $V^*$ is the sum of the defect volumes and is given by

$$V^* = \Sigma V' = \frac{4\pi}{3}\int_0^t V_o G^3 \dot{N}(t - \tau)d\tau = \frac{\pi V_o G^3 \dot{N} t^4}{3}. \tag{4.32}$$

The fractional volume that is damaged is $F = V^*/V_o$, or

$$F = \frac{\pi}{3}\left(G^3 \dot{N} t^4\right). \tag{4.33}$$

This equation applies only to short times or small $F$ values. Later, however, the defective regions impinge on one another in much the same way that circular water wavelets do when raindrop nuclei randomly strike a pond and spread outward. Therefore, $F$ stops increasing according to Eqn (4.33) and decreases with time as impingement of damaged regions occurs. Eventually, $F$ levels off. Equation (4.34) accounts for both short- and long-time behavior:

$$F(t) = 1 - \exp\left[-\left(\frac{\pi}{3}G^3 \dot{N} t^4\right)\right]. \tag{4.34}$$

At short times a Taylor expansion yields Eqn (4.33), while at long times $F$ approaches 1.

This well-known relationship is the celebrated Avrami equation. It has been used for well over a half century to describe the fractional amount of solid-state diffusional transformation of a matrix from which new phases nucleate and grow. A readable derivation and simple application of this equation is given in Ref. [21]. The most general form of the Avrami equation, valid for arbitrary defect geometries and failure processes, is commonly written as

$$F(t) = 1 - \exp[-K(T)t^n], \tag{4.35}$$

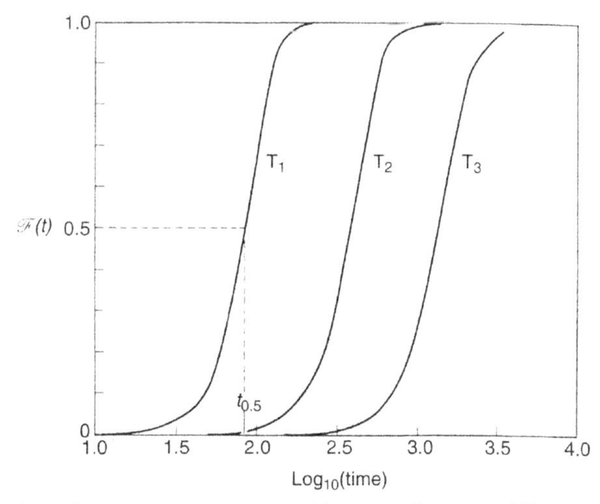


**Figure 4.22** Plot of Avrami equation as $F(t)$ versus log $t$ at different temperatures ($T_1 > T_2 > T_3$).

and plotted in Figure 4.22. Here $K(T)$ is a thermally activated, temperature-dependent constant ($K(T) = K_o \exp[-E/kT]$), and $n$ is a constant typically varying between 1 and 4. In this form, constant $K(T)$ has absorbed both time- and temperature-dependent nucleation and growth terms. One suspects that rapid damage growth follows a long period during which defects incubate; alternatively, failure nuclei can form readily, but subsequent growth of damage is sluggish. These behavioral extremes model nucleation- and growth-controlled damage regimes, respectively. The critical question of what value of $F$ is necessary for failure depends on the particular damage mechanism.

There are two publications on issues related to reliability that make use of an Avrami-like equation. The first deals with void nucleation and growth damage during electromigration, an important reliability problem treated at length in Chapter 5. Resulting increases in resistance ($R$) with time were fitted to an equation of the form $\exp[-K(T)t^n]$ [22]. A second application, interestingly, involves modeling the time-dependent learning curve and yield improvement in processing integrated circuits. The theory developed involved an Avrami-like formula [23].

Comparison with Eqn (4.14) shows that $F$ in Eqn (4.35) has exactly the same form as the Weibull CDF $F$; thus $n = \beta$, and $K(T) = (1/\alpha)^\beta$. As suggested by Eqn (4.17), a plot of $\ln\{-\ln[1 - F(t)]\}$ versus $\ln t$ yields a line with a slope of $n$. There are other noteworthy similarities. For example, $n$, like $\beta$, is usually a small number, while $K(T)$ like $\alpha$, assumes a wide range of values.

### 4.5.4 Physics of the Lognormal Distribution

The lognormal distribution is characterized by a failure rate $\lambda(t)$ that has a single maximum, with $\lambda$ equal to zero at both zero and infinite time. This is distinct from the Weibull distribution, which simply shows a monotonic increase with time. Electronic components have wear-out hazard functions that do not rise monotonically, and this partially accounts for the utility of the lognormal distribution. Physical models of failure generally invoke the central limit theorem (Section 4.2.1) to explain that specific energy levels $E_i$ are nominally the same in all components and are distributed in a Gaussian fashion. Since degradation is directly related to the occupancy of such levels according to the Maxwell–Boltzmann factor, exp $[-E_i/kT]$, the logarithm yields $E_i/kT$, which is normally distributed; hence the lognormal distribution. Service environments of electronic products are relatively benign and vary little among populations. Since the spread in product life is primarily a function of manufacturing variability, it is possible to predict lifetimes on the basis of reliability testing.

In a specific application, Jordan [18] has discussed the aging degradation of GaP red LEDs in terms of diffusion and accumulation of point defects in the p–n junction region. Explicit expressions for the reduction of the electroluminescent efficiency were derived and had the form of the lognormal function.

There is an interesting inequivalence between the lognormal description of electronic components and the increasing hazard function for human lifetimes [3,12]. For living species, internal aging processes apparently generate species-specific biological clocks with increasing hazard functions as $t \rightarrow \infty$. Furthermore, the presence of redundancy at many levels complicates the form of the hazard function. For example, loss of a kidney does not cause death in humans. In addition, random external variables, e.g., bacteria, viruses, accidents, addiction to cigarettes, alcohol, or drugs, become increasingly life-threatening as the recuperative powers of the body wear out. In contrast to electronic devices, the longevity of humans cannot be predicted with any accuracy through a physical examination.

### 4.5.5 Acceleration Factor

In recent years electronic devices and components have become very reliable, so that failure rates are now quite low when tested under ambient conditions. However, in order to make more meaningful lifetime projections, many failures, or data points, are required. This can be achieved experimentally through accelerated testing. Since many aging mechanisms are governed by the Maxwell–Boltzmann factor, raising the temperature of the product under test is usually the most effective way to hasten failure.

The AF, an important term in the subject of reliability, was previously defined as the ratio of the degradation or failure time at the use temperature $T_1$ relative to that at an elevated, accelerated-test temperature $T_2$. In using AF concepts it is assumed that the mechanism of damage does not change in the process. This caveat is an important one. After all, when a hen incubates eggs at room temperature, chicks will hatch; but if the process is accelerated in boiling water, hard boiled eggs result! True acceleration simply means speeding up the action without distorting it in any way. Essentially reproducing Eqn (1.10), the AF is given by.

$$\mathrm{AF} = \frac{\mathrm{Rate}\ (T_2)}{\mathrm{Rate}\ (T_1)} = \frac{\mathrm{MTTF}\ (T_1)}{\mathrm{MTTF}\ (T_2)} = \exp\left[\frac{E}{R}\left(\frac{1}{T_1} - \frac{1}{T_2}\right)\right], \tag{4.36}$$

where the simple symbol $E$ replaces the previously used $\Delta G^*$ in all subsequent Arrhenius-type formulas. The activation energy ($E$) is normally positive, so that when $T_2 > T_1$, AF $> 0$. More complex expressions for AF arise when driving forces differ at $T_1$ and $T_2$, as noted in Eqn (1.11). For example,

$$\mathrm{AF} = \exp\left\{\frac{E}{R}\left(\frac{1}{T_1} - \frac{1}{T_2}\right) + C(V_2 - V_1)\right\}, \tag{4.37}$$

with $C$ a constant, is commonly used to describe accelerated testing and burn-in under the combined influence of temperature and applied voltage $V$.

More generally, the question arises as to how to incorporate AF values obtained from accelerated testing under some set of stress-test conditions (s), into reliability functions that are valid under use conditions (u). In the sense used, stresses include temperature, voltage, current, and humidity, taken singly or in combination. *True*, or *linear*, acceleration factors apply when every test failure time and distribution function is multiplied by the same constant value to obtain projected results during use. Therefore, $t_f(\mathrm{u}) = \mathrm{AF} \times t_f(\mathrm{s})$, and the PDF transforms as

$$f_{\mathrm{u}}(t) = (1/\mathrm{AF}) \times f_{\mathrm{s}}(t/\mathrm{AF}). \tag{4.38}$$

Similarly, formulas for the cumulative distribution and hazard functions transform as

$$F_{\mathrm{u}}(t) = (1/\mathrm{AF}) \times \{\mathrm{AF} \times F_{\mathrm{s}}(t/\mathrm{AF})\} = F_{\mathrm{s}}(t/\mathrm{AF}) \tag{4.39}$$

and

$$\lambda_u(t) = (1/\mathrm{AF}) \times \lambda_s(t/\mathrm{AF}), \tag{4.40}$$

respectively. Substitution shows that there is no simple relation between the hazard function at use relative to test conditions for the lognormal case, but explicit formulas for the other two distributions are

$$\text{Exponential} \quad \lambda_u = (1/\mathrm{AF}) \times \lambda_s, \tag{4.41}$$

$$\text{Weibull} \quad \lambda_u(t) = (1/\mathrm{AF})^{\beta} \times \lambda_s(t) \tag{4.42}$$

---

### Example 4.5

Tantalum capacitors were tested at elevated temperature, and the failure times were Weibull distributed with scale parameter $\alpha = 2.5 \times 10^5$ h and a shape parameter $\beta = 0.5$. If the AF is estimated to be $1 \times 10^4$, what is the probability that the capacitors will last 1 year?

**Answer**

The failure probability is given by $F(t) = 1 - \exp\left[-\left(\frac{t}{\mathrm{AF} \times \alpha}\right)^{\beta}\right]$. In 1 year there are 8760 h.

Therefore, $F(8760) = 1 - \exp\left\{-\left[-\frac{8760}{1 \times 10^4 \times 2.5 \times 10^5}\right]^{1/2}\right\} = 0.00187$. Thus the survival probability is equal to $1 - F(8760)$ or 0.998.

---

## 4.5.6 The Arrhenius Model

Rooted in the very nature of atoms and molecules is their thermally activated response to increasing temperature expressed by the Maxwell–Boltzmann or Arrhenius factor (Eqn (1.2)). Because of the intimate linkage of reaction rates, failure times, and AFs, it is natural to incorporate the Arrhenius factor into the probability functions and their derived parameters. However, this involves more than a simple substitution of $\exp\left[\frac{E}{R}\left(\frac{1}{T_1} - \frac{1}{T_2}\right)\right]$ for AF; while AF applies to only two temperatures, the Arrhenius factor is more generally applicable to a range of temperatures.

The activation energy for degradation and damage processes at the atomic level critically affects the macroscopic MTTF through the formula

$$\text{MTTF} = A \exp\left[\frac{E}{RT}\right], \tag{4.43}$$

where $A$ is a constant. A list of activation energies for common failure mechanisms in silicon devices is included in Table 4.4.

To illustrate how temperature influences failure times, consider the data of Figure 4.23(a), which are plotted in lognormal fashion. In obtaining the activation energy that characterizes these data, lifetimes are picked off the abscissa axis at the ordinate value corresponding to 50% cumulative failures. These times ($t_{50}$) are defined as MTTF and are plotted in the Arrhenius manner, i.e., log MTTF versus $1/T(K)$, as shown in Figure 4.23(b). From the equation of the resulting straight line,

$$\log \text{MTTF} = \log A + E/(2.303RT) \tag{4.44}$$

the slope represents $E/(2.303\ R)$. Calculation yields a value of $E = 1.03$ eV. Note that in this case the same activation energy would be obtained when selecting times corresponding to other percentage cumulative failures (e.g., 10%), because the lines are parallel. When this is true it means that the same failure mechanism is probably operative at different temperatures and times.

The same procedure can be followed for Weibull data consisting of a set of parallel failure-distribution lines as a function of time, when plotted at different temperatures. As noted earlier in Eqn (4.17), the slopes yield the value for β; by then reading off times to reach a given $F$ $\left(\text{usually } F = \frac{1}{2}\right)$ and plotting them versus $1/T$ in the Arrhenius manner, the activation energy can be extracted.

Complications arise when cumulative failure distributions at different temperatures are not parallel to one another. This is what has occurred in Figure 4.24 where thin-film dielectric breakdown data are plotted. Outwardly, it appears that the activation energy is dependent on electric field. An admixture of failure mechanisms may be involved to account for the nonparallel failure distributions. In such cases, variations of Eyring-type relationships, which are considered next, may be more appropriate than the simple Arrhenius expression.

### 4.5.7 The Eyring Model

While the Arrhenius model is based solely on temperature stressing, the so-called "Eyring model" [25] takes into account the action of multiple

**Table 4.4** Time-dependent failure mechanisms in silicon devices

| Device association | Failure mechanism | Relevant variables | Accelerating factors | Activation energy (eV) |
|---|---|---|---|---|
| $SiO_2$ and Si-$SiO_2$ | Surface charge | Mobile ions | V, $T$ | 1.0–1.05 |
| | Dielectric breakdown | $E$, $T$ | $E$, $T$ | 0.2–1.0 |
| | Charge injection | $E$, $T$, $Q_{ss}$ | $E$, $T$ | 1.3 (Slow trapping) |
| Metallization | Electromigration | $T$, $j$, $A$, gradients of $T$ and $j$, grain size | $T$, $J$ | 0.5–1.2 |
| | Corrosion (chemical, galvanic, electrolytic) | Contamination, $H$, $V$, $T$ | $H$, $V$, $T$ | ~0.3–0.6 (for electrolysis $V$ may have thresholds) |
| | Contact degradation | $T$, metals, impurities | $T$ | |
| Bonds and other mechanical interfaces | Intermetallic growth | $T$, impurities, bond strength | $T$ | In Al–Au $E = 1.05$ eV |
| | Fatigue | Bond strength, temperature cycling | $T$ extremes | |
| Metal penetration hermeticity | Aluminum penetration into Si seal leaks | $T$, $j$, $A$ | $T$, $j$ | 1.4–1.6 |
| | | Pressure differential, atmosphere | Pressure | |

$V$ = voltage, $E$ = electric field, $A$ = area, $Q_{ss}$ = interfacial fixed charge, $T$ = temperature, $j$ = current density, $H$ = humidity.
After Ref. [24].

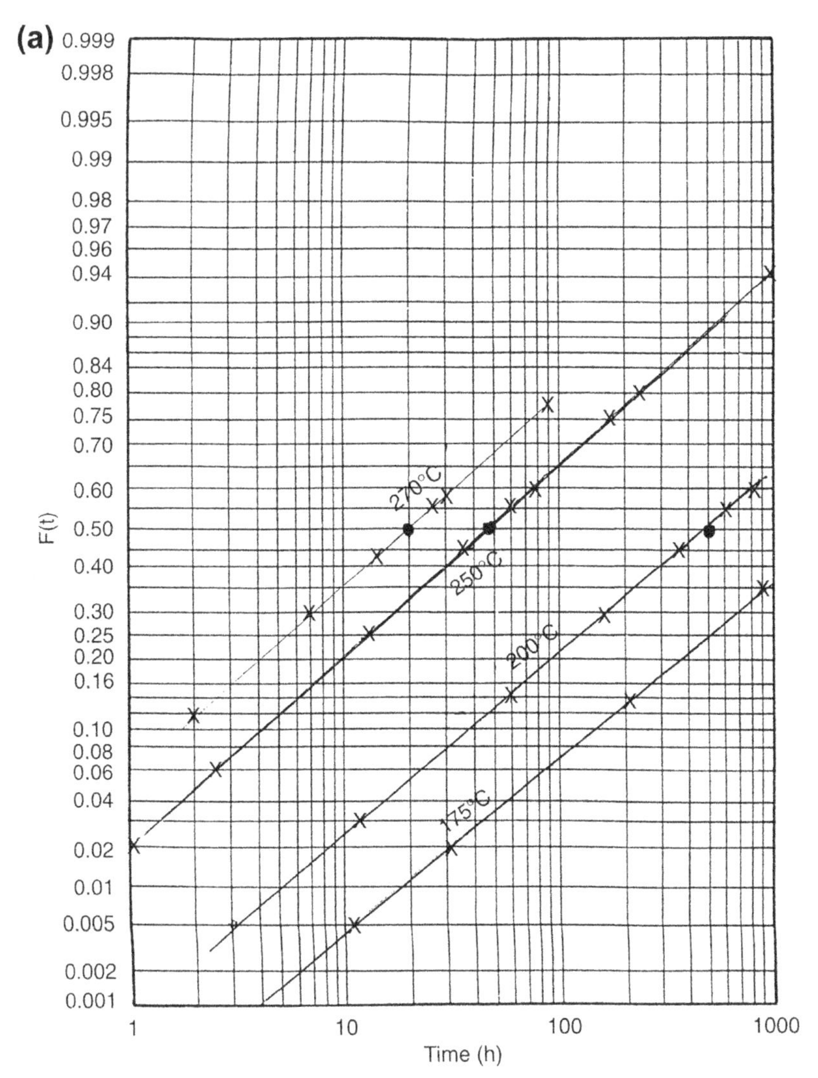


**Figure 4.23** (a) Simulated device failure data at different temperatures plotted as log-normal distributions. (b) Lifetimes (MTTF) corresponding to 50% cumulative failures plotted as log MTTF versus $1/T$ (K).

stresses. Based on the Nernst–Einstein approach to reactions and processes introduced in Section 1.3.3, it is not difficult to incorporate such effects into an expression for predicted times to failure. The stresses may exponentially affect lifetime, serve to reduce the effective activation energy for failure, be a multiplicative factor, or appear as some combination of these in

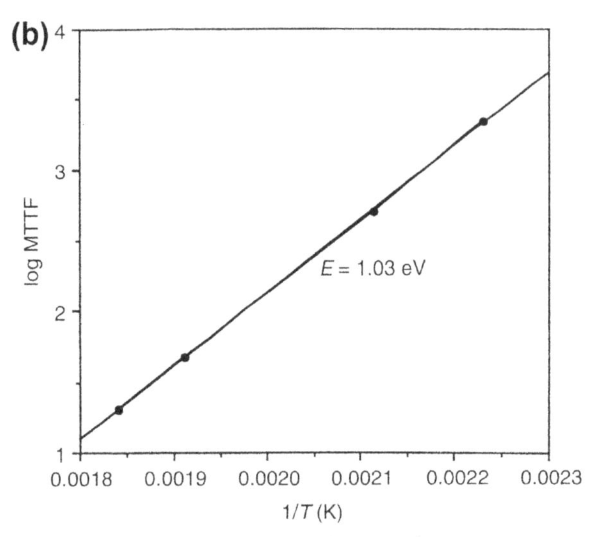


**Figure 4.23** —Continued

expressions for MTTF. A form of the Eyring model for the influence of an additional stress in addition to temperature is thus

$$\text{MTTF} = AT^a \exp\left[\frac{E}{RT}\right] \exp\left\{\frac{1}{R}\left(B + \frac{C}{T}\right)\right\}, \tag{4.45}$$

where $a$, $A$, $B$, and $C$ are constants. Often the influence of the exponential factor swamps that of the $T^a$ term, so that it is sometimes omitted, in which case we are back to Eqn (4.43). Reliability practitioners have often invoked assorted variants of this phenomenological equation in the literature. One example of the form

$$\text{MTTF} = AV^{-b} \exp\left[E/RT\right] \tag{4.46}$$

is used for situations where voltage ($V$) acceleration is operative. To use this formula, three constants, $A$, $b$, and E, must be known or estimated.

## 4.5.8 Is Arrhenius Erroneous?

Among the issues involving reliability prediction, the exact role of temperature in accelerating failure is one of the most controversial. Traditionally thought to be the prime variable influencing the failure of microelectronics through thermal activation of damage mechanisms, Hakim [27] raises the question, "Is Arrhenius erroneous?" This critic

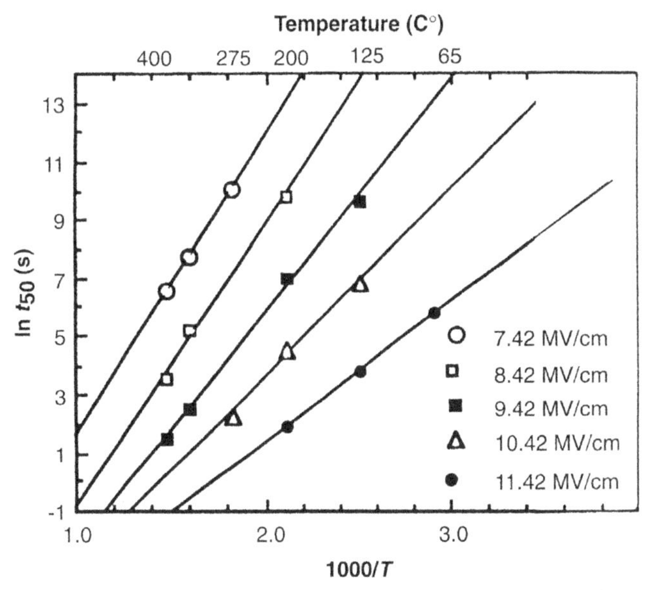


**Figure 4.24** Arrhenius plots of mean dielectric breakdown times ($t_{50}$) for 15 nm thick $SiO_2$ films. Data points are derived from lognormal failure distributions obtained at the indicated electric fields. Note that in Figure 6.13 the activation energy for breakdown is essentially independent of electric field. *From Ref. [26].*

contends that lowering the operating temperature of equipment will not necessarily improve system reliability. If true, the size of cooling systems in electronic equipment could be reduced, with with no loss of reliability an added bonus. In general, temperature AFs in MIL-HDBK-217 may work well for individual devices and components. But a large proportion of electronic systems failures today are not caused by parts failures but by other events within the system such as electromagnetic interference, timing problems, electrostatic discharge, and mishandling [28]. This again raises questions as to the utility of MIL-HDBK-217.

Pecht et al. [29] have addressed the role of temperature, with an emphasis on those integrated circuit failure mechanisms operative in the range −55 to 125 °C. They suggest that most mechanisms have not quantitatively distinguished the roles of steady-state temperature, temperature transients, rate of temperature rise or drop, and temperature gradients. This assessment is probably correct simply because the admixture of these effects with those of the simultaneously operative nonthermal stresses, all within the complex geometry and environment of devices, makes failure modeling practically intractable. A consequence of this is the wide variability of activation energies reported in Table 4.4. Questions related to whether failures at elevated temperatures are the same as those at use temperatures, and whether mechanisms and activation energies change, have already been raised; they will be addressed again in Section 4.6.3.

Blanks [30] in a similar vein has questioned the validity of the Arrhenius relation when nonconstant failure rates prevail. He suggests that the temperature-dependent rate of equipment failure is less than that predicted by the Arrhenius AF when the hazard rate decreases with time. This implies that failure lifetimes would be underestimated in the case of common lognormal failure distributions. Conversely, when the hazard rate increases with time, the Arrhenius AF would overestimate equipment failure rates.

The Arrhenius truth of increased atom movements and reaction rates with rising temperature is not being challenged in these collective criticisms. Our ignorance lies in how to unambiguously isolate and characterize individual damage mechanisms when an admixture or distribution of activation energies may exist. Nevertheless, blaming steady-state temperatures for failures and blindly invoking Arrhenius AFs probably shifts emphasis away from design and manufacturing deficiencies in many cases.

### 4.5.9 The Bathtub Curve Revisited

In Section 1.3.8 the famous bathtub curve, so named for its shape, was introduced as a way to plot failure rates over product life cycles. The early failure, or "infant mortality," regime has, for understandable reasons, been the subject of considerable scrutiny for many years. During short times the weak or marginally functional members of the population fail according to a decreasing failure rate. In a model developed at AT&T [31] based on life testing under simulated field conditions, the bathtub curve is plotted as the hazard function, or failure rate (in FITs) versus time, as shown in Figure 4.25(a). Early failures are modeled by a Weibull hazard function of the form $\lambda(t) = \lambda(0)t^{-(1-\beta)}$, where $1 - \beta < 1$, and $\lambda(0)$ is basically the initial failure rate. On the log $\lambda$ – log $t$ plot the resulting declining failure rate line is defined by an $\sim$20 FIT data point after a period of $10^4$ h, a time somewhat longer than 1 year.

The temperature dependence of infant failures is important because of thermal burn-in screening. Assorted bipolar, MOS, and electro-optical device families have surprisingly exhibited a common low activation energy ranging from 0.25 to 0.4 eV despite the fact that many different failure mechanisms are involved [33]. Following elevated temperature stressing, hazard rates at service temperatures can be estimated by calculating the

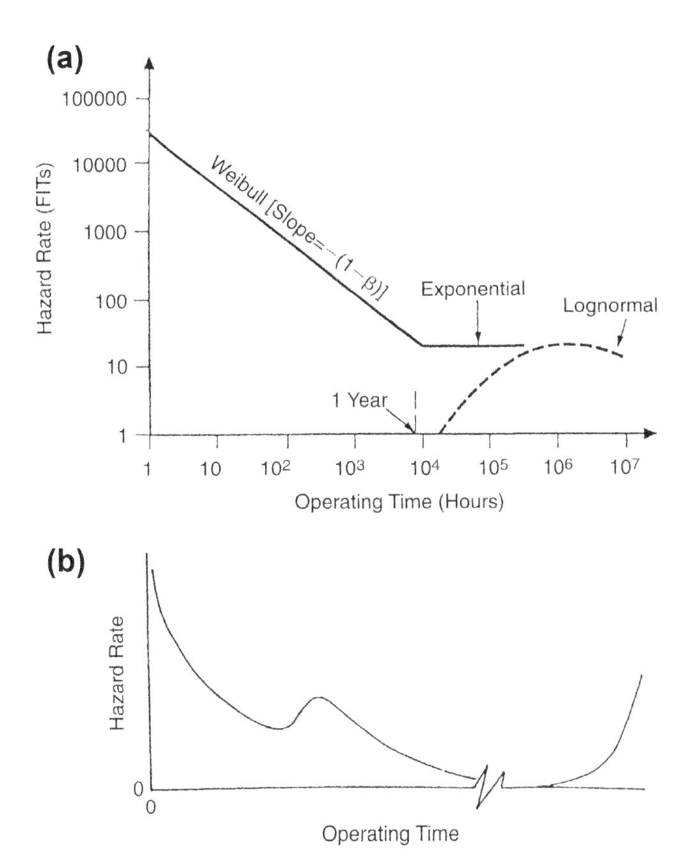


**Figure 4.25** (a) Bathtub curve plotted as the hazard function or rate (in FITs) versus log time. The failure rates are typical but do not correspond to any particular device. (b) Roller coaster $\lambda(t)$, which starts high initially, falls to a local minimum (a valley), and then rises up and over a hump finally to merge with flatter portions. *From Ref. [32].*

appropriate AF and substituting it into Eqn (4.40). Failures on the left-hand edge of the bathtub curve reflect workmanship defects that include faulty design, omitted inspections, and mishandling. Particulates and contamination during the processing of IC chips are also leading causes of early failures.

Beyond the infant mortality regime is the long, roughly flat portion known as the *intrinsic failure* period. Failures occur randomly in this region, and the failure rate is approximately constant. Mathematically, the constant hazard rate of the exponential distribution is admirably suited to model such random failures. Most of the useful life of a component is spent here, so that much reliability testing is conducted to determine values of $\lambda$ in this region.

Finally, there is the *wear-out* failure regime. Here components degrade at an accelerated pace, so the failure rate increases in this region. This is the regime where lognormal and Weibull statistics enjoy a dominant role. A manufacturing goal for high-reliability components is to extend the life of components prior to the onset of wear out. Remarkable strides have been made in this direction. It should be noted that while experimental failure data for some devices and components may reflect the entire shape of the bathtub curve, this is not true of the mathematical distribution functions. These distributions fit only one or, at most, two regions of the curve reasonably well.

What has been said thus far about the bathtub curve is the traditional wisdom. For more than a decade this view has been disputed in criticism to the effect that "the bathtub does not hold water any more" [34] and "the roller coaster curve is in" [32]. Very low infant mortality failure rates of devices nowadays is the basis for the former observation, while the latter characterization is due to the sometimes-observed short time hump in the hazard rate. Thus $\lambda(t)$, which starts high initially, falls to a local minimum (a valley) and then rises up and over a hump to finally merge with flatter portions of the curve, as suggested in Figure 4.25(b). As far back as 1968, Peck [35] attributed the humps to what he called "freak" failures and suggested that they could be eliminated by stress screening such as burn-in. Later, Peck [36] discerned a considerable lengthening of the infant mortality region for integrated circuits to times approaching 100,000 h. After that, the hazard rate apparently meets the main failure distribution. At this stage it becomes an academic question; devices do not actually fail but become obsolete due to design changes and introduction of new technology.

Note $\lambda(t)$ for a system on a chip with $10^8$ transistors is possibly an admixture of a few of the mechanisms in Table 4.4, all competing or behaving to give a low or high reliability, respectively. Rarely do all chips fail due to one and only one failure mechanism. If they do then it is quite possibly wear out, and the root cause of failure may not have been determined or removed from the design and fabrication process.

## 4.6 PREDICTION CONFIDENCE AND ASSESSING RISK

### 4.6.1 Introduction

Thus far we have been been blessed with a generous amount of failure data to enable us to readily perform statistical analyses of the kind discussed in this Chapter Indeed, much of the electronics reliability physics literature is devoted to accelerated testing that generates such data in abundance. In practice, however, there are many important situations where there are very few, if any, failures. Military systems, nuclear reactors, airport traffic control systems, undersea communication links, and surgical electronic equipment spring to mind as examples where catastrophe may result from a single failure. In such cases, how are statistical predictions of failure made, and how confident can we be of reliability estimates? What do we do when zero failures are seen in laboratory failure-testing experiments? Questions involving statistically insignificant numbers of failures, and the risks involved in drawing conclusions from them, are addressed in this section [37].

Measurement error associated with the statistical counting of % failure is fairly straight forward and essentially a discrete parameter. For example, 1 fail in 10, 10%. An analogue parameter such as temperature or threshold voltage has an error associated with the technique of measurement and the accuracy and precision of the device used for the measurement. The interested reader is referred to Yardley Beers [38], "Theory of Error" for a thorough and complete discussion of experimental error. A careful analysis of the data generated by any experiment must be a prerequisite to what follows below.

### 4.6.2 Confidence Limits

The degree of confidence we have in the estimates of $\mu$, $\sigma$, $\alpha$, $\beta$, $\lambda$, or any other statistical quantity mentioned in this chapter is dependent on the number of samples tested and quantity of data, as well as on the measurement error of the statistic under consideration. Since the average value for a specific sample will differ from the true value of the population, the latter should be quoted to lie within a range in which we have confidence. The limits that bound this range of values on the high and low ends are known as confidence limits. To be meaningful, reliability projections should be accompanied by a statement of the confidence level. Statistical inference will not inform us of the fate of a particular specimen, but it could, for example, state that 97% of the components in a population will have an MTTF of 1500 h. When the population is very large, then its standard deviation will also serve for the specimen itself. There are standard mathematical techniques for evaluating confidence limits, but a discussion of them would take us too far afield. For this information the reader is referred to the indicated Refs [3,4,39].

More relevant is what happens when very few failures are available for analysis. Does it mean that traditional reliability modeling is precluded?

Surprisingly, all is not lost! An approach is available to estimate an upper bound for the probability of failure and even to attach a level of confidence to it. Following Nash [3,37], let us consider a subpopulation of $N$ devices where $n$ failures are seen in time $t = t_a$. The probability of such an outcome is given by the binomial formula (Eqn (3.23)), now written in the form

$$P(n, t_a) = \frac{N!}{n!(N-n)!} F(t_a)^n [1 - F(t_a)]^{N-n}. \quad (4.47)$$

In this equation $F(t_a)$ is the failure function or probability of failure in the time interval $t = 0$ to $t = t_a$. In the extreme case where there are no failures, $n = 0$. Since $F(t) = 1 - R(t)$ by Eqn (4.4),

$$P(0, t_a) = R^N(t_a), \quad (4.48)$$

where $R$ is the reliability or survival function. Because, $P(0, t_a) + P(n > 1, t_a) = 1$ by definition, it follows that the probability of 1 or more failures is given by

$$P(n \geq 1, t_a) = 1 - R^N(t_a). \quad (4.49)$$

The problem now reduces to selecting an appropriate functional form of $R(t_a)$. As we shall see, $R(t_a)$ will contain an undetermined parameter related to a characteristic lifetime. Since no failures were observed, a conservative bound on this characteristic lifetime will require that the chance of this occurrence be small. If, for example, $P(n > 1, t_a) = 0.95$ or 95%, then a lower limit on the characteristic lifetime is such that the actual observation of zero failures is very unlikely, i.e., only 5%. A confidence level $C(0)$, for the case of no failure, is defined to be

$$C(0) = P(n \geq 1, t_a). \quad (4.50)$$

This value of $C(0)$ is a known quantity that we impose in the analysis.

In what follows, expressions for the failure rate will be developed for two different failure models. In the case that failures are described by a Weibull function,

$$R(t) = \exp\left[-(t/\tau)^{\beta}\right], \quad (4.51)$$

the characteristic lifetime, $\tau$, is used in place of $\alpha$ (see Eqn (4.14)). When no failures occur in testing $N$ components for a time $t_a$, then

$$C(0) = 1 - \exp\left[-N(t_a/\tau)^{\beta}\right]. \quad (4.52)$$

Suppose we wish to calculate the hazard rate at a specific time, say 1 h. Then by Eqn (4.15),

$$\lambda_1 = \beta \tau^{-\beta}. \tag{4.53}$$

It is usually desired to include temperature effects, so that inclusion of the AF (Eqn (4.42)), coupled with elimination of $\tau$ between the last two equations, yields the following expression for the failure rate at use temperature $T_u$.

$$\lambda_1(T_u) = \frac{\beta \ln[1/(1 - C(0))]}{N(\mathrm{AF}t_a)^{\beta}}. \tag{4.54}$$

---

## Example 4.6

Suppose that ICs are used in a critical application at 40 °C. The manufacturer selects a sample population of 1000 chips, ages them at 150 °C for 2 years ($t_a = 1.75 \times 10^4$ h) and no failures are observed. Typical data for other similar products suggest that the activation energy for failure is 0.4 eV and $\beta = 0.25$. What upper bound value for the failure rate, $\lambda_1$ (40 °C), can be predicted with a confidence of 90% ($C = 0.9$)?

**Answer**

First the AF is evaluated from Eqn (4.36): $\mathrm{AF} = \exp\left[\frac{0.4}{8.62\times10^{-5}}\left(\frac{1}{313} - \frac{1}{423}\right)\right] = 47.2$. After substituting in Eqn (4.54), $\lambda_1(T_u) = \frac{0.25 \ln\frac{1}{(1-0.9)}}{10^3\times(47.2\times1.75\times10^4)^{0.25}} = 1.91 \times 10^{-5}/\mathrm{h}$ or 19,094 FITs

Compared to the decreasing hazard rate of the Weibull model, the use of the exponential distribution is easily derived. Under conditions where no failures are observed during testing, $\lambda(t) = \lambda_1(T_U) =$ const, and it is easily shown that.

$$\lambda = (N \cdot \mathrm{AF} \cdot t_a)^{-1} \times \ln[1/(1 - C(0))]. \tag{4.55}$$

Using the same data above in Example 4.6, the assumption of an exponential distribution would result in $\lambda = 2.7$ FITs, quite a difference.

---

### 4.6.3 Risky Reliability Modeling

Making reliability estimates is often fraught with pitfalls, as the following example illustrates [40]. In a hypothetical but realistic scenario, suppose that GaAs high-electron mobility transistors (HEMTs) are needed for a state-of-the-art communications satellite that costs $100,000,000. The desired life is 25 years, and failure of even a single transistor will degrade the performance of the satellite. Engineers at the company contracted to design, manufacture, test, and specify the reliability of the HEMTs conduct a testing program involving accelerated aging of the devices under bias. Ten transistors were tested at each of four aging temperatures, 225, 234, 240, and 260 °C, and the MTTF were found to be 980, 380, 190 and 33 h, respectively, as depicted in the plot of Figure 4.26. Significantly, there were no infant-mortality failures. Longer time tests at lower temperatures were planned, but an answer was needed as soon as possible, and it was based on these four data points.

Employing the familiar relation MTTF $\approx$ exp $[E/(RT)]$ (Eqn (4.43)), an Arrhenius plot (Figure 4.27) of the failure times versus $1/T$(K) yields an activation energy ($E$) for failure of 2.2 eV. Projecting the 225 °C data to the use temperature results in MTTF(298 K) = MTTF(498 K)exp $\left[\frac{-2.2}{8.62\times10^{-5}}\left(\frac{1}{498}-\frac{1}{298}\right)\right]$ or $1.18 \times 10^{15}$ MTTF(498 K); this is equivalent to an AF of over $10^{15}$. It means that the predicted life at 25 °C is $1.2 \times 10^{18}$ h $\left(1.3 \times 10^{14}\right)$ years, or equivalently, more than $5 \times 10^{12}$ desired lifetimes. Based on this calculation the overjoyed engineers enthusiastically endorsed the reliability of these transistors. But, can this predicted life, which is some 30,000 times the age of the earth, be believed? After all, Evans [42] has commented that "Acceleration factors of 10 are not unreasonable. Factors much larger than that tend to be figments of the imagination and

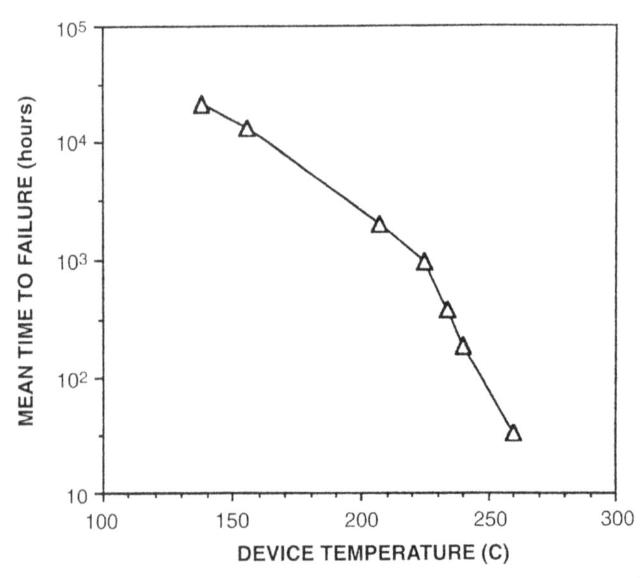


**Figure 4.26** Failure of high-electron mobility transistor transistors as a function of time. *Data reported by Ref. [41].*

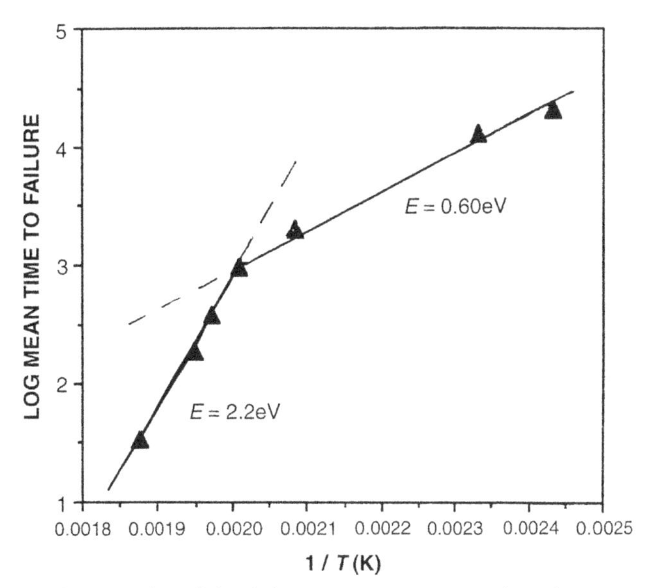


**Figure 4.27** Arrhenius plot of the failure times versus 1/*T* (K). High temperature data yield an activation energy for failure of 2.2 eV. Low temperature data yield an activation energy for failure of 0.60 eV.

of lots of correct, but irrelevant, arithmetic." If so, is there a more reasonable approach to this problem?

One difficulty with the analysis is the assumption that elevated-temperature data can be simply extrapolated to the use temperature. Making an ~200 °C extrapolation is risky for semiconductor devices and requires caution; accelerated aging should preferably be done closer to the use temperatures, if possible. What if a change in failure mechanism having a smaller activation energy occurred at these lower temperatures? To check such a possibility, it would have been a good idea to estimate the activation energy ($E$) that would be required to give a 25 year ($2.19 \times 10^5$ h) life. Under the above assumptions, $E = \frac{8.62\times10^{-5}}{(1/298-1/498)}\ln\left(\frac{2.19\times10^5}{0.98\times10^3}\right) = 0.34$ eV. The decrease in activation energy from 2.2 to 0.34 eV is perhaps fortuitous, but significant because the latter value is characteristic for infant mortality failures of many semiconductor products, as indicated in Section 4.5.9. In fact, as Figures 4.26 and 4.27 reveal, longer-time testing of these HEMTs at three lower temperatures indicated failures with an activation energy of 0.60 eV.

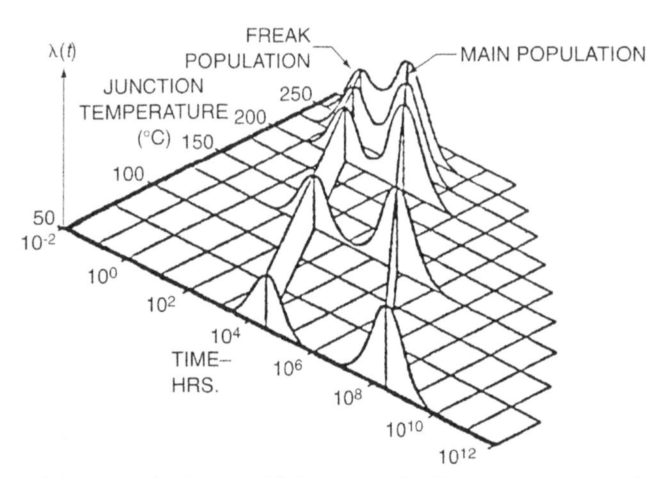


**Figure 4.28** Schematic depiction of failure rate distribution consisting of large main and small freak populations. At low temperatures the two populations diverge, with freak failures occurring at short times. *After Ref. [43].*

It is instructive to view graphically what happens to the hazard rate when two failure populations, a large main and small freak, are exhibited, as shown in Figure 4.28. At elevated temperatures both failure behaviors merge into what appears to be a main (bimodal) distribution. However, at lower temperatures these distributions diverge according to their respective activation energies. Now the freak population poses a particular threat at low temperatures.

A similar methodology for early-life reliability prediction in integrated circuits was advanced in a recent publication [44]. It was suggested that early failures are not intrinsic, but rather premature or infant failures. Therefore, the common practice of extrapolating intrinsic life-test distributions to estimate early-life reliability is incorrect and often leads to unduly optimistic predictions. This all underscores the necessity for screening weak products, some of which may have the potential to cause freak failures.

## 4.6.4 Freak Failures

The subject of freak failures is perhaps the most difficult topic addressed in reliability. Valuable insights into this problem have been provided by Nash [3]. Assuring reliability against freak phenomena is open-ended and unstructured because statistically insignificant numbers of freak failures occur. In order to design strategies against freak failures it is instructive to first distinguish between the *normal* and *freak* products of manufacture. The reliability of normal products has the following characteristics:

1. Failure mechanisms are few in number, e.g., dielectric breakdown, electromigration, etc.
2. These mechanisms are distributed among the entire population of products.
3. Failures are independent of lot number or time of manufacture.
4. The normal or statistical variability of manufacturing processes gives rise to the defects that cause normal failures.
5. Normal failure mechanisms are characterized by well-known activation energies and AFs. Product lifetimes are therefore statistically predictable.
6. Importantly, normal products do not fail prematurely. Screening prior to shipment generally weeds out the weak products that would have failed by the normal failure mechanisms.

Abnormal or freak failures, on the other hand, differ from normal failures in all respects. Specifically, they have the following characteristics:

1. There are many causes and mechanisms of freak failures. Each one is different and occurs with low frequency. Thus freak failures cannot be statistically modeled. If they could, they would cease to be freak and become normal failures.
2. Freak failures are found only in small subsets of the overall population of products.
3. Such failures are dependent on lot as well as time of manufacture.
4. Failure times are often very short and frequently unrelated to operating times. Freak failures are not wear-out failures!
5. In contrast to normal failures, there are very few studies of freak failures in the literature.

Two observations of failures summarize the distinction between normal and abnormal products. Normal products do not fail under anticipated use conditions; field failures are of the freak variety.

### 4.6.5 Minimizing Freak Failures

Is there any hope for a workable strategy to deal with abnormal products? Before answering the question it is helpful to better understand what are thought to be the sources of freak failures. These include:

1. Random one-of-a-kind accidents (e.g., electrostatic discharge or electrical overstress due to mishandling).
2. Manufacturing processes that are poorly controlled (e.g., particulate contamination) or prone to yield low-quality results.
3. Changes in the quality or properties of raw materials used (e.g., due to new suppliers).

Based on the last two considerations, the remedies required to minimize the possibility of freak failures are those that continuously improve product quality. Specific measures to improve products for high-reliability applications include:

1. Additional and stricter testing to identify and eliminate products that exhibit signs of being in the least atypical. Tighter pass-fail limits, aging at higher temperatures and bias voltages, and more stages of inspection represent some of the new quality hurdles that should be instituted.
2. Laboratory analysis of each and every service failure to establish fundamental causes of failure. Such information should result in corrective redesign and manufacturing changes aimed at improving the product.
3. Implementation of more robust designs with higher safety margins.
4. Rejection of products that have been reworked.
5. Institution of traceability to pinpoint faulty processes and identify suspected lots.
6. Stricter supplier monitoring.

The remainder of this book is primarily concerned with normal failures because virtually the entire reliability physics literature is concerned with them. Unfortunately, this same literature offers little insight into the uncharted territory of freak failures; we therefore travel the more comfortable and well illuminated paths of normal behavior.

## 4.7 A SKEPTICAL AND IRREVERENT SUMMARY

Now that rather elaborate statistical methods have been presented to analyze degradation and failure data, we may ask how correct and reliable the conclusions are. Skepticism about statistics being able to justify any particular point of view is not solely limited to the social sciences; examples were given in the chapter questioning the validity of the predictions made. I am not the only one to raise such doubts, but rather have humbly followed more eloquent analysts and critics of reliability methodology and its philosophical underpinnings. One such critic is R.A. Evans, an editorial writer for the *IEEE Transactions on Reliability,* whose column appears in each issue. In these often memorable editorials, many issues related to reliability, statistics, and improving product quality are addressed in a witty and perceptive manner. Several excerpts from this column will give the reader a sense of his concerns and provide a perspective on this chapter.

## STATISTICS AND IGNORANCE

*We spend too much time in our reliability courses on probability and statistical inference ... that show us how to quantify our ignorance. We do not spend enough time on removing that ignorance ... the engineering, physics and chemistry of why things fail and why things don't fail.*

**from Vol. 39, p. 257, Aug. 1990**

## SUPERSTITION, WITCHCRAFT, PREDICTION

*The title is a parallel to Disraeli's phrase, "Lies, Damned Lies, and Statistics". ... A main use of engineering reliability prediction is for the marketplace. That is, a customer wants to know the reliability of a product, before buying the product. The vagaries of such numbers* (predictions *for the marketplace) leads to the infamous numbers games.*

**from Vol. 37, p. 257, Aug. 1988**

## STATISTICS VERSUS PHYSICS

*Statistics describes only populations, not individuals. In particular, the hazard rate (failure rate) of a population does not say anything at all about the behavior of an individual. ... An individual part is likely to degrade with time, and thus get worse as time goes by; it can do that and still be a member of a constant-hazard-rate population. ... Reliability-statistics is not physics; don't ever confuse them. ... The quality and reliability disciplines use statistics as a last resort. If we were smarter about the world, we wouldn't have to.*

**from Vol. 38, p. 273, Aug. 1989**

## WHERE DO I BEGIN?

*... you are placed in charge of creating a cost-effective reliability program ... tell me what I should do? ... Most corrective action will be a combination of management and engineering changes. Some will require scientific advances. A very few will require mathematical advances. Very rarely will mathematical statistics be a bottleneck, although some statistical principles are always important. ... Don't get a bunch of experts to come and lecture everyone on reliability statistics.*

**from Vol. 38, p. 513, Dec. 1989**

## RELIABILITY PREDICTION AND MIL-HDBK-217

*One should use MIL-HDBK-217, not because it is so great, but in spite of its weaknesses. When it is used like any other engineering tool, it can be a great help. If it is confused with holy writ, then there is no end of trouble. The search for certainty often ends up in the acceptance of superstition.*

**from Vol. 37, p. 257, Aug 1988**

## 4.8 EPILOGUE—FINAL COMMENT

Catastrophic failure, device degradation, and AFs are some of the most important and the most difficult aspects of working as a reliability engineer. The AF for a given test depends on test parameters, samples, and stress conditions. If we first take a thermally activated mechanism, we will necessarily assume a Maxwell–Boltzmann thermal activation for the rate of degradation. But the degradation may not always be outright catastrophic failure. The degradation may be a % change in some parameter, say threshold voltage or resistance, such that the circuit is slower than it needs to be. A case in point, which is counter intuitive comes up with hot electron induced threshold voltage shift of MOSFETs. The threshold voltage shift is less at higher temperature because it slowly anneals above room temperature. This gives the appearance that it is inverse with temperature. This comes about because the observation is a competition of two separate effects. First, the hot electrons are generated in the channel and some of them surmount the barrier at the oxide interface and are trapped in the gate oxide at the drain end of the channel. Second, the electrons trapped in the gate oxide anneal out as they are released from the traps in the gate oxide. The annealing effect is thermally activated and causes less observed shift at higher temperatures. The complexity of this effect is further complicated by the generation of hot electrons which also increases with applied drain voltage and shorter channel length, both of which increase the electric field, which is the driving force producing the hot electrons. In order to determine an AF, one must first determine how much threshold voltage shift will constitute failure. This shift is usually called an end of life shift. The trick is to put this end of life shift in the context of the mean and standard deviation of the expected values of the threshold voltage values expected in a chip and a wafer. Notice we are talking about the tail of the onset of failure, which although not early life failure, is somewhat like early life failure. Chip failure will be determined by the MOSFET device which is the closest to the end of life value. Six sigma will bear quite heavily on our decision as to whether a process is well controlled or not, both from a yield and a reliability standpoint. In very sophisticated designs one may even have to consider the duty cycle of the device to come to a final decision. Such considerations often lead to design groundrules for dimensions and guidelines for maximum currents and voltages in devices. Violation of these rules will result in premature failure. However the rules are design, process and technology dependent, which is another way of saying they are dependent upon the company making the chip, and even possibly the organization within the company.

## EXERCISES

**4.1** Calculate the mean and standard deviation for the fuse data listed in Table 4.1. What percentage of the measurements fall within the interval between $\mu - \sigma$ and $\mu + \sigma$? How does this percentage compare with the predicted area under a normal distribution curve?

**4.2** During the cyclic stress or fatigue testing of a large number of solder joints, the following failures were recorded in the indicated intervals.

| Number of cycles | Percentage of failed joints during given interval of testing |
|---|---|
| 0–50 | 1.21 |
| 50–100 | 0.12 |
| 100–150 | 0.13 |
| 150–200 | 0.62 |
| 200–250 | 0.36 |
| 250–300 | 0.95 |
| 300–350 | 4.51 |
| 350–400 | 6.73 |
| 400–450 | 3.24 |
| 450–500 | 13.2 |
| 500–600 | 4.65 |
| 600–700 | 8.58 |
| 700–800 | 2.24 |

Calculate the failure rates per cycle and plot the results as a function of the number of cycles in order to obtain a "bathtub" curve.

**4.3** A failure mechanism that follows the exponential distribution function will not benefit from burn-in. Why?

**4.4** "Grandfather's clock ... stood ninety years on the floor ... but it stopped short, never to go again when the old man died (at exactly 90)." Even more reliable was "The Deacon's Masterpiece," the wonderful one-horse shay that lasted "one hundred years and a day." Each of these fabled marvels of American craftsmanship had a reliability of unity until failure, at which time it became zero. Assuming Weibull statistics can be applied in these cases, calculate the characteristic lifetimes ($\alpha$) and shape parameters ($\beta$) in each case.

**4.5** The extreme value distribution (due to Gummel) has a probability density function given by $f(t) = (1/b)\exp[(t-\tau)/b]\exp(-\exp[(t-\tau)/b])$, where $b$ and $\tau$ are constants. What are the mathematical forms of $F(t)$ and $\lambda(t)$?

**4.6** Demonstrate the explicit relationship connecting $F(t)$ and $\lambda(t)$, namely,

$$F(t) = 1 - \exp\left[-\int_0^t \lambda(t)\mathrm{d}t\right].$$

**4.7** A LED has a MTTF of 250,000 h.

**a.** Based on the exponential distribution, predict the probability of failure within a year.

**b.** At what time would half of the devices fail?

**c.** Suppose failures are modeled with a lognormal distribution function with a standard deviation $\sigma = 1.5$. What is the probability of LED failure in one year?

**4.8** Fifteen 75-W light bulbs were tested for 1000 h, and the failure times in hours were 890; 808; 501; 760; 490; 658; 832; 743; 993; 812; 378; 576; 899; 910; 959.

**a.** What is the MTTF?

**b.** Assuming constant failure-rate modeling. What is the reliability after 400 and after 800 h?

**4.9** The failure fraction of a laser transmitter is 3% over a period of 25 years. Express the hazard rate in FITs.

**4.10** Certain device failures obey lognormal statistics with $\sigma = 4.5$.

**a.** What is the maximum value of the normalized failure rate and at what normalized time does it occur?

**b.** How long does it take to reach a maximum failure rate of 10 FITs?

**4.11 a.** Provided that lognormal behavior is obeyed, what minimum median lifetime ($t_{50}$) is required for devices to function 40 years and suffer only 100 FITs if $\sigma = 1.0$?

**b.** What maximum standard deviation should be targeted if no more than 10 FITs can be tolerated in devices with a median lifetime of $1 \times 10^8$ h?

**4.12** A reliability engineer using the MIL-HDBK-217 calculates a failure rate of 225 FITs for a particular device using exponential distribution function statistics.

**a.** What is the CDF for 20 years of operation?

**b.** As a result of accelerated testing it was found that wear-out phenomena were governed by Weibull failure functions. The Weibull parameters were $\alpha = 5 \times 106$ h, $\beta = 3$. What is the predicted CDF after 20 years?

**4.13** A bathtub curve consists of three linear regions as a function of time ($t$), namely,

**1.** an infant mortality failure rate that decreases as $\lambda(t) = C_1 - C_2 t$;

**2.** a zero random failure rate;

**3.** a wear-out region that varies as $C_3(t - t_o)$.

**a.** Sketch the bathtub curve.

**b.** Calculate $f(t)$ and $F(t)$ values in each of the three regions.

**4.14** The Rayleigh distribution is defined by a single parameter $k$, and the hazard rate increases linearly with time such that $\lambda = kt$. From this information calculate $f(t)$ and $F(t)$ for Rayleigh distributions. What is the connection between the Rayleigh and Weibull distribution functions?

**4.15** A missile guidance system contains 100 transistors each of which is 99.9% reliable. The failure of any transistor will cause the guidance system to fail. How reliable is the guidance system? Suppose there are 1000 transistors. What is the system reliability?

**4.16** For a particular telephone IC circuit pack, Weibull infant mortality failure rates are characterized by $\lambda(t) = 34{,}100\ t^{-0.728}$ FITs.

**a.** If the failure rate at the end of a year is 25 FITs, calculate the percentage of devices failing during the first 6 months.

**b.** Suppose the circuit pack contained 250 such components. How many would fail after 6 months?

**4.17** A communications system consists of a transmitter, a receiver, and an encoder such that the failure of any of these components will cause failure of the system. If the individual reliabilities are 0.93, 0.99, and 0.95, respectively, what is the system reliability?

**a.** If the exponential distribution governs failure in each of the components, what is the system failure rate?

**4.18** Consider a control system that contains subsystems having the following reliability characteristics.

| | |
|---|---|
| Central processing unit | Weibull $\beta = 1.05$, $\alpha = 100$ h |
| I/O card | Exponential $1/\lambda_o = 750$ h |
| Actuator | Lognormal $\mu = 6.4$, $\sigma = 1.5$ |

It is necessary for all subsystems to function for the system to operate. What is the reliability of the system at 300 h? From Ref. [45]

**4.19** Sketch what happens to the bathtub curve if

**a.** component dimensions shrink further.

**b.** devices are powered under successively higher current–voltage stressing.

**4.20** Consider the bathtub curve for humans. Specify examples of the causes of death that are operative in each of the three time domains.

**4.21** A 16-position connector is subjected to five different degradation mechanisms. The reliability for each mechanism is $R_i (i = 1-5)$.

**a.** Write an expression for the overall reliability $R$ of the connector.

**b.** If the $R_i$ are all equal to $R_o$, what is value of $R$?

**c.** What contact reliability is required to produce a connector reliability of 0.9999?

**4.22** In a population of photodiodes, 20% display freak ($f$) failures at a MTTF of 12,000 h, while the main (m) population has an MTTF of 28,000 h.

**a.** Sketch the probability density function, $f(t)$, for this population if they are normal distributions.

**b.** Suppose $\sigma(f) > 2\sigma(m)$. What is the ratio of the peak values of $f(t)$, i.e., $f_f(t)/f_m(t)$?

**4.23** In GaAs transistors were tested at 180, 196, and 210 °C, and the times to failure in hours were:

180 °C: 250, 420, 1000, 1300, 1300, 1300, 1300, 1500, 1500, 1600

196 °C: 190, 200, 330, 330, 400, 400, 420, 600, 800

210 °C: 105, 110, 120, 130, 210, 230, 290, 300, 320, 400

**a.** What are the MTTF at each temperature if the data follow lognormal statistics?

**b.** What is the activation energy for failure?

**c.** What is the mean life at 110 °C?

**4.24** While testing a 10,000 FIT device, 23 failures were observed within the first hour of burn-in. What was the total number of devices in the burn-in lot?

**4.25** The overall reaction rate that leads to failure of a component is the result of two mechanisms having different activation energies, $E_1$ and $E_2$. In failure mode A these mechanisms occur in series (sequentially), while for failure mode B they add in parallel. Sketch the overall reaction rate versus temperature in the form of an Arrhenius plot for both failure modes. What are the reliability implications of the thermal dependencies of these two failure modes with respect to predicting low-temperature failure?

## REFERENCES


[1] M.G. Pecht, F.R. Nash, Proc. IEEE 82 (1994) 992.
[2] P.A. Tobias, D.C. Trindade, Applied Reliability, Van Nostrand Reinhold, New York, 1986.
[3] F.R. Nash, Estimating Device Reliability: Assessment of Credibility, Kluwer Academic, Boston, 1993.
[4] P.D.T. O'Connor, Practical Reliability Engineering, second ed., Wiley, Chichester, 1986.
[5] R.A. Evans, Tutorial notes, in: Annual Reliability and Maintainability Symposium, 1994.
[6] P.E. Fieler, Tutorial notes, in: IEEE International Reliability Physics Symposium, 2.1, 1990.
[7] E.E. Lewis, Introduction to Reliability Engineering, second ed., Addison Wesley Publishing, 1996.
[8] J.A. Carr, Electron. Technol. J.
[9] K.L. Wong, Qual. Reliab. Eng. Int. 7 (1991) 489.
[10] J.A. McLinn, Qual. Reliab. Eng. Int. 6 (1990) 237.
[11] S.F. Morris, J.F. Reilly. Proc. Ann. Reliability and Maintainability Symposium, 1993, p. 503.
[12] W.B. Joyce, P.J. Anthony, IEEE Trans. Reliab. 37 (1988) 299.
[13] F.R. Nash, Estimating Device Reliability: Assessment of Credibility, Kluwer Academic, Boston, 1993.
[14] L.R. Goldthwaite, in: Proc. 7th National Symposium on Reliability and Quality Control in Electronics, 1961, p. 208.
[15] L.R. Goldthwaite, Proceedings of the Seventh National Symposium on Reliability and Quality Control in Electronics, 1961, p. 208.
[16] W.J. Bertram, in: S.M. Sze (Ed.), VLSI Technology, second. ed., McGraw Hill, New York, 1988.
[17] K.D. Hong, S.S. Lee, J.Y. Kim, S. Daniel, C.K. Yoon, 33rd Annual Proceedings of the IEEE Reliability Physics Symposium, 1995, p. 85.
[18] A.S. Jordan, Microelectron. Reliab. 18 (1978) 267.
[19] D.S. Peck, in: 9th Annual Proceedings of the IEEE Reliability Physics Symposium, 1971, p. 69.
[20] R. Ramakumar, Engineering Reliability—Fundamentals and Applications, Prentice Hall, Englewood Cliffs, NJ, 1993.
[21] K.N. Tu, J.W. Mayer, L.C. Feldman, Electronic Thin Film Science for Electrical Engineers and Materials Scientists, Macmillan, New York, 1992.
[22] K.P. Rodbell, M.V. Rodriguez, P.J. Ficalora, J. Appl. Phys. 61 (1987) 2844.
[23] W. Hillberg, Microelectron. Reliab. 20 (1980) 337.
[24] W.J. Bertram, in: S.M. Sze (Ed.), VLSI Technology, second ed., McGraw Hill, New York, 1988.
[25] H. Eyring, S.H. Lin, S.M. Lin, Basic Chemical Kinetics, J. Wiley, New York, 1980.
[26] J.S. Suehle, P. Chaparala, C. Messick, W.M. Miller, K.C. Boyko, 32nd Annual Proceedings of the IEEE Reliability Physics Symposium, 1994, p. 120.
[27] E.B. Hakim, Solid State Tech. 33 (8) (1990) 57.

[28] P.D.T. O'Connor, Solid State Tech. 33 (8) (1990) 59.
[29] M. Pecht, P. Lall, E.B. Hakim, Qual. Reliab. Eng. Int. 8 (1992) 165.
[30] H.S. Blanks, Qual. Reliab. Eng. Int. 6 (1990) 259.
[31] D.J. Klinger, Y. Nakada, M.A. Menendez, AT&T Reliability Manual, Van Nostrand Reinhold, 1990.
[32] K.L. Wong, Qual. Reliab. Eng. Int. 5 (1989) 29.
[33] D.S. Peck, in: 16th Annual Proceedings of the IEEE Reliability Physics Symposium, 1978, p. 1. F.S. Beltramo. Proceedings of the Annual Reliability and Maintainability Symposium, vol. 327, 1988.
[34] K.L. Wong, Qual. Reliab. Eng. Int. 4 (1988) 279.
[35] D.S. Peck, in: 6th Annual Proceedings of the IEEE Reliability Physics Symposium, 1968.
[36] D.S. Peck, Trans. Electron. Devices 26 (1979) 38.
[37] F.R. Nash, in: Materials Research Society Symposium Proceedings, vol. 184, 3, 1990.
[38] Yardley Beers, Introduction to the Theory of Error, Addison Wesley Publishing, 1957.
[39] G.W.A. Dummer, N.B. Griffin, Electronics Reliability—Calculation and Design, Pergamon Press, Oxford, 1966.
[40] F.R. Nash, Seminar at Stevens Institute, September 27, 1995.
[41] A. Christou, in: A. Christou (Ed.), Reliability of Gallium Arsenide MMICs, Wiley, 1992.
[42] R.A. Evans, IEEE Trans. Reliab. 40 (1991) 497.
[43] F.R. Nash, Lucent Technologies, Bell Laboratories Innovations.
[44] S.S. Menon, K.F. Poole, Microelectron. Reliab. 35 (1147) (1995).
[45] T.D. Thomas, P.B. Lawler, in: Electronic Materials Handbook, vol. 1 Packaging, ASM International, Materials Park, Ohio, 1989, p. 899.

# CHAPTER 5

# Mass Transport-Induced Failure

## 5.1 INTRODUCTION

The preceding chapters have introduced the devices, processing defects, and yields of electronic products that will now see service. Chapter 4 prepared us to expect product failures and provided mathematical tools that enable future performance to be predicted. From now until the end of Chapter 10, we shall be primarily concerned with the detailed microscopic mechanisms that cause these products to degrade in use. We have already noted (Section 1.3.1) that virtually all failures in electronic materials and devices are the result of the physical movement of atoms or charge carriers from benign locations associated with normal behavior to other sites where they contribute to creating or enlarging defects that lead to component malfunction. Because solids are often not in thermodynamic equilibrium, atoms feel compelled to move or react. Thus, silicon and aluminum should theoretically revert to $SiO_2$ and $Al_2O_3$ in air at room temperature, while dopants at junctions should intermix in response to thermodynamic imperatives. But fortunately, for the most part, these and other scenarios for change and reaction in chemically unstable materials, interfaces, and structures do not occur. The low operating temperatures involved essentially reduce reaction rates to negligible levels. Nevertheless, there are other material combinations and forces present that do pose threats to reliability because the chemical and physical changes they induce are on a scale comparable to device or component feature dimensions.

In this chapter we primarily deal with phenomena attributable to atomic diffusional motion on the chip level. This should be distinguished from other transport effects treated elsewhere, i.e., electron (Chapter 6), ionic (Chapter 7), and stress-induced flow of solder (Chapter 9). Effects of atomic diffusion include compound formation, precipitation reactions, void formation, alloying, contact and interconnect reactions, and general degradation of the metallization. Related examples involve environmentally induced ion migration during corrosion and the permeation of water through plastic packages.

*Reliability and Failure of Electronic Materials and Devices*
ISBN 978-0-12-088574-9
http://dx.doi.org/10.1016/B978-0-12-088574-9.00005-7

It is a primary objective of this chapter to predict the time dependence of the visible manifestations of failure based on the mass transport models that describe them. Metals are the major culprits associated with mass transport-induced reliability problems at both chip and packaging levels. Degradation and failure phenomena due to interdiffusion, compound formation, electromigration (EM), and contact reactions are some of the effects modeled in this chapter.

## 5.2 DIFFUSION AND ATOM MOVEMENTS IN SOLIDS

### 5.2.1 Mathematics of Diffusion

Diffusion is defined as the migration of atoms (or molecules) in a matrix under the influence of a concentration-gradient driving force. Through sequential motion of countless numbers of atoms, each jumping from site to site, an observable macroscopic change in atomic concentration, $C$, results. In most cases the diffusing and matrix atoms differ, but they may be the same, as during EM in Al conductors. Fick's law, the phenomenological equation that defines diffusion, is written as

$$J_{\mathrm{m}} = -D\mathrm{d}C/\mathrm{d}x. \tag{5.1}$$

Concentration gradients ($\mathrm{d}C/\mathrm{d}x$) in the positive $x$ direction cause transport of a mass flux of atoms ($J_{\mathrm{m}}$) in the negative $x$ direction, signifying that the vectors defining these terms are oppositely directed; hence the negative sign. Flux units mirror those of concentration. For $C$ in atoms/cm$^3$ or g/cm$^3$, the corresponding units of $J_{\mathrm{m}}$ are atoms/cm$^2$-s or g/cm$^2$-s. Irrespective of concentration units, values of $D$, the diffusion coefficient or diffusivity, are always expressed in units of (distance)$^2$/time, i.e., cm$^2$/s. The diffusion coefficient, a measure of the extent of mass transport, depends on a number of factors, including the nature of the diffusing atoms and matrix, the specific transport path (i.e., lattice, grain boundary (GB), dislocation, surface, interstitial, etc.), temperature, and the concentration of the diffusing species. Temperature, the most important factor influencing $D$, was already noted in Eqns (5.1)–(5.3), while dopant concentration effects in semiconductors were sketched in Section 3.2.2.1.

During nonsteady-state diffusion, more atoms enter a region of the matrix than leave, and therefore, their concentration builds in time ($\mathrm{d}C/\mathrm{d}t > 0$). This is normally the case in a homogeneous matrix. But there are situations where more atoms leave a region than enter it. In time, this loss of matter ($\mathrm{d}C/\mathrm{d}t < 0$) leads to mass depletion or even voids that can adversely affect reliability. Such effects may occur at interfaces between different phases, or when the matrix is heterogeneous and contains numerous GBs and structural defects. In either case, nonsteady-state diffusion can be accounted for by equating the negative divergence of the mass flux (i.e., $-\nabla \times J_m$) to $dC/dt$. When proper account is taken of instantaneous mass conservation at each value of $x$, the equation governing diffusion in one dimension is

$$\frac{\partial C(x,t)}{\partial t} = D\frac{\partial^2 C(x,t)}{\partial x^2} \tag{5.2}$$

if $D$ is constant.

There is a simple solution to this equation when diffusion occurs in a semi-infinite matrix, and atoms are supplied at the $x = 0$ boundary from an inexhaustible or *continuous* surface source of concentration $C_s$. Further, it is assumed that the initial concentration in the matrix is $C_o$, or constant throughout, and remains $C_o$ at infinite distance. These initial and boundary conditions may be written as $C(x, t = 0) = C_o$, $C(x = 0, t) = C_s$ ($C_s$ = constant), and $C(x = \infty, t) = C_o$. Employing standard boundary-value methods for solving partial differential equations, the solution to Eqn (5.2) is

$$[C(x,t) - C_o]/[C_s - C_o] = \text{Erfc}\left[x/(4Dt)^{1/2}\right]. \tag{5.3}$$

Known as the complementary error function, Erfc $[x/(4Dt)^{1/2}]$ is mathematically defined as

$$\text{Erfc}[x/(4Dt)^{1/2}] = 1 - \frac{2}{\pi^{1/2}}\int_0^{\frac{x}{2(Dt)^{1/2}}} \exp - z^2 dz. \tag{5.4}$$

An even simpler solution to Eqn (5.2) results if a *finite* or *instantaneous* rather than continuous surface concentration $S$ (in units of atoms/cm$^2$) is involved. The Gaussian solution has the form

$$C(x,t) = \frac{S}{(\pi Dt)^{1/2}} \exp - \frac{x^2}{4Dt} \tag{5.5}$$

For example, if dopant atoms diffuse into a Si wafer, the amount left on the surface decreases with time; i.e., at $x = 0$, $C(0, t) = S/(\pi Dt)^{1/2}$. Graphic solutions for continuous and finite-source concentration profiles

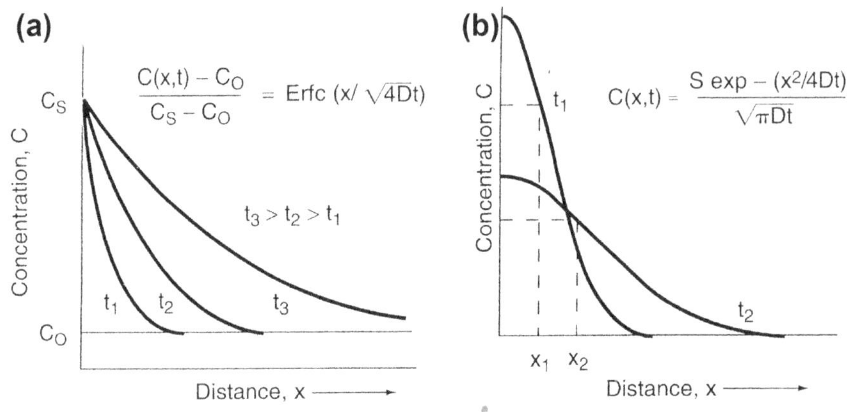


**Figure 5.1** Sequence of concentration profiles at different diffusion times. (a) Continuous (constant) source at surface yields complementary error function solutions. (b) Instantaneous (finite) source at surface yields Gaussian solutions. As the time quadruples from $t_1$ to $t_2$ ($t_2 = 4t_1$), the diffusional penetration distance doubles from $x_1$ to $x_2$.

are shown in Figure 5.1. Like trigonometric functions, values for the complementary error function and Gaussian are tabulated as a function of the argument $[x/(4Dt)^{1/2}]$ in mathematical references and texts on statistics. These same functions are used in the probability and reliability mathematics of Chapter 4.

When back-of-the-envelope estimates for the extent of diffusional penetration are sought, the equation

$$x^2 = 4Dt, \tag{5.6}$$

based on the practical implications of Eqns (5.4) and (5.5), serves rather well.

## 5.2.2 Diffusion Coefficients and Microstructure

Together with Eqn (5.6), the other important thing to remember about diffusion is that the diffusivity is thermally activated (Eqns (5.1)–(5.3)). In the expression for diffusivity, both $D_o$ and $E_D$ are dependent on the nature of the diffusant and matrix, and the specific path taken by the atoms (see section 5.2.1). Thus for diffusion through grains of a polycrystalline matrix, atoms can either exchange places with matrix atoms (vacancies), hop within the confines of a dislocation core, migrate along GBs, or skip along the surface. Short-circuit atom transport via the latter three paths requires successively less energy expenditure than for motion through the bulk. The reason is due to a correspondingly less confining atomic environment that presents fewer physical obstacles to diffusion. These ideas have been quantitatively confirmed in face-centered cubic metals. The following expressions typically describe the diffusivity of atoms along the indicated paths [1]. To obtain these values it is assumed that both the GB width and dislocation radius are 0.5 nm in size.

$$\text{Lattice or bulk diffusion} \quad D_{\mathrm{L}} = 0.5 \exp - 17.0 T_{\mathrm{M}}/T \ \mathrm{cm}^2/\mathrm{s} \tag{5.7a}$$

$$\text{Dislocation} \quad D_{\mathrm{d}} = 0.7 \exp - 12.5 T_{\mathrm{M}}/T \ \mathrm{cm}^2/\mathrm{s} \tag{5.7b}$$

$$\text{Grain boundary} \quad D_{\mathrm{GB}} = 0.3 \exp - 8.9 T_{\mathrm{M}}/T \ \mathrm{cm}^2/\mathrm{s} \tag{5.7c}$$

$$\text{Surface} \quad D_{\mathrm{s}} = 0.014 \exp - 6.5 T_{\mathrm{M}}/T \ \mathrm{cm}^2/\mathrm{s} \tag{5.7d}$$

Note should be taken of the different activation energies for each diffusion mechanism. Thus bulk, dislocation, GB, and surface diffusion mechanisms have activation energies that vary as 17.0 $RT_{\mathrm{M}}$, 12.5 $RT_{\mathrm{M}}$, 8.9 $RT_{\mathrm{M}}$, and 6.54 $RT_{\mathrm{M}}$, respectively, where $T_{\mathrm{M}}$ is the melting point and $R$ is the gas constant.

The regimes of dominant diffusion mechanisms are outlined in Figure 5.2 as a function of grain size ($l$) and dislocation density ($\rho_{\mathrm{d}}$) for different temperatures. This diffusion map is the outcome of allowing the various transport processes to freely compete, where the lines represent the conditions for equal mass fluxes from neighboring regime mechanisms. Although strictly intended for self-diffusion, we shall assume that the map applies to impurity diffusion as well. In the thin-film regime (shown dotted) at low temperatures, GB diffusion dominates, and therefore GB diffusion coefficients must be used for the purposes of calculation. Even if driving forces other than concentration gradients are operative, microscopic atomic migration occurs along these same paths.

Effective diffusivity values vary as a function of grain size, necessitating some sort of averaging of the individual contributions from lattice and GB mechanisms. An effective diffusivity, $D_{\mathrm{eff}} = D_{\mathrm{L}} + fD_{\mathrm{GB}}$ has been suggested, where $f$ is the fraction of atoms in short-circuit paths. A measure of $f$ is the ratio of the GB thickness to grain diameter. In general, grain size variations will correspondingly cause differing diffusivities.

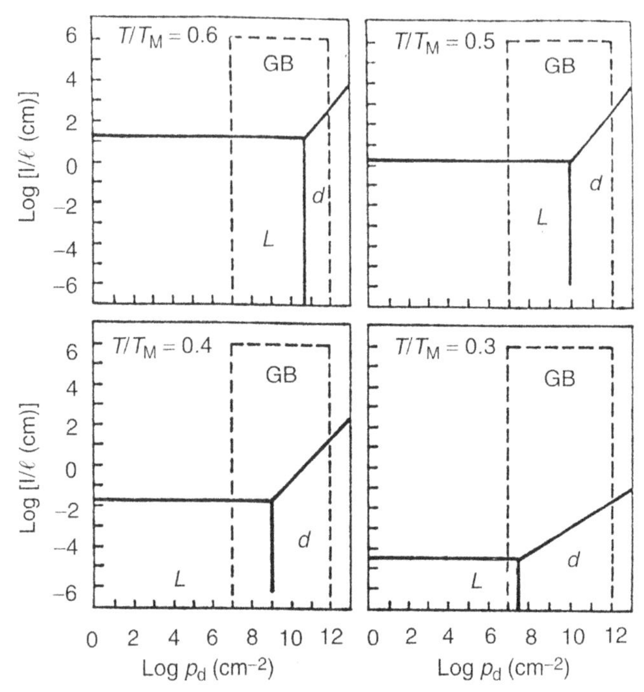


**Figure 5.2** Regimes of dominant diffusion mechanism in FCC metals as a function of grain size (ℓ) and dislocation density ($p_d$) at different temperatures. *From Ref. [2].*

## 5.3 BINARY DIFFUSION AND COMPOUND FORMATION

### 5.3.1 Interdiffusion and the Phase Diagram

Many reliability problems stem from interdiffusion effects at the interface between two dissimilar materials e.g., solder joint, semiconductor contact. To better understand the nature of reaction, let us consider a couple consisting of two elements $A$ and $B$, in contact at a planar interface. What will happen if the couple is now heated to temperature $T_o$, where diffusion of atoms normally occurs? The answer depends on the nature of $A$ and $B$, their grain structures, and the magnitude of the involved diffusivities. A good-first step is to look at the phase diagram for $A$ and $B$. Among the possible reactions are those shown in Figure 5.3, where solid solution (Figure 5.3(a)) and eutectic (Figure 5.3(b)) behaviors are depicted. Below each of these, typical interdiffusion profiles are depicted (Figure 5.3(c) and (d)) at $T_o$ and time $t$. These representations emphasize the connection between phase-concentration distributions and phase diagram-imposed solubility limits. For complete solid solution systems

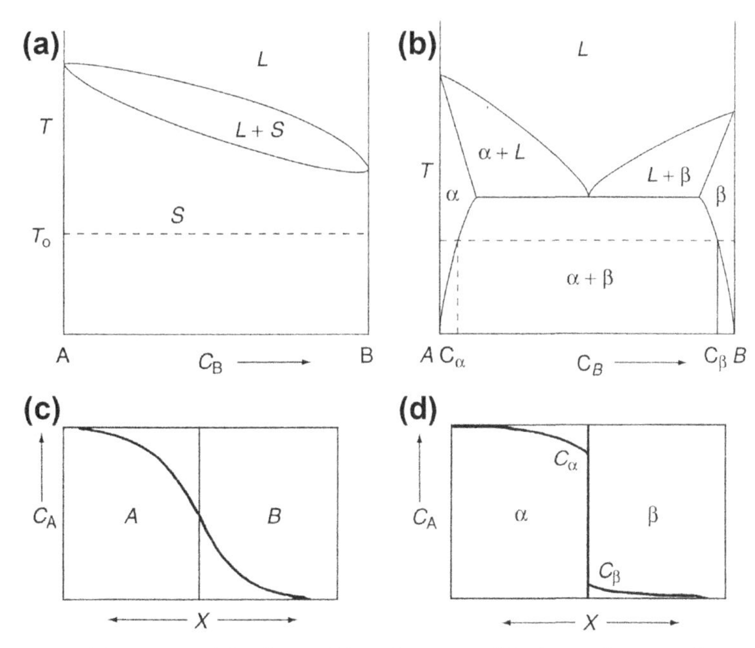


**Figure 5.3** Schematic solid solution (a) and eutectic (b) phase diagrams for *A* and *B* components. Interdiffusion profiles at temperature $T_o$ (dotted) corresponding to solid solution (c) and eutectic (d) phase diagrams.

(e.g., Ge–Si), phases can theoretically assume any binary composition. But in eutectic systems (e.g., Pb–Sn), interdiffusion occurs only up to the solubility limit, resulting in the formation of a range of terminal solid solutions.

Concentration profiles in real systems are generally not error-function-like or symmetric across the couple interface, because the diffusivity usually depends on composition and hence position. The rather involved graphic Boltzmann–Matano analysis can yield values for the overall diffusivity, $D(C)$, in such couples, provided that the diffusional profiles are known [3]. For many solid solutions (e.g., Cu–Ni, In–Pb) the first of the two Darken equations (the second is Eqn (5.11)) yields the (chemical) diffusivity in the couple, namely,

$$D(C) = C_A D(B) + C_B D(A) \tag{5.8}$$

This equation suggests that a kind of weighted average of individual diffusivities $D(A)$ and $D(B)$ determines the magnitude of $D(C)$. In turn, the resulting nonlinear diffusion equation

$$\frac{\partial C(x,t)}{\partial t} = \frac{\partial}{\partial x}\left[D(C)\frac{\partial C(x,t)}{\partial x}\right],$$

yields complex diffusion profiles.

## 5.3.2 Compound Formation

Many interdiffusion-related reliability problems are due to compound formation effects at solder–metal interfaces or semiconductor contacts. These applications will be addressed in later sections, e.g., 8.3.3.3 and 5.4, but the underlying roles of phase diagrams, transport mechanisms, and kinetics are considered here. Compound-forming systems generally have more complex phase diagrams than those we have been considering. An idealized version is shown in Figure 5.4, where the resulting concentration profile across a couple composed of components $A$ and $B$ is also indicated. The growth kinetics of the ($AB$) compound phase is of particular interest. Atoms from both $\alpha$ and $\beta$ phases must be transported to the interface for the compound to thicken. Conservation of mass requires that the two shaded areas on the $\alpha$-phase side be equal, and similarly on the $\beta$-phase side.

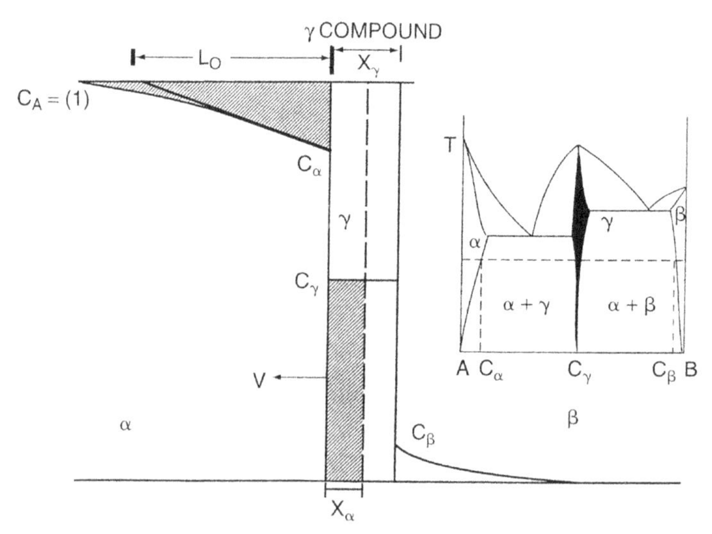


**Figure 5.4** Depiction of intermediate compound (γ) formation arising from reaction between components $A$ and $B$. The reaction temperature is dotted in on phase diagram.

Analysis [1] shows that reaction at the $\alpha/\gamma$ interface results in compound growth kinetics given by

$$X_\alpha = \left[\frac{D_\alpha(C_A - C_\alpha)^2}{(C_\alpha - C_\gamma)C_\gamma}\right] t^{1/2}. \tag{5.9}$$

A similar expression holds for the $\beta/\gamma$ interface, and therefore the final compound layer thickness, the sum of both, is expected to grow parabolically in time i.e.,

$$X_\gamma = \text{constexp}\left[-\frac{E_C}{RT}\right] t^{1/2}. \tag{5.10}$$

Parabolic, thermally activated growth with activation energy $E_C$ should come as no surprise when Eqn (5.10) is compared with Eqn (5.6), which expresses the fundamental connection among variables for diffusional processes.

Examples of parabolic growth occur during the elevated-temperature *diffusion-controlled* oxidation of many metals (e.g., Al, Cu) and the formation of intermetallic compounds and metal silicides. When the oxides are adherent to the base metal, parabolic growth dominates, but this is not so when the oxide is porous and flakes off. In such a case oxygen (or the metal) does not have to diffuse through the oxide; it can directly access the metal surface from the gas phase and form oxide through direct chemical reaction. Such *reaction-controlled* oxidation frequently occurs only for short times in continuous or protective oxides (e.g., $SiO_2$) but persists indefinitely in nonprotective (porous) oxides (e.g., MgO). Reaction-controlled growth is characterized by linear, rather than parabolic, kinetics. Other time dependencies have also been observed during oxidation.

### 5.3.3 The Kirkendall Effect

In this section we address the implications of a binary alloy system whose individual components have different diffusivities. Let us consider a diffusion couple composed of equal volumes of $A$ and $B$ bonded across a planar interface that contains a row of minute, inert markers as shown in Figure 5.5(a). For simplicity, the $A$–$B$ solid solution system of Figure 5.3(a) is assumed. Suppose $A$-vacancy exchanges are less prevalent than those between $B$ atoms and vacancies. Therefore, in a given time less $A$ is transported to the right (Figure 5.5(b)) than $B$ to the left (Figure 5.5(c)). If the couple suffers no volume change, then the marker has effectively shifted

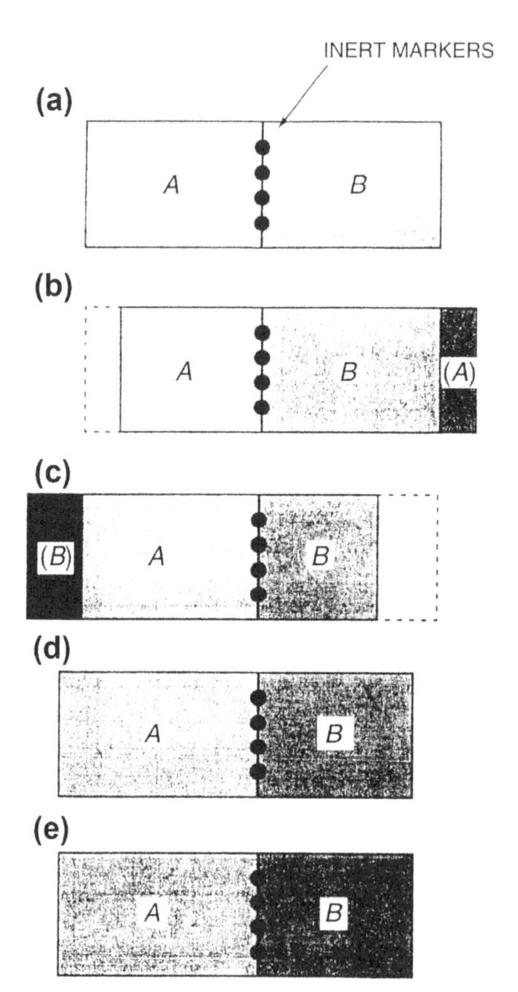


**Figure 5.5** (a) Illustration of Kirkendall effect in a diffusion couple composed of *A* and *B* components bonded across a planar interface containing a row of small, inert markers. (b) Diffusion of *A* atoms to the right. (c) Diffusion of *B* atoms to the left. (d) Final position of marker, which effectively shifts to the right (e) if the couple suffers no volume change.

to the right (Figure 5.5(d) and (e)). This effect is known as the Kirkendall effect, and it dispelled early notions that both atoms in an alloy interdiffused at the same rate. The second Darken equation predicts that the effective velocity ($v$) of the marker-plane motion is

$$v = (D_A - D_B)\frac{\mathrm{d}C_A}{\mathrm{d}x}, \tag{5.11}$$

where the concentration gradient is evaluated at the interface [3].

Not only is there an effective interfacial migration, but the possibility also exists that voids may form. To see why, we note that the difference between the large flux of $B$ atoms and the smaller counterflux of $A$ atoms must be equal to the flux of vacancies. The basic diffusion equation (Eqn (5.2)), derived by setting the negative divergence of the mass flux equal to the time rate of change of concentration, also applies to vacancies. Therefore, there will tend to be a vacancy buildup on the $B$ side balanced by a corresponding depletion on the $A$ side. When a critical vacancy supersaturation develops, there is always the possibility that they may condense into voids. The big question, of course, is what level of vacancy supersaturation is critical. This depends on generally unknown vacancy trapping and emission probabilities, void/matrix interfacial properties, and local driving forces. There is scarcely a more alarming occurrence for reliability prospects than void or incipient crack formation, but this is just what happens in the Al–Au contact system considered next.

### 5.3.4 The Purple Plague

This colorful term refers to the purple $AuAl_2$ intermetallic phase formed when Al and Au reacts. The compound was known for at least 70 years before it surfaced as a reliability concern in wire bonds to transistors and hybrid ICs some 45 years ago. Various bonding configurations are susceptible to purple-plague formation and the damage associated with it. These include Au wires or wedges bonded to thin-film pads of Al on Si, Al wires attached to an Au-plated film on Si or a contact post, and Au wire/Au balls bonded under heat and pressure (thermal compression) to Al films. At elevated temperatures and long times the $AuAl_2$ compound forms in the reaction zone, and the bond fails as a result of annular microcrack formation.

The Al–Au phase diagram reveals the existence of five intermetallic compounds. Each of the bond types above can be modeled by a planar diffusion couple between the involved elements. But depending on the relative initial metal thicknesses ($d$), different final phase mixtures result if equilibrium is attained. For example, when $d_{Al} > d_{Au}$, then the formation of $AuAl_2$ and excess Al is predicted. Similarly, in an excess of Au, the $AlAu_4$ compound will eventually form. In these as well as other Al–Au combinations, some $AuAl_2$ will also form. What is so alarming about its presence is the Kirkendall porosity that accompanies it (Figure 5.6). Since voids form on the Au side of the interface, Au apparently migrates more

**Figure 5.6** Extensive $AuAl_2$ growth and Kirkendall porosity at the interface between a thermosonically formed gold ball bonded to an aluminum pad. *From Ref. [4].*

rapidly than Al; a fact confirmed independently in the diffusion literature.

While $AuAl_2$ alone has an acceptable electrical conductivity and is mechanically strong, Kirkendall voids degrade both of these bond properties. The rate at which $AuAl_2$ thickens is given by $x^2_{AuAl_2}/t = 1.23\ \exp[-1.2\ \mathrm{eV}/(kT)](\mathrm{cm}^2/\mathrm{s})$ and follows parabolic kinetics [1].

## 5.4 REACTIONS AT METAL–SEMICONDUCTOR CONTACTS

### 5.4.1 Introduction to Contacts

Achieving the requisite ohmic or Schottky electrical behavior of contacts poses challenges in processing first, which are then followed by the reliability concerns of property change in service. Metal–semiconductor interdiffusion effects at contacts are primarily responsible for reliability problems manifested as short circuits at junctions, increases in contact resistance, and Schottky barrier height variations. These effects are aggravated in devices that pass large currents and heat appreciably. In silicon devices prior to about 1980, aluminum was the contact metal of choice. However, as device dimensions shrank, the Al–Si interdiffusion damage described in the next section led to the adoption of alternative metal–silicide contacts.

New contact metallurgies for Si are less susceptible to reliability problems than metal contacts to GaAs and compound semiconductor devices. Among the reasons for this are [5]:

1. GaAs decomposes into Ga and $As_2$ gas at 580 °C.
2. The Schottky barrier height ($\Phi_B$) is typically ~0.8 eV for most metals. An implication of this is intrinsically high contact resistance. (In contrast, $\Phi_B$ for Si is ~0.6 eV.)
3. It is difficult to dope *n*-GaAs to levels higher than $5 \times 10^{18}/cm^2$.
4. Annealing up to 850 °C is required to electrically activate dopants following ion implantation. Therefore, special precautions must be taken to avoid decomposition or melting.
5. Because two elements comprise compound semiconductors, there are more possibilities for reaction with contact metals.

### 5.4.2 Al–Si Contacts

The problems considered in this section arise during processing rather than service. Nevertheless, they are instructive and when extrapolated to low temperatures and long times may have reliability implications. We first consider the simple Al–Si contact, which until the VLSI regime was the common method for making electrical connection to Si devices. To form the contact, Al is deposited on the Si, which is covered by a native, discontinuous thin film of oxide, as shown in Figure 5.7(a) and (b). The contact structure essentially consists of a diffusion couple of the two elements. Reference to the accompanying Al–Si phase diagram (Figure 5.8) reveals that the solubility of Si in Al increases as the temperature is raised, reaching ~1 wt% at 500 °C. At this common processing temperature, Si readily diffuses in Al in order to satisfy its solubility requirements (Figure 5.7(c)); the loss of Si leaves pits behind in the semiconductor. Meanwhile, Al counter-migrates into the Si. However, very small amounts of Al dissolve in Si. Excess Al fills the Si-depleted regions and precipitates in the form of conducting filaments that can "spike," or short, shallow junctions (Figure 5.7(d)).

Damage at Al–Si contacts shown in Figure 5.9 takes the form of surface pits, which can be made visible by etching off the aluminum. The Si-pit size ($\delta_{Si}$) has been shown [6] to grow in a thermally activated way with parabolic kinetics; i.e., $\delta_{Si} \approx \exp(-\frac{E}{kT})t^{1/2}$. Furthermore, as the contact area ($A_C$) increases, the pit size was observed to decrease according to $\delta_{Si} \approx A_C^{-0.4}$. In order to explain this effect we note that increasing the contact area enlarges

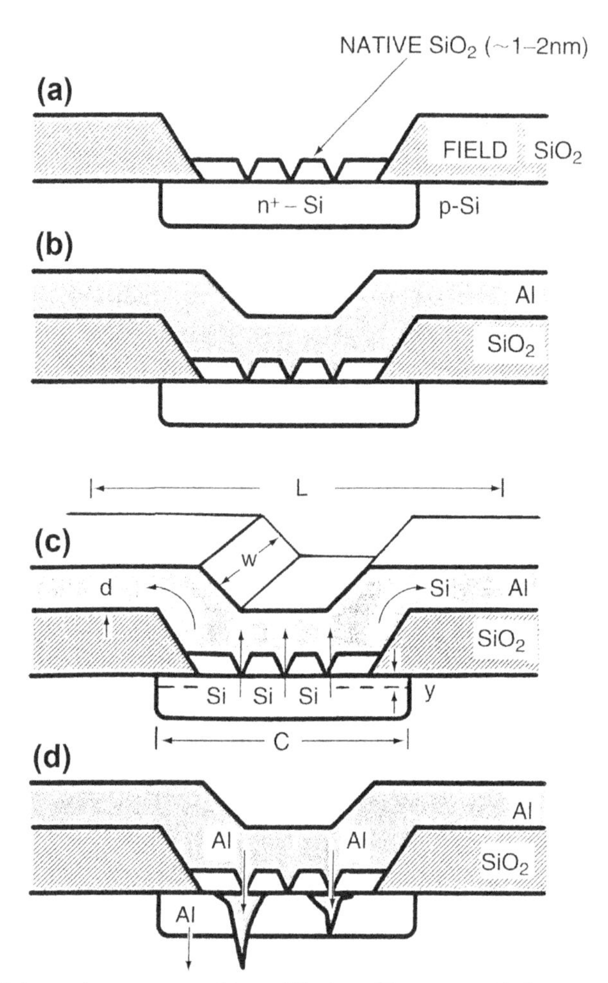


**Figure 5.7** Schematic sequence of interdiffusion effects at an Al–Si contact, leading to junction spiking. (a) Native oxide in contact window. (b) Deposition of aluminum. (c) Silicon diffusion into aluminum upon sintering treatment at 450 °C. See Example 5.1. (d) Counter diffusion of aluminum into silicon shorts junction.

the Si to Al ratio. But the volume of Al that can be saturated at any diffusion temperature remains fixed. More locations and avenues for interdiffusion means less reaction at each site, and hence a smaller $\delta_{Si}$.

---

## Example 5.1

Consider the contact structure of Figure 5.7(c) consisting of a pure Al film of indicated geometry over Si. Calculate the depth ($y$) of Si depletion in the wafer during a processing heat treatment of 1 h at 450 °C, where the solubility ($S$) of Si in Al is 0.5 wt%. The respective densities for Si and Al are $\rho_{Si} = 2.33$ and $\rho_{A1} = 2.7$.

**Answer** The amount of Si needed to saturate the metal contact structure depends on the Al volume as well as the solubility of Si in Al. Conservation of mass requires that the amount of Si lost beneath the contact window must equal that which enters the metal. For (polycrystalline) Al–(single crystal)Si couples, the diffusivity of Si in Al is given by $D = 25 \times 10^{-4} \exp[-0.79 \text{ eV}/(kT)] \text{ cm}^2/\text{s}$ [5]. By Eqn (5.6), the total diffusion length ($L$) along both sides of the contact at 450 °C is calculated to be $L = 4 \times \{(25 \times 10^{-4} \exp[-0.79 \text{ eV}/8.62 \times 10^{-5} \times 723]) \times 3600\}^{1/2} = 1.07 \times 10^{-2}$ cm (107 μm). Therefore, $Lwd\rho_{Al}S = wCy\rho_{Si}$. Solving, $y = \frac{Ld\rho_{Al}S}{(C\rho_{Si})}$, substituting $\rho_{Al} = 2.7$, $\rho_{Si} = 2.33$, and assuming $w = 2$ μm, $d = 1$ μm, and $C = 5$ μm, $y$ is calculated to be 0.123 μm.

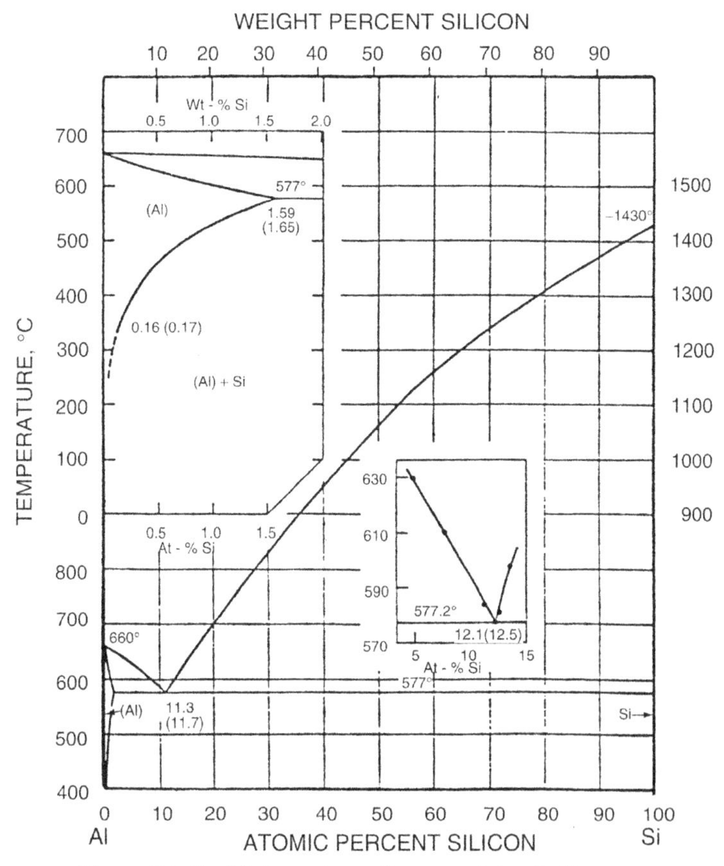


**Figure 5.8** Equilibrium phase diagram of the Al–Si system.

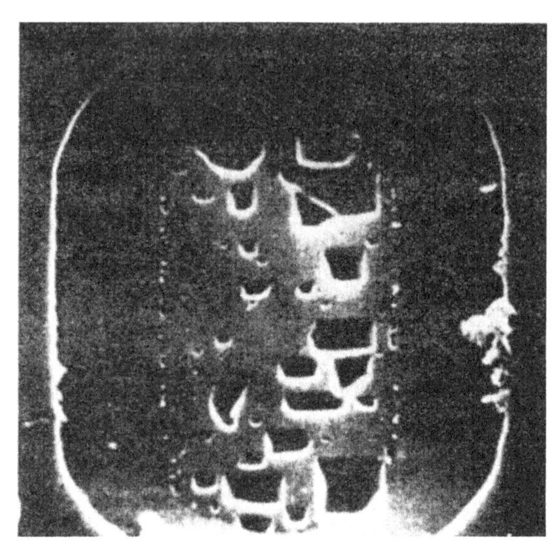

**Figure 5.9** Alloy penetration pits formed by interdiffusion of aluminum and silicon. Contact reaction occurred during sintering at 400 °C for 1 hour. *From C.M. Bailey, D.H. Hensler, AT&T Bell Laboratories.*

This example shows that direct Al–Si contact cannot be tolerated under any circumstances, because the reaction dimensions are uncomfortably large compared to device junction depths. The use of Al metallizations saturated with greater than 1 wt% Si eliminates the driving force for these interdiffusion effects. But when this alloy cools, Si precipitates as its solubility declines, leaving nodules behind on the semiconductor surface. These nodular islands are *p*-type Si (Al) doped and generally raise the contact resistance. Similar but generally more severe interdiffusion effects result from reaction between Al and polycrystalline Si (polysilicon) films. Diffusion and reaction lengths are expected to be larger for polysilicon.

### 5.4.3 Metal Silicide Contacts to Silicon

To protect submicron-deep junctions against the intrusion of the Al metallization, diffusion barriers composed of metal silicides, or Ti–W alloys are commonly interposed to limit interdiffusion between Al and Si. Silicides are metals with resistivities of ~15–50 μΩ-cm, which is considerably higher than the value for Al (2.8 μΩ-cm). They are formed by reacting refractory metal films deposited on Si wafers. Rather than the pitted Al–Si interface, the resulting silicide layers yield very stable planar and uniformly thick contacts. Silicides that have seen use in microelectronics applications are listed in Table 2.1 together with some of their properties. These contacts are relatively immune to degradation in normal service. Problems do, however, arise at elevated temperatures during EM and such effects in $TiSi_2$ and $TaSi_2$ contacts will be discussed in Section 5.6.1.1.

Silicides not only provide necessary electrical contact functions, but also serve as diffusion barriers. Occasionally, Al and the silicide chemically combine so that the fundamental problem of curbing Al reactivity remains. As an example of the potential degradation of metal silicides consider the case of PtSi [7]. This intermetallic has been used in microelectronics because it makes nearly ideal ohmic and Schottky barrier contacts. However, the reaction between Al interconnects and PtSi at a Si contact yields the $PtAl_2$ intermetallic compound and attendant electrical instabilities in the operation of Schottky diodes. Compound formation was reported to obey reaction kinetics given by

$$f_{PtAl_2} = 1 - \exp(-[K(T)t^n]. \tag{5.12}$$

Here $f_{PtAl_2}$ is the fractional amount of PtSi transformed to $PtAl_2$, and $K(T)$ is a temperature-dependent constant proportional to the ubiquitous Boltzmann factor ≈ exp[−E/kT]. The constant $n$ had a value close to 3, and activation energy $E$ was equal to 2.2 eV. As a gage of the reaction, $f$ reached a value of 0.5 at 340 °C after 1000 h.

This study is interesting for a couple of reasons. First, Eqn (5.12) is the Avrami equation (see Eqns 4.35), which describes the kinetics of phase transformations that occur by nucleation and growth processes. Secondly, such kinetics have the form of the Weibull distribution function.

## 5.4.4 Contacts to GaAs Devices

### *5.4.4.1 Scope*

Just as in Si devices, the seeds of GaAs contact reliability problems are often planted during processing. A good example is reproduced in the transmission electron microscopic cross-sectional image of a high-electron-mobility transistor contact structure (Figure 5.10). Due to a 1–2 nm thick oxide barrier between the metallization and underlying AlGaAs, metal alloying occurred nonuniformly. In reducing contact quality, conducting projections that threatened to short to the GaAs layer, formed. The resemblance to the "spiking" of shallow junctions in Si, discussed in Section 5.4.2, is evident.

Service failures in GaAs devices can occur at surfaces and interfaces, in the substrate or active layers, at Schottky and ohmic contacts, and by EM

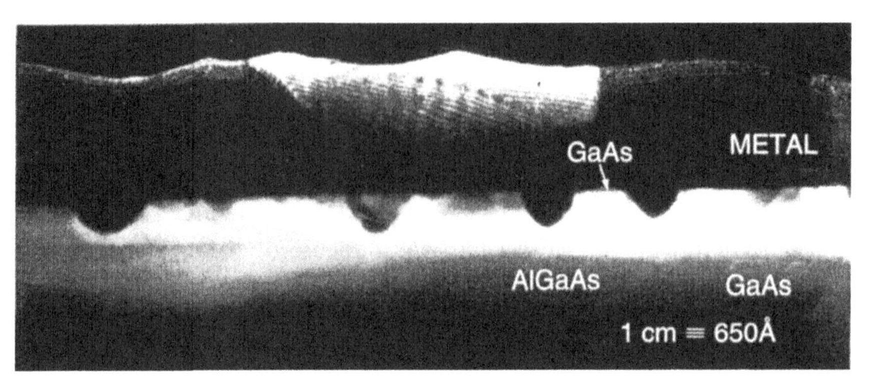


**Figure 5.10** TEM micrograph of Ni/Au-Ge/Ag/Au alloyed ohmic contact (420 °C, 20 s) on a GaAs/AlGaAs HEMT structure. *From Ref. [8]. Courtesy F. Ren, Lucent Technologies, Bell Laboratories Innovations.*

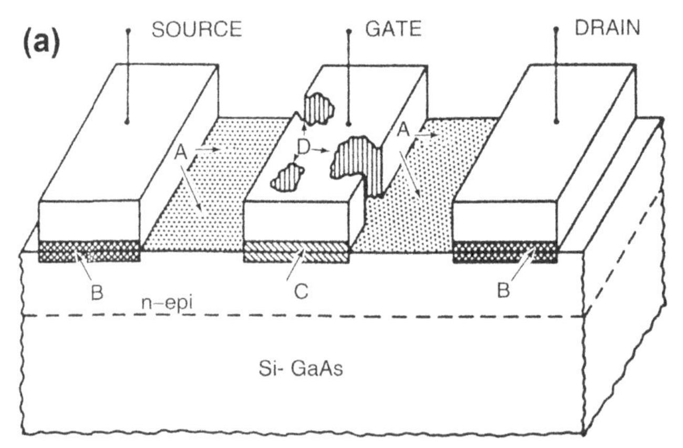


**Figure 5.11** Views of damage to metal contacts in GaAs transistors. (a) Localization of damage in a MESFET at A. surface region between gate and source (and drain) metallization, B. source and drain ohmic contacts, C. Schottky gate contact and underlying channel, and D. gate metallization. *(From Ref. [9].)* (b) Damage due to a high-power pulse applied to the gate. A. interdiffusion of Au through Cr to active channel, B. thermal runaway, C. metal–GaAs interdiffusion. *(From Ref. [9].)* (c) Interelectrode short-circuit bridge. A. material depletion, B. initiation of migration, C. drain (AuGeNi), D. edge of gate recess, E. gate (Al), F. traces of material migration, G. droplets due to material accumulation. *(From Ref. [10].)*

and corrosion damage of the metallization. Damage manifestations of several of these are schematically shown in Figure 5.11 indicating mass transport effects between and around the contact metals. Lifetimes of all GaAs devices (e.g., MESFETs, MODFETs, LEDs) depend strongly on the

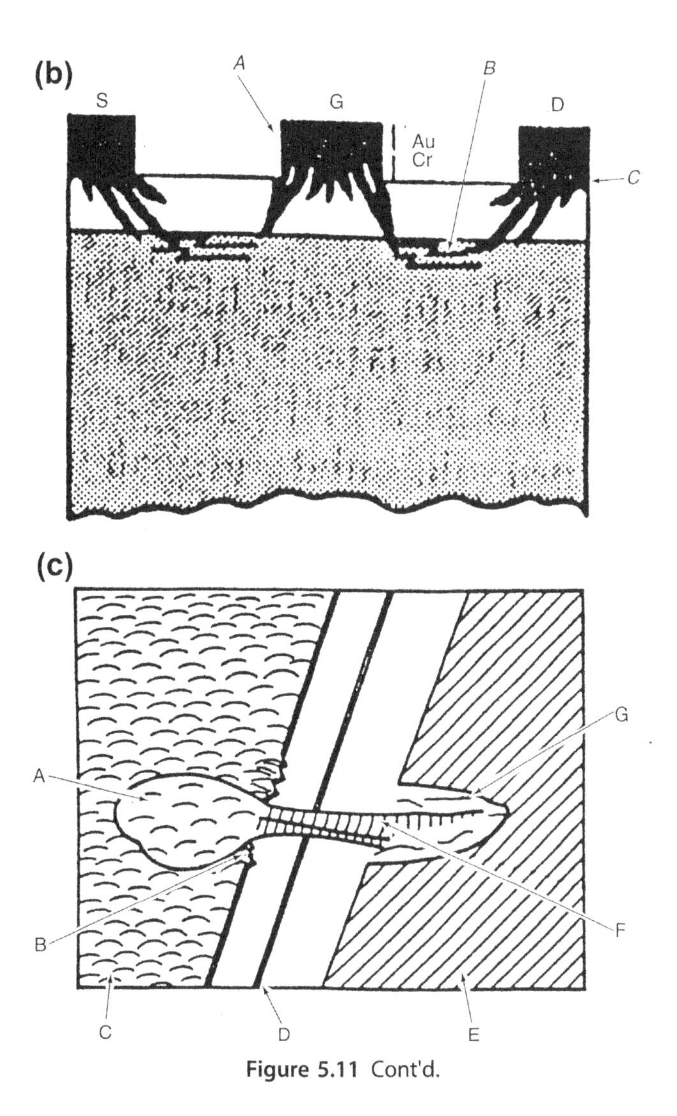


**Figure 5.11** Cont'd.

stability of the metals employed in ohmic, Schottky, and heat-sinking contacts [9–12]. The three main contact-degradation effects include:

1. Long-term interdiffusion within metal sandwich structures;
2. Electric field-induced metal migration between closely spaced neighboring electrodes; and
3. Metal migration through narrow conductors carrying large current densities (EM).

#### *5.4.4.2 Schottky Contacts*

These contacts must adhere well to GaAs, have a high conductivity (low sheet resistance), be amenable to lithographic processing, and be stable with respect to interdiffusion, corrosion, and EM. Only a few metals possess this complement of properties, with Al and Ti being the most commonly employed. In the case of Al, two specific failure mechanisms have been identified, namely Al/GaAs interdiffusion and EM. Accompanying the degradation at the Al/GaAs interface is an undesirable increase in electron barrier height, typically from 0.7 to 0.9 eV. While Al/GaAs contacts are thermally stable, the same is not true for Ti, which readily oxidizes. This makes wire bonding difficult, a problem solved by an Au overlayer. However, the Au/Ti/GaAs contact structure is unstable at temperatures as low as 200 °C. The solution here as in Si technology is to interpose diffusion barriers that minimize, but do not totally eliminate, reactions between Au and Ti. Metals such as Pt, W, Pt/Cr, and Pd have been used for this purpose.

Diffusion of Au into GaAs after penetrating through metal-barrier layers is a primary failure mechanism in Au-based gate metallizations. When this happens, the metal/semiconductor interface advances into the channel, narrowing its effective thickness. In addition, diffused metal alters the net donor concentration in the channel and modifies device behavior.

Typical activation energies for degradation of Al contacts range from 0.8 to 1.0 eV. Reflecting greater stability, the corresponding activation energies for Au-based contacts are larger, and values of 1.5–1.8 eV have been reported.

#### *5.4.4.3 Ohmic Contacts*

Parasitic contact resistance at source and drain ohmic contacts influences a number of FET device parameters (transconductance, saturation current, etc.). The most widely used metallization for ohmic contacts to GaAs is derived from an Au–Ge eutectic alloy covered by a Ni layer to improve wettability. The resulting AuGeNi possesses low contact resistance required for high-frequency and high-power applications. Increases in contact resistance have been attributed to four effects:

1. Outdiffusion of Ga and As from GaAs and subsequent reaction with the contact metal. A nonstoichiometric defect-rich region is then left behind under the contact.
2. Indiffusion of Ga.
3. The doping density in the semiconductor is reduced by indiffusion of Au and Ni.

**4.** Formation of NiAsGe, AuGa, and $Ni_2GeAs$ compound phases. This reduces the effective contact area.

Among the interesting manifestations of damage in unpassivated GaAs MESFETs are $As_2O_3$ polyp growths over ohmic contacts containing Au [13]. A model for their growth suggests that As is liberated as a by-product of GaAs dissociation by Au, a reaction that results in the formation of an AuGa alloy. Room temperature Kirkendall-like interdiffusion between Ga and Au facilitates the simultaneous migration of As to the contact surface, where it oxidizes. Passivation protection against moisture minimizes the occurrence of such damage.

#### *5.4.4.4 Gate Sinking*

Diffusion of gate metals into the underlying GaAs effectively reduces the thickness of the active layer, an effect known as "gate sinking." The change in device pinch-off voltage ($V_p$), a degradation mode suffered by GaAs MESFETS, is a consequence of gate sinking [14]. A diffusion model to explain this effect is based on the metal migration that occurs. For one-dimensional diffusion of the Schottky contact metal into GaAs we have for constant surface concentration, $C_s$ (Eqn (5.3)),

$$C(x,t) = C_s \mathrm{Erfc}\left[\frac{x}{(4Dt)^{1/2}}\right]. \tag{5.13}$$

Assuming that the metal front is at depth $X_d$ with a concentration of $\sim 10^{-4}$, then $X_d = 2.75 \times 2\,(Dt)^{1/2}$. Theory shows that the pinch-off voltage in an active layer of thickness $W$, reduced in time $t$ by $X_d$, is given by

$$V_p(t) = -\frac{qN(W - X_d)^2}{2\varepsilon} + V_o, \tag{5.14}$$

where $N$ is the effective layer doping level, $q$ is the electronic charge, $\varepsilon$ is the effective dielectric constant, and $V_o$ is the built-in voltage. Relative to the initial pinch-off voltage, $V_p(0) = \frac{-qNW^2}{2\varepsilon} + V_o$, the change is given by

$$\Delta\left[V_p\right] = -qNX_d W/\varepsilon, \tag{5.15}$$

where it is assumed that $V_o$ is constant and $X_d/(2W) < 1$.

Therefore, it is predicted that the pinch-off voltage drifts parabolically with time according to

$$\Delta[V_p(t, T)] = \frac{5.5qNW}{\varepsilon}(D_o t)^{1/2} \exp - \frac{E_d}{2kT} \qquad (5.16)$$

by combining previous equations and decomposing the temperature-dependent diffusivity into constants $D_o$ and $E_d$. Comparison of Eqn (5.16) with the measured behavior of MESFETs is made in Figure 5.12(a) and lends support to the proposed diffusion model.

#### *5.4.4.5 Failure in MODFETs*

Mechanisms for MODFET failure have been summarized by Anderson [11] and include those mentioned earlier. Contact degradation and gate interdiffusion are the most common. Thus gate sinking from Schottky contacts, in addition to effectively thinning the AlGaAs layer, alters Schottky barrier heights. Interdiffusion of ohmic contact metal into GaAs apparently causes electron trapping, establishing a barrier to current flow. Deconfinement of the two-dimensional electron gas (Section 2.7.4), together with reduction of carrier mobility and saturation velocity, are other suggested consequences of the diffusion of either contact metal, dopants, or semiconductor constituents. The parabolic kinetics of change in barrier height and source resistance shown in Figure 5.12(b) and (c) lend further credence to diffusion-based degradation mechanisms. These effects are not solely limited to GaAs-based HEMTs. A recent study on 0.1 μm gate length InP HEMTs [15] revealed a transconductance decrease as a result of gate sinking. The kinetics of degradation were parabolic with time, and defined by an activation energy of 1.26 eV.

## 5.5 EM PHYSICS AND DAMAGE MODELS

### 5.5.1 Introduction

Electromigration, a phenomenon not unlike electrolysis, is characterized by the migration of metal atoms in a conductor through which large direct-current densities ($j_e$) pass. It occurs in pure metals as well as in alloys and has been suggested as a mechanism for failure of tungsten filaments in light bulbs. Lamp filament failures will be addressed at the end of the chapter. EM was first identified as a reliability problem in IC metallizations more than 30 years ago. Interest in EM has persisted to the present day, where it still remains a troublesome reliability issue, as reflected by the many publications and conference sessions devoted to it each year.

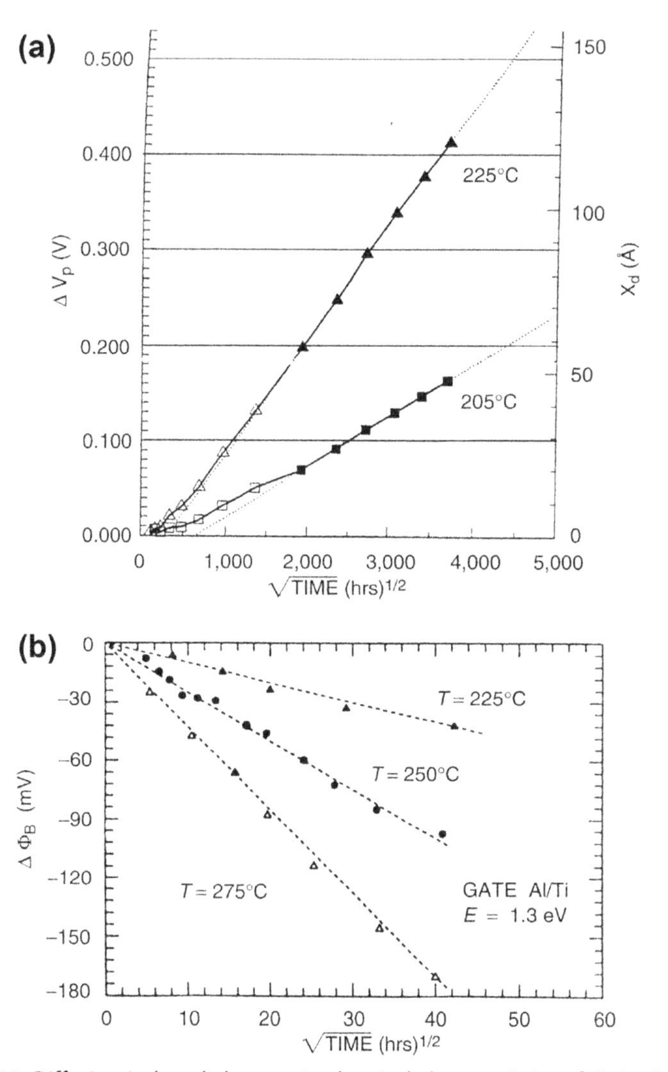


**Figure 5.12** Diffusion-induced changes in electrical characteristics of GaAs devices. In all cases, data plotted versus square root of time reveals parabolic kinetics. (a) Pinch-off voltage ($V_p$) drift for MESFETs aged at 205 and 225 °C. *(From Ref. [14].)* (b) Decrease in barrier heights ($\Phi_B$) of AlGaAs/GaAs HEMTs aged at 225, 250, and 275 °C. (c) Increase in parasitic source resistance ($R_s$) of AlGaAs/GaAs HEMTs aged at 225, 250 and 275 °C. *(From Ref. [9].)*

Of the order of hundreds to several thousand meters of narrow, thin-film interconnections of Al less than 1 μm in width are now present on typical IC chips to connect devices to one another and to contact pads. For a 700 million-transistor microprocessor chip, average interconnect lengths

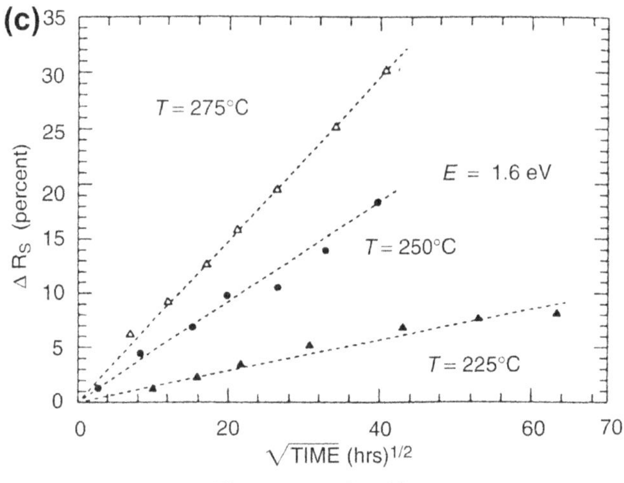


**Figure 5.12** Cont'd.

of ~30 μm are suggested, but in actuality a distribution of both shorter and longer lines is present. Critical interconnections such as clock, control, and data lines between the processor and cache may run the entire length of a chip and be 1–2 cm long. Optimum operation of devices frequently requires high signal currents. Because conductor cross-sectional areas are now quite a bit less than one square micron, current densities can reach high levels even at modest applied voltages. In fact, Al interconnects are designed to carry about $10^5$ A/cm$^2$ and can sustain up to ~$10^7$ A/cm$^2$ without immediately opening. This is a very large current density and would cause an isolated bulk Al conductor to rapidly melt, as we shall see in the calculation of Section 5.5.3. However, catastrophic failure does not occur immediately in thin-film IC interconnections because they are effectively bonded to a massive heat sink that efficiently conducts away the joule heat that is generated.

Passage of high current densities through interconnects causes time-dependent mass-transport effects that are manifest as surface morphological changes. Resulting conductor degradation includes mass pileups in hillocks and whiskers, void formation and thinning, localized heating, and cracking of passivating dielectrics; examples of the effects of EM are shown in Figure 5.13. In bootstrap fashion these damage processes accelerate to the point where open circuiting terminates conductor life. There is thus a basis for a corollary of one of Murphy's laws: "A million-dollar computer will protect a 25-cent fuse by blowing first."

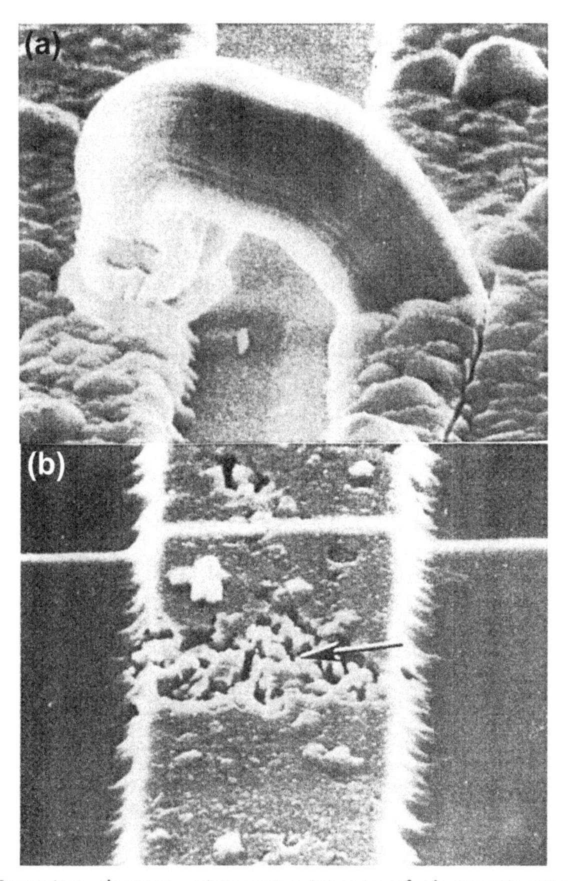


**Figure 5.13** Scanning electron microscope images of electromigration damage in aluminum interconnects. (a) Hillock extruded through a hole in overlying glass layer causing a short. (b) Void formation across the stripe. *From Ref. [16]. For other examples of electromigration damage, see Figure 5.26.*

As the feature size of dimensions has shrunk the need for a higher conductivity metal has resulted in the choice of copper as a replacement for Al–Cu. Copper has unique requirements, namely it must be encased in a diffusion barrier on all sides to keep it from diffusing into $SiO_2$ and the Si devices, as well as prevent corrosion and EM extrusions. The use of lower dielectric constant (k) interlayer dielectrics (ILD), which are softer than $SiO_2$, has further complicated the occurrence of stress and voiding due to these material properties. As we discussed in Chapter 2, the dual damascene process is used to fill prelined trenches with copper, and the excess copper is removed using chemical mechanical polishing (CMP). We will present the details for copper electromigration (EM) and stress voiding (SV) in Section 5.8, after discussing Al–Cu, which is still used in all manufacturing facilities which are not practicing state-of-the-art technology. A key point is that as technology evolves it takes a very long time for older technology to be replaced in existing facilities. In particular, new technology is usually placed in new facilities and the old facilities are used to continue making older products, until the cost of manufacture will no longer sustain the selling price. This is called end of life for a product made in a given older technology and is the death knell for availability of that product. Often firms that depend on that product are given the option of a one time final buy to help them transition to a replacement.

## 5.5.2 Physical Description of EM

### *5.5.2.1 Atomic Model*

At a fundamental level, EM involves the interaction between current carriers and migrating atoms. The nature of the interaction is not entirely understood, but it is generally accepted that in an electron conductor, electrons (of charge $q$) streaming toward the anode can impart sufficient momentum to atomic ion cores upon impact to propel them into neighboring vacant sites. Depicted in Figure 5.14 is the sequence of events. The electric field, which exerts an electrostatic pull on positively charged ion cores toward the cathode, is simply insufficient to withstand the oppositely (anode) directed "electron-wind" force at high current densities. Therefore,

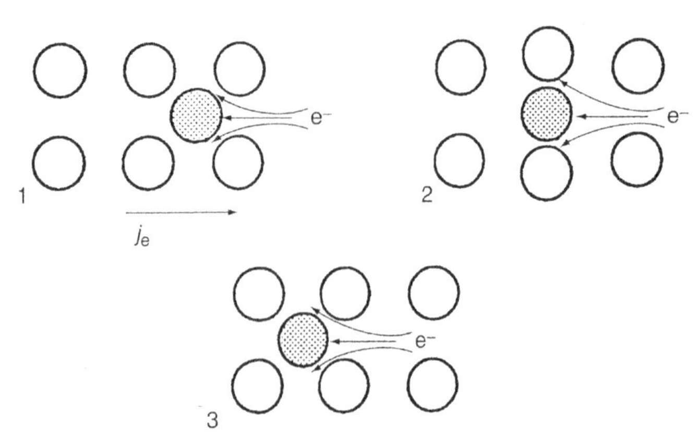


**Figure 5.14** Atomic model of electromigration. The electron wind at large current densities imparts sufficient momentum to ion cores to propel them into neighboring vacancies.

there is a *net* force ($F$) that biases the normally random motion of an atom precariously perched in a saddle-point configuration between two flanking half-vacancies. This force, given by

$$F = Z^{*}qE = Z^{*}q\rho j_{\mathrm{e}}, \tag{5.17}$$

is sufficient to push the ion over the energy barrier, so that it preferentially migrates toward the anode. In this equation the electric field $E$ is related to the current density through the resistivity $\rho$, i.e.,$E = \rho j_{\mathrm{e}}$, and $Z^{\star}$ is the *effective* ion valence. Experiment reveals $Z^{\star}$ to be negative in sign and has a magnitude far in excess of typical chemical valences. There is wide scatter in reported magnitudes of $Z^{\star}$ in different metals. For Al a typical value is $-10$, while values several times larger have been reported in Au and Sn.

### *5.5.2.2 Macroscopic Equations of EM*

Even though there are no concentration gradients present in pure metals, diffusional atomic movement by EM does in fact occur. How is this possible, and how is the apparent paradox resolved? Earlier, in Section 1.3.3, notions of atom transport were enlarged to include the effect of generalized forces such as electric fields, and the very important Nernst–Einstein equation (Eqn 1.6) was derived. We start here with an operating definition of the field-induced mass flux that results in a diffusional drift of migrating atoms with velocity v, namely.

$$J = C_{\mathrm{v}}, \tag{5.18}$$

where $C$ is the atomic concentration (atoms/cm$^3$). The total mass-flux $J$ (atoms/cm$^2$-s) through the conductor due to both atomic diffusion (Eqn (5.1)) and drift (Eqns. (5.18) and (5.16)) is

$$J = -D\frac{\mathrm{d}C}{\mathrm{d}x} + \frac{CDF}{RT}, \tag{5.19}$$

where $F$ is the generalized force and $RT$ has the usual meaning. For our particular application, $F = Z^{\star}q\,E$ (Eqn (5.17)), and if the concentration-gradient term is neglected,

$$J = \frac{CDZ^{*}q\rho j_{\mathrm{e}}}{RT}. \tag{5.20}$$

EM is thus characterized at a fundamental level by the material constants $Z^{\star}$, $D$, and $\rho$.

The passage to nonsteady-state mass transport is made via the equation of continuity; i.e., $\frac{\partial C}{\partial t} = -\nabla \cdot J$. Film damage effects are caused by a nonvanishing divergence of $J$ given by $\frac{\partial C}{\partial t} = -\frac{\partial J}{\partial x} - \left(\frac{\partial J}{\partial T}\right)\left(\frac{\mathrm{d}T}{\mathrm{d}x}\right)$, or

$$\frac{\partial C}{\partial t} = -\frac{\partial}{\partial x}\left\{\frac{CDZ^* q\rho j_e}{RT}\right\} - \frac{\partial}{\partial T}\left\{\frac{CDZ^* q\rho j_e}{RT}\right\}\frac{\partial T}{\partial x}. \tag{5.21}$$

Thus when $\mathrm{d}C/\mathrm{d}t < 0$, mass depletion occurs and voids form. On the other hand, mass accumulates in growths and hillocks when $\mathrm{d}C/\mathrm{d}t > 0$. However, when $\mathrm{d}C/\mathrm{d}t = 0$ there is no change in atomic density, and no damage occurs. If alloys are involved, the appropriate diffusion term $\left(D\frac{\partial^2 C}{\partial x^2}\right)$, normally absent in the case of a pure conductor, would be included.

Many of the time-dependent failure models that we will consider later are based either on portions of Eqn (5.21) or on modifications of it. The first term on the right-hand side arises from local structural and property gradients at constant temperature. Temperature gradients and thermally dependent properties that exist along powered film stripes contribute to the second term on the right. The resultant damage is schematically depicted in Figure 5.15.

## 5.5.3 Temperature Distribution in Powered Conductors

Before addressing material transport during EM it is instructive to consider the separate issue of heat transfer in powered metal wires. Calculation of the

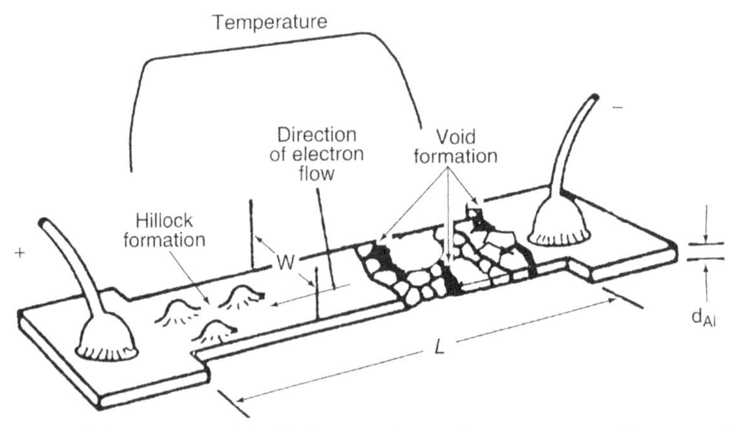


**Figure 5.15** Schematic model of damage in an interconnect subject to electromigration. Hillocks and voids develop in response to temperature gradients superimposed on the directed electron current flow.

temperature attained is a useful prelude to estimating the current passed prior to failing. Such an analysis [17,18] is applicable to the behavior of fuses that cause an open circuit by melting.

Let us first consider current $i$ (A) passing through a wire of length $L$ and cross-sectional area $A$ whose electrical resistivity is $\rho$ ($\Omega$-cm) and thermal conductivity is $\kappa$ (W/cm-K). Assuming no heat loss occurs by radiation or convection, the nonsteady-state heat flow equation describing the temperature ($T$) at any position ($x$) may be written as

$$\delta c \frac{\partial T}{\partial t} = \kappa \frac{\partial^2 T}{\partial x^2} + \frac{i^2 \rho}{A^2}, \tag{5.22}$$

with $\delta$ and $c$ the wire density and heat capacity, respectively. (Note that $\kappa/\delta c$ is the thermal diffusivity, which has units of $cm^2/s$.) In addition to the two familiar terms in Eqn (5.22), the rate of heat generation must be considered. In a short length of resistor wire d$x$, this amounts to $i^2\ \rho\ dx/A$. Thermal and electrical conductivities ($\sigma$, where $\sigma = 1/\rho$) of metals are theoretically connected by the well-known Weidemann–Franz law, expressed by $\sigma = \kappa/LT$, where $L$ is the Lorenz number, a constant equal to $2.44 \times 10^{-8}$ W-$\Omega$/$K^2$. In the steady state, $dT/dt = 0$, and therefore Eqn (5.22) now assumes the far simpler form of the harmonic oscillator equation,

$$\frac{dT^2}{dx^2} + B^2 T = 0, \tag{5.23}$$

where $B^2 = i^2 L/(A^2 \kappa^2)$.

The physical problem we now address is a conductor whose ends are maintained at ambient temperature $T_o$. Wirebonds that connect silicon chips to metal lead frames approximate this geometry. By solving Eqn (5.23), it is not difficult to show that the maximum temperature ($T_{max}$) reached, which occurs at the midspan (L/2), is given by

$$\frac{T_{max}}{T_o} = \frac{1}{\cos\frac{BL}{2}} = \frac{1}{\cos\left(\frac{iL^{1/2}L}{2A\kappa}\right)}. \tag{5.24}$$

Wire failure occurs when $T_{max}$ reaches the melting point $T_M$, in which case the critical current density $j_{crit}$ ($i/A$) is given by

$$j_{crit} = \left(2\kappa/L^{1/2}L\right)\cos^{-1}\left(T_o/T_M\right). \tag{5.25}$$

Substituting the following values for Al, $\kappa = 240$ W/m-K, $L = 10^{-2}$ m, $T_M = 933$ K, and $T_o = 298$ K, yields $j_{crit} = 3.83 \times 10^8$ A/m$^2$ or $3.83 \times 10^4$A/cm$^2$. As noted earlier, thin-film stripes are able to sustain passage of considerably larger current densities because the substrate serves as a massive heat sink for the conductor.

More involved nonsteady-state heat flow analyses that include radial heat loss from powered wires have also been published [19]. Such calculations of transient temperature behavior have proven useful in designing operating conditions for wire bonding processes that avoid excessive heat damage due to passage of high currents.

What are the implications of temperature gradients on EM in thin-film stripe metallizations? Assuming that the above calculation applies, we may expect a sine-like temperature distribution, i.e., $T_o$ at the ends, $T_{max}$ at the center. Therefore, atoms originating at the negative end of the stripe migrate not only in the direction of electron flow, but also into a hotter region of the stripe. They accelerate because of the strong thermal activation of the ion mobility and in moving ever faster leave mass-depleted regions behind. At the anode end the reverse happens. Now the atoms are driven from a hot region, where they move rapidly, to the cooler end, where they decelerate. Atoms pile up, and mass tends to accumulate there.

These expectations are borne out by the overall negative sign of second term in Eqn (5.21). If at the cathode end $dT/dx$ is negative, the temperature derivative of the quantity in parentheses is also negative; i.e., $Z^\star < 0$, and the rise in D with temperature swamps any countertrend in the other terms. Therefore, $(\partial C/\partial t) < 0$, and mass depletions are anticipated. The heated stripe is like a highway that goes from one lane to two lanes and then back to one lane. Cars that initially travel slowly pick up speed in the two-lane portion only to slow down again. In concert, the intercar spacing is small initially, then widens (as gaps or voids develop), and finally decreases as cars enter a jam once again (where hillocks form). Some of these features are illustrated in the model of void and hillock damage depicted in Figure 5.15.

## 5.5.4 Role of Stress

Metal thin films are invariably deposited in a state of stress. Therefore, we may well ask what role stress plays in EM. Just as an electric field causes atomic drift, so too a gradient of stress ($\sigma$) acts as a generalized force to induce atomic motion. Thus, atoms preferentially migrate from compressively ($\sigma$ more negative) stressed regions and accumulate at locations stressed in tension ($\sigma$ more positive), while vacancies diffuse the other way. The resulting stress gradient causes a backflow of matter, and, as we shall see, this effect has significance in short conductors. Thermodynamics argues that the free energy per atom in a stress field, or chemical potential ($\mu$) of an atom, depends on the stress and is equal to $\mu = -\sigma\Omega$, where $\Omega$ is the atomic volume ($cm^3$/atom). Furthermore, the force per atom in a stress field is the negative gradient of the chemical potential, or

$$F = \Omega \frac{d\sigma}{dx}. \tag{5.26}$$

Blech [20] demonstrated the role of stress in a clever experiment involving electrotransport in short Au lines deposited on longer more resistive molybdenum film conductors. Current through the latter shunted to the Au as shown in Figure 5.16, causing atoms to electromigrate from the cathode to anode side and pile up there. The displaced cathode edge enabled a direct measurement of the atomic drift velocity. A surprising result was the cessation of edge migration when stripes were shorter than a critical length $L_c$. The effect arises from a stress-induced backflow of atoms

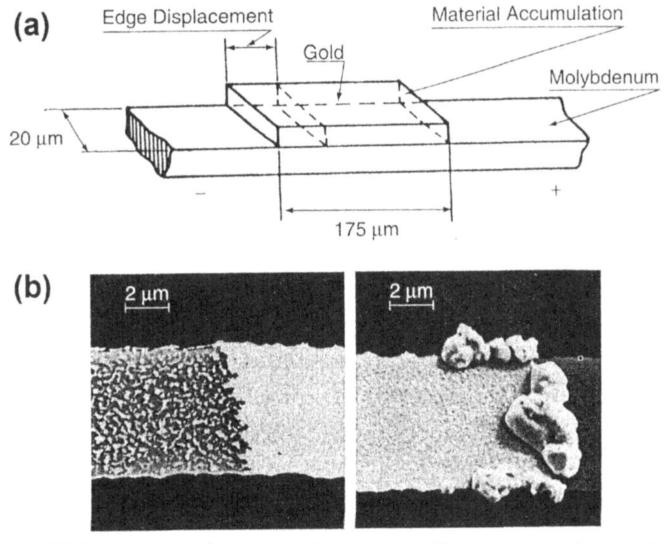


**Figure 5.16** (a) Experiment demonstrating stress effects during electromigration in short conductors. (b) Scanning electron micrographs of gold stripe powered at $10^6$ A/$cm^2$. Cathode end at left is depleted of gold, while anode end at right contains hillock growths. *From Ref. [20].*

that counters the primary EM flux. Under these conditions, the two mass-transport driving forces balance, or

$$Z^* qE = \Omega \frac{d\sigma}{dx}. \tag{5.27}$$

Substituting, $d\sigma/dx \approx \sigma_c/L_c$ and $E_c = \rho j_e(c)$, where critical values are denoted by c, the result that emerges is

$$L_c j_e(c) = \frac{\Omega \sigma_c}{Z^* q \rho}. \tag{5.28}$$

Significantly, the shorter the stripe length, the larger the current density required to sustain electrotransport. This effect must be accounted for during lifetime testing because otherwise-unduly pessimistic life predictions may result for short stripes.

Later, in Section 5.7, the role of stress in voiding phenomena, a reliability problem of current concern, will be explored.

## 5.5.5 Structural Models for EM Damage

### *5.5.5.1 Electromigration in Grain Boundaries*

Thus far we have tacitly assumed that powered interconnects have a homogeneous structure. But we know that the grain structure of thin films can vary widely depending on deposition conditions and that structure has a profound effect on EM damage. Powered single-crystal Al stripes, for example, have been shown to exhibit virtually "infinite" life. The primary reason for this is that diffusion and drift of atoms in GBs far exceeds transport through the bulk of the grains. In fact, the activation energy for the former is only about half that of the latter (compare Eqn 5.7(a) and (c)). At low temperatures, in particular, it is an excellent assumption to view the stripe as an interconnected GB network and neglect mass transport everywhere except within the network channels. Thus, fine-grained films have many channels, or paths, for atomic migration, while coarse-grained films have many fewer such avenues for transport.

### *5.5.5.2 Structural Barriers*

Interfaces between adjoining regions of an interconnect where properties differ may be regarded as structural barriers. The difference can be very large or barely perceptible; it can take the form of variations in grain orientation, grain size, or differences in chemical composition, such as at a contact between the metallization and semiconductor. In addition, differences in atomic diffusivity, effective valence and vacancy generation, and trapping characteristics can serve as barriers capable of creating local mass-flux divergences. Furthermore, the situation is not static. With EM-induced heating, boundaries may move, grains may recrystallize, and subtle structural and chemical evolution may occur in the conductor.

To appreciate structural barriers, consider what happens when a fine-grained stripe segment abuts one of larger grain size. For one direction of current flow, atoms that migrate readily through numerous GBs suddenly come upon a region with a dearth of GB paths that lead away. Such a barrier results in a pileup of matter. Similarly, void formation can be expected in regions where more atoms leave than enter. It is not difficult to model what might be expected at the barrier interface ($x = 0$) between stripe region 1, extending from $x = -\infty$ to $x = 0$, and adjoining region 2, ranging from $x = 0$ to $x = \infty$ [21]. Suppose differing vacancy GB diffusivities, effective valences, current densities, resistivities, vacancy lifetimes ($\tau_1, \tau_2$), temperatures, etc., distinguish each region. Under isothermal conditions, and neglecting stress or other effects that influence vacancy motion.

$$\frac{\partial C_i}{\partial t} = D_i \frac{\partial^2 C_i}{\partial x^2} - \left(\frac{DZ^* q \rho j_e}{RT}\right)_i \frac{\partial C}{\partial x} + \frac{C_{io} - C_i}{\tau_i}, \tag{5.29}$$

where $i = 1, 2$, $C$ is the vacancy concentration, and $C_o$ is the equilibrium vacancy concentration. The first two terms on the right express the diffusional and field-driven mass contributions; only the last term is unfamiliar and represents the vacancy generation (or annihilation) rate. Steady-state conditions typically prevail within minutes, so $\partial C_i / \partial t$ can be safely dropped. Imposing the conditions of a continuous concentration ($C_1 = C_2$) and flux $\partial C_1 / \partial x = \partial C_2 / \partial x$ at $x = 0$ enables $C$ to be calculated at the interface. The results of a sample calculation for the vacancy supersaturation $S$ ($S = (C - C_o)/C_o$) versus distance are shown in Figure 5.17 for the case where grain sizes ($d_1$, $d_2$) differ across the interface.

Although a vacancy supersaturation (or subsaturation) may be predicted, the crucial question is whether it is sufficient to cause nucleation of voids (or growths). This is a very complicated issue that depends on the chemical and physical nature of the heterogeneous void nucleation site. Unfortunately, it is not known with any certainty what level of vacancy supersaturation is required to generate damage. The problem of void nucleation and its time dependence during EM has been addressed [22] employing concepts of classical or capillarity theory. A significant feature

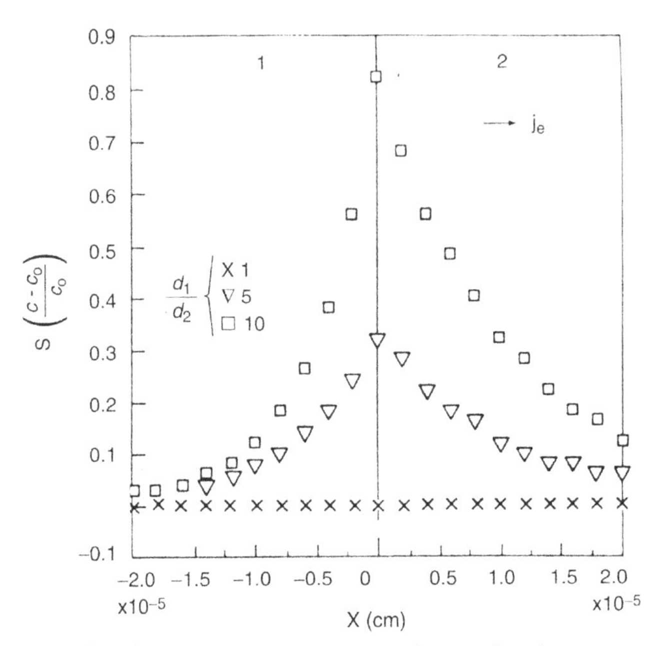


**Figure 5.17** Calculated vacancy supersaturation at the interface between regions with a grain-size discontinuity. This condition is mathematically approximated by a difference in number of GB transport paths within the two regions. Note: $\bar{x}$ is the vacancy sink distance ($\bar{x} = (2D\tau)^{1/2}$), $j_e = 1 \times 10^6$ A/cm$^2$ and $T = 400$ K. *From Ref. [21].*

of the analysis is the prediction of a void incubation time that has, in effect, been observed [23].

With increasingly shrinking interconnect stripe widths, broad-area or blanket metallizations leave greater numbers of GB "triple points" and grains lined up in a bamboo-like structure after patterning. The confluence of three GBs at a triple point is schematically depicted in Figure 5.18(a), and shown in a portion of the Al interconnect reproduced in Figure 5.18(b). Divergences in mass transport can occur, assuming that migrating atoms are solely confined to GB channels. It is apparent that when fewer atoms leave the triple point than enter it, a mass accumulation is favored. But, when more atoms leave the triple point than enter it, voids form. Less EM damage is anticipated when fewer GBs are oriented parallel to the axis of current flow. Thus we may expect conductor life to sequentially increase as grains change from a bimodal size distribution to small, large, near-bamboo and hyperbamboo, structures.

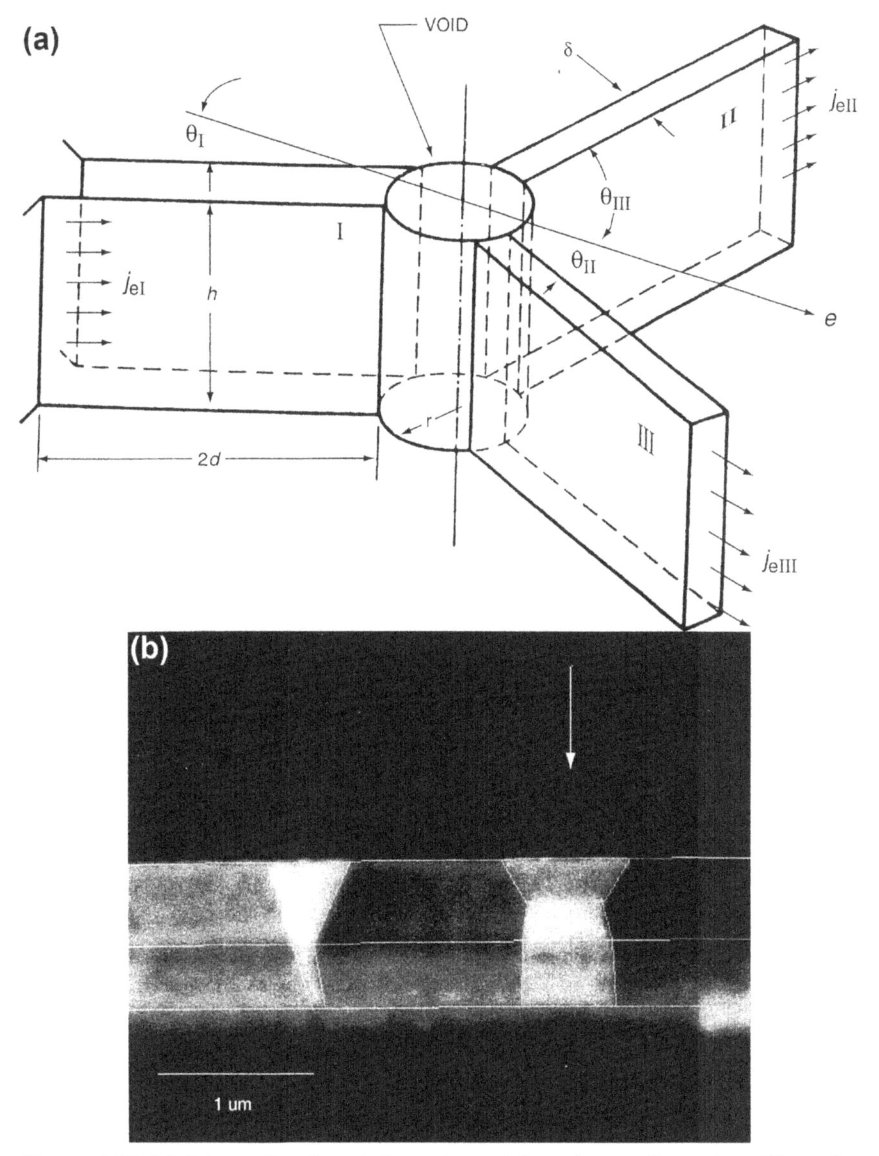


**Figure 5.18** (a) Schematic of grain-boundary triple point configuration. When the electron current polarity is reversed, the electromigration damage in the form of mass accumulation and voids at triple points also reverses. *(From Ref. [21])*. (b) Actual triple point grain configurations in an aluminum interconnect (800 nm wide, 400 nm thick) imaged by focused ion beam (FIB) microscopy. *Courtesy of D. Barr, Lucent Technologies, Bell Laboratories Innovations.*

#### *5.5.5.3 GB Grooving*

As another example of a structural barrier that may generate void damage we consider the phenomenon of GB grooving [24]. In Figure 5.19(a) the cross-section of a film at a junction between two grains is shown. Although such a structure may be stable for all practical purposes, it is, nevertheless, not in thermodynamic equilibrium. The reason is that the surface ($\gamma_s$) and GB ($\gamma_b$)

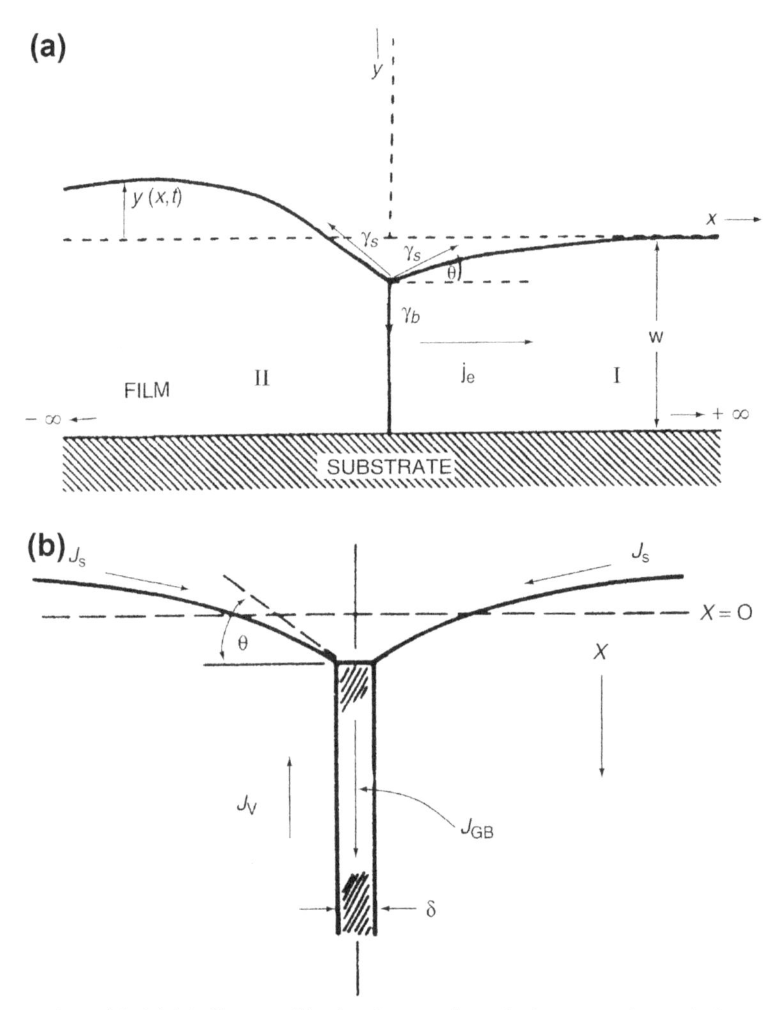


**Figure 5.19** (a) Initial film profile is shown dotted. Asymmetric grain-boundary grooving as a result of electromigration. *(From Ref. [24].)* (b) Schematic of electromigration-induced vacancy flux at grain-boundary triple point. The electromigration-induced surface ion flux ($J_S$) compensates for the vacancy flux ($J_V = -J_{GB}$) to preserve the dihedral angle. *(From Ref. [21].)*

interfacial tensions are not mechanically balanced in the vertical direction. But, through thermal GB grooving processes as a result of mass transport, equilibrium between the involved tensions is achieved such that $2\,\gamma_s \sin\theta = \gamma_b$. If a unidirectional surface flux of atoms now flows past the junction as a result of EM, mechanical equilibrium at the groove is upset. As atoms preferentially accumulate on one side of the boundary, an asymmetry in film height develops and the dihedral angle sharpens. In order to reestablish equilibrium the groove widens and deepens. In the next instant of time the equilibrium is upset again, and so on. This perpetual drive to equilibrium punctuated by intermediate moments of instability ultimately causes the groove to reach the substrate where a void or open circuit is produced. Such a mechanism occurs at any level of vacancy supersaturation, in contrast to the damage model presented in Section 5.5.5.2. Analytical expressions for the kinetics of grooving and surface topography of the grains have been published.

Finally, we may consider grooving to cause damage at a GB triple point by an EM-induced vacancy flux, as shown in Figure 5.19(b). This case has been modeled [21], and the length ($y$) of the GB hole produced as a function of time ($t$) is given by

$$y(t) = \frac{M(Bt)^{3/4}}{\sqrt{2}\Gamma(7/4)}, \tag{5.30}$$

where $M = C_b \delta D_b Z^\star qE/2\nu_s D_s \Omega \gamma_s$, $B = D_b\, \gamma_s\, \nu_s \Omega^2/kT$, and $\Gamma$ is the gamma function. In these expressions b refers to the GB, s to the surface; $\nu_s$ is the effective surface ion concentration, and $\Omega$ is the atomic volume. Times calculated to produce damage are of the order of tens of hours for reasonable choices of the involved constants.

## 5.6 EM IN PRACTICE

### 5.6.1 Manifestations of EM Damage

All powered metals in ICs are potentially susceptible to EM effects. Interconnects, contacts, and vias are the three locations that are particularly prone to damage. Much of Section 5.5 was devoted to degradation and failure in interconnections; this discussion is by no means over and will continue. We start by discussing EM-damage manifestations in contacts and vias.

#### *5.6.1.1 EM at Contacts*

Large mass divergences are expected at powered contacts because any metal that leaves the semiconductor interface cannot be replenished. The subject of diffusional degradation at contacts in both Si and GaAs devices was treated earlier. However, in the presence of simultaneous EM, the damage accelerates. Ondrusek et al. [25] have studied the wear-out of both Al–Si and $TiSi_2$–Si contacts and found the latter to be considerably less susceptible to EM degradation. For the case of Al–Si contacts, previously introduced in Section 5.4.2 in connection with junction spiking, the following was observed:

1. Current-induced migration of Si through Al–1%Si metallization results in contact failure having activation energy of 0.9 eV.
2. Either accumulations or depletions of Si in the contact window occur depending on the direction of current flow. Electron flow into the contact causes a Si buildup at the contact; when the polarity of current flow is reversed, Si is driven into the Al–1%Si metallization. These effects are shown respectively in Figure 5.20.
3. Large reverse-bias $p-n$ junction leakage currents occur during Si depletion and the contact resistance increases when Si accumulates.
4. The mean time for contact failure depends on current density as MTTF $\approx j_e^{-n}$; $n$ is found to range from 5 to 11 as the current increases.

In contrast, the failure activation energy was 1.1 eV for the Si accumulation mode of failure in $TiSi_2$–Si contacts. However, for the depletion mode of failure the activation energy was 1.5 eV. Similar beneficial life extensions occur with the use of $TaSi_2$ contacts to Si [26]. Activation energies for EM failures ranged from 1.1 to 1.4°eV, suggesting damage and open circuiting in the matrix rather than GBs of the Al. Transition to a high-resistance, nonohmic state is a manifestation of failure.

EM damage is not limited solely to silicon devices but occurs in compound semiconductors as well. High-power, broad-band MMICs are

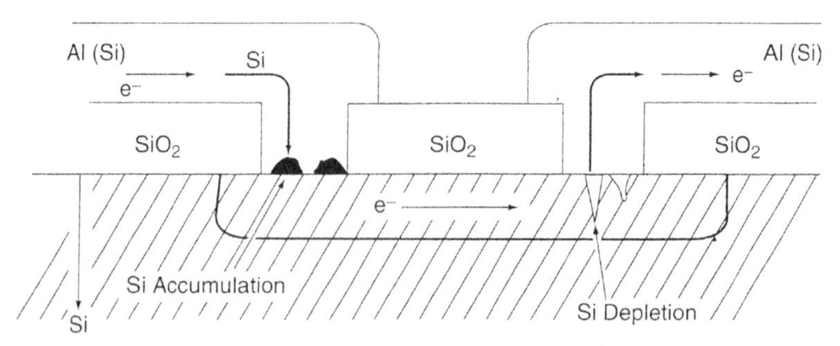


**Figure 5.20** Accumulation- and depletion-mode contact failures caused by silicon electromigration are dependent on direction of electron flow. Depletion-mode failures occur by aluminum spiking. *From Ref. [25].*

susceptible under conditions of accelerated testing. Failure analysis of these devices reveals that EM of the AuCr ohmic contact metallization occurs in the manner schematically depicted in Figure 5.11(c).

#### 5.6.1.2 EM at Vias

It has already been noted (Section 2.9.8) that IC architecture makes extensive use of successive interconnect levels and vias that enable current to flow between levels. Typical structures employ Al–Cu vias sandwiched between Al–Cu interconnects, but for higher reliability, tungsten plug vias are employed. In addition, thin diffusion-barrier films of TiN are frequently used to line vias. While the Al metallization tends to migrate, or "walk" off the via, W exhibits negligible diffusion or transport. Therefore, like contacts, via–interconnect interfaces are sites of potentially large mass divergences where Al depletion due to EM cannot be replaced. By carrying the shunted current when the Al has migrated, the TiN layer plays an important role in forestalling failure.

Degradation at vias is not only dependent on the composition and grain structure of the metals employed and the uniformity of film coverage, but also on the direction of current flow. Several models for EM damage at vias are shown in Figure 5.21 illustrating the numerous opportunities for mass divergence at contact interfaces within the complex via–conductor geometry [27,28]. Simple electrotransport of Al leaves voids at the bottom (Figure 5.21(a)) and top (Figure 5.21(b)) plug interfaces when electrons flow into or away from them. In addition, thinned metal conductors in the via sidewalls (Figure 5.21(c)) and corners induce current crowding and excessive joule-heating effects that lead to accelerated degradation. For these reasons it should not be surprising that failure is significantly more frequent in via structures than in interconnect stripes as the comparison in Figure 5.22 reveals.

### 5.6.2 EM in Interconnects: Current and Temperature Dependence

Irrespective of the kind of damage produced, the analysis of more than a quarter century of EM reliability testing of interconnections has distilled the following general relationship between mean time to failure (MTTF) and operating current density and temperature, namely

$$\mathrm{MTTF}^{-1} = Bj_{\mathrm{e}}^{n} \exp\left[-\frac{E_{\mathrm{e}}}{kT}\right]. \tag{5.31}$$

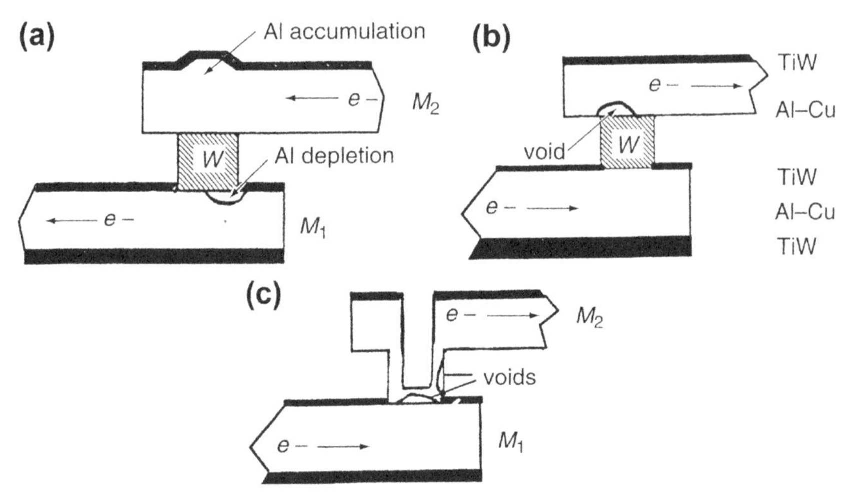


**Figure 5.21** Electromigration damage at vias between metallization levels $M_1$ and $M_2$. (a) Void formation in metallization at bottom of via plug. *(From Ref. [27].)* (b) Void formation in metallization at top of via plug. (*From Ref. [27].*) (c) Thinned metallizations at via bottom and sidewalls. *(From Ref. [28].)*

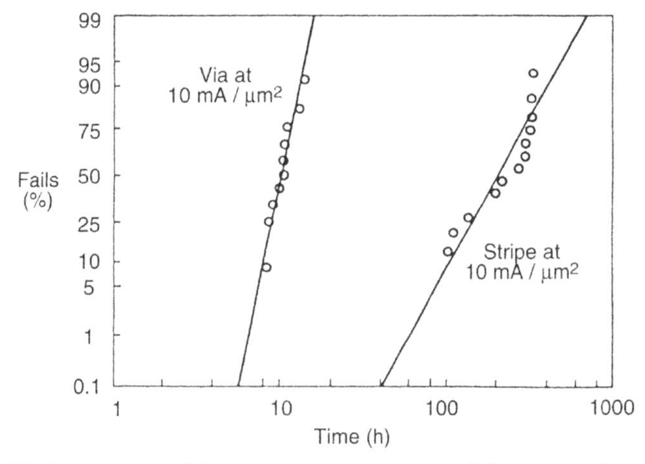


**Figure 5.22** Comparison of via and interconnect stripe failure rates. *From Ref. [29].*

This equation, where $B$ is a constant and $E_e$ is the activation energy for EM, is attributed to Black [30], who found that $n = 2$. In this or other versions, the equation plays a useful role in designing metallizations.

An argument that supports the $n = 2$ current density–dependence suggests that $\text{MTTF}^{-1}$ is proportional to the product of the number of voids in a stripe and the rate at which voids grow. Each of these factors depends on $j_e$, so that a $j_e^2$ dependence is predicted. An additional theoretical justification for this formula, based on stress effects, will be derived later, in Section 5.7.3.2. Finally, a recent experimental study in which temperature was measured over a 1-μm diameter spot size supports the Black equation for low values of current density [31]. However, at high current densities larger values of $n$ reflect synergistic effects beyond simple isothermal transport, e.g., due to thermal gradients, current crowding, stripe narrowing, etc. The data of Figure 5.23 display these trends and indicate that values as high as $n \approx 10$ occur when $j_e$ approaches $1 \times 10^7$ A/cm$^2$.

The thermally activated nature of EM damage is underscored by Eqn (5.31). It is important to realize, however, that $E_e$ and $n$ values extracted from lifetime testing, resistance monitoring, or other indirect measurements typically reflect a complex admixture of mass transport and damage processes: void nucleation, migration and coalescence, GB diffusion, surface diffusion, stress-assisted diffusion, thermomigration, grooving, extrusion, and so on. Increasing current densities enhance these processes in a

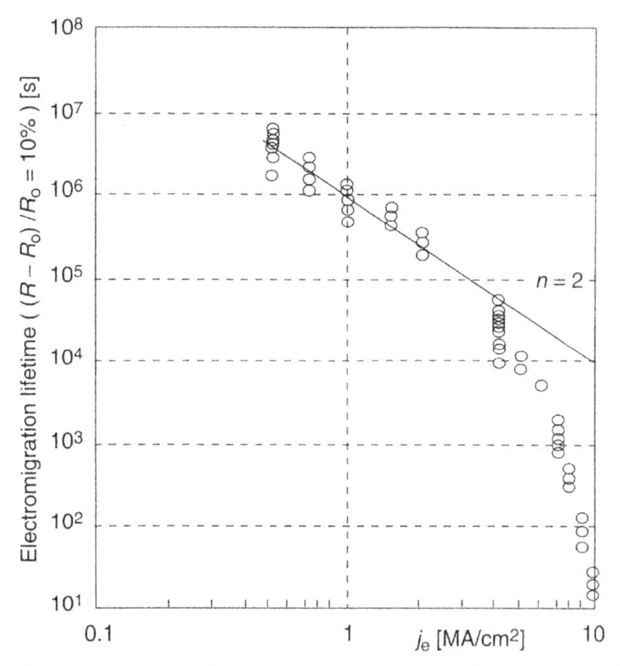


**Figure 5.23** Electromigration lifetime of aluminum metallizations as a function of current density. Criterion for life is a 10% change in resistivity. Below, a current density of $2 \times 10^6$ A/cm$^2$, $n = 2$. Larger values of $n$ are attained as $j_e$ approaches $1 \times 10^7$ A/cm$^2$. *From Ref. [31].*

nonlinear fashion. It is little wonder, then, that hot spots develop to further accelerate failure.

A range of activation energy values typically measured in assorted metal films used in electronics technology are entered in Table 5.1. Corresponding composition and grain size ranges employed are also listed. For more detailed information on the more than 60 studies from which the representative composite values of $E_e$ were selected, the reader should consult the indicated reference. The overwhelming bulk of the data pertains to Al and Al alloy metallizations, where the following observations apply:

1. The activation energy for bulk diffusion in pure Al is 1.4 eV. For films with large grain size there are few avenues for defect diffusion paths, and thus $E_e$ is found to be 1–1.2 eV. However, measured $E_e$ values in the neighborhood of 0.4–0.6 eV for the usual fine-grained films are indicative of GB mass transport-induced damage. Observed grain size-dependent activation energies are the apparent consequence of weighting admixtures of low bulk and high GB diffusivity contributions.
2. Additions of Cu to Al raise $E_e$ by a few tenths of 1 eV and extend MTTF lifetimes of Al accordingly. This important finding is the basis for the use of Al alloy metallizations containing approximately 0.3–5 wt% Cu. The effect of Cu is not completely understood, but

**Table 5.1** Electromigration in metal films

| Metal | Grain size (μm) | Failure criteria | $E_e$ (eV) | $n$ |
|---|---|---|---|---|
| Pure Al | 0.1–2 | $\Delta R/R$, open circuit | 0.35 to ~0.6 | 2–4 |
| | 6–8 | | 1.0–1.2 | |
| Al–0.3 to 5 wt% Cu | 1–6 | $\Delta R/R$, open circuit | 0.5–0.8 | 2 |
| Al–0.3 to 3 wt% Si | 0.3–7.6 | $\Delta R/R$, open circuit | 0.31–0.8 | 1.7–5.4 |
| Al–Cu–Si | 0.25–0.6 | $\Delta R/R$, open circuit | 0.25–0.86 | 1.7 |
| Au | 0.2–0.5 | $\Delta R/R$, open circuit, tracer, | 0.7–0.9 | 3.3–4 |
| Ag | | | | 0.3–0.95 |
| Cu | | | 1.1–1.4[a] | 1.1 |

[a]= added by authors.
Primarily adapted from Ref. [32].

the $Al_2Cu$ precipitates and dissolved Cu appear to interact with Al atoms in GBs and impede their motion. (In commercial Al–Cu precipitation hardening alloys these same $Al_2Cu$ precipitates strengthen the Al matrix.) It is also common to add 1 wt% Si as well to prevent Al–Si interdiffusion at semiconductor contacts.

3. The activation energy values are in general agreement with the diffusivity systematics presented in Eqn (5.7). Thus $E_e$ scales directly with higher melting point and larger grain size.

### 5.6.3 Effect of Conductor Geometry and Grain Structure on EM

The dependence of conductor length and width on EM life has been a subject of continuing interest. Assuming constant width, longer stripes simply have more potential failure sites (e.g., triple points) than shorter conductors, and therefore, they are observed to exhibit reduced lifetimes. But, when stripes are unusually short, stress-induced mass backflow effects (Sect. 5.5.4) surface to counter EM damage.

From the earlier discussion of structural barriers (Section 5.5.5.2) the desirability of having equiaxed grains is evident. Even better than equiaxed grains are interconnects with no GBs, i.e., a single-crystal Al film. Since it is impractical to deposit such a film, the best alternative is an interconnect with a bamboo grain structure that effectively behaves like a series stack of single crystals. The GBs are then oriented normal to the current flow. In such a case the only path available for mass transport is through the bulk grains, an unlikely avenue at low temperature. From a geometric standpoint, shrinking line dimensions make bamboo structures more probable. Interestingly, the effect of decreasing film linewidths on EM failure time reveals an unexpected beneficial twist, as revealed in Figure 5.24. Film life decreases as the linewidth is reduced from 4 to 2 μm, in accord with intuitive expectations that the narrower the stripe, the easier it is for damage to span across it. However, an encouraging increase in life occurs below 2 μm because of the bamboo grain development [33].

From the standpoint of processing, the film lifetime seems to be affected by method of film deposition; evaporated films are apparently longer lived than sputtered films, a fact perhaps attributable to a more random crystallographic texture in the case of the latter. Multiple GB orientations thus appear to encourage electrotransport damage effects. A suggested formula that neatly summarizes the dependence of mean lifetime on grain size ($S$),

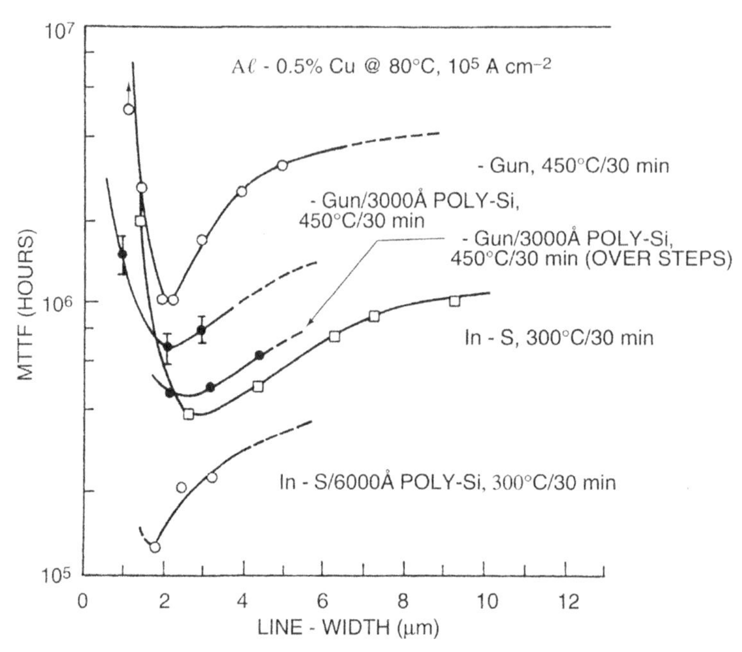


**Figure 5.24** Mean time to failure as a function of interconnect linewidth for evaporated (E-gun) and sputtered (In–S) Al films. *From Ref. [33].*

the standard deviation statistical spread in grain size ($\sigma_{gs}$), and the preferred crystallographic orientation is

$$\mathrm{MTTF} = \frac{BS}{\sigma_{gs}^2} \cdot \log \left[ \frac{I_{111}}{I_{200}} \right]^2 . \tag{5.32}$$

Here the $I$'s are the corresponding X-ray intensities for planes of indicated indices and $B$ is a proportionality constant. Clearly, more nearly equiaxed large grains that have a strong [111] orientation (texture) in the film plane are desirable.

A significant experimental advance in characterizing the role of film texture on EM has recently emerged. By capturing the backscattered electron-diffraction pattern from patterned interconnects with the use of a scanning electron microscope, the crystallographic orientation, or microtexture, of each individual grain has been mapped (see Section 11.3.2.1). Intriguing findings using this so-called electron backscatter diffraction technique include the disappearance of certain off [111] texture grains as well as grain rotation as a result of EM [34].

### 5.6.4 EM Lifetime Distributions

EM failure times are invariably modeled as lognormal distributions. Examples of EM failure data plotted in lognormal form are shown in Figure 5.25. There are two physical reasons that explain why this particular statistical function is applicable to EM failures. The first has to do with profusion of damage mechanisms exhibited. As a result of the suggestion in Section 4.5.4 we may assume that EM exhibits a corresponding number of randomly distributed activation energies. But since failure times are directly proportional to the Maxwell–Boltzmann factor, exp $[-E_e/kT]$, the logarithm yields $E_e/kT$, which is normally distributed. Hence the overall use of the lognormal distribution.

A second explanation justifying the applicability of the lognormal distribution is based on the temperature dispersion of conductors during life testing [36]. Due to the different thermal resistance values, individual stripe temperatures are normally distributed. Given the Black equation's exponential dependence on temperature, a lognormal distribution of failure times would be expected.

Another approach to the statistics of interconnect reliability adopts the view that a metal stripe is a chain of $N$ independent links such that $F(t)$ is the cumulative distribution function (CDF) of a single link [37]. The scaled CDF of the chain, $F_N(t)$, is then the strength distribution of the "weakest link," or

$$F_N(t) = 1 - [1 - F(t)]^N. \tag{5.33}$$

Simple substitution of the Weibull form of $F(t)$ reveals that $F_N(t)$ is also a Weibull distribution. Nevertheless, attempts to fit large failure populations to such a self-reproducing Weibull function have proven unsuccessful [38]. But if $F(t)$ is lognormal, $F_N(t)$ is not lognormal. The conclusion is that lognormal distributions can approximate true failure times only in a finite percentile interval. Furthermore, what characterizes the length of the weak link is not clear. Is it, for example, the grain size or perhaps the threshold length, $L_c$?

### 5.6.5 EM Testing

Irrespective of degradation mechanism, it is always a desideratum to accelerate testing in order to rapidly screen production lots prone to failure. In the case of EM several test methods have gained popularity. Each has individual advantages and drawbacks, but in common fast response (minutes

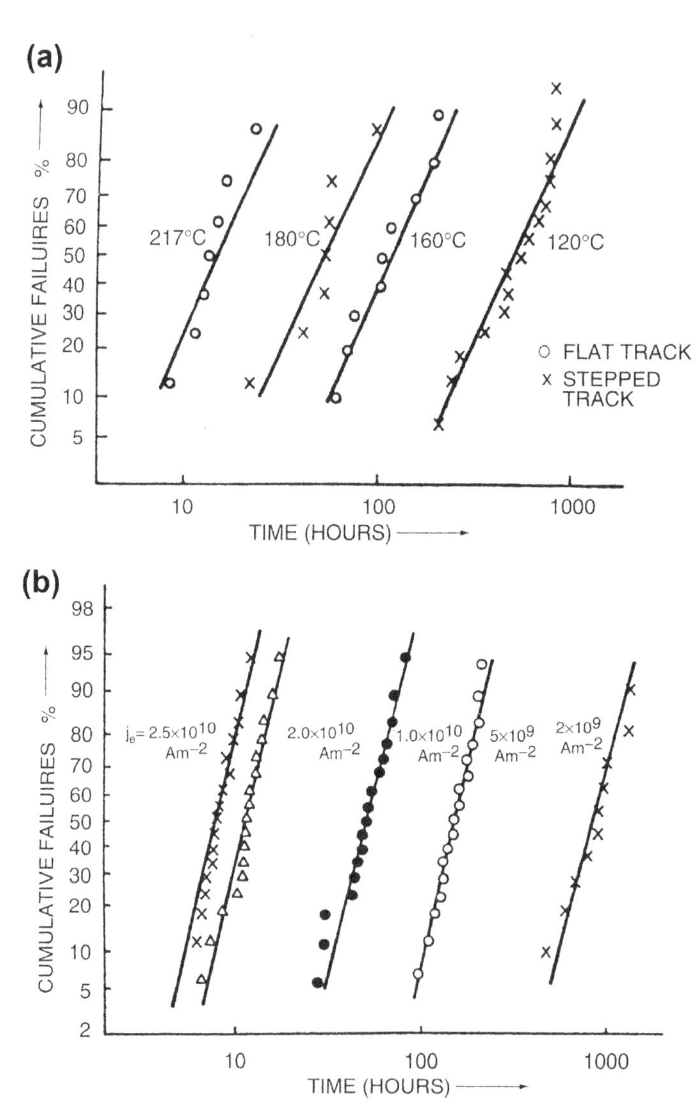


**Figure 5.25** Mean electromigration failure times of E-beam evaporated aluminum conductors plotted in lognormal form. (a) Effect of temperature on electromigration failure. Data pertain to unstepped (flat) and stepped specimens powered at $1 \times 10^6$ A/cm$^2$. (b) Effect of current density on electromigration failure. Data pertain to stepped track specimens powered at 180 °C. *From Ref. [35].*

to a few hours at most) and scaleup to the wafer level are sought-after features. By wafer level we mean placing a pair of probes across any interconnect prior to separation into dies. In this section three of these tests are briefly described.

The first test is known as the temperature-ramp resistance analysis to characterize EM, or TRACE [39]. In this test a temperature ($T$) versus time ($t$) ramp is applied to a packaged, nonsealed metal track while current simultaneously flows in the stripe. Resistance ($R$) change is monitored for about an hour. The kinetic parameters $B$ and $E_e$ that best fit the equation $(1/R)\, dR/dt = B \exp[-E_e/kT]$ are then extracted and used to model failure times.

A highly accelerated wafer testing technique known as breakdown energy of metal was introduced in 1985 [40]. The test consists in forcing EM open-circuit failures by means of a computer-controlled current staircase. At each current step the test line resistance is measured, and the temperature is calculated from the known temperature coefficient of resistivity for the metallization. With current densities of $1 \times 10^7\ \mathrm{A/cm^2}$, failures typically occur within minutes at generally elevated interconnect temperatures. When open circuiting occurs, the total energy transferred to the interconnect is calculated. Values so determined have been shown to obey lognormal statistics. The 50% point of the distribution is a failure parameter that enables comparison of the EM performance in different conductors.

Finally, there is the standard wafer-level electromigration acceleration test, or SWEAT [41]. This very rapid test is supposedly capable of yielding information about the metallization in less than 15 s. In the SWEAT test acceleration occurs solely by joule heating. First, relationships are established between the metallization temperature and resistance change. Then the dependence of the power dissipated in the interconnect and the corresponding temperature rise are experimentally evaluated. A minimum acceptable failure lifetime (e.g., $t_m = 20$ years) is chosen at the operating current and temperature, thus defining an acceleration factor (AF) for a particular test time to failure (e.g., MTTF = 15 s). The current ramps up and stops when a value $j_s$ corresponding to $T_s$, is calculated from Black's equation, or

$$\frac{\mathrm{MTTF}}{t_m} = \frac{1}{\mathrm{AF}} = \left(\frac{j}{j_s}\right)^2 \exp\left[\frac{E_e}{k}\left(\frac{1}{T_s} - \frac{1}{T}\right)\right]. \tag{5.34}$$

If the wafer under test lasts longer than $t_m/\mathrm{AF}$, it is accepted; otherwise, it fails during the test and is rejected.

These fast tests have the disadvantage of employing large current densities that generate steep thermal gradients. Because of this, measured failure times may not physically extrapolate well to situations where smaller current densities prevail.

### 5.6.6 Combating EM

Concurrent with the discovery of EM as a reliability problem in aluminum interconnects were strategies devised to cope with it. But serendipity played a role and intervened on two separate occasions in combating EM. In the mid-1960s a misfocused electron beam evaporated some hearth (Cu) rather than intended charge (Al) metal. The resultant Al–Cu alloy films were found to be significantly resistant to EM. Because Cu raises the activation energy for Al migration, it is now universally used in metallizations. Figure 5.24 is the basis for the second instance of unexpected luck. While the deleterious effects of GBs have not been totally eliminated, shrinking film widths have, nevertheless, decreased their influence. Over the years the strategy of using multilayers, or sandwiches, of Al between more EM-resistant refractory metals (e.g., Ti, W, TiN) has evolved. Thus, in the emergency of Al open circuiting, the outer metals act as more resistive shunts. Rather than immediate failure, the metallization still functions, albeit with minor degradation of electrical behavior.

## 5.7 STRESS VOIDING

### 5.7.1 Introduction

One of the more unsettling reliability problems associated with metallizations is known as stress voiding [42,43]. It was first reported around 1984–1985 by a number of investigators [44,45], who recognized that film stress played a role in the phenomenon. More than ~~a~~ two decades later SV continues to be an active area of reliability research. The basic manifestations of failure include wedge or slit-shaped voids or cracks that initiate at the interconnect edge, as shown in Figure 5.26. Some of the characteristics of SV are the following:

1. It occurs even when there is no current flowing or EM-induced damage. Simultaneous EM accelerates SV.
2. Void formation can occur during thermal aging.
3. A passivation layer such as $SiO_2$ or silicon nitride is required for voiding to occur.
4. Voiding generally develops in line widths less than 2 μm and appears to be a greater hazard the narrower the stripe.
5. The void density varies with composition and grain structure.
6. In Al, SV has been observed in stripes aged at room temperature; the rates of damage are thermally activated.

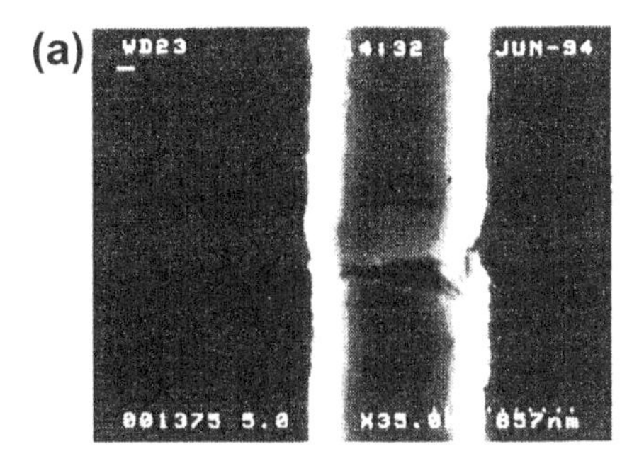


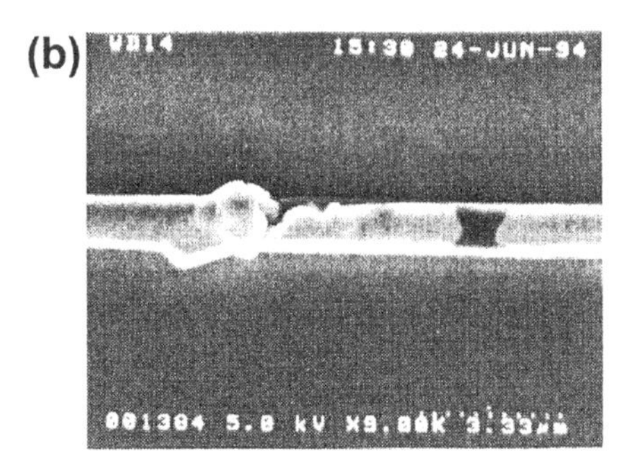


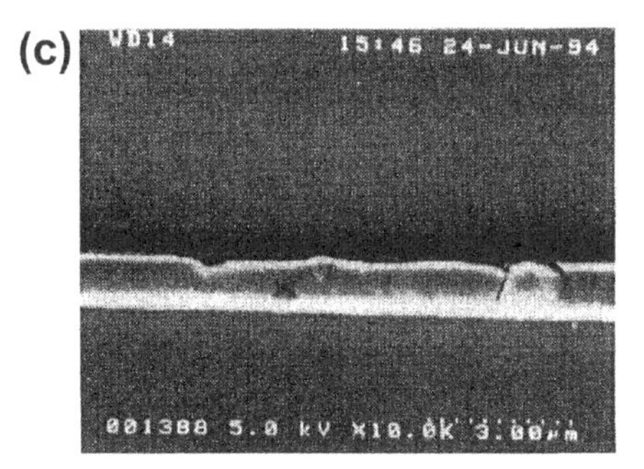


**Figure 5.26** SEM images of stress-related electromigration damage in Al interconnects. (a) Wedge or slit void in a 0.8 μm wide stripe. (b) Void and hillock formation in a 1 μm wide stripe. (c) Edge void. *Courtesy of M. Marcus, Lucent Technologies, Bell Laboratories Innovations.*

Among the pertinent issues we will explore in connection with SV are the origin and role of stress, the effect of conductor confinement, mechanisms for mass transport–induced damage, and the optimization of materials and processing for improved reliability. We start with stress effect considerations.

## 5.7.2 Origin of Stress in Metallizations

It is instructive to start with Figure 5.27, which summarizes the state of stress in metal films for different material and geometric configurations [29]. The blanket, unpassivated thin film deposited on a substrate in Figure 5.27(a) is under a balanced biaxial stress distribution, i.e., $\sigma_x = \sigma_y$. But, because of the stress-free surface, $\sigma_z = 0$. As noted in Section 3.3.2, film stress stems from the thermal mismatch between film and substrate as well as from the detailed mechanics of growth. For a wide stripe, a balanced biaxial stress distribution still prevails for the most part. However, if the film is in the form of a narrow stripe (Figure 5.27(b)), then near the lateral edges of the film $\sigma_x$ does not equal $\sigma_y$, but one still expects that $\sigma_z = 0$. If now this narrow line is encompassed by an overlying passivating thin film (Figure 5.27(c)), an unbalanced triaxial state of stress (i.e., $\sigma \neq \sigma_y \neq \sigma_z$) will develop.

The thermal origins of these stresses allow for a rough estimate of their magnitudes. For the case of Al, passivated by $Si_3N_4$ deposited at elevated temperature, the metal contracts more than the nitride upon cooling. Therefore, the Al is stressed in tension and the $Si_3N_4$ in compression. Provided that a stress-free state exists at the passivation deposition

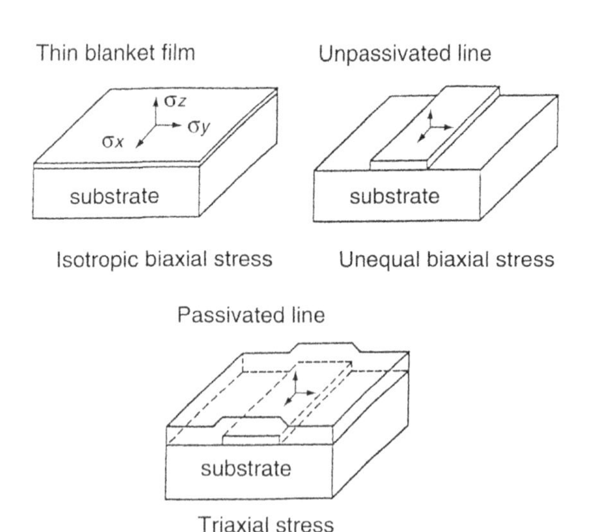


**Figure 5.27** States of stress in metal films. *(From Ref. [29].)* (a) Blanket, unpassivated thin metal film ($\sigma_x = \sigma_y$, $\sigma_z = 0$). (b) Patterned, unpassivated narrow metal stripe ($\sigma_x \neq \sigma_y$, $\sigma_z = 0$). (c) Metal stripe is passivated by blanket insulator film ($\sigma_x \neq \sigma_y \neq \sigma_z$, $\sigma_z \neq 0$).

temperature ($T$), use of Eqn (3.8) enables the stress in the Al to be calculated at the use temperature $T_o$. Assuming $T - T_o = 300$ °C, $E$(Al) = 69 GPa, ν (Al) = 0.33, $\alpha_f$ (Al) = 25 × $10^{-6}$/C, and $\alpha_s$ ($Si_3N_4$) = 2.5 × $10^{-6}$/C, the calculated thermal stress in the metal is $\sigma = +\,0.695$ GPa. Such a level is well above the tensile yield stress of Al alloys, and therefore we may expect stripes to be plastically deformed. In addition, the elastic strain ($\sigma/E$) is 1% in each direction, accounting for a 3% greater metal volume than in the unconstrained state.

The mathematical form of the stress distribution in stripes is complex but has been modeled by finite element analysis. An example is shown in Figure 5.28 for a 1-μm wide Al–Si stripe [46]. Stress levels approaching 450 MPa are evident at film corners. In vias the conductors are small in all three dimensions. When confined by overlying dielectrics, a hydrostatic

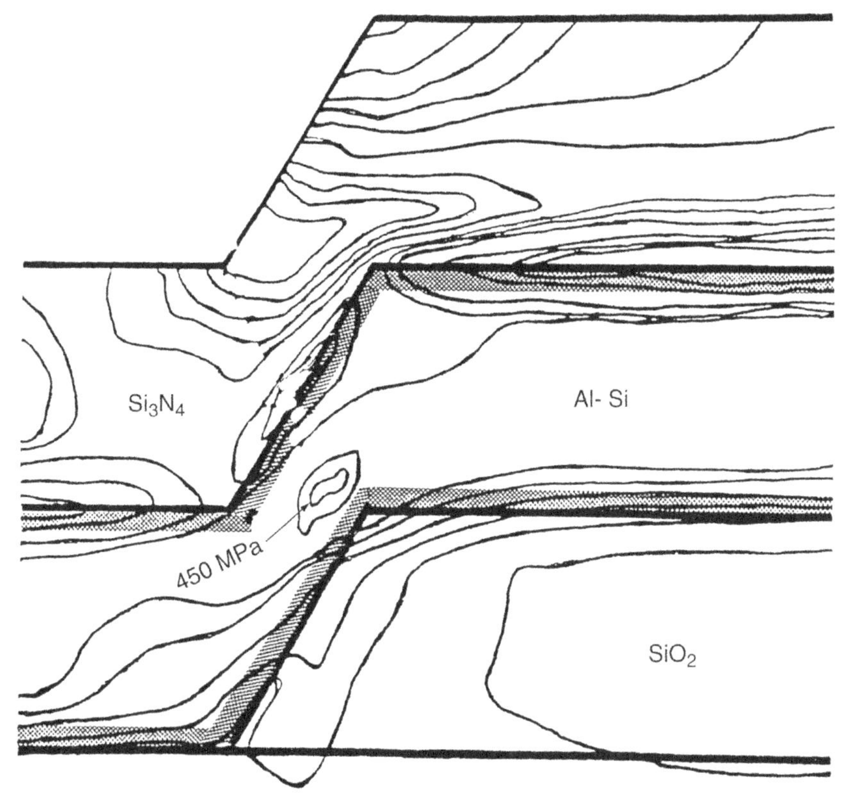


**Figure 5.28** Finite element analysis of stress distribution in 1 μm wide Al–Si stripe passivated by $Si_3N_4$. Stress levels of 450 MPa are sufficient to nucleate voids. (Contours at every 50 MPa) *From Ref. [46].*

or triaxial tensile stress-state develops, leading to cavitation effects and void nucleation.

## 5.7.3 Vacancies, Stresses, Voids, and Failure

### *5.7.3.1 Void Nucleation*

This section attempts to explore the suggested linkage among vacancies, stresses, and voids, as well as to forge connections between EM, SV, and failure. We start this ambitious undertaking by noting that for reasons given in Section 5.7.2, passivated interconnects are observed to contain GB voids when cooled to room temperature. According to classical nucleation theory, a spherical void of radius $r$ is stable when the strain (free) energy reduction in forming it exceeds the energy increase associated with creating new void surface area. Furthermore, according to thermodynamics, atoms have different "effective" atomic concentrations depending on their chemical potentials ($\mu$). Thus, when atoms lie within stressed regions, $\mu = -\sigma\Omega$ (see Section 5.5.4), while on curved (void) surfaces, $\mu = 2\gamma\Omega/r$, where $\gamma$ is the surface energy (J/m$^2$). If the potentials are equal at equilibrium, $r = 2\gamma/\sigma$. For $\gamma = 1$ J/m$^2$ and $\sigma = 400$ MPa, a void of radius 5 nm is stable. Importantly, stressed regions and voids modify vacancy concentrations (Eqn (3.1)) by altering their (free) energy of formation.

### *5.7.3.2 Atom Transport under EM and Stress Forces*

Directed atom (or vacancy) transport from stressed GBs, dislocations, interfaces, etc. to voids, or vice versa, under gradients in the chemical potential ($\nabla\mu$) is responsible for enlargement or shrinkage of voids. Atomic drift preferentially occurs from locations of higher chemical potential to those of lower chemical potential. To develop a quantitative macroscopic model of stress-driven atom (vacancy) flow, we recall that the force on an atom in a stress field is given by $F = \Omega\, \mathrm{d}\sigma/\mathrm{d}x$ (Eqn (5.26)). When this stress-induced force is added to the primary electric field driving force, substitution in the Nernst–Einstein equation yields

$$J = \frac{CD}{RT}\left\{Z^* q\rho j_\mathrm{e} + \Omega\frac{\mathrm{d}\sigma}{\mathrm{d}x}\right\}. \tag{5.35}$$

Interconnect damage occurs under isothermal conditions when the divergence of the mass flux,

$$\frac{\partial C}{\partial t} = -\frac{\partial}{\partial x}\left[\frac{CD}{RT}\left\{Z^* q\rho j_\mathrm{e} + \Omega\frac{\partial\sigma}{\partial x}\right\}\right], \tag{5.36}$$

is nonzero. Changes in atomic concentration cause stress variations and vice versa. Therefore, $\partial C/\partial t$ is proportional to $\partial\sigma/\partial t$, and it has been specifically shown that $\partial C/\partial t = -(C/B)\ \partial\sigma/\partial t$, where $B$ is an appropriate constant with a value equal to ~0.6 times Young's modulus [47]. Substituting yields

$$\frac{\partial \sigma}{\partial t}=\frac{\partial}{\partial x}\left[\frac{DB\Omega}{RT}\left\{\frac{\partial \sigma}{\partial x}+\frac{Z^{*}q\rho j_{\mathrm{e}}}{\Omega}\right\}\right]. \tag{5.37}$$

These last two equations form the basis of interrelated mass or vacancy transport and change in film stress in the presence of EM. According to them the current-induced vacancy flux creates a stress-directed counterflux that retards EM damage.

We now consider the Fick-like diffusion portion of Eqn (5.37).

$$\frac{\partial \sigma}{\partial t}=\frac{BD\Omega}{RT}\frac{\partial^{2} \sigma}{\partial x^{2}}, \tag{5.38}$$

and note that stress changes can be expected over a long period with a characteristic time given by an effective diffusion coefficient equal to $DB\Omega/(RT)$. It is instructive to obtain a solution to this equation for a semi-infinite long stripe where there is a vanishing EM flux of vacancies at $x=0$. This means $J=0$ and the condition that $\partial\sigma/\partial x = -Z^{\star}q\rho\, j_{\mathrm{e}}/\Omega$ (Eqn (5.27)). Such a situation is physically approximated in a stripe with a "bamboo" grain structure because the current direction is normal to the GB paths. As a result, no voids form. Upon solving Eqn (5.38) with the use of Laplace transforms, the value of stress at $x=0$ is readily given by

$$\sigma=\frac{2Z^{*}q\rho j_{\mathrm{e}}}{\Omega}\left(\frac{DB\Omega t}{\pi RT}\right)^{1/2}. \tag{5.39}$$

If we suppose that the time to failure (MTTF) occurs when a critical stress ($\sigma = \sigma_{\mathrm{c}}$) effectively pulls the conductor open at time $t = t_{\mathrm{c}}$, then

$$\mathrm{MTTF}=t_{\mathrm{c}}=\frac{\pi RT}{4DB\Omega}\left[\frac{\sigma_{\mathrm{c}}\Omega}{(Z^{*}q\rho j_{\mathrm{e}})}\right]^{2}. \tag{5.40}$$

The result that the failure time varies inversely as the square of the current density is commonly observed in practice (Eqn (5.31)).

#### *5.7.3.3 Void Growth*

In this section expressions for the growth rate of voids located on GBs will be given, largely without proof [48–50]. The geometry is shown in

Figure 5.29, which consists of an internally stressed interconnect with a transverse GB that has a wedge shaped void at one end. Physically, the tensile stress provides the driving force for atoms to diffuse from the void root, where the chemical potential is larger, into the void surface where it is smaller. In doing so, the void opens and dimension $a$, advances into the stripe interior. Simultaneously, the stress relaxes, as indicated in Figure 5.29. The key questions are, What is the velocity of void growth, and how long will it take $a$ to span the conductor width?

Suggested equations for the time ($t$) dependence of $a$ are [50]

$$\frac{da}{dt} = M\sigma^3, \tag{5.41}$$

which applies to slit-like voids, and

$$\frac{da}{dt} = N\sigma^{3/2}. \tag{5.42}$$

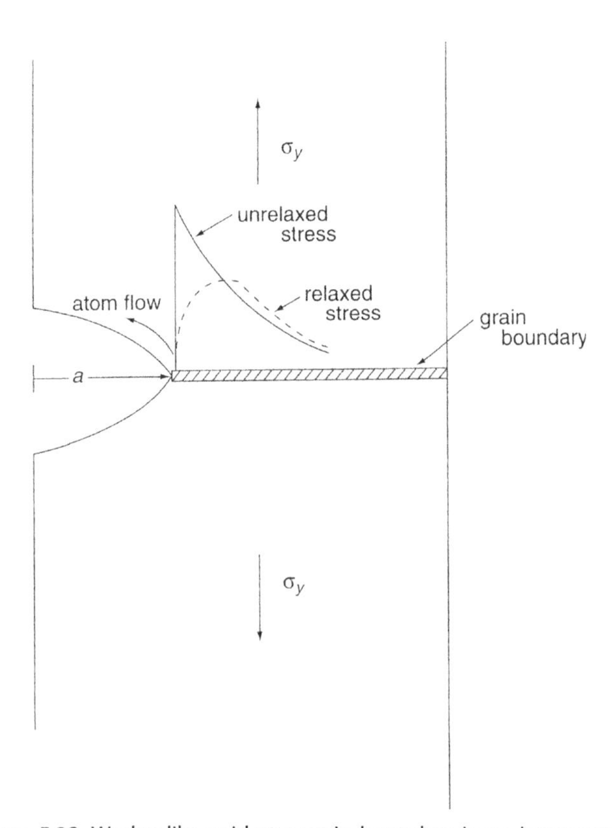


**Figure 5.29** Wedge-like void at a grain boundary in an interconnect.

which better describes the growth of wedge-shaped voids. These creep-like formulas (see Chapter 9) should not be surprising, considering that stress-induced diffusion is involved. The terms $M$ and $N$ are constants that depend on void geometry, surface ($\gamma_s$) and GB ($\gamma_{gb}$) tensions, and the ubiquitous Boltzmann factor. When GB diffusion is faster than surface diffusion on the void surface, Eqn (5.41) is applicable, but when the reverse is true, Eqn (5.42) applies.

## 5.7.4 Role of Creep in SV Failure

It is well known in elasticity theory that any three-dimensional state of stress at a point in a body can be replaced by an equivalent set of three mutually perpendicular principal stresses ($\sigma_x$, $\sigma_y$, and $\sigma_z$). Furthermore, the effective stress at the point can be estimated. For example, in the case of the complex three-dimensional stress distribution in Al conductors, the von Mises formula for combined stresses [51]

$$\sigma_{VM} = 2^{-1/2}\left\{\left(\sigma_x - \sigma_y\right)^2 + \left(\sigma_y - \sigma_z\right)^2 + \left(\sigma_z - \sigma_x\right)^2\right\} \tag{5.43}$$

provides a criterion for plastic deformation. Thus plastic deformation ensues if the calculated value of the von Mises stress, $\sigma_{VM}$, exceeds the yield stress in tension, $\sigma_y$.

We have just seen in the previous section that SV, and the time-dependent deformation that causes damage, is suggestive of creep phenomena. This is the reason that creep models (e.g., GB sliding) and formulas have been invariably used to describe SV rates ($R_{SV}$). The simplest expression proposed for $R_{SV}$ is basically a combination of Eqns (3.8) and (5.41) given previously. A MTTF associated with $R_{SV}$ is predicted [52] to be.

$$\mathrm{MTTF}^{-1} \approx R_{SV} = A(T - T_o)^n \exp\left[\frac{-E_{SV}}{RT}\right], \tag{5.44}$$

where $R_{SV}$ is associated with the steady-state creep strain rate, and $T - T_o$ is the difference between the zero stress (deposition) and service temperatures. Constant $A$ includes the factor $[E(\alpha_f - \alpha_s)/(1-\nu)]^n$ for the (von Mises) stress value. Two opposing trends that are evident in this formula can be visualized in Figure 5.30. At elevated operating temperatures the Boltzmann factor is large, thus fostering enhanced atomic motion with activation energy $E_{SV}$. Simultaneously, the induced thermal stress is relaxed, so that

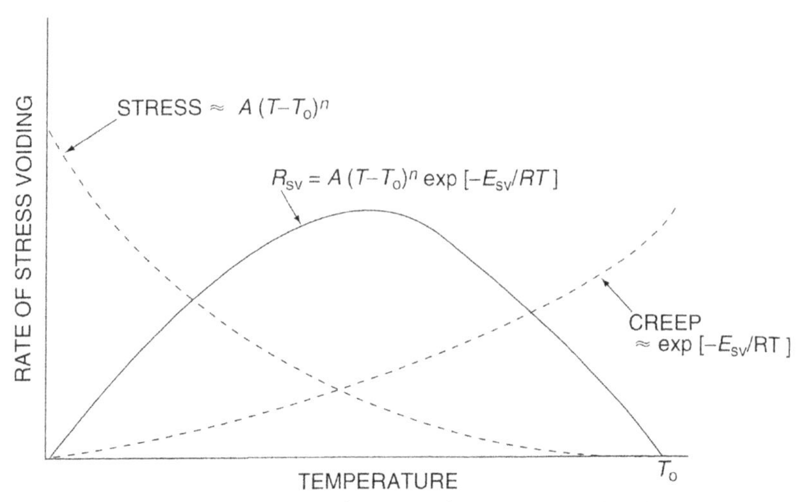


**Figure 5.30** Stress voiding rate as a function of temperature. The shape of the curve reflects the competition between creep and stress relaxation effects. *After Ref. [52].*

overall, $R_{SV}$ is small. At low temperatures the reverse is true, and again the SV rate is minimal. In between, $R_{SV}$ reaches a maximum value. Such a trend in stress-voiding failures has been observed by some investigators [53,54]. In these studies a typical value of $n = 4$, and maximum stress-voiding effects occurred at about 180 °C. This very unusual result that damage does not increase monotonically in the expected Arrhenius manner should be further verified before being accepted as law.

## 5.7.5 SV at Vias

Stress voiding is manifested not only along the sides of isolated aluminum interconnects, it occurs also at vias [55]. After the via hole is etched open it is usually annealed to ~450 °C to remove impurities that promote corrosion or adversely affect surface migration and nucleation of the subsequently deposited plug metals of Al or W. This creates the reliability problem depicted in Figure 5.31. Upon heating, the Al expands more than the $SiO_2$, compressing the metal over the central portion of the stripe ($L_s$). At the via hole the stress-free state results in the development of a stress gradient that forces Al atoms to be extruded into the via. This leaves vacancies behind at the $L_s/L_c$ interface and they condense into voids. The damage has been done and it is not reversed on cooling. Apparently, the excess high-temperature vacancies now precipitate and further stabilize the voids. In support of this mechanism, experimental studies have revealed

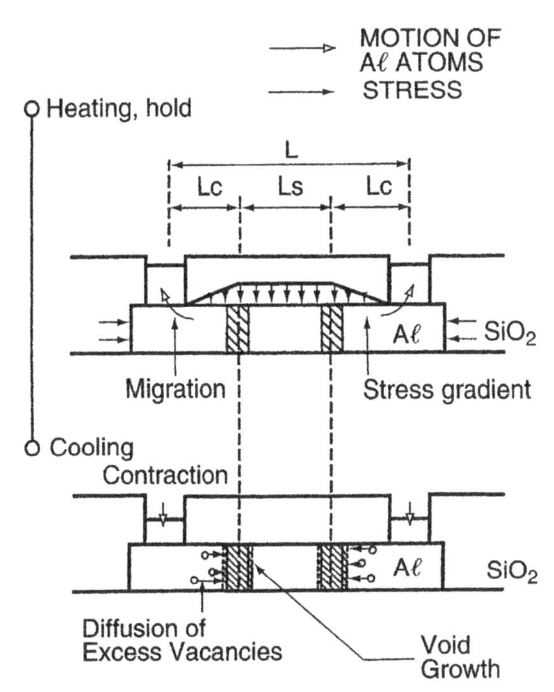


**Figure 5.31** Model of aluminum extrusion and void formation at vias during thermal processing. *From Ref. [55].*

greater volumes of extruded metal with increasing distance between vias. Greater void densities are produced during both higher temperature and longer time anneals.

In this as well as other comparable situations, the lesson is to avoid deposition of high-compressive-strength dielectrics.

## 5.8 MULTILEVEL COPPER METALLURGY—EM AND SV

### 5.8.1 Introduction

A need to reduce signal delay in ICs has resulted in the selection of copper as a replacement for Al/Cu. Dual damascene is the new technique used to form interconnect structures based on conductive copper metal lines inlaid into an oxide or low dielectric constant (k) nonconductive layer. The dual damascene technique forms trenches and vias (hence dual or twice used) into which copper is eventually electroplated. CMP is used to remove excess metal from the wafer surface. The process has been adopted for forming copper interconnects because traditional plasma etch techniques cannot be used for patterning copper films. Copper/low k interconnects formed by the dual damascene technique are ubiquitous in state-of-the-art manufacture of IC devices.

## 5.8.2 Technology

Since 1997, copper multilevel metallization has been used to combat the signal delay caused by the resistivity of Al and Al–Cu [56]. The first use was with six levels of metallization [57] as described in Chapter 2. The additional delay, caused by the capacitance of the insulation between metallization levels, was further reduced by lowering the dielectric constant of the dielectric material used to isolate levels. Today, in IC chips with 700 million FETs, copper has essentially replaced Al–Cu. Copper has about a 40% lower resistivity, than Aluminum, see Table 2.1 in Chapter 2. Copper, however, has a much higher propensity for corrosion, unlike Al, which develops a self-limiting oxide. The copper must be protected on all sides, essentially encapsulated to prevent corrosion. The diffusivity of copper in Si and $SiO_2$ is high and as a result a diffusion barrier is also needed to stop copper diffusion into dielectrics and Si devices. Copper encapsulation not only prevents corrosion but also helps eliminate EM extrusions, which in turn increases stress in the copper. The use of copper therefore presents a different set of trade-offs and requirements, namely, the copper is typically covered with TaN/Ta, with TaN/Ta barriers surrounding vias and an additional barrier between levels of metallization. The dual damascene process starts with trenches, which are etched into the ILD. The ILD is used for electrical isolation between the metallization levels. After trenches are formed, the vias are etched at the ends of the trenches to form the contact to the metal level below. The trenches and vias are first lined with TaN/Ta and then lined with copper seed. The seed helps initiate electroplating of copper into the vias and trenches. Copper outside the trenches (on top of the ILD) is removed using CMP. CMP is used to planarize the surface such that only the copper in the trenches remains. A barrier layer (usually SiN, SiCN, SiC, or TiSiN) is deposited and then a dielectric (ILD) layer is deposited and the process starts all over again with trench and via formation. The resulting structure consists of wiring levels that are isolated by layers of dielectric and connected by via structures between wiring levels. The complexity of this structure has prompted developers to require stress testing not only for EMEM and SV (also known as stressmigration), but also for time-dependent dielectric breakdown, thermal cycling (T/C), temperature humidity bias, and functional stressing.

An example of an eight-level microprocessor with copper low k technology is shown in Figure 5.32(a), [58].

This technology can be used with 10 levels of metallurgy [59], see Figure 5.32(b).

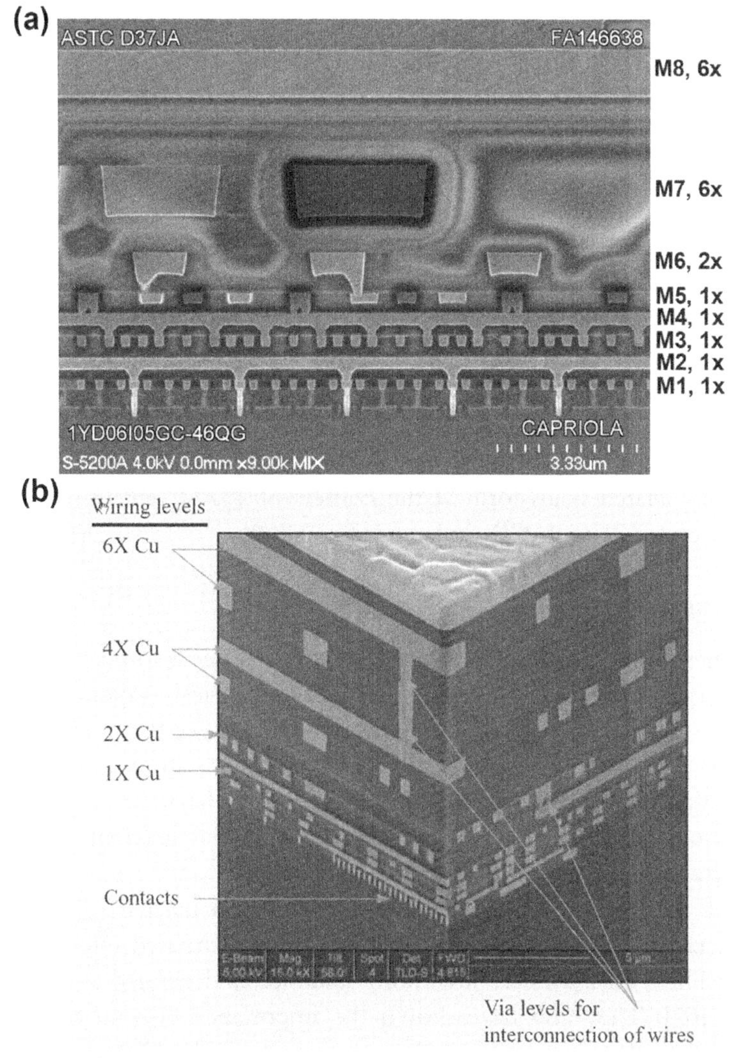


**Figure 5.32** (a) SEM x-section of 8-level microprocessor BEOL. Levels M1-M6 are in low k SiCOH, M7-M8 are in FTEOS. Hierarchical scaling factors are indicated (1×,2×,6×). *(From Ref. [58].)* (b) SEM x-section of 10-level microprocessor BEOL. *(From Ref. [59].)*

All metal levels are dual damascene except M1, which is W, and one Al(Cu) terminal level. To reduce the capacitance the technology has no polish stop and no trench etch stop. The ILD is low k SiCOH, except for the top two levels which are SiOF. The barrier cap over copper levels is PECVD (plasma enhanced chemical vapor deposition) SiC(N,H), which leads to improved hermiticity. The copper diffusion barrier (liner) is TaN/Ta, shown in Figure 5.33 [59].

## 5.8.3 Copper Processing

Grain size and orientation of the electroplated copper can also play a role in EM. However, annealing to form Copper grain structure appears rather ineffective. Self-annealing takes place at room temperature after deposition and texture is not affected by subsequent heat treatment. GB grooving appears to take place around random grains (see Figure 5.34 [60]).

The texture of the electrodeposited copper films is generally (111) and does not appear to depend on barrier material. The texture further seems to be inherited from the copper seed and Ta. Self-annealing results in many twins with no apparent effect on texture as a result of subsequent heat treatment.

Twin related voids form at the corners of {322} twins or the intersection of {322} with GBs and other twins [60].

## 5.8.4 Manufacturing

As dimensions get smaller the liner in the trench needs to become thinner and resputtering during TaN/Ta, PVD (physical vapor deposition —sputtering) has been shown to significantly improve the liner properties and EM life as it is made thinner, see Figure 5.35 [61]. Proper adjustment of liner resputter intensity and liner thickness can affect the degree of via anchoring and reduce stressmigration due to a stress gradient for a given liner thickness, as shown in Figure 5.36 [61].

This technology claims "The SiCOH and cap mechanical, chemical, and electrical strengths are increased, as well as associated interface adhesion." These attributes are just those described and defined as needed by others [62]. They also have shown the importance of removing plasma damaged material from the dual damascene sidewalls as shown in Figure 5.37.

The tree structure into which a given copper line and via are inserted can significantly influence the chance of a void and the growth rate of voids

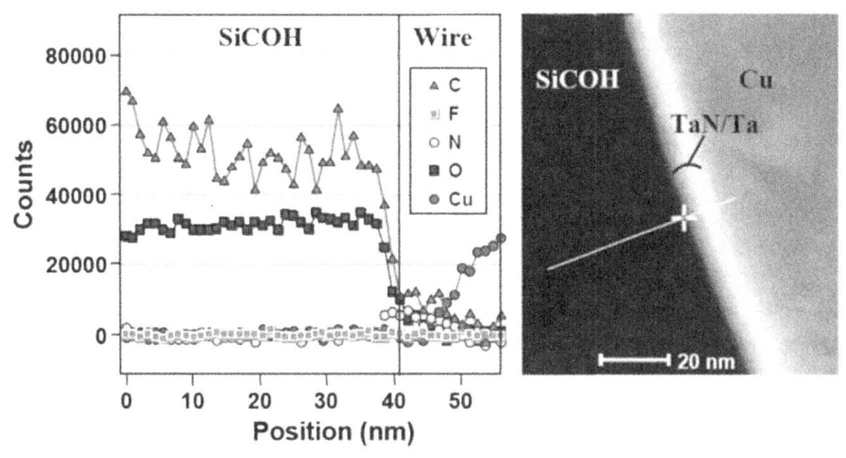


**Figure 5.33** TEM/PEELS line scans across a Cu/SiCOH sidewall interface showing no plasma damaged layer in the SiCOH. *From Ref. [58].*

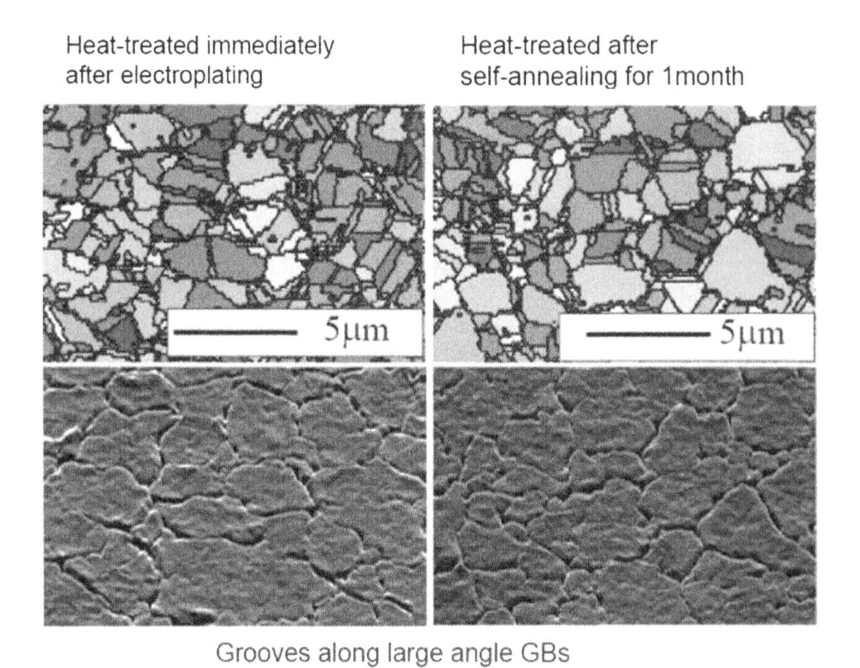


**Figure 5.34** Texture and surface morphology after heat treatment. *From Ref. [60].*

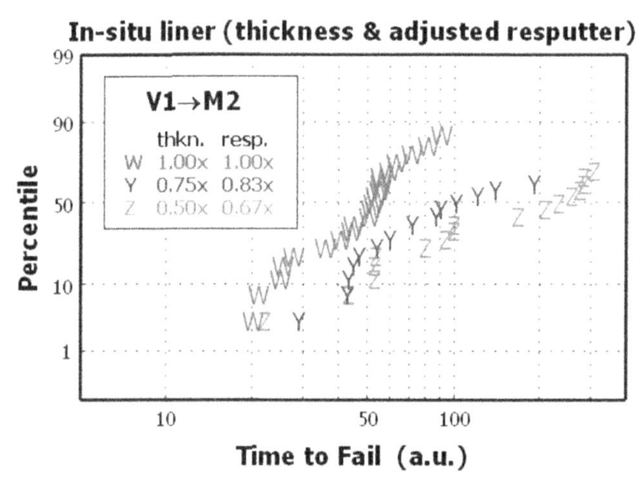


**Figure 5.35** Electromigration behavior of upstream-stressed via-line structures. *From Ref. [61].*

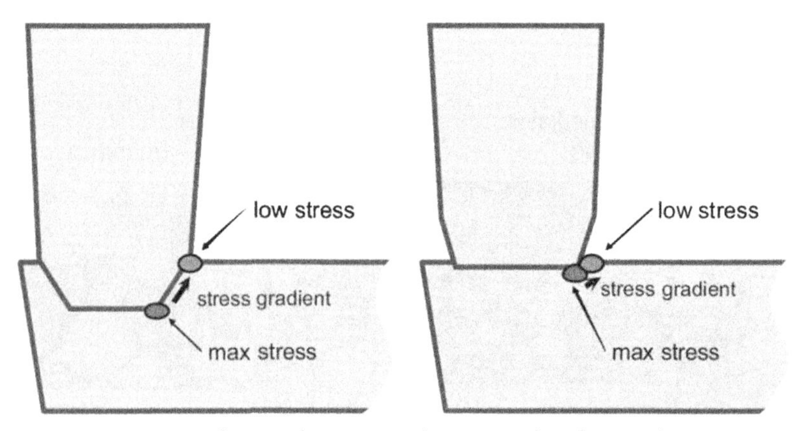


**Figure 5.36** Increasing distance between via bottom and triple point liner/cap/copper yields lower stress gradients on vias with deeper anchoring. *From Ref. [61].*

due to the effect of being either upstream or down stream from the apparent copper reservoir in the interconnect system [63a,63b]. This dependence on reservoir location is shown quite well in Figure 5.38(a) [61].

Maekawa et al. have reported dramatic improvement in time to failure for EM as shown in (Figure 5.39), if 1–2% Aluminum is used in the copper seed (Figure 5.40) [64].

Vairagar et al. have treated Copper surfaces with $NH_3$, $H_2$ or Silane ($SiH_4$) prior to dielectric cap, and shown (Figure 5.41) that the formation of

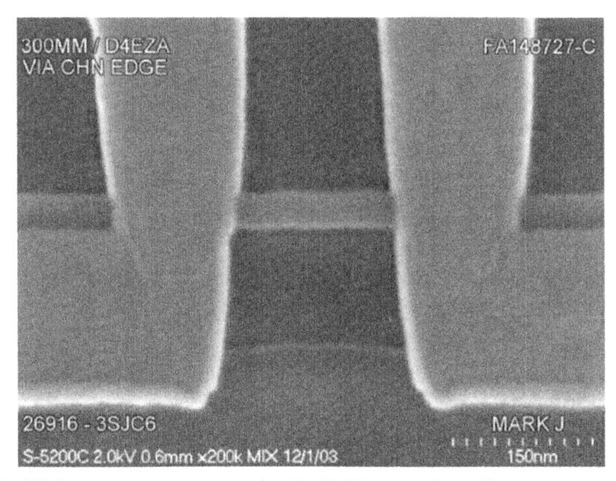


**Figure 5.37** SEM cross-section of Cu/SiCOH sidewall showing no plasma damage. *From Ref. [58].*

a thin layer of copper silicide improves EM life compared to copper nitride. $H_2$ and $SiH_4$ treatment gave about 2× improvement in $t_{50}$ [65].

ILD also plays a key role in MTTF because the reliability is very dependent upon contamination, adhesion, strength, leakage, and (coefficient of thermal expansion) CTE of the ILD.

Low k dielectrics such as SiOC and FSG are softer and reduce adhesion and release stress, and thus promote voiding.

The manufacturing process can also cause a new class of defects, which include poor adhesion due to contamination, incomplete copper fill, and residue from CMP. The low k dielectric is also subject to cracking and scratching during CPM which promotes failures.

Some of these issues are exacerbated by the dissimilar CTEs for the low k dielectric, barrier and cap. Voids can also be formed during manufacture because of copper underfill and poor adhesion Copper to barrier.

## 5.8.5 Copper EM and SV

EM and SV occur in copper but have slower kinetics, which extend time to failure compared to Al/Cu. The melting point of copper is 1083 °C compared to 660 °C for Aluminum. Also, lattice diffusion in copper has an activation energy of 2.2 eV compared to 1.4 eV for Aluminum. The GB diffusion activation energy in Copper is 0.7–1.2 eV, compared to 0.4–0.8 eV for aluminum. Luce [66] has shown 100x improvement in $t_{50}$ for copper compared to Al. Copper technology, however, has a bimodal

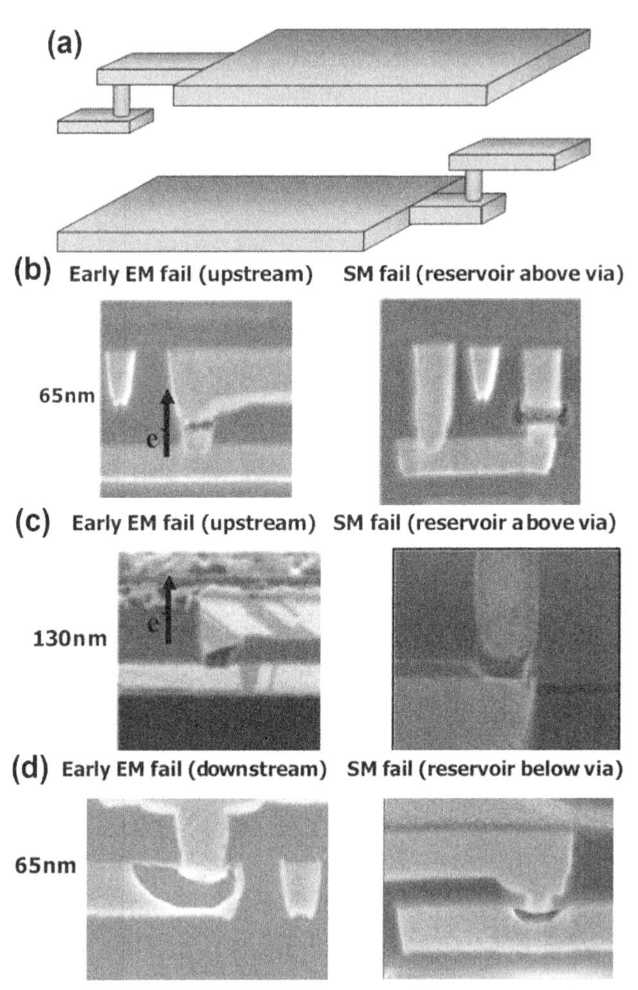


**Figure 5.38** (a) Schematics of the test structure configurations used for SM tests, with reservoir either above or below. *(From Ref. [61].)* (b) Typical early EM-fail in upstream-stressed via (left) and SM-fail in single via connected to large metal reservoir above. For both mechanisms the voids are characterized by slit like voids at mid-half of the via. *(From Ref. [61].)* (c) Typical early EM-fail in upstream-stressed via (left) and SM-fail in single via connected to large metal reservoir above observed in previous technologies. For both wear-out mechanisms the voids are located at the very bottom of the via. *(From Ref. [61].)* (d) EM and SM failure scenarios for downstream stressed vias and vias connected to large metal reservoirs below. Despite the use of liner deposition process with resputtering, voids appear in the same way as in previous technologies with moderate or no resputter. *(From Ref. [61].)* The impact of upstream or down stream mass transport on EM and SV (stressmigration) will be at least in part driven by the location of the reservoir as indicated in Figure 5.38(b)–(d) [61].

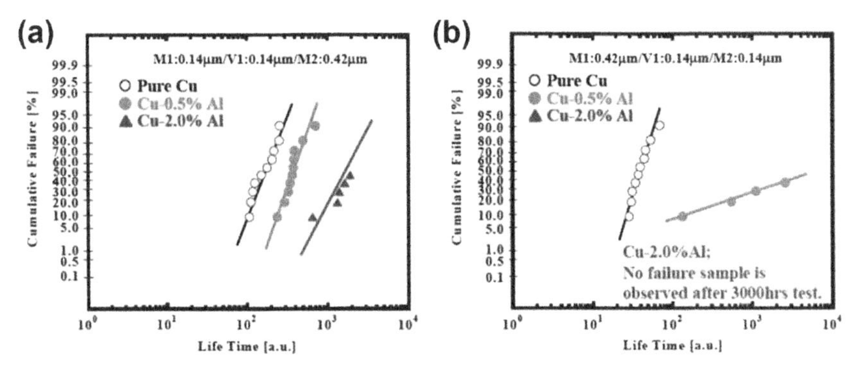


**Figure 5.39** Cumulative failure time distribution of the EM lifetime at a current of 0.6mA/via and a temperature of 250C (a) and 300C (b): alloy seed applied to (a) M1 and (b) M2. *From Ref. [64].*

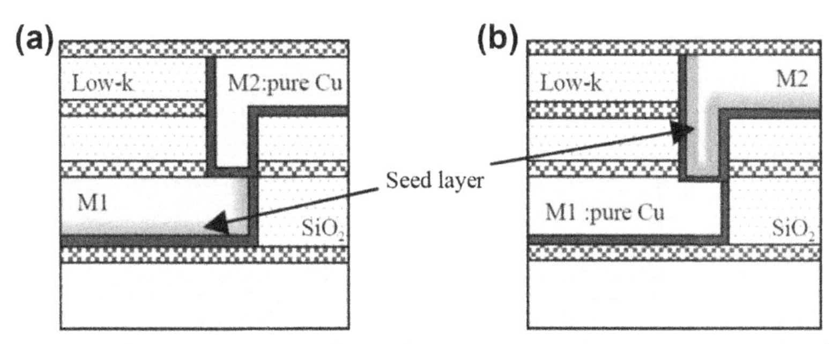


**Figure 5.40** Schematic diagrams of sample: (a) M1 is Cu-alloy or pure Cu seed and M2 is pure Cu seed (M1 alloy sample), (b) M1 is pure Cu seed and M2 is Cu-alloy or pure Cu seed (M2 alloy sample). *From Ref. [64].*

failure distribution. Failures, which occur early (EM) tend to be voids at the bottom of the via, whereas long-term fails (SV) tend to develop as voids in the trench. Voiding can also be related to adhesion issues between the Copper and the barrier which is a direct result of the copper electroplating process and the copper seed used to initiate the plating at the bottom of the via.

A review of the literature leads one to conclude that the reliability is very process and material dependent. However, EM in Copper is primarily driven by interface diffusion and surface diffusion [67], not by GB diffusion as in aluminum. EM appears to be a bit slower in copper with typical activation energy in the range of 0.8–1.1 eV [56].

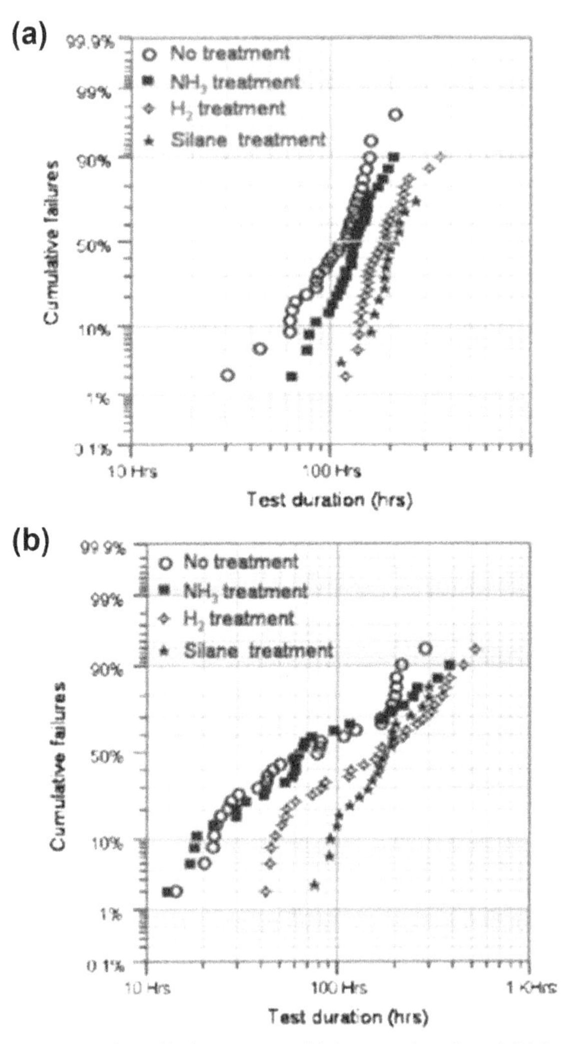


**Figure 5.41** Lognormal plot of failure times of (a) upper (M-2) and (b) lower (M-1) layer test structures with different surface treatment. *From Ref. [65].*

So the MTTF can be treated as before using Eqns (5.31) and (5.40), with appropriate change of parameters and constants.

The low k dielectric materials used with copper have higher coefficients of expansion than the oxides used with aluminum. The attendant cohesion between layers is different as is the elastic modulus and the hardness. This makes it much more difficult to deal with the release of stress build up.

The CMP itself can be a source of voids in copper. Poor CMP can reduce adhesion between the layers of copper, cap, barrier and dielectric.

CMP can also induce film cracking, scratching and residues, which impact reliability. Inspection is key to keeping the process well controlled. Insufficient or excessive CMP can cause intralevel shorting if gross, or contribute to leakage between lines in a given level [68]. The use of TaN/Ta as a cap on top of the Copper lines has exhibited an activation energy for EM of 1.4 eV compared with other caps ($SiN_x$ or $SiC_xN_yH_z$) with 1.0–1.1 eV and no cap with 0.87 eV [69].

The multilayer encapsulation may be part of the reason the electromigration life is longer than for Al–Cu. However the ILD used can also reduce the MTTF if it is too soft. Lower dielectric constant materials are not as rigid as $SiO_2$ and usually have less adhesion, which in turn can cause delamination and extrusion. Low-k dielectric materials tend to have a lower electrical breakdown field and somewhat higher leakage current, which we will discuss in Chapter 6. Finally, poor Copper adhesion to the barrier cap can also seriously degrade MTTF.

### 5.8.6 Future Technologies

The relentless reduction of feature size and increased device count on IC chips will undoubtedly continue for the foreseeable future. The optimization of wiring in chips will pose difficult decisions and tradeoffs. Guidelines and groundrules can help but their implementation on 12–15 wiring levels will become difficult if not unwieldy. Rule based wiring may eventually be necessary because of the complexity of current entering and leaving connection nodes between levels. Optimization of line length based on the current it carries may not be as viable as with Al/Cu. In particular, F. Wei. et al. [70] report that line length dependence of EM appears somewhat similar to Aluminum, namely short lengths have improved reliability. But there does not appear to be a deterministic current-density line-length product (jL) for which all lines are immortal, rather there is a sub-population at long lengths, which are expected to be immortal.

The selection of low k ($k \sim 2.3$) dielectric constant insulating layers between metallization levels, to further reduce delay, has continued to the point that a new technology called "air gap" (actually a vacuum, $k = 1$) insulation of wiring has been published and may be available in the not too distant future [71]. The microprocessor cross-section in Figure 5.42 shows empty space in between the chip's wiring. Wires are usually insulated with a glass-like material. In this process IBM uses integrated self-assembly techniques confined to laboratories, with its manufacturing lines, to create a test

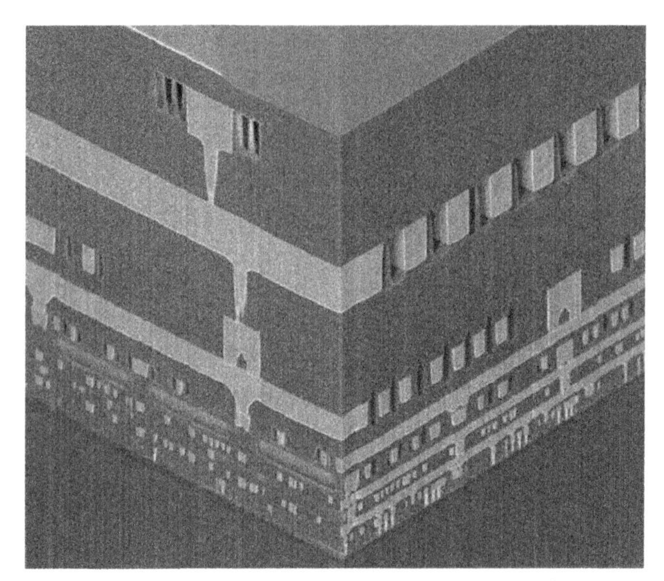

**Figure 5.42** IBM Airgap microprocessor. *From Ref. [71].*

version of its latest microprocessors that use vacuum gaps to insulate the miles of nano-scale wire that connect hundreds of millions of transistors. The technology reduces electrical interference, raises processor performance and lowers energy consumption.

## 5.9 FAILURE OF INCANDESCENT LAMPS

As a change of pace, the chapter ends with a discussion of the failure of incandescent light bulbs. Now that vacuum tubes are just a memory, tungsten filament failures in light bulbs are probably the most common and annoying reliability problem of an electrical nature that we normally encounter; happily, it is easily and cheaply dealt with. The "hot spot" theory of failure maintains that the temperature of the filament is higher in a small region because of some local inhomogeneity in the tungsten. Wire constrictions or variations in resistivity or emissivity constitute such inhomogeneities. With time, the temperature difference between the hot spot and the rest of the filament, at temperature $T$, increases. Preferential evaporation of W not only darkens the inner surface of the bulb, but it thins the filament, making the hot spot hotter, resulting in yet greater evaporation. In bootstrap fashion the filament eventually melts or fractures. In this model, applicable to filament *metals* heated in vacuum, the life of the bulb is exponentially dependent on filament temperature, or correspondingly, inversely proportional to the metal vapor pressure, $P$. For tungsten, $\mathrm{MTTF}^{-1} \approx P_{\mathrm{w}} = P_{\mathrm{o}} \exp[-\Delta H_{\mathrm{vap}}/RT]$, where $P_{\mathrm{o}}$ is a constant, $RT$ has the usual meaning, and $\Delta H_{\mathrm{vap}}$, the heat of vaporization, is 183 kcal/mol. Any residual oxygen in the bulb helps to transport W in the form of volatile $WO_3$.

Aside from hot spots, tungsten filaments experience vibration and tend to sag due to elevated temperature creep. The chief operative mechanism for creep damage in lamp filaments is GB sliding. When this occurs, the filament thins, and the local electrical resistance increases, raising the temperature in the process. Higher temperature accelerates creep deformation (sagging) and leads to yet hotter spots. The cycle feeds on itself, and the filament is well on its way to failure. Making nonsag tungsten wire is a way to attain longer-lasting bulbs. Drawn tungsten wire with minute additions of aluminum and potassium has an interlocking GB structure that is particularly resistant to sliding and creep during high-temperature operation. The GBs make acute angles with respect to the wire axis and are additionally pinned by alloying additions. Nonsag wire is the result.

Common light bulbs are filled with argon. While the main parameter affecting bulb life is the tungsten vapor pressure, Ar alters the gas-phase transport of W. Tungsten atoms no longer fly off in linear paths, but may swirl around in the Ar ambient due to diffusion and convection effects. Bulb dimensions as well as $P_{\mathrm{w}}$ play important roles in influencing MTTF. Filaments are also coiled. When straight wires are coiled, the *effective* surface area responsible for mass loss is reduced by a factor of about two. While this may be expected to enhance life proportionately, other W transport mechanisms are operative. A high concentration gradient of tungsten in the coil promotes gas-phase diffusion, while severe temperature gradients foster surface diffusion along the filament. The latter is responsible for the undesirable crystallographic faceting of filament grains shown in Figure 5.43.

Tungsten halogen lamps emit considerably more light than ordinary incandescent bulbs because the filament is operated at higher temperatures. The reason they don't fail is due to the chemical-vapor transport of the evaporated W back to the filament. Small amounts of iodine and bromine added to the argon enable the formation of volatile $WI_6$ and $WBr_6$ compounds, which circulate and regenerate W upon decomposition.

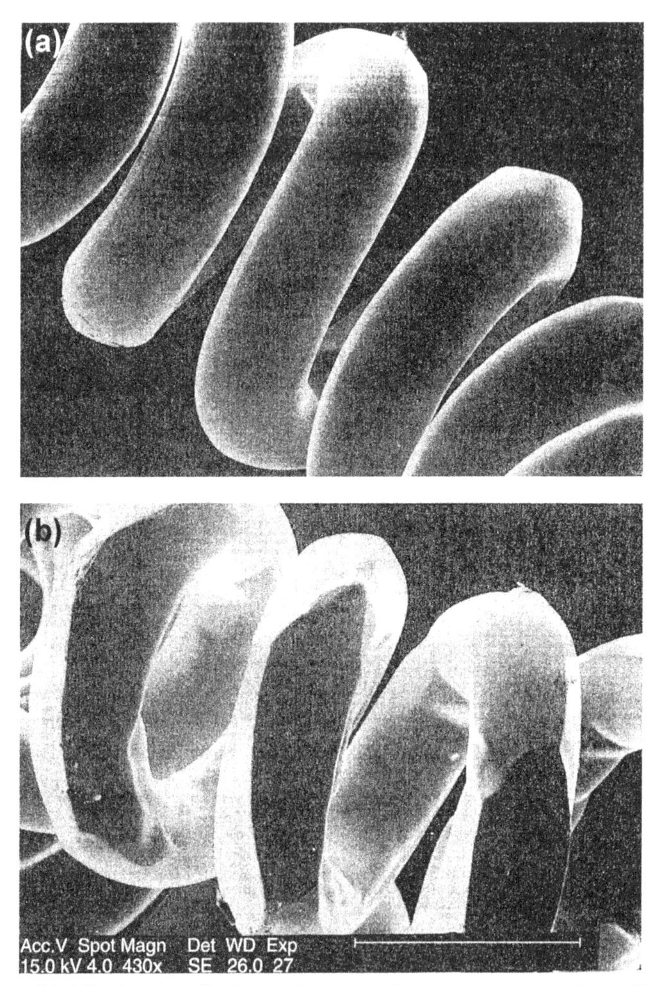


**Figure 5.43** (a) SEM image of unheated 50 μm diameter tungsten lamp filament. (b) Crystallographic faceting of heated filament. *Courtesy of R. Anderhalt, Philips Electronic Instruments Company, a Division of Philips Electronics North America Corporation.*

## EXERCISES

**5.1** A finite-source layer of $A$ deposited onto the surface of matrix $B$ diffused into it after heating to time $t_1$ and the concentration profile versus distance was obtained.

**a.** How would you plot the concentration—distance data to obtain a straight line?

**b.** If the slope of the line is $m$, what is the value of the diffusion coefficient?

**c.** What additional information would be required to obtain the activation energy for diffusion?

**5.2.** In a couple composed of semi-infinite materials $A$ and $B$ bonded at $x = 0$, the concentration profile can be curve fit by the equation $C(x, t) = 1/2 \text{ Erfc } x/(4Dt)^{1/2}$ for $\infty > x > -\infty$. The Matano analysis is designed to extract concentration-dependent diffusivity values from arbitrary diffusion-couple concentration profiles. Key to the analysis is a graphical integration that is made at any particular value of $C$ (e.g., $C'$), such that $D(C') = -\frac{1}{2t}\left(\frac{dx}{dc}\right)_C = C' \int_0^{c'} dxC$. Using this formula, show that the diffusivity is concentration-independent for the given profile.

**5.3.** According to the systematics of diffusion in FCC metals, what are the activation energies for bulk and GB diffusion in Al, Au, and Cu thin films?

**5.4** In considering the reliability of a thin-film Cu—Ni bilayer contact structure for solar cells, it was feared that these metals would interdiffuse and degrade electrical resistance. To estimate contact lifetimes, the available high-temperature lattice-diffusion data for Cu in Ni yielded a $D$ value of $3 \times 10^{-24}$ cm$^2$/s when extrapolated to 300 °C.

**a.** Using this value, how long will it take for these metals to interdiffuse through 200 nm at 300 °C?

**b.** Experiment showed that interdiffusion occurred in an hour. What happened?

**5.5** To prevent silicon and aluminum films from reacting, a diffusion barrier is imposed. What properties are required for optimal function of this barrier?

**5.6** Gold bonding wire 25 μm in diameter is attached to an aluminum film pad 2 μm thick.

**a.** How long would it take a 0.1 μm thick $AuAl_2$ compound layer to form if the contact is exposed to a 120 °C environment?

**b.** Suppose the length of the Au wire is 1000 μm, the Al pad has a diameter of 25 μm, and the electrical resistivity of $AuAl_2$ is 10 times that for Au. Write an equation for the time dependence of overall resistance change measured between the metal ends.

**5.7** In MESFETs the saturation drain-source current is given by $i_{DSS} = qN\, V_{sat}\, W\, Z$, where $V_{sat}$ is the saturation value of electron drift velocity and $Z$ is the gate width. Predict how $i_{DSS}$ will change with time if gate sinking occurs.

**5.8** Suppose good pure Al contacts to Si could be produced at low temperatures instead of 450 °C. Would spiking of junctions 50 nm deep become a reliability problem if such devices operate at 50 °C? Make any assumptions you wish.

**5.9** Derive the result indicated in Eqn (5.24).

**5.10** Consider the steady-state heat flow in a metal stripe where heat loss to the ambient temperature $T_o$ occurs by convection. If the equation describing the temperature ($T$) at any position ($x$) is written as

$$0 = \kappa \frac{dT^2}{dx^2} + \frac{i^2 \rho}{A^2} + h(T - T_o),$$

where $h$ is the (constant) heat transfer coefficient, obtain a solution for $T(x)$ using the same boundary conditions as in Section 5.5.3.

**5.11** Under simultaneous EM and diffusional transport show that $\frac{\partial C}{\partial t} = D\frac{\partial^2 C}{\partial x^2} - V\frac{\partial C}{\partial x}$, where $v$ is given by $DZ^{\star}q\rho\, j_e/RT$. Prove by direct substitution that

$$C(x, t) = \frac{C_o}{2}\left\{\exp\frac{vx}{D}\,\mathrm{Erfc}\frac{(x + vt)}{(4Dt)^{1/2}} + \mathrm{Erfc}\frac{(x - vt)}{(4Dt)^{1/2}}\right\}$$

is a solution.

**5.12.**

**a.** It has been determined that the mean life of pure aluminum stripes of width $w$ (in cm) and thickness $d$ (in cm) is given by MTTF (h) $= 4.4 \times 10^{12}\, w\, d\, j^{-n}$ exp $[E_e/kT]$, where $n = 2$, $E_e = 0.49$ eV, and $j$ is in A/cm$^2$. Which has a greater influence in changing the value of MTTF at 50 °C, a 1% change in $n$ or a 1% change in $E_e$?

**b.** What is the mean life of an interconnect 0.4 μm wide and 0.5 μm thick powered with $1 \times 10^5$ A/cm$^2$? How does it compare to the mean life of an Al–Si interconnect of the same dimensions for which MTTF (h) $= 2.2 \times 10^{15}\, w\, d\, j^{-2}$ exp $[0.54\ \mathrm{eV}/kT]$?

**5.13** A 0.5 μm thick aluminum stripe is powered with $1 \times 10^6$ A/cm$^2$. Assuming typical constants for Al, whose surface energy is $10^{-4}$ J/cm$^2$, predict how long it would take a GB groove-hole to reach the substrate.

**5.14** Electromigration failure data for pure Al and Al–4wt.% Cu–1.7 wt% Si interconnects obey lognormal statistics. For pure Al, $t_{50} = 85$ h and $\sigma = 0.32$, while for alloyed Al, $t_{50} = 5600$ h and $\sigma = 0.22$. Sketch the behavior of both metallizations to scale on lognormal plots.

**5.15** Show that after substitution of the Weibull cumulative distribution function $F(t)$, $F_N(t)$, defined in Eqn (5.33), is also a Weibull distribution.

**5.16** A third-level IC interconnect consists of a distribution of Al–Cu stripes deposited over $SiO_2$ as well as over the sides and bottoms of via contacts. On the $SiO_2$ the stripe width ($w$) and thickness ($d$) dimensions are each $5 \times 10^{-5}$ cm, but in the via $w$ is thinned to $0.25 \times 10^{-5}$ cm. The conducting length of the stripe is three times that of the via, and the mean life of the interconnect metallization is given by MTTF (h) $= 2.4 \times 10^{15}\, w\, d\, j^{-2} \exp\,[0.65 \text{ eV}/kT]$.

**a.** What is the ratio of MTTF for the involved interconnect relative to a linear (unthinned) stripe of the same length if $j = 4 \times 105$ A/cm$^2$?

**b.** For an overall conducting length of 1 mm, what is the voltage drop across the interconnect?

**5.17** From the plots of Figure 5.25 evaluate the constants $B$, $E_e$, and $n$ in the Black equation (Eqn (5.31)).

**5.18** Solve the boundary-value problem

$$\frac{\partial \sigma}{\partial t} = \frac{DB\Omega}{RT} \frac{\partial^2 \sigma}{\partial x^2}$$

under the following initial and boundary conditions:

At $t = 0$, $x > 0$; $\sigma = 0$.

At $x = 0$, t; $d\sigma/dx = -Z^\star/q\rho\, j_e/\Omega$.

At $x = \infty$, t; $\sigma = 0$.

Derive an expression for $\sigma(x,t)$.

**5.19** Suppose it is desired to passivate aluminum metallization with surrounding layers of $SiO_2$, phosphosilicate, and borophosphosilicate glass, as well as silicon nitrides. In each case the insulators are deposited at ~400 °C and cooled to 25 °C, a process that thermally stresses the Al. Average values for the Young's modulus, thermal expansion coefficient, and Poisson ratio for a number of involved materials are listed below.

| Material | E (GPa) | $\alpha$ ($10^{-6}$/K) | v |
|---|---|---|---|
| Al alloy | 60 | 27 | 0.33 |
| $SiO_2$ | 94 | 0.5 | 0.25 |
| PSG | 79 | 2 | 0.2 |
| BPSG | 120 | 2 | 0.2 |
| Silicon oxynitride | 150 | 2 | 0.25 |
| Silicon nitride | 158 | 2 | 0.25 |

Employing Eqn (3.8), calculate the thermal stresses induced in Al in contact with each of the dielectrics and insulators. How do the values compare with the average yield stress of 80 MPa for Al–Si alloys?

**5.20** Compare the critical Blech length in unpassivated as well as passivated aluminum interconnects at 100 °C through which a current density of $2 \times 10^5$ A/cm$^2$ flows. Assume that $Z^\star$ is $-10$ and passivation is accomplished with $SiO_2$ deposited at 425 °C. Typical stresses in thin films can be found in Table 3.5.

**5.21** In aluminum films passivated by SiN, SV occurs with an $n$ value equal to 4. Sketch the trends for SV as a function of temperature. Suggest ways to minimize $R_{SV}$ in Al.

**5.22** Consider a thin-film interconnect with a bamboo grain structure. One of the GBs containing a small void of radius $r$ is stressed in tension along the stripe axis.

**a.** Write expressions for the concentration of vacancies on the void and in the GB.

**b.** Describe the motion of vacancies and the nature of damage they produce.

**5.23** A bi-layer metal-film line of aluminum on titanium is deposited having a length $L$ and a width $W$. The Al film is $d_{Al}$ thick, while $d_{Ti}$ is the thickness of the Ti film. During powering, a void of length $L_v$, width $W$, and thickness $d_{Al}$ totally displaces the original aluminum in a central portion of the line, leaving only Ti to carry the current. Derive an expression for the increase in electrical resistance of the voided region relative to the original line in terms of the resistivities of Al and Ti ($\rho_{Al}$, $\rho_{Ti}$) and conductor geometries.

**5.24** Future metallizations may very well utilize copper instead of aluminum. Discuss the likely implications of such a processing change with respect to the reliability issues of EM, SV, contamination of silicon devices, and corrosion.

**5.25** During testing of 100 W light bulbs a manufacturer measured a tungsten filament temperature of 1900 °C and claimed that the bulbs will last 1000 h. However, the average electric power drawn by the bulbs in service is greater than that used during testing. Measurements show that the electrical resistance of the bulb under use conditions is 5% higher than during testing. What is a more realistic value for bulb life under these conditions?

## REFERENCES


[1] M. Ohring, The Materials Science of Thin Films, Academic Press, Boston, 1992.
[2] R.W. Balluffi, J.M. Blakely, Thin Solid Films 25 (1975) 363.
[3] D. Gupta, in: D. Gupta, P.S. Ho (Eds.), Diffusion Phenomena in Thin Films and Microelectronic Materials, Noyes Publications, Park Ridge, NJ, 1988.
[4] H.K. Charles, in: Electronic Materials Handbook, Packaging, ASM International, Materials Park, Ohio, vol. 1, 1989, p. 224.
[5] J.W. Mayer, S.S. Lau, Electronic Materials Science for Integrated Circuits in Si and GaAs, Macmillan, New York, 1990.
[6] S. Vaidya, J. Electron. Mater. 10 (1981) 337.
[7] C.C. Goldsmith, G.A. Walker, N.J. Sullivan, in: 16th Annual Proceedings of the IEEE Reliability Physics Symposium, 1978, p. 64.
[8] F. Ren, A.B. Emerson, S.J. Pearton, T.R. Fullowan, J.M. Brown, Appl. Phys. Lett. 58 (1991) 1030.
[9] F. Magistrali, C. Tedesco, E. Zanoni, C. Canali, in: A. Christou (Ed.), Reliability of Gallium Arsenide MMICs, Wiley, Chichester, 1992.
[10] H.L. Hartnagel, in: A. Christou, B.A. Unger (Eds.), Semiconductor Device Reliability, Kluwer, 1990.
[11] W.T. Anderson, Tutorial Notes, IEEE International Reliability Physics Symposium, 7.1, 1992.
[12] A. Christou, in: A. Christou (Ed.), Reliability of Gallium Arsenide MMICs, Wiley, Chichester, 1992.
[13] H. Goronkin, D. Convey, IEEE T. Electron Dev. 36 (1989) 600.
[14] A. Bensoussan, P. Coval, W.J. Roesch, T. Rubalcava, in: 32nd Annual Proceedings of the IEEE Reliability Physics Symposium, 1994, p. 434.
[15] D.J. LaCombe, W.W. Hu, F.R. Bardsley, in: 31st Annual Proceedings of the IEEE Reliability Physics Symposium, 1993, p. 364.
[16] J.R. Devaney, in: Electronic Materials Handbook, Packaging, ASM International, Materials Park, Ohio, vol. 1, 1989, p. 1006.
[17] J.T. May, M.L. Gordon, W.M. Piwnica, S.M. Bray, in: ASM International Symposium for Testing and Failure Analysis (ISTFA), 1989, p. 121.
[18] M. Coxon, C. Kershner, D.M. McEligot, IEEE T. Compon. Hybr. 9 (1986) 279.
[19] J.R. Lloyd, M. Shatzkes, in: 26th Annual Proceedings of the IEEE Reliability Physics Symposium, 1988, p. 216.
[20] I.A. Blech, E. Kinsbron, Thin Solid Films 25 (1974) 327.
[21] M. Ohring, R. Rosenberg, J. Appl. Phys. 42 (1971) 5671.
[22] M. Ohring, Mat. Sci. Eng. 7 (1971) 158.

[23] M.H. Wood, S.C. Bergman, R.S. Hemmert, in: 29th Annual Proceedings of the IEEE Reliability Physics Symposium, 1991, p. 70.
[24] M. Ohring, J. Appl. Phys. 42 (1971) 530.
[25] J.C. Ondrusek, C.F. Dunn, J.W. McPherson, in: 25th Annual Proceedings of the IEEE Reliability Physics Symposium, 1987, p. 154.
[26] S.D. Steenwyk, E.F. Kankowski, in: 24th Annual Proceedings of the IEEE Reliability Physics Symposium, 1986, p. 30.
[27] J. Tao, K.K. Young, N.W. Chung, C. Hu, in: 30th Annual Proceedings of the IEEE Reliability Physics Symposium, 1992, p. 338.
[28] T. Yamaha, S. Naitou, T. Hotta, in: 30th Annual Proceedings of the IEEE Reliability Physics Symposium, 1992, p. 349.
[29] P.S. Ho, D. Bouldin, Tutorial Notes, in: IEEE International Reliability Physics Symposium, vol. 5.1, 1995.
[30] J.R. Black, IEEE T. Electron Dev. ED-16 (1969) 338.
[31] M. Sakimoto, T. Itoo, T. Fujii, H. Yamaguchi, K. Eguchi, in: 33rd Annual Proceedings of the IEEE Reliability Physics Symposium, 1995, p. 333.
[32] A. Scorzoni, B. Neri, C. Caprile, F. Fantini, Mater. Sci. Rep. 7 (1991) 143.
[33] S. Vaidya, T.T. Sheng, A.K. Sinha, Appl. Phys. Lett. 36 (1980) 464.
[34] D.L. Barr, Electron Backscatter Diffraction and Focused Ion Beam Studies of VLSI Aluminum Interconnect Microstructure (Ph.D. thesis), Stevens Institute of Technology, 1996.
[35] S.P. Sim, Microelectron. Reliab. 19 (1979) 207.
[36] A. Bobbio, O. Saracco, Thin Solid Films 17 (1973) S13.
[37] J.R. Lloyd, J. Kitchin, J. Appl. Phys. 69 (1991) 2117.
[38] C.V. Thompson, J.R. Lloyd, MRS Bull. XVIII (12) (1993) 18.
[39] R.W. Pasco, J.A. Schwarz, in: 21st Annual Proceedings of the IEEE Reliability Physics Symposium, 1983, p. 10.
[40] C.C. Hong, D.L. Crook, in: 23rd Annual Proceedings of the IEEE Reliability Physics Symposium, 1985, p. 108.
[41] B.J. Root, T. Turner, in: 23rd Annual Proceedings of the IEEE Reliability Physics Symposium, 1985, p. 100.
[42] C.-K. Hu, K.P. Rodbell, T.D. Sullivan, K.Y. Lee, D.P. Bouldin, IBM J. Res. Develop. 39 (1995) 465.
[43] P. Borgesen, M.A. Korhonen, C.-Y. Li, Thin Solid Films 220 (1992) 8.
[44] J. Curry, G. Fitzgibbon, Y. Guan, R. Muollo, G. Nelson, A. Thomas, in: 22nd Annual Proceedings of the IEEE Reliability Physics Symposium, 1984, p. 6.
[45] J.T. Yue, W.P. Funsten, R.V. Taylor, in: 23rd Annual Proceedings of the IEEE Reliability Physics Symposium, 1985, p. 126.
[46] S.F. Groothuis, W.H. Schroen, in: 25th Annual Proceedings of the IEEE Reliability Physics Symposium, 1987, p. 1.
[47] M.A. Korhonen, P. Borgesen, K.N. Tu, C.-Y. Li, J. Appl. Phys. 73 (1993) 3790.
[48] F.G. Yost, F.E. Campbell, IEEE Circuits Devices 6 (5) (1990) 40.
[49] F.G. Yost, D.E. Amos, A.D. Romig, in: 27th Annual Proceedings of the IEEE Reliability Physics Symposium, 1989, p. 193.
[50] Q. Guo, L.M. Keer, Y.-W. Chung, in: J.H. Lau (Ed.), Thermal Stress and Strain in Microelectronics Packaging, Van Nostrand Reinhold, New York, 1993.
[51] W.A. Backofen, Deformation Processing, Addison-Wesley, Reading, Mass, 1972.
[52] J.W. McPherson, C.F. Dunn, J. Vac. Sci. Technol. B.5 (1987) 1321.
[53] J. Klema, R. Pyle, E. Domangue, in: 22nd Annual Proceedings of the IEEE Reliability Physics Symposium, 1984, p. 1.
[54] A. Tezaki, T. Mineta, H. Egawa, in: 28th Annual Proceedings of the IEEE Reliability Physics Symposium, 1990, p. 221.

[55] H. Shibata, T. Matsuno, K. Hashimoto, in: 31st Annual Proceedings of the IEEE Reliability Physics Symposium, 1993, p. 340.
[56] H.S. Rathore, Tutorial Notes, in: IEEE International Reliability Physics Symposium, vol. 3.1, 1997.
[57] T.N. Theis, IBM J. Res. Develop 44 (3) (2000) 379.
[58] D. Edelstein, H. Rathore, et al., Comprehensive reliability evaluation of a 90 nm CMOS technology with Cu/PECVD low-k BEOL, in: 42th Annual Proceedings of the IEEE Reliability Physics Symposium, 2004, p. 316.
[59] D.J. Poindexter, S.R. Stiffler, et al., Optimizaion of silicon technology for the IBM system z9, IBM J. Res. Dev. 52 (2007) 5–18.
[60] J. Koike, Tutorial Notes, in: IEEE International Reliability Physics Symposium, vol. 243.1, 2006.
[61] A.H. Fischer, O. Aubel, et al., Reliability challenges in copper metallizations arising with the PVD resputter liner engineering for 65 nm and beyond, in: 45th Annual Proceedings of the IEEE Reliability Physics Symposium, 2007, p. 511.
[62] M.W. Lane, E.G. Liniger, J.R. Lloyd, J. Appl. Phys. 93 (2003) 1417–1421.
[63] [a] C.L. Gan, C.V. Thompson, et al., Mat. Res. Soc. Proc. 766 (2003). E1.5.1–6; [b] C.L. Gan, C.V. Thompson, et al., Mat. Res. Soc. Proc. 766 (2002). B8.13.1–6.
[64] K. Maekawa, K. Mori, et al., Improvement in reliability of Cu dual-damascene interconnects using Cu-al Alloy seed, in: Advanced Metallization Conference Asian Session (ADMETA2004), 2004.
[65] A.V. Vairagar, Z. Gan, Wei Shao, et al., Improvement of electromigration lifetime of submicrometer dual-damascene Cu interconnects through surface engineering, J. Electrochem. Soc. 153 (2006) G840–G845.
[66] S. Luce, in: IEE ITC, 1998.
[67] E.T. Ogawa, J.W. McPherson, et al., Stress-induced voiding under vias connected to wide Cu metal leads, in: 42th Annual Proceedings of the IEEE Reliability Physics Symposium, 2002, p. 312.
[68] L. Peters, Exploring advanced interconnect reliability, Semiconductor Int. (8) (2002).
[69] C.-K. Hu, L. Gignac, B. Herbst, D.L. Rath, Comparison of Cu electromigration in Cu interconnects with various caps, Appl. Phys. Lett. 83 (2003) 869–871.
[70] F. Wei, C.L. Gan, C.V. Thompson, et al., Mat. Sci. Soc. Proc. B13 (3.1–6) (2002) 716.
[71] EE Times, IBM Airgap Microprocessor, IBM Brings Nature to Computer Chip Manufacturing 05.03.07.

# CHAPTER 6

# Electronic Charge-Induced Damage

## 6.1 INTRODUCTION

To the extent that the degradation and failure effects are the consequence of matter transport, this time of electrons and holes, we may view this chapter as a companion to the previous one. Whereas metal–metal, metal–semiconductor, and electron–metal interactions were largely the focus of Chapter 5, here our interest is primarily in insulators and the corresponding electron (hole)–electron or electron–atom interactions. The physical laws that govern the motion of neutral atoms and electronic carriers differ, and the damage wreaked by their motion also differs markedly. Insulators, and to a lesser extent semiconductors, rather than conductors, are the primary victims of electric field-induced accumulation and transport of electronic charge. The failure mechanism depends on how the charge was generated, and where it ultimately resides or is dissipated. This process of generation, transport, accumulation, and discharge can be quite brutal, to say nothing of how destructive. Nonsemiconductor electrical components, such as transformer oil and high-voltage ceramic and polymer insulation, are also susceptible to electron charging and discharging damage.

The phenomena of breakdown, hot electron transport, charge trapping, and electrostatic discharge and all serious reliability issues in integrated circuit devices, are the primary focus of this chapter. Central to an understanding of these specific charge-induced damage phenomena are the electrical conduction mechanisms in insulators [1]. In all cases, damage is directly proportional to the amount of accumulated charge, as well as, the way it is subsequently redistributed. We start by considering the nature of charge transport in insulators, materials we do not normally think of as being able to conduct electricity. Some feel for the rapidity of these phenomena that can be obtained by considering the dimensions and velocities involved. A typical event takes place in the space of a few to tens of atoms, which implies maybe 1–10-nm distance. Typical drift velocities for electrons in insulators and semiconductors range from $10^5$ to $10^7$ (cm/s). Hence

*Reliability and Failure of Electronic Materials and Devices*
ISBN 978-0-12-088574-9
http://dx.doi.org/10.1016/B978-0-12-088574-9.00006-9

these phenomena occur in less than a picosecond ($10^{-12}$ s). The result is often very subtle and cumulative over time, resulting in parametric drift, or it can be explosive and seem almost instantaneous.

## 6.2 ASPECTS OF CONDUCTION IN INSULATORS

### 6.2.1 Current–Voltage Relationships

When sufficiently high electric fields, $E$(V/cm), are applied to insulators, the response is a passage of a current of density $j$ (A/cm$^2$). If $E$ is small, the response is generally linear or ohmic, and the conductivity ($\sigma$), given by $j/E$, is independent of $E$. For larger applied fields the $j$–$E$ relationships defining charge transport through insulators are more complex. Depending on mechanism, the conduction characteristics are described by one or more of the equations listed in Table 6.1. In most cases the magnitude of $j$ is strongly temperature dependent according to the usual Maxwell–Boltzmann factor.

Charge injection into insulators from either metal or semiconductor contacts is usually modeled in terms of band diagrams. The case of electron injection from the gate electrode into the oxide of an MOS structure is schematically indicated in Figure 6.1. Even modest voltages ($V$) can generate extremely high electric fields across thin oxides of thickness ($d_o$) because $E = dv/dx$ or $V/d_o$. The electric field supported by the insulator is represented by the sloped potential curve; steeper slopes mean higher $E$ fields. Where local fields are highest, electrons from the gate enter the insulator conduction band by either climbing over the electron energy barrier (Figure 6.1(a)) or by burrowing or tunneling through the triangular barrier shown in Figure 6.1(b). The first of these processes, known as Schottky emission, resembles thermionic emission and typically has the $T^2$ exp $(-\Phi_B/kT)$ dependence, where $\Phi_B$ is the barrier height. However, the applied electric field causes a continuously varying electron potential-energy variation at the interface rather than an abrupt change. When an electron enters the insulator, it produces a so-called image field that adds to the barrier field, effectively lowering it. The result is a reduced barrier($\Phi_b$, $\Phi_w$)height (Figure 6.1(c)) and an enhanced current given by the second exponential factor in Eqn (6.1). Overall, the $j - E$ characteristics reveal that current injection depends strongly on the barrier height, temperature, and electric field.

The electron-tunneling mechanism is quantum-mechanical in origin and depends on there being empty states, in the insulator, available for

**Table 6.1** Conduction processes in insulators

| Process | $j$–$E$ characteristics | Voltage (V), temperature (T) dependence |
|---|---|---|
| **1.** Schottky emission | $j = A^* T^2 \exp\left[\frac{-q\left[\Phi_B - (qE/(4\pi\varepsilon))^{1/2}\right]}{kT}\right]$ | $\approx T^2 \exp\left[\frac{+aV^{1/2} - b}{T}\right]$ (6.1) |
| **2.** Frenkel–Poole emission | $j = B \exp\left[\frac{-q\left[\Phi_B - (q\varepsilon/(\pi\varepsilon))^{1/2}\right]}{kT}\right]$ | $\approx V \exp\left[\frac{+d'V^{1/2} - b}{T}\right]$ (6.2) |
| **3.** Tunneling | $j = \frac{CE^2}{\Phi_B} \exp - \left[\frac{8\pi(2m)^{1/2}(q\Phi_B)^{3/2}}{3hqE}\right]$ | $\approx V^2 \exp(-c/V)$ (6.3) |
| **4.** Space charge limited | $j = \frac{4\varepsilon_o(2q/m)^{1/2}}{9} \frac{E^{3/2}}{d^{1/2}}$ | $\approx V^{3/2}$ (6.4) |
| **5.** Ohmic (intrinsic) | $j \approx ET^{3/2} \exp[-E_g/2kT]$ | $\approx VT^{3/2} \exp(-eE_g/T)$ (6.5) |
| **6.** Ionic conduction (low field) | $j \approx \frac{E \exp[-E_i/kT]}{T}$ | $\frac{\approx V \exp(-E_i/T)}{T}$ (6.6) |
| **7.** Ionic conduction (high field) | $j \approx \sinh[qaE/kT]\exp[-E_i/kT]$ | $\approx \sinh(gV/T)\exp(-hE_i/T)$ (6.7) |

$A^\star$ = Richardson constant, $\Phi_B$ = barrier height, $E$ = electric field, $\boldsymbol{\varepsilon}$ = insulator dielectric constant, $\boldsymbol{\varepsilon}_o$ = permittivity of free space, $m$ = effective electron mass, $d$ = insulator thickness, $h$ = Planck's constant, $E_g$ = insulator energy gap, $E_i$ = activation energy for ion creation and motion. Constants $B$, $C$, $a$, $d'$, $b$, $c$, $e$, and $f$ are independent of $V$ and $T$.
After Ref. [1].

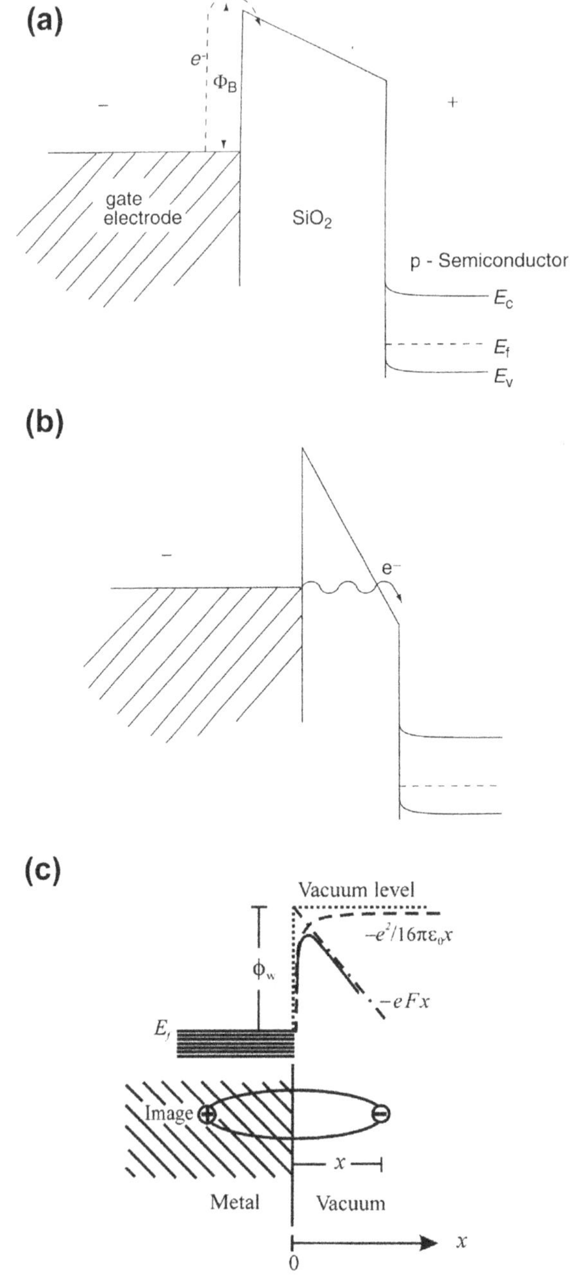


**Figure 6.1** Band diagram of MOS structure with negative voltage applied to the gate electrode. Electrons are injected into the oxide conduction band by: (a) climbing over the electron energy barrier; (b) Fowler–Nordheim tunneling (at large gate bias); (c) Barrier height lowering.

electrons to occupy. Resulting currents depend very strongly on $E$, but unlike other conduction mechanisms, tunneling characteristics (Eqn (6.3)) are temperature-independent. Tunneling phenomena play important roles in dielectric breakdown failure and hot-electron degradation. In view of this, tunneling current-field characteristics are shown in Figure 6.2, where $\ln j/E^2$ is plotted versus $1/E$ [2]. Shrinking device features make tunneling a favored mechanism for charge injection into both semiconductors and insulators.

Once inside the insulator, injected charge often resides in traps for times that are either very short, very long, or periods in between, relative to typical mean times to failure for electronic equipment. Traps maybe thought of as quantum-mechanical states in the forbidden energy gap having energies somewhere between the valence and conduction band edges. They possess a cross section or probability for capturing charge (either positive or negative). Traps are generated at broken silicon-oxygen bonds, defects, dangling bonds, impurity atoms (e.g., hydrogen, nitrogen); they are located within the bulk of the insulator, but often near a semiconductor or metal contact interface.

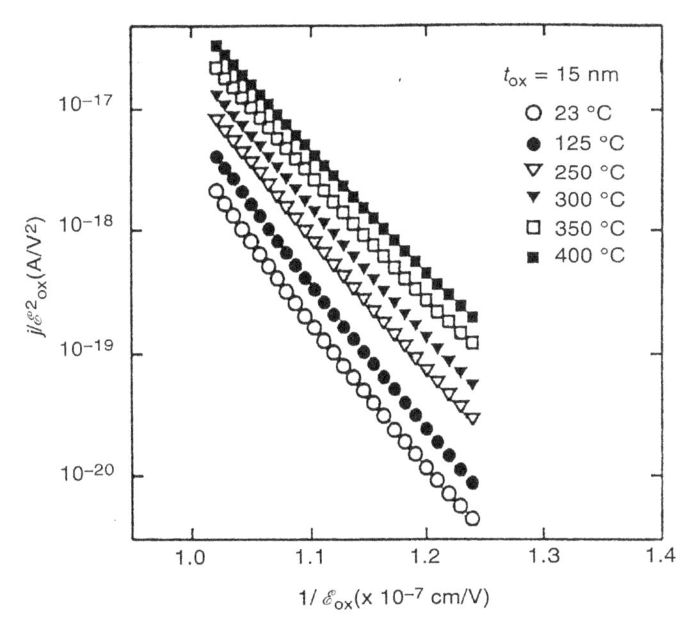


**Figure 6.2** Fowler–Nordheim tunneling current-field characteristics measured in 15 nm thick $SiO_2$ films at temperatures ranging from 25 to 400 °C. Theoretically, Fowler–Nordheim tunneling is temperature-independent. The actual characteristics reveal a slight temperature dependence of the barrier height. *From Ref. [2].*

Once charge is trapped, it takes energy to expel it. Electron or hole charge motion then often means hopping from trap to trap. Frenkel–Poole emission (Eqn (6.2)) and intrinsic conduction (Eqn (6.5)) are two such transport mechanisms. Whereas the former emission process is exponentially accelerated by $E$, the conduction by the latter is linearly dependent in a Boltzmann manner. Often, hopping electrons encounter holes and recombine, thus reducing the overall charge density.

Instead of electron or hole transport, charged ions ($Na^+$, $H^+$, $Ag^+$) often carry current in insulators, especially if moisture is present. Ionic conduction also occurs in glasses and polymers that contain intentionally added ionic compounds ($P_2O_5$). It is found that ionic currents are linearly dependent on $E$ and thermally activated. Unlike the other conduction mechanisms, ionic transport is accompanied by mass transport as well, because matter is depleted at anodes and transferred to cathodes. This may take the form of projections of metal known as whiskers or dendrites that can cause short circuits when they bridge conductors. Such effects are discussed in Chapter 7.

### 6.2.2 Leakage Current

In contrast to the above conduction through the bulk of insulators, *leakage current* is said to flow when charge strays from the intended metals and semiconductors and migrates through bulk defects or along surface layers. It is usually the presence of condensed water on insulators that makes their surfaces conductive. As a result of such surface ion effects, corrosion, electrical arcing, and partial short circuiting, overall circuit characteristics often degrade progressively over a long time, but sometimes with catastrophic abruptness. The surface current is ohmic and generally expressed by a formula of the type

$$j = AE \exp(bxRH) \exp\left(\frac{-E}{kT}\right), \tag{6.8}$$

where $RH$ is the relative humidity, $E$ is the activation energy for conduction, and $A$ and $b$ are constants. Values of $E$ for a number of insulators range from 0.4 to 1.1 eV, while $b$ is approximately 0.1–0.3 [3]. An analysis shows that $b$ depends on the extent of water-vapor adsorption and is thus strongly temperature dependent.

### 6.2.3 $SiO_2$—Electrons, Holes, Ions, and Traps

Silicon technology owes much to the remarkable properties of $SiO_2$ in its crucial function as the gate oxide, for fulfilling passivating, insulating, and dielectric needs and for its important role in masking and processing. However, the demands made on the gate oxide simultaneously make it the target of several important degradation effects in MOS transistors. Thus, dielectric breakdown, hot electron transport [4], radiation damage, and some modes of electrostatic discharge failure are phenomena intimately associated with charge transport in thin $SiO_2$ films. Although the microscopic details are controversial and sometimes confusing [5], the picture of the four different kinds of charge associated with the unbiased $Si-SiO_2$ system and their location within it, is shown in Figure 6.3.

1. Mobile $Na^+$ and $K^+$ *ionic charge*, may be present in $SiO_2$. In the early 1960s p-channel MOSFETs were used because n-channel devices could not be controlled due to the presence of ionic contamination in the gate oxide, in particular $Na^+$. The positive gate bias used in n-channel devices drove the positive ions to the $Si–SiO_2$ interface, causing threshold voltage shifts. Once this was realized, methods (phosphosilicate glass and $Si_3N_4$ barriers) were developed to prevent these ions from entering the gate oxide, and thus allowed n-channel devices to become prevalent. This was an important step forward because n-channel devices were faster than p-channel due to the higher electron mobility compared to hole mobility in p-channel devices. Introduced as unwanted contamination during processing, this source of oxide charge has been virtually eliminated as a source of reliability problems. However a lack of vigilance can still cause ionic contamination in processing.

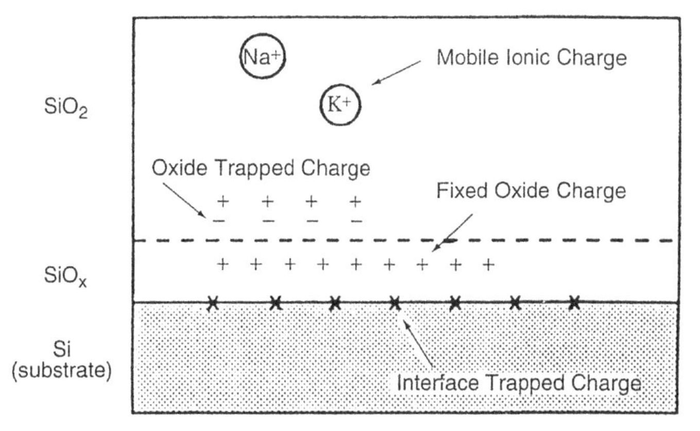


**Figure 6.3** Location and identification of electronic and ionic charge associated with $SiO_2$ films.

2. Secondly, there is the *oxide trapped charge*, which is distributed throughout the bulk of the $SiO_2$ film. A portion of this charge ($Q_b$) stems from processing and can be removed through low-temperature annealing. X-ray radiation and high-energy electrons are additional sources of $Q_b$. Oxide-trapped charge is associated with shallow traps that are not located near the $Si{-}SiO_2$ interface. Electron and hole traps apparently contain Si dangling bonds $(Si{-}O)_3{\equiv}Si^{\bullet}$ and oxygen dangling bonds $(Si{-}O)_3{\equiv}Si{-}O{-}$ to which charge attaches. (The symbol "−" denotes an electron-pair bond, while "•" signifies an unpaired or dangling electron bond.)
3. Next we consider the positive *fixed oxide charge* density ($Q_f$), which is located some 3–5 nm from the $Si{-}SiO_2$ interface. Because such charge is fixed, it does not usually charge or discharge and is largely uninfluenced by the normal operating voltages of the MOS transistor. Regarded as a sheet of charge near the $Si{-}SiO_2$ interface, $Q_f$ is associated with incomplete Si−Si and Si−O bonds. It is present in densities of $10^{10}/cm^2$ on (100) Si surfaces and $\sim 5 \times 10^{10}/cm^2$ on (111) Si surfaces. In addition to silicon orientation, the distribution and density of $Q_f$ is related to the thermal oxidation process and depends on the oxidation temperature and subsequent cooling rate, as well as the oxidizing species ($O_2$, $H_2O$, $OH^-$)
4. Lastly, there is *interfacial oxide charge* density, $Q_i$, that is localized within $\sim$0.2 nm of the $SiO_2{-}Si$ interface. This charge is associated with interface traps that are also called interface, surface, and fast states. $Q_i$ arises from oxidation-induced structural defects, metal impurities, and broken bonds due to charge injection (e.g., radiation, hot electrons). The Si wafer orientation dependence of $Q_i$ is similar to that for trapped interfacial charge. To borrow a term from the Civil War, these states have been aptly called "Border States" [6] or traps because of their mixed allegiance to bulk and interfacial character. Furthermore, the interfacial states are amphoteric; they are acceptor-like (negative when filled) in the upper half of the Si band gap, and donor-like (positive when empty) in the lower half of the band gap.

### 6.2.4 Consequences of Charge in $SiO_2$—Instability and Breakdown

How did oxide charge get there in the first place? The first oxidation process and all subsequent oxidation and annealing process steps contribute to this charge. As we discussed in Chapter 2 (Sections 2.5.2 and 2.5.3) the total charge and its distribution, in the gate dielectric of a MOSFET, determines its threshold voltage ($V_t$). So the device, as fabricated, has a certain $V_t$, which may change if the charge in the gate oxide changes. One model suggests that, under the influence of an applied field, either the available electrons and holes recombine immediately, or they begin to migrate through the oxide. When the gate electrode is positively biased, electrons are rapidly swept toward the gate and collected there. The holes and positive ions slowly diffuse to the $SiO_2$—Si interface, where some fraction of them are trapped in deep hole (fixed) oxide traps some 5 nm from the interface. Holes bound to shallow traps give rise to bulk oxide-trapped charge that is present farther from the interface. Importantly, some holes make it to the $SiO_2$—Si interface and are trapped there. This interfacial charge can then rapidly exchange places with vacant states in the adjoining Si band gap. The objective, of a final anneal of semiconductor devices, is to stabilize the device parameters so that charges do not move around during the service life of the device. If the processing and anneal are successful the device will have a long and stable life. However, if the operation of the device causes any charges to move around, then particular instability mechanisms will cause the device parameters, e.g., $V_t$, to change. Later in this chapter we will discuss these mechanism ($\pm V_t$ drift, hot carrier injection, negative or positive Bias temperature instability (NBTI or PBTI)). Another consequence of the movement of charges in the gate oxide is possible dielectric breakdown, which destroys the device. The injection mechanisms in Figure 6.1 are of particular importance to these breakdown mechanisms. Shorting of devices or shorting between metal levels of an integrated circuit are catastrophic events that may cause permanent device failure.

It is, the dynamic generation of new and/or the alteration of existing charge states, under unusually high electric fields, or local defects where fields are large, that ultimately causes dielectrics to breakdown. This is the subject we now consider.

## 6.3 DIELECTRIC BREAKDOWN

### 6.3.1 Introduction

Short-circuit failure of dielectric materials, used in capacitors, insulators, and encapsulants, that stems from electric fields applied across them is known as dielectric breakdown. Ceramic oxides, glasses, ionic compounds, polymers, and dielectric fluids are all susceptible to such failure.

Manifestations of the damage include the pitting, cratering, and melted regions shown in Figure 6.4. Breakdowns are often classified as *electronic* or *thermal*. Electronic breakdown mechanisms serve as the catalyst for the thermal damage that invariably follows to destroy the dielectric. *Intrinsic* and *defect related* are two other descriptors for breakdown. As the term implies, intrinsic breakdown refers to failure in the absence of defects. During electrical stressing of dielectrics, the largest electric field sustainable maybe assumed to be the intrinsic breakdown field. It is impractical to entirely rid dielectrics of defects during manufacture. As such then defects reduce the breakdown field relative to the intrinsic level in ways discussed in Section 6.3.3.1. Table 6.2 lists the dielectric breakdown fields of assorted ceramics and glasses near the intrinsic limit.

## 6.3.2 A Brief History of Dielectric Breakdown Theories

Attempts to explain dielectric breakdown phenomena date at least to the 1930s, when von Hippel calculated the breakdown field by equating the energy gained by electrons in the field to the energy lost through interaction with the atoms of the dielectric. A fuse-like breakdown was assumed to occur by joule heating. It is apparent that the field strength applied and the current it produces are the crucial parameters that define dielectric breakdown.

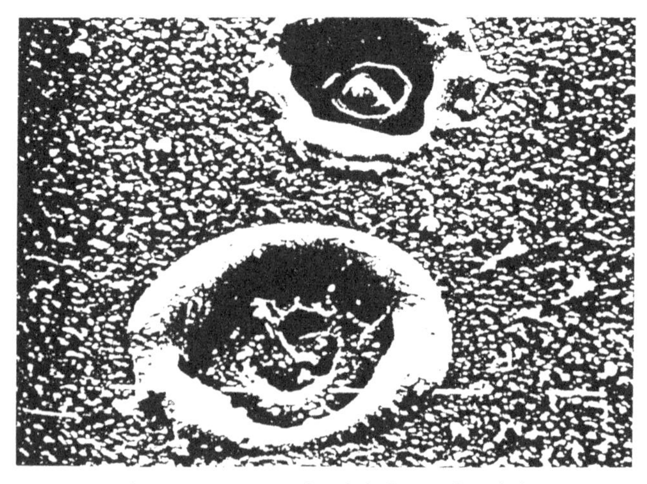

**Figure 6.4** Scanning electron micrograph of dielectric breakdown in a 40 nm thick $SiO_2$ film. The crater diameter is ~3 μm. *From Ref. [7].*

**Table 6.2** Dielectric breakdown strengths for ceramics and glasses

| Material | Form | Thickness (cm) | Temperature °C | Breakdown Field ($10^6$ V/cm) |
|---|---|---|---|---|
| $Al_2O_3$ | Anodized film | $3 \times 10^{-6}$ | 25 | 7.0 |
| 99.5% $Al_2O_3$ | Polycrystalline bulk | 0.63 | 25 | 0.18 |
| Alumina porcelain | | 0.63 | 25 | 0.15 |
| High-voltage porcelain | | 0.63 | 25 | 0.15 |
| Low-voltage porcelain | | 0.63 | 25 | 0.03 |
| Lead glass | | 0.02 | 25 | 0.25 |
| Lime glass | | 0.004 | 25 | 2.5 |
| Borosilicate glass | | 0.0005 | 20 | 6.5 |
| Quartz crystal | | 0.005 | 20 | 6.0 |
| Quartz fused | | 0.005 | 20 | 6.6 |
| $TiO_2$ (‖ optic axis) | | 0.01 | 25 | 0.02 |
| (‖ optic axis) | | 0.01 | 25 | 0.12 |
| $BaTiO_3$ | Single crystal | 0.02 | 0 | 0.040 |
| | Polycrystal | 0.02 | 25 | 0.12 |
| $SrTiO_3$ | Single crystal | 0.046 | 25 | 0.41 |
| $PbZrO_3$ | Dense polycrystal | 0.016 | 20 | 0.079 |
| $PbZrO_3$ | (10% porosity) | 0.016 | 20 | 0.033 |
| Mica | (Muscovite crystal) | 0.002 | 20 | 10.1 |
| $SiO_2$ | Thin film | | | 11 |
| $Si_3N_4$ | Thin film | | | 10 |
| SiN (H) | Thin film | | | 6 |

Data from Ref. [8].

Subsequent treatments [7,9] and theories of dielectric breakdown considered the following factors.

1. The collective energy of the entire electron distribution, not just individual electrons.
2. Thermal-runaway phenomena. In bootstrap fashion, joule heating as a result of high localized currents makes the dielectric matrix more conductive, so that it passes more current, heating it further, making it even more conductive, etc., until permanent damage occurs.

3. Electron multiplication and avalanche occur in the high field due to electron-matrix atom collisions. Such impact ionization can rapidly build the charge density. For example, it has been suggested that approximately 40 such impacts are required for breakdown if two electrons are liberated in a collision. As a result, $2^{40}$, or $1.1 \times 10^{12}$, electrons would be generated. Theories based on electron avalanches are readily applicable to thin films.
4. Field emission from the cathode as the charge-injection mechanism. Then, as in case 3, impact ionization proceeds to generate an avalanche.
5. Space charge alteration of the effective field at the injecting electrode. Mobile space charge generated by ionic contaminants is responsible for time-dependent conduction and breakdown effects.
6. Field enhancement due to asperities on the dielectric surface.
7. Internal field distortion as a result of successive multiplication of charge carriers by impact ionization.
8. Interface-state generation during constant current injection.

The list is incomplete, but it gives a sense of what has been considered and modeled.

### 6.3.3 Current Dielectric Breakdown Theories

It is now generally agreed that the electrical breakdown of thin oxides proceeds by a linked, two-step process involving *wear out* followed by *thermal runaway* failure [10,11]. During wear out the applied voltages and currents generate an assortment of defects and charge traps or states that accumulate both within the bulk oxide and at either oxide interface. With time, charge accumulation raises the density of defects and traps to a critical level, where sufficiently high local electric fields and currents are generated to cause a thermal runaway and melting of microscopic regions. Thus, whereas wear out occurs globally, breakdown occurs locally. The specific operative mechanisms, as well as the equations that define them and are used for managing failure data, are matters of controversy. The reciprocal and linear electric field models have emerged in recent years as the two most popular ways of describing dielectric breakdown.

#### *6.3.3.1 Reciprocal Electric Field Model*

In the approach by Hu and coworkers [12,13], a hole-induced oxide breakdown model is assumed. Electrons injected from the gate metal cathode into the oxide undergo impact ionization events that generate holes in the process. The slow-moving holes become trapped in the oxide near the cathode, distorting the band diagram and increasing the electric field there, as shown in Figure 6.5(a). (Recall that the slope of the inclined insulator band lines reflects the magnitude of the electric field.) Electron tunneling is enhanced in the high field, resulting in greater current injection. An alternative mechanism for breakdown stresses the role of anode hole injection [14] into the oxide valence band. In the model of Figure 6.5(b) electrons drop down to the Fermi level of the anode, where the ~3.1 eV energy released breaks bonds at the $SiO_2$–metal interface. This bond breaking proceeds from anode to cathode, forming a convenient conductive path for the discharge that causes failure.

In either case the injected oxide charge is generally accommodated in traps. Early during wear out, shallow traps that weakly bind charge form; but with time, additional traps are generated, while former ones deepen. More and more charge is captured, until a critical hole charge density ($Q_{BD}$), having a typical value of 1–10 C/cm$^2$, eventually causes breakdown and destruction of the dielectric [15]. The dependence of the mean time to failure (MTTF) on $Q_{BD}$ is plotted in Figure 6.6 for two different gate oxides. For dielectric breakdown we shall denote MTTF by $t_{BD}$. Intrinsic breakdown occurs when

$$Q_{BD} = \text{const} = j \times t_{BD}. \tag{6.9}$$

We may thus expect $t_{BD}$ to vary inversely with the tunneling current ($j$), which in turn depends very strongly on $E$ (Eqn (6.3)). Thus, the so-called *reciprocal field* expression for failure time takes the form (for hole trapping)

$$t_{BD} = t_{OR} \exp\left(\frac{G_R}{E}\right) = t_O \exp\left(\frac{G_R X_o}{V_o}\right) \tag{6.10}$$

where $V_o$ and $X_o$ are the oxide voltage and thickness, respectively. Constants $t_{OR}$ and $G_R$ are temperature dependent and given by $t_{OR} = 5.4 \times 10^{-7} \exp(-0.28 \text{ eV}/kT)$ (sec), and $G_R(T) = 120 + \frac{5.8}{kT}$ MV/cm, where $kT$ has the usual meaning. Included in these constants are, among other things, the hole generation and trapping efficiencies. Reference to the Fowler–Nordheim tunneling equation, which reveals that the inverse $E^2$ term has been neglected compared to the exponential dependence on $1/E$.

Early dielectric failures have been attributed to defects at or near the Si–$SiO_2$ interface. Suggested defect sources include Si stacking faults, surface roughness, metallic impurities, and precipitates. The presence of

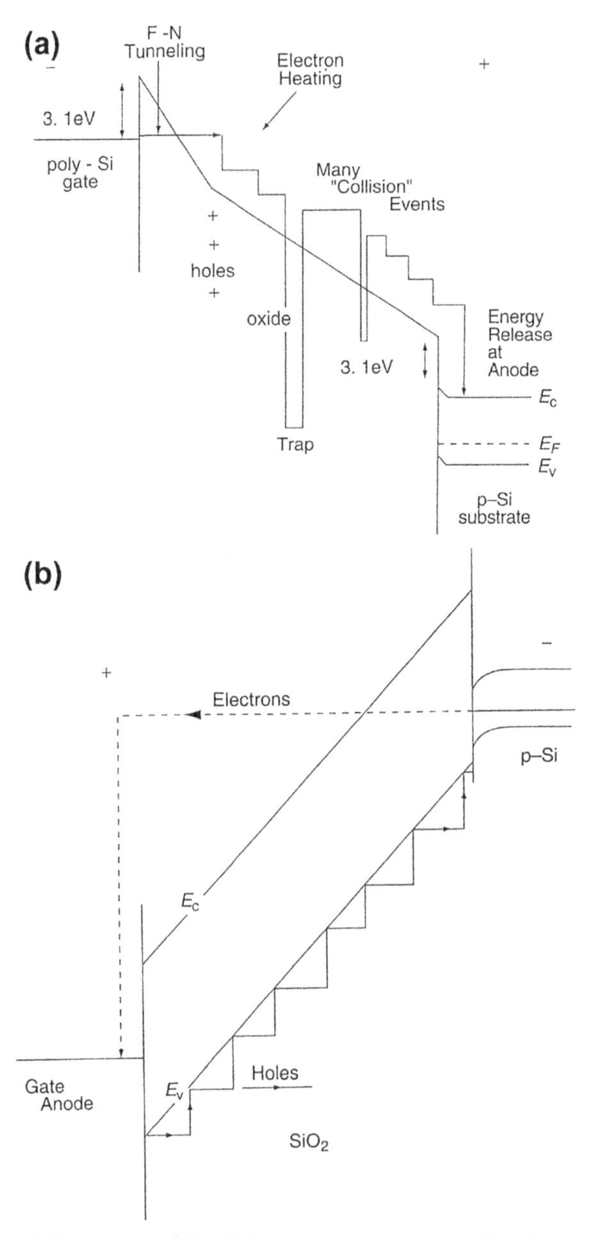


**Figure 6.5** Band diagram models of electron injection into $SiO_2$. (a) Negative bias on the gate electrode traps holes near the cathode. As a result, the local electric field is increased and electron emission into the oxide is enhanced. Electron traps in the oxide are schematically shown. (b) Negative bias on the p-type semiconductor causes holes to travel from the anode gate to the cathode.

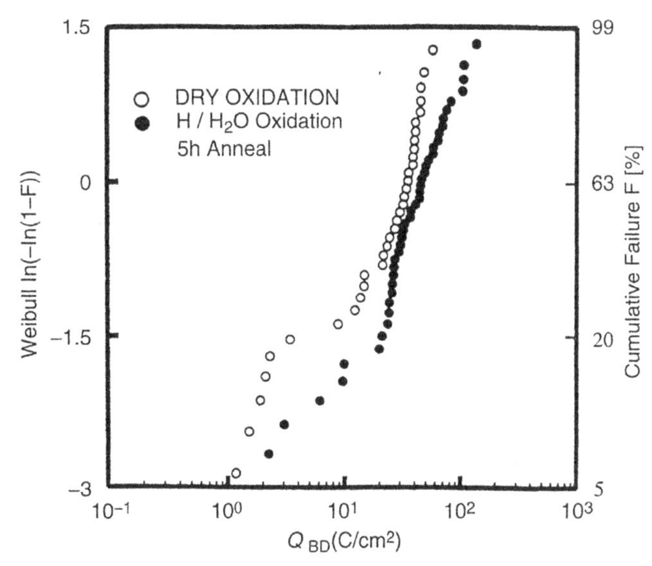


**Figure 6.6** Dielectric breakdown characteristics in 6.3-nm thick dry and wet oxides on p-type silicon as a function of injected charge ($Q_{BD}$). Better $Q_{BD}$ characteristics are observed in wet oxides that are hydrogen annealed after oxidation. The stress current density is 0.1 A/cm$^2$, and the gate area is $5 \times 10^{-5}$ cm$^2$. *From Ref. [15].*

such defects effectively thins the oxide from $X_o$ to $X_{eff}$, where the latter is the effective thickness at the weakest spot in the dielectric. It has been observed that increased microroughness of silicon wafer surfaces lowers the dielectric breakdown field [16]. Under such conditions

$$t_{BD} = t_{OR} \exp\left[\frac{G_R\ X_{eff}}{V_o}\right]. \tag{6.11}$$

The concept of thinning is also applicable to electrode asperities that raise the interfacial electric field because of a "lightning-rod effect." In addition, it is a mathematical way to deal with local regions of modified chemical composition that may promote charge trapping or reduce the $Si–SiO_2$ barrier height (lower $G_R$).

### *6.3.3.2 Linear Electric Field Model*

In contrast to the reciprocal field theory (Eqn (6.10)), the linear field model [17,18] suggests that dielectric breakdown times are given by the expression

$$t_{BD} = t_{OL} \exp(G_L[E_{BD} - E]). \tag{6.12}$$

Both $t_{OL}$ and $G_L$ are also temperature dependent and vary in the same manner as $t_{OR}$ and $G_R$ above, but with different values for the constants. Here $E$ is the applied electric field and $E_{BD}$ is the breakdown field. Unlike the physical basis underlying the reciprocal field model, justification for this equation is largely empirical. Nevertheless, we discern the concept of "distance-to-fail" in the factor ($E_{BD} - E$). This means that dielectric defects with breakdown fields less than $E$ will fail in very short times, so they maybe considered yield failures. On the other hand, those defects with breakdown fields $E < E_{BD}$ will fail in times determined by the magnitude of the difference between the two fields. The greater the difference $E_{BD} - E$, the greater the distance-to-fail. And if that distance is too great, the defect may not fail during the life of the product.

A comparison between the two dependencies for breakdown is made in Figure 6.7, where it appears that plotting the data as a function of the inverse field provides a better fit.

## 6.3.4 Dielectric Breakdown Testing: Ramp Voltage

Two common tests are generally performed to evaluate dielectric breakdown and the resulting reliability of MOS capacitors. The first treated in this section is known as the ramp-voltage test. It consists in stressing the

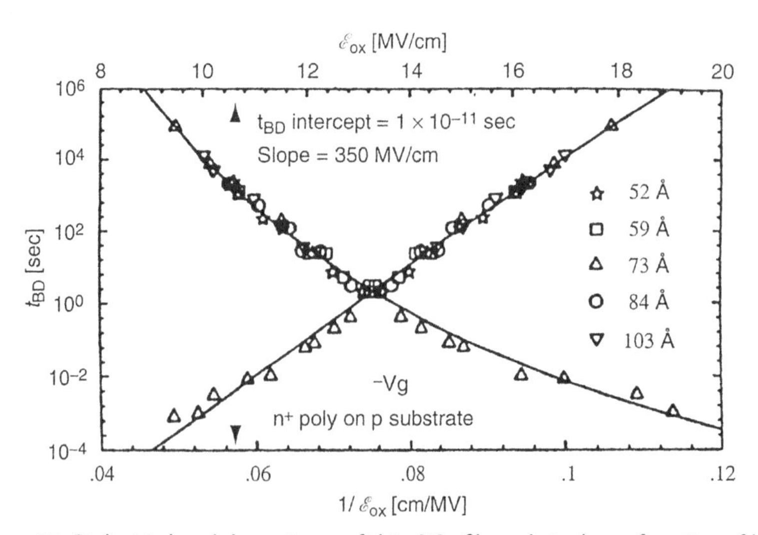


**Figure 6.7** Dielectric breakdown times of thin $SiO_2$ films plotted as a function of both $E$ and $1/E$. The polysilicon gate electrode is negatively biased. For data shown the reciprocal field dependence, rather than linear field dependence more accurately describes breakdown. *From Ref. [19].*

oxide films to breakdown by application of a voltage that linearly increases with time. Speed of acquiring data is an advantage in this testing method, enabling it to be used in assessing wafer-level reliability. Voltage-to-breakdown data are usually modeled by Gaussian or extreme-value distributions. Histograms that plot the number of failed capacitors or devices versus the breakdown field are the common way to display the raw data [20]. As shown in Figure 6.8, two distinct distributions typically appear. One is centered on high fields and represents the main population of failures; while the other, spread over a wide range of low breakdown fields, corresponds to freak behavior and failures of weak devices. These latter populations are normally screened out and do not see service. Data included in Table 6.2 were essentially obtained by ramping the voltage and presumably they correspond to the mean value of the main failure population.

### 6.3.4.1 Ramp-Voltage Breakdown Statistics

Due to its potential connection to weakest-link processes (Section 4.5.2), dielectric breakdown data from ramp-voltage tests are usually plotted as ln (–ln(1 – $F$)) versus ln $E$. This method of plotting emphasizes the distinction between intrinsic and defect-related failures [21]. For example, MOS capacitors often fail at weak spots or defects in the oxide film. These defects are introduced during processing and maybe due to the incorporation of

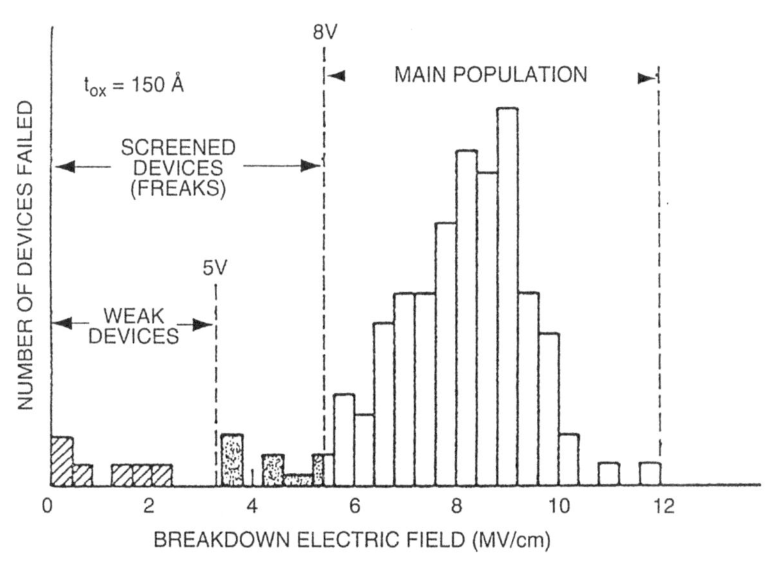


**Figure 6.8** Histogram of oxide breakdown strength. *From Ref. [20].*

contaminants (particularly metals), or to fine cracks, pin-holes, and asperities. Often defects arise during deposition of the gate electrode or subsequent processing. Importantly, these defects cause breakdown to occur at lower electric fields than is the case in intrinsic or defect-free dielectrics. If there are many defects, breakdown will occur at the weakest defect in much the same way a chain breaks at the weakest link.

How do we distinguish between intrinsic and defect-related failures, and how many of each will there be? Answers to these important questions can be obtained by statistically analyzing the failures of large numbers of capacitors. The results of testing some 12,000 capacitors, each with an area of 0.02 $mm^2$, are shown in Figure 6.9. Readers will recognize this as a Weibull plot of the data in the form of percent failures ($F$) versus the electric breakdown field. Plotting the data as $\ln(\ln(1-F)^{-1})$ rather than $F$ accentuates the transition from defect-related failures below $\sim 9 \times 10^6$ V/cm to intrinsic failures. Approximately 7% of the failures are attributed to defects, while the remaining 93% are intrinsic failures, all of which occur at virtually the same breakdown field of $9 \times 10^6$ V/cm. The intrinsic-oxide failure mechanism is generally related to the robustness of oxide processing.

More information can be extracted from these results through further statistical analysis. For one thing, the defect density $D$ (number/$cm^2$) can be determined. If defects are randomly distributed over surface area $A$ and a

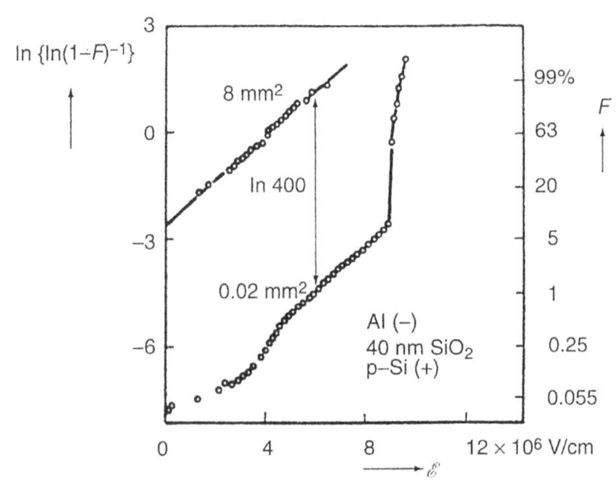


**Figure 6.9** Results of testing some 12,000 MOS (Al-40 nm $SiO_2$-p-type Si) capacitors, each with an area of 0.02 $mm^2$. The Weibull breakdown distribution is plotted as a function of electric field. *From Ref. [21].*

Poisson distribution can be assumed, we can build on our knowledge of yield theory (Eqns (6.3)–(6.24)). Thus, the probability that there will be $n$ defects in the dielectric is

$$P = \frac{(DA)^n}{n!}\exp[-DA], \tag{6.13}$$

and the probability of finding dielectrics without any defects (i.e., $n = 0$) is therefore

$$P = \exp[-DA]. \tag{6.14}$$

By taking probability, $P = 1 - 0.07$ or 0.93, and $A = 0.02\ \text{mm}^2$, the calculated value of the defect density is $D = 3.7 \times 10^2/\text{cm}^2$.

The interesting connection between defect yield modeling in Chapter 3 and the analysis of dielectric breakdown here should be noted. A test of Eqn (6.14) was made by enlarging the capacitor area to 8 mm$^2$, a 400-fold increase over the original contact dimensions. This is physically equivalent to testing groups of 400 of the original capacitors connected in parallel. As predicted, the straight line representing defect-related failures was shifted vertically by a factor of ln 400 in Figure 6.9, assuming the same value of $D$.

## 6.3.5 Dielectric Breakdown Testing: Constant Voltage

In constant-voltage tests the oxide is exposed to higher than designed operating voltages, and the time to breakdown is measured. Known as time-dependent dielectric breakdown (TDDB) life tests, they are essential in order to estimate voltage acceleration factors and model reliability. A great disadvantage of this test is that it takes a long time; months of testing are not uncommon. Failure times have been modeled by both lognormal and Weibull distributions. Data plotted in the Weibull form are shown in Figure 6.10. This figure is interesting because like Figure 6.9, it demonstrates that dielectric breakdowns occur with greater frequency as the electrode area is enlarged [22,23]. In addition, extrinsic failure times are apparently electrode size dependent.

Just as in ramp-voltage testing, intrinsic and extrinsic breakdown failures also occur in constant-voltage tests. The lognormal plots of Figure 6.11 with two linear portions of different slope also clearly show the transition from defect-controlled extrinsic to intrinsic breakdown. On the bathtub curve, intrinsic failures would be characterized by an increasing failure rate in the wear-out portion. There is no need to worry, however, because for

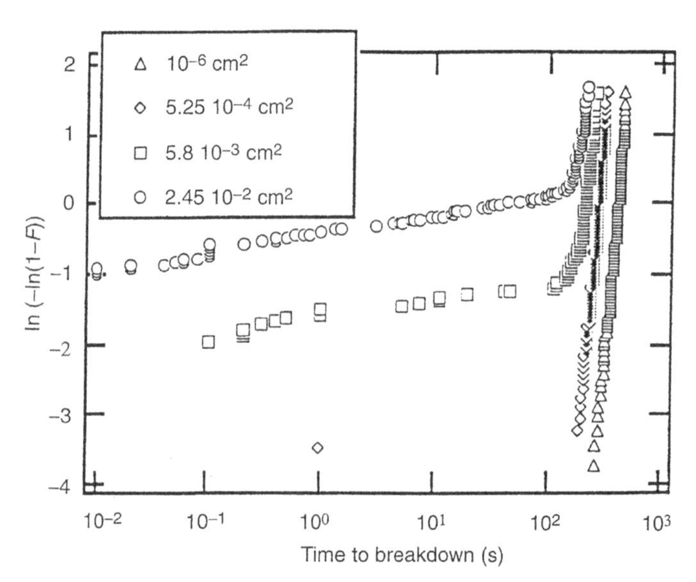


**Figure 6.10** Weibull distributions for time to breakdown in capacitors stressed with a current density of 0.1 A/cm$^2$. Both extrinsic and intrinsic breakdown modes are electrode area dependent. *From Ref. [22].*

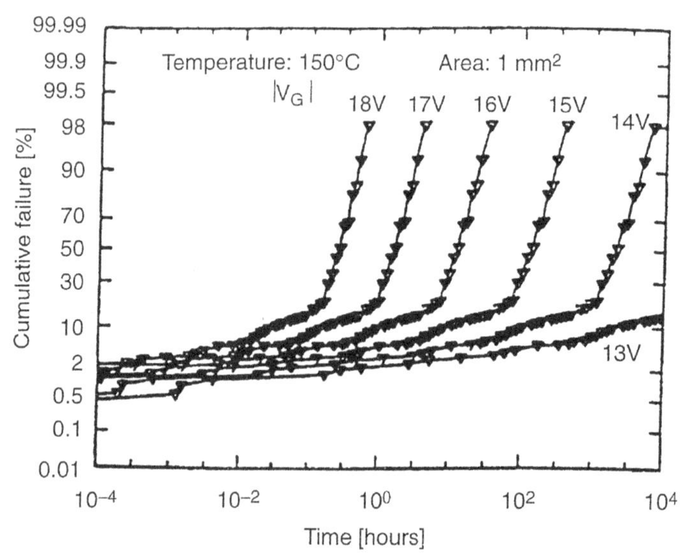


**Figure 6.11** Lognormal plots of dielectric breakdown failures as a function of time. The transition from defect-controlled (extrinsic) failure at short time to intrinsic breakdown at long time is illustrated. *From Ref. [11].*

most mature processes, intrinsic breakdown is not a major issue; extrinsic breakdown will terminate the operational life of the product early.

To completely characterize dielectric life, TDDB tests are generally conducted to span a desired matrix of electric field and temperature values. Thus, the lifetime dependence on $E$ is measured at a particular $T$, while its variation with $T$ is measured at a fixed value of $E$. An example of the former case is shown in Figure 6.12 for $SiO_2$ films ranging in thickness from 2.5 to 22 nm [10]. The results demonstrate the relative independence of failure time on film thickness; instead, the value of $E$ is the significant variable.

In Figure 6.13 an Arrhenius plot of median failure times for 22.5 nm oxides stressed at different electric fields is plotted versus $1/T$. Unlike the data displayed in Figure 4-20 we may conclude that the activation energy for breakdown is virtually independent of electric field in this case.

## 6.3.6 Ramp-Voltage and Constant-Voltage Tests: Is There a Connection?

Driven by the need for faster reliability tests to assess dielectric quality, there have been many attempts [24–26] to reconcile voltage-to-breakdown and time-to-breakdown data from respective ramp-voltage and constant-voltage tests. The task is not always an easy one and generally requires assumptions as to which failure distribution (e.g., lognormal or Weibull)

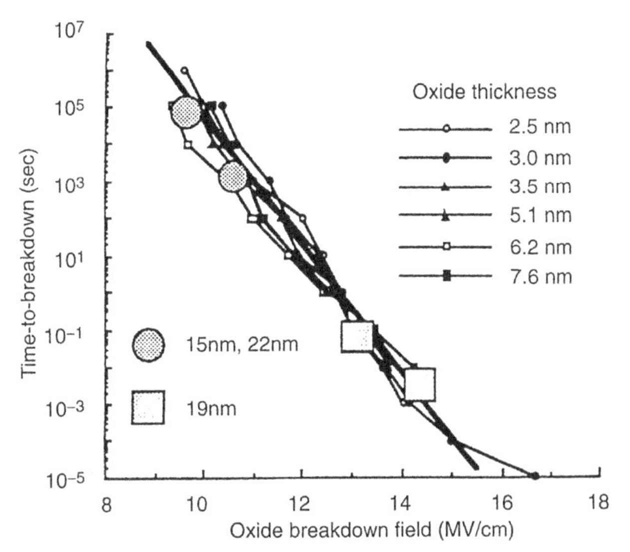


**Figure 6.12** Time to breakdown as a function of electric field for $SiO_2$ films ranging in thickness from 2.5 to 22 nm. *From Ref. [10].*

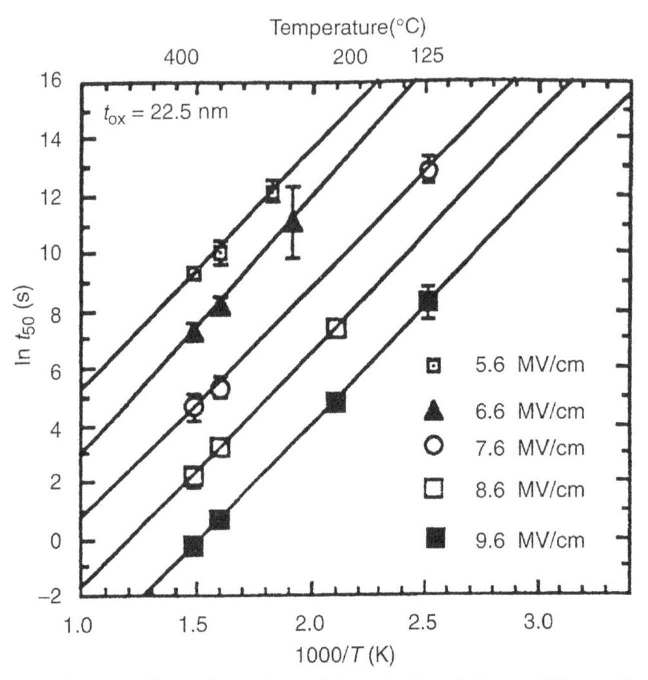


**Figure 6.13** Arrhenius plot of median dielectric breakdown failure times versus *1/T. From Ref. [2].*

applies, and what electric field dependence (e.g., reciprocal or linear) better fits the breakdown data. When correlations between tests are derived, questions of validity and utility always arise; the former because of the assumptions made and the latter because of the scatter and uncertainty inherent in breakdown testing.

The physical approach to the problem by Chen et al. [25] is not difficult conceptually and will be sketched. Breakdown in either test is assumed to occur by the model given above based on electron tunneling. A key assumption supposes that the critical magnitude of trapped hole charge ($Q_{BD}$) necessary to induce thermal runaway and breakdown is the same in both tests. In the constant-voltage test, an integral of the tunneling current with respect to time between $t = 0$ and $t = t_{BD}$, gives $Q_{BD}$ a quantity that depends on $t_{BD}$ and the constant oxide field, $E$. From Eqn (6.9) we have.

$$Q_{BD} = t_{BD} \times K_1 \exp\left(\frac{K_2}{E}\right), \tag{6.15}$$

where $K_1$ and $K_2$ are constants.

In the ramp-voltage test $V$ or $E$ is no longer constant but depends on time ($t$) as $V = Rt$, where $R$ is the voltage ramp rate. Now, $Q_{BD}$ is given essentially by the integral

$$Q_{BD} = \int_{0}^{t_{BD}} K_1 \exp - \left(\frac{K_2}{Rt}\right) dt. \tag{6.16}$$

Upon equating the two values of $Q_{BD}$, the sought-after expression linking $t_{BD}$, $V$(or $\varepsilon$) and $R$ is obtained. A specific formula for room-temperature testing is [19]

$$t_{BD} = t_o(T)\left\{\frac{3.6 \times 10^8 V_{BD}^2}{R}\right\}^{\left[\frac{G(T)V_{BD}}{350V_o}\right]}, \tag{6.17}$$

where $V_{BD} = Rt_{BD}$, and terms $V_o$ $G$, and $t_o$ were defined earlier. Therefore, in a TDDB test a lifetime of 10 years at $V_o = 5.5$ V and $T = 125$ °C corresponds to $V_{BD} = 11.5$ V at $R = 1$ V/s in a ramp-voltage test. However, much testing of both types would be necessary before sufficient confidence develops in the credibility of this correspondence.

### 6.3.7 Electric Field and Temperature-Acceleration Factors

The electric field acceleration factor, $\gamma$, is defined as the negative slope of the log $t_{BD}$ versus $E$ plot derived from constant voltage testing. As a consequence of Eqn (6.10), the reciprocal electric field model yields

$$\gamma = -\frac{d[\log t_{BD}]}{dE} = \frac{G_R}{\ln 10E^2}(\text{decades/MV/cm}), \tag{6.18}$$

while for the linear field model yields,

$$\gamma = G_L. \tag{6.19}$$

Therefore, while $\gamma$ is predicted to be dependent on the electric field according to Eqn (6.18), it assumes a constant value, irrespective of the field, in linear field modeling. The results of investigations over the years [19,25] typically show that $\gamma \approx 6$ decades per MV/cm when $E \approx 5$ MV/cm, and roughly 1.5 decades per MV/cm when $E \approx 10$ MV/cm. Thus, the reciprocal breakdown-field model better accounts for the wide variation in $\gamma$ values reported in the literature.

The temperature acceleration factor, AF($T$), is field dependent in both models. For fixed electric field testing

$$\mathrm{AF}(T) = \frac{t_{\mathrm{BD}}(T_1)}{t_{\mathrm{BD}}(T_2)} = \frac{t_o(T_1)\exp[G(T_1)/E]}{t_o(T_2)\exp[G(T_2)/E]}. \tag{6.20}$$

Accounting for the temperature dependence of both $t_o$ and $G$ at temperatures $T_1$ and $T_2$, the result for the reciprocal field case is

$$\mathrm{AF}(T) = \exp\left[-\frac{E_{\mathrm{BD}}}{k}\left(\frac{1}{T_2}-\frac{1}{T_1}\right)\right]\left\{\exp\frac{[G_{\mathrm{R}}(T_1)-G_{\mathrm{R}}(T_2)]}{E}\right\}. \tag{6.21}$$

Similarly, for the linear field model

$$\mathrm{AF}(T) = \exp\left[\frac{-E_{\mathrm{BD}}}{k}\left(\frac{1}{T_2}-\frac{1}{T_1}\right)\right]$$

$$\{\exp[G_{\mathrm{L}}(T_1)\,(E_{\mathrm{BD}}(T_1)-E) - G_{\mathrm{L}}(T_2)\,(E_{\mathrm{BD}}(T_2)-E)]\}. \tag{6.22}$$

In this case the temperature acceleration factor depends on $T$ and $E$ in a complex way.

If one assumes that the exponential term within each second pair of brackets is weakly temperature dependent relative to the first exponent, then we are left with the simple Boltzmann acceleration factor

$$\mathrm{AF}(T) = \exp\left[-\frac{E_{\mathrm{BD}}}{k}\right]\left(\frac{1}{T_2}-\frac{1}{T_1}\right). \tag{6.23}$$

Older studies quoted by Anolick and Nelson [17] on breakdown in relatively thick $SiO_2$ sandwiched between an assortment of electrodes, e.g., Al–Si, Si–Si, have yielded $E_{\mathrm{BD}}$ values ranging from 0.6 to 2.0 eV. The more recent data of Figure 4-20 on 15 nm $SiO_2$ yield values spanning 0.41–0.86 eV. Lower temperature-dependent $E_{\mathrm{BD}}$ values have also been reported. This broad distribution in measured $E_{\mathrm{BD}}$ values stems from the complex dependence of this energy on oxide thickness, $E$ and $T$, and the inability to isolate the effects of these variables in analyzing dielectric breakdown data.

### 6.3.8 Plasma Charging Damage to Gate Oxides

To combat the danger of gate oxide breakdown it is important to understand what the potential sources of the harmful charge are. Perhaps the most important charge source stems from VLSI plasma etching and deposition processes as previously noted in Section 3.4.2.3; this subject has been recently reviewed by Cheung [27]. Key to the damage induced in the gate oxide is the antenna effect shown schematically in Figure 6.14. The antenna is a conductor (e.g., polysilicon, aluminum) of large effective area connected to a much smaller gate. Upon exposure to the plasma, the antenna collects and channels the ion current to the gate. To dissipate the charge an electric field is established that promotes an electron tunneling current to flow through the oxide. In effect, the plasma acts as a current-limited voltage source. The damage produced in the oxide depends on the antenna ratio ($A_r$), or area of antenna divided by gate area. As an example, let us assume $A_r = 1000$ and that a high-density plasma supports an ion-current density of 20 mA/cm$^2$. Under these conditions, 20 A/cm$^2$ of ion current is channeled to the gate. An electric field of perhaps 15 MV/cm may then be established to provide the required tunneling current. This larger than breakdown field would cause rapid oxide failure. Antenna ratio-dependent damage is depicted in Figure 6.14, where the resemblance to electrode area-dependent breakdown effects is immediately apparent. In practice, values of $A_r$ range from several hundred to a thousand or more. In designing circuit layouts, great efforts are made to reduce $A_r$, since this is the route to a diminished oxide-failure probability.

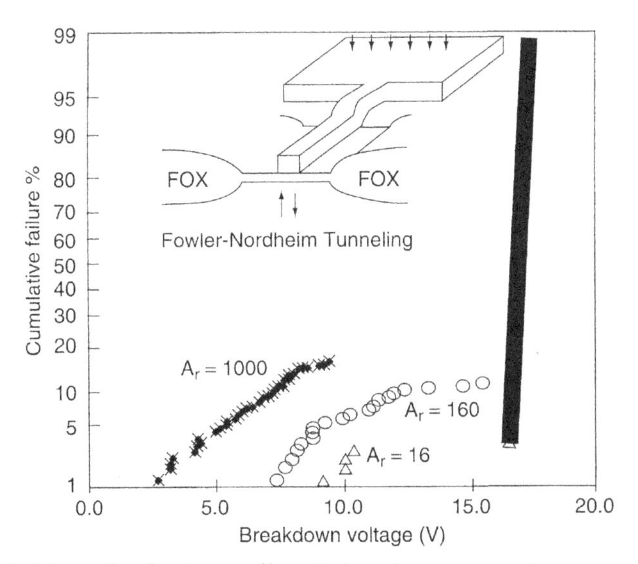


**Figure 6.14** Schematic of antenna effect during plasma processing superimposed on Weibull failure distribution data as a function of gate-oxide breakdown voltage for different antenna ratios. *From Ref. [28].*

Plasma damage to gate oxides occurs primarily during overetching, not during the main etching [29]; it is manifested by capacitor breakdown yields (as measured by $Q_{DB}$ and TDDB), trap densities (as measured by capacitance–voltage curves), enhanced leakage currents and noise, and transistor degradation (as measured by transconductance and threshold voltage). Fortunately, high-temperature annealing is generally effective in emptying traps of the plasma-injected charge. But during final wafer-processing steps only less effective, low-temperature annealing can be tolerated. One must then be vigilant of potential latent-charge degradation effects.

### 6.3.9 Analogy between Dielectric and Mechanical Breakdown of Solids

It is an interesting pedagogical exercise to view dielectric breakdown phenomena within the broader context of failure in materials by pointing out some remarkable similarities to mechanical breakdown of solids [7,21]. In both cases the stimulus (stress $\sigma$, or voltage $V$) produces a flow (elongation strain $e$, or charge $Q$). Mechanical energy of amount $\frac{1}{2}\sigma e$ or $\frac{1}{2}Ee^2$ can be stored elastically ($E$ is Young's modulus) through stretching of interatomic bonds (Figure 6.15(a)). Correspondingly, the electrical energy stored in a capacitor of capacitance $C$ is given by $\frac{1}{2}CV^2$ (Figure 6.15(b)). As long as there is no charge leakage, the applied energy polarizes the dielectric. When the elastic rangers is exceeded, the material flows plastically, and work hardens; mechanical energy is converted to heat as dislocation defects glide in the process. Similarly, above a certain electric field strength, charge flows, or is injected, into the capacitor, and electrical energy is converted into heat. Just as work hardening inhibits further deformation, so too, dielectric trapping of injected charge prevents further injection through the evolution of space charge fields.

When the strains or injected charge levels exceed critical values, failure occurs, and the material suddenly loses part or all of the stored energy. In mechanics we speak of fracture, while in dielectrics, breakdown occurs. Instead of statically applied loads, dynamic stresses of low amplitudes are known to cause fatigue. The analogous phenomenon in dielectrics is known as treeing, a damage process briefly discussed in the next section.

There is yet another parallel between the effects of straining and charge injection rates on failure. Slow solids straining generally exhibit greater elongations to failure than when strained very rapidly. Fractures are often ductile below and brittle above a roughly defined transition strain-rate level. In the dielectric analogue, the charge stored to breakdown ($Q_{BD}$)

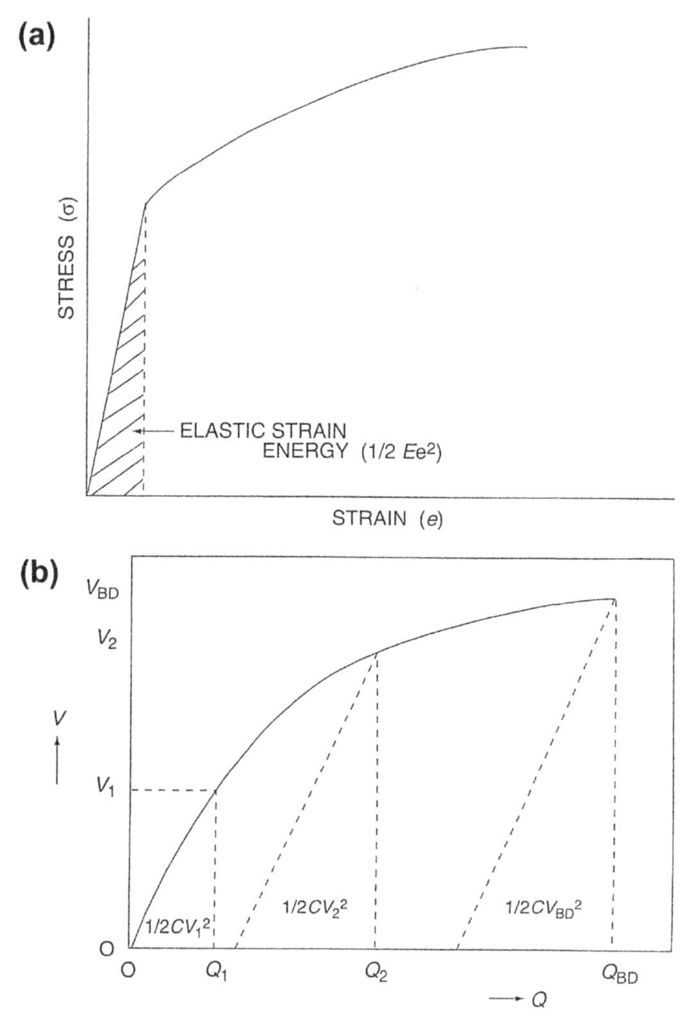


**Figure 6.15** (a) Stress-strain curve of a deformed solid. Mechanical energy expended in deformation is represented by the area under the curve. (b) Schematic representation of the energy content of a charged capacitor of capacitance $C$ in terms of the voltage ($V$)-charge ($Q$) characteristics. Below $V_1$, voltage increases linearly with charge, and the electrical energy stored is given by $\frac{1}{2}CV^2$. Above $V_1$, charge leakage causes a less marked increase in $V$. At an injected charge level of $Q_{BD}$ the energy dissipated is so great that breakdown occurs.

decreases abruptly beyond a certain injection rate. Above the transition injection rate, charge apparently accumulates more rapidly than it can be dissipated. In the mechanical case we would say that a ductile-to-brittle transition occurs because the system is straining faster than it can relax.

In summary, we may regard dielectrics as "ductile" materials when $Q_{BD}$ is large. In such a case the matrix can dissipate a large amount of energy, and $Q_{BD}$ is orders of magnitude larger than the displacement charge $D(q)$. For "brittle" dielectrics, in contrast, $Q_{BD}$ is not much larger than $D(q)$.

### 6.3.10 Discharges and Water Trees

Realizing that electrical breakdown of insulation is not limited to thin films and microelectronics, it is instructive to end our formal discussion of dielectric breakdown with examples drawn from bulk glasses and polymers. Consider the dramatic lightning-like discharge in glass captured in the photograph of Figure 6.16. This so-called Lichtenberg discharge occurred as the result of exposing the glass block to high-energy radiation and grounding it. Prior to breakdown, injected electrons accumulate in the solid, causing a kind of dielectric wear. Like rivers eroding the earth, the damage spreads as electrons preferentially lose their energy at previously damaged sites. Just as small streams form broad rivers that leave deep channels in the landscape, so too, leaking charge wears away the insulator until a low resistance path between electrodes ultimately causes destruction of the capacitor. Erosion wear is apparently not related to the rate of water

**Figure 6.16** Lichtenberg discharge pattern of dielectric breakdown in glass. *Courtesy of Corning Inc.*

(charge) flow, but only to the accumulated volume (amount). Such a damage model applies, in principle, to thin gate oxides, but such tree-like discharge patterns have apparently not been reported.

It has been more than a quarter century since the morphologically similar phenomenon of *water treeing* was first observed in cable insulation [30]. Water trees grow in a wide range of hydrophobic–polymeric insulating materials, e.g., extruded polyethylene, when exposed to both moisture and high voltage. The bush- or tree-like growths within the polymer result in lower dielectric breakdown strengths, threatening the huge investment electrical utilities have in polymeric distribution cables. Many studies have, therefore, been conducted in order to understand the chemistry of polymer degradation.

Oxidation theories seem to dominate the explanations for water treeing [31]. Growth of trees is apparently driven by electro-oxidation of polar amorphous regions of polymer in the direction of the local electric field. Electro-oxidation produces two effects. First, polymer chains are broken, and a "track" or "tree path" is formed. Local oxidation then fosters the conversion of the original hydrophobic track to a more hydrophilic nature. This encourages water molecules dispersed in the polymer matrix to condense out as liquid water in the track. In turn, this water promotes transport of ions in the track to help mediate further electro-oxidation of the polymer at the tip of the track. Thus, track extension becomes self-propagating in a manner similar to the electrical tree-like discharges noted above.

## 6.4 HOT-CARRIER EFFECTS

### 6.4.1 Introduction

Charge that is injected into the gate oxide is a concern because the switching characteristics of a MOSFET can degrade and exhibit instabilities as a result. When a MOSFET acts as a good switch, the drain-source current is as large as possible in the on state and very low in the off state. The typical effect of hot-carrier, or commonly hot-electron, degradation is to reduce the on-state current in an n-MOSFET and increase the off-state current in a p-MOSFET [32] (see Figure 2–17). A measure of transistor degradation or lifetime is commonly defined in terms of some percentage shift of threshold voltage, change in transconductance, or variation in drive or saturation current. To understand the origin of these effects we start with a discussion of the nature of hot electrons. The observed degradation of device characteristics will then be discussed and interpreted in terms of models for hot-carrier effects.

## 6.4.2 Hot Carriers

For semiconductors in thermal equilibrium, electrons and holes continually absorb and emit acoustical phonons (low-frequency lattice vibrations), resulting in an average energy gain of zero. Such electrons have kinetic energies ($E$) that are normally slightly higher than that of the conduction band edge ($E_c$) by an amount $\sim kT_r$. Similarly, for holes, $E$ is slightly less than the valence band edge ($E_V$), again by $\sim kT_r$. At room temperature ($T_r$), $kT_r$ is only 0.025 eV, which is small compared to the carrier kinetic energy corresponding to $E_c$ and $E_v$. This is the case in low electric fields, where the carrier velocity is field-independent (Ohm's law). Under nonequilibrium conditions, however, carriers with far larger kinetic energies contribute to the current flow. In electric fields of 20 kV/cm, for example, the carriers lose energy by scattering with optical or high-frequency phonons and their velocities saturate at roughly $10^7$ cm/s. But at fields of 100 kV/cm, carriers gain more energy than they lose by scattering. Such accelerated electrons have energies of $E_C + kT_e$, where $T_e$ is an effective temperature such that $kT_e > kT_r$. With effective temperatures ($\sim E_C/kT$) of tens of thousands of degrees Kelvin, these electrons are at the very top of the Fermi distribution. Known as *hot electrons*, they pose serious reliability problems in MOS transistors [4].

Consider now the MOS structure with a current that flows through the silicon channel parallel to the Si–$SiO_2$ interface. As long as they remain in the Si, hot carriers cause no damage. But during scattering they may acquire a transverse component of velocity that enables them to enter the adjacent oxide. Such charge injection occurs when the electron energy exceeds the Si–$SiO_2$ barrier height of $\sim$3.5 eV, or typically three times the gap energy. Similarly, hot-hole injection over an even larger 4.6-eV Si–$SiO_2$ barrier leads to their occupation of accommodating oxide traps, with cross sections orders of magnitude larger than those of electron traps. Not only do hot electrons occupy traps, they also generate interface traps. It is this excess charge in the oxide that generates the device damage collectively known as hot-carrier degradation.

## 6.4.3 Hot Carrier Characteristics in MOSFET Devices

In MOS transistors we expect hot-carrier effects to occur when energetic electrons are catapulted from the Si lattice into traps within the $SiO_2$. The widely accepted picture of this process is shown in Figure 6.17(a). Depicted is an n-MOSFET where channel hot-electron injection occurs when the gate voltage ($V_G$) is comparable to the drain voltage ($V_D$). Typical gate current–voltage characteristics for this case are depicted in Figure 6.18 [33]. Two factors cause the gate current ($i_G$) to rise as $V_G$ initially increases. First, the inversion charge in the channel increases, so that more electrons are present for injection into the oxide. Second, the stronger influence of the vertical electric field in the oxide prevents electrons in the oxide from detrapping and drifting back into the channel. However, with large gate voltages the effective channel electric field decreases, reducing the number of hot electrons that can surmount the Si–$SiO_2$ barrier. Therefore, $i_G$ peaks when $V_G$ is roughly equal to the drain-source potential $V_D$, and drops thereafter.

When $V_D$ exceeds $V_G$, we speak of drain avalanche hot carrier injection, a regime schematically depicted in Figure 6.17(b). This mechanism first depends on an impact-ionization avalanche to create carriers. These secondary electrons then become hot, and they, not the primary electrons, are responsible for causing degradation. In the case of high substrate-bias voltages, additional secondary hot electrons generated from deeper Si substrate regions can also be injected into the oxide. However, there is little reason to believe that secondary electrons produce more damage than the primary hot electrons. Irrespective of the origin of the hot electrons, the dominant damage mechanism is creation of surface traps at the Si–$SiO_2$ interface.

### 6.4.4 Models for Hot-Carrier Degradation

A widely used model to quantitatively account for time-dependent hot-carrier degradation is due to Hu et al. [35]. The basic assumption of the model hinges on there being a supply of hot electrons that are "lucky." According to the "lucky electron" concept [36], the probability that a channel electron will travel a distance $d$ or more without suffering any collision is equal to exp $[-d/\lambda]$, where $\lambda$ is the mean free path between scattering events. Now consider an electron of charge $q$, traveling a distance $\lambda$ in the channel electric field $E_c$. The probability that it will reach energy $\phi$ without suffering a collision is given by exp $(-\phi/(\lambda\ q\ E_c))$, because $d = \phi/(q\ E_c)$. For channel electrons to reach the gate oxide, two "lucky" processes are involved. The first requires that electrons gain sufficient kinetic energy from the channel field to become hot. Secondly, the electron momentum must be redirected perpendicularly, so that hot

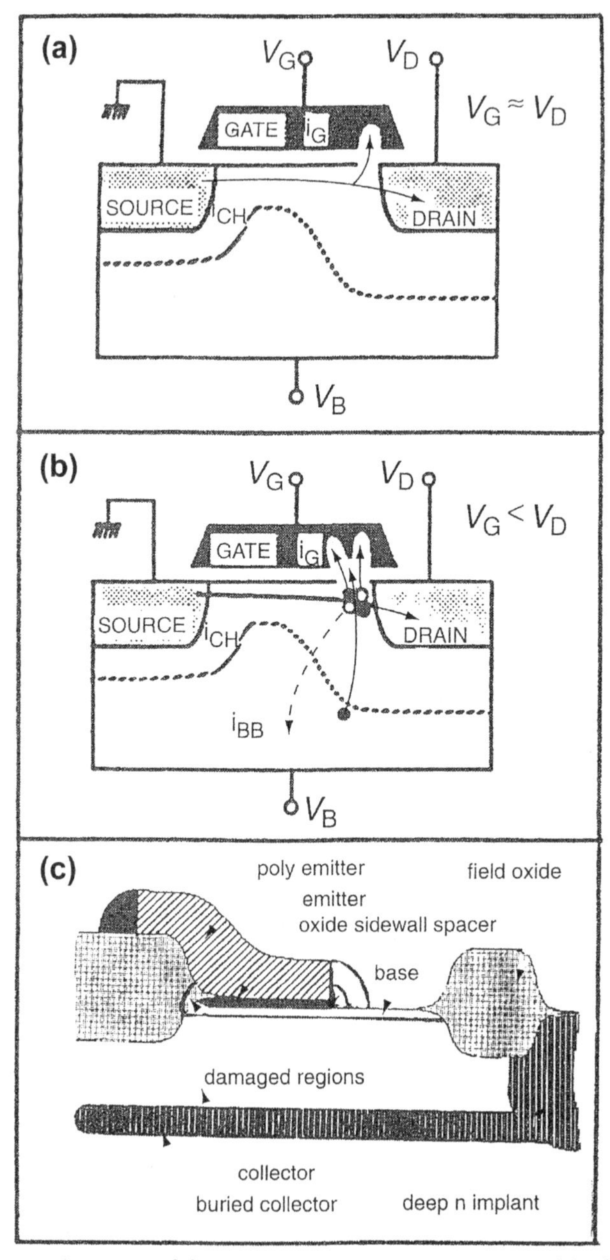


**Figure 6.17** Mechanisms of hot-carrier injection in transistors. (a) Channel hot-electron injection occurs when the gate voltage ($V_G$) of n-MOSFET is comparable to the drain voltage ($V_D$). (b) Drain avalanche hot-carrier injection occurs when $V_D > V_G$. Secondary hot electrons generated within deeper Si substrate regions maybe injected into the oxide when $V_B > 0$ V. *From Ref. [33].* (c) Damage to field and sidewall oxides due to hot-carrier injection in reverse-biased n-p-n transistors. *From Ref. [34].*

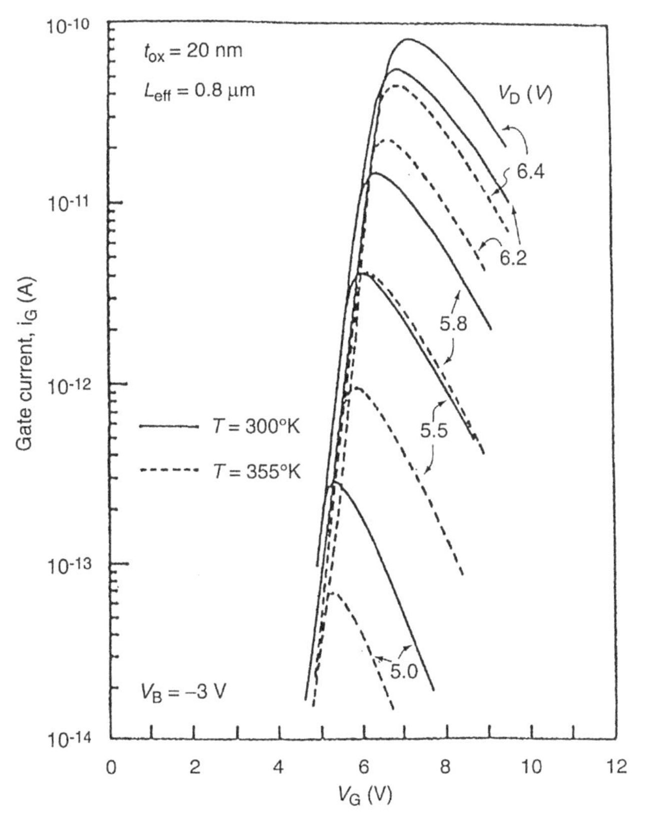


**Figure 6.18** Gate current–voltage characteristics for hot-carrier injection by the channel-hot-electron injection mechanism. *From Ref. [33].*

electrons can enter the oxide. The explicit consequences of these processes are substrate ($i_S$) and gate currents ($i_G$) whose magnitudes depend on electron energies reaching those required for impact ionization ($\phi_i$), and for surmounting the Si–$SiO_2$ energy barrier ($\phi_b$), respectively. In particular

$$i_S = C_1 i_D \exp\left[\frac{-\phi_i}{\lambda q E_c}\right] \tag{6.24}$$

and

$$i_G = C_2 i_D \exp\left[\frac{-\phi_b}{\lambda q E_c}\right], \tag{6.25}$$

where $C_1$ and $C_2$ are constants. In these equations, $i_D$ is the (cold) drain current flow that supplies some of the eventually "lucky" electrons. Eliminating $\lambda\ qE_c$ from Eqns (6.24) and (25) yields

$$\frac{i_G}{i_D} = C_2\left(\frac{i_S}{C_1\ i_D}\right)^m, \tag{6.26}$$

where $m = \phi_b/\phi_i$. This equation, which applies to the case where $V_G$ exceeds $V_D$, has been verified in n-channel transistors. It is found that $m$ is approximately 3, which is roughly consistent with values of $\phi_i \approx 1.3$ eV and $\phi_b \approx 3.5$ eV.

In normal MOSFET operation the gate current is negligible and barely measurable. When hot electrons are generated, $i_G$ rises rapidly, as seen in the characteristics of Figure 6.18, thus providing a good measure of the ensuing damage. To link the gate current to time-dependent degradation, let us denote the shift, change, or variation in electrical parameter due to hot-carrier effects by $\Delta$. If the time ($t$) rate of change in $\Delta$ is proportional to $i_G$, then

$$\frac{d\Delta}{dt} \approx i_G = \frac{A(\Delta)}{W} i_D \left(\frac{i_S}{i_D}\right)^m. \tag{6.27}$$

Here $W$ is the MOSFET width, and $A(\Delta)$ has absorbed $C_2$ and $C_1$ and additionally accounts for the dependence of degradation on existing damage. When $d\Delta/dt$ reaches a certain level, we say that hot-carrier failure has occurred. Since the MTTF depends on the reciprocal of $d\Delta/dt$.

$$\mathrm{MTTF}^{-1} = Bi_D\left(\frac{i_S}{i_D}\right)^m, \tag{6.28}$$

where $B = A(\Delta)/W$. A test of this equation is made in Figure 6.19 [37] for different devices operating under a wide variety of static and dynamic operating conditions. Here device lifetime is defined as the time it takes for the drain current to decrease by 5%. In this case a slope of $-1$ is obtained.

The model presented thus far tacitly assumes static, or dc, voltages and currents. As device operation involves ac time-dependent wave forms, it is in our interest to correlate static and dynamic degradation. This is simply achieved by integrating or time averaging over the substrate and drain currents, so that

$$\mathrm{MTTF}^{-1} = \frac{B}{T_c}\int_0^{T_c} i_D\left(\frac{i_S}{i_D}\right)^m dt, \tag{6.29}$$

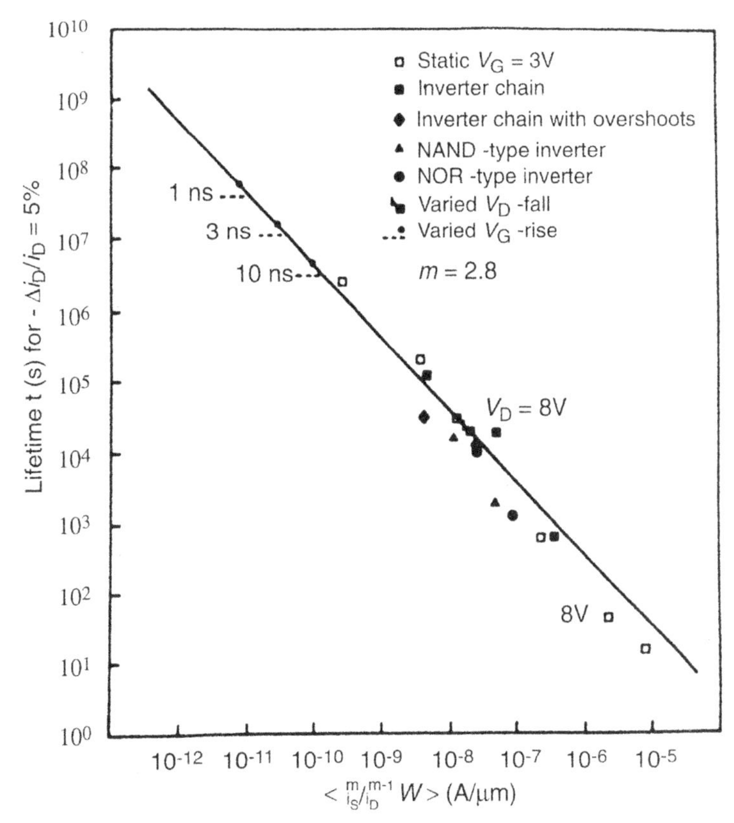


**Figure 6.19** Lifetime (5% change in drain current) of an n-MOSFET versus the time average function of substrate and drain current. Various switching devices show good agreement with this plot of Eqn (6.28). *From Ref. [37].*

where $T_c$ is the full cycle time.

It is common practice [38,39] to plot the change in Δ according to either a log time,

$$\Delta \sim \log \frac{t}{\tau} (\tau = \text{const}), \tag{6.30}$$

or power law dependence,

$$\Delta \sim t^n. \tag{6.31}$$

The increase in drain current of p-MOSFETs, shown in Figure 6.20(a), is an example obeying logarithmic kinetics. On the other hand, the time-dependent threshold-voltage increase and transconductance decrease for

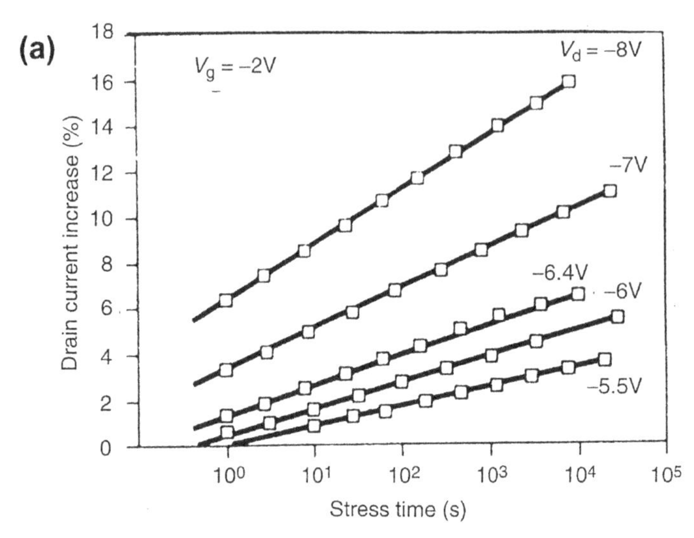


**Figure 6.20** Changes in electrical characteristics of field effect transistors due to hot-electron effects. (a) Drain current increase in p-MOSFETs. From Ref. [38]. (b) Transconductance decrease in n-MOS transistors from different prophecies depicted by different symbols. From Ref. [38]. (c) Time-dependent threshold-voltage shifts as a function of temperature. *From Ref. [39].*

the indicated n-MOSFETs seem to better obey power-law kinetics, as seen in Figure 6.20(b) and (c).

The fact that greater threshold voltage shifts occur at lower, rather than higher, temperatures in Figure 6.20(c) [40] should not go unnoticed. Similarly, larger changes in transconductance occur at 77 K relative to 300 K for the same gate voltage. Hot-electron degradation is unique because it is the only example in this book where more damage is produced, the lower the temperature. In essence, this implies that the activation energy for failure is negative. The reason has to do with the higher population of hot electrons and their enhanced trapping at lower temperatures.

## 6.4.5 Hot-Carrier Damage in Other Devices

Hot-carrier effects are not limited to MOSFETs. They also occur in n-p-n as well as p-n-p Si bipolar transistors. In particular, bipolar transistors used in read/write amplifiers suffer loss in current gain when subjected to reverse emitter-base voltage stress. Apparently the oxide regions adjacent to the emitter-base become damaged due to electron trapping at interface states, an effect that raises the base current, thus lowering the transistor gain.

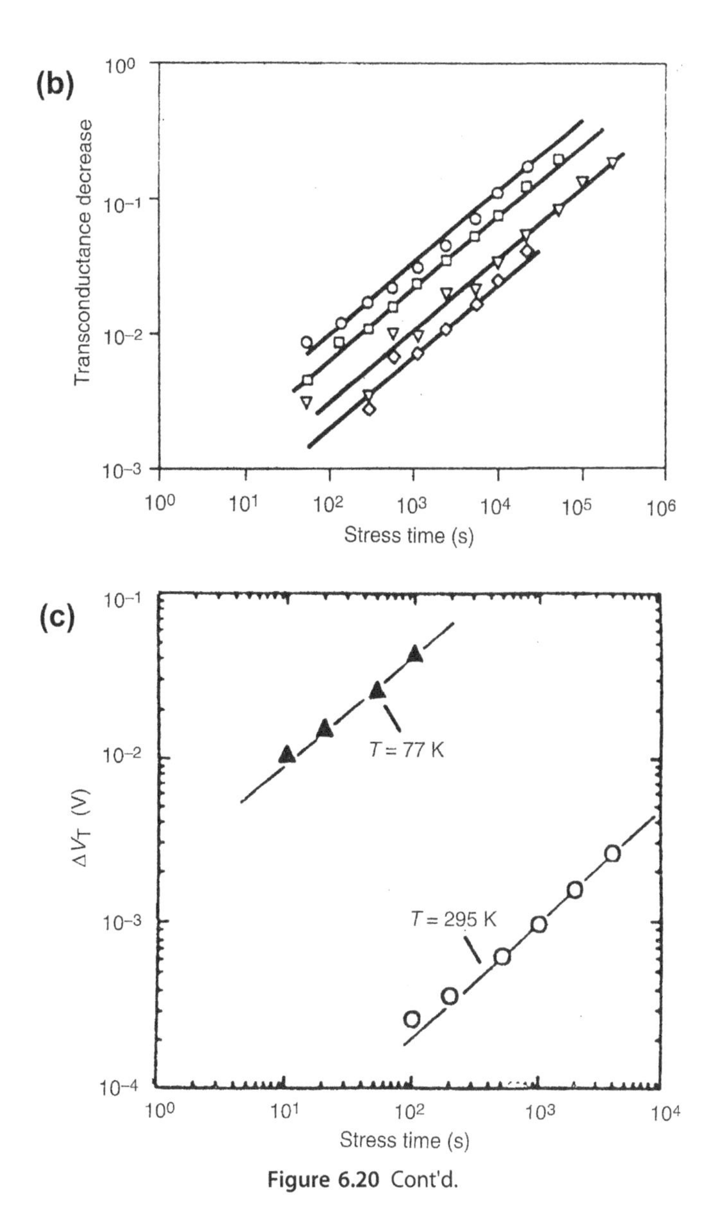


**Figure 6.20** Cont'd.

Models for the leakage-current paths through the oxide sidewall spacer and field-oxide ("birdsbeak") structures are depicted in Figure 6.17(c). It has been observed that the change in base current ($\Delta i_{\mathrm{B}}$) with time ($t$) is linear for short times, but $\Delta i_{\mathrm{B}} \approx t^{0.2}$ for longer times, indicating partial saturation [41]. The transition in degradation kinetics occurs after a characteristic time $\tau$ given by $\tau = \tau_{\mathrm{o}} \exp(-V_{\mathrm{EB}}/V_{\mathrm{o}})$, where $V_{\mathrm{EB}}$ is the reverse emitter-base voltage; values for $\tau_o$ and $V_o$ depend on the particular device and are typically $5 \times 10^{-11}$ s and 0.25 V, respectively.

Degradation of GaAs-based MESFET and (high-electron mobile transfer) HEMT devices due to hot electrons has also been reported [41]. In the case of HEMTs these effects show up at large drain-source voltages that enable carriers to reach high energies and cause impact-ionization effects. As a result, the drain current decreases, the parasitic drain-resistance increases, and there is a change in the transconductance frequency response.

### 6.4.6 Combating Hot-Carrier Damage

Basic to a number of processing strategies for minimizing hot-carrier effects in n-MOSFETs is the introduction of ion implanted regions of lighter doping between the channel and heavily doped drain regions. This doped-substrate engineering is accomplished through double diffused drain, lightly doped drain, and large angle tilt implanted drain structures. As a consequence of these creatively altered doping profiles, the electric field is spread out, reducing its peak value.

Additional approaches to increasing immunity to hot-carrier damage involve thermochemical processing. In an effort to reduce the Si–$SiO_2$ interfacial trap density, postmetallization anneals at ~400 °C are commonly carried out in hydrogen. This improves device function by passivating interface states that are otherwise active. However, under the stimulus of hot-electron impact it is thought that hydrogen desorbs from the Si–$SiO_2$ interface and, in effect, causes the observed degradation. To minimize this possibility and further combat hot-carrier injection, the gate oxide is now lightly nitrided. This dielectric is more resistant to interface-trap generation, presumably because weaker silicon–hydrogen bonds possessing an energy of 0.5 eV are replaced by stronger silicon–nitrogen bonds of 1.6 eV energy. A final strategy to limit the driving force for hot-carrier phenomena is to simply reduce the transistor operating voltage. One would imagine that at voltages below 3.5 V, electrons could not access the oxide. This is not the case, however, and probably means that tunneling effects are involved.

## 6.5 ELECTRICAL OVERSTRESS AND ELECTROSTATIC DISCHARGE

### 6.5.1 Introduction

Two additional failure mechanisms of very great importance are *electrical over-stress* (EOS) and *electrostatic discharge* (ESD); they are reviewed in Refs

[42–46]. One difference between the two is the voltage magnitude involved. The dividing line is blurry, but in EOS, tens to hundreds of volts are typical: while in ESD, voltage pulses one to two orders of magnitude higher may develop. Current flow is basic to both EOS and ESD, but the source of the charge is generally different in each mechanism. In EOS large currents flow through device junctions because of excessive applied fields that arise from poor initial circuit design, mishandling, or voltage pulses. Local hot spots at junctions and parasitic transistor behavior between closely spaced devices are some of the features associated with EOS. The destruction of a diode due to EOS is depicted in Figure 6.21.

Electrostatic-discharge phenomena are part of a broader class of electrical overload environments that cause damage to sensitive electronics. In this category we include high voltages generated by static charge, signal-switching transients, electromagnetic pulses, spacecraft charging, high-power electromagnetic and radio-frequency interference, and lightning. Common to all of these ESD sources are high-voltage pulses that commonly generate large currents (>1 A) of short duration (risetime ~10 ns, decay time ~150 ns). These effects can occur during device processing, assembly into systems, and use of the product by the consumer.

An important cause of ESD is static-charge buildup as a result of *triboelectric* (rubbing) effects. Peak potentials ranging from tens of volts up

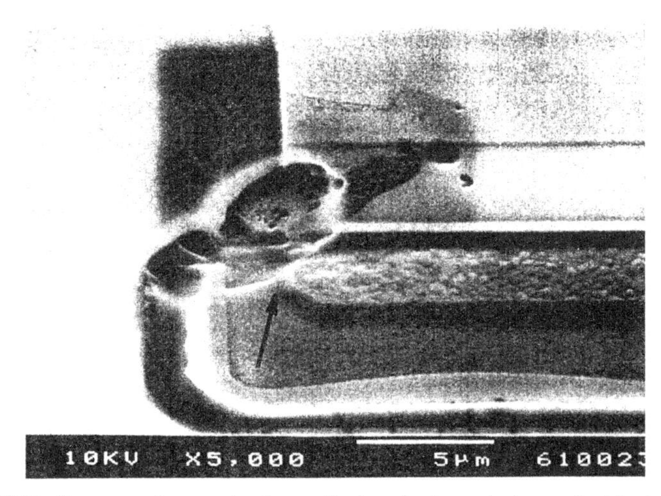


**Figure 6.21** Damage to protection diode device during electrical-overstress testing. *Courtesy of M. C. Jon, Lucent Technologies, Bell Laboratories Innovations.*

to ~30 kV can be generated this way. Simply walking across vinyl or carpeted floors can result in electrostatic voltages ranging between approximately 10 and 30 kV at relative humidities of 20%. Frictional contact with plastic packaging or furniture can generate as much as 20 kV. Pulse currents of more than 150 A have been drawn from metal furniture; typical discharge energies are of the order of several to tens of millijoules. At relative humidities of 60–90%, charge leaks away, and the tribopotentials are 10–100 times smaller, but enough, in some cases, to cause damage. The ordering of triboelectric effects in different materials is illustrated in Table 6.3. When two materials in the series are rubbed together, the one that is higher in the series acquires a positive potential relative to the lower one.

In discharges of tens of kV the resulting currents cause indiscriminate damage; i.e., failures are not specific to any particular component or device. Oxides, interconnections, contacts, and semiconductor junctions of integrated circuits and electro-optical devices are vulnerable. The sensitivity of a device to ESD generally scales inversely to the minimum-size feature. This can be appreciated with reference to Table 6.4 where discrete devices such as high-speed MOS and compound semiconductor field effect transistors are most sensitive to damage because of their small dimensions.

Depending on the voltage level, speed of approach, and shape of charged bodies, signals can exceed 1 GHz and may reach 5 GHz. At these frequencies equipment cables and stripes on printed-circuit boards act as efficient receiving antennas. In general, the voltage and current necessary to cause damage is one to two orders of magnitude greater than that to cause a temporary upset (e.g., causing a memory change of 0–1 or 1–0). Thus, conductive coupling through an ESD spark will destroy circuit lines, whereas radiation coupling may only cause upset.

### 6.5.2 ESD Models

The two most popular models used to characterize and simulate ESD events are known as the human-body model (HBM) and charged device model (CDM). Circuits for each of these models are shown in Figure 6.22. In the HBM case (Figure 6.22(a)) a person simulated by a 100-pf capacitor, $C_B$, is charged to voltage $V_{HBM}$. Upon touching an uncharged device with the body simulated by a 1500-Ω resistor, $R_B$, discharge occurs at a single lead, while a second lead is grounded. In evaluating the ESD vulnerability of a device, $C_B$ is first charged to $V_{HBM}$

**Table 6.3** Electrostatic triboelectric series

| **Most positive (+)** |
|---|
| Air |
| Human skin |
| Asbestos |
| Fur (rabbit) |
| Glass |
| Mica |
| Human hair |
| Nylon |
| Wool |
| Silk |
| Aluminum |
| Paper |
| Cotton |
| Steel |
| Wood |
| Sealing wax |
| Hard rubber |
| Nickel, copper |
| Brass, silver |
| Gold, platinum |
| Acetate fiber (rayon) |
| Polyester (mylar) |
| Celluloid |
| Polystyrene (styrofoam) |
| Polyurethane (foam) |
| Polyethylene |
| Polypropylene |
| Polyvinyl chloride |
| Silicon |
| Teflon |
| Silicone rubber |
| Most negative (−) |

From Ref. [47].

with switches $S_1$ closed and $S_2$ open. Switch $S_2$ is then closed after opening $S_1$, and the capacitor charge ($Q = C_B V_{HBM}$) is discharged into the device. If the device being tested has resistance $R_T$, then Ohm's law gives for the time ($t$)-dependent ac current,

$$i(t) = i_o \exp\left[\frac{-t}{\tau}\right], \tag{6.32}$$

**Table 6.4** Susceptibility of various devices to ESD ddamage

| Device | Range of damage ESD susceptibility (V) |
|---|---|
| MOSFET | 10–100 |
| GaAs FET | 100–300 |
| EPROM | 100–500 |
| JFET | 140–7000 |
| Surface acoustic wave | 150–500 |
| Operational amplifier | 190–2500 |
| CMOS | 250–3000 |
| Schottky diodes | 300–2500 |
| Thin and thick film resistors | 300–3000 |
| Bipolar transistors | 380–7000 |
| Silicon controlled rectifiers | 680–1000 |
| 8085 Microprocessor | 500–2000 |

After Ref. [43].

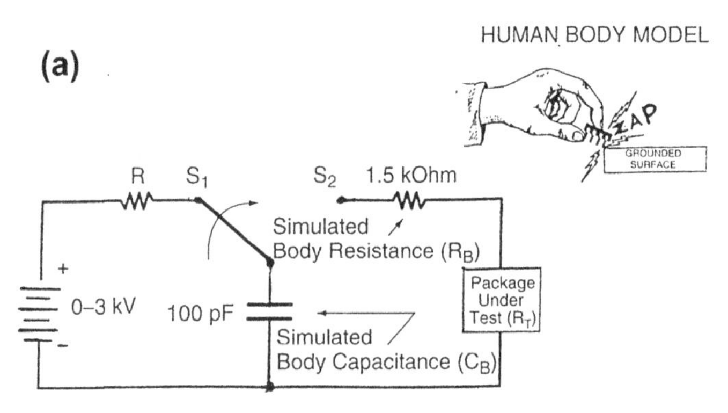


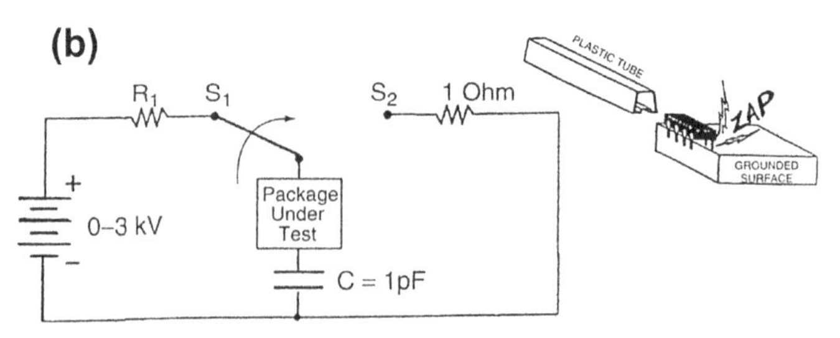


**Figure 6.22** Testing configurations for evaluating electrostatic discharge susceptibility of devices. (a) Human body model. Body resistance = 1500 Ω, Body capacitance = 100 pF; (b) Charged device model. *From Ref. [48].*

where $i_o = V_{HBM}/(R_T + R_B)$, and the time constant $\tau = C_B\,(R_T + R_B)$. The power $P$ dissipated in resistor $R_T$ is simply

$$P = i(t)^2 (R_T). \tag{6.33}$$

Since we are eventually concerned with the resistor temperature rise, the energy discharged is of importance. This energy is partitioned between $R_B$ and $R_T$, and that dissipated by $R_T$ alone is given by

$$E(R_T) = \frac{E_o R_T}{(R_T + R_B)}, \tag{6.34}$$

where $E_o$ is the energy originally stored in $C_B$, namely, $\frac{1}{2} C_B V_{HBM}^2$.

The CDM considers the situation where charge initially resides on a device as shown in Figure 6.22(b). This charge may have arrived via triboelectric rubbing or sliding in shipping tubes. During simulation testing the device is charged to a potential $V_{CDM}$ by touching its leads with a power supply probe (switch $S_1$ closed). When $S_2$ is closed, the charge stored on the device is quickly discharged into the 1-$\Omega$ resistance. Because the device normally has resistive, capacitive, and inductive character of an unknown nature, the resulting RLC circuit response is more complex than for the HBM. Rather than an exponentially decaying current signal, a damped sine wave maybe involved. For our purposes we will consider only the human body model.

### 6.5.3 Thermal Analysis of ESD Failures

Junction burnout in diodes and bipolar and MOS transistors due to ESD effects has been a subject of considerable interest over the years. The first to model the problem were Wunsch and Bell [49], who derived the following relation for the MTTF of a p-n junction of area $A_o$ that absorbs power $P$:

$$\text{MTTF} = \frac{\pi c \rho \kappa (T_M - T_o)^2}{\left(\frac{P}{A_o}\right)^2}. \tag{6.35}$$

In this equation $c$, $\rho$ $\kappa$, and $T_M$ are the heat capacity (J/g °C), density (g/cm$^3$), thermal conductivity (W/cm °C), and melting point (°C) of the semiconductor, respectively, and $T_o$ is the ambient temperature. The Wunch–Bell relation predicts that the MTTF of the component or device varies as the inverse square of the power density dissipated in it. Failure ensues when the device temperature $T$ rises to the melting point. These investigators subjected diodes and transistors of varying junction areas to both forward- and reverse-bias electric power pulses that destroyed them; the functional dependence between device lifetimes and power densities expressed by Eqn (6.35) was generally obeyed.

Thermal burnout of devices has been theoretically modeled more recently [48,50,51]. Because the results have relevance to ESD damage, it is worthwhile to present aspects of the analysis. Following Lai [48] we assume the human body model for characterizing ESD damage to a resistor $R_T$. Provided that the electrical energy is dissipated *adiabatically* in the resistor volume v, no thermal energy is lost externally, and the temperature rise ($\Delta T$) is easily calculated to be $E(R_T)/(c\rho \nu)$. Collecting terms, and using Eqn (6.34), the final resistor temperature ($T$) is

$$T = T_o + \frac{C_B V_{HBM}^2 \ R_T}{2c\rho\nu(R_T + R_B)}. \tag{6.36}$$

We now consider the time dependence of heating. In the case of *nonadiabatic* energy absorption the equation governing the resistor temperature is given by the familiar nonsteady-state heat-flow equation

$$\frac{\partial T}{\partial t} = \frac{\kappa}{c\rho}\frac{\partial^2 T}{\partial x^2}. \tag{6.37}$$

Assuming that a constant heat flux, $Q(R_T)$, (in units of J/cm$^2$-s or W/cm$^2$) flows into one end of an infinitely long resistor, the initial and boundary conditions are, respectively,

$$T(x > 0, \ t = 0) = T_o \tag{6.38a}$$

$$T(x = \infty, t > 0) = T_o \tag{6.38b}$$

$$Q(R_T) = -\kappa(\mathrm{d}T/\mathrm{d}x)_{x=0} \quad \text{for } t > 0. \tag{6.38c}$$

The solution to the above boundary value problem is readily obtained using Laplace transforms and is given by

$$T = T_o + \frac{2Q(R_T)}{\kappa}\left\{\left(\frac{Kt}{\pi}\right)^{1/2}\exp\left(-\frac{x^2}{4Kt}\right) - \frac{x}{2}\mathrm{Erfc}\left(\frac{x}{2(Kt)^{1/2}}\right)\right\}, \tag{6.39}$$

where the thermal diffusivity $K$ (cm$^2$/s) $= \kappa/(c\rho)$.

Normally we are only interested in the temperature at the resistor surface ($x = 0$), and therefore, Eqn (6.39) reduces to

$$T = T_o + \frac{2Q(R_T)}{\kappa}\left(\frac{Kt}{\pi}\right)^{1/2} = T_o + 2Q(R_T)\left[\frac{t}{(\pi c\rho\kappa)^{1/2}}\right]. \tag{6.40}$$

Assuming that the heat flux $Q(R_T)$ is equal the power density $P/A_o$,

$$T = T_o + \left(\frac{2P}{A_o}\right)\left[\left(\frac{t}{\pi c\rho\kappa}\right)^{1/2}\right], \tag{6.41}$$

where $A_o$ is the resistor surface area through which heat flows. Aside from numerical constants the result is essentially the Wunsch–Bell relationship (Eqn (6.35)) after associating $t$ with MTTF.

The solution of Eqn (6.39) directly applies to square wave testing where the heat flux is constant during pulse duration $t$. However, in an ESD test, the pulse decays exponentially such that over 99% of the power is dissipated in the first five time constants. An *average* power, $<P>$ is dissipated for $t = 5\tau$, and therefore,

$$T = T_o + \left(\frac{2<P>}{A_o}\right)\left[\left(\frac{5\tau}{\pi c\rho\kappa}\right)^{1/2}\right]. \tag{6.42}$$

The value of $<P>$ can be readily calculated over this time period through the use of Eqn (6.32)

$$\begin{aligned} <P> &= \frac{1}{5\tau}\int_0^{5\tau}\left[i_o \exp\left(-\frac{t}{\tau}\right)\right]^2 (R_T)\ dt \\ &= \frac{R_T i_o^2}{10}[\exp(-0) - \exp(-10)] \approx \frac{R_T i_o^2}{10} \end{aligned} \tag{6.43}$$

A couple of examples will illustrate the use of these equations.

---

## Example 6.1

A pulse of 4500 V is discharged from a person into a p–n junction. The junction area is $5 \times 10^{-7}$ cm$^2$ and the resistance is 20 Ω. Assume that the voltage across the device is negligible compared with that of the ESD pulse.

a. Over what time span does the discharge occur?
b. What is the maximum current?

c. Calculate the power density discharged through the diode.

**Answers:**

a. For the human body model the time constant is $\tau = C_B (R_T + R_B)$. Substituting, $\tau = 100 \times 10^{-12} (20 + 1500) = 0.152$ μs. Therefore, $5\tau = 0.76$ μs.
b. Ohm's law simply states that $i_o = V_{HBM}/(R_T + R_B) = 4500/(20 + 1500) = 2.96$ A.
c. The power density is $\langle P \rangle /A_o$, or $R_T\, i_o^2/(10\, A_o)$. Substituting, $\langle P \rangle /A_o = 20\, (2.96)^2/(10\, (5 \times 10^{-7})) = 35$ MW/cm$^2$.

---

## Example 6.2

What human-body voltage pulse is necessary to cause a 1 cm long, 20 μm diameter gold wire to melt under adiabatic conditions? Gold has a melting temperature ($T_M$) of 1064 °C, a density ($\rho$) of 19.3 g/cm$^3$, a heat capacity ($c$) of 0.13 J/g-K, and a resistivity given by $2.4 \times (1 + 0.0034(T°C - 20))$ μΩ-cm.

**Answer** According to Eqn (6.36), the human-body voltage is given by $V_{HBM} = \{(T_M - T_o)[2\, c\rho v(R_W + R_B)]/(C_b R_W)\}^{1/2}$, where $R_W$ is the wire resistance. In calculating $R_W$, it is assumed that the average temperature is 550 °C, so that $R_W = 2.4 \times 10^{-6}\, [1 + 0.0034(550 - 20)]\, [1/\pi (10^{-3})^2] = 2.1$ Ω. The resistor volume is $3.1 \times 10^{-6}$ cm$^3$. Therefore, upon substitution, $V_{HBM} = \{[550 - 20][2 \times 0.13 \times 19.3 \times (3.1 \times 10^{-6})\, (2.1 + 1500)]/(100 \times 10^{-12} \times 2.1)\}^{1/2} = 243$ kV. Such a large pulse magnitude indicates why wires will not open from human body ESD.

---

The Wunsch–Bell-like failure model has been extended to both earlier and later times by considering shorter and longer ESD pulse widths as shown in Figure 6.23. Failure power-density thresholds vary with failure time $t_f$ as

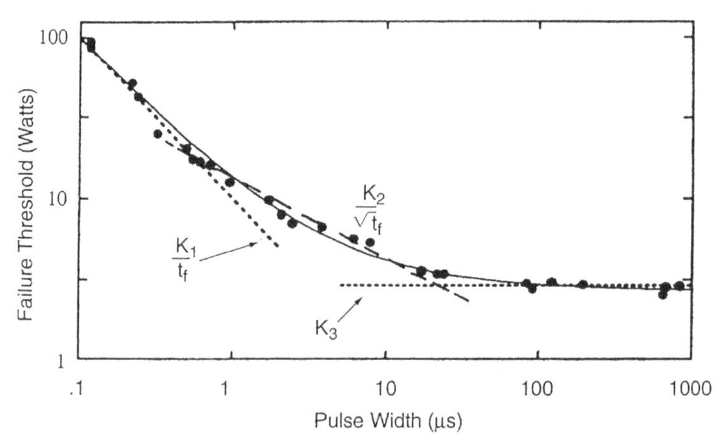


**Figure 6.23** Failure threshold power versus pulse width, illustrating three regimes of behavior indicated by Eqn (6.44). *From Ref. [42]. Experimental data points are due to Ref. [52].*

$$\frac{P}{A_o} = \frac{K_1}{t_f} + \frac{K_2}{t_f^{1/2}} + K_3 \tag{6.44}$$

where the first and last terms correspond to very short and long pulse widths, and $K_1$, $K_2$, and $K_3$ are constants. Because heat does not travel far in a very short time, we may view the temperature rise as occurring adiabatically. Conversely, failure thresholds are constant for long pulse widths, and steady-state (time-independent) conditions prevail. Sandwiched in between at intermediate pulse widths ($\sim$ 1–10 μs) is the Wunsch–Bell regime. In addition to assorted diodes and transistors, these equations have been used to model failure times in thin-film resistors and capacitors.

## 6.5.4 ESD Failure Mechanisms

The destructive power of ESD in integrated circuits is shown in Figure 6.24. Post mortem analysis of such failures has revealed essentially three basic mechanisms for permanent damage to semiconductor devices [42]. They are shown schematically in Figure 6.25 and include junction burnout, oxide punch-through, and metallization burnout. All of these mechanisms are thermal in nature, with damage generally occurring when the temperature of the affected region dissipating the ESD pulse energy rises to the melting point. Each mechanism is briefly considered in turn, starting with junction burnout.

### *6.5.4.1 Junction Burnout*

While the macroscopic modeling of junction burnout was treated in Section 6.5.3, we now explore the microscopic aspects of such failures. Junction burnout is commonly manifested by alloy spiking or shorting. Unlike the very slow solid-state diffusion-controlled junction spiking discussed in Section 6.5.4.2, damage occurs at a far greater speed under ESD conditions. Local melting and the creation of a shorting metal filament is the reason. For example, if a junction depth of $10^{-5}$ cm is shorted during a $10^{-6}$ s ESD pulse, the metal filament growth velocity is estimated to be $10^{-5}/10^{-6} = 10$ cm/s.

Pierce [53] has modeled the complex electrothermomigration-induced mass transport as a result of electrical overstress. The mass flux is similar to that which governs electromigration (Section 6.5.5), but there is now an added thermomigration contribution proportional to the temperature gradient. What makes filament growth so rapid are the very large current densities (e.g., $10^7$ A/cm$^2$), and the fact that the melting point has been

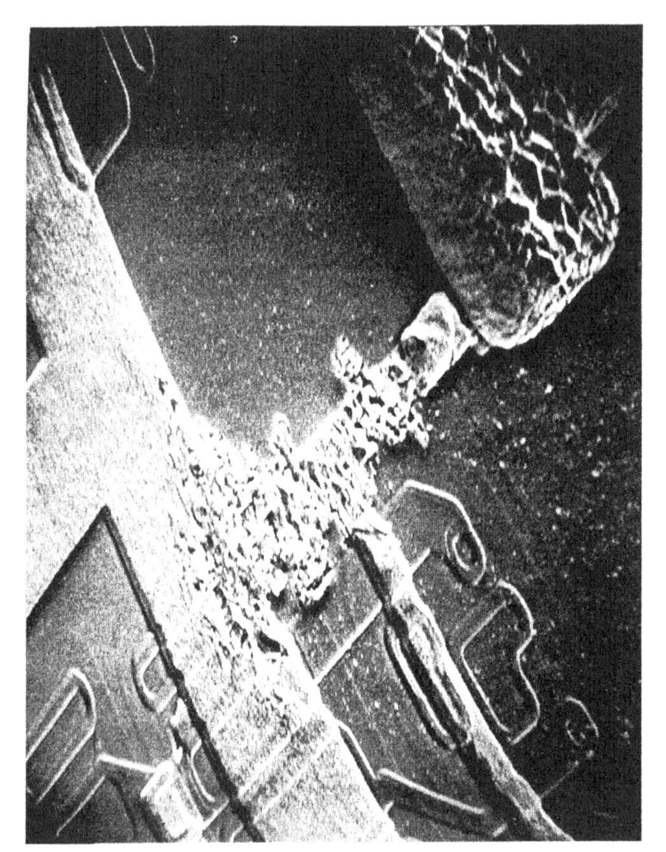

**Figure 6.24** Damage to an integrated circuit due to ESD. *Courtesy of G. Steiner, Lucent Technologies, Bell Laboratories Innovations.*

exceeded. Liquid-state diffusivities of $10^{-4}$ cm$^2$/s can now be expected instead of the much smaller solid-state values. In the case of aluminum metallizations on silicon, the eutectic temperature of 577 °C figures in the calculation of failure (Eqn (6.35)). In this case the failure threshold power would be reduced relative to the case of Al in contact with a high-melting barrier metal.

The junction structure itself contributes to ESD failure tendencies. For example, junction corners, defects, and small sidewall areas cause current crowding and reduced damage thresholds. Lastly, it should be noted that for equivalent power ($P$) levels, reverse-biased p-n junctions are more susceptible to ESD failure than forward-biased junctions. In the former, with breakdown voltage $V_{\mathrm{BD}}$ applied, power ($P = iV_{\mathrm{BD}}$) is concentrated in the relatively narrow depletion width ($\delta$). However, during forward

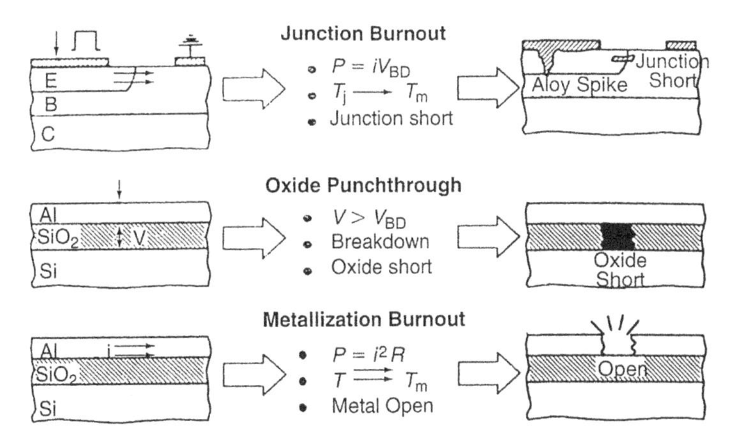


**Figure 6.25** Schematic depiction of ESD failures. Junction shorting occurs when the critical power causes the semiconductor junction temperature $T_j$ to reach the melting point. Oxide punch-through is essentially dielectric shorting. Metallization burnout occurs when the critical power causes the melting point of the metal to be reached. *From Ref. [42].*

bias, the joule power ($i^2$ $R$, $R$ = resistance) is dissipated through the entire device of size $D$. Through comparison of power densities, $iV_{BD}/\delta > i^2$ $R/D$, because $D > \delta$.

#### 6.5.4.2 Metallization Burnout

It was determined in Section 5.5.3 that critical dc current densities required to melt aluminum wires were of the order of $4 \times 10^4$ A/cm$^2$, but thin-film metallizations can tolerate far more current without immediately melting. Zapping a metal interconnect with a sufficiently large ESD power pulse accelerates melting failure in a way not unlike that experienced by a blown fuse. The process has been modeled by Pierce [54] and results of the analysis are shown in Figure 6.26(a) and (b) for the case of constant power pulses and variable pulse widths, respectively. In the first of these figures, temperature–time histories reveal when the melting point of the aluminum is reached. The data of Figure 6.26(b) are in good agreement with the adiabatic and steady-state heating regimes defined by Eqn (6.44). In accord with our previous electromigration experience, current densities in excess of $10^7$ A/cm$^2$ cause failure within about 1 μs.

#### 6.5.4.3 Oxide Punch-Through

The last of the materials that suffer ESD damage is the oxide. Whether insulating a metal runner or serving as the gate oxide, rupture of the oxide

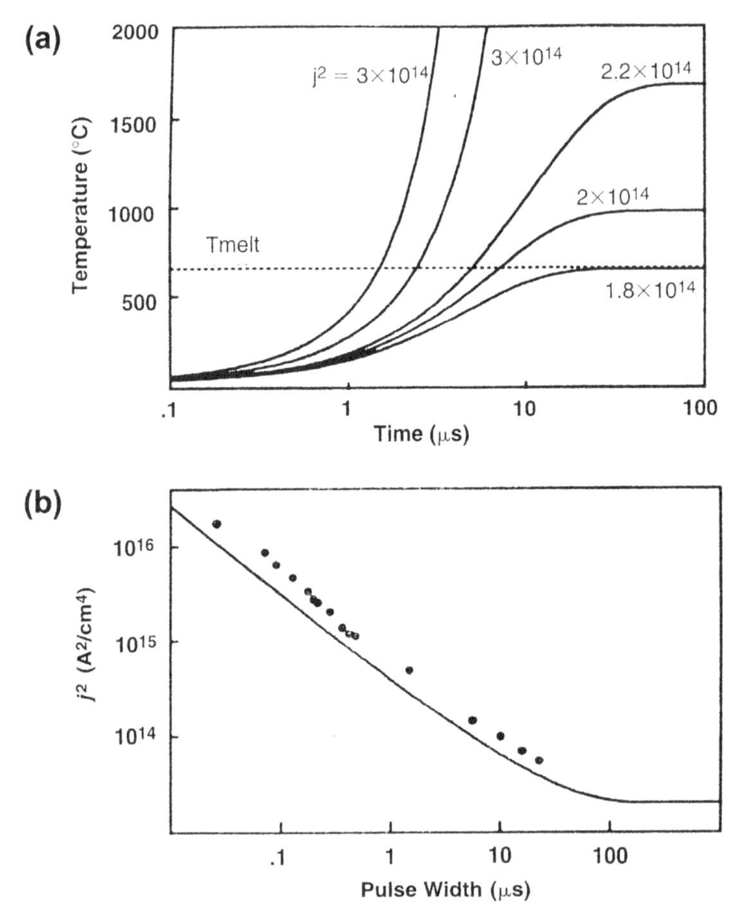


**Figure 6.26** (a) Burnout model for 0.5 μm Al metallization on 1 μm $SiO_2$ with constant power pulses. The failure current density (A/cm$^2$) and time are given by the intersection with the melting temperature of Al ($T_M$). (b) Comparison of theory with metallization failure data generated by ESD power pulses of variable widths. *From Ref. [42].*

(punch-through) or dielectric breakdown has been previously dealt with at length in this chapter. Studies of ESD failures in oxides have revealed their adiabatic nature.

## 6.5.5 Latent Failures

We know that high-energy ESD failures convincingly destroy the unfortunate devices involved. But, what about low-level ESD transients that produce a spectrum of "soft," or latent, effects that wound, rather than catastrophically destroy. Depending on their robustness, devices can sustain a lesser or greater number of discharges without apparent loss of function. Such latent defects may slightly reduce hot-carrier lifetimes, shorten dielectric breakdown times, and enhance electromigration, but not to the point of failure. One might expect, however, that recurrent low-voltage zap wounds would ultimately result in a transition from the dormant defect stage to actual malfunction or even failure.

A summary of the results of latency testing reveals that,

1. Life-test studies yield inconclusive results.
2. Most devices degraded by ESD actually improve or heal during life testing.
3. Many latent failures are simply leaky pins operating within specifications.
4. Gate oxides are probably prone to latent defects.
5. On-chip protection structures are generally such that further degradation is unlikely during normal device operation.

On the basis of this collection of observations, latent damage and failure is difficult to model with any certainty; its impact on reliability is likewise uncertain.

### 6.5.6 Guarding against ESD

Two basic strategies are employed to protect sensitive electronics from ESD damage, namely, preventing it from occurring and modifying devices and circuits to make them immune [4,45,55]. The first approach deals with sources of charge and voltage external to the electrical hardware and attempts to neutralize potential harmful effects arising from personnel, workstations, packaging and transportation, and the workplace environment. For example, by wearing metal wrist straps, antistatic gloves, and smocks, ESD damage from handling can be minimized. Equipping workbenches with tabletops and mats that dissipate charge, and using grounded tools (e.g., soldering iron tips, solder suckers), stools, and chairs helps to insure against ESD damage at workstations. Similarly, the use of grounded carts, antistatic bags, and tote boxes, as well as shielded or protective containers for transporting integrated circuit chips, minimizes damage arising during movement of electronic products. Lastly, the work environment can be made safer through humidity control and suppressing ionization of air.

It is virtually impossible to eliminate ESD from occurring, so the second, and most widely used, strategy is to improve device immunity to destructive damage. One way to do this is to connect solid-state clamping elements between critical points such as the input and ground of devices. If the ESD-induced voltage exceeds the clamping voltage, the element turns on and becomes a low-impedance shunt to divert currents to ground. Incorporation of either shunt capacitors or series inductor filters are other approaches to minimizing ESD damage. Further discussion of circuits to protect electronics against ESD is well beyond the scope of this book.

### 6.5.7 Some Common Myths about ESD

Common to all poorly understood damage phenomena is an accumulated "wisdom" invested with differing proportions of fact and fiction. Sometimes this wisdom is scientifically based, but in other cases it maybe little more than a collection of myths. Because the subject of ESD has its share of them, it is prudent to be aware of the mythology before proceeding to devise countermeasures. McAteer [45] has enumerated the following myths about static electricity and ESD.

1. Conductive components are never involved in ESD.
2. Antistatic materials always protect components that are susceptible to static electricity.
3. ESD is primarily a MOS problem.
4. Devices with input-protection circuitry are always immune to ESD.
5. Components are insensitive to ESD after they are installed in an assembly.
6. Devices or printed-circuit boards with connector pins shorted together are immune to ESD.
7. Antistatic, static-dissipative, and conductive materials are equivalent.
8. ESD damage to printed-circuit boards can be prevented by handling them by the edges.
9. Controlling the relative humidity to greater than 40% eliminates static problems.
10. Static-electricity problems can be eliminated by controlling room ionization.

Clearly, separating fact from fiction, evaluating myths, and successfully combating ESD will necessitate a more advanced understanding of the complex interaction among electrostatics, materials, and circuit analysis. If this is not achieved, one myth that will persist is, We are not having ESD problems.

## 6.6 BIAS TEMPERATURE EFFECTS

### 6.6.1 Negative-Bias Temperature Instability

NBTI is a key reliability issue in MOSFETs. NBTI manifests as an increase in the threshold voltage and consequent decrease in drain current and transconductance of a MOSFET. The degradation exhibits logarithmic dependence on time. It is of immediate concern in p-channel MOS devices (pMOS), since they almost always operate with negative gate-to-source voltage; however, the very same mechanism also affects nMOS transistors when biased in the accumulation regime, i.e., with a negative bias applied to the gate [56–58].

In sub-micrometer devices, nitrogen is incorporated into the silicon gate oxide to reduce the gate leakage current density and prevent boron penetration. However, incorporating nitrogen enhances NBTI.For new technologies (32 nm and shorter nominal channel lengths), high-K metal gate stacks are used as an alternative to improve the gate current density for a given equivalent oxide thickness. Even with the introduction of new materials like hafnium oxides, NBTI remains.

It is possible that the interfacial layer composed of nitrided silicon dioxide is responsible for those instabilities. This interfacial layer results from the spontaneous oxidation of the silicon substrate when the high-*K* dielectric is deposited. To limit this oxidation, the silicon interface is saturated with *N* resulting in a very thin and nitrided oxide layer.

It is commonly accepted that two kinds of trap contribute to NBTI: first, interface traps are generated. Those traps cannot be recovered over a reasonable time of operation. Some authors refer to them as permanent traps. Those traps are the same as the one created by channel hot carrier. In the case of NBTI, it is believed that the electric field is able to break Si–H bonds located at the Silicon-oxide interface. H is released in the substrate where it migrates. The remaining dangling bond Si– (Pb center) contribute to the threshold voltage degradation.

In addition to the interface state generation, second, some preexisting traps located in the bulk of the dielectric (and supposedly nitrogen related) are filled with holes coming from the channel of pMOS. Those traps can be emptied when the stress voltage is removed. This degradation in threshold voltage can be recovered over time.

The existence of two coexisting mechanisms created a large controversy, with the main controversial point being about the recoverable aspect of interface traps. Some authors suggest that only interface traps are generated and recovered; today this hypothesis is ruled out. The situation is clearer but not completely solved. Some authors suggest that interface traps generation is responsible for hole trapping in the bulk of dielectrics. A tight coupling between two mechanism may exist but nothing has been definitely demonstrated.

### 6.6.2 Positive-Bias Temperature Instability

With the introduction of high-*K* dielectrics and metal gates, a new degradation mechanism appeared. The PBTI for positive bias temperature instabilities affects nMOS transistors (negative channel MOSFET) when positively biased. In this particular case, no interface states are generated and 100% of the Vth degradation maybe recovered. Those results suggest that there is no need to have interface state generation to trapped carrier in the bulk of the dielectric.

## EXERCISES

1. Large, conically shaped ceramic insulators with a fluted-surface profile are often used to isolate high electric potentials from ground in high-voltage transmission lines. Give a reason for the wavy profile in these insulators.
2. Capacitors composed of 11 nm thick $SiO_2$ gates sandwiched between p-type Si substrates and polysilicon were stressed to breakdown either through application of a constant current of 40 μA or a constant voltage bias of 12.866 V. For electrode areas measuring 0.02 mm$^2$, 50% of the capacitors failed at a breakdown charge of $Q_{BD} = 2\ C/cm^2$, independent of testing method.
   a. What is the MTTFbased on constant-current testing?
   b. Can the resistivity of the oxide be calculated from this information? If so, how does the apparent value compare with typical resistivities of $SiO_2$?
3. The gate oxide of an MOS transistor has an area of 1 μm$^2$ and it is effectively connected to a square metal antenna with an area of 1200 μm$^2$, as shown in Figure 6.14.
   a. During a plasma-etching step, a current density of 10 mA/cm$^2$ is sustained. What current is supplied to the oxide?
   b. Suppose a nonconducting mask 0.2 μm thick sits above the metal antenna so that only the antenna film edges are exposed to the plasma. How much current is supplied to the gate now?

4. At the instant of dielectric breakdown, the discharge current density is much larger than typically applied external densities resulting from currents of microamperes. Demonstrate this fact by assuming that the discharge current arises from flow of displacement charge in $\sim 0.1\ \mu s$. Note that the electric displacement $D(q)$ and field ($E$) are related by $D(q) = \varepsilon\varepsilon_o E$, where $\varepsilon$ is the dielectric constant and $\varepsilon_o$ is the permittivity of free space. The electrode area is 0.01 $mm^2$, and the discharge damage area is 1 $\mu m^2$.
5. Based on the results of the previous problem calculate:
   a. the total energy density at breakdown for a capacitor of unit area and thickness, and estimate
   b. the radius of the breakdown crater of melted material assuming it is either Si or $SiO_2$.

   How does the estimated value compare to the actual crater size?
   Assume that the average heat capacities of Si and $SiO_2$ are 6 and 14.5 cal/mol-K, respectively.
6. A 10 nm thick gate oxide suffers dielectric breakdown at an electric field of 10 MV/cm. It was observed that a 100 nm diameter pinhole formed within 0.12 ms. What current was probably responsible for breakdown if the $SiO_2$ melted? Assume that the melting point of $SiO_2$ is 1700 °C.
7. Gate oxides are now approximately 8 nm or less in thickness. Speculate what happens to the distribution and magnitude of the ionic, oxide trapped, fixed, and interfacial oxide charge. In particular, comment on the magnitudes of the electrostatic interactions between these charge distributions.
8. It has been found that TDDB failure times of $SiO_2$ capacitors remain relatively independent of oxide thickness when polysilicon contacts are used. However, for aluminum contacts the failure times decrease greatly in thinner oxides. Why?
9. It is desired that a 12.5 nm thick dielectric last 5 years within a capacitor charged to 5 V. What breakdown electric field would be probably be measured if the voltage were ramped at a rate of 1 V/m?
10. TDDB failure times are commonly plotted as log MTTF versus $1/E$ rather than $\log[(\text{MTTF})E^2]$ versus $1/E$.
    a. Why is this latter way of plotting data more physically appealing?
    b. At high fields, does a plot of log MTTF versuss $1/E$ over- or underestimate the time of failure if $\log[(\text{MTTF})\ E^2]$ versus $1/E$ is the correct dependence?

**c.** Set up equations that would enable the correspondence between voltage-ramping and constant-voltage tests to be determined assuming MTTF $\approx 1/E^2 \exp(1/E)$.

**11.** Ultrathin oxides grown on silicon wafers having longer carrier lifetimes are observed to be more reliable (as measured by injected charge to breakdown) than oxides grown on wafers with shorter carrier lifetimes. Give a plausible reason for this finding.

**12.** It has been suggested [59] that iron-precipitate contaminants of size $d$ (cm) increase the effective local electric field in $SiO_2$ linearly according to $E_{eff} = (V_o/X_o)\,(1 + 7 \times 10^6 d)$, where $V_o$ is the applied voltage and $X_o$ is the oxide thickness. What failure time can be expected in 10 nm thick oxides containing 0.5 nm precipitates? How does this compare to the MTTF when precipitates are 2 nm in size? Assume $t_{OR} = 10^{-11}$ s and $G_R = 320$ MV/cm.

**13.** Dielectric breakdown in mica capacitors appears to be governed by ion transport. Suggest a relationship between the failure time, material constants, temperature, and applied electric field.

**14.** Estimate the largest leakage current a gate oxide measuring 0.1 $\mu m^2$ can tolerate if it must survive for 5 years if $Q_{BD}$ is 10 $C/cm^2$.

**15.** **a.** Derive a single comprehensive equation for the MTTFof $SiO_2$ films as a function of applied electric field, temperature, and thickness, if the oxide properties are as depicted in Figures 6.12 and 6.13.

**b.** What testing temperature was used to obtain the data shown in Figure 6.12?

**16.** **a.** If the quantity $m$ in Eqn (6.28) is changed by an amount $dm$, by what factor does the mean lifetime of a MOSFET subject to hot-electron damage change?

**b.** What is the apparent activation energy for hot-electron degradation in devices whose behavior is shown in Figure 6.20(c)?

**17.** **a.** Mention three strategies to limit hot-electron damage in field effect transistors.

**b.** Annealing devices in deuterium instead of hydrogen appears to reduce hot-electron degradation. Suggest a possible reason why.

**18.** Provide physical arguments for the following observations:

**a.** Breakdown of dielectrics has similarities to the plastic deformation of metals.

**b.** Dielectric breakdown can be caused by either electron or hole injection processes.

**c.** Electronic components subjected to narrow voltage-pulse widths can sustain higher power levels prior to failure.
**d.** Hot electrons can reach temperatures of tens of thousands of degrees.

**19.** Consider a discrete silicon diode whose p-n junction area is 1 $\mu m^2$. An electric power pulse destroyed the device in 10 ns. Roughly estimate the threshold power discharged in the diode.

**20.** The lateral extent of the heat-induced discoloration surrounding a failure site has been used as a means of crudely estimating the time duration of an EOS event. Assume that the temperature in the damage zone is given by $T = T_M \text{Erfc} \frac{x}{(4Kt)^{1/2}}$, and the discoloration zone extends to points where temperatures of $\frac{1}{2} T_M$ are reached.

In a certain integrated circuit chip failure, silicon was discolored for a distance of ~5 μm. What pulse time is suggested? What is the reason for the appearance of colors in thermally damaged silicon?

**21.** Sketch the critical power density versus pulse time behavior for GaAs devices as in Figure 6.23. Are GaAs devices more or less sensitive to ESD than Si devices? Note for GaAs: $\rho = 5.32$ g/cm$^3$, $T_M = 1236$ °C, $\kappa = 0.81$ W/cm-°C, $c_p = 0.35$ J/g-°C.

## REFERENCES


[1] S.M. Sze, Physics of Semiconductor Devices, second ed., Wiley, New York, 1981.
[2] J.S. Suehle, P. Chaparala, C. Messick, W.M. Miller, K.C. Boyko, 32nd Annual Proceedings of the IEEE Reliability Physics Symposium, IEEE, 1994, p. 120.
[3] R.B. Comizzoli, in: P.J. Singh (Ed.), Materials Developments in Microelectronic Packaging: Performance and Reliability, ASM, 1991, p. 311.
[4] Y. Leblebici, S.M. Kang, Hot Carrier Reliability of MOS VLSI Circuits, Kluwer, Boston, 1993.
[5] S.M. Sze, Semiconductor Devices—Physics and Technology, Wiley, New York, 1985.
[6] D.M. Fleetwood, IEEE Trans. Nucl. Sci. NS-39 (1992) 269.
[7] D.R. Wolters, J.J. van der Schoot, Philips J. Res. 40 (1985) 115.
[8] G. C. Walther and L. L. Hench, eds of Physics of Electronic Ceramics, Part A, Dekker, New York.
[9] D.R. Wolters, J.J. van der Schoot, Philips J. Res. 40 (1985) 164.
[10] R.S. Scott, N.A. Dumin, T.W. Hughes, D.J. Dumin, 33rd Annual Proceedings of the IEEE Reliability Physics Symposium, IEEE, 1995, p. 131.
[11] J.S. Suehle, B.W. Langley, C. Messick, Tutorial Notes, IEEE International Reliability Physics Symposium, IEEE, 1995, p. 131.
[12] J. Lee, I.-C. Chen, C. Hu, 26th Annual Proceedings of the IEEE Reliability Physics Symposium, p. 131, 1995.
[13] R. Moazzami, J.C. Lee, C. Hu, IEEE Trans. Electron Devices 36 (1989) 2462.
[14] J.F. Verweij, J.H. Klootwijk, Microelectron. J. 27 (1996) 611.

[15] T. Ohmi, K. Nakamura, K. Makihara, 32nd Annual Proceedings of the IEEE Reliability Physics Symposium, p. 161, 1994.
[16] T. Ohmi, Proc. IEEE 81 (5) (1993) 716.
[17] J.W. McPherson, D.A. Baglee, 23rd Annual Proceedings of the IEEE Reliability Physics Symposium, p. 1, 1985.
[18] E.S. Anolick, G.R. Nelson, IEEE Trans. Reliab. R-29 (1980) 217.
[19] R. Moazzami, C. Hu, Tutorial Notes, IEEE International Reliability Physics Symposium, 5.1, 1992.
[20] A.G. Sabnis, VLSI Electronics Microstructure Science, in: VLSI Reliability, vol. 22, Academic Press, San Diego, 1990.
[21] D.R. Wolters, Philips Tech. Rev. 43 (1987) 330.
[22] R. Degraeve, J.L. Ogier, R. Bellens, Ph Roussel, G. Groesseneken, H.E. Maes, 34th Annual Proceedings of the IEEE Reliability Physics Symposium, p. 44, 1996.
[23] D.J. Dumin, Tutorial Notes, IEEE International Reliability Physics Symposium, 2.1, 1995.
[24] C.K. Chan, IEEE Trans. Reliab. 39 (1990) 147.
[25] J.C. Chen, S. Holland, C. Hu, 23rd Annual Proceedings of the IEEE Reliability Physics Symposium, p. 24, 1985.
[26] A. Berman, 19th Annual Proceedings of the IEEE Reliability Physics Symposium, p. 204, 1981.
[27] K.P. Cheung, Tutorial Notes, IEEE International Reliability Physics Symposium, 4.1, 1997.
[28] S. Fang, A.M. McCarthy, J.P. McVittie, Proceedings of 3rd International Symposium on ULSI, vol. 473, 1991.
[29] Y. Uraoka, K. Eriguchi, T. Tamaki, K. Tsuji, IEEE Trans. Semicond. Manuf. 7 (3) (1994) 293.
[30] T. Miyashita, Boston, Proc. IEEE-Nema Electr. Insul. Conf. 131 (1969).
[31] J.J. Xu, S.A. Boggs, IEEE Electr. Insul. Mag. 10 (5) (1994) 29.
[32] K. Mistry, B. Doyle, IEEE Circuits Devices 11 (1995) 33.
[33] E. Takeda, Tutorial Notes, IEEE International Reliability Physics Symposium, 1b1, 1993.
[34] J. Scarpulla, J. Dunkley, S. Lemke, E. Sabin, M. Young, 35th Annual Proceedings of the IEEE Reliability Physics Symposium, p. 34, 1997.
[35] C. Hu, S.C. Tam, F.-C. Hsu, P.K. Ko, T. Chan, K.W. Terrill, IEEE Trans. Electron Devices ED-32 (1985) 375.
[36] S. Tam, P.K. Ko, C. Hu, IEEE Trans. Electron Devices ED-31 (1984) 1116.
[37] W. Weber, M. Brox, T. Kuenemund, H.M. Muehlhoff, D. Schmitt-Landsiedel, IEEE Trans. Electron Devices ED-38 (1991) 1859.
[38] W. Weber, M. Brox, MRS Bull. XVIII (12) (1993) 36.
[39] R. Bellens, P. Heremans, G. Grosseneken, H.E. Maes, 26th Annual Proceedings of the IEEE Reliability Physics Symposium, p. 8, 1988.
[40] H.E. Maes, P. Heremans, R. Bellens, G. Grosseneken, Qual. Reliab. Eng. Int. 7 (1991) 307.
[41] C. Tedesco, C. Canali, F. Magistrali, A. Paccagnella, E. Zanoni, Qual. Reliab. Eng. Int. 9 (1993) 371.
[42] D.G. Pierce, Tutorial Notes, IEEE International Reliability Physics Symposium, 8.1, 1995.
[43] N. Sclater, Electrostatic Discharge Protection for Electronics, TAB Books, McGraw Hill, Blue Ridge Summit, PA, 1990.
[44] A. Amerasekera, J. Verwey, Qual. Reliab. Eng. Int. 8 (1992) 259.
[45] O.J. McAteer, Electrostatic Discharge Control, McGraw Hill, 1990.

[46] W.D. Greason, Electrostatic Discharge in Electronics, Research Studies Press, Wiley, New York, 1992.
[47] R.Y. Moss, IEEE Trans. Compon. Hybrids Manuf. Tech. 5 (512) (1982).
[48] T.T. Lai, IEEE Trans. Compon., Hybrids Manuf. Tech. 12 (1989) 627.
[49] D.C. Wunsch, R.R. Bell, IEEE Trans. Nucl. Sci. 15 (1968) 244.
[50] V.M. Dwyer, A.J. Franklin, D.S. Campbell, IEEE Trans. Electron Devices 37 (1990) 2381.
[51] H.H. Choi, T. DeMassa, Solid State Electr. 36 (1993) 1511.
[52] H. Domingos, Proc. EOS/ESD Symp (1980). EOS 2.6.
[53] D.G. Pierce, EOS 7, 67, Proc. EOS/ESD Symp. (1985).
[54] D.G. Pierce, EOS 4, 55, Proc. EOS/ESD Symp. (1982).
[55] W. Boxleitner, IEEE Spectr. 26 (8) (1989) 36.
[56] K. Dieter Schroder, Jeff A. Babcock, J. Appl. Phys. 94 (July 2003) 1–18.
[57] M. Alam, S. Mahapatra, Microelectronics Reliab. 45 (1) (January 2005) 71–81.
[58] D.K. Schroder, Microelectronics Reliab. 47 (June 2007) 841–852.
[59] W.B. Henley, et al., in: 31st Annual Proceedings of the IEEE Reliability Physics Symposium, 1993, p. 22.

# CHAPTER 7

# Environmental Damage to Electronic Products

## 7.1 INTRODUCTION

The sources of integrated circuit (IC) damage considered up to this point originated within the materials and substances used in manufacturing. But now we confront degradation and damage to electronic products from outside sources such as humid atmospheres, airborne contaminant particles, and ionizing radiation. Even in the very early years of microelectronics it was recognized that moisture and atmospheric particulates, often ionic in nature, posed serious reliability problems in devices, components, and packages. Thin-film metallizations and fine wires composed of corrodible metals (e.g., Al, Ag) and insulation (plastic, ceramics) prone to moisture absorption resulted in degradation products that were uncomfortably large in size compared to the small feature dimensions involved. Today, humidity-induced corrosion of exposed or unprotected metals and water pickup, as well as release from polymers, are among the major failure mechanisms in electronic products.

In this chapter, we explore a number of issues related to environmental damage starting with the sources of atmospheric contamination. Then the kinetics of permeation as well as leakage of harmful environmental gases (e.g., water vapor) and contaminants into electronic packages will be discussed. Later we shall see how particulates and humid environments act in synergy to promote corrosion and metal migration damage.

Quite different from the environmental damage to electronic products from atomic, molecular, and particulate matter is that wrought by exposure to ionizing radiation. The damage from this source normally causes instabilities in MOS transistor function through charging effects in the gate oxide. Errors in stored information due to impingement by cosmic rays and alpha particles emitted by radioactive elements is another common form of radiation damage that will be discussed at the end of the chapter.

*Reliability and Failure of Electronic Materials and Devices*
ISBN 978-0-12-088574-9
http://dx.doi.org/10.1016/B978-0-12-088574-9.00007-0

## 7.2 ATMOSPHERIC CONTAMINATION AND MOISTURE

### 7.2.1 Sources of Airborne Contamination

Important constituents of the outdoor environment include corrosive gases and particles [1]. The compositions and concentrations of these harmful species vary with geographic location in the United States due to the distributions of humidity and air pollution, shown in Figure 7.1. Included under air pollution is acid rain composed of raindrops (200–5000 μm in size) as well as fog droplets (0.5–50 μm), which in turn contain $NH_4^+$, $SO_4^{-2}$, and $NO_3^-$ ions. These and other ionic species are found in airborne particulates; a breakdown of their concentrations within Newark, New Jersey, indoor and outdoor ambients is given in Table 7.1.

The size distribution of atmospheric particles typically ranges from 0.1 to 15 μm. There are very many particles smaller than 0.1 μm, but their total mass is small. Within the 0.1–15 μm size range are coarse particles (>2.5 μm) of mineralogical or biological origin that are usually formed by wind-induced abrasion. Their residence time in the atmosphere is typically hours to days, and they deposit by gravitational settling or wash out by the action of rain. As Table 7.1 notes, their concentration indoors is but a fraction of that outdoors, a fact explained by the use of modest filtration systems. On the other hand, particles smaller than 2.5 μm stem from natural sources such as volcanic ash and fossil fuels. In urban areas where fossil fuels are the dominant pollution source, small carbonaceous/oxide particles 0.05 μm in size are the seed nuclei for $SO_2$ and $N_2O$, which then oxidize to $H_2SO_4$ and $HNO_3$. Ammonia released from fertilization and biological decay then neutralizes the acidified solid, all the while enlarging the particles to the stable 0.1–2.5 μm range. Such particles can remain airborne for months and even years through diffusional and convective transport. Importantly, the concentration ratio of fine particles within buildings to that outdoors is 0.2–0.5, depending on the filtration efficiency, the rate of air exchange, and the indoor sources of particles, e.g., due to smoking.

The particulate and gaseous contaminants that manage to bypass filtration barriers in buildings land on the vertical and horizontal surfaces of electronic components and systems, where they have the potential to cause corrosion. In wafer fabrication facilities, some finer particles even penetrate clean-room defenses and become the defects that lower yields.

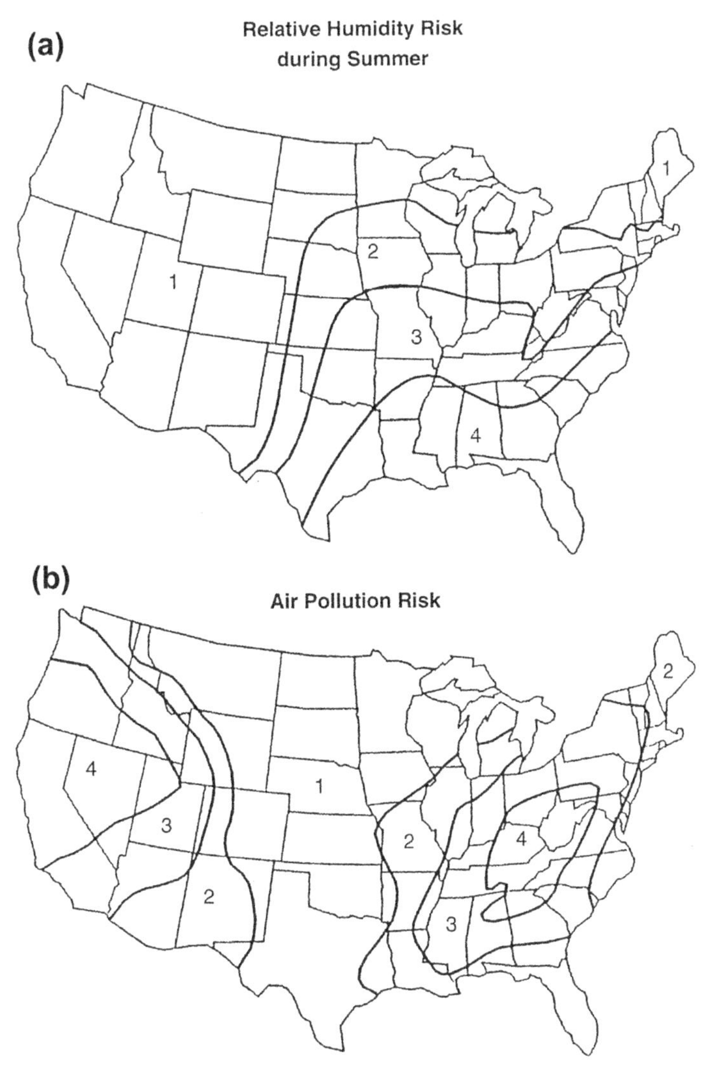


**Figure 7.1** Relative summer humidity (a) and air pollution (b) risks in different US geographic locations rated from 1 (least severe) to 4 (most severe). *After D. J. Sinclair, Lucent Technologies, Bell Laboratories Innovations.*

It is instructive to predict the surface accumulation rate ($A$) of a given pollutant particle assuming the known outdoor concentration ($C_o$) and filtration efficiency ($F$). Filters are rated in terms of the ratio of contaminant level indoors ($C_i$) after passing through the filter, to the outdoor level, i.e., ($F = C_i/C_o$). Typically $C_i/C_o = 0.1$ for high-efficiency filters. An

**Table 7.1** Average concentrations of ionic species and corrosive gases in the air around Newark, NJ (ng/m$^3$)

| **Particles** | **Indoor** | | **Outdoor** | |
|---|---|---|---|---|
| | **Fine** | **Coarse** | **Fine** | **Coarse** |
| $Cl^-$ | 4 | 4 | 66 | 502 |
| $SO_4^{2-}$ | 721 | 25 | 5213 | 725 |
| $Na^+$ | 14 | 3 | 103 | 446 |
| $NH_4^+$ | 168 | 0 | 1631 | 1 |
| $K^+$ | 19 | 2 | 53 | 30 |
| $Mg^{2+}$ | 2 | 1 | 14 | 90 |
| $Ca^{2+}$ | 10 | 20 | 25 | 297 |
| Total mass | 2880 | 6100 | 18,090 | 12,340 |
| ***Gases*** | | | | |
| Sulfur dioxide | 5400 | | | |
| Nitrogen oxides | 24,000 | | | |
| Chlorine gases | 510 | | | |
| Reduced sulfur | 800 | | | |
| Ammonia | 40,000 | | | |

From Ref. [1].

additional quantity that must be known is the particle terminal settling, or deposition, velocity, $v$ (Eqn (3.13)). We now simply note that the accumulation rate is the product of the indoor concentration and deposition velocity, and in typical units,

$$A(\mu g/cm^2 - yr) = 0.526C_i(\mu g/m^3)v(cm/s). \tag{7.1}$$

As an example, consider the accumulation of $SO_4^{2-}$ for which $C_o = 20$ μg/m$^3$. In passing through a filter for which $C_i/C_o = 0.12$, and assuming $v = 0.005$ cm/s, substitution yields a surface accumulation of $6.31 \times 10^{-3}$ μg/cm$^2$-year. Measurements in many cities across the United States typically reveal that annual sulfate and chloride accumulations on outdoor zinc and aluminum surfaces range between 0.1 and 0.8 μg/cm$^2$.

## 7.2.2 Moisture Damage

Moisture-induced damage (e.g., swelling of IC packages, electrical shorting) and corrosion of metal components (e.g., conductors, contacts) and auxiliary hardware (e.g., closed-loop laser cooling system [2]) associated with electrical and electronic equipment have traditionally been major reliability concerns. An annual monetary toll of a few billion ($10^9$) dollars is not an exaggeration of the magnitude of moisture/corrosion problems in electrical equipment. Although our focus later will be corrosion in microelectronics, larger electrical and electronic systems are also quite vulnerable [3]. Similarly, a review of corrosion damage in disk recorder heads, aircraft avionic equipment, stepping motors, klystron tubes and other microwave equipment, printed wiring boards, antennas, contacts, and so on, can be found in reference [4]. Since a necessary condition for corrosion is the presence of water, we start by exploring aspects of moisture accumulation on surfaces and permeation of electronic packages [5,6].

#### *7.2.2.1 Moisture Accumulation on Surfaces*

According to the kinetic theory of gases, a surface immersed in an environment containing a gas at pressure $P$ will be impacted by a flux ($\Phi$) of its molecules given by

$$\Phi = \frac{P}{(2\pi MRT)^{1/2}}$$

or

$$\Phi = \frac{3.5 \times 10^{22} P}{(MT)^{1/2}} (\text{molecules/cm}^2 - \text{s}). \tag{7.2}$$

In these expressions, $M$ is the molecular weight, the pressure is expressed in millimeters of mercury or torr (1 torr = 1 mm Hg = 133.3 Pa), and $RT$ has the usual meaning. Before this formula is used, let us consider relative humidity (RH), which is defined as the ratio of the actual pressure of water vapor to the equilibrium vapor pressure at the given temperature. For water at 25 °C, the equilibrium vapor pressure is 24 torr; if, for example, RH = 42%, the value of $P$ is $0.42 \times 24$, or 10 torr. Substitution in Eqn (7.2) yields a flux of $4.8 \times 10^{21}$ $H_2O$ molecules per square centimeter-second at this temperature. Since a solid surface contains some $10^{15}$ atomic sites per square centimeter, a monolayer of condensed water will form in approximately 0.2 μs, assuming that all impinging vapor molecules stick.

Although a monolayer film forms almost instantaneously on a metal surface, it does not grow much thicker than a few monolayers with time. Furthermore, the pure water film does not generally destroy the thin surface oxide layer that passivates and protects the metal. However, when airborne pollutants of reactive chlorides, sulfides, sulfates, and oxygen and nitrogen compounds simultaneously impinge on the surface, the trouble begins. As the latter dissolve in the thin moisture films, high concentrations of corrosive ions arise to initiate and perpetuate damage by the mechanisms listed in the next section.

Moisture poses different reliability problems in nonmetallic materials. Continuous water films containing ions (electrolytes) that coat ceramics and glasses are paths for surface leakage currents and may cause insulators to break down in applied electric fields.

#### *7.2.2.2 Moisture Permeation of Bodies*

In porous electronic products, permeation of moisture often creates reliability problems. Semiconductor chips are encased within packages that occasionally leak. Thus, moisture and contaminants are able to permeate plastic, ceramic, and metal packages through microscopic interconnected pores, capillaries, channels, flaws, and even cracks in these materials and the glues, sealants, and solders used to seal them. In addition, moisture in packages arises from entrapped water due to poor bakeout procedures and is generated internally by decomposition of polymers. For example, moisture ingress in plastic packages has been identified to be the cause of "popcorn" cracking during soldering, a subject discussed later, in Section 8.2.3.2.

How large is package permeation or leakage, and how is it quantitatively characterized? Fortunately, package leaks typically do not pose a significant reliability hazard. A quantitative measure of the size of a leak is the leak flow rate ($Q_s$), defined to be the quantity of dry air that passes through the leak when there is atmospheric pressure on one side and vacuum on the other and expressed in units of atmosphere cubic centimeters per second or torr-liters per second. Leak rates can actually be calculated for certain leakage path geometries. Such calculations are done for vacuum system components and can be located in the vacuum science literature [7]. They are based on either kinetic theory of gases or compressible fluid flow, depending on whether the molecular or viscous gas-flow regime, respectively, is involved. Molecular flow occurs when the mean free path of gas molecules ($\lambda$) is much larger than the minimum dimensions ($d_{min}$) of the leakage channel, i.e., $\lambda > d_{min}$; molecules then collide with the channel walls more often than with each other. The leak rate for a gas of molecular weight $M$, at temperature $T$ (**K**), in a cylindrical channel of length $L$ (cm) and radius $r$ (cm), is given by

$$Q_s(\text{molecular}) = \frac{30.48r^3}{L}\left(\frac{T}{M}\right)^{1/2}(P_1 - P_2). \tag{7.3}$$

In this equation, the units of $Q_s$ are torr-liters per second, while $P_1$ and $P_2$ are the upstream and downstream pressures in torr, respectively.

The more complex case of viscous flow occurs when $d_{min} > \lambda$. Now molecules collide with each other more frequently than with the walls of the channel. For gases of viscosity $\eta$ (in Poise), the flow rate in torr-liters per second through a long circular channel of length $L$ and radius is

$$Q_s(\text{viscous}) = \left(\frac{\pi r^4}{16\eta L}\right)\left(P_1^2 - P_2^2\right). \tag{7.4}$$

Transitional flow occurs when $\lambda = d_{min}$ under conditions intermediate between molecular and viscous flow.

Now we consider the rate of pressure change caused by transport of a gaseous species into a package that does not contain gas initially. Assuming that ingress occurs by molecular flow, then

$$\text{Rate of ingress} = \frac{P_e Q_s}{P_o}\left(\frac{M_{air}}{M}\right)^{1/2}, \tag{7.5}$$

where $P_o$ is the atmospheric pressure, $P_e$ is the external pressure, and $M$ and $M_{air}$ are the molecular weight of the species and air, respectively. Gas that entered through the leak can also escape via the leak. Therefore, if the partial pressure has reached $P_i$,

$$\text{Rate of loss} = \frac{P_i Q_s}{P_o}\left[\frac{M_{air}}{M}\right]^{1/2}. \tag{7.6}$$

The net time ($t$) rate of gas pressure buildup or throughput is given by the difference, or

$$V^{\frac{dP_i}{dt}} = (P_e - P_i)\frac{Q_s}{P_o}\left[\frac{M_{air}}{M}\right]^{1/2}. \tag{7.7}$$

Integration of this simple differential equation for a package of volume $V$ yields

$$P_i(t) = P_e\left\{1 - \exp\left(-\frac{Q_s t}{P_o V}\left[\frac{M_{air}}{M}\right]^{1/2}\right)\right\}. \tag{7.8}$$

Similarly, it is easily shown that if the initial gas pressure within a package surrounded by a vacuum is $P_i(0)$, leakage will cause the pressure to decay according to

$$P_i(t) = P_i(0)\exp - \left[\frac{Q_s t}{P_o V}\left(\frac{M_{air}}{M}\right)^{1/2}\right]. \tag{7.9}$$

Diffusion of water through packages is a different way to view moisture ingress. Rather than molecular or viscous flow of water vapor through leakage channels, the diffusion equation (Eqn (5.2)) models transport of $H_2O$ through the package matrix with an "effective" diffusivity $D$. As an example, consider an epoxy-encapsulated package with a continuous concentration of moisture $C_c$ at its surface $x = 0$. The moisture concentration buildup $C(x, t)$ at the chip surface located $x = h$ deep is given by

$$\frac{C(x,t) - C_i}{C_c - C_i} = \frac{4}{\pi}\sum_{n=0}^{\infty}\frac{1}{n}\sin\left(\frac{n\pi x}{2h}\right)\exp\left[-\left(\frac{n\pi}{2h}\right)^2 Dt\right], \tag{7.10}$$

where $C_i$ is the initial moisture content. Note that this approach incorporates the geometry of the package, which is not true of Eqn (7.8). In this case, the diffusivity depends on the geometric array of pores, capillaries, channels, and flaws in the epoxy.

## 7.3 CORROSION OF METALS

### 7.3.1 Introduction

Chief among the mechanisms of degradation in metals are the electrochemical reactions that lead to corrosion or loss of metal. The potential for metallic corrosion exists whenever two metals, one acting as a cathode, the other as an anode, are surrounded by an aqueous electrolyte and connected electronically so as to complete an electrical short circuit. These requirements can be physically manifested in many varied ways. Galvanic corrosion can often be easily diagnosed because two different metals are involved. But corrosion can also occur when two regions of a single metal are somehow different (e.g., due to differential deformation, defects, local compositional inhomogeneities). This makes it difficult to recognize all of the corrosion disguises whose ominous faces are responsible for the large toll of damage to electrical and electronic equipment in humid environments.

Extensive treatments of the science and engineering of corrosion can be found in a number of excellent texts [8,9]. As an introduction to the subject, let us consider a metal M immersed in an acidic electrolyte. As the metal dissolves or corrodes, two reaction zones can be identified. The anodic reaction can be written as

$$M^{o} - ne^{-} \rightarrow M^{+n} \tag{7.11}$$

where the loss of electrons signifies chemical oxidation. The electrons are then consumed in cathodic regions where ions are reduced or discharged. In acidic environments, the cathodic reduction reaction

$$2H^{+} + 2e^{-} \rightarrow H_2 \tag{7.12}$$

occurs. However, in basic electrolytes, the production of hydroxyl ions via the reaction

$$H_2O + 1/2O_2 + 2e^{-} \rightarrow 2OH^{-} \tag{7.13}$$

is favored.

In the dissolution process above, both anode and cathode reactions occurred in the same metal. Each of these so-called half-cell electrode reactions can be isolated, however, under ideal conditions. Associated with each is a potential $E_{emf}$ that depends on the metal, nature of the surrounding electrolyte, ions present, and their concentration and temperature. For the anode reaction of Eqn (7.11), the Nernst equation yields

$$E_{emf} = E^{o} + \frac{RT}{nF_a}\ln(M^{+n}) = E^{o} + \frac{0.059}{n}\log(M^{+n}) \text{ at } 25^{\circ}\text{C}, \tag{7.14}$$

where $E^{o}$ is the standard potential and $F_a$ is the Faraday constant. A similar expression defines the electric potential at the other electrode. If the two half-cells are electrically connected, the individual emfs are added to yield the overall cell or reaction potential. Standard potentials ($E^{o}$) and reactions for a number of metals used in electrical applications are entered in Table 7.2. Relative positioning of the

**Table 7.2** Standard electrode potentials at 25 °C

| Electrode reaction | Electrode potential $E^o$ (volts) (relative to standard hydrogen) |
|---|---|
| $Au \rightarrow Au^{3+} + 3e^-$ | +1.498 |
| $H_2O + Cl^- \rightarrow H^+ + HOCl + 2e^-$ | +1.49 |
| $2Cl^- \rightarrow Cl_2(g) + 2e^-$ | +1.36 |
| $2H_2O \rightarrow O_2 + 4H^+ + 4e^-$ | +1.229 |
| $Pt \rightarrow Pt^{2+} + 2e^-$ | +1.200 |
| $Ag \rightarrow Ag^+ + e^-$ | +0.799 |
| $Fe^{2+} \rightarrow Fe^{3+} + e^-$ | +0.771 |
| $4(OH)^- \rightarrow O_2 + 2H_2O + 4e^-$ | +0.40 |
| $Cu \rightarrow Cu^{2+} + 2e^-$ | +0.337 |
| $Sn^{2+} \rightarrow Sn^{4+} + 2e^-$ | +0.150 |
| $\mathbf{H_2 \rightarrow 2H^+ + 2e^-}$ | **0.000** |
| $Pb \rightarrow Pb^{2+} + 2e^-$ | −0.126 |
| $Sn \rightarrow Sn^{2+} + 2e^-$ | −0.136 |
| $Ni \rightarrow Ni^{2+} + 2e^-$ | −0.250 |
| $Co \rightarrow Co^{2+} + 2e^-$ | −0.277 |
| $Cd \rightarrow Cd^{2+} + 2e^-$ | −0.403 |
| $Fe \rightarrow Fe^{2+} + 2e^-$ | −0.440 |
| $Cr \rightarrow Cr^{3+} + 3e^-$ | −0.744 |
| $Zn \rightarrow Zn^{2+} + 2e^-$ | −0.763 |
| $Al \rightarrow Al^{3+} + 3e^-$ | −1.662 |
| $Na \rightarrow Na^+ + e^-$ | −2.714 |

Cathodic behavior or reduction occurs more readily the higher the electrode reaction is in the table. Anodic behavior or corrosion occurs more readily the lower the electrode reaction is in the table.

involved metals in this table determines the tendency to react chemically and corrode. The driving force for galvanic corrosion directly depends on the magnitude of the algebraic difference between the two half-cell potentials.

### 7.3.2 Pourbaix Diagrams

A useful way to graphically depict thermodynamic aspects of aqueous metallic corrosion is through Pourbaix diagrams [8]. These diagrams, so named after their developer, map the oxidizing power, or potential, ($E$) of electrolyte solutions versus their acidity or alkalinity (pH). The result is a kind of metal–water phase diagram with boundaries defining regions of stable phases (e.g., ions, oxides, compounds) that either promote the tendency of metals to corrode or remain passive. Construction of Pourbaix diagrams is straightforward in principle, but involved in practice. The reactions of metals with water are pertinent, and there are essentially three kinds, defined by Eqn (7.11), by

$$M^{o} + nH_2O \rightarrow M(OH)_n + nH^{+} + ne^{-}, \tag{7.15a}$$

and by

$$M^{o} + nH_2O \rightarrow MO_n^{n-} + 2nH^{+} + ne^{-}. \tag{7.15b}$$

Let us consider the case of a metal undergoing the reaction defined by Eqn (7.15b). By the Nernst equation and the definition of pH as $-\log(H^{+})$,

$$\begin{aligned} E_{emf} &= E^{o} + \frac{0.059}{n}\log\left\{\left(MO_n^{n-}\right)\left(H^{+}\right)^{2n}\right\} \\ &= E^{o} + \frac{0.059}{n}\log\left(MO_n^{n-}\right) - 0.118\,\text{pH}. \end{aligned} \tag{7.16}$$

A plot of $E_{emf}$ versus pH yields a line with a slope of $-0.118$.

As a further example, we consider aluminum, arguably the most vulnerable metal in microelectronics from a corrosion standpoint. Reactions in aqueous solutions include

$$2Al + 6H^{+} \rightarrow 2Al^{+3} + 3H_2, \tag{7.17}$$

$$2Al(OH)_3 + 6H^{+} \rightarrow 2Al^{+3} + 6H_2O, \tag{7.18}$$

and

$$Al + 3(OH)^{-} \rightarrow Al(OH)_3 + 3e^{-}, \tag{7.19}$$

$$Al_2O_3 + H_2O \rightarrow 2AlO_2^{-} + 2H^{+}. \tag{7.20}$$

The first two reactions occur in acidic solutions. On the other hand, the latter two reactions are favored when the solutions are basic and when $Na^{+}$ and $K^{+}$ impurities are present to raise the local pH. Importantly, the potential in each of these reactions can be cast into the form of Eqn (7.16) and plotted as $E$ (relative to the standard hydrogen electrode) versus pH lines once it is recalled that pH + pOH = 14.00. Proceeding in this way, the resultant Pourbaix diagram for aluminum is shown in Figure 7.2, where the following items are noteworthy:

1. Aluminum is amphoteric, and the protective oxide film dissolves at both low and high pH values. In acids, aluminum dissolves forming

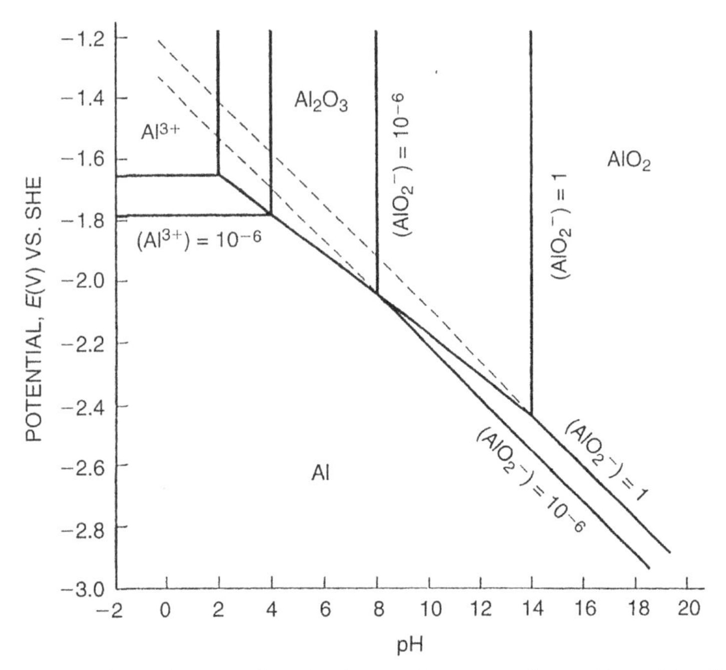


**Figure 7.2** Pourbaix diagram for aluminum. Lines are shown for arbitrarily, but frequently chosen, ionic concentrations of 1 and $10^{-6}$. *From Ref. [8].*

$Al^{3+}$ ions, while in basic solutions corrosion occurs with the generation of $AlO_2^-$ ions.

2. There is a region of passivity extending between pH 4 and 8, where aluminum is protected by $Al_2O_3$. Passivation also requires a potential above $\sim -1.8$ V.
3. Lower concentrations of ions containing Al compress the field of oxide stability.
4. Aluminum metal is immune to corrosion at potentials that lie below those indicated by the lowest lines on the Pourbaix diagram.

### 7.3.3 Corrosion Rates

In assessing reliability, it is the rates of corrosion that are of interest. Corrosion rates are not predictable from thermodynamic considerations and cannot be determined from Pourbaix diagrams. Rather complex processes at the metal–electrolyte interface generally govern the speed of corrosion reactions, and these proceed at finite rates. If the discharge of hydrogen is not rapid enough to accommodate all of the electrons provided, the cathode surface charges negatively. The negative potential that develops is known as cathodic polarization. Similarly, anodic polarization, or a positive potential, develops at the anode because of a depletion of electrons. And with greater electron deficiencies, higher anodic polarizations are produced. This, in turn, favors more dissolution of metal or corrosion. Thus, we may view anodic polarization as the driving force for corrosion. What happens is that the surface reaches a steady-state potential that depends on the rate at which electrons are exchanged between anode and cathode reactions. Without polarization, the smallest driving force would result in infinitely large electrochemical reaction rates.

Irrespective of the specific metals or aqueous environment, Faraday's law determines the extent of damage done. Thus, the passage of 96,500 Coulombs (C) of charge will remove a gram equivalent weight of anode metal and discharge a like weight of ions at the cathode. Because corrosion occurs heterogeneously, corrosion currents are usually unknown. This makes it difficult to predict the extent of corrosion damage and the life of involved components.

### 7.3.4 Damage Manifestations

Corrosion damage in microelectronics is similar to that which occurs in materials used in mechanically functional applications. Although the morphology of corrosive attack can vary appreciably, there are several basic mechanisms, and these are presented in turn.

#### *7.3.4.1 Uniform Corrosion*

In this case, corrosion damage occurs uniformly over large areas of metal or alloy surfaces. Low-carbon steels and copper alloys are subject to such attack. In the passive state, the rate of corrosion is essentially negligible. Protection is typically provided by a thin oxide film. However, if passivity is destroyed by puncturing the oxide, the metal becomes active, with effective anodic and cathodic sites constantly shifting, so that metal loss is uniform. Although it is relatively easy to protect against uniform corrosion, examples of failures by this mechanism are generally rare.

#### *7.3.4.2 Galvanic Corrosion*

Two different metals (alloys) are required for galvanic action, perhaps the easiest form of corrosion to recognize. When dissimilar metals in electrical contact are immersed in an electrolyte, an electrochemical cell (i.e., a battery) is produced. The more active metal becomes the negatively charged electrode, or anode, while the more noble metal becomes the positively charged electrode, or cathode.

Galvanic corrosion is magnified when the anode is small and the cathode large. Examples of such a size effect occur commonly in electronics when electrodeposited noble metal coatings are used to protect less noble substrates, e.g., Au on Ni. If the Au coating is porous or punctured at a point, Ni may rapidly corrode. The reason has to do with conservation of current $i(A)$ in both electrode reactions. Thus, if $j$ is the current density (amperes per square centimeter) and $A_o$ is the electrode area, the condition that $i(\mathrm{Au}) = i(\mathrm{Ni})$ means that $j(\mathrm{Au})A_o(\mathrm{Au}) = j(\mathrm{Ni})A_o(\mathrm{Ni})$. Because $A_o(\mathrm{Au}) > A_o(\mathrm{Ni})$, $j(\mathrm{Ni}) > j(\mathrm{Au})$; the high current density causes the small unprotected Ni region to rapidly dissolve, while the cathode reaction is collectively borne by the large coating area.

### *7.3.4.3 Pitting Corrosion*

As the description implies, pit or hole formation is the form of attack during pitting corrosion. Localized pitting attack generally occurs while the bulk of surface is passive. Chloride, and other halogen ions to a lesser extent, are notorious for initiating pitting attack in otherwise passive metals and alloys. A suggested mechanism for the pitting of stainless steel in the presence of $Cl^-$ ions is shown in Figure 7.3. Corrosion begins with the loss of passivity

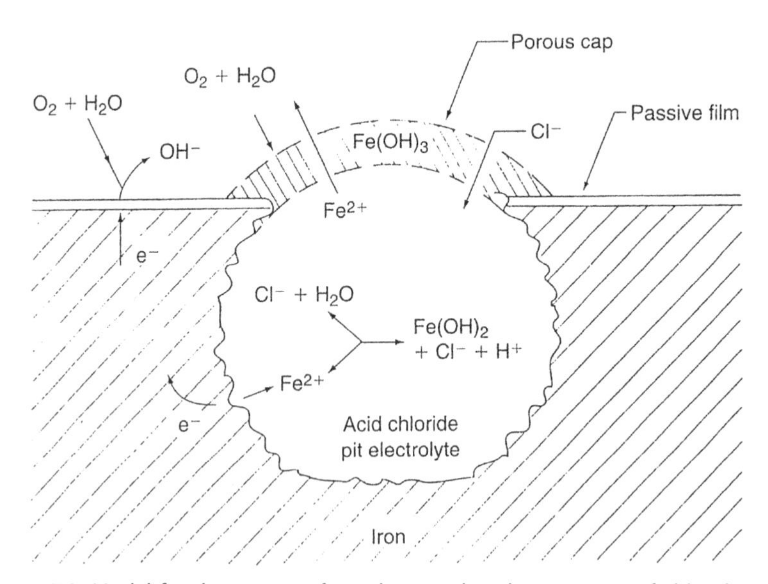


**Figure 7.3** Model for the pitting of stainless steel in the presence of chloride ions.

caused by the local destruction of the thin chromium oxide passive film. A copious production of $Fe^{2+}$ ions attracts $Cl^-$, where hydrolysis occurs by the reaction

$$Fe^{2+} + 2H_2O + 2Cl^- \rightarrow Fe(OH)_2 + 2HCl. \tag{7.21}$$

In the process, $Fe(OH)_2$ forms, and the pH is lowered at the pit initiation site. The acid chloride solution further accelerates anodic dissolution, which in turn concentrates more chloride at the pit, etc., in this so-called self-propagating autocatalytic reaction. In addition, $Fe^{2+}$ oxidizes to $Fe^{3+}$, so that a porous cap of rust ($Fe(OH)_3$) covers the pit hole. This encourages $Cl^-$ ions to migrate in order to balance the excess positive charge generated by metal oxidation. All the conditions are now fulfilled for continued pitting corrosion. Similar pitting reactions occur for the nickel and chromium atoms in stainless steel.

Crevice corrosion also causes local attack of metal that is similar to pitting damage. Metal surface regions hidden under washers, gaskets, sealing rings, surface scale, switch contacts, rivets, etc., serve as crevices. Restricted oxygen supply to the crevice diminishes the cathode reaction of Eqn (7.13).

#### 7.3.4.4 Fretting Corrosion

Tin, plated solder, and other electrical contact materials often suffer corrosion-accelerated wear when the contacting metals undergo small-amplitude cyclic rubbing or fretting. Particles chemically dissolve and mechanically detach, leaving a surface geometry behind that generally increases the contact resistance. Fretting effects at electrical contacts are addressed again in Chapter 9.

#### 7.3.4.5 Stress Corrosion Cracking

Corrosion in the presence of tensile stress accelerates the damage to the metal surface. The mechanisms are not well understood, but it is thought that applied or internal stress causes atoms to leave the anodic tips of incipient cracks, thereby sharpening them. Usually, particular environment–metal combinations are associated with stress corrosion cracking (SCC), e.g., chlorides for stainless steel, ammonia for copper, and $H_2S$ for low-carbon steel. Residual thermal stresses, or those induced by bending electrical lead wires, in combination with moisture serve to promote such cracking. When the loading is cyclic rather than static, the stress-corrosion damage is known as corrosion fatigue.

#### *7.3.4.6 Atmospheric Corrosion*

This form of corrosion attack stems from the environment, but without the presence of a bulk electrolyte. Instead, the electrolyte is usually a thin film of moisture that condenses from the atmosphere when the RH is sufficiently high. The electrolytic process of metal migration on nominally insulating substrates occurs in this manner (Section 7.4).

## 7.4 CORROSION IN ELECTRONICS

### 7.4.1 Overview

After having just introduced some of the classic manifestations of corrosion, we now focus on corrosion failure in electronic components. A couple of corrosion case histories are reproduced in Figure 7.4 and discussed in the figure captions. In microelectronic devices, the scale of damage is far smaller but no less destructive. Vulnerable metal structures are thin film contacts, interconnect lines, and wire or solder bonding pads [11,12]. These structures are often coated with glassy inorganic and/or organic polymers, which not only provide electrical insulation but mechanically protect and chemically passivate devices as well. However, defective coatings and packages render unprotected metals vulnerable to chemical attack from moisture as well as atmospheric and encapsulant contaminants.

An old but still instructive review by Schnable et al. [13] of corrosion in IC chips encased within hermetically sealed and plastic-encapsulated packages has usefully classified corrosion as

1. chemical (where no bias voltage is present),
2. electrochemical (where a dc bias voltage is applied and currents flow), and
3. galvanic (involving dissimilar metals).

The role of the passivating layers is of crucial significance in the corrosion of underlying metals. Pinholes and cracks in these brittle dielectric films ($SiO_2$, silicon nitride) allow the ingress of moisture and contaminants, while their residual internal stresses may promote SCC of the metallizations. Corrosion on the chip level is largely a processing problem because metallizations are exposed then. Once metal structures are passivated, corrosion largely ceases to be a reliability issue. On the packaging level, ionic contaminants (e.g., $Cl^-$) provided a substantial corrosion hazard for thin-line metallizations in early molded plastic packages. However, today's epoxy molding materials have a lower concentration of corrosive ions, so that failures are infrequent.

**Figure 7.4** Examples of corrosion-related problems in electronics. (a) Core of failed nichrome-film resistor showing missing metal. The resistor was damaged by excessive heat applied during the lead retinning operation. This opened the resistor epoxy end-cap seal and allowed entry of chlorides from flux and cleaning solvents. Because chlorine was detected in the region of nichrome loss, the role of this element in fostering corrosion attack was confirmed. *(From Ref. [10].)* (b) Corrosion of copper on a printed-circuit board plated through hole. Prior to soldering, the copper was mechanically brushed, generating metal flow and surface crevices. Flux, entrapped in the crevices, remained after the Sn–Pb solder covered the Cu (see Section 8.3.3.2). The combination of different metals in contact, salts in the flux, and a humid environment resulted in galvanic action with the formation of blue ($CuSO_4$) corrosion products surrounding the lower half of the hole. *Courtesy of Manko Associates.*

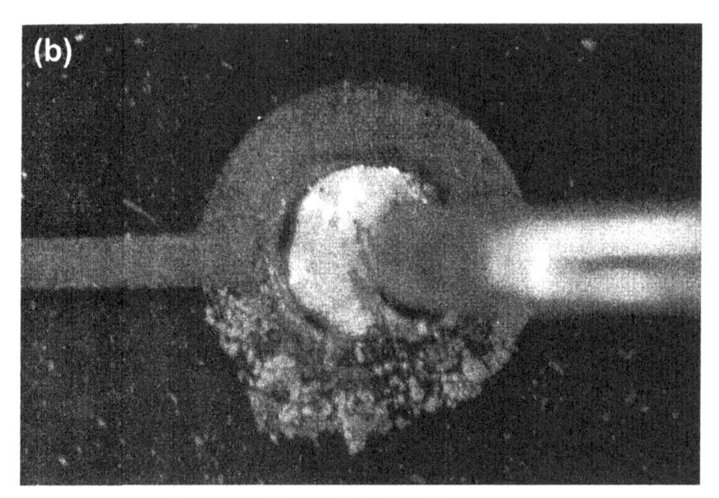


**Figure 7.4** Cont'd.

## 7.4.2 Accelerated Corrosion Testing

The observation that aluminum metallization on ICs is prone to corrosion has led to accelerated testing at elevated temperatures and RH levels. Results distilled from such testing reveal that mean failure times can be described by

$$\mathrm{MTTF} = \mathrm{A}\mathrm{V}^{-m}(\mathrm{RH})^{-n} \exp\left[\frac{E_{\mathrm{c}}}{kT}\right], \tag{7.22}$$

where $V$ is the applied voltage, $E_{\mathrm{c}}$ is an effective activation energy, $A$, $n$, and $m$ are constants, and $kT$ has the usual meaning. If the test voltages, circuit designs, and specimen geometry and spacings are constant, Peck [14] has suggested that

$$\mathrm{MTTF} \approx (\mathrm{RH})^{-3} \exp\left[\frac{0.8\mathrm{eV}}{kT}\right]. \tag{7.23}$$

Like similar formulas that describe electromigration (Eqn (5.31)) and dielectric breakdown (Eqn (6.10)), Eqn (7.22) is quite useful in design and for making life projections. Unfortunately, however, little insight into the kinetics of the degradation process can be gleaned from such phenomenological formulas.

While early test conditions varied widely, much testing is presently carried out at 85 °C and 85% RH. Highly accelerated stress tests are popular

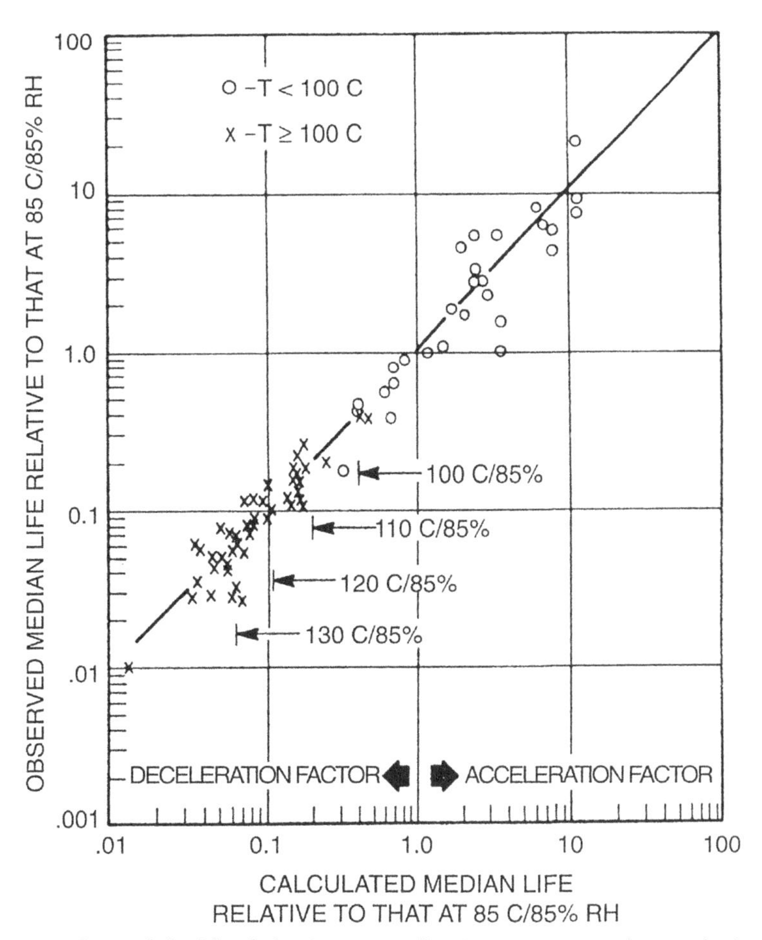


**Figure 7.5** Ratio of the life of aluminum metallization in epoxy packages obtained in humidity tests relative to 85°C/85% relative humidity (RH) conditions. *From Ref. [14].*

in order to accentuate the adverse roles humidity and temperature have on corrosion; otherwise, testing times are impracticably long. Results of a wide survey of available corrosion-failure data for aluminum in epoxy packages are shown in Figure 7.5. Relative to the standard 85°C/85% RH test conditions, higher temperatures shorten failure times according to Eqn (7.23).

## 7.4.3 Corrosion of Specific Metals

### *7.4.3.1 Aluminum*

Corrosion effects have been widely studied in aluminum and Al base alloy metallizations. As we have already seen, Al suffers corrosion damage in the presence of water. Even the passive, amorphous native-oxide layer can undergo hydrogen blistering damage by the reaction sequence depicted in Figure 7.6 [15]. Most commonly, however, Al corrodes at both anodic and cathodic sites in the presence of contaminant ions. For example, chloride ions from the atmosphere, solder fluxes, chlorinated hydrocarbons (Freon, trichloroethylene), solvents, etc., accelerate a

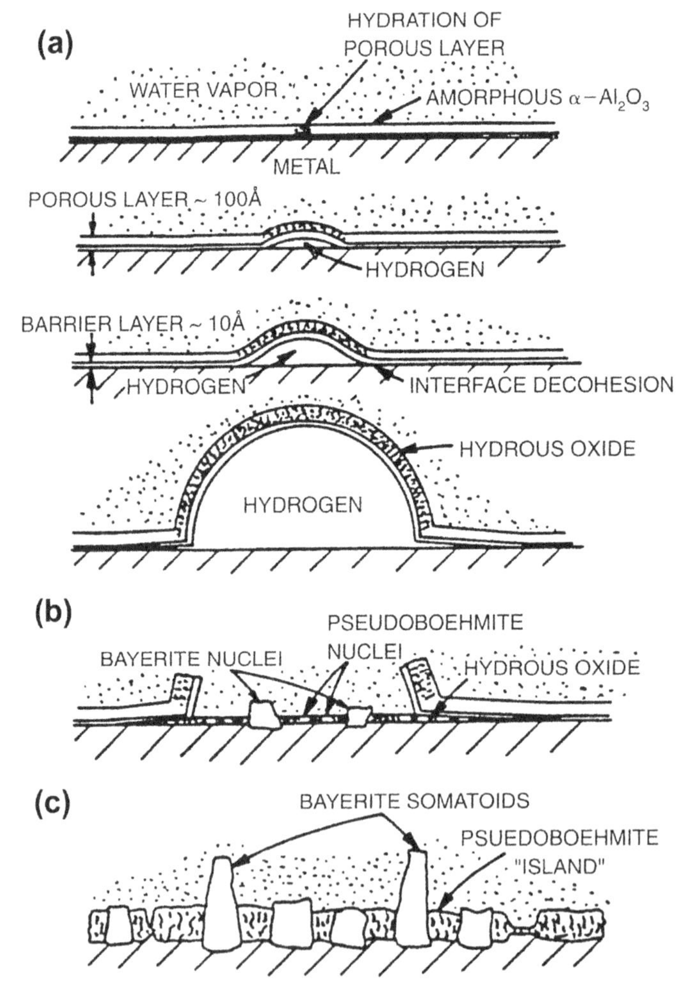


**Figure 7.6** Schematic model of reaction occurring at aluminum metal/oxide-water vapor interface. (a) Initiation of hydrogen attack; (b) blister fracture and nucleation of aluminum-oxygen-hydrogen phases; (c) hydroxide growth. *From Ref. [51].*

pitting-type anodic corrosion reaction that is autocatalytic via the following cycle:

1. Dissolution of protective hydrated oxide upon adsorption of $Cl^-$ ions

$$Al(OH)_3 + Cl^- \rightarrow Al(OH)_2Cl + OH^- \tag{7.24a}$$

2. Chloride attack of Al

$$Al + 4Cl^- \rightarrow AlCl_4^- + 3e^- \tag{7.24b}$$

3. Regeneration of $Cl^-$

$$AlCl_4^- + 3H_2O \rightarrow Al(OH)_3 + 3H^+ + 4Cl^- \tag{7.24c}$$

Corrosion problems in IC chips can start during processing. For example, during patterning of thin-film interconnects by reactive plasma etching, gases containing chlorine are used, which can leave halide residues behind to contaminate photoresists and metallizations. Exposure to humidity then sets the above chemical reactions in motion. In addition, cathodic corrosion occurs when Al interconnects contact phosphosilicate glass (PSG) films that contain excessive phosphorus levels. For reasons explained in Chapter 5, aluminum metallizations contain copper, and sometimes silicon. This enhances the prospects for galvanic corrosion, which takes the form of intergranular corrosive attack and loss of Al [16] in Al–Cu metallizations during wet etching and use of chlorinated solvents. Similarly, microcorrosion of Al–Cu bonding pads has been found to be a barrier to high-quality wire bonding [17]. Examples of accelerated corrosion in aluminum metallizations exposed to chlorides and humid atmospheres are shown in Figure 7.7.

The specter of even more catastrophic corrosive attack exists in different regions of the chip because the Al metallization is additionally in contact with other metals and alloys. These serve as semiconductor contacts (e.g., PtSi), diffusion barriers (Ti–W, Cr), bonding pads (Au wire), vias (W, TiN), and flip-chip solder pads (Au, Cu–Cr). To expose the extent of potential corrosion between assorted metal pairs, Griffen et al. [18] have recently experimentally determined the galvanic series for thin-film metallization and diffusion barriers displayed in Table 7.3. Corrosion tests revealed that the loss of Al from Al–0.5 wt% Cu–1 wt% Si on CVD tungsten occurred some 25 times more rapidly than from Al–0.5 wt% Cu–1 wt% Si on sputtered W–10 wt% Ti, despite the ordering of metals in the table; passivation of W–Ti is the probable cause.

Exposure of these metal couples to an aqueous environment would unleash a galvanic corrosion nightmare, but for the most part metallizations

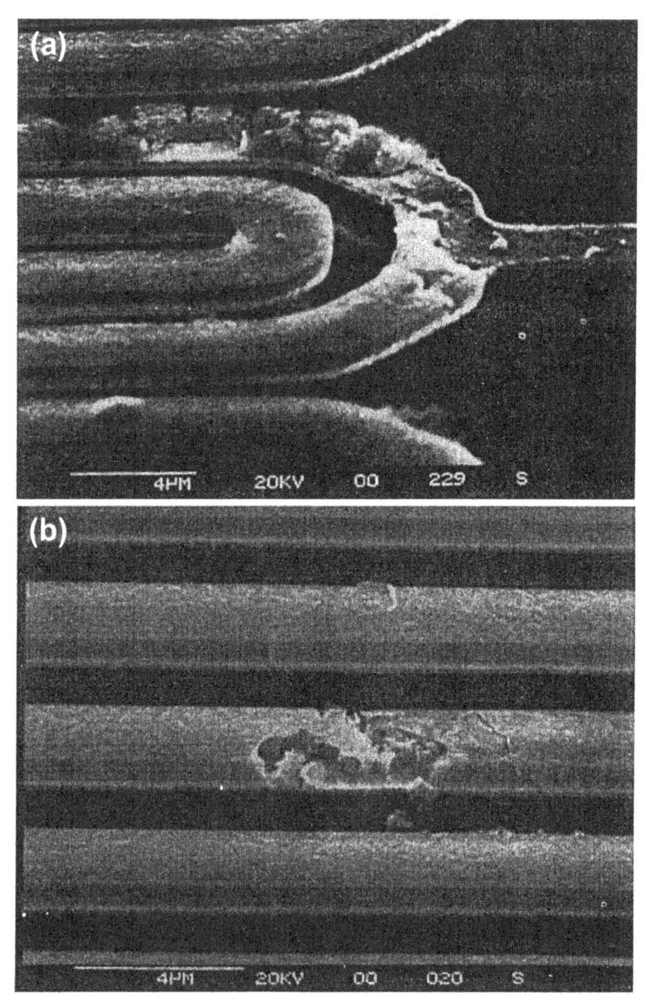


**Figure 7.7** Corrosion of aluminum metallizations exposed to 1000 ppm solutions of $NH_4Cl$. (a) Metal corrosion extending from bonding pad into an array of interconnections. Once corrosion of the aluminum is complete, current flows in the more resistive Ti:W layer. (b) Localized corrosion where the native oxide has been penetrated. *From Ref. [52].*

are not susceptible to such attack because they are passivated with dielectric overcoats. Lastly, it should be appreciated that Al, as well as other metal films, is normally stressed in tension following oxide passivation. Since bulk Al alloys exhibit stress-corrosion cracking, the role of stress in accelerating corrosion failures of Al metallizations is always a possibility.

**Table 7.3** Galvanic series for thin-film metallizations and diffusion barriers

| Metal | EMF (mV)[a] |
|---|---|
| Cu (1 μm) | +70 |
| Al-0.5 wt% Cu-1 wt% Si (0.75 μm)on CVD W (0.55 μm) | −6 |
| CVD W (0.75 μm) | −23 |
| W-10 wt% Ti (0.75 μm) | −92 |
| Al-2 wt% Cu (1 μm) | −368 |
| Al-2 wt% Cu-1 wt% Si (1 μm) | −686 |
| Al-1 wt% Si (0.75 μm) on CVD W (0.55 μm) | −783 |
| Al (0.75 μm) on CVD W (0.55 μm) | −800 |
| Al (0.75 μm) on sputtered W-10 wt% Ti (0.55 μm) | −948 |
| Al (1 μm) | −1030 |

Electrolyte: Stirred solution of 2000 ppm $NH_4Cl$ at 25°C.
Reference electrode: Saturated calomel (+0.242 mV).
[a]Potentials relative to standard hydrogen electrode.
From Ref. [18].

#### 7.4.3.2 Gold

Even the most noble of metals, gold, corrodes. For the most part, damage requires application of a dc bias voltage to the metal. In the presence of moisture and chloride ions, gold migration occurs by an anode dissolution–cathode plating process known as metal migration, which is described more fully in Section 7.5. The anode reaction is thought to be

$$Au + 4Cl^- \rightarrow AuCl_4^- + 3e^-, \tag{7.25}$$

while the reverse reaction at the cathode results in dendritic Au deposits that often cause leakage currents or shorting of components. This type of electrolytic corrosion has also been observed in gold-containing metallizations, e.g., Ti–Pd–Au, Ni–Au–Cr, Ti–W–Au, and Ti–Pt–Au.

Polymer-encapsulated gold wire bonds to aluminum pads have been reported to corrode when there is sufficient moisture permeation of the package [19]. Not only is there the $AuAl_2$ (purple plague—see Section 5.3.4) compound to contend with, but the epoxy flame retardants used contain halogens, which provide the $Br^-$ necessary to autocatalyze local pitting corrosion.

Gold is commonly employed in GaAs MESFET contact and metallization schemes. Accelerated temperature–humidity–ionic contamination testing not surprisingly revealed that Al gate electrodes were more susceptible to corrosion damage than Au/refractory metal gates [20]. Corrosion of Al, migration of Au between source and gate, shorting metallic bridges, and formation of Al–Au compounds correlated with degradation of device characteristics.

#### *7.4.3.3 Other Metals*

A number of other metals are used in electronic applications, and a brief survey of corrosion damage in these follows.

(a) Nickel. Nickel-plated components and hardware are widely used in electronics. An example of nickel corrosion in solid-state microwave hybrid-blocking and bypass capacitors has been reported by Torkington and Lamoreaux [21]. Failure in Ni, a normally corrosion-resistant metal, was attributed to the sealed-in moisture derived from the cover gas used during package solder sealing. With condensed moisture, the galvanic voltage ($-1.75$ V between Au and Ni) added to the larger applied voltage (5 V). In the presence of chlorine and $CO_2$, $NiCl_2$ and $NiCO_3$ corrosion products were observed to short the capacitor electrodes. Interestingly, the 50:1 Au:Ni ratio is an example of the corrosion-prone, small anode–large cathode configuration (Section 7.3.4.2). In this case, it greatly increased the current density and attack of nickel.

(b) Copper. For fear of contaminating and poisoning silicon, copper was sparingly used in microelectronics, until recently, when Cu-based, thin-film metallizations (see Chapter 2) have been used in 10 level metallurgy systems [22]. Additionally, copper and its alloys play a very major role in electrical components such as wire, lead frames, printed-circuit board circuitry, heat sinks, vacuum tube electrodes, and RF gaskets, as well as in systems that cool electronics. Much of the corrosion damage in these items is attributable to attack by atmospheric pollutants, in particular $(NH_4)_2SO_4$ [1]. In the presence of sufficient moisture, copper dissolves in ammonium sulfate solution to form $Cu(NH_3)_2^+$; i.e.,

$$Cu + 2NH_3 \rightarrow Cu(NH_3)_2^+ + e^-. \tag{7.26a}$$

The corrosion products $Cu_2O$ and hydrated copper sulfates then form via the following typical reactions:

$$2Cu(NH_3)_2^+ + H_2O \rightarrow Cu_2O + 2H^+ + 4NH_3 \tag{7.26b}$$

and

$$3Cu^{2+} + 4(OH)^{-1} + SO_4^{2-} \rightarrow Cu_3(SO_4)(OH)_4. \tag{7.26c}$$

Copper is also susceptible to corrosion in humid sulfide atmospheres [23]. Dendrites of corrosion products were observed to grow laterally from interconnections on printed-circuit boards in the absence of applied electric potentials. However, the processes of corrosion, dendrite growth, and filament shorting are accelerated by several orders of magnitude when electric fields are present.

#### *7.4.3.4 Corrosion in Magnetic Disks*

Among the metal-containing electronic products sensitive to corrosion are computer hard disks [5,24]. Magnetic recording media typically consist of a layer of either $Fe_2O_3$ particles or a thin metal-alloy (Co–Ni) film sandwiched between an aluminum-substrate pad on the bottom, and metal oxide plus lubricant film on top. Common corrodents were found to be chlorine and sulfur. In the form of $SO_4^{2-}$, the latter attacked the Co, which was found on the disk surface.

### 7.4.4 Modeling Corrosion Damage

Over the years there have been numerous attempts to model corrosion damage with the objective of predicting failure times. This is an extraordinarily difficult task because

1. the chemical reactions are often complex;
2. ion movements occur in an electrolyte fluid under the influence of spatially complex electric fields, concentration gradients, and convective flows; and
3. the metal surface structure, chemistry, and heterogeneous nature of damage are not well known, particularly under conditions of pitting or stress corrosion.

For these reasons, it has been common to use familiar, phenomenological formulas like Eqn (7.22) to predict failure times. However, these have obvious shortcomings, as noted earlier. The approach taken by Pecht [5] and reproduced here is more satisfying, particularly because it connects uniform corrosion rates to reliability functions. To see how, we consider degradation of microelectronic packages that is caused by the two-step process of moisture permeation of packages followed by uniform corrosion of the metallization within.

An equation describing permeation of moisture was given earlier (Eqn (7.8)):

$$\frac{P_i(t)}{P_e} = 1 - \exp[-\beta t], \tag{7.27}$$

where $\beta = Q_S/(P_o V)\ [M_{air}/M]^{1/2}$. It is further assumed that uniform corrosion is essentially described by a first-order chemical reaction of the form

$$\frac{dC}{dt} = -\alpha(C - C_o). \tag{7.28}$$

In this equation, $C$ may be thought of as a corrosion density per unit volume or concentration, with $C_o$ being the maximum, or self-limiting, value of $C$; i.e., $0 \leq C \leq C_o$. Thus, the rate of corrosion is greatest initially when $C = 0$ and declines thereafter. Rate constant $\alpha$ combines a product of terms including the RH in the package and the ubiquitous Boltzmann factor, $\exp[-E_c/kT]$, where $E_c$ is the activation energy for corrosion and $kT$ has the usual meaning. Explicitly, $\alpha = \alpha_o \cdot (\text{RH}) \cdot \exp \cdot [-E_c/kT]$, where factor $\alpha_o$ is dependent on the nature of the materials and voltage bias level.

The effect of temperature is often not what one might expect. When the internal package temperature is higher than the ambient temperature, moisture cannot condense. Thus, devices that dissipate a lot of heat may actually suffer less corrosion damage than devices that consume and dissipate little power, e.g., CMOS devices. Moisture condenses more readily in the latter, and there is greater corrosion risk for metallizations, wire bonds, lead frames, etc.

Since a measure of RH is $P_i(t)/P_e$, for humidity conditions that allow for condensed moisture on metallic components, Eqn (7.28) yields

$$\frac{dC(t)}{dt} = -B(1 - \exp[-\beta t])(C - C_o), \tag{7.29}$$

where constant $B = \alpha_o \cdot \exp \cdot [-E_c/kT]$. Upon separation of variables and integration,

$$\frac{C(t) - C_o}{C_i - C_o} = \exp\left[-\frac{B}{\beta}\{\beta t - (1 - \exp -\beta t)\}\right]. \tag{7.30}$$

The constant of integration $C_i$ satisfies the initial condition that at $t = 0$, $C = C_i$. If we now expand the term $\{\beta t - (1 - \exp[-\beta t])\}$ as $\beta t - 1 + 1 - \beta t + \frac{1}{2}(\beta t)^2 + \ldots$, or $\frac{1}{2}(\beta t)^2$ for times of order $t^3$, then

$$\frac{C(t) - C_o}{C_i - C_o} = \exp\left[-\frac{1}{2}B\beta t^2\right]. \tag{7.31}$$

If it is assumed that $C_i = 0$, then $(C(t) - C_o)/(C_i - C_o)$ is a measure of the reliability $R$, so that

$$R(t) = \exp\left[-\frac{1}{2}B\beta t^2\right]. \tag{7.32}$$

This association is consistent with full reliability initially (no corrosion) and total loss of reliability (fully corroded) at very long time. A test of experimental corrosion failure data plotted as $R$ versus $t$ is shown in Figure 7.8 and compared with a fit of Eqn (7.32). Evidently, there is satisfactory agreement between the data and the predicted form of this equation.

From Eqn (4.4) we recall that $F(t) = 1 - R(t)$, and therefore,

$$F(t) = 1 - \exp\left[-\frac{1}{2}B\beta t^2\right]. \tag{7.33}$$

This cumulative distribution function is of the Weibull form, and it suggests the use of such distributions for physically modeling corrosion failures. Finally, the hazard function, $\lambda(t)$, defined by $F'(t)/R(t)$, is simply calculated to be

$$\lambda(t) = \frac{B\beta t \exp\left[-\frac{1}{2}B\beta t^2\right]}{\exp\left[-\frac{1}{2}B\beta t^2\right]} = B\beta t. \tag{7.34}$$

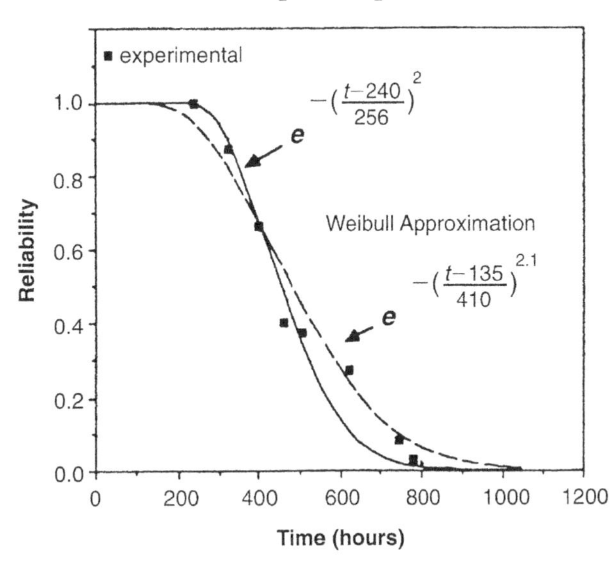


**Figure 7.8** Reliability of an encapsulated package in a corrosive environment. The experimental failure data are fit to a Weibull distribution. *From Ref. [5].*

This result corresponds directly to the Rayleigh failure rate (see Section 4.3.3).

## 7.5 METAL MIGRATION

### 7.5.1 Introduction

The phenomenon of metal migration, also referred to as "metallic," "wet," and "electrolytic" electromigration, has caused many catastrophic microcircuit failures [23,25]. Unlike solid-state electromigration (Section 5.5), which generally requires elevated temperatures and high current densities ($\sim 10^5$ A/cm$^2$), metal migration can occur at room temperature with passage of less than $1 \times 10^{-3}$ A/cm$^2$. Like corrosion, it is an electrochemical phenomenon that can occur both within and outside of IC packages. The effect was first observed in polymer (phenol fiber) circuit boards containing silver lead wires and contacts. In essence, Ag metal was ionically removed from its initial location, then Ag ions migrated in the electric field, and finally Ag metal redeposited elsewhere [26]. Water absorbed on the insulator surface served as the electrolyte. At the anode, Ag formed a colloidal oxide, while at the cathode Ag metal assumed the dendritic morphology. Examples of dendritic growth in silver and copper conductors are shown in Figure 7.9(a) and (b).

An extensive review of metal migration–accelerated testing [27] suggests that the effects are largest in silver. In addition to Ag, metals that migrate in a distilled water electrolyte include Bi, Cd, Cu, Pb, Sn, and Zn. When distilled water is contaminated with sodium or potassium chloride (0.001–0.01 M), Au, Pd, and Pt migrate. Other metals, e.g., Ni, Cr, and Ag–Au–Pd alloys, also migrate under certain conditions. When liquid water is present, electromigration effects increase greatly in speed and severity.

For metal migration to occur, the following conditions are generally required:

1. The anode metal must easily form ions.
2. Nearby there must be a metal that serves as the cathode.
3. A dc voltage must exist between the anode and the cathode. Depending on the interelectrode spacing, as little as 1–2 V up to 100 V or more may be required.

**Figure 7.9** Examples of wet electromigration. (a) Silver dendrites growing in deionized water on glass–epoxy printed circuit board under applied voltage (24×). (b) Copper sulfide dendritic filaments growing from copper conductor between polyester sheets without application of voltage, but after exposure to a humid $H_2S$ atmosphere (24×). *From Ref. [53]. Courtesy of S. J. Krumbein, AMP Incorporated.*

4. Moisture or a water film must be present on the insulator substrate surface to provide the electrolyte needed for ionic conduction. Porous and fibrous substrates, e.g., epoxy resin/glass fiber, polyesters, laminates, and fabrics, serve to retain the electrolyte.

Metal ions also migrate on oxide surfaces, and these effects are discussed in Sections 7.5.3 and 7.5.4.

## 7.5.2 Metal Migration on Porous Substrates

The need to accumulate moisture within the porosity of the insulator was emphasized above in order for metal migration to become a reliability problem. In this regard, the model developed by DiGiacomo [28] is instructive. The analysis assumes that the metal ion current density that flows is directly proportional to the fraction of substrate pores containing condensed water. For a lognormal distribution of pore radii, the fractional area ($f$) on which condensation occurs is given by

$$f = \frac{1}{2}\,\mathrm{Erfc}\left[\frac{1}{2}\sigma \ln\left(\frac{r_{\mathrm{avg}}}{r}\right)^2\right], \tag{7.35}$$

where $\sigma$ (see Section 4.3.5) is $\ln(r_{50}/r_{16})$, and $r_{\mathrm{avg}}$ is the average pore radius. Water condensation in pores from a saturated vapor pressure ($p$) is basically driven by a chemical reaction whose free energy change per unit volume $\Delta G_V$ is given by $-kT \ln\frac{p}{p_\infty}$ [29]. The ratio of the vapor pressure above the pore radius, or meniscus, $r$, relative to that over a flat pore surface $p_\infty$, may be thought of as the RH. In this context, elementary nucleation or capillarity theory establishes the connection between $r$ and $\Delta G_V$ as $r = \frac{2\gamma\Omega}{kT\ln(\mathrm{RH})}$, where $\gamma$ and $\Omega$ are the water surface tension and atomic volume, respectively, and $kT$ has the usual meaning.

The ionic mass flux ($J_i$) is the product of the ion concentration ($C_i$ in grams per cubic centimeter) and its velocity, where the latter is given by the Nernst–Einstein equation (Eqn (1.6)); i.e., $v_i = DF/RT$. As in electromigration (Section 5.5.2), the force $F$ is equal to $ZqE$, where $Z$ is the ion valence, $q$ is the electronic charge, and $E$ is the electric field. Combining all of the above terms, the equation

$$J_i = \left(\frac{ZqC_iDE}{RT}\right) f = ZqC_iDE\ \mathrm{Erfc}\left[\frac{\sigma}{2}\ln\left(\frac{r_{\mathrm{avg}}kT\ln(\mathrm{RH})}{2\gamma\Omega}\right)^2\right] \tag{7.36}$$

results. The goal is to obtain the time to failure (MTTF), which is defined by how long it takes a dendritic filament to span conductors separated by a length $L$ when a voltage $V$ is applied. To produce and extend a dendrite, the current density at its tip must be orders of magnitude higher than its average value. Dendritic growth occurs by a focusing of the ion current at its tip of radius $r_d$, an effect that magnifies the local field to $E = V/r_d$. The critical mass of metal that is transported to create the shorting dendrite filament is assumed to be $Lr_d \cdot \delta \cdot d$, where $\delta$ is the electrolyte thickness and $d$ is the metal density. Failure occurs in time $t = \text{MTTF}$, and by definition the corresponding mass flux (mass/area-time) is $J_i = Lr_d \cdot \delta \cdot d / r_d\ \delta\ \text{MTTF} = L \cdot d/\text{MTTF}$. Solving for MTTF yields the desired result

$$\text{MTTF} = \frac{r_d RTLd}{ZqC_iDV\ \text{Erfc}\left[\frac{\sigma}{2}\ln\left(\frac{r_{avg}kT\ \ln\text{RH}}{2\gamma\Omega}\right)^2\right]}. \quad (7.37)$$

In this closed-form equation, the effects of voltage, humidity, nature of the substrate, dendrite geometry, and temperature (through $D$ and $RH$) are all linked.

### 7.5.3 Metal Migration on Ceramic Substrates

In this section, a simple model [30] for metal migration based on Figure 7.10(a) will be developed. A parallel pair of metal stripes with indicated dimensions has a potential $V$ applied across it. Due to moisture condensation, a thin electrolyte layer coats the ceramic substrate and spans the metal electrodes. The electrolyte contains ionic contaminants that make it sufficiently conductive to cause metal migration. Corrosion will proceed until a length of electrode, approximately equal to its width, is corroded to an open condition (Figure 7.10(b)). Dendrite growth, which can short conductors, is often a secondary effect of corrosion. Using Ohm's law ($i = V/R$) and the definition of resistance $R(\Omega) = \rho S/(Lt_o)$, we obtain

$$i = \frac{VL}{(S\rho/t_o)}, \quad (7.38)$$

where $\rho$ (ohm centimeters) is the resistivity of the electrolyte. The volume of corroded metal, $\nu$ (cubic centimeters), is proportional to the total charge Q (coulomb) passed, or $i \times \tau$, where $\tau$ is the time. Faraday's law essentially states that $Q = nd\nu F_a/M$, where $F_a$ is the Faraday constant and $n$, $d$, and $M$ are the metal valence, density, and atomic weight, respectively. Therefore, combining terms, the effective time to failure ($\tau = \text{MTTF}$) when $\nu = lwh$, is given by

$$\text{MTTF} = \frac{nd\nu F_a S}{MVL}\left(\frac{\rho}{t_o}\right) = \text{constant} \times \rho/t_o, \quad (7.39)$$

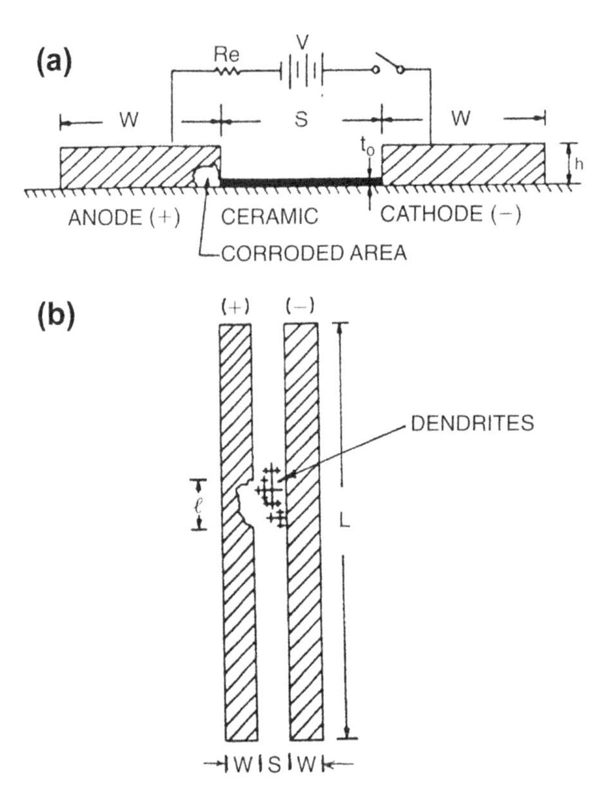


**Figure 7.10** (a) Cross-section of corrosion process involving metal migration between conducting stripes on a ceramic substrate covered with a water film. (b) Plan view of corrosion process. *From Ref. [30].*

where the constant depends on the metal in question and the amount lost, the corrosion geometry, and the applied voltage.

This model predicts that corrosion failure times are directly proportional to the resistivity of the electrolyte divided by its thickness. The quantity $\rho/t_o$ is known as the sheet resistance and has the peculiar units of ohms per square. This definition stems from the fact that the electrical resistance of a thin square of material is independent of the size of the square. In thin films or layers, the sheet resistance is measured by probing the wide surface area with a four-point resistivity probe. In aqueous solutions, the equivalent conductances ($\Lambda$) are given by $\Lambda = 1000/(\rho C)$, and are tabulated for concentrations ($C$) of different ionic species [31]. This makes it possible to estimate the sheet resistance.

---

**Example 7.1** After introduction of 1.7 ppm of $Cl_2$ into an 85°C/80% RH ambient, the corrosion leakage current across metal stripes on a ceramic substrate increased. If the moisture layer is 100 nm thick, what is the electrolyte sheet resistance?

**Answer** Chlorine dissolves in water through the reaction

$$Cl_2(g) + H_2O \rightarrow H^+ + Cl^- + HOCl,$$

which is the difference of the half reactions

$$H_2O + Cl^- \rightarrow H^+ + HOCl + 2e^-, \quad E^o = +1.49V,$$

and

$$2Cl^- \rightarrow Cl_2(g) + 2e^-, \quad E^o = +1.36V.$$

Applying Eqn (7.14),

$$0 = (1.49 - 1.36) + 8.314(358)/2(96,500)\ln(H^+)(Cl^-)/(Cl_2).$$

Substituting for the chlorine concentration, $(Cl_2) = 1.7 \times 10^{-6}$ yields $(H^+)$ $(Cl^-) = 3.36 \times 10^{-10}$. Thus, there are $1.83 \times 10^{-5}$ equivalents per liter for each ion. The equivalent conductance of $H^+$ and $Cl^-$ is 775 at 85 °C [31]. Since $\rho = 1000/(C\Lambda)$, the sheet resistance, $\rho/t_o = 1000/(C\Lambda t_o)$. Substitution yields $\rho/t_o = 1000/1.83 \times 10^{-5}$ $(775)(10^{-5}) = 7.1 \times 10^9$ $\Omega$/sq.

---

## 7.5.4 Mobile Ion Contamination in CMOS Circuits

Related to the problem of ion migration in liquid electrolytes is ion transport through a solid ionic medium like $SiO_2$. Though not a corrosion problem, the fact that ionic motion is involved makes it of interest here. One of the most serious reliability problems exhibited by early MOS transistors was instability of the threshold voltage ($V_T$) causing operation at more negative voltages than desired. With positive bias, $V_T$ drifted with time to more negative values. At elevated temperatures device behavior was very erratic. The cause of these problems was drift of positively charged mobile ions in the gate oxide. A particular culprit was identified to be sodium, hence the appellation "the sodium problem." Sodium entered processing lines from the quartzware used as well as from the salt introduced by winter footwear. Despite the progress made in tracking down sources of impurities and reducing instabilities due to positive mobile ion contamination, failures due to this cause still occur [32].

Ion drift velocities ($v$) can be calculated from the measured mobilities ($\mu$ in units of square centimeter per volt-second) through the formula

$$v = \mu E, \tag{7.40}$$

where $\varepsilon$ is the electric field. For $Na^+$ and $K^+$ in $SiO_2$ the respective values of mobility are

$$\mu(Na^+) = 1.0 \exp\left(-\frac{0.66 \text{ eV}}{kT}\right) \text{cm}^2/\text{V} - \text{s} \tag{7.41a}$$

and

$$\mu(K^+) = 0.03 \exp\left(-\frac{1.09 \text{ eV}}{kT}\right) \text{cm}^2/\text{V} - \text{s}. \tag{7.41b}$$

This means that $Na^+$ ions will traverse a 15-nm-thick oxide in $10^{-5}$ s at 200 °C in fields of 1 MV/cm. On the other hand, the larger $K^+$ ions require about 10 s. Maximum tolerable $Na^+$ and $K^+$ concentrations are $1 \times 10^{10}$ $\text{cm}^{-2}$. When positive ions drift through the oxide of an MOS structure, they also shift the effective flat-band voltage. Thus, a positive gate bias voltage applied to NMOS transistors drives ions toward the Si interface and displaces the flat-band voltage by $-Q/C_o$ (Section 2.5.3), where $Q$ is the ion charge and $C_o$ is the oxide capacitance. Equivalently, transistor threshold-voltage shifts occur as a result of charge transport (Eqn (2.6)). Such instabilities are not as evident in PMOS transistors, which is the reason these devices were the first to be commercially developed.

NMOS transistors required taming "the sodium problem," which was accomplished initially by using PSG whereby the sodium was immobilized, and later $Si_3N_4$, on top of the gate $SiO_2$, as a barrier to sodium.

## 7.6 RADIATION DAMAGE TO ELECTRONIC MATERIALS AND DEVICES

### 7.6.1 A Historical Footnote

It was a "big bang" on July 9, 1962, that catapulted the reliability of radiation-damaged solid-state electronics to the center stage of world politics [33]. On that day the US high-altitude nuclear device Starfish was detonated. This was followed by several similar Russian tests later that year.

The resulting nuclear contamination of the exoatmosphere generated sufficient electronic perturbation to the Van Allen belt to cause failure of the Telstar 1 communications satellite. Not long after this, Nikita Khrushchev, premier of the Union of Soviet Socialist Republics, announced that this radiation vulnerability would be considered a possible strategy to destroy future US military space systems. This threat provided an incentive and challenge to develop survivable space electronics and acquire an understanding of the nature of radiation damage to semiconductor devices. Fortunately, the damage to Telstar was reversible. In time, enough was learned so that damaged p–n junctions could be repaired from the ground by altering their bias voltages.

Even though the cold war is now history, interest persists in the subject of radiation damage to semiconductor devices and electronic equipment used in the peaceful exploitation of nuclear energy. A number of recent books on this and related topics [34–36] amply demonstrate this interest and importance.

## 7.6.2 Some Definitions

The radiation we are primarily concerned with is composed of energetic particles and photons possessing energies ranging from tens to millions of electron volts. While the former include neutrons, electrons, and alpha particles, X-rays, gamma rays, and ultraviolet light are examples of photons. The radiation energy density that materials are exposed to is known as the dose ($D$). If $\mathrm{d}E$ is the mean energy absorbed per unit mass $\mathrm{d}m$ of irradiated material, the dose is defined by

$$D = \frac{\mathrm{d}E}{\mathrm{d}m}. \tag{7.42}$$

Although the absorbed dose in a small region, i.e., a gate oxide, can be rigorously defined, it is impossible to know the actual energy imparted. The common unit of dose is the rad, defined as 1 rad = 100 erg/g. In the SI system, the gray (Gy) is the unit of dose, where 1 Gy = 1 J/kg, so that 1 Gy = 100 rad. Since the absorbed dose depends on the material in question, the latter should be referenced directly following the dose unit, e.g., rads (Si), rads ($SiO_2$), Gy (GaAs).

In many applications, the dose rate $\mathrm{d}D/\mathrm{d}t$, expressed in rads per second, is important. For example, during exposure to long-lived radioactive isotopes, $D$ is assumed to be constant, and the dose rate is zero. If radiation is pulsed, e.g., flash X-ray machines or a nuclear blast, then the dose rate also becomes a significant variable. Examples will be given of cases where $D$ and $dD/dt$ are reported.

### 7.6.3 Radiation Environments

The primary radiation environments that degrade or damage electronic materials and devices include outer space, nuclear reactors, radiation processing activities, nuclear weapon, and controlled fusion facilities [35]. Electronic equipment is exposed to these radiation environments during interplanetary space flight, in power-generating plants, in factories, and in defense applications.

In outer space, particle and photon radiation have typical energies ranging from kiloelectronvolt to gigaelectronvolt and consist of

1. a broad spectrum of energetic particles trapped in the earth's magnetic field, forming radiation belts;
2. galactic cosmic rays, which are the fluxes emanating from energetic heavy ions. Energies can extend even beyond teraelectronvolt;
3. solar winds and flares, which are eruptions of energetic protons and smaller amounts of alpha particles, heavy ions, and electrons with energies ranging to hundreds of megaelectronvolt.

The nuclear-reactor environment depends on reactor type. Within the core of pressurized-water, boiling-water, and fast-breeder reactors, neutron fluxes range from $10^{12}$ to $10^{14}/\text{cm}^2/\text{s}$, while gamma radiation levels between $10^5$ to $10^{10}$ rad/h are common. This contrasts with neutron fluxes of $3 \times 10^5/\text{cm}^2/\text{s}$ and gamma ray doses of $10^{-3}$–$10^2$ rad/h in the containment area where electronic equipment and components operate.

Radiation processing is employed in a number of industrial activities, which are noted together with the doses involved; retardation of food spoilage (30–500 krad), sterilization of medical supplies (2.5 Mrad), sewage processing ($\sim$1 Mrad), and modification of polymers (10–15 Mrad). The processing of spent nuclear fuels generates maximum dose rates of $10^5$ rad (Si)/h for gamma radiation and 0.1 rad (Si)/h for neutrons.

In contrast, a 1-megaton nuclear weapon produces prompt radiation dosages of 2100 rad neutrons and 62,000 rad gamma at a distance of 1 mile from the detonation point. If and when controlled nuclear fusion reactors are a reality, radiation levels of $10^{12}$–$10^{16}$ equivalent neutrons-cm$^{-2}$ and ionizing doses of $10^3$–$10^7$ rad (Si) s$^{-1}$ will have to be dealt with. The primary source of this radiation is 14 MeV neutrons from the deuterium–tritium reaction.

In each of the above environments, electronic equipment is used to sense and monitor radiation as well as to control robots and other machinery remotely. To appreciate the damage to solid-state and electro-optical devices that the above radiation magnitudes can wreak, we must first understand how energetic particles and ionizing radiation interact with materials.

## 7.6.4 Interaction of Radiation with Matter

We are primarily concerned with the consequences of atomic displacements and ionization in materials caused by radiation. This broad subject, which has been researched for a half century, does not easily lend itself to a brief review. Nevertheless, in flow-chart style, Figure 7.11 neatly summarizes radiation damage into particle- and photon-induced categories. Of particular interest is the long-lived, as well as transient, damage produced by the particular radiation.

(a) Atomic displacement. Starting with particles, we note that physical displacement of atoms from their original positions is the major form of damage created. The particle trajectory, and the damage produced, depends on its energy, mass, and charge, as well as on the nature of the matrix. What happens during ion implantation doping of silicon illustrates these effects. Dopant ions with kinetic energies of 50–100 keV, or thousands of times larger than the binding energies of lattice atoms, knock atoms off lattice sites in collision cascades, creating vacancy and interstitial defects in the process. It is usually difficult to estimate just how many vacancy–interstitial pairs are created. For example, if the displacement energy is 10 eV, a 12-MeV proton is predicted to generate only 17 vacancy–interstitial pairs [35]. Each incoming ion executes a different trajectory and comes to rest at a different depth from the surface. The ion range is statistically distributed, and lighter particles penetrate further than heavier ones of the same energy. Heating the radiation-damaged matrix heals it because atoms again resume their equilibrium positions by diffusional processes.

Materials exposed to an energetic-particle environment may also become radioactive. Protons, neutrons, and nuclei are all capable of transforming stable nuclei into radioactive ones by removal of nucleons, or in the case of low-energy neutrons by neutron capture. Depending on the particular atom, such induced radioactivity decays with a characteristic

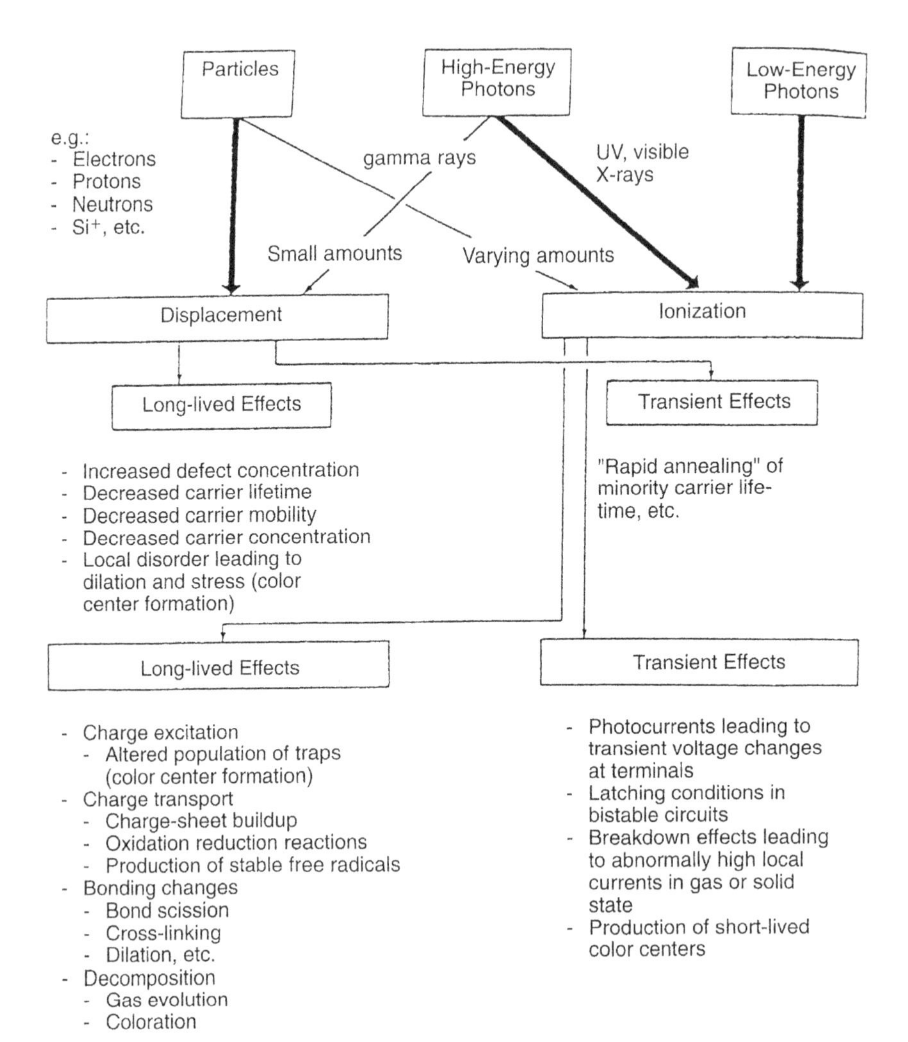


**Figure 7.11** Summary of radiation-induced degradation effects in solid-state materials and devices. *From Ref. [35].*

half-life through emission of gamma rays, beta rays (electrons), and alpha particles.

It should be realized that different classes of solids suffer different levels of particle-induced damage. Most vulnerable are semiconductors and insulators because significant changes in electrical properties occur. Metals, on the other hand, suffer some mechanical hardening and embrittlement but no appreciable change in electrical behavior. Furthermore, damage to metals from gamma and X-rays is negligible, unlike their effect in nonmetals.

(b) Electronic charge displacement (ionization). The interactions between energetic radiation and the electronic structure of atoms is every bit as varied as the particle momentum transfer to atoms described above. Radiation energies are usually large enough to cause complex core-electron transitions in individual atoms. These involve a train of events that begins when electrons are ejected from low orbital levels by incoming energetic photons (X-rays, gamma rays) or particles (electrons, ions). Electrons from outer levels fill these now vacant levels, and X-ray emission, characteristic of the excited atom, results. More complex internal electron excitations and transitions ensue in the ejection of Auger electrons. Such processes cause no damage; they are capitalized upon to fingerprint atoms. Applications of X-ray fluorescence and Auger electron spectroscopy in failure mode analysis will be discussed in Chapter 11.

In addition to localized atomic excitation, energy loss in semiconductors and inorganic insulators is ultimately converted into electron–hole pair production. Both energetic particles and photons are capable of causing such ionization damage. The excitation of valence-band electrons to the conduction band creates charge carriers that are mobile in electric fields. Simultaneously, slower moving holes remain behind in the valence band. As we have already seen, trapped holes in thin dielectrics enhance the probability of degrading dielectric behavior.

Depending on whether particles or photons of the same energy are involved, there is a difference with respect to the amount of ionization produced. Because some of the particle energy is partitioned to momentum transfer to atoms, less is available for ionization than is the case with photons. Thus, ionization effects produced by megavolt particles are equivalent to those induced by much less energetic photons, electrons, or ultraviolet radiation. Unlike alpha particles and protons, which can be effectively shielded by thin metal foils, cosmic, gamma, and X-rays easily penetrate housings of electronic equipment.

### 7.6.5 Device Degradation Due to Ionizing Radiation

The subject of primary and secondary damage mechanisms in various devices and components has been reviewed by several authors [34–39] and summarized in Table 7.4. In this section, a number of these mechanisms are

**Table 7.4** Primary and secondary damage mechanisms in various devices, circuits, and components in the space radiation environment

| | Permanent | | | | | | | Temporary |
|---|---|---|---|---|---|---|---|---|
| **Devices, circuits, and components** | **Lifetime degradation** | **Carrier removal** | **Trapping** | **Mobility degradation** | **Charge buildup** | **Latchup** | **Absorption** | **Single event upset** |
| Si bipolar transistors and integrated circuits | P | S | | | S | P | | P |
| Si MOS transistors and integrated circuits | | | | S | P | P | | P |
| JFETs | | P | P | S | S | | | |
| p–n junction diodes | P | P | P | | S | | | |
| LEDs, laser diodes | P | | | | | | | |
| Charge-coupled devices | P | | P | | P | | | |
| Photodetectors | P | | | | P | | | |
| Microwave devices and circuits | P | P | | S | | | | |
| GaAs transistors and integrated circuits | | P | | S | S | | | |
| Capacitors | | | | | P | | | P |
| Diffused resistors | | P | | S | | | | |
| Optical components (fibers, windows, mirrors) | | | | | | | | P |

P = Primary; S = Secondary.
Charge buildup can be a primary failure mechanism in some types of bipolar integrated circuits.
Adapted from Ref. [37].

described; the important subject of single-event upset, however, merits a discussion on its own and is treated in Section 7.6.6.

### *7.6.5.1 MOS Devices*

The basic effect of ionizing radiation in $SiO_2$ films can be understood in terms of dielectric charging effects previously introduced in Chapter 6. Damage begins in the gate oxide in the form of electron–hole pairs. Because it requires 17 eV to produce electron–hole pairs in $SiO_2$, such charge can be generated with radiation of modest energy. In comparison, carrier pair production is energetically easier in semiconductors, requiring only 3.6 eV in Si and 4.8 eV in GaAs. For an exposure of a rad, approximately $10^{13}$ pairs/cm$^2$ are generated in $SiO_2$. While the more mobile electrons are removed, some holes tend to be trapped at the Si/$SiO_2$ interface. This additional radiation-induced oxide charge ($Q_r$) reduces the threshold voltage ($V_T$) by an amount $Q_r/C_o$, where $C_o$ is the oxide capacitance (see Section 2.5.3). Thus, the MOSFET may be turned on even without any applied gate voltage.

Threshold-voltage shifts are both dose and dose rate dependent. This is illustrated in Figure 7.12 for *n*-channel MOS devices. At first, $V_T$ is unchanged up to doses of approximately $10^5$ rad ($SiO_2$), then it goes

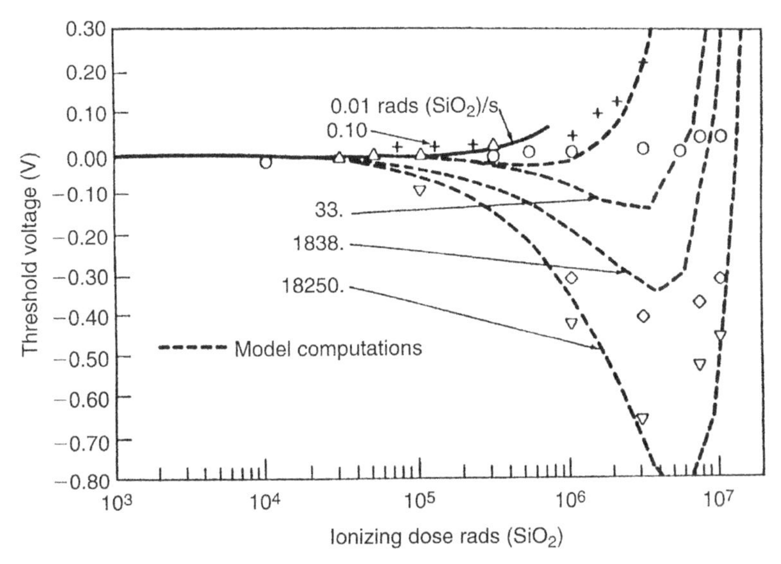


**Figure 7.12** Threshold voltage for *n*-channel CMOS transistors as a function of dose for various dose rates. *From Ref. [54].*

negative, where it reaches a minimum and finally swings up toward more positive values as the dose rises. This behavior, which is accentuated by increasing dose rates, is a superposition of two effects. Gate oxide-trapped holes cause a monotonic decrease in $V_T$ with dose, while negative interface states induce a rise in $V_T$. Other deleterious effects of radiation in MOSFETs include channel-mobility degradation and increased leakage currents at high dose levels.

#### 7.6.5.2 GaAs Devices

In some environments, GaAs devices are far more radiation tolerant than Si devices, but in other radiation environments, the damage is comparable. For example, GaAs FETs have no gate insulators to experience charge buildup and, therefore, are relatively immune to radiation. In bulk GaAs, like Si, ionizing radiation generates electron–hole pairs, introduces defect levels and scattering centers in the energy gap, removes carriers, and reduces mobility. The latter two effects are responsible for transconductance or gain degradation in transistors. For this to occur in GaAs, dose levels of $\sim 10^8$ rad are typically required when channel doping levels are relatively high.

#### 7.6.5.3 Electro-Optical Devices

Laser diodes, light-emitting diodes, and photodetectors all suffer degradation of operating characteristics due to radiation damage [37]. In lasers, permanent increases in threshold currents result from ionizing radiation, which generates nonradiative recombination centers. The latter are also primarily responsible for degradation of LEDs. Silicon-based photodetectors and charge-coupled devices are prone to the same degradation mechanisms, i.e., charge buildup, surface leakage currents, and reduction in carrier mobility and lifetime, that plague MOS devices.

Optical fiber, often used with these devices, offers advantages relative to conventional transmission lines because it is insensitive to electromagnetic noise. However, optical fiber is more sensitive to radiation damage in the form of luminescence and increased absorption. Spurious light signals and attenuation of light pulses can then be expected.

In general, the damage and degradation discussed to this point occur only at high radiation levels that are created artificially. We now consider the often more serious radiation damage that can occur in circuits powered in what would be considered natural, everyday environments.

## 7.6.6 Soft Errors

### 7.6.6.1 Introduction

Computer circuits function by identifying small packets of charge as elemental bits of information. Any noise that modifies this charge also may change the information stored. This can happen, for example, when radiation, a source of this noise, impinges upon and alters the electrical state of an internal node, causing a false logic transition, e.g., a bit flop. Such errors may mean data corruption, execution of wrong commands, or even a "locking up" that requires reinitialization. In digital electronics, errors that are not caused by permanent damage to circuits are referred to as soft errors, soft fails, or single event upsets (SEUs). Soft fails, happily, do not necessarily affect the computer user, because the system may be turned off or the incorrect memory bits may be overwritten before they are used. In microelectronic circuits employed in satellites, digital avionics, high-flying aircraft, as well as interrestrial electronics and computers, soft errors are a major concern. Over the years from 1970 to 1982, there have been over 40 recorded upsets in space satellites (e.g., Voyager, Pioneer Venus, Galileo) that affected NMOS, CMOS, and bipolar memories, as well as assorted flip-flop circuits [40]. Occasionally, however, the more serious effect of single event latchup (SEL) has been triggered, resulting in circuit destruction.

Soft errors and their reliability implications are sufficiently important to be the subject of an entire recent issue of the *IBM Journal of Research and Development* [41]. In it there is an interesting history of the IBM experience from 1978 to 1994 with terrestrial cosmic rays and soft errors, including experimental measurements to detect them, results of accelerated testing, and models used to analyze the response of ICs. Much of what follows is derived from this excellent survey of the subject.

### 7.6.6.2 Sources of Radiation That Cause Soft Errors

In 1978, the first evidence of sea-level soft fails from energetic particle impact was published by May and Woods [42]. Radioactive alpha particles emanating from ceramic packages were determined to be the cause of damage. The source was the water used in the factory, built downstream from the tailings of an old uranium mine that contained high levels of radioactive elements. A year later, Ziegler and Lanford [43] suggested that sea-level cosmic rays could also cause upsets in electronics. Thus, the nuclear particles that create electronic noise and soft fails stem from two sources, namely, the decay of radioactive atoms and extraterrestrial cosmic ray particles.

(a) Nuclear particles from radioactive atoms. A number of materials employed in IC manufacture contain trace levels of radioactive atoms that emit gamma rays (high-energy photons), beta rays (electrons), and alpha particles (helium ions), or some combination of these. From the standpoint of producing soft errors, alpha particles are the most "upsetting." These alpha particles, with energies ranging from 3.8 to 8.8 MeV, stem from the decay of naturally occurring radioactive heavy actinides like uranium ($U^{238}$) and thorium ($Th^{232}$) and their daughter nuclei. Because these elements are part of the earth's crust, trace amounts are present in metallization alloys, gold, alumina, and lead frame alloys used in packaging; they can also be introduced during processing, as we shall see.

Alpha particle ($\alpha$P) activity is often quoted in terms of a flux, or rate of alpha emission from the surface per unit area (square centimeters) per unit time (h). As a rule of thumb, an alpha flux of 1 alpha/$cm^2$-h corresponds to one part per million (ppm) of $U^{238}$ nuclei. The following calculation shows us how to estimate the flux more exactly [44]. First we note that $U^{238}$ has an activity of $6.85 \times 10^{-7}$ Ci/g, where 1 curie (Ci) $= 2.2 \times 10^{12}$ disintegrations/min (dpm). Furthermore, the decay of $U^{238}$ to $Pb^{206}$ releases 8 alpha particles.

---

**Example 7.2** Estimate the alpha particle flux from an $Al_2O_3$ package containing 1 ppm $U^{238}$ per gram of ceramic. Assume that only 25% of the alphas generated escape the package from a depth of 14 μm or less, and that the density of $A1_2O_3$ is 3.9 g/$cm^3$.

**Answer** In 1 $cm^3$ of $Al_2O_3$ there are $1 \times 10^{-6} \times 1\ cm^3 \times 3.9\ g/cm^3 = 3.9 \times 10^{-6}$ g of $U^{238}$. This corresponds to $6.85 \times 10^{-7}$ Ci/g $\times$ $3.9 \times 10^{-6}$ g/$cm^3$ $= 2.67 \times 10^{-12}$ Ci/$cm^3$, or an equivalent radiation flux of $2.67 \times 10^{-12}$ Ci/$cm^3$ $\times$ $14 \times 10^{-4}$ cm $= 3.74 \times 10^{-15}$ Ci/$cm^2$. Collecting terms, 8 $\alpha$P/dis $\times$ 0.25 $\times$ ($3.74 \times 10^{-15}$ Ci/$cm^2$) $\times$ ($2.2 \times 10^{12}$ dpm/Ci) $\times$ 60 min/h $=$ 0.98 $\alpha$P/$cm^2$-h.

---

Typical $\alpha$P flux levels in IC chips without tungsten silicides rarely exceed 0.005 $\alpha$P/$cm^2$-h. Ceramic packaging components (lids, lead frames) may exhibit flux levels ten to as much as a thousand times larger. In both cases, there is considerable scatter, depending on processing conditions and involved vendors. Nevertheless, such $\alpha$P fluxes are sufficient to cause troublesome SEUs. Details of the $\alpha$P damage mechanism and energy required in silicon will be treated in Section 7.6.6.4.

An interesting example of chip contamination by such radioactive elements involves $Po^{210}$, an isotope of polonium that occurs naturally as a daughter of radon gas. Hera random-access memory chips manufactured by IBM in 1987 were found to be contaminated by $Po^{210}$, an alpha emitter, such that hot chips suffered failure rates a thousand times higher than clean chips. Following a frantic search for the source of contamination, the culprit was found to be radioactive nitric acid bottles. In producing these, the manufacturer had to meet stringent semiconductor standards of cleanliness. Therefore, a bottle-cleaning machine employing radioactive $Po^{210}$ was used (as in smoke alarms) to ionize an air jet that dislodged electrostatic dust left behind after washing. The jets leaked radioactivity only intermittently, confounding the problem. This example is illustrative of the painstaking efforts that periodically surface in the industry to track down and eliminate sources of SEUs.

(b) Nuclear particles from cosmic rays. The second, and in many ways more troubling, source of nuclear particles stems from cosmic rays that bombard Earth from the recesses of outer space. Primary sources of cosmic rays are energetic galactic particles and solar wind particles. The former are far more energetic than the latter, which typically have energies of ~1 GeV. Secondary cosmic rays are generated when the primary cosmic rays hit atmospheric atoms and create a shower, or cascade, of secondary particles. Finally, there are the terrestrial cosmic rays that hit Earth. These are composed of assorted nuclear particles, e.g., neutrons, pions, protons, and muons, that have energies ranging from 1 MeV to 1 GeV. About 97% of the nucleon flux at sea level is due to neutrons, and fluxes are typically $10^5/cm^2$-year. It is interesting to note that cosmic ray fluxes vary strongly with solar cycle activity and altitude [45]. In going from New York (0 ft) to Denver (5280 ft), the neutron flux increases 3.4-fold, and at Leadville, Colorado (10,200 ft), the flux is 12.8 times higher than at sea level. This trend continues and accounts for the fact that the soft fail rate of electronics at airplane altitudes is 100 times the magnitude recorded at terrestrial altitudes.

A relative comparison of the radioactive and cosmic ray contributions to soft error rates (SERs) depends on various factors, including levels of nuclear contamination, location, and altitude of devices. Data for $Po^{210}$-contaminated 288-Kb DRAM chips in Figure 7.13 depict measured SERs

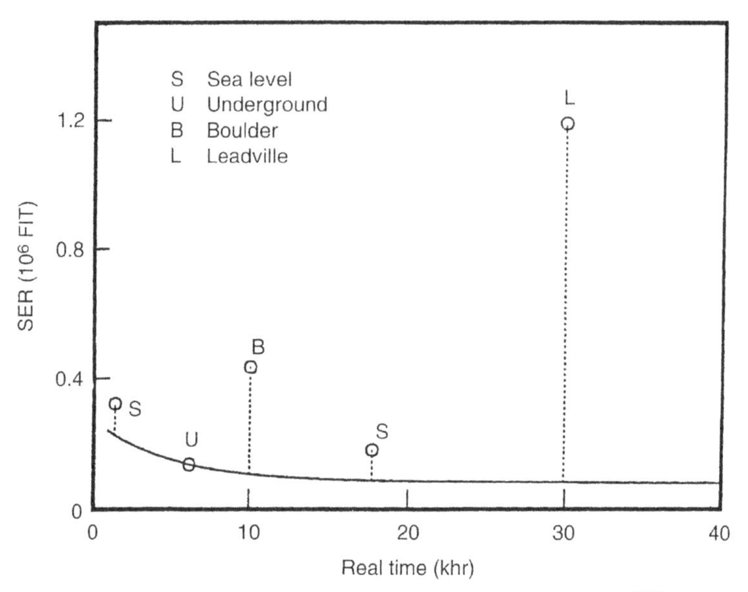


**Figure 7.13** Soft error rates for 288-Kb DRAMs contaminated with $Po^{210}$ measured at various altitudes. The solid line is the alpha-particle component of soft error rate (SER), while the vertical distance above (dotted lines) corresponds to the cosmic ray contribution to the SER. *From Ref. [55].*

as a function of time and altitude. The results can be described by the equation

$$\mathrm{SER} = A \exp\left(-\frac{t}{\tau_1/2}\right) + \mathrm{B} + \mathrm{C}, \tag{7.43}$$

where $A$ is the initial SER due to $Po^{210}$ contamination, $\tau_{1/2} = 138\ d$ half-life, $B$ is the contribution due to longer lived alpha emitters, and $C$ is the cosmic ray contribution.

The roles of these radiation sources may also be coupled. For example, cosmic ray activation of boron in borophosphosilicate glass has recently been suggested as a dominant source of alpha particles that cause soft errors in Si [46].

#### *7.6.6.3 Sample Calculations of Neutron–Silicon Interactions*

Since energetic neutrons are the primary particles involved in cosmic ray SEUs, it is instructive to reproduce Ziegler's [45] thumbnail calculations of their interactions with Si circuits. Silicon contains 28 nucleons (protons and neutrons), each of which has a strong interaction radius of $1.3 \times 10^{-13}$ cm;

therefore, for the Si nucleus, the radius scales to about $4 \times 10^{-13}$ cm, corresponding to a cross-sectional area of $5 \times 10^{-25}$ cm$^2$. This circular area for Si is the cross-section for interaction with neutrons. If the nuclei are uniformly distributed, they present a black wall, or total absorption cross-section, to a neutron flux given by

$$1/(\text{area}/\text{atom}) = 1/(5 \times 10^{-25}\ \text{cm}^2) = 2 \times 10^{24}\ \text{atoms}/\text{cm}^2. \quad (7.44)$$

For Si with an atomic density of $5 \times 10^{22}$ atoms/cm$^3$, the absorption length is $(2 \times 10^{24}$ atoms/cm$^2)/(5 \times 10^{22}$ atoms/cm$^3) = 40$ cm. Assuming that Si–neutron interactions must occur within 10 μm of the surface to be important, the probability of a nuclear "hit" is

$$(\text{target depth})/(\text{absorption length}) = 10^{-3}\ \text{cm}/40\ \text{cm} = 0.25 \times 10^{-4}. \quad (7.45)$$

Thus, one out of 40,000 incident neutrons will interact within 10 μm of the surface.

The key question we now address is how many of these hits will cause a soft error. Assuming bipolar transistor ICs whose typical active device area is 0.04 cm$^2$, the number of active atoms/chip is (active area)(electrical depth)(atom density), or $(0.04\ \text{cm}^2)(10^{-3}\ \text{cm})(5 \times 10^{22}\ \text{atoms/cm}^3) = 2 \times 10^{18}$ atoms. Further, the active cross-section = (active atoms)/(absorption cross section) = $(2 \times 10^{18}\ \text{atoms})/(2 \times 10^{24}\ \text{atoms/cm}^2) = 10^{-6}\ \text{cm}^2$. Accelerated testing on a 4096-bit bipolar memory chip revealed a soft-fail cross-section of $40 \times 10^{-12}$ cm$^2$/bit. Thus, the SER cross-section = $(4096\ \text{bits})(40 \times 10^{-12}\ \text{cm}^2/\text{bit}) = 1.6 \times 10^{-7}\ \text{cm}^2$. The ratio of the active cross-section to the SER cross-section = $(1 \times 10^{-6}\ \text{cm}^2)/(1.6 \times 10^{-7}\ \text{cm}^2) = 6$. This means that one in six neutron-silicon "hits" will result in a soft fail.

Consequently, the problem increases with smaller dimensions. If the active circuit thickness is reduced to 1 micron, a soft fail will result from almost every hit!

### *7.6.6.4 Energetic Particle Damage in Silicon*

When a high-energy cosmic ray or alpha particle penetrates a reverse-biased p–n junction, it leaves a track of ions and electrons in its wake, as illustrated in Figure 7.14. In the DRAM cell shown, the plasmalike track collapses the depletion layer and distorts the equipotentials there into the shape of a cylindrical funnel. This "funneling" produces a very large

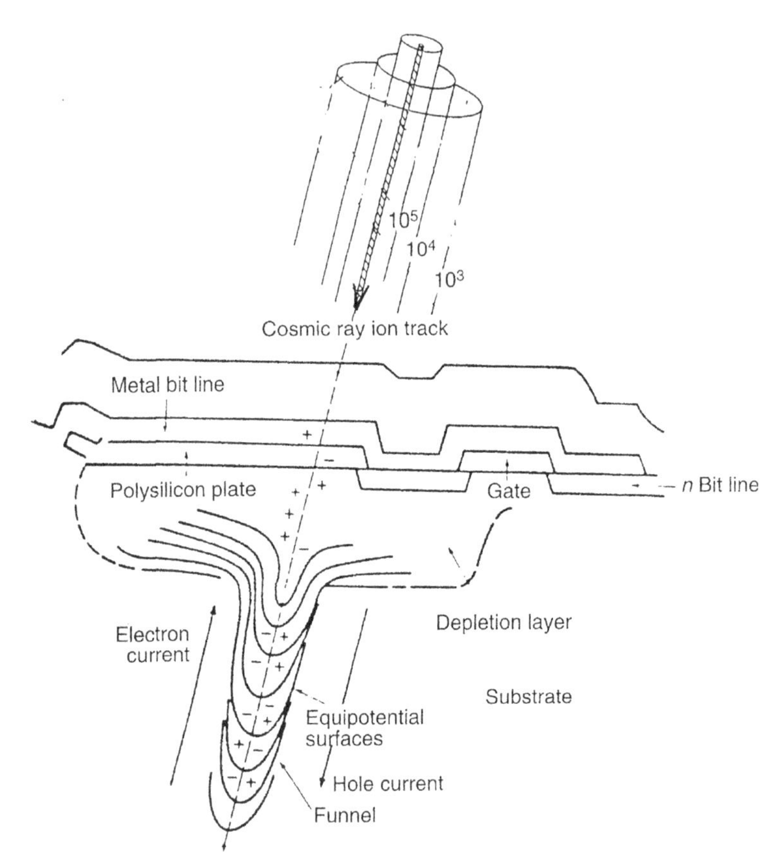


**Figure 7.14** Collapse of DRAM cell depletion layer along cosmic ray track produces a funnel in the device bulk. Representative dose profiles around the single ionizing particle are indicated. *After Ref. [36].*

electric field that propels electrons into the depletion layer while holes flow down into the bulk Si. If the particle track impales or passes close to a memory charge-storage node, an upset error may occur within nanoseconds. Subsequent diffusion in the next microsecond may distribute charge to neighboring nodes, causing upset there. Finally, the depletion-layer damage relaxes back to its original state without any permanent damage to devices. The funneling phenomenon has been numerically modeled [47] and experimentally verified through charge collection measurements [48].

It has been found that a critical charge ($Q_c$) is required to produce upsets, and that $Q_c$ depends on the particular device and feature dimensions

($l$) as shown in Figure 7.15. The linear relationship between the variables is well described by

$$Q_c = 0.023l^2, \tag{7.46}$$

with $Q_c$ in picocoulombs and $l$ in micrometers. Interestingly, this equation is independent of whether Si or GaAs devices are involved. In 1-μm DRAM cells, the critical charge to produce upset is about $2.3 \times 10^{-14}$ C, or close to $1.5 \times 10^5$ electronic charges. If hit by a 5-MeV alpha particle, it has been estimated that the particle range is a few tens of microns in Si; because the energy to produce electron–hole pairs in Si is 3.6 eV, about

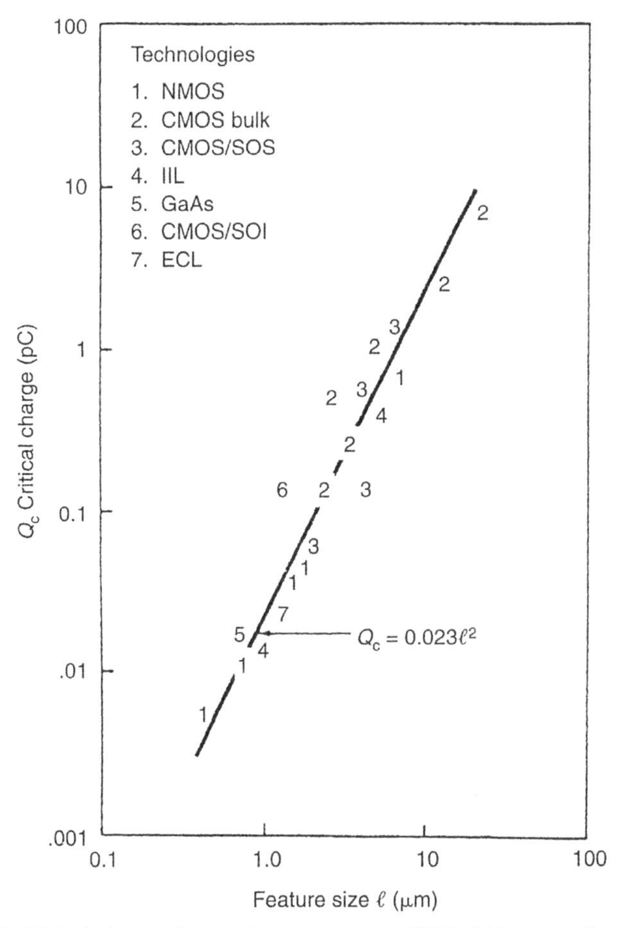


**Figure 7.15** Critical charge for single event upset (SEU) ($Q_c$) versus feature size in different devices. *From Ref. [56].*

$1.4 \times 10^6$ electronic charges can be expected. Thus, it is easy to see why alpha particles can trigger soft memory errors. All that is needed is a sufficient flux of them. Further, at feature dimensions of 10 nm, $Q_c$ is in the range of $1.2 \times 10^{-18}$ C, or close to 10 electronic charges. Hence, there is great interest in radiation hardening, which we discuss below.

### 7.6.7 Radiation Hardening of Devices

We close the chapter by commenting on strategies employed to protect or harden devices against the effects of radiation, and in particular, against soft errors. The increasing amount of critical spaceborne electronics in satellites, deep space probes, and strategic weapons systems has driven the need for increased performance and reliability of circuits employed in the space radiation environment. Perhaps the primary method to harden both silicon and GaAs devices against radiation has been through circuit design. This subject is well beyond the scope of this book, but has been reviewed by Kerns and Schafer [49]. In the processing area, a favorite strategy has been to isolate devices from electrical continuity with the base semiconductor wafer. Thus, dielectric isolation, or the imposition of an oxide layer or moat around individual MOSFETs or bipolar transistors, has been extensively practiced in silicon technology. Numerous silicon on insulator approaches, including growth of Si on sapphire, separation by ion implantation of oxygen, zone melt recrystallization, porous oxidized Si, and epitaxial lateral overgrowth, have been attempted [35]. In general, dielectric isolation has removed latchup paths and reduced volumes around junctions and oxides where carrier generation and trapping can occur.

With respect to soft errors, it is instructive to consider the techniques adopted to minimize their influence in 16-Mb DRAMs subjected to $\alpha$ particles [50]. In addition to error-correcting codes, the $\alpha$ particle sensitivity was reduced by the following means:

1. The use of a deep-trench capacitor. Charge stored on this capacitor was designed to be greater than that transferred by three $\alpha$ particles traversing the longest diagonal through the memory-cell volume.
2. Minimizing charge collected by p–n junctions of the MOS transfer device source and drain implants. By fabricating the transfer device in an *n*-well, charge generated in the substrate is collected by the well supply.
3. Reducing charge collection efficiency by storing memory data "inside" rather than "outside" the dielectrically isolated storage trench. In the

latter, charge is collected over a larger potential substrate volume and source of radioactive atoms.

## EXERCISES

**7.1** An electronic vacuum tube 20 $cm^3$ in volume is pumped to $10^{-11}$ atm prior to glass–metal sealing. For a moisture penetration rate of $1 \times 10^{-8}$ atm-$cm^3$/s, how long will it take for the pressure in the tube to rise an order of magnitude?

**7.2** The mean time to failure of plastic packages as a function of temperature and RH is given by $MTTF = B(RH)^{-n}\exp[E_c/kT]$, where $B$ is a constant.

**a.** If $n = 2.66$ and $E_c = 0.79$ eV, does a 10% increase in RH have a greater effect in reducing mean life than a 10% increase in temperature?

**b.** For RH and temperature variables to have an equal influence in degrading plastic packages, as measured by MTTF, what relationship must connect RH, $T$, $n$ and $E_c$?

**7.3** An aluminum contact film that is 1 μm thick and 0.1 × 0.1 mm in area carries a current of 1 μA. Under a worst case corrosion scenario, how long would it take the metal to be totally converted into ions?

**7.4** Write possible balanced chemical reactions for the formation of $NiCl_2$ and $NiCO_3$ corrosion products when Ni is exposed to the atmosphere and water containing chlorides.

**7.5 a.** Is "purple plague" formation likely to occur during aqueous corrosion of a gold-wire bond to an aluminum contact pad?

**b.** Suppose a condensed moisture layer containing a $10^{-6}$ M concentration of $Al^{+3}$ surrounds the pad, while a $10^{-8}$ M concentration of $Au^{+3}$ coats the wire. What maximum potential can develop between these metals? What minimum potential can develop?

**7.6** It has been claimed that the $Au_4Al$ intermetallic compound corrodes in the presence of bromine ions according the reaction $Au_4Al + 3Br^- \rightarrow 4Au + AlBr_3 + 3e^-$. Suggest a subsequent reaction that regenerates bromine ions. What evidence is there that such a reaction sequence occurs?

**7.7** The contact metal $WSi_2$ is subject to electrolytic corrosion in the presence of moisture and voltage bias. Write a chemical reaction that would tend to convert this conducting silicide to insulating oxide by-products. If the corresponding free energies of formation

($\Delta G$) from elements at 298 K are $\Delta G(\text{WSi}_2) = -22$ kcal/mol, $\Delta G(\text{WO}_3) = -204$ kcal/mol, and $\Delta G(\text{SiO}_2) = -182$ kcal/mol, what is the overall free energy change of the corrosion reaction? Note: You may have to consult a table of thermodynamic data.

**7.8** The use of aluminum electrical wiring in the home has been known to cause fires that apparently initiate at plug outlets; in contrast, copper wiring appears to be immune from this problem. Give a possible reason for these observations.

**7.9** Suppose corrosion of metal within a plastic package occurs by moisture permeation followed by a chemical reaction described by second-order kinetics such that $dC/dt = -k(C - C_o)^2$ where $k$ is the rate constant. Derive an expression for the reliability function.

**7.10** What similarities and differences in matrix environment, mass-transport mechanisms, current densities, damage morphology, temperature dependence, and degradation kinetics exist between wet migration and solid-state electromigration?

**7.11** During electrolytic refining of copper, the impure anode sheet metal is plated onto pure Cu cathodes in cells that employ aqueous $CuSO_4$ electrolytes. In this process, as with wet metal migration, dendrites often span the anode–cathode spacing and short the electrodes. Suggest reasons why the formation of dendrites is so heterogeneous and why they grow with such variable velocity. In the case of copper refining, what practical measures can be taken to minimize cell shorting by dendrites?

**7.12** The 100-nm-thick $SiO_2$ film of a capacitor was uniformly contaminated with a concentration of $1 \times 10^{15}$ sodium ions/cm$^3$. Application of 5 volts swept all of the sodium to the negative electrode.

**a.** How long does it take the $Na^+$ ions to electromigrate across the oxide at 300 K?

**b.** The voltage is reduced to zero. Sketch resultant time-dependent concentration profiles of $Na^+$ as a function of time.

**c.** How long will it take the $Na^+$ to diffuse back across the oxide?

**7.13** As a result of exposure to radiation

**a.** Metals harden but become less ductile and suffer damage in the form of creep, void swelling, and hydrogen embrittlement.

**b.** Ceramics undergo color changes and exhibit loss of thermal conductivity but enhancement of electrical conductivity.

**c.** Polymers generally lose strength and ductility, discolor, and evolve gas (e.g., hydrogen).

Select one of these material classes and write a report explaining reasons for the radiation-damage effects manifested. (The April 1997 issue of the *MRS Bulletin* is devoted to materials performance in radiation fields.)

**7.14** How many electron–hole pairs are created in Si and GaAs by exposing each to a radiation dose of 1 Gy?

**7.15** Nuclei of $Th^{232}$ have an activity of $1.1 \times 10^{-7}$ Ci/g and release 6 alpha particles per disintegration when they decay to $Pb^{206}$. If these alpha particles can penetrate 50 $\mu$m of Si, what $\alpha$P flux can be expected?

**7.16** A source of neutrons impinges on both silicon and germanium. What ratio of subsurface interaction depths will enable the probability of neutron–Ge atom hits to equal the number of neutron–Si atom hits?

**7.17** Soft-error fail rates per chip in 288-Kb DRAMS (1986) were measured to be approximately 1 fail per year. More recent data for 4-Mb DRAM (1993) memory chips from different manufacturers reveal SER values of 0.0026–0.000,46 fails per year.

**a.** Suggest possible reasons for the differences in these values.

**b.** How many FITS does 0.000,46 fails per year correspond to?

**7.18** A 64-Mb bipolar cache memory chip containing $5 \times 10^8$ bits in the memory is exposed to an energetic nucleon flux equal to $10^5/cm^2$-year. The probability of a soft error is given in terms of the cross-section, which has a value of $4 \times 10^{-12}$ $cm^2$/bit. How many soft errors can be expected if this chip operates for a year?

## REFERENCES

[1] R.B. Comizzoli, R.P. Frankenthal, R.E. Lobnig, G.A. Peins, L.A. Psota-Kelty, D.J. Siconolfi, D.J. Sinclair, Electrochem. Soc. Interface 26 (1993). Fall.
[2] D.J. Schneider, D.C. Williams, Lasers Optronics 25 (October 1994).
[3] J.D. Guttenplan, ASM Handbook, vol. 13, 1987, p. 1107.
[4] E. White, G. Slenski, B. Dobbs, ASM Handbook, vol. 13, 1987, p. 1113.
[5] M. Pecht, IEEE Trans. Compon. Hybrids Manuf. Tech. 11 (3) (1990) 383.
[6] Tutorial Notes R.P. Merrett, IEEE Int. Reliab. Phys. Symp. 1 (2.1) (1984).
[7] A. Roth, Vacuum Technology, North-Holland, Amsterdam, 1976.
[8] D.A. Jones, Principles and Prevention of Corrosion, Macmillan, New York, 1992.
[9] M.G. Fontana, Corrosion Engineering, third ed., McGraw Hill, New York, 1978.
[10] J.D. Guttenplan, Mater. Perform. (April 1990) 76.
[11] M. Tullmin, P.R. Roberge, IEEE Trans. Reliab. 44 (1995) 271.

[12] X. Shan, M. Pecht, in: J.H. Lau (Ed.), Thermal Stress and Strain in Microelectronics Packaging, Van Nostrand Reinhold, New York, 1993.
[13] G.L. Schnable, R.B. Comizzoli, W. Kern, L.K. White, RCA Rev. 40 (1979) 416.
[14] D.S. Peck, in: 24th Annual Proceedings of the IEEE Reliability Physics Symposium, 1986, p. 44.
[15] M. Iannuzzi, IEEE Trans. Compon. Hybrids Manuf. Tech. CHMT-6 (1983) 181.
[16] P. Totta, J. Vac. Sci. Tech. 13 (1976) 26.
[17] S. Thomas, H.M. Berg, in: 23th Annual Proceedings of the IEEE Reliability Physics Symposium, 1985, p. 153.
[18] A.J. Griffen, S.E. Herna'ndez, F.R. Brotzen, J.D. Lawrence, J.W. McPherson, C.F. Dunn, in: 31st Annual Proceedings of the IEEE Reliability Physics Symposium, 1993, p. 327.
[19] K.H. Ritz, W.I. Stacy, E.K. Broadbent, in: 25th Annual Proceedings of the IEEE Reliability Physics Symposium, 1987, p. 28.
[20] W.T. Anderson, A. Christou, IEEE Trans. Reliab. R-29 (1980) 222.
[21] R.S. Torkington, D. Lamoreaux, in: ASM International Symposium for Testing and Failure Analysis (ISTFA), 1990, p. 225.
[22] T.N. Theis, IBM J. Res. Develop. 44 (3) (2000) 379.
[23] S.J. Krumbein, IEEE Trans. Reliab. 44 (4) (1995) 539.
[24] T. Saigusa, T. Ohtaki, M. Nakajima, Y. Nakao, Materials Developments in Micro-electronic Packaging Conf. Proc., Quebec, Canada (August 1991) 345.
[25] B. Rudra, D. Jennings, IEEE Trans. Reliab. 43 (1994) 354.
[26] G.T. Kohman, H.W. Hermance, G.H. Downes, Bell Syst. Tech. J. XXXIV (6) (1955) 1115.
[27] J.J. Steppan, J.A. Roth, L.C. Hall, D.A. Jeanotte, S.P. Carbone, J. Electrochem. Soc. 134 (1) (1987) 175.
[28] G. DiGiacomo, in: A.H. Landzberg (Ed.), Microelectronics Manufacturing Diagnostics Handbook, Van Nostrand Reinhold, New York, 1993.
[29] M. Ohring, The Materials Science of Thin Films, Academic Press, Boston, 1992.
[30] R.T. Howard, IEEE Trans. Compon. Hybrids Manuf. Tech. CHMT-4 (1981) 520.
[31] Handbook of Chem and Physics, 59th ed., CRC Press, Cleveland, 1978–79.
[32] Tutorial Notes R.L. Hance, J.W. Miller, K.B. Erington, M.A. Chonko, IEEE Int. Reliab. Phys. Symp. 4.1 (1995).
[33] H. Hughes, in: T.P. Ma, P.V. Dressendorfer (Eds.), Ionizing Radiation Effects in MOS Devices and Circuits, J. Wiley and Sons, New York, 1989.
[34] T.P. Ma, P.V. Dressendorfer (Eds.), Ionizing Radiation Effects in MOS Devices and Circuits, J. Wiley and Sons, New York, 1989.
[35] A. Holmes-Siedle, L. Adams, Handbook of Radiation Effects, Oxford University Press, Oxford, 1993.
[36] G.C. Messenger, M.S. Ash, The Effects of Radiation on Electronic Systems, second ed., Van Nostrand Reinhold, New York, 1992.
[37] J.R. Srour, J.M. McGarrity, Proc. IEEE 76 (1988) 1443.
[38] K.F. Galloway, R.D. Schrimpf, Microelectron. J. 21 (1990) 67.
[39] G.J. Papaioannou, in: A. Christou, B.A. Unger (Eds.), Semiconductor Device Reliability, Kluwer, Amsterdam, 1990.
[40] L. Adams, Microelectron. J. 16 (1985) 17.
[41] J.F. Ziegler, G.R. Srinivasan (Eds.), IBM J. Res. Develop. 40 (1) (1996).
[42] T.C. May, M.H. Woods, IEEE Trans. Electron Devices ED-26 (1979) 2.
[43] J.F. Ziegler, W.A. Lanford, Science 206 (1979) 776.
[44] L. Lantz, IEEE Trans. Reliab. 45 (2) (1996) 174.
[45] J.F. Ziegler, IBM J. Res. Develop. 40 (1) (1996) 19.

[46] R. Baumann, T. Hossain, S. Murata, H. Kitagawa, in: 33rd Annual Proceedings of the IEEE Reliability Physics Symposium, 1995, p. 296.
[47] H.L. Grubin, J.P. Kreskovsky, B.C. Weinberg, IEEE Trans. Nucl. Sci. NS-31 (1984) 1161.
[48] G.C. Messinger, IEEE Trans. Nucl. Sci. NS-29 (1982) 2024.
[49] S.E. Kerns, B.D. Shafer, Proc. IEEE 76 (1988) 1470.
[50] C.H. Stapper, W.A. Klassen, in: 30th Annual Proceedings of the IEEE Reliability Physics Symposium, 1992, p. 3.
[51] G. M. Scamans and A. S. Rehal, J. Mater. Sci. 14 (1979) p. 2459.
[52] J.D. Lawrence, J.W. McPherson, in: 29th Annual Proceedings of the IEEE Reliability Physics Symposium, IEEE, 1991, p. 102.
[53] S.J. Krumbein, in: 33rd Meeting of the IEEE Holm Conference on Electrical Contacts, 1987.
[54] D.B. Brown, W.C. Jenkins, A.H. Johnston, IEEE Trans. Nucl. Sci. 36 (1989) 1954.
[55] T.J. O'Gorman, J.M. Ross, A.H. Taber, J.F. Ziegler, H.P. Muhlfeld, C.J. Montrose, H.W. Curtis, J.L. Walsh, IBM J. Res. Develop. 40 (1) (1996) 41.
[56] E.L. Peterson, P. Marshall, J. Rad. Eff. Res. Eng. (January 1989).

# CHAPTER 8

# Packaging Materials, Processes, and Stresses

## 8.1 INTRODUCTION

As R.E. Gomory of IBM has so aptly noted, "Far from being passive containers for microelectronic devices, the packages in today's advanced computers pose at least as many engineering challenges as the chips that they interconnect, power, and cool" [1]. This is not an idle exaggeration, because packaging technology encompasses an integrated bundle of interdisciplinary concerns. It requires the expertise of scientists as well as mechanical, electrical, chemical, and materials engineers engaged in processing and assembling a host of different materials for a variety of functions. At least three levels of packaging architecture depicted in Figure 8.1 are required to interconnect, power, distribute signals, and cool chips in electronic equipment [2]. Chip-level interconnections may be thought of as zero-level packaging. Individual dies encased in a single package then constitute the first level.

Types of single-chip packages include the plastic encapsulated and ceramic enclosures shown in Figure 8.2. These packages may be conveniently divided into through-hole (Figure 8.2(a)) and surface-mount (Figure 8.2(b)) varieties depending on how they are attached to second-level packaging substrates. Using the dual in-line plastic package (DIP) as an example, a cutaway view reveals the integrated circuit (IC) chip, its mechanical attachment to the lead frame, and its electrical connection by means of wire bonds. Typical materials employed, number of input/output (I/O) pins, and lead spacing, or pitch, in both types of packages are indicated.

The second level of packaging involves assembling two or more of the first-level chip packages onto polymer-base printed circuit boards (PCBs) (also known as printed wiring boards (PWB)), ceramic substrates, or in assorted multichip modules (MCMs). Characteristic of this level of packaging is the almost universal use of low-melting-point solders to extend the interconnections between levels. Like small brooks that feed narrow streams

*Reliability and Failure of Electronic Materials and Devices*
ISBN 978-0-12-088574-9
http://dx.doi.org/10.1016/B978-0-12-088574-9.00008-2

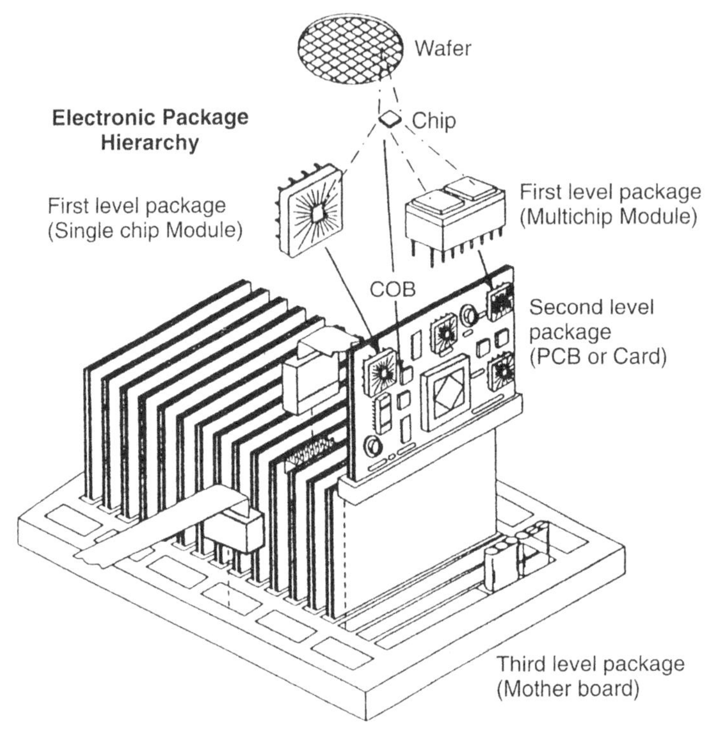


**Figure 8.1** Electronic package hierarchy. *From Ref. [2].*

that swell into rivers, so too the tiny currents through interconnections on the chip are collected by successively larger conductors. First, there are chip contact pads, next bonded wires, then metal frames whose leads are inserted into or soldered to the circuit board, and finally mechanical connectors. At each level along the way, insulating packages, housings, and electrical connections are needed. System performance is adversely affected by increasing the number of packaging levels and stretching the length of signal paths. At the chip level, signals are processed at speeds approaching $10^{-12}$ s, but computers run almost a thousand times slower. Thus, the primary packaging challenge is to preserve chip performance without compromising system reliability.

In Table 8.1, some of the manufacturing processes used to produce first- and second-level packages are noted. The first part of this chapter is concerned with a number of these packaging issues and begins with a discussion of how single-chip carriers are processed. As was the case for the chip level in Chapter 3, attention will focus on the defects introduced in these

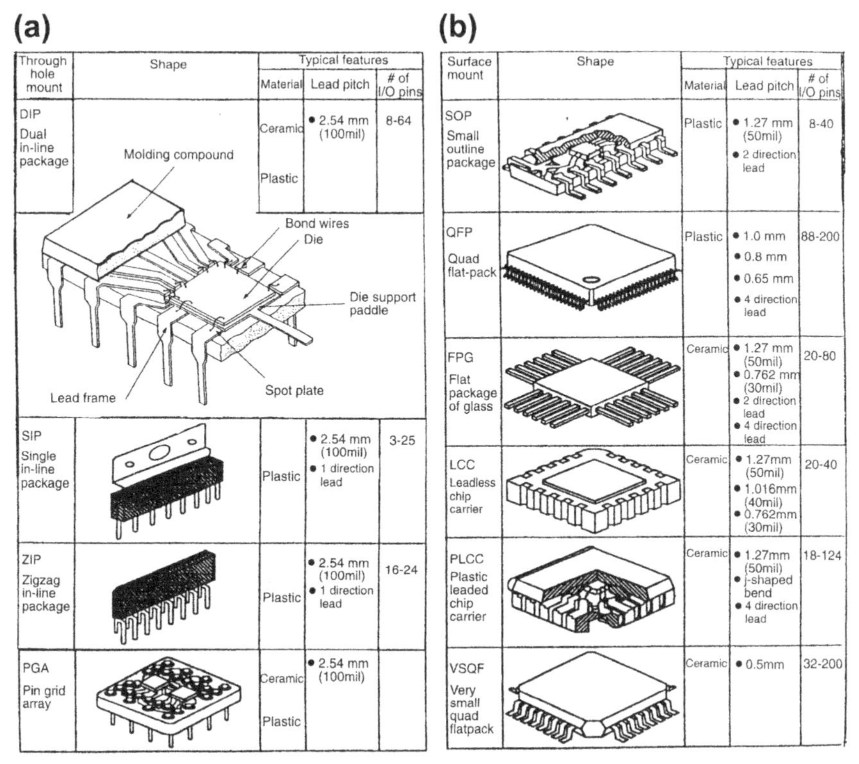


**Figure 8.2** Package profiles and characteristics. (a) Through-hole mounted packages. The components of a typical DIP are shown. (b) Surface mounted packages. *From Ref. [6].*

processing steps. All too often, later degradation and failure of the "passive containers" have limited or even halted the functioning of sophisticated IC chips without otherwise damaging them. In many cases, stress, in particular thermal stress, is the culprit.

Much of the remainder of this chapter is broadly devoted to two topics. The first deals with the properties of solders, their reactions, and use in joining operations. Analysis of the elastic thermal stresses that develop in an assortment of first-level and second-level packaging structures is the second topic. Unfortunately, we rarely know the magnitude of the stresses present. One reason for this is that internal stresses are involved, and there are no measurable external forces. In Section 3.3.2.4, it was noted that stress could be determined in films or coatings by measuring the curvature, or bow, of large wafer substrates. But this is not practical in the small structures present on either the chip or the packaging level. In addition,

**Table 8.1** Classification of packaging manufacturing processes

| ***Packaging manufacturing processes*** | |
|---|---|
| Die attachment | Brazing |
| Wire bonding | Controlled-collapse |
| Plastic encapsulation | Chip connection |
| Thin-film metal deposition | Cleaning |
| Thin-film resistor deposition | Coating processes |
| Thin-film dielectric deposition | Visual Inspection |
| Photoresist patterning | Package sealing |
| Image transfer | Leak testing |
| Etching | Electrical testing |
| Screen printing | Branding |
| Thermal treatment | |
| Resistor trimming | |
| ***Printed-circuit board manufacturing processes*** | |
| Cutting | Contact plating |
| Punching | Solder Masking |
| Drilling | Bare board testing |
| Screen printing | Component Inspection |
| Image transfer | Soldering |
| Etching | Inspection |
| Cleaning | Electrical testing |
| Electroplating | |
| Electroless plating | |
| Laminating | |

From Ref. [50].

the geometries are too complex. The situation is not hopeless, however, because stresses can often be calculated under special circumstances from elasticity theory or determined using finite-element methods. Occasionally, the thermal stress levels are high enough to induce degradation of packaging materials and plastic deformation of solder joints that causes them to fail by creep and thermal fatigue. These latter time-dependent failure mechanisms and associated reliability issues will be addressed in Chapter 9.

Many of the above and related packaging issues are treated in a growing number of edited handbooks and texts [1–9]. Readers requiring a more detailed knowledge of the engineering analysis and design, materials selection, cooling, and reliability of electronic packaging materials and structures should consult them.

## 8.2 IC CHIP PACKAGING PROCESSES AND EFFECTS

### 8.2.1 Scope

By the time the final step of chip processing has been completed, some wafers are worth a few hundred times their weight in gold. However, fabrication of the wafer does not necessarily mean that we have a finished product. Only when the chip is tested and safely enclosed within one of the packages of Figure 8.2 is this phase of manufacturing completed. These packaging and assembly steps, schematically outlined in Figure 8.3, are fraught with processing pitfalls that can seriously limit yield. Wafers produced with such great care and cost must now be separated into chips or dies, interconnected, and packaged. There is a great contrast between processing at the chip and package levels, the defects created, and the reliability implications involved. Distinguishing features worth noting include the following.

#### *8.2.1.1 Size*

If ~1 μm or less is the characteristic process feature size on chips, tens to thousands of micrometers and larger are typical packaging-component dimensions.

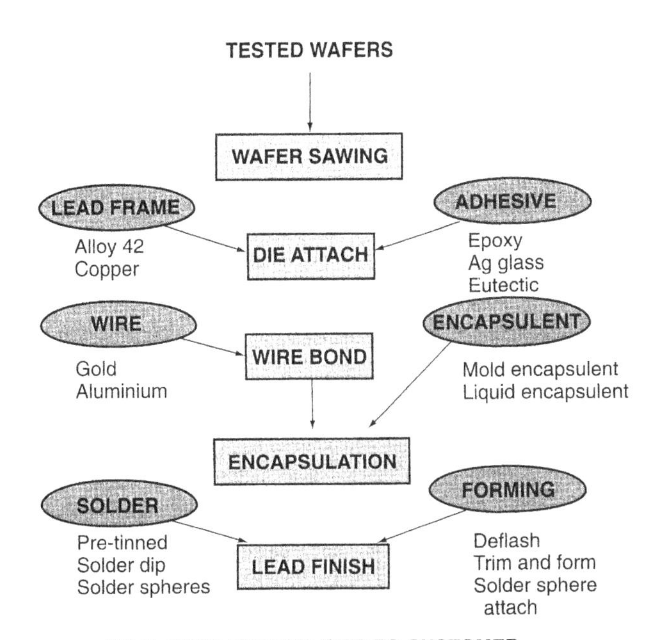


**Figure 8.3** Integrated circuit package assembly steps. *After Ref. [55].*

#### 8.2.1.2 Materials

Packaging materials and structures are chosen to meet the demands of ICs. Thus, metals are selected according to how well they conduct current into and out of chips, and encapsulants on their ability to encase and protect the precious chip. A wider spectrum of materials is employed in packages than in chips. In addition to solely electrical conduction functions, metals are used in packages as mechanical supports, to conduct heat away (heat sinks), and to seal the contents. Ceramics like alumina also serve as containers for chips and are often the substrates for mounting semiconductor chips. Polymers are employed to hermetically encase the chips and are the substance of PCBs. Even silicon is used as a substrate in the silicon-on-silicon multichip module packaging technology. Relevant electrical and mechanical properties of a number of packaging materials are listed in Tables 2.1 and 8.2, respectively.

Dielectric materials, which include ceramics, polymers, glass, and composites, have an important influence on system performance. From an electrical standpoint, packages contain transmission line structures in which it is essential to reduce signal propagation delay times. The latter are proportional to $\varepsilon^{1/2}L$, where $\varepsilon$ is the insulator dielectric constant and $L$ is the wiring length. Small values of both $\varepsilon$ and $L$ are therefore desirable for high-speed signal transmission.

#### 8.2.1.3 Processes

Unlike chip-level processing, packaging is generally carried out in low-level clean rooms. Materials and components (wire, sheet metal parts, ceramic plates, etc.) are usually purchased from fabricators and assembled. In this sense, packaging resembles traditional secondary-manufacturing operations. Molding of polymers, soldering, joining, and bonding are important packaging processes.

#### 8.2.1.4 Processing Temperatures

Compared to chip-level processing, temperatures employed in packaging are relatively low. Instead of 1000 °C and above during epitaxy, diffusion, and especially during crystal growth, eutectic die-attach processes only reach ~425 °C. Other than die attachment, packaging materials rarely experience temperatures above a few hundred degrees centigrade. Note that these comments refer only to processes where the IC chip is present; excluded are prior processing of components, e.g, firing glass or ceramic–metal hermetic seals and conductive inks in hybrid circuits.

**Table 8.2** Properties of packaging materials

| | CTE $10^{-6}\,°C^{-1}$ | Thermal conductivity W/m K | Tensile strength[a] MPa | Modulus of elasticity GPa |
|---|---|---|---|---|
| ***Semiconductors*** | | | | |
| Si | 3.0–3.5 | 84 | | 131 |
| GaAs | 5.7 | 46 | | 86 |
| InP | 4.8 | 68 | | |
| ***Metals*** | | | | |
| Cu | 16.7 | 393 | 310 | 116 |
| Ag | 19.7 | 418 | 145 | 76 |
| Au | 14.2 | 297 | 340 | 57 |
| Al | 23.1 | 247 | 147 | 68 |
| Mo | 5.2 | 138 | 690 | 325 |
| Ni | 14 | 92 | 760 | 221 |
| Pd | 9.0 | 76 | 152 | 145 |
| W | 4.5 | 160 | 621 | 388 |
| Alloy 42 (42Ni–58Fe) | 5.3 | 15 | | 145 |
| Invar (36Ni–64Fe) | 2 | 11 | | |
| Kovar | 5.5 | 17.1 | 520 | 138 |
| Au–Sn (eutectic) | 16 | 251 | 198 | 69 |
| Au–Si (eutectic) | 13 | 293 | 270 | 69.5 |
| Au–Ge (eutectic) | 12 | | 233 | 70 |
| Eutectic solder | 24.7 | 50.6 | 39 | 15 |
| 95 Pb–5Sn | 29 | 35.5 | 23 | 24 |
| ***Ceramics/Glasses*** | | | | |
| AlN | 2.9–3.8 | 195–319 | 400 | 345 |
| $Al_2O_3$ (96%) | 5.4–7.2 | 31–46 | 400 | 303 |
| BeO | 5.9–7.2 | 184–275 | 240 | 312 |
| Cordierite | 7.9 | 4 | 30 | 34 |
| Mullite | 5 | 6.7 | 200 | 175 |
| $SiO_2$ | 0.57 | 0.1 | 98 | 94 |
| Si (110) | 2.6–3.25 | 98.9–148 | 185 | 169 |
| SiC | 3.4 | 120 | 450 | 439 |
| SiN | 1.1–2.17 | 40.2 | 106 | 150 |
| Diamond | 1.1 | 2000 | | 1050 |
| ***Polymers*** | | | | |
| FR4 | 13.6–15.4 | 0.29 | 276 | 71 |
| Polyimide | 40–50 | 0.155 | 103 | 1.37 |
| PTFE | 75 | 0.26 | 24.5 | 0.4 |
| Epoxy resin | 60–80 | 0.2 | 43–85 | 2.7–3.4 |

Values are for room temperature.
[a]Flexural strengths for ceramics and glasses.
Data taken from a variety of sources.

#### *8.2.1.5 Defects and Reliability*

Instead of crystallographic and other microscopic defects on the chip level, packaging defects are macroscopic in size. They consist, for example, of voids in polymers and die-attach materials, and cracks in chips, dies, and solders. Packaging failures often stem from thermal-stress effects, but environmental damage due to corrosion and humidity is not uncommon. A compilation of defects arising from package manufacturing is listed in Table 8.3.

Although this table specifically targets the reliability of IC packages, very similar issues arise in the packaging of related electronic products. For example, Figure 8.4 depicts the reliability concerns in a micromachined silicon pressure sensor and its associated package. Fabrication of the miniature mechanical sensor borrows heavily from silicon lithography and processing technology to create the diaphragm and associated electrical connections.

We now briefly survey some of the important first-level packaging processes in a descriptive way, emphasizing some of the defects introduced and reliability issues raised along the way.

### 8.2.2 First-Level Electrical Interconnection

#### *8.2.2.1 Die Attachment*

The first step in the packaging process is scribing and breaking the wafer into dies. As noted in Section 3.3.2, even this step can be troublesome by producing serious losses due to die cracking. Next, the die is bonded to a conducting paddle or substrate that is mechanically strong and makes provision for electrical connection to subsequent packaging levels. Sheet metal (e.g., Cu, alloy 42 (42Ni/58Fe), alloy 50 (50Ni/50Fe) lead (pronounced "leed") frames, metallized ceramic ($Al_2O_3$, AlN) plates), and even Si wafers have served as substrates in different packaging technologies. Most commonly, however, metals are the substrates. The Si chip is bonded to them either by means of Pb–5 wt%Sn and more popularly, by Au–Si eutectic alloy solders, or by metal (Ag)-filled conductive organic and glass adhesives. In eutectic bonding, a thin preform layer of Au is interposed between die and substrate and then heated to approximately 425 °C. Some backside silicon dissolves in the gold, and a melt layer containing 3.6 wt.% Si forms at the eutectic temperature of 370 °C. This eutectic liquid serves to wet the die and securely bond it to the substrate upon solidification. The high cost of Au has provided the driving force to use Ag-filled epoxies or glass-frit adhesives with additional low bonding and curing-temperature

**Table 8.3** Defects associated with packaging manufacturing processes
**A. Chip packaging**

| ***Die bonding*** |
|---|
| Alkali ion contamination; chip fracture; epoxy creep; excessive electrical resistance; excessive moisture source; excessive thermal resistance; halide contamination; horizontal chip cracks; inadequate bond area; incorrect bond-line thickness; incorrect die selection; incorrect positioning; ionic contamination; lead frame oxidation; low die shear; mechanical damage to chip; nonwetting; outgassing source; poor adhesion; poor flow of adhesive; poor wetting; resin bleed; silver migration; source of particles; stress effects; too little or too much adhesive; vertical chip cracks; voids; wrong alloy. |
| ***Wire bonding*** |
| Chip scratches; chip fracture; contamination; cratering; incorrect wire dress; incorrect bond placement; loose particles; low bond strength; overbonding; purple plague; smeared metallization; substrate metal too hard; underbonding; wire too hard or too soft. |
| ***Plastic encapsulation*** |
| Cracks; damage by filler particles; excessive cure; excessive flash; excessive source of alpha particles; excessive coefficient of thermal expansion (CTE) mismatch; inadequate cure; inadequate moisture resistance; inadequate plastic flow; inadequate thermal conductivity; ionic contamination; poor adhesion to lead frame; voids; wire sweep. |
| ***Screen printing*** |
| Cracks; excessive amount of screened paste; excessive bleed; excessive pinholes; inadequate amount of screened paste; missing features; poor adhesion; poor alignment; poor resolution. |
| ***Package sealing*** |
| Cavitation; contamination; deformed metallization; excessive deformation; excessive moisture in package; excessively high leak rate; lack of hermeticity; loose particles; mechanical damage; oxide cracks in multilevel metallized ICs; passivation cracks; poor wetting; weld splash; wrong ambient. |
| ***Cleaning*** |
| Electrostatic discharge (ESD) effects; inadequate removal of surface impurities; ionic contamination; organic residues; particulate contamination; radiation damage to semiconductor devices; surface pitting. |

*Continued*

**Table 8.3** Defects associated with packaging manufacturing processes—Cont'd
**A. Chip packaging**

| *Controlled-collapse chip connection* |
|---|
| Bridging; dewetting; flux residues; incorrect final spacing; poor reflow; poor solder wetting; too little or too much lead; too little or too much tin. |
| **Printed-circuit board manufacturing processes** |
| *Drilling* |
| Burrs; debris; delamination; drill wander; extra holes; foreign matter in holes; holes not drilled through; loose fibers; mechanical damage; misregistered holes; missing holes; nailheading; oblong holes; plowing; rifling; smears; voids; wrong size holes. |
| *Electroplating* |
| Burning; edge feathers; grainy deposits; inadequate substrate preparation; poor adhesion; poor reflow; poor throwing power; roughness; treeing; whiskers; wrong alloy composition; wrong thickness. |
| *Soldering* |
| See Table 8.6. |

From Ref. [50].

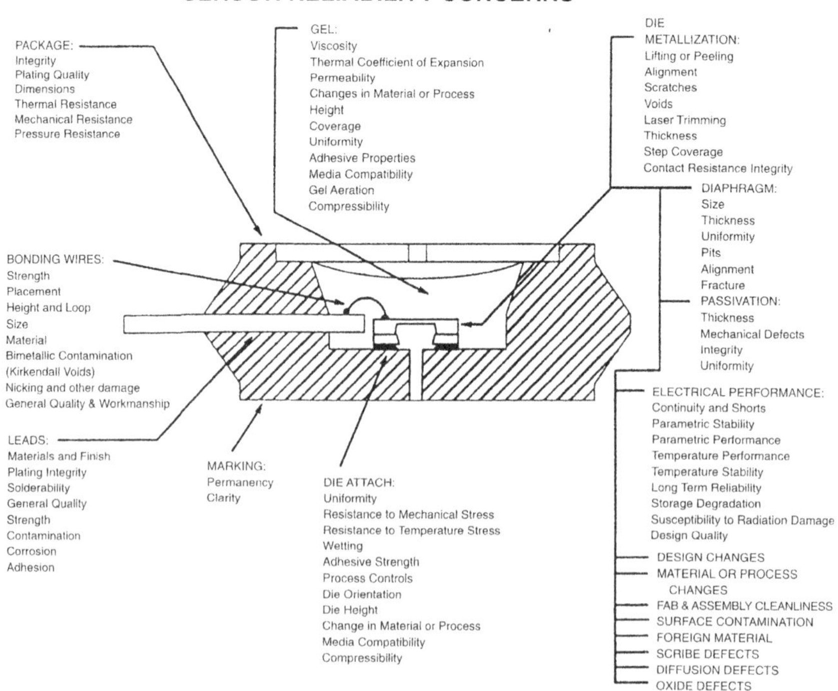


**Figure 8.4** Reliability concerns in a micromachined sensor and package. *From Ref. [56].*

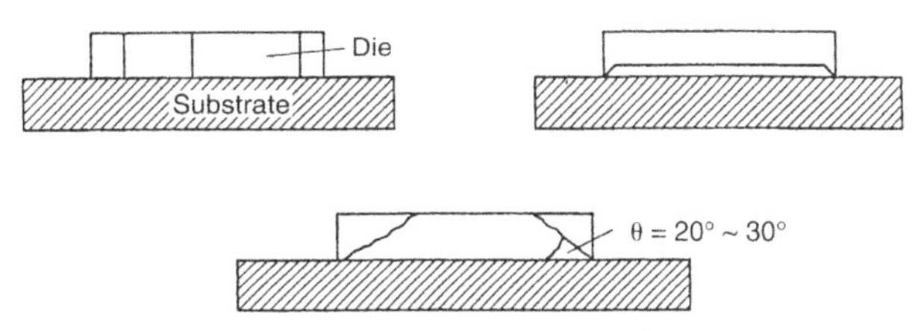


**Figure 8.5** Vertical, horizontal, and corner patterns of die cracking. *From Ref. [57].*

advantages. Dendrite filament formation during bias-induced migration of Ag is always a short-circuit threat in packages, and must be guarded against when conducting epoxies are used.

The worst thing that can happen in bonding operations is die cracking. Vertical, horizontal, and corner nucleated cracking, illustrated in Figure 8.5, sometimes occurs. Cracking appears to be induced by thermal expansion ($\alpha$) differences between the chip (*c*) and substrate (*s*) as mediated by the adhesive. For example, let us assume that a stress-free bond forms at die-attach processing temperatures. At lower temperatures, the thermal stress is given by the well-known expression (Eqns (8.3)–(8.8))

$$\sigma = \frac{E}{1-\nu}\left[\alpha_c - \alpha_s\right]\left(T - T_o\right). \tag{8.1}$$

If $\alpha_c > \alpha_s$, upon cooling to room temperature ($T_o$) the chip will contract more than the substrate, which restrains the full extent of die shrinkage. This places the die in tension, an always dangerous stress state for a brittle material. The thermal-stress magnitude developed in the chip is given by Eqn (8.1), and it is clear that it can be minimized through a close thermal-expansion match of the involved materials, coupled with low processing temperatures. A more exact quantitative stress analysis of this trilayer die-attach geometry, subject to temperature changes, will be deferred to Section 8.5.

Further examination of die bonds has revealed the presence of voids in the adhesive layer. Voids arise in eutectic bonds as a result of poor wetting due to incomplete oxide or contamination removal from the bond interfaces. In hermetic polymer packages, evolution of gas and moisture during high-temperature sealing are the suspected sources of voids. Moisture absorption in die-attach adhesives remains a source of delamination and cracking problems that are unleashed when heat is applied during soldering operations. Examples of cracking in die-attach materials are reproduced in Figure 8.6.

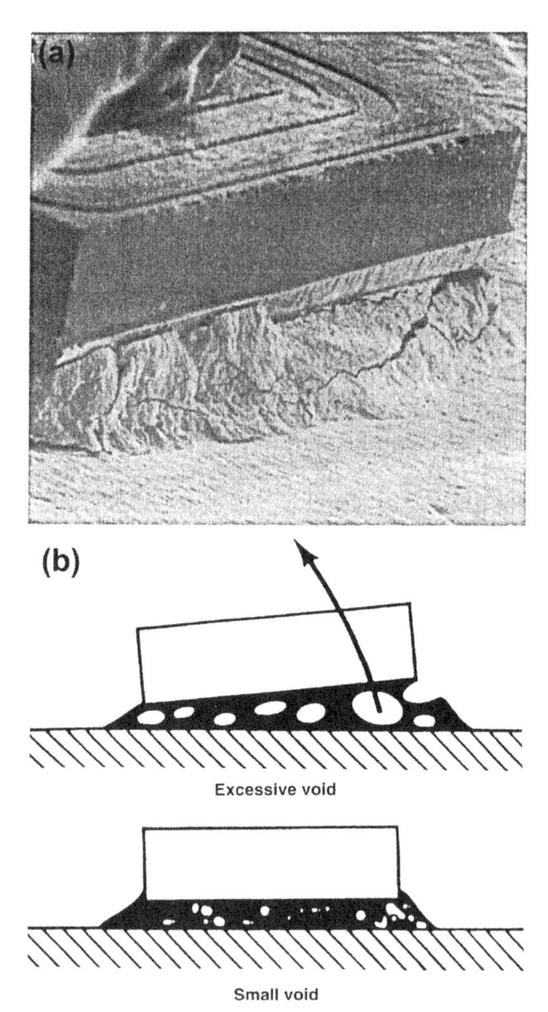


**Figure 8.6** Degradation of die-attach materials. (a) Cracking of the die-attach solder due to excessive power cycling. *(From Ref. [10].)* (b) Voiding in die bond adhesive results in low shear strength and die "popping." *(From Ref. [11].)*

### *8.2.2.2 Wire Bonding*

After the bottom of the chip is bonded to the substrate, the remaining contacts are made to the bond pads on the top, or processed, surface of the silicon die. Wire bonding employing fine Au wire, approximately 25 μm in diameter, is the most widely practiced method. During thermocompression bonding, specially designed die-bonder machines feed the Au wire to the Al die pad in a carefully controlled time, temperature, and applied-pressure cycle. In the process, the wire acquires a ball shape after heating and pressing it onto the metallized chip bond pad. Sometimes ultrasonic vibration is simultaneously applied to facilitate the removal of surface oxides and ensure better metal-to-metal contact, plastic deformation, and adhesion. The other end of the wire is then looped in a prescribed arc and bonded at the lead frame as the wire is severed. This automated cycle repeats in a rapid sewing-machine-like fashion until all the connections to the die are completed.

The photos in Figure 8.7 indicate that wire bond failures can occur in the wire, at the ball–wire interface, through the ball, and along the chip bonding pad. Among the factors that weaken Au wires during thermosonic wire bonding are recrystallization and grain growth, necking, wire scratching, and even cracking. "Purple plague" formation (see Section 5.3.4) sometimes causes decline in bond strength and wire lifting, literally leaving a footprint behind on the Al pad (Figure 8.7(c)).

#### 8.2.2.3 Tape Automated Bonding

Fast as it appears to be, the bonding of 8–10 wires per second is relatively slow. Tape automated bonding (TAB) speeds up matters by simultaneously "gang" bonding one end of all the leads first to the IC chip (inner lead bonding (ILB)) and then the other lead ends to the substrate bond pads (outer lead bonding (OLB)). In TAB, which is reviewed by Lau et al. [12], a prefabricated metal (typically copper) interconnection pattern is supported on a polyimide film. Like a motion picture film, the tape contains sprocket aligning holes that ensure registry with the die I/O pads as it unwinds from reels. When centered over the chip, a single thermocompression-like stroke gang bonds all contacts at once. Some 400–700 I/O pads can be TAB bonded with minimum pitch values of 100 μm; these characteristics compare with 250–500 I/Os and ~150 μm pitch for wire bonding.

Both OLB and ILB TAB leads are subject to processing and reliability problems. The former leads usually consist of Cu coated with a thin Sn or Au layer (0.5–2 μm thick). These are joined to the substrate via metallized pads selectively coated with eutectic solder (63 wt% Sn—37 wt% Pb). Upon melting of the solder, a number of different Au—Sn intermetallic binary and ternary compounds ($Au_5Sn$, $AuSn_4$, $AuSn_2$, AuSn, $Cu_5AuSn_5$, $Cu_4Au_2Sn_5$) form. Some of these compounds tend to be brittle. Together with Kirkendall porosity, which is often observed when the Au content is too high [13], the mechanical integrity of the bond is threatened. With moisture and chloride ions, the collection of metals comprising bumps on the IC chip (see Section 8.2.2.4) and the inner TAB leads plated with Sn presents a plethora

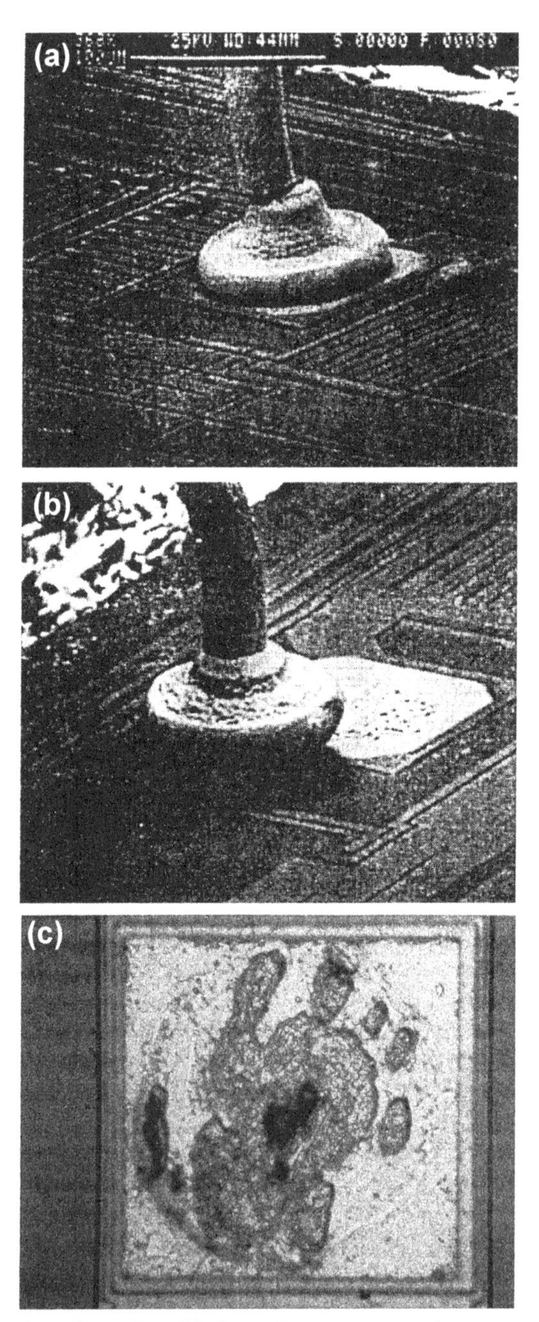

**Figure 8.7** Examples of wire bond failures: (a) A scanning electron microscope (SEM) image of wire fracture during elevated temperature cycling. *(From S.J. Kelsall, Texas Instruments.)* (b) SEM image of a bond lifted from a substrate pad. *(From T.M. Moore, Texas Instruments.)* (c) Lifted ball bond leaves trace in the shape of a footprint on the Al pad. *(Courtesy of Carl Dorsey, Lucent Technologies, Bell Laboratories Innovations.)*

of potential galvanic-corrosion couples. Under these circumstances, corrosion of Al interconnects and the bond pad-bump interface, as well as debonding of Cu/Sn inner leads from bumps, occurs [14]. In addition, TAB leads are fragile and susceptible to handling damage in the form of bending, loss of coplanarity, and fracture. TAB is often used along the edges of displays and small handheld devices. It relies upon anisotropic conductive film, which is compressed under temperature to achieve contact between metal contacts. These contacts provide mechanical as well as electrical contact and strain relief. These shortcomings have led to implementation of other lead configurations such as pin and ball grid arrays (BGAs).

#### *8.2.2.4 Flip-Chip Bonding*

A flip chip is broadly defined as a die mounted on the substrate using various interconnect materials and methods, e.g., fluxless solder bumps, wire, TAB, conductive adhesives, and compliant bumps, so that the chip active area faces the substrate. Solder bump flip-chip bonding, pioneered by IBM in the mid-1960s for the manufacture of computer modules and shown in Figure 8.8, is still very much used in high-performance electronic packages. An excellent book reviewing the attributes of this, as well as other flip-chip technologies, has been edited by Lau [2].

##### 8.2.2.4.1 Package Structure

In Figure 8.8, passivated chips are processed to contain solder bumps at opened, metallized contact areas on the active device surface. This structure is known as BLM, or ball limited metallurgy. The chip is then flipped 180° or turned to face the next level of assembly, typically a ceramic substrate. Complementary top surface metallurgy (TSM) on the substrate consisting of a collection of metallized microsockets and interconnections enables alignment with the bumps on the chip. Interconnects also extend through interior layered levels of circuitry within the substrate. When the solder is heated until it is molten and reflows (see Section 8.4.3), an array of solder balls forms to attach the chip above to the substrate below. When properly done, the two metallurgies are merged, and a controlled-collapse chip connection (or C-4) is achieved between the BLM and the TSM. Similarly, on the bottom substrate surface there are metal multilayers at the I/O pads that enable solder balls or brazed pins to mate with the PCB below.

Trilevel thin-film structures of Au, Cu, and Cr on both the IC pads and microsocket surfaces sandwich the reflowed solder between them. Each metal plays a distinct role in ensuring the required reliability of the structure.

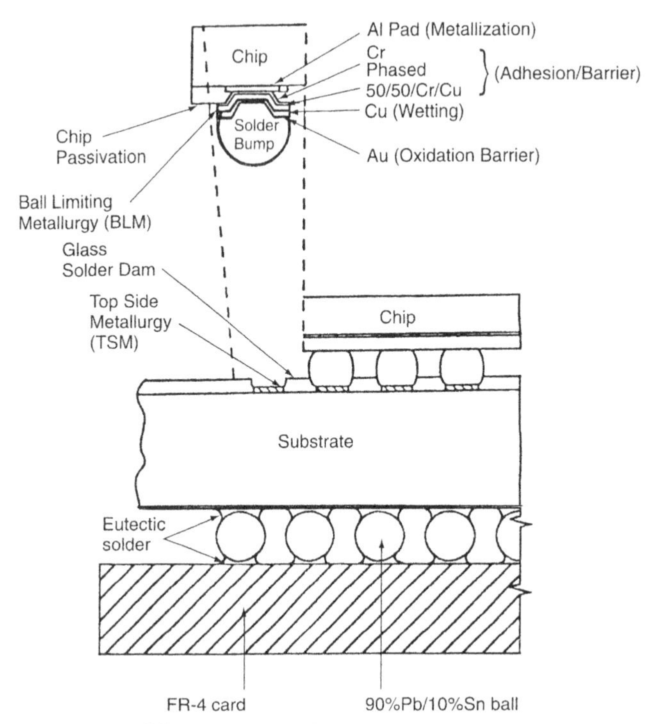


**Figure 8.8** Schematic of flip-chip technology employed between chip and substrate (first-level packaging). Similar flip bonding is utilized between substrate and PCB (second-level packaging).

The Cr film is needed to bond to the Al pads, which cannot be directly soldered. Next, Cu is the actual base metal for soldering. But Cu oxidizes on storage, and when it does, solder will not wet it. Therefore, a Au film is needed to protect it. When reflow occurs, a number of complex mass transport and intermetallic compound formation processes occur throughout the molten solder during reaction with the metal films (Section 8.3.3.4). Brittle cracking of intermetallic compounds is a potential source of solder defects.

#### 8.2.2.4.2 Stress

It is normally assumed that joints become stressed when they cool to low temperatures after soldering. If the chip and substrate are of the same material, e.g., silicon, then the BLM–TSM solder joints will be unstressed. However, when these materials differ, we may expect the solder post or column to suffer a shear distortion of magnitude directly proportional to its distance from the zero displacement axis. The extent and sense of column tilt depends on the difference in chip and substrate thermal expansion coefficients and temperature excursion. In addition, thermal expansion differences between the solder ($\alpha_{so}$) and intersolder encapsulant ($\alpha_e$) can alter the chip–substrate separation. For example, if $\alpha_e > \alpha_{so}$, the chip and substrate are drawn toward each other, but if $\alpha_e < \alpha_{so}$, stresses develop to push them apart. When superimposed on the compositional changes in the solder, one can appreciate how the shear stresses might cause fracture at compound interfaces. Furthermore, these effects are accelerated by thermal cycling. The subject of thermal stress and its role in the reliability of solder connections will be quantitatively dealt with in both this (Section 8.5.2) and the next chapter (Section 9.5.1).

#### 8.2.2.4.3 Corrosion

The collective array of more than a half dozen metals that are either in close proximity or directly in contact presents a potential cornucopia of galvanic couples and corrosion problems. Like inner lead TAB bonding, moisture and impurity ($Cl^-$, $Na^+$)-induced corrosion of positive and negatively biased Al tracks, reaction at solder bump interfaces, passivation cracking due to corrosion products, and debonding are degradation processes that can affect flip-chip bonded structures [13]. Similar studies by Frankel et al. [15] found that corrosion susceptibility of Cr/Cu, with attendant crack initiation and loss of adhesion at the pad edge, was reduced by interposing a thin Ni film between these metals.

Flip-chip bonding is also used in attaching many chips to a single substrate. The advantages of doing this, including packaging of the entire MCM, are discussed in Section 8.2.3.4. Now, however, we turn to the problem of enclosing individual IC chips.

## 8.2.3 Chip Encapsulation

The flip-chip structure, with its passivation everywhere except at the solder, naturally protects the silicon device surface and needs no further enclosure. But the wire-bonded chips must be sealed within containers or packages in such a way that the chip and first-level contact pads are protected. Furthermore, leads must pass through the package and be available for second-level interconnection.

### *8.2.3.1 Processing Plastic Packages*

Enclosing chips within polymer molding compounds is the most popular way to package ICs [7,16]. In fact, more than 90% of all packages produced are of the plastic variety; they enjoy a virtual monopoly in the civilian sector and have been increasingly accepted in military applications. Annually, close to $100 \times 10^9$ plastic DIPs are produced worldwide. Epoxies containing curing agents, fillers, flame retardants, mold release compounds, and coloring agents are typically employed and are molded into packages via a sequence of steps that includes (1) placement of lead frames in molds; (2) conditioning, preheating, and loading of molding compound; (3) transfer or injection molding; and (4) curing and ejection. Extreme care must be taken to minimize molding defects and yield losses in this relatively low-tech (and cheap) packaging step to preserve the fruits of high-tech (and expensive) chip-level processing.

Many defects found in plastic packages originate from the thermal stresses of processing. For example, the thermal expansion coefficients of four key components within packages, i.e., chip ($4.4 \times 10^{-6}\,°C^{-1}$), alloy lead frame ($\sim 4.0$–$4.7 \times 10^{-6}\,°C^{-1}$), copper ($17 \times 10^{-6}\,°C^{-1}$), and molding compound ($\sim 20 \times 10^{-6}\,°C^{-1}$), differ. This implies that stresses will develop between different pairs of contacting materials and foster interfacial delamination and even component cracking. Delaminated regions between the molding polymer and metal leads can be the conduits for moisture and chloride ion ingress to foster corrosion, as Figure 8.9 suggests. Aided and abetted by ionic species (e.g., $Br^{-1}$) and oxygen contained within epoxy encapsulant flame retardants, corrosion can lead to $Al_2O_3$ formation and wire bond liftoff [17].

Stresses in lead frames [18] and wire bonds [19] arising from the fluid flow of the polymer during the transfer molding process are additional sources of defects. Calculations by Suhir and Manzione have shown that the maximum stress in the wire is proportional to the square of the ratio of the wire bond span to the wire diameter. Stresses could be large enough to lift

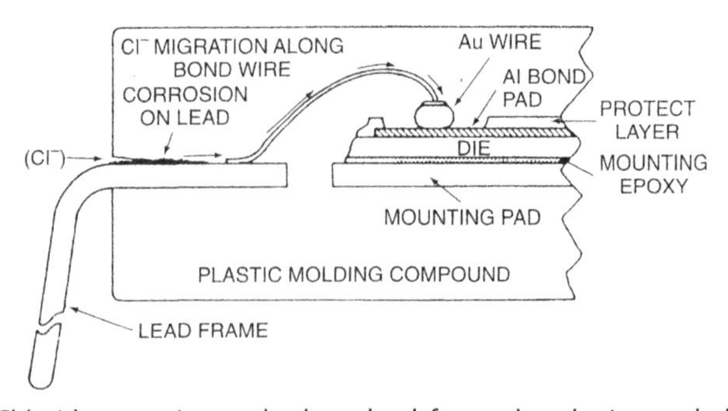


**Figure 8.9** Chloride corrosion path along lead frame, bond wire, and chip pad in a plastic IC package. *From Ref. [58].*

or sweep the wire from the bonding pad. In the case of lead frames, flow-induced forces can cause lead deflections. The forces arise because the lead frame edge splits the flow of the molding compound into two mismatched flow fronts, one above and one below the sheet-metal plane. This asymmetric flow exerts a net pressure that deforms the cantilever leads. Analysis has predicted that lead deflection varies directly as the fifth power of the lead length and inversely as the third power of the lead thickness.

### 8.2.3.2 Popcorn Cracking

After the apparently successful development of plastic encapsulation, it was observed that the frequency of so-called popcorn cracking of packages increased during the summer months. Furthermore, these failures occurred considerably after molding and usually during solder reflow of surface-mounted packages. These two observations strongly suggested that humidity, an overlooked yet ever present factor, was responsible. Package swelling and popcorn cracking develop according to the sequence shown in Figure 8.10. First, moisture permeates the package, most likely along the

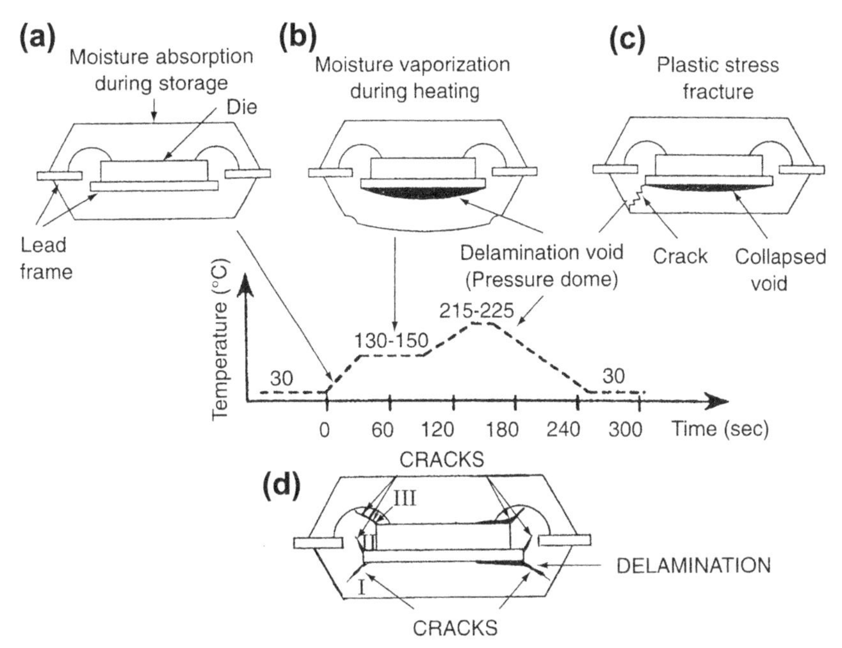


**Figure 8.10** Sequence of events leading to package swelling and popcorn cracking keyed to the solder heating cycle. (a) Moisture absorption during storage. (b) Vaporization of moisture, doming, and delamination during soldering. (c) Package cracking. (d) Locations of delamination and crack propagation. *From Ref. [59].*

metal–plastic interfaces, and is absorbed in the molding compound during storage (Figure 8.10(a)). Moisture ingress is driven by thermodynamic imperatives and essentially implemented by Fickian laws of diffusion (Section 7.2.2.2).

During reflow soldering of the package to the circuit board, the condensed moisture at internal polymer interfaces rapidly expands. This phase change and expansion of the confined vapor leads to a pressurized dome of steam (Figure 8.10(b)). Coupled with the strength reduction in the molding compound as its temperature rises, the void collapses and the package ultimately bursts (Figure 8.10(c)). Such a mechanism of moisture-induced damage is operative even in the absence of preexisting cracks or delaminated interfaces [20,21]. The critical condition for package cracking is simply $\sigma_P > \sigma_S$, where $\sigma_P$ is the maximum stress at the edge of the die pad and $\sigma_S$ is the strength of the molding compound at the soldering temperature. Furthermore, larger die pad ($L_{DP}$) and thinner encapsulant thickness ($L_P$) dimensions both increase $\sigma_P$ and promote the tendency to cracking; larger pads allow for more potential damage sites, while thinner encapsulants enable moisture to penetrate more readily. A suggested formula for the onset of cracking is thus

$$\sigma_P = K\frac{(L_{DP})^2}{(L_P)^2}P_{H_2O}, \tag{8.2}$$

where $(P_{H_2O})$ is the water vapor pressure and $K$ is a constant of proportionality.

Three types of popcorn cracks are observed, as shown in Figure 8.10(d). Types 1 and 11 nucleate as a result of delamination at the paddle–encapsulant interface. Cracks initiate below and above the paddle and propagate downward or upward, respectively, to the package surface. Type 111 cracks nucleate at the die corner and propagate upward. Once cracking of the plastic occurs, weakened or lifted wire bonds and die-surface delamination threaten package integrity. Further ingress of moisture and contaminants is then facilitated, raising prospects for corrosion.

The primary way to prevent package popcorning is baking prior to solder reflow in order to remove moisture. Other methods include desiccant bagging of all moisture-sensitive devices, mechanically strengthening and toughening molding compounds to resist fracture, and enhancing the ability of these materials to prevent water diffusion and absorption. Groothuis et al. [22] have recently discussed the mechanical behavior of molding polymers and suggested metrics for selecting them.

Finally, the introduction of lead-free solders for packaging connections has exacerbated this situation because the lead-free solders invariably require a temperature higher than lead solders. This ideosyncracy and others are very well presented in Refs [23] and [24].

#### *8.2.3.3 Hermetic Packages*

Due to their perceived reliability advantages over plastic packages, hermetically sealed packages have found favor in military and high-risk applications. Hermetic packages encase the IC chips within boxes and enclosures made of ceramics and metal alloys and allow for interconnection to the next packaging level, e.g., a PCB. Alumina is the most widely used ceramic in packages known as CERDIPs and CERPACs. Its low thermal conductivity makes $Al_2O_3$ unsuited for power packages, and therefore AlN and BeO have also been employed. Either flip chip or chips on lead frames are essentially installed within the open ceramic boxes, interconnected to the package pins, and then sealed with a ceramic lid. Glass frits that must be heated, as well as epoxies and solders, are used as sealant materials in both lids and pins. One has only to recall the disaster caused by the spaceship *Challenger's* gaskets to realize that sealants and the technology of sealing are common causes of leakage and unreliability in all kinds of systems; ceramic packages are no exception. Prior to sealing, contaminants are removed and protective silicone encapsulants are used to prevent exposure of the chip to possible moisture in the package environment. Among reliability problems in these packages are pin and lid leakage, corrosion of embedded bond wires, cracking of ceramic cases, and delamination. In an interesting case, void defects (Section 5.7) were produced in Al–Cu interconnects as a result of annealing the glass seals during packaging [25]. Evidently, thermal stresses developed in the IC chip as a result of this heat treatment.

Metal packages are similar in function to their ceramic counterparts. They are mechanically tougher but suffer from corrosion phenomena and problems associated with plating. Shorting to the case is a constant danger, not only from wire bonds or lead frames, but from whiskers that sprout and grow from tin-plated surfaces [6].

#### *8.2.3.4 Multichip Modules*

Although we have focused on single IC chip packages, there are other ways to organize electronic components and circuits [1–4]. Hybrid microcircuits are an example, and they are used in virtually every commercial and military electronic product from computers to radar systems. These consist of combinations of IC chips and additional discrete active and passive add-on devices and components, usually mounted on ceramic substrates.

In contrast to the idea of one IC chip per package, much can be gained by mounting several chips on a single substrate and encapsulating them within a single package to produce MCMs [26,27]. In principle, MCMs allow for integration of a mix of different technologies, e.g., analog, digital, CMOS, GaAs, and hybrid. In addition, MCMs enable a higher ratio of silicon or semiconductor device area to substrate area than is possible in single-chip packages. In the latter, this ratio is typically less than 0.1, whereas it may be greater than 0.5 in MCMs, signifying great efficiency in substrate or board-area usage. Both of these advantages are immediately evident if we look at Figure 8.11. Shown is the circuitry within a cellular phone that performs speaker-dependent voice recognition and telephone answering. Ten separate packages containing mixed IC technologies (e.g., microprocessor, memory) are essentially replaced by one MCM that contains the 10 IC dies, flip chip bonded to a silicon substrate. In the process, the circuit size shrank from 43.2 to 6.5 $cm^2$ with a corresponding weight reduction from 32 to 2.3 g.

Three popular multichip packaging configurations known as MCM-C, MCM-D, and MCM-L are distinguished in Table 8.4. Wire bonding, TAB, and flip chip (C-4) options exist for the first-level interconnection between the IC chip and the substrate. Of these, flip-chip bonding offers the highest packing density because the "footprint" on the substrate is smallest. Once a number of chips have been so bonded to a substrate, the entire assembly may, for example, be molded within a single plastic package.

Improvement in system performance and reliability are the chief advantages of a MCM, now acting as a unit, relative to the traditional one-chip packages. The latter may have some 10 times the number of connections and 6 times the wiring length relative to the MCM package [27].

#### 8.2.3.4.1 Performance

Higher system-operating speeds occur because there are shorter chip-to-chip interconnects with lower overall capacitance. In addition, there are lower noise characteristics due to the short signal paths, low dielectric constant media for chip-to-chip signal routing, controlled impedance in interconnects, and low inductance.

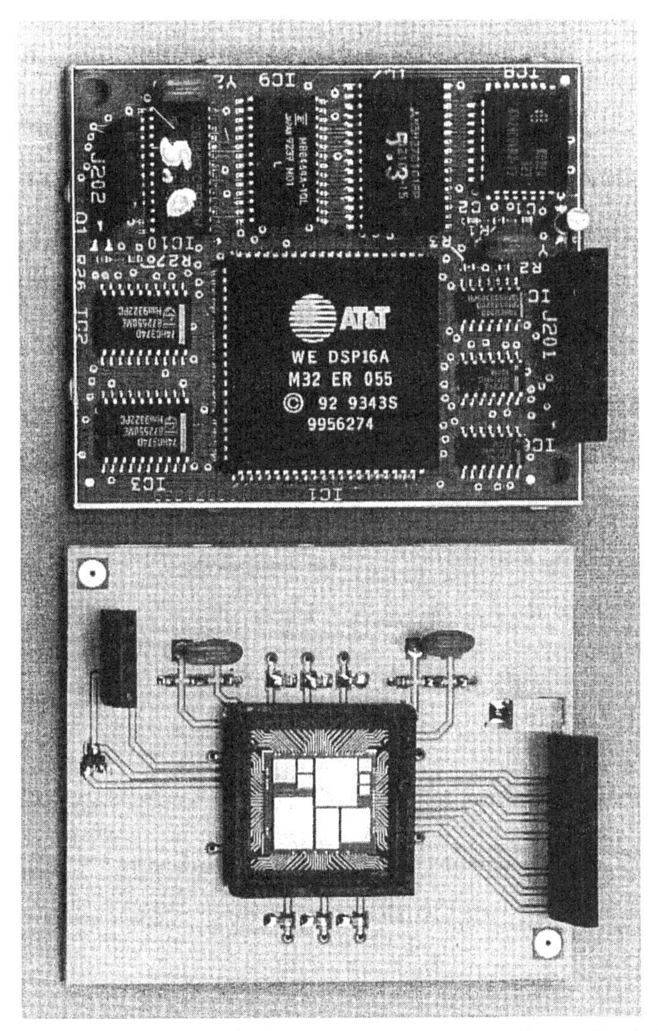


**Figure 8.11** Voice recognition/telephone answering device. Above: Circuit board assembly module. Below: Multichip module on a test board. *From Ref. [60]. Courtesy of K. L. Tai and R. C. Frye, Lucent Technologies, Bell Laboratories Innovations.*

### 8.2.3.4.2 Reliability

The reliability and failure issues in MCMs are similar to those for the single-chip modules discussed earlier. Nevertheless, greater system reliability is achieved because MCM technology essentially eliminates one whole packaging level. As a result, there is a dramatic decrease in the number of fatigue-prone leads and solder joints, and amount of

**Table 8.4** Characteristics of MCM technologies

| Property or characteristic | MCM-C (Thick film) | MCM-D (Thin film) | MCM-L (PCB) |
|---|---|---|---|
| Substrate | ceramic ($Al_2O_3$, glass-ceramic) | Various (organic, metal, ceramic) | Organic (FR-4 (epoxy glass), polyimide, polyester) |
| Metallization | Thick film Mo–Mn, W | Thin film Al, Cu | Laminated Cu |
| Number of layers | Up to ~60 | Up to ~8 | Up to ~50 |
| Linewidth (μm) | 75 | <25 | ~65 |
| Via size (μm) | ~150 | ~60 | ~300 |
| Via density ($cm^{-2}$) | 120 | 5000 | 60 |
| Interconnect density ($cm/cm^2$) | 20–30 | 160–320 | 12–16 |
| Dielectric constant | 2.5–3 | 8–9 | 4.5–5 |
| Propagation delay (ps/cm) | ~120 | 60–80 | 80–100 |
| Technology maturity | high | Limited | high |
| Cost | Medium | high | Low |

From Refs [51,52,53]

troublesome encapsulant materials and sealants. Heat removal from MCMs is often a concern, however, particularly in the high-density interconnect (HDI) MCMs; for this reason, they can be prone to reliability problems.

In HDI modules, the substrate, which is usually silicon, is often covered with a dielectric layer of polyimide that is permeated by multilevels of interconnects. These terminate in bonding pads to which ICs are flip-chip bonded. Except for scale, the interior of the polyimide layer is reminiscent of the IC chip, where the overlying $SiO_2$ is permeated by the stack of metallization levels and vias that connect them. Reliability issues of such polyimide structures involve metal–polyimide interactions, difficulty in wire bonding to Al pads, polyimide decomposition, and cracking. While the polyimide–Al interface is inert, Cu interconnects are chemically unstable, and if used require diffusion barriers. In addition, polyimide can carbonize during processing, causing loss of metal adhesion and surface leakage currents. Simultaneous volume shrinkage during curing can cause polyimide cracking as well.

## 8.3 SOLDERS AND THEIR REACTIONS

### 8.3.1 Introduction

For making permanent interconnections, solders have the great virtue of enabling low-temperature joining of metallic conductors to establish low-resistance ohmic electrical performance at all packaging levels. Solders also provide mechanical support of components and are, in some cases, a conduit for heat transfer. However, together with their excellent electrical and joining characteristics, they have attributes that make them a source of processing defects, and they are susceptible to a number of reliability problems. The former issues will be addressed in this chapter, while the latter concerns will be primarily reserved for Chapter 9. In addition to inclusion in the references dealing with packaging, there are a number of books [28–30] devoted solely to solder materials, processes, and reliability issues that are worth consulting.

Let us start by reviewing some elementary aspects of solder metallurgy. The phase diagram for Pb–Sn, a system that exhibits eutectic solidification, is shown in Figure 8.12. In addition to Pb–Sn, three other eutectic systems play a role in this book. Within the Au–Si system, a solderlike eutectic alloy is used to bond or attach ICs to the metal frame during the packaging of IC chips. Those with good memories will also recall that similar eutectic

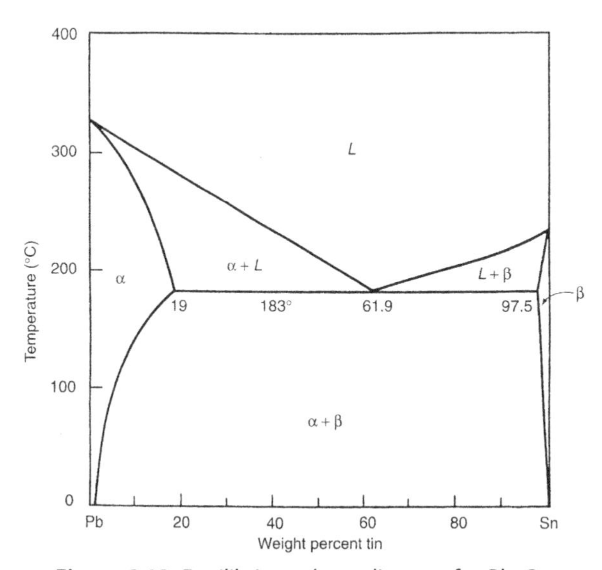


**Figure 8.12** Equilibrium phase diagram for Pb–Sn.

systems consisting of Al–Cu and Al–Si have relevance to electromigration and contact reliability, respectively (see Chapter 5).

Upon solidification of the eutectic solder alloy (61.9 wt% Sn–38.1 wt% Pb), a eutectic transformation occurs at the critical isotherm 183 °C according to the reaction

$$\text{Eutectic } L(61.9 \text{ wt\% Sn}) \xrightarrow{183\ \mathrm{C}} \alpha(19 \text{ wt\% Sn}) + \beta(97.5 \text{ wt\% Sn}) \quad (8.3)$$

Since the alloy undergoes eutectic solidification, the proeutectic phase is embedded within the two-phase eutectic microstructure. Equilibrium at 183 °C implies that phase compositions given by the tie lines are $\alpha$ (19 wt% Sn) and $\beta$ (97.5 wt% Sn). Just below 183 °C the lever rule predicts a eutectic mixture composed of 45.4 wt% $\alpha$ and 54.6 wt% $\beta$.

While this standard equilibrium analysis provides information on phase compositions and amounts, it says nothing about phase morphologies; these are dependent instead on prior processing history. For example, the cooling rate during electronic soldering operations is often too fast for thermodynamic equilibrium conditions to prevail. Rapid cooling of solders yields fine microstructures, i.e., those with a large number of small features. Finer grained cast structures, characterized by small interdendritic spacings, are not only generally stronger, but also metastable. Provided that sufficient thermal energy is available for the necessary atom movements, there will be a gradual solid-state transformation toward equilibrium structures and solubilities dictated by the phase diagram. Because of their low melting points, room temperature for solders is actually about 0.65 of $T_M$ ($T_M$ is the absolute melting point in K). At such high homologous temperatures, atoms diffuse and metals creep at observable rates. And during higher temperature operation, even more thermal energy is available to accelerate transformations. Thus, the structure that forms during solidification is gradually altered, as shown in Figure 8.13. Simultaneously, the solder loses strength over a period of a few days or weeks, a phenomenon known as age softening. The cause of the softening is the relieval of supersaturation within the eutectic matrix. Continued coarsening due to diffusion of atoms continues for years but has a much lesser effect on room-temperature mechanical properties since the solder strength stabilizes after a few weeks. It takes little imagination to grasp the potential reliability implications of time-dependent structural change in solder contacts.

Chemical and structural change can also occur during thermal cycling, a common service environment. Solders will then undergo cycles of phase

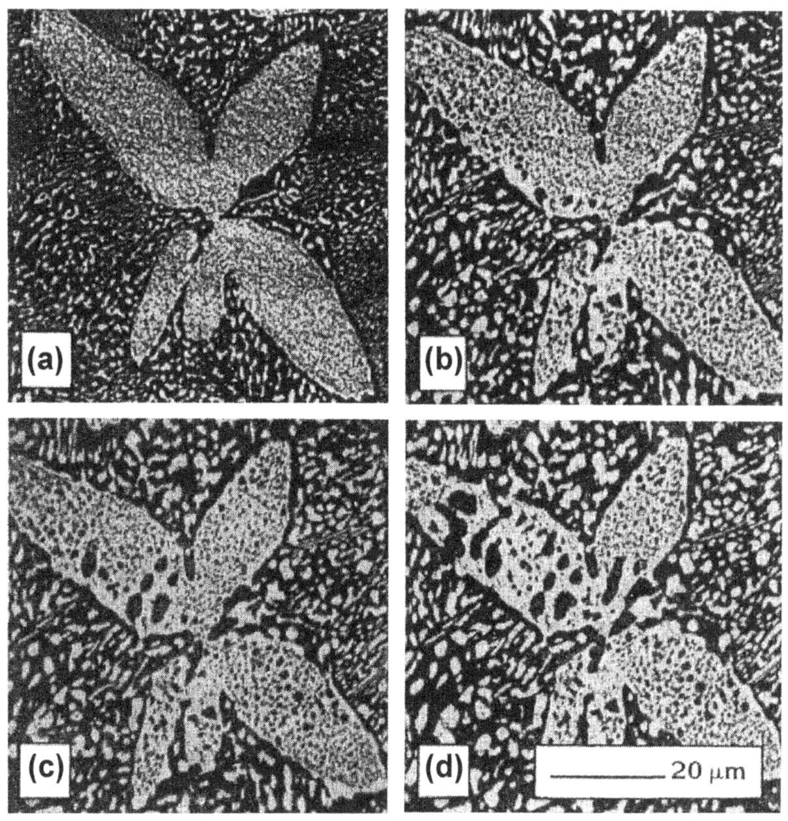


**Figure 8.13** Coarsening of Pb–Sn solidified solder structure as a function of room temperature aging. (a) 2 h; (b) 17 days; (c) 34 days; (d) 63 days. Lead and tin appear white and black, respectively. *From Ref. [61].*

precipitation and dissolution in concert. If the thermal cycle frequency is low, both the precipitation and dissolution processes may go to completion at low and high temperatures, respectively. However, if the thermal cycle frequency is high, neither precipitation nor dissolution will approach completion during a cycle; some steady-state structure, different from that at equilibrium, may then evolve. The influence of chemical and morphological change must always be borne in mind when assessing solder reliability, even at room temperature.

## 8.3.2 Solder Properties

The composition of a broad number of electrical solder alloys and their relevant properties are tabulated in Table 8.5. Binary eutectic, and

**Table 8.5** Properties of solder alloys (in order of increasing liquidus temperature)

| Alloy composition wt% | L °C | S °C | CTE $10^{-6}$ °C$^{-1}$ | Electrical resistivity μΩ-cm | Tensile stress ksi | 0.2% offset yield stress ksi | Elongation % | Creep resistance |
|---|---|---|---|---|---|---|---|---|
| 50Sn−50In | 117 | | | 14.7 | 1.72 | | | |
| 40Sn−60In | 122 | 113 | | | 1.10 | 0.67 | 5.5 | Low |
| 42Sn−58Bi | 138 | 138 | 14.9 | 34.5 | 9.71 | 6.03 | 1.3 | Mod |
| 43Sn−43Pb−14Bi | 163 | 144 | | | 5.60 | 3.60 | 2.5 | Low−Mod |
| 30In−70Sn | 175 | 117 | | | 4.67 | 2.54 | 2.6 | Low |
| 62Sn−36Pb−2Ag | 179 | 179 | | | 4.50 | 2.57 | 17.9 | High |
| 63Sn−37Pb | 183 | 183 | 24.7 | 15.0 | 5.13 | 2.34 | 1.38 | Mod |
| 60In−40Pb | 185 | 174 | | | 4.29 | 2.89 | 10.7 | Mod |
| 60Sn−40Pb | 190 | 183 | 23.9 | 15.0 | 4.29 | 2.89 | 10.7 | Mod |
| 80Sn−20Pb | 199 | 183 | | | 6.27 | 4.30 | 0.82 | Mod |
| 96.5Sn−3.5Ag | 221 | 221 | | 12.3 | 8.36 | 7.08 | 0.69 | High |
| 85Sn−10Pb−5Sb | 230 | 188 | | | 6.45 | 3.63 | 1.40 | Mod |
| 95Sn−5Ag | 240 | 221 | | 13.7 | 8.09 | 5.86 | 0.84 | High |
| 95Sn−5Sb | 240 | 235 | 27 | | 8.15 | 5.53 | 1.06 | High |
| 30In−70 Pb | 253 | 240 | | | 4.83 | 3.58 | 15.1 | Mod |
| 5Sn−85Pb−10Sb | 255 | 245 | | | 5.57 | 3.67 | 3.50 | High |
| 25Sn−75Pb | 266 | 183 | | | 3.35 | 2.06 | 8.4 | Low |
| 15Sn−82.5Pb−2.5Ag | 280 | 275 | | | 3.85 | 2.40 | 12.8 | Mod |
| 10Sn−88Pb−2Ag | 290 | 268 | | | 3.94 | 2.25 | 15.9 | Mod−high |
| 95Pb−5Sb | 295 | 252 | | | 3.72 | 2.45 | 13.7 | |
| 5Sn−93.5Pb−1.5Ag | 301 | 296 | | | 6.75 | 3.85 | 1.09 | Mod |
| 10Sn−90Pb | 302 | 268 | | | 3.53 | 2.02 | 18.3 | Mod |
| 1Sn−97.5Pb−1.5Ag | 309 | 309 | | | 5.58 | 4.34 | 1.15 | Mod |
| 5Sn−95Pb | 312 | 308 | | | 3.37 | 1.93 | 26.0 | Mod−high |
| 95 Pb−5In | 314 | 292 | | 33.8 | 3.66 | 2.01 | 33.0 | |

L = liquidus temperature, S = solidus temperature, CTE = coefficient of thermal expansion, creep resistance (low, moderate, high).
From Refs [28,54]

near-eutectic tin–lead alloys have traditionally dominated soft-solder usage in both electrical and plumbing applications. It is not surprising, therefore, that both of these elements have served as the basis for developing improved multicomponent alloys containing varying additions of indium, bismuth, silver, antimony, etc. Lead, however, is toxic, and in recent years environmental regulations have mandated reduced levels of Pb exposure and total elimination of this metal in various products. Although legislation has not yet curtailed the use of Pb in electronic solders, such action will probably occur in the not-too-distant future, with potentially significant implications for joining and packaging reliability [31].

Solders tend to be more electrically resistive than the base metals they interconnect. However, because the length-to-area ratio of solder joints is low, negligible resistance is added to the electrical circuit. Resistivity, of course, depends on temperature, and the data in Table 8.5 are for 20 °C. Interestingly, many solder alloys are superconducting at temperatures near 4 K. Thermal expansion coefficients for solder alloys are usually higher than those of the metals they interconnect. Thus, a solder droplet that solidifies and wets a metal surface is expected to be stressed in residual tension while the underlying metal is compressed.

Solders are mechanically weak materials with strengths typically an order of magnitude less than those for structural aluminum and copper alloys [32]. Importantly, solder alloys are susceptible to low-temperature creep deformation. Young's modulus for these materials ranges between 21 and 55 GPa ($3–8 \times 10^6$ psi) [33]. Tin is harder, stronger, and stiffer than lead; e.g., Young's moduli for Sn and Pb are 50 GPa and 14 GPa, respectively. If one thinks of Pb–Sn solders as composites whose overall properties are weighted averages of those of the two constituent phases, it is easy to see why both the tensile strength and Young's modulus increase directly with Sn content.

## 8.3.3 Metal–Solder Reactions

### *8.3.3.1 Wetting*

In considering the interaction of solder with a base metal surface, the question of wetting is the primary issue. Solder has a reasonably high surface tension ($\gamma_s$), so that it, like mercury, tends to ball up when molten. If a solder droplet is placed on a substrate, the mechanical equilibrium of forces indicated in Figure 8.14 prevails between the surface tensions exerted by the involved materials. The surface tensions between the solder and vapor ($\gamma_{sv}$), and the base metal and vapor ($\gamma_{bv}$), are essentially fixed. Thus, for

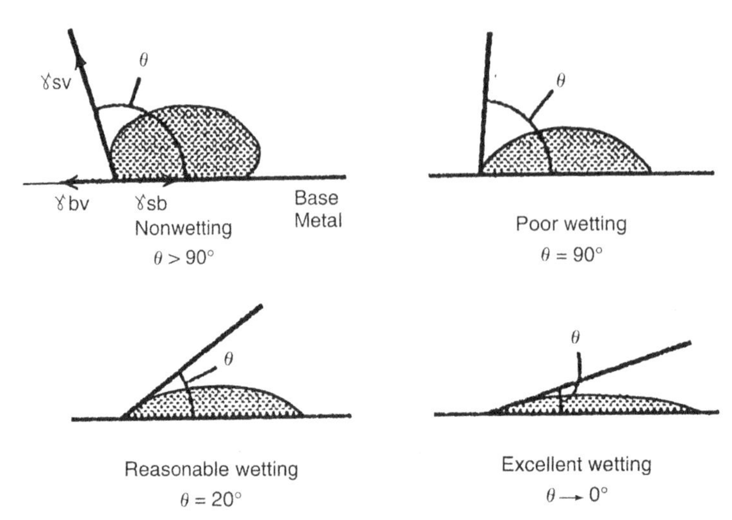


**Figure 8.14** Interfacial forces acting on a molten solder drop in mechanical equilibrium with a substrate, resulting in various wetting conditions.

solder to effectively spread on or wet the base metal, the interfacial tension ($\gamma_{sb}$) between the solder and the base metal must be reduced. Applying flux enables this to happen; the solder then flows over and metallurgically coats the surface at the desired low contact or wetting angle ($\theta$). Good wetting occurs for $\theta$ between 0 and 20°, while for angles greater than ~60° the wetting is poor.

In addition to ensuring wetting, there is the problem of avoiding dewetting, a condition where $\theta$ exceeds 90°. Visually, dewetting is most often manifested by the solder on the surface pulling back into irregular mounds, exposing base metal or intermetallic compounds in between as the contact angle increases. Dewetting has been shown [34] to be caused by gas evolution during molten solder contact with the base metal. The gas apparently stems from solder-flux decomposition, in particular, the thermal breakdown of contaminant organics and the release of water vapor from hydrated inorganics.

### *8.3.3.2 Solder Fluxes*

Fluxes are employed in virtually all soldering operations and play important chemical, thermal, and physical roles in ensuring the quality and reliability of the soldering process. The chemical function of fluxes is to react with surface tarnish oxide and sulfide films, so that reaction products can be displaced by the molten solder alloy. Not only do fluxes expose clean metal surfaces, but they also maintain a chemical blanket over them during soldering, preventing their reoxidation during the heating period. The thermal function of fluxes is to promote heat transfer from the molten solder to the joint area so that base metals reach temperatures high enough to be wetted by the solder. Finally, fluxes facilitate the physical need to reduce the interfacial surface tension between the solder and the base metal.

Traditional rosin fluxes contain only three types of ingredients, namely, rosin, an ammonium hydrohalide salt that serves as the activator, and lastly a solvent, often isopropanol. During soldering, the activator decomposes to release hydrogen chloride (or bromide), which reacts with the oxide present to form small quantities of metallic chlorides. Most of the chloride ions are dissipated in gaseous form, but some may remain undecomposed, especially on cooler surfaces that are not in direct contact with the solder. Minute residues of chlorides, hydrohalides, and metallic rosin salts cause potential reliability problems due to ionic leakage currents. For example, NaCl contamination levels of 1 $mg/cm^2$ have been shown to be sufficient to initiate corrosion on PWBs [35].

On an electronics assembly, especially under humid conditions, ionic conduction can have several deleterious effects. The four most serious are [36] as follows:

1. Malfunction of the circuit due to current flow between adjacent non-equipotential tracks,
2. Electrolytic corrosion due to the formation of reactive substances at the electrodes and their subsequent chemical reaction with the conductor metal (e.g., see Figure 7.4(b)),
3. Metal-dendrite formation due to the passage of current through an electrolyte containing metal salts, and
4. Electrical noise. An ionic current is never sensibly constant, as is a purely electronic current; due to their low mobility, ions often "clump" in the electrolyte.

Thus, it is important either to remove as much of the ionic contamination as possible by efficient and controlled cleaning, or to render it as harmless as possible.

#### *8.3.3.3 Dissolution and Compound Formation Reactions*

There are important compositional changes associated with molten solder wetting of base-metal surfaces. In the case of copper, for example, a small amount first dissolves upon contact with solder. Then, reaction between Cu and Sn atoms produces an intermetallic-compound layer of $Cu_6Sn_5$ at the interface. Although $Cu_6Sn_5$ is hard and brittle in nature, it essentially fosters the adhesion required for reliable soldered joints. However, the thick $Cu_6Sn_5$ layers that form at elevated temperatures are brittle and induce premature failure. Thus, thin layers of intermetallic compounds are beneficial, but thicker layers are detrimental.

In reflow soldering processes (Section 8.4.4), a limited amount of solder remains molten for almost a minute. This results in a high level of dissolved copper and contributes to a dull joint luster because $Cu_6Sn_5$ precipitates upon solder solidification. In contrast, wave soldered joints are generally shiny and bright because different operating temperature and time variables are involved. Although traditional wisdom attributes imperfection to dull solder surfaces, there is little evidence that they are less reliable than shiny joints. The kinetics of metal dissolution reactions in solder are generally observed to be linear such that metal thickness loss ($X_d$) upon submersion into the melt follows a time ($t$) dependence of the form

$$X_d = A \exp\left[-\frac{E_d}{RT}\right] t. \tag{8.4}$$

In this relation, $A$ is a constant, $E_d$ is the activation energy for dissolution, and $RT$ has the usual meaning. Values of $E_d$ for dissolution in molten 60Sn-40Pb are observed to scale with the melting point of the metal [37], as seen in Figure 8.15. In contrast, de Kluizenaar has essentially reported a single value of $E_d$ equal to $\sim 13.1$ kcal/mol, irrespective of base metal [38]; however, dissolution rates varied inversely with melting point.

After the joint has been made, potential reliability problems may stem from the solid-state interdiffusion between base metal and solder. In the case of Cu, a second intermetallic compound, $Cu_3Sn$, now forms between the $Cu_6Sn_5$ and Cu base. Both intermetallic phases then slowly thicken with time, even at room temperature, according to the parabolic kinetics described below in Eqn (8.5). Depletion of Sn by this mechanism is one of the reasons for solderability deterioration of stored Sn or Sn–Pb coated components. Solid-state diffusional growth of intermetallic compounds proceeds more slowly than dissolution and is described by parabolic kinetics with a dependence given by

$$X_c = B \exp\left[-\frac{E_c}{RT}\right] t^{1/2}. \tag{8.5}$$

Here, $X_c$ is the compound layer thickness, $B$ is a constant, and $E_c$ is the activation energy for compound growth.

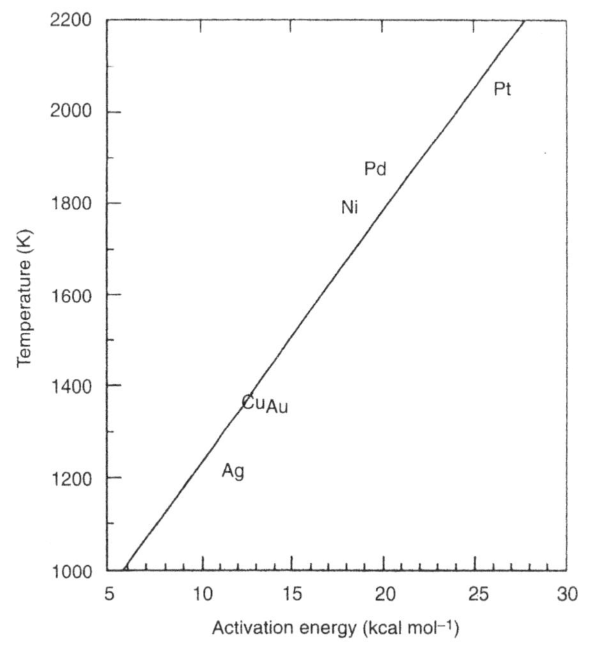


**Figure 8.15** Correlation between melting temperature and activation energy for dissolution of various metals in molten 60Sn–40 Pb solder. *From Ref. [37].*

A study [38] of both the dissolution kinetics of platinum in 60Sn-40 Pb solder and the subsequent solid-state growth of the $PtSn_4$ compound provides some notion of the kind of temperature–time relationships involved. For Pt, $E_d$ and $E_c$ were found to be 20.4 kcal/mol and 11.5 kcal/mol, respectively. Faceted crystals of $PtSn_4$ form as a result of dissolution during solder reflow (Figure 8.16). Similar kinetics apply to other base metallizations used in electrical applications such as gold with compounds $AuSn_4$, $AuSn_2$, and AuSn; silver with the compound $Ag_3Sn$; palladium with the compound $PdSn_4$; and so on. These intermetallic compounds can sometimes compromise solder-joint performance; for example, when gold-plated terminations are soldered, the $AuSn_4$ plates that form are particularly brittle, and cracks readily nucleate and propagate through them.

### *8.3.3.4 Compound Formation in Flip-Chip Structures*

In flip-chip technology, control of soldering is a more critical issue than in any other joining application. Metal dissolution and compound formation reactions are unleashed during reflow processes, leaving a residue of products frozen within the solidified solder ball. Details of these reactions

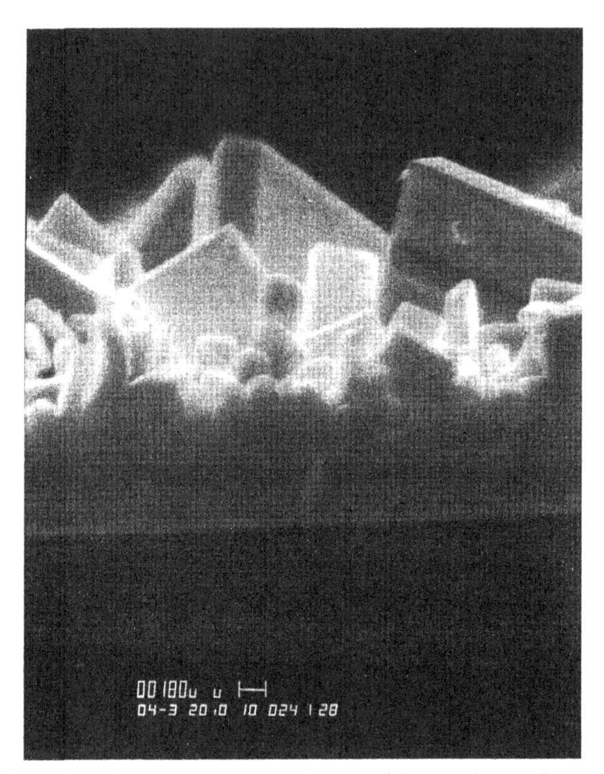

**Figure 8.16** Scanning electron microscope image of the products of reaction between Pt and Pb–Sn solder during reflow heating at 253 °C for 2500 s. The $PtSn_4$ crystals were revealed by etching away the surrounding solidified solder melt. *Courtesy of D. Schwarcz.*

depend intimately on the particular metals and solder. The thermal history is compressed within a few tens of seconds, but considerable compositional change can occur within that time. To estimate the expected extent of reaction, we note that dissolution is followed by diffusion in the melt. Diffusion lengths ($X$) are given by $2(Dt)^{1/2}$ (Eqns (5) and (6)), and the typical diffusivity ($D$) of elements in solder melts is $10^{-5}$ cm$^2$/s. For $t = 30$ s, $X = 49 \times 10^{-3}$ cm, or 490 μm. When added to dissolution and melt-convection effects, compositional change essentially spans the entire solder melt dimensions, with potentially alarming reliability implications.

These concerns are in fact substantiated in Figure 8.17, where reactions in both Pb–Sn and Pb–In solders are schematically depicted [39]. Since the reaction products lack ductility, cracks sometimes initiate in Pb–Sn solders at points that are closer to either the BLM or TSM interface, especially

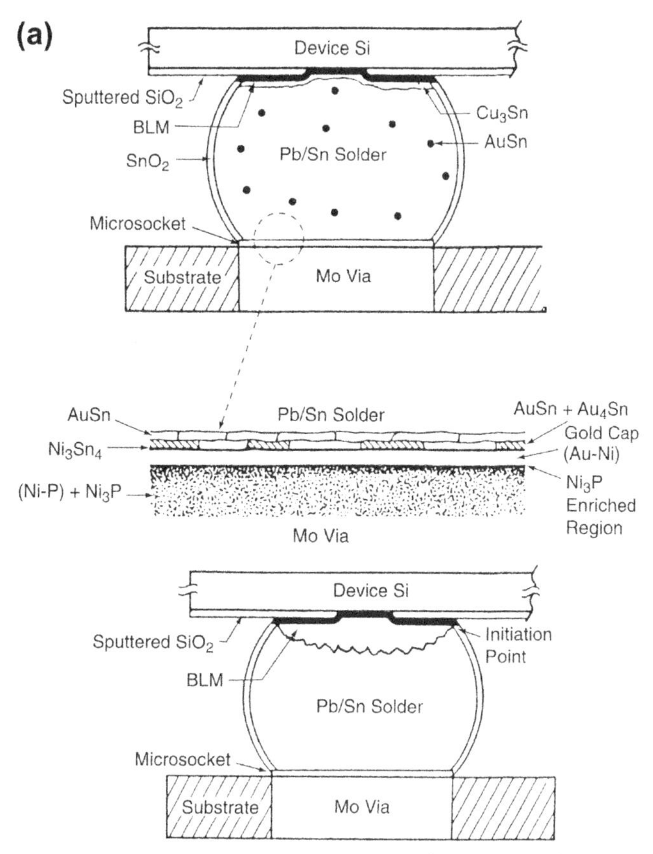


**Figure 8.17** Solder reactions with assorted metals in flip-chip bonding. (a) Reactions in Pb–Sn solder. (b) Reactions in Pb–In solder. *From Ref. [39].*

when thermally fatigued by repeated reflow cycling. Intermetallic-compound particles of AuSn are also distributed through the solder while a layer of $Cu_3Sn$ forms near the chip pad. In the case of Pb–In, there is a richer assortment of reactions and compounds. For example, $AuIn_2$, $Au_3In$, and $Cu_7In_4$ phases have been identified, and an $In_2O_3$ layer coats the solder. Since the substrate may contain electroless-plated $Ni_3P$ films, additional nickel–tin and nickel–indium compounds form. Irrespective of composition, intermetallic compounds are potential sites of crack nucleation during thermal cycling. Molten tin and indium are particularly reactive metals, but lead is relatively inert, and lead-rich compounds do not form. Because of this tendency, 95Pb–5Sn solders have been used in flip-chip bonding to prevent thin base-metal (Cr, Cu, and Au) films from dissolving in the

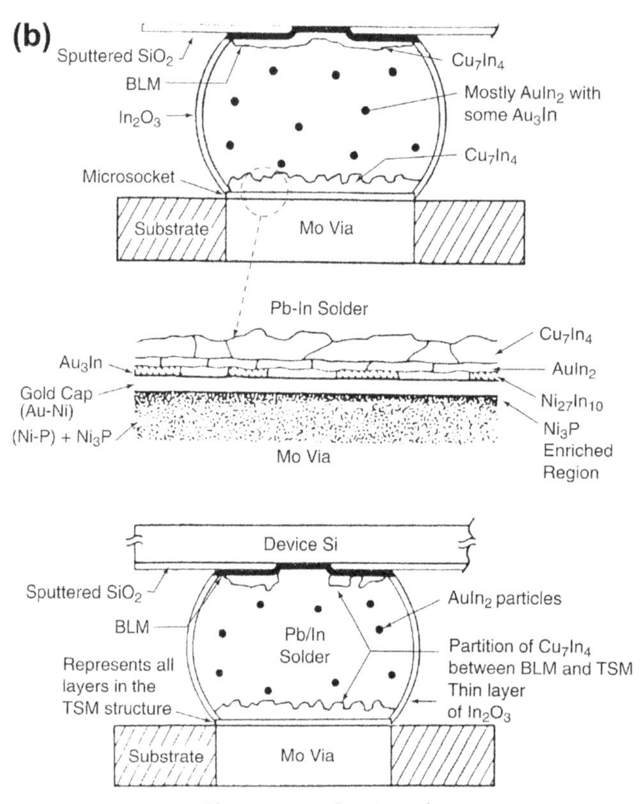


**Figure 8.17** Continued.

"massive" molten solder bumps. Lead's limited reactivity will be sorely missed if (when) it is banned for use in electronic solders.

## 8.4 SECOND-LEVEL PACKAGING TECHNOLOGIES

### 8.4.1 Introduction

After first-level packages are manufactured, they are connected to higher level assemblies, e.g., substrates, cards, boards. This step has an important influence on overall system performance, cost, and reliability [40]. There are two basic methods of interconnection. In the first, a nonpermanent, pluggable pin–socket arrangement is used to allow for easy disconnect or repair. However, for high performance and reliable, permanent interconnections, soldering is almost universally employed. We have already noted the importance of solders in flip-chip bonding of IC chips to substrates, and TAB of solder bumps to dies and substrates. Now the soldering operations that lead to the completed circuit boards we all recognize in electronic products are considered.

## 8.4.2 Through Hole and Surface Mounting

The difference between these two types of mounting packages or passive components on PCBs is shown in Figure 8.18. In 1995, approximately 90 billion IC packages were mounted by these two methods, with through-hole methods accounting for ~20% of this market. By the year 2000, the number of packages processed will double, but the through-hole mounting share is projected to shrink to perhaps only 5%.

In through-hole mounting (Figure 8.18(a)), the package or component leads are inserted through holes in the PCB. These holes are electroplated with copper on the cylindrical wall, and the package leads are soldered to effect both electrical and mechanical interconnection. For the past 30 years, the chief through-hole product processed has been the DIP shown in Figure 8.2. In pin-type packages, the primary alternative to the DIP is a pin grid array (PGA), where the entire package bottom is covered with pins. Significantly, PGA packages outperform DIPs and occupy a smaller area, but they are more expensive.

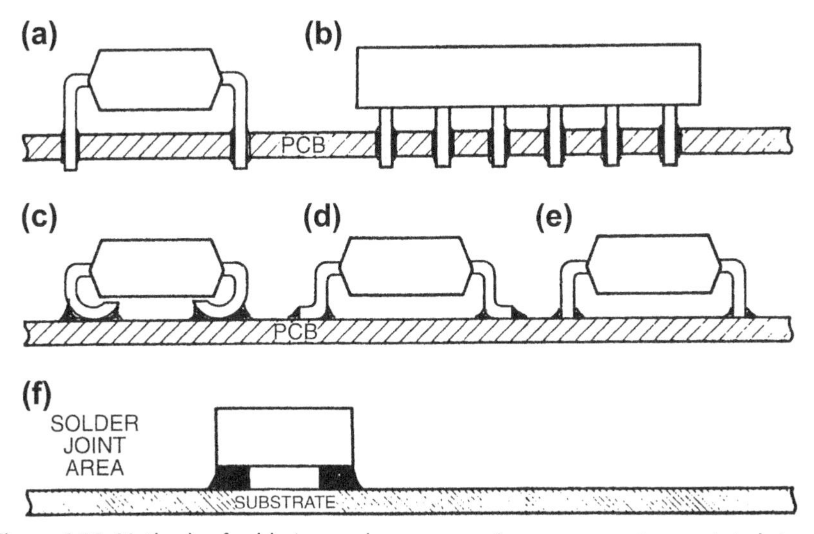


**Figure 8.18** Methods of soldering packages or passive components on printed-circuit boards. Mounting configurations and typical package examples. Through-hole mounting (a) (DIP). (b) PGA. Surface mounting (c) J (PLCC). (d) Gullwing (SOP, QFP). (e) Butt leaded (f) leadless (LCC).

Surface mount technology (SMT), introduced in the early 1980s, provides many attractive advantages relative to through-hole packaging, including

1. an approximately eight fold smaller metallized area for soldering,
2. capability of mounting on both sides of the board,
3. reduced package size, increased device density, and greater PCB utilization,
4. better electrical performance and improved repairability,
5. fewer and smaller PCB holes, and
6. increased line density.

These attributes have enabled SMT to become the dominant method for chip-carrier attachment to circuit boards. Two principal SMT packages that have evolved are leaded and leadless, and examples of each are shown in Figure 8.18(b). (Note that lead is pronounced here as "leed.") Leaded packages have compliant leads that extend from the package to the PCB where they are soldered. Two popular lead configurations are the "J" and "gullwing." The former are tucked under the package in a J shape. Compliant leads flex under applied loads, reducing stresses and strains in the solder joint. On the other hand, in leadless packages, solder forms a more rigid bridge between the metallized areas of the package and the PCB. The basic difference between leaded and leadless SMT is the elastic compliance of the lead. As we shall see in Section 9.5.2, this factor becomes significant when the response of leadless and leaded solder joints to thermal cycling is treated.

### 8.4.3 Ball Grid Arrays

Increasing transistor-gate counts have perpetually driven the development of smaller packages with ever more I/O connections. These I/O demands have driven peripherally leaded surface-mounted components to smaller lead-to-lead pitches. As a result of the added complexity, circuit board assembly yields and costs have been adversely affected. To address these problems, one of the newer packaging techniques, which will increasingly replace the leaded and leadless packages discussed above, makes use of BGAs. Except for the fact that BGAs are used in second-level packaging, they are basically similar to flip-chip solder connections. This can be seen in Figure 8.8, where the solder balls that connect the substrate (containing the previously attached flip chip) to the PCB are part of a BGA. In general, BGA solder balls are larger than chip solder bumps, but they continue to shrink in size in the unceasing quest for miniaturization. In contrast to leaded connections, the overall footprint of the BGA on the circuit board is only slightly larger than that of the IC chip itself. Basically, all of the reliability and failure experience with flip-chip (e.g., C-4) soldering is translatable to BGA technology.

Although BGA packages exhibit good electrical and thermal performance and enhanced package and assembly yields, they are, like all new technologies, not without problems. Some stem from the lack of infrastructure and acceptance. Because they have hidden solder connections, BGA packages cannot be easily inspected or reworked. Latent reliability defects may be problematical, and such plastic packages are also prone to popcorn cracking.

### 8.4.4 Reflow and Wave Soldering Processes

There are two basic soldering processes to connect chip packages and components to PCBs, and they differ depending on the methods of solder application and subsequent heating [41]. In reflow soldering, the solder is first electrolytically deposited, applied as a preform, or most often selectively screened on to the printed-circuit board in the form of a paste. After placement of the chip packages and related components, heat is usually applied by radiation and convection methods in order to dry, melt, and reflow the solder to effect the desired interconnections. Concurrent with these stages are the processing problems of thermal shock (during preheating); solder splatter and ball formation (during drying); and burnt boards, dewet solder, and cold solder-joints (during melting and reflow). Temperatures in the range of 160–230 °C are accessed during the ~1-min heating reflow and cooling portion of the cycle. Maintaining low temperatures is quite important when soldering heat-sensitive plastic packages. As noted earlier, carefully controlled reflow soldering is practiced during flip-chip bonding; surface-mounted assemblies are also reflow soldered.

In wave soldering, the PWB is passed over a jet of molten solder that exits from a nozzle and wets its underside. For through-hole PCBs, the steps include automatic insertion, flux application, preheating, and wave soldering. On the other hand, surface-mounted components require glue application to hold the packages and components in place while the PCB is turned upside-down for flux application and wave soldering. Good solder joints require raising the temperature of the base metal to allow wetting, capillary action, and intermetallic formation, and finally to drain excess solder off unwetted surfaces.

### 8.4.5 Defects in Solder Joints

A number of the common soldering defects found in through-hole and surface-mount technologies are itemized separately in Table 8.6. To better appreciate these, defects representative of both soldering processes are depicted in Figures 8.19 and 8.20 .

**Table 8.6** Defects in reflow and wave soldering processes

| Description | Possible causes | Counter measure |
|---|---|---|
| **Reflow soldering** | | |
| **a.** Solder bridging conductors | Paste formed single meniscus | Correct printing, dispensing or placement errors, smearing of paste. |
| **b.** Discolored PCB laminate | Overheating | Reflow profile wrong, use long-wavelength infrared, PCB not aligned. |
| **c.** Component placement, displacement | Incorrect positioning, component drifted due to wetting and surface tension | check tolerances, vibration, and wetting forces. |
| **d.** Grainy joint | Contamination, too much base metal in joint | Shorten solder time, cool board. |
| **e.** Open solder joint | No contact between paste and lead | check paste deposition process, solder sucked into via. |
| **f.** Solder balls | Flux carried solder away from joint | Higher metal content in paste, use higher viscosity paste, check flux activity. |
| **g.** Tombstone (lifting of a component) | Unequal surface tension wetting forces and masses, nonsymmetric metallization | Dispense paste equally, check heat sinking. |
| **h.** Wicking (solder draws up lead leaving little or none on pad) | lead wets before pad | use hot solder-dip leads, improve pad solderability. |

| Wave soldering | | |
|---|---|---|
| **a.** Bare metal | lead cut in solder-cut-solder operation | use stronger flux. |
| **b.** Blow hole | gas escaping from hole during solder solidification | check organic contamination from flux, PCB. |
| **c.** Cold solder joint | temperature of base metal too low | Increase preheat and solder temperature. |
| **d.** Dewetting | solder withdraws from base metal | check contamination, prevent $Cu_3Sn$ formation by reducing solder temperature and time. |
| **e.** Disturbed joint (rough solder surface) | Mechanical vibration during solidification | check vibration in conveyor pallets. |
| **f.** Excess solder | Large wetting angles, joint surface is not concave | bad drainage at wave exit, flux not controlling surface tension, check flux. |
| **g.** Hole not filled | No solder meniscus on top of board, no wetting of eyelet | Hole-lead ratio incorrect. |
| **h.** Icicles | Conical, icicle-like solder adhering to metal on PCB | Increase solder temperature, correct drainage of wave. |
| **i.** Insufficient solder | Excessive solder drainage | surface tension of solder too low, reduce solder or PCB temperature. |
| **j.** Nonwetting | Partial wetting with high contact angle | Improve solderability of PCB or component. |

From Ref. [29].

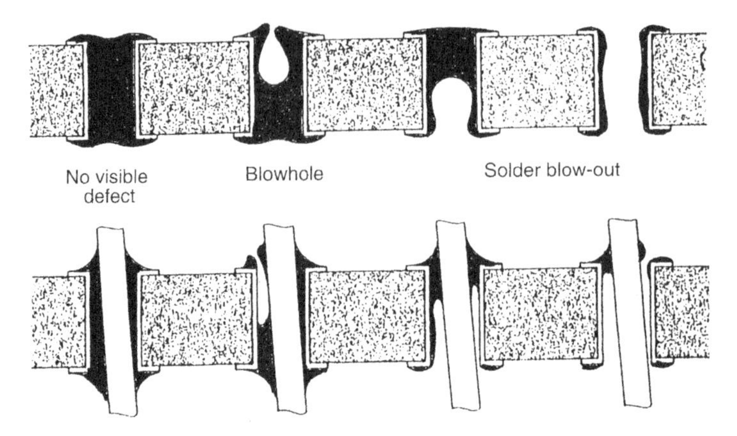


Figure 8.19 Schematic of through-hole soldering defects. *After Ref. [62].*

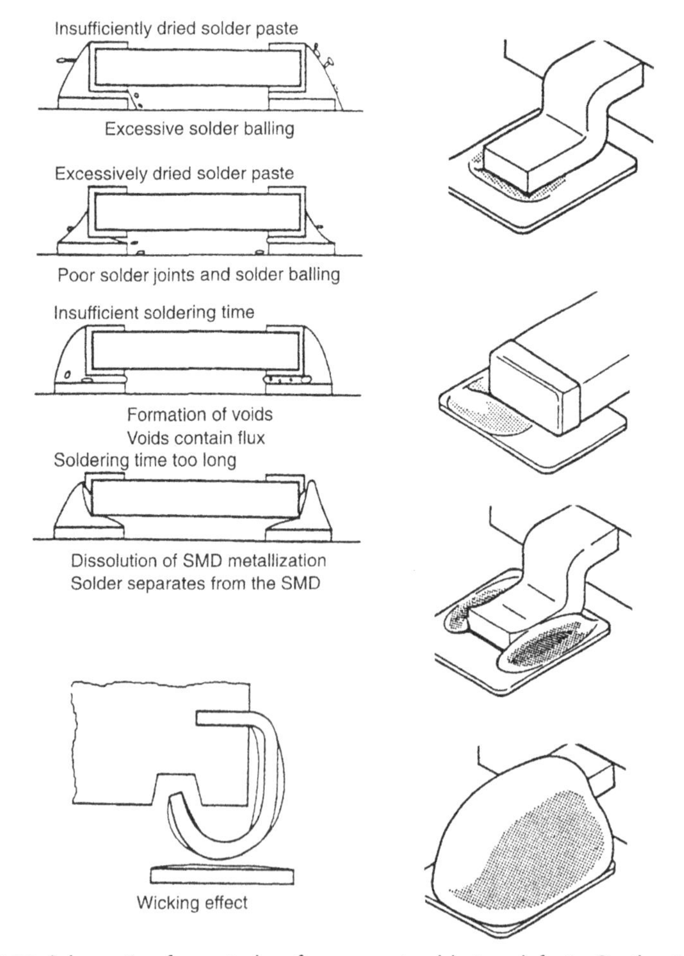


Figure 8.20 Schematic of assorted surface-mount soldering defects. On the right, the effects of too little or too much solder are evident. *From Ref. [63].*

# 8.5 THERMAL STRESSES IN PACKAGE STRUCTURES

## 8.5.1 Introduction

The source of many packaging processing defects and reliability problems is thermal stress. In our first encounter with internal stress we saw (Section 3.3.2.2) how thermal stress could develop in a single material when constrained. Thermal stresses can also arise in so-called thermostat or bimetallic strip structures, consisting of two different materials in intimate contact. Normally it is assumed that the thermostat is unstressed at some reference temperature. Stress then develops in the thermostat as it is either cooled or heated to the use temperature because the thermal expansion coefficients of the two materials differ. A film of $SiO_2$ on silicon is such a thermostat structure that was discussed in connection with oxidation (Section 3.3.2.4). As noted on earlier occasions, Eqn (8.1) (or Eqn (3.8)) yields the stress magnitude.

Stress can also develop in a bilayer structure by other than thermal means. Epitaxial mismatch and film-growth processes are examples of intrinsic sources of internal stress in film/substrate bilayers. The Stoney formula (Eqn (3.10)) describes the bowing of the bilayer in terms of the stress. Thermal stresses will also cause bilayers to bow. Thus, by combining Eqns (8.1) and (3.10), the predicted radius of curvature, $R$, of the combination is given by

$$R = \frac{(1 - \nu_f) E_s d_s^2}{6(1 - \nu_s) d_f E_f [\alpha_s - \alpha_f] (\Delta T)}, \tag{8.6}$$

where f and s refer to film and substrate, respectively, and $\Delta T$ is the temperature difference.

The packaging structures we are about to consider differ from thermostats in that they contain three, not two, contiguous layers. Even though packaging layers are thick compared to thin films, the same principles of mechanics apply when considering response to loading. Thermal stress problems are generally complex, requiring knowledge of elasticity theory and finite element analysis. Two of the leading practitioners of their application to microelectronic packaging are Suhir [42,43] and Lau [9]; their publications are highly recommended.

## 8.5.2 Chips Bonded to Substrates

### *8.5.2.1 Shear Stresses (Continuous Attachments)*

There are a number of relatively simple yet rather important thermal stress problems that involve layers of different components that are bonded by means of an intermediate adhesive. A specific example of such a sandwich structure consists of a Si die and substrate (the adherents) bonded by an epoxy adhesive in between. In other cases, solder acts as the adhesive.

Such structures can be modeled in terms of three layers, representing the die, denoted by [1]; substrate, denoted by [2]; and an adhesive of thickness $d_o$, as shown in Figure 8.21(a). Thermal stresses in such structures have been treated rigorously by Suhir [42]. A simplified approach by Johnson et al. [44] will be presented here, however. It employs linear stress analysis to calculate the thermal shear stress that develops in the adhesive when the temperature of the assembly is changed by an amount $\Delta T$. A considerable simplification occurs if a two-dimensional geometry is considered. It is further assumed that

1. the adhesive is linearly elastic and the shear stress does not vary through its thickness,
2. both top and bottom layers are uniformly stressed through the thickness,
3. bending of the structure is neglected, and
4. the temperature is uniform in all layers.

In the equations that follow, $G$ is the shear modulus of the adhesive, and other symbols have been defined earlier. Separating the assembly into

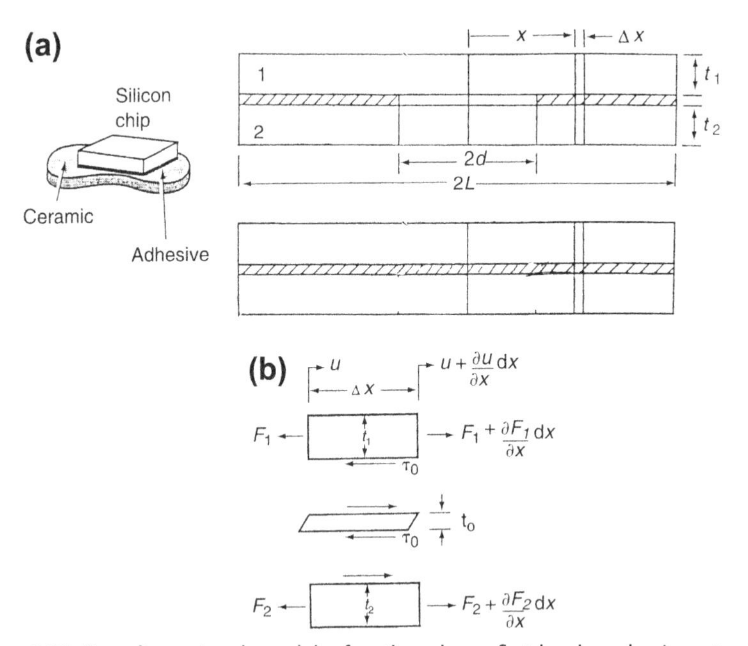


**Figure 8.21** Two-dimensional model of a three-layer first-level packaging structure subjected to thermal stresses. (a) Discontinuous attachment (above); continuous attachment (below). (b) Free body forces on the die, adhesive, and substrate. *From Ref. [44].*

individual free bodies, the equations for axial force ($F$, per unit width) equilibrium in the $x$ direction of the indicated elements are

$$\frac{dF(1)}{dx}dx - \tau\, dx = 0 \quad \text{and} \quad \frac{dF(2)}{dx}dx + \tau\, dx = 0 \tag{8.7}$$

The strain in the top and bottom layers arises both from internal forces and thermal expansion and may be expressed in terms of displacements ($u$) as

$$\frac{du(1)}{dx} = \frac{F(1)}{E(1)d(1)} + \alpha(1)\Delta T \quad \text{and} \quad \frac{du(1)}{dx} = \frac{F(2)}{E(2)d(2)} + \alpha(2)\Delta T. \tag{8.8}$$

Hooke's law gives the shear strain in the adhesive layer as

$$\frac{\tau}{G} = \frac{u(1) - u(2)}{d_o}. \tag{8.9}$$

After double differentiation of Eqn (8.9), the result is

$$\frac{d^2\tau}{Gdx^2} = \frac{1}{d_o}\left[\frac{d^2u(1)}{dx^2} - \frac{d^2u(2)}{dx^2}\right]. \tag{8.10}$$

The right-hand side is evaluated after differentiating Eqn (8.8) once and noting that $dF/dx = \tau$ (Eqn (8.7)). Finally, the basic equation

$$\frac{d^2\tau}{dx^2} = \beta^2\tau \tag{8.11}$$

emerges, where $\beta^2 = \frac{G}{d_o}\left[\frac{1}{E(1)d(1)} + \frac{1}{E(2)d(2)}\right]$ This ordinary differential equation has the solution

$$\tau = A \sinh \beta(x) + B \cosh \beta(x), \tag{8.12}$$

which applies in the region $L > x > 0$. In determining constants $A$ and $B$, the following physical conditions hold at $x = L$ and at $x = 0$:

1. At $x = 0$ the shear stress is expected to vanish; because of symmetry there is no displacement at the center, so $\tau$ must rise from zero (antisymmetrically) toward maxima at $x = L$ and $x = -L$. Thus,

$$\tau = 0 \ \text{at} \ x = 0. \tag{8.13}$$

This condition implies that $B = 0$.

2. At end $x = L$, the normal stresses and forces must vanish, i.e., $F(1) = 0$, $F(2) = 0$. Combining Eqns (8.8) and (8.9), the boundary condition is

$$\frac{d\tau}{dx} = \frac{G[\alpha(2) - \alpha(1)]\Delta T}{d_o} \quad \text{at} \quad x = L. \tag{8.14}$$

Therefore, $\frac{G[\alpha(2)-\alpha(1)]\Delta T}{d_o} = A\beta \cosh \beta L$, enabling $A$ to be determined. Finally, the general solution is given by

$$\tau(x) = \frac{G[\alpha(2) - \alpha(1)]\Delta T}{d_o\beta} \frac{\sinh \beta x}{\cosh \beta L}, \tag{8.15}$$

and its distribution is plotted in Figure 8.22. Normally, for reliability applications we are only interested in the maximum shear stress, which occurs at $x = L$. Therefore,

$$\tau(\max) = \frac{G}{d_o\beta}[\alpha(2) - \alpha(1)]\Delta T \tanh \beta L. \tag{8.16}$$

It should be noted that the shear stress at the interface depends directly on the thermal mismatch strain, the modulus of the adhesive, and the length of the assembly joint. But surprisingly, the shear stress is independent of the coefficient of thermal expansion of the adhesive material. Based on this solution one can imagine a number of different approximations based on the magnitudes of the material constants and geometry involved.

1. When the adherents are much more rigid or stiff than the adhesive, $1 \gg \beta L$, and $\tanh \beta L \approx \beta L$. Therefore,

$$\tau_{\max} = \frac{L}{d_o} G[\alpha(2) - \alpha(1)]\Delta T. \tag{8.17}$$

This formula is essentially the equation for thermal stress (Eqn (8.1)) magnified by the ratio of the joint length to thickness. It applies to the common situation of a silicon die ($E = 62 \times 10^6$ psi) attached to a ceramic substrate ($E = 50 \times 10^6$ psi) via an adhesive, where $G$ may be as low as 1000 psi.

2. If the chip is thin and flexible compared to the substrate, then $\beta = \{G/[d_o\, E(1)\, d(1)]\}^{1/2}$ and substitution in Eqn (8.15) yields

$$\tau(\max) = \left\{\frac{d(1)E(1)G}{d_o}\right\}^{1/2} [\alpha(2) - \alpha(1)]\Delta T \tanh \beta L. \tag{8.18}$$

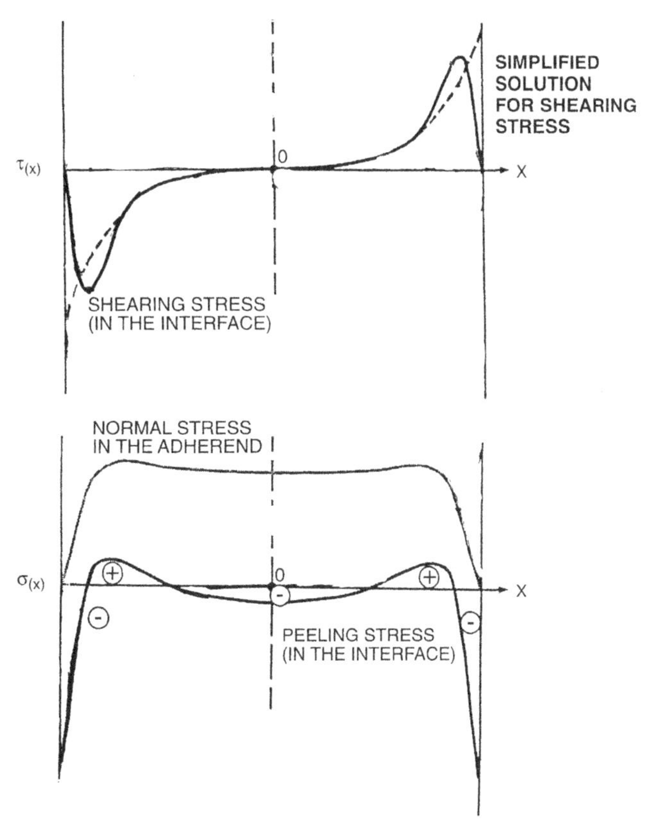


**Figure 8.22** Above: General solution for $\tau(x)$ (Eqn (8.14)) plotted as dotted line. The solid line represents the exact solution. Below: Plots of normal and peeling stresses. *Courtesy of E. Suhir, Lucent Technologies, Bell Laboratories Innovations.*

3. If the assembly is large, $\beta L \gg 1$, tanh $\beta L$ is approximately equal to unity. In this case,

$$\tau(\max) = \frac{G}{d_o\beta}[\alpha(2) - \alpha(1)]\Delta T. \tag{8.19}$$

The importance of the shear strength of the adhesive is evident in these cases. Die bond adhesives are either soft or hard, and their shear strengths should be compared to the values of $\tau(\max)$ in these formulas. Soft adhesives include the Pb–Sn-based solders and the organics consisting mostly of epoxies and polyimides that are filled with conductive materials. Thermal mismatch stresses between the die and substrate are borne primarily by mechanically weak bonds. This makes them susceptible to creep, fatigue fracture, or debonding during thermal stressing while transmitting little stress to damage the die.

The situation is reversed in the hard adhesives. These materials primarily consist of gold-based eutectic alloys, e.g., Au–Si, Au–Ge, and Au–Sn. Because they are mechanically strong, these hard adhesives transmit thermal mismatch stresses to the die, with the danger of cracking it. Materials with intermediate mechanical properties have been proposed to strike a balance between these extreme failure risks. In the next chapter, we shall consider the reliability issues associated with the use of adhesives.

#### *8.5.2.2 Shear Stresses (Discontinuous Attachment)*

Rather than a continuous attachment over the entire bond length, it is often the case that segments of adhesive bond a chip to the substrate, as shown in Figure 8.21. This problem is physically more complex, because the boundary conditions at $x = d$ are not obvious. Nevertheless, the general solution for the shear stress in the adhesive is

$$\tau = \frac{G[\alpha(2) - \alpha(1)]\Delta T}{d_o \beta} \frac{[\sinh \beta(x - d) + \beta d \cosh \beta(x - d)]}{[\cosh \beta(L - d) + \beta d \sinh \beta(L - d)]}, \qquad (8.20)$$

where $d$ is the gap in the adhesive. Note that when the gap vanishes, i.e., $d = 0$, this solution immediately reduces to Eqn (8.15). Finally, the shear stress dependence on adhesive length varies in the manner shown in Figure 8.23. Also plotted in this figure is the variation of the normal stress, a quantity addressed next.

#### *8.5.2.3 Normal and Peeling Stresses*

In addition to interfacial shear stresses in the adhesive, the adherents are subject to normal and peeling stresses. The normal, or tensile stress acts along the die axis and arises from $F(1)$; it can be simply obtained by integrating $\tau(x)$ in Eqn (8.7). As shown in Figure 8.22, these stresses are large in the center and vanish at the outer edges of the chip. In the case of a film attached to a substrate, $d(2) \gg d(1)$, a simple formula for the maximum tensile stress in the film is given by Suhir [42]:

$$\sigma(\max) = -E(1)[\alpha(2) - \alpha(1)]\Delta T \left[1 - \frac{1}{\cosh \beta L}\right], \qquad (8.21)$$

while stresses in the substrate are only $d(1)/d(2)$ times as large.

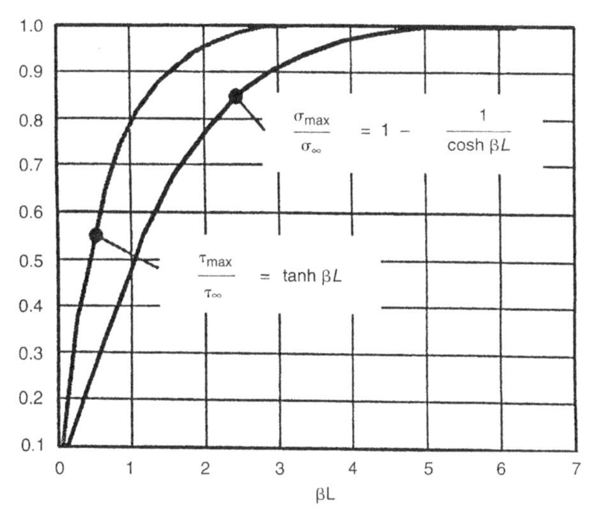


**Figure 8.23** Shear stress distribution as a function of joint length in adhesively bonded joints. Note: $\sigma_\infty = E(1)[\alpha(2) - \alpha(1)]\Delta T$, $\tau_\infty = (1/(d_o\beta))G[\alpha(2) - \alpha(1)]\Delta T$, and $2L =$ joint length. *Courtesy of E. Suhir, Lucent Technologies, Bell Laboratories Innovations.*

The transverse normal, or peeling, stresses arise from the constraint that forces the assembled components to bend jointly and have the same curvature at all sections despite differences in their flexural rigidities (i.e., $E\,d^3/12$). An explicit simplified formula for the peeling stress ($\sigma_p$) distribution is given by

$$\sigma_p(x) = -\frac{\mu G[\alpha(2) - \alpha(1)]\Delta T}{d_o}\frac{\cosh \beta x}{\cosh \beta L}. \tag{8.22}$$

In this formula,

$$\mu = \frac{\dfrac{d(2)E(1)d(1)^3}{12[1-\nu(1)^2]} - \dfrac{d(1)\ E(2)\ d(2)^3}{12[1-\nu(2)^2]}}{\dfrac{E(1)d(1)^3}{6[1-\nu(1)^2]} + \dfrac{E(2)d(2)^3}{6[1-\nu(2)^2]}}$$

reflecting the difference in adherent thicknesses and flexural rigidities. Peeling stresses are small everywhere except at the extreme edges of the attachment (Figure 8.22).

As an illustrative example, the three relevant stresses were calculated for the case of a GaAs MMIC power amplifier device [45], and the results are plotted in Figure 8.24. Using a value of $\Delta T$ equal to 345 °C, the maximum

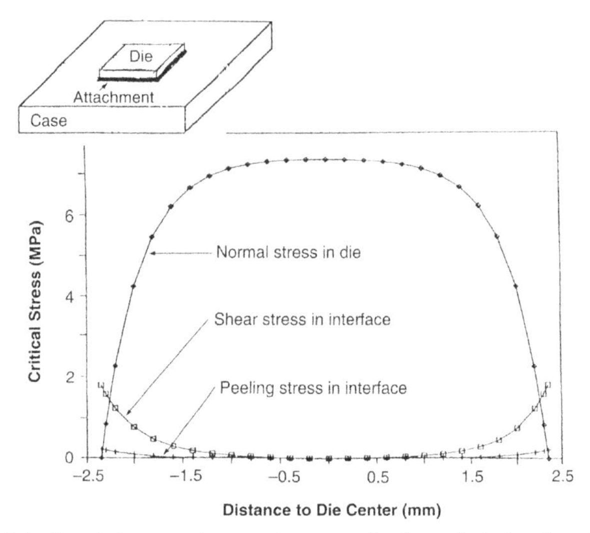


**Figure 8.24** Calculated shear and normal stresses in the substrate of an MMIC power amplifier device composed of GaAs substrate with dimensions $4.7 \times 2.7 \times 0.1\ mm^3$, Au–Ge eutectic attachment with dimensions $4.7 \times 2.7 \times 0.0254\ mm^3$, and CM 15 (15% Cu–85% Mo) case with dimensions $14.4 \times 12.7 \times 2.8\ mm^3$. *From Ref. [45].*

tensile stress in the GaAs die is found to be 7.36 MPa, the maximum shear stress in the Au–Ge adhesive is 1.8 MPa, and the peeling stress at the edge of the substrate (CM 15 or 15%Cu–85%Mo) is only 0.22 MPa.

## 8.5.3 Thermal Stress in Other Structures

With slight modification, the basic analysis of thermal stress has been applied to other than planar semiconductor/adhesive/substrate assemblies. Stress in various wire bond configurations and attachments used in ceramic packages are among the examples briefly described next.

### 8.5.3.1 Wire Bonds

In the case of wire bonds, the Al bonding pad on a chip was assumed by Hu et al. [46] to act something like an adhesive between the wire above and the semiconductor substrate beneath. The equation derived for the shear stress in the bond pad is given by

$$\tau(x) = \frac{G_p \Delta T}{b_p \gamma} \left\{ [\alpha(w) - \alpha(s)] - \frac{[\alpha(s) - \alpha(p)]}{1 + \frac{E(s)A(s)}{E(p)A(p)}} \right\} \frac{\sinh \gamma x}{\cosh \gamma L(w)}. \tag{8.23}$$

The symbols $w$, $p$, and $s$ refer to wire, pad, and substrate quantities, respectively, and $SS$ is defined by $\gamma^2 = (G_p/b_p)\ \{r/[E(w)\ A(w)] + w(p)/[E(s)\ A(s)]\}$; in addition, $r$ is the wire radius, and $A$, $L$, and $w$ are the indicated cross-sectional area, length, and width dimensions, respectively. A more complex expression than Eqn (8.15) results because the geometry is more involved. Nevertheless, the spatial distribution of the stress in this solution is essentially the same.

### *8.5.3.2 Sealing Ceramic Packages*

Low-melting solder glass is used to seal metal-to-ceramic and ceramic-to-ceramic joints in high-performance ceramic packages. There are two critical areas that must be sealed for the package to be hermetic. The first is the metal lead frame, which is sealed on both sides to the ceramic base with a layer of solder glass. Similarly, at the top of the package, a metal–metal welded or soldered seam joins the ceramic lid to the base. Glass is inherently brittle, and therefore the thermal expansion mismatch between the glass and lead frame as well as between glass and ceramic is of concern. However, since the lead frame is thin, its contribution to the stress of the glass layer is small, and thus only the ceramic–glass–ceramic structure is of interest.

A formula derived [47] for the maximum normal stress in a very wide glass layer is

$$\sigma_{max} = \frac{E(g)[\alpha(c) - \alpha(g)](T - T_o)}{\left[1 + \frac{d(g)E(g)}{d(c)E(c)}\right]} \tag{8.24}$$

where $g$ and $c$ refer to the glass and ceramic, $d$ represents the total thickness of the involved glass or ceramic parts, and all moduli are reduced by the appropriate $(1 - \nu)$ factor; i.e., $E(i) = E/[1 - \nu(i)]$. Frequently, $d(c) > d(g)$, and

$$\sigma = E(g)[\alpha(c) - \alpha(g)](T - T_o) \tag{8.25}$$

is the simple and conservative solution (Example 8.1).

## Example 8.1

Consider a ceramic package where the solder glass used has the following properties; $E(g) = 0.7 \times 10^6$ kg/cm$^2$, $\nu(g) = 0.27$, $\alpha(g) = 6.8 \times 10^{-6}$ °C$^{-1}$. Similarly, for the alumina ceramic, $E(c) = 3.5 \times 10^6$ kg/cm$^2$, $\nu(c) = 0.25$, $\alpha(c) = 7.2 \times 10^{-6}$ °C$^{-1}$. If the seal processing temperature is 450 °C, what stress can be expected in the glass at 25 °C?

**Answer** It is assumed that all stresses are relieved at 450 °C. Therefore, by Eqn (8.25) the thermal stress produced in the glass at 25 °C is $\sigma = 0.7 \times 10^6/(1 - 0.27)[7.2 \times 10^{-6} - 6.8 \times 10^{-6}](450 - 25) = 163$ kg/cm$^2$. This value should be compared to the 5500 kg/cm$^2$ compressive (fracture) strength of the glass. The undesirable tensile stress that develops underscores the need for high-toughness solder glasses.

#### *8.5.3.3 Thermal Stresses in Solder Joint Structures—Case Histories*

As final examples of thermal stresses in packaging let us consider a couple of case histories. In the first, strains were experimentally measured in a solder BGA employing Moiré interferometry techniques [48]. The overall structure shown in Figure 8.25(a) consisted of a heat sink bonded to a ceramic package, which was in turn soldered to a PCB via the BGA. Upon cooling this structure from 60 °C to room temperature, all of the associated components contracted. As a result, the fringe pattern reproduced in Figure 8.25(b) and (c) was obtained, corresponding to relative material displacements (strains) in the horizontal ($x$) and vertical ($y$) directions. Since the deformation magnitude and direction are essentially proportional to the fringe density and orientation, it can be seen that the ceramic module suffers little thermal strain due to its low coefficient of thermal expansion (CTE) ($6.5 \times 10^{-6}$ °C$^{-1}$) and high elastic modulus. Interestingly, the circuit board (bottom layer of fringes) shows greater deformation in the $y$ than in the $x$ direction, largely because of the anisotropy in its thermal expansion. The shear deformation of the solder balls (third layer of fringes) is our primary concern, however, and it clearly increases from the module center to either outer edge. Furthermore, the tensile deformation ($y$ direction) of the solder balls shows a complex behavior that also increases with $x$. Although we have a segmented, discontinuous attachment, calculated shear and tensile strains show similar variations to those displayed in Figures 8.22 and 8.24 for continuous attachments.

Finally, we consider the case of a corner lead and solder joint in a 256-pin fine-pitch, plastic quad flat-pack surface-mount assembly [48]. The

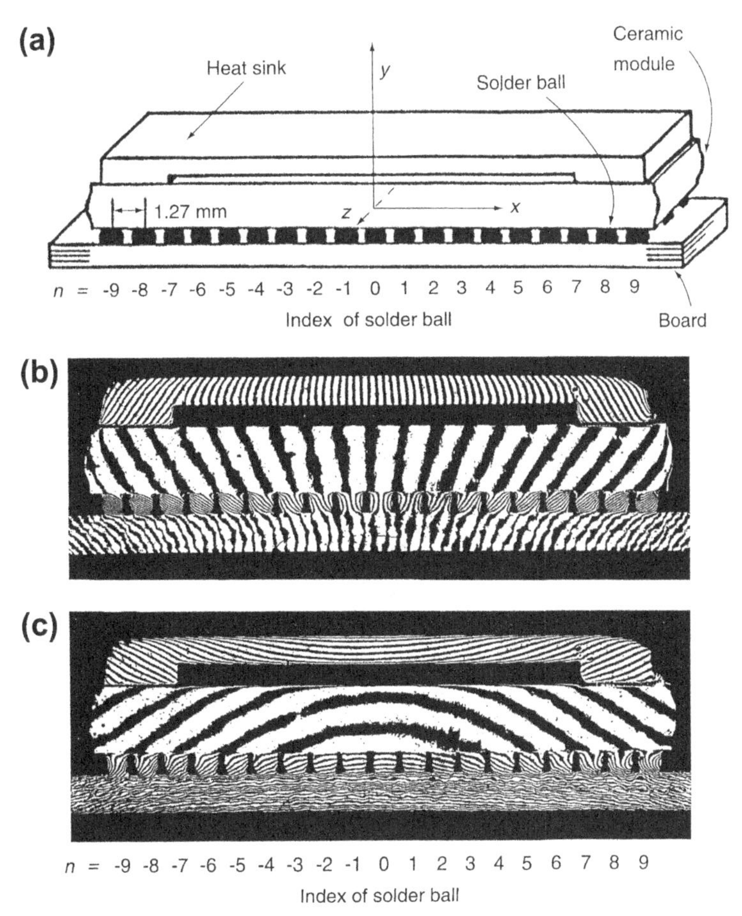


**Figure 8.25** Cross-section of a solder ball grid array with 19 joints in a row (a). Moiré fringe pattern of the module cross-section showing the *x* displacement field (b) and *y* displacement field (c). *From Ref. [48].*

geometry of this structure, shown in Figure 8.26, is far too complex for thermal stresses to be calculated by the exact methods we have employed for layered thermostat-like structures. Instead, three-dimensional finite-element methods of calculation were used. First, both lead (Figure 8.26(a)) and solder (Figure 8.26(b)) geometries are subdivided into a network of numerous prismlike elements. Then the concepts of elasticity theory are applied to each element, and the solutions are stitched together to yield the stresses and strains throughout the structure.

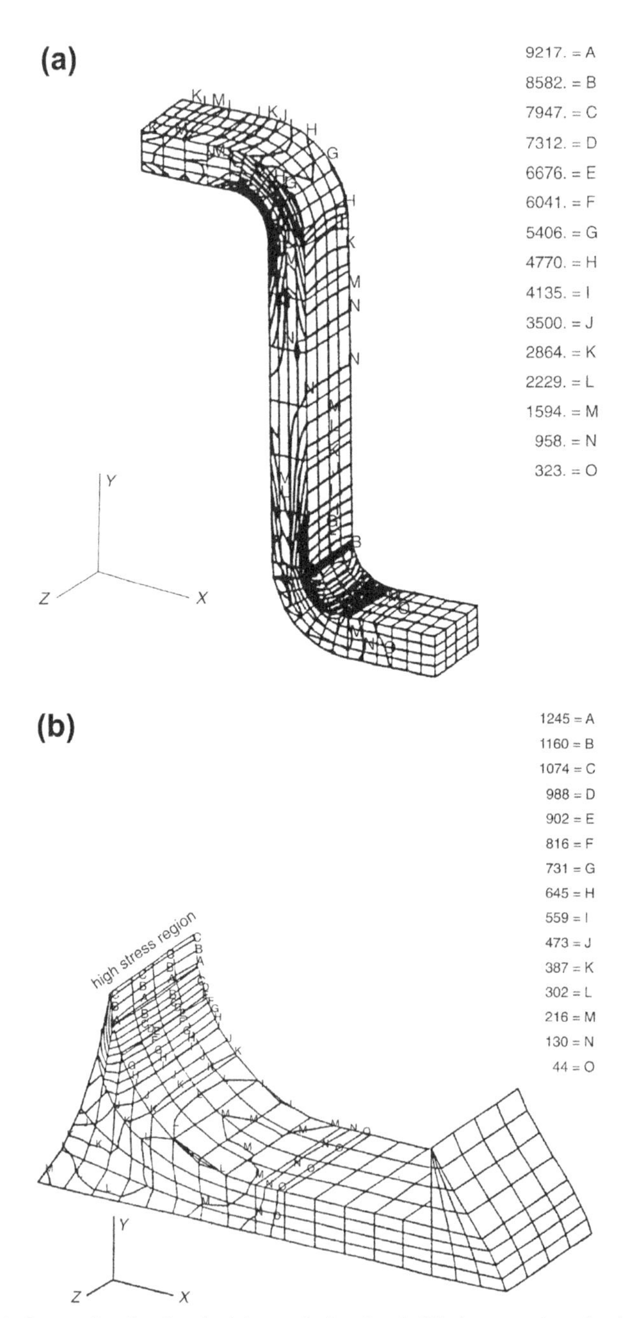


**Figure 8.26** Stress distribution in (a) a gullwing lead, (b) the associated solder joint of a plastic quad flat pack surface-mount assembly. Units of stress are in pounds per square inch for a unit displacement of 0.0001 inch in one direction. An example of thermal distortion in the joint is shown in (c). *From Ref. [64].*

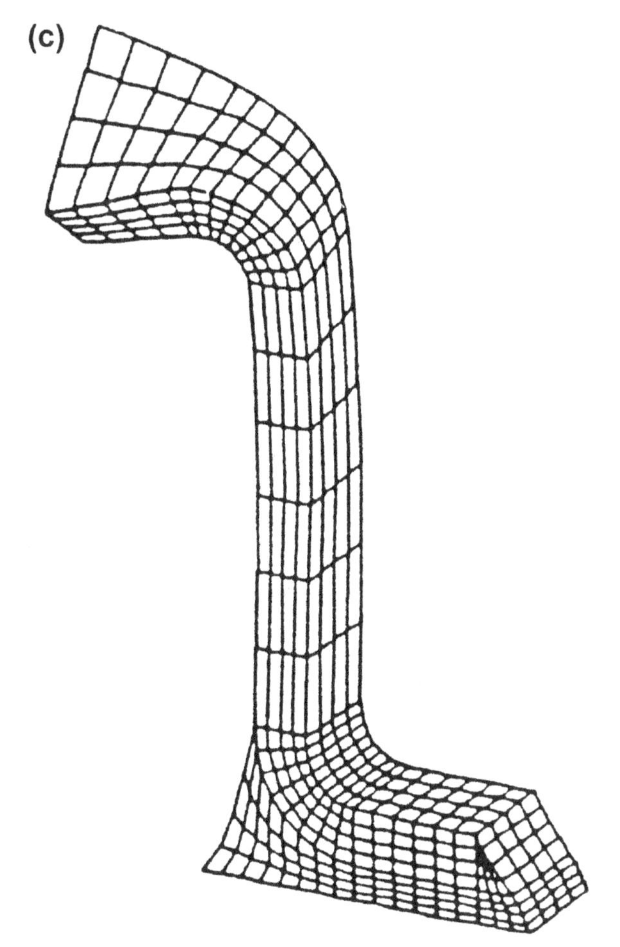


**Figure 8.26** Continued.

Of interest is the response to an 85 °C temperature difference impressed on the assembly. Not only does the lead deflect noticeably, but the solder is also stressed and deformed. More severe deformations involving twisting of the leads can also occur due to stressing modes that cause rotational displacements, as shown in Figure 8.26(c). In view of this, one can very well imagine that the reliability of the solder might be problematical during repeated thermal cycles. This is one of the subjects treated in the next chapter.

## EXERCISES

**8.1.** A 5-chip multichip module consists of one microprocessor, two ASIC (application-specific IC) and two memory chips with respective yields of 92%, 83%, and 97%. What is the expected yield of the MCM?

**8.2.** Both large and small dielectric constant materials are required for various electrical functions discussed in this book. Give an example of each.

**8.3.** Alloy 42 and Kovar were developed to provide metallic electrical feedthroughs in hermetic ceramic packages.

**a.** Why have these materials proven to be reliable in this application?

**b.** Due to their success in hermetic packaging, these metals have found their way into plastic packages for lead-frame-die-attach applications. Here they have been less successful. Why?

**8.4.** **a.** Explain the differences between fracture of IC chips (Figure 8.5) and die-attach materials (Figure 8.6). In particular, contrast brittle and ductile fracture phenomena.

**b.** Analyze the wire bonding failures shown in Figure 8.7 in terms of the applied tensile and shear stresses, and the mechanical properties of the materials involved in the attachment.

**8.5.** What are the packaging advantages of flip-bonding IC chips to a silicon rather than alumina substrate?

**8.6.** Assume that swelling of a package of thickness $2L$ due to external moisture in-diffusion obeys the following one-dimensional boundary value problem:

$$\frac{\partial C(x,t)}{\partial t} = D\frac{\partial^2 (x,t)}{\partial x^2}$$

$$C = C_o \text{ at } x = \pm L, t > 0;$$

$$\frac{\partial C}{\partial x} = 0 \text{ at } x = 0, t > 0$$

$$C = 0 \text{ at } x > 0, t = 0.$$

Here $C(x, t)$ is the moisture concentration in the package at position $x$ and time $t$, $C_o = 5$ mg/cm$^3$, $D$ is the diffusion coefficient of water $= 5.5 \times 10^{-9}$ cm$^2$/s, and the chip-lead frame is located at $x = 0$.

**a.** Derive an expression for moisture concentration as a function of $x$ and $t$.

**b.** Write an expression for the time-dependent moisture weight gain of the package. If you cannot derive the mathematical solutions, sketch the approximate moisture profile and package weight gain as a function of time. Hint: See Section 7.2.2.2.

**8.7.** Sketch a map that schematically outlines safe and unsafe processing regimes with respect to popcorn cracking. Use water concentration and die pad dimensions as the ordinate and abscissa.

**8.8.** Many a phase diagram has been consulted in the intense search to uncover environmentally safe lead-free solders, and the path appears to be converging on tin and bismuth-based alloy compositions. After perusing common references such as the *Metals Handbook* (ASM) and collections of binary phase diagrams, comment on the reliability prospects for Sn-rich and Bi-rich solders. Electrical conductivity, thermal expansion, mechanical strength, wettability, and corrosion resistance are some of the issues of concern.

**8.9.** **a.** Within the family of tin–lead solders predict whether Sn-rich or Pb-rich alloys are more prone to be metastable after solidification, and thus undergo phase-precipitation reactions during aging.

**b.** Precipitation reactions in the solid state tend to follow a kinetics that obey the Avrami formula, $f = 1 - \exp[-K(T)]t^n$, where $f$ is the fractional amount precipitated, $K(T)$ is a constant dependent on temperature, $t$ is the time, and $n$ is a constant. Sketch the progress of the solder aging reactions.

**8.10.** In mechanical equilibrium a molten droplet of solder on a substrate assumes a spherical cap shape such that $\gamma_{sv} = \gamma_{ls} + \gamma_{lv}\cos\theta$. The wetting angle is $\theta$, $\gamma_{ij}$ represents the particular interfacial surface energy (tension), and $l$, $s$, and $v$ refer to the liquid, substrate, and vapor, respectively.

**a.** Prove this relationship among the involved surface tensions.

**b.** What tensions are likely to be affected and how by (1) fluxing, (2) compound formation, (3) high soldering temperatures, (4) solder oxidation? How do these processes affect the wetting angle?

**8.11.** An indium-rich solder is in contact with the gold film of a TSM microsocket. Sketch a concentration profile across the In–Au diffusion couple and indicate where you would expect the $AuIn_2$ and $Au_3In$ compounds to appear.

**8.12.** Suggest a reason why cracking in a Pb–Sn solder occurs closer to the BLM interface after exposure to low temperatures and few reflows, and near the TSM surface at higher reflow temperatures.

**8.13.** Analysis of intermetallic-compound formation in an In–Sn solder on copper revealed that the growth rate at temperatures above 90 °C was governed by an activation energy of 5.3 kcal/mol. At this temperature the growth rate was $1 \times 10^{-9}$ cm/s, while below 90 °C an activation energy of 18.6 kcal/mol controlled compound growth. (From Ref. [49].)

**a.** Sketch the compound growth rate on an Arrhenius plot spanning temperatures from 25 to 120 °C. What compound thickness can be expected after a year at a temperature of 30 °C?

**b.** Is it common for high-temperature reactions to occur at lower activation energies than low-temperature reactions?

**c.** What are the reliability implications of such behavior at low temperatures?

**d.** Explain possible mechanisms for the two regimes of compound growth.

**8.14.** Two different sets of metal dissolution data in 40 Pb–60Sn solder have been reported in the literature. Meagher et al. [38] claim that the activation energy increases directly with the melting point of the dissolving metal, while de Kluizenaar [39] suggests a constant activation energy of 13.1 kcal/mol independent of metal.

Try to explain either set of conclusions based on the physics of the dissolution process, potential complications due to compound formation, diffusion in the melt, and the nature of the metals involved.

**8.15.** Prove that the velocity of compound growth during soldering is infinite initially, whereas dissolution of the metal in solder proceeds at a finite velocity initially. Similarly, demonstrate that precipitation reactions in solder during aging start off slowly, reach a maximum rate, and then slow to imperceptible rates at long times.

**8.16.** **a.** Aluminum is bonded to alumina using a rigid epoxy with a $\text{CTE} = 45 \times 10^{-6}\ ^\circ\text{C}^{-1}$, and $E = 10 \times 10^6$ psi. If the joint is cured at 150 °C, calculate the stress in the $Al_2O_3$ at 25 °C.

**b.** Suppose the adhesive had a $\text{CTE} = 45 \times 10^{-6}\ ^\circ\text{C}^{-1}$ and $E = 0.02 \times 10^6$ psi. For the same curing temperature, what is the stress in the $Al_2O_3$?

**8.17.** A 0.5-μm-thick $YBa_2Cu_3O_7$ ceramic film is deposited on a 0.5-mm-thick silicon wafer maintained at 600 °C. To capitalize on its superconducting properties the film must be cooled to 90 K. What sign and radius of wafer bow can be expected under these conditions? For $YBa_2Cu_3O_7$, $E = 96$ GPa, CTE $= 11.8 \times 10^{-6}\ °C^{-1}$, and $\nu = 0.25$.

**8.18.** **a.** Calculate the maximum shear stress in a Au–Si (eutectic) die-attach adhesive that is used to bond a silicon chip to an alloy 42 lead frame after heating to 425 °C and cooling to 25 °C. The thicknesses of the chip, lead frame, and eutectic are 0.5, 0.3, and 0.15 mm, respectively, and the length of the continuous attachment is 1.5 cm.

**b.** Compare this stress with that produced using conducting epoxy in a package containing the same component geometry. Assume that curing occurs at a temperature of 175 °C.

**c.** In which package is the chip less likely to suffer cracking?

**8.19.** Consider a flip-chip soldering operation involving five solder balls in a line. Schematically sketch the shape of the connection if

**a.** the chip and substrate differ in CTE.

**b.** the contact registration is slightly misaligned.

**c.** the two outer solder balls are larger than the three inner ones.

**8.20.** Derive a formula for the normal stress as a function of distance along a continuous chip–adhesive–substrate structure of length $2L$ subjected to a processing temperature difference $\Delta T$.

## REFERENCES


[1] R.R. Tummala, E.J. Rymaszewski (Eds.), Microelectronics Packaging Handbook, Van Nostrand Reinhold, New York, 1989.

[2] J.H. Lau (Ed.), Flip Chip Technologies, McGraw Hill, New York, 1996.

[3] Electronic Materials Handbook, vol. 1, Packaging, ASM International, Materials Park, Ohio, 1989.

[4] C.A. Harper (Ed.), Electronic Packaging and Interconnection Handbook, McGraw-Hill, New York, 1991.

[5] D.P. Seraphim, R. Lasky, C.-Y. Li (Eds.), Principles of Electronic Packaging, McGraw-Hill, New York, 1989.

[6] M. Pecht (Ed.), Integrated Circuit, Hybrid, and Multichip Module Package Design Guidelines—A Focus on Reliability, Wiley, New York, 1994.

[7] M. Pecht, L. T Nguyen, E.B. Hakim (Eds.), Plastic-encapsulated Micro-electronics-materials, Processes, Quality, Reliability, and Applications, Wiley, New York, 1995.

[8] M. Pecht, A. Dasgupta, J.W. Evans, J.Y. Evans (Eds.), Quality Conformance and Qualification of Microelectronic Packages and Interconnects, Wiley, New York, 1994.

[9] J.H. Lau, Thermal Stress and Strain in Microelectronics Packaging, Van Nostrand Reinhold, New York, 1993.

[10] T.W. Lee, in: T.W. Lee, S.V. Pabbisetty, ASM International (Eds.), Microelectronic Failure Analysis, Materials Park, Ohio, 1993, p. 344.
[11] L. Perkins, Electronics Materials Handbook, vol 1, Packaging, ASM International, Materials Park Ohio, 1989, 1047.
[12] J.H. Lau, S.J. Erasmus, D.W. Rice, Circuit World 16 (2) (1990) 5.
[13] E. Zakel, G. Azdasht, H. Reichl, IEEE Trans. Components Packaging Manuf. Tech. B17 (4) (1994) 569.
[14] R. Padmanabhan, IEEE Trans. Components Hybrids Manuf. Tech. CHMT-8 (4) (1985) 435.
[15] G.F. Frankel, S. Purushothaman, T.A. Peterson, S. Farooq, S.N. Reddy, V. Brusic, Trans. Components, Hybrids Manuf. Tech. B 18 (1995) 709.
[16] L.T. Manzione, Plastic Packaging of Microelectronic Devices, Van Nostrand Reinhold, New York, 1990.
[17] K.N. Ritz, W.T. Stacy, E.K. Broadbent, 25th Annual Proceedings of the IEEE Reliability Physics Symposium, 1987, p. 28.
[18] E. Suhir, L.T. Manzione, J. Electron. Packaging Trans. ASME 113 (1991) 421.
[19] E. Suhir, L.T. Manzione, J. Electron. Packaging Trans. ASME 113 (1991) 16.
[20] A.A. Gallo, R. Munamarty, IEEE Trans. Reliability 44 (3) (1995) 362.
[21] R. Shook, T. Conrad, Tutorial Notes, IEEE Int. Reliability Phys. Symp. (1995), 1.1.
[22] S.K. Groothuis, K.G. Heinen, L. Rimpillo, 33rd Annual Proceedings of the IEEE Reliability Physics Symposium, 1995, p. 76.
[23] Dongkai Shangguan, Lead-free solder interconnect reliability, 292 p. ASM Int. (December 5, 2005)
[24] K. Subramanian, Lead-free Solders: Materials Reliability for Electronics, Wiley, 2012.
[25] M.R. Lin, J.T. Yue, 25th Annual Proceedings of the IEEE Reliability Physics Symposium, 1987, p. 164.
[26] J.H. Lau (Ed.), Chip on Board Technologies for Multichip Modules, Van Nostrand Reinhold, New York, 1994.
[27] N. Sherwani, Q. Yu, S. Badida, Introduction to Multichip Modules, Wiley, New York, 1995.
[28] H.H. Manko, Solders and Soldering, 2nd. ed, McGraw Hill, New York, 1979.
[29] A. Rahn, The Basics of Soldering, Wiley, New York, 1993.
[30] J.H. Lau (Ed.), Solder Joint Reliability, Van Nostrand Reinhold, New York, 1991.
[31] W.B. Hampshire, Soldering Surf. Mount Tech. 14 (6) (1993) 49.
[32] H.J. Frost, in: J.H. Lau (Ed.), Solder Joint Reliability, Van Nostrand Reinhold, New York, 1991.
[33] R. Darveaux, K. Banerji, IEEE Trans. Components Hybrids Manuf. Tech. 15 (1992) 1013.
[34] J.A. DeVore, J. Metals (July 1984) 51.
[35] L.J. Turbini, D. Cauffield, Soldering Surf. Mount Tech. 17 (6) (1994) 30.
[36] B.N. Ellis, Circuit World 20 (4) (1994) 26.
[37] B. Meagher, D. Schwarcz, M. Ohring, J. Mat. Sci. 31 (1996) 5479.
[38] E.E. de Kluizenaar, Philips Electron Opt. Bull. 131 (1991) 3.
[39] K.J. Puttlitz, IEEE Trans. Components Hybrids Manuf. Tech. 13 (4) (1990) 647.
[40] R. Prasad, Semicond. Int. 19 (4) (1996) 116.
[41] A. Malhotra, in: M. Pecht (Ed.), Soldering Processes and Equipment, Wiley, New York, 1993.
[42] E. Suhir, in: A. Bar-Cohen, A.D. Kraus (Eds.), Advances in Thermal Modeling of Electronic Components and Systems, vol. 1, Hemisphere, New York, 1988.
[43] E. Suhir, MRS Symp. Proc. Electron. Packaging Mat. Sci.-II. (1986) 133.
[44] E.A. Johnson, W.T. Chen, C.K. Lim, in: D.P. Seraphim, R. Lasky, C.-Y. Li (Eds.), Principles of Electronic Packaging, McGraw-Hill, New York, 1989.

[45] J.M. Hu, in: A. Christou (Ed.), Reliability of Gallium Arsenide MMICs, J. Wiley and Sons, Chichester (, 1992.
[46] J.M. Hu, M. Pecht, A. Dasgupta, J. Electron. Packaging Trans. ASME 113 (1991) 275.
[47] E. Suhir, B. Poborets, J. Electron. Packaging Trans. ASME 112 (1990) 204.
[48] Y. Guo, C.K. Lim, W.T. Chen, C.G. Woychik, IBM J. Res. Develop. 37 (1993) 635.
[49] D.R. Frear, F.G. Yost, MRS Bull. XVIII (12) (1993) 49.
[50] G.L. Schnable, in: A.H. Landsberg (Ed.), Microelectronics Manufacturing Diagnostics Handbook, Van Nostrand Reinhold, New York, 1993.
[51] B.R. Livesay, R.K. Shukla, IEEE IRPS Tutorial Notes (1993).
[52] J.E. Sergent, Semicond. Int. (June 1996) 263.
[53] K. Losch, K. Allaert, S. Smernos, J. Novotny, Electron. Commun. 3rd quarter (1993) 260.
[54] J.S. Hwang, R.M. Vargas, Soldering Surf. Mount Tech 5 (1990) 38.
[55] C. Bauer, Semicond. Int. 19 (4) (1996) 127.
[56] R. Frank, Understanding Smart Sensors, Artech, Boston, 1996.
[57] L.G. Feinstein, Electronics Materials Handbook, vol 1, Packaging, ASM International, Materials Park Ohio, 1989, 215.
[58] L. Gallance, M. Rosenfeld, RCA Rev. 45 (1984) 256.
[59] H. Lee, Y.Y. Earmme, IEEE Trans. Components Hybrids Manuf. Tech. 19A (1996) 168.
[60] B.J. Han, S. Das, R.C. Frye, K.L. Tai, M.Y. Lau, IEEE Multi-Chip Module Conference, California, Santa Cruz, January 31, 1995, p. 45.
[61] R.D. Denman, Tin and Its Uses, No. 165, 1991, 16.
[62] C. Lea, Circuit World 16 (4) (1990) 23.
[63] T.L. Landers, W.D. Brown, E.W. Fant, E.M. Malstrom, Electronics Manufacturing Processes, Prentice Hall, Englewood Cliffs, 1994. NJ.
[64] J.H. Lau, in: J.H. Lau (Ed.), Thermal Stress and Strain in Microelectronics Packaging, Van Nostrand Reinhold, New York, 1993.

# CHAPTER 9

# Degradation of Contacts and Package Interconnections

## 9.1 INTRODUCTION

The unifying theme of this chapter encompasses problems that arise in maintaining electrical continuity between conductors. We have all instinctively diagnosed poor contacts as the source of malfunction in some piece of electrical equipment or other, and often been correct. In the days before transistors, the simple act of making a local telephone call activated hundreds of electro-mechanical contacts, whose inefficient operation or malfunction meant signal distortion, noise, or even lack of signal transmission. Whether they are "static," as in permanent soldered interconnections, or are "dynamic," sliding, and rotating, as in switches and motors, electrical contacts pose a variety of interesting and important reliability problems.

In this chapter we will discuss both types of contacts, starting with a classical treatment of the nature of contacting, current-carrying conductors. Dynamic contact problems arising from mechanical wear, corrosion, and stress relaxation are among the first reliability issues we will address. Much of the chapter's later focus, however, will be devoted to the reliability of solder materials used in interlevel packaging connections. In this sense the chapter continues the discussion that began in Chapter 8 dealing with packaging configurations, joining techniques, and solder defects.

Building on the previously developed thermal-stress analyses, failure modeling of contacts and solder interconnections is possible, at least in principle. Nevertheless, it still remains a formidable task to obtain estimates for contact failure times because:

1. stress or strain levels required to cause failure are often unknown;
2. the applied stresses are usually cyclic in nature;
3. component contact and structural geometries are complex;
4. contacting surfaces may wear due to the relative motion between them; and
5. the elevated temperature plastic and viscoelastic behavior of the involved materials often displays a complex temperature–time–environment history.

*Reliability and Failure of Electronic Materials and Devices*
ISBN 978-0-12-088574-9
http://dx.doi.org/10.1016/B978-0-12-088574-9.00009-4

When applied to solder interconnections, a logical failure-modeling approach first assumes the operative thermal stress levels expected as calculated previously. The cyclic nature of the stressing must then be considered and this information folded together with material properties into the relevant expressions that describe the deformation response. Creep, fatigue, and fracture are the operative mechanisms that cause packaging components and structures to fail. These important subjects are basically the substance of two of the later chapter sections.

Finally, there is a discussion of vibration and dynamic loading effects. In addition to static and cyclic differential thermal-expansion effects, dynamic loading and vibration can also be quite harmful from the standpoint of equipment reliability. Because of potential fatigue damage to soldered leads due to vibratory flexure of circuit boards and components, the subject is treated in this chapter.

## 9.2 THE NATURE OF CONTACTS

### 9.2.1 Scope

The purpose of electrical contacts is threefold: to make or complete a circuit, to maintain current flow through the circuit, and to break or open the circuit when required. Contacts must reliably repeat these functions over specified time intervals that may be as short as a fraction of a millisecond. The electrical spectrum of contacting devices and methods includes switches of all kinds, relays, pin, and socket connectors, as well as solder contacts. Although there is considerable overlap, applications may be broadly classified according to their power usage as follows:

1. power engineering, where high voltages and currents are involved;
2. medium power, with moderate voltages and currents, as in home appliances; and
3. low voltage–low current applications, as in telecommunications and computers.

In these applications, electrical continuity is often achieved mechanically through impact, sliding, or rotational motion of metal surfaces that are forced together to make and sustain contact. Different mechanical phenomena are involved in each of these domains of usage, and the corresponding contact-failure mechanisms reflect this. In the first category, for example, electrical discharges and arcs may be operative, causing extensive erosion and wear of the contacting metal-surfaces. Contacts also degrade by mechanical wear in medium and low-power applications, where in addition, contact opening and closing causes insulating surface films to form as a result of (dry) oxidation or (wet) corrosion. In such cases, contact unreliability arises through the destruction of the initially smooth contacting and conducting surfaces. The legacy is open and noisy or intermittent electrical behavior.

There is considerable literature on the subject of contacts [1,2]. Investigations continue to be carried out on contact problems, where for over 40 years there has been an annual IEEE Holm Conference on Electrical Contacts, so named for R. Holm, a prominent researcher in this field. In these conference proceedings, also published in the *IEEE Transactions on Components, Hybrids and Manufacturing Technology*, many papers are devoted to the damage of high-power contacts, a subject we will not discuss further. Instead, the subsequent treatment will be limited to low- and medium-power contacts, the involved materials, and the degradation and reliability issues that surround them.

## 9.2.2 Constriction Resistance

Consider the physics of what happens when two nominally identical smooth metals contact one another. Although there appears to be contact over a substantial area, actual contact is made at a limited number of microscopically small areas. (If the involved metals were infinitely hard, there would be only three points of contact.) With analytical tools such as the scanning electron microscope and atomic force microscope, we now know that even very finely polished metal surfaces exhibit arrays of protruding surface asperities separated by valleys, with peak-to-peak height differences of perhaps 10 nm.

When a potential difference is applied across touching asperity-free cylindrical conductors, the current flow lines are parallel, as shown in Figure 9.1(a). The contact voltage ($V_c$) that develops between points $A$ and $B$ is simply given by Ohm's law; i.e., $V_c = i \cdot R_o$, where $i$ is the current and $R_o$ is the metal ohmic resistance. This situation is approximated by a soldered joint because the solidified solder effectively fills the spaces between asperities. However, when contact occurs only at asperities, current flows in the constricted manner of Figure 9.1(b), distorting the flow lines. The contact voltage ($V_c$) developed is now larger in order to maintain the same current and is given by:

$$V_c = i(R_o + R_c), \qquad (9.1)$$

where $R_c$ is the so-called constriction resistance. Analogy to fluid flow through a constricted pipe under a fixed pressure head (constant $V_c$) means

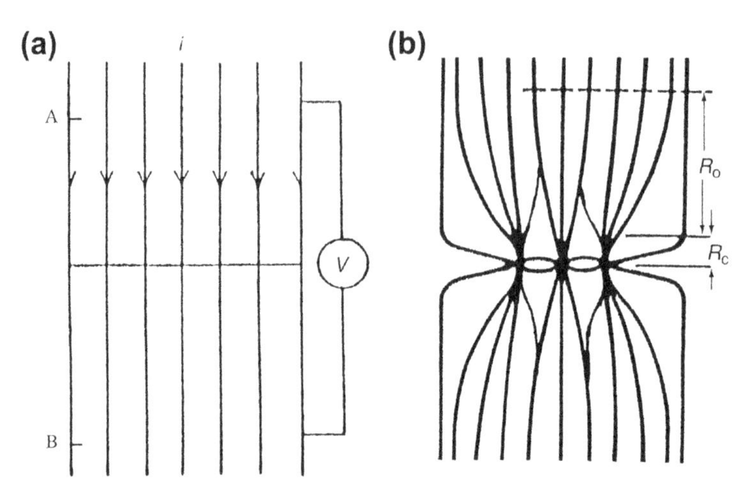


Figure 9.1 (a) Parallel current flow lines over a large asperity-free area of contact. (b) Constricted current flow lines at contact asperities.

a smaller throughput ($i$), which is caused by the added resistance ($R_c$). If, in addition, there is a resistance contribution $R_f$ due to a surface film at the contact, it is added to $R_o + R_c$.

Using concepts of electromagnetic theory it has been shown that when two plane surfaces make contact over an asperity area of radius $a$, the total constriction resistance is:

$$R_c = \frac{\rho}{2a}, \tag{9.2}$$

where $\rho$ is the resistivity. For metals we recall that $\rho \approx 10^{-6}$ to $10^{-4}$ $\Theta$-cm, and, we note that if $a$ is large, $R_c$ is negligible; but if $a$ is sufficiently small, $R_c$ can rise to unacceptable levels and cause circuit malfunction.

### 9.2.3 Heating Effects

One might imagine currents to cause appreciable local heating effects at the asperity–asperity contact welds, known as **a-spots**, because of the small areas involved. Then the contact resistance would rise from $R_c(1)$ at temperature $T(1)$ to $R_c(2)$ at $T(2)$ according to the relation $R_c(2) = R_c(1)\{1 + \alpha\,[T(2) - T(1)]\}$, where $\alpha$ is the temperature coefficient of resistivity. Assuming that no heat is lost from the hot contact constriction and that steady-state temperatures are quickly reached, it is not hard to roughly estimate the temperature rise, $T(2) - T(1)$. The power generated from Joule heating $(V_c^2/R_c)$ is converted into an equivalent heat (power) flow $(\kappa[T(2) - T(1)]\, A_c/\Delta x)$, where $A_c$ and $\Delta x$ are the constriction area and length, respectively, and $\kappa$ is the thermal conductivity. Noting that $R_c = \rho \cdot \Delta x/A_c$, and equating the two power contributions, the approximate expression

$$V_c^2 = \rho\kappa\left[T(2) - T(1)\right] \tag{9.3}$$

emerges. In metals the electrical conductivity ($\sigma = 1/\rho$) and thermal conductivity are proportional to each other through the formula, $\rho\kappa = LT$, where $L$ is the Lorenz number (Section 5.5.3). If $T$ is assumed to be some mean temperature between $T(1)$ and $T(2)$, then $V_c^2 = LT[T(2) - T(1)]$. A more accurate expression [3] is

$$V_c^2 = 4L\left[T^2(2) - T^2(1)\right]. \tag{9.4}$$

To see the implications of this formula, let us assume $V_c$ = IV and $T(1) = 300$ K. Taking $L = 2.44 \times 10^{-8}$ W-$\Omega$/K$^2$, $T(2)$ is calculated to be 3200 K, a value much higher than the melting point of almost all metals. The mere hint of such high temperatures implies sources of reliability problems. If, for example, the contact area at asperities was $10^{-8}$ cm$^2$, the current density would be sufficiently high for electromigration damage to occur; and if contact tarnish films existed, they would probably not survive the heat generated.

### 9.2.4 Mass Transport Effects at Contacts

The unusually high temperatures and electric current densities present at contact spots mean that at least three different atomic mass transport mechanisms may be operative to transfer material from one contact metal to the other [4]. Diffusion, electromigration, and thermomigration are the mechanisms, and they are explored in this section.

1. **Diffusion** The diffusional flux of atoms ($J_a$) in a concentration gradient ($dC/dx$) driving force, with $x$ normal to the contact surface, is a subject treated in this book on at least two prior occasions. Because a discontinuous concentration gradient exists for atoms at a contact junction between two different metals, the mass flux given by

$$J_m = -D\frac{dc}{dx}, \tag{9.5}$$

is potentially large. For appreciable interdiffusion to occur, the diffusivity given by $D = D_o \exp[-E_D/RT]$ must be sufficiently large, and this requires a high temperature. Even if both contacting metals are identical, atomic diffusion is possible. In this case the driving force is the surface-energy gradient that stems from differences in surface curvature. Atoms at asperity tips will tend to diffuse into the valleys and lower the overall surface area. Essentially the same mechanism that is operative during the sintering of metal powder compacts is active here. Sintering or welding at hot spots effectively widens the contact area and reduces the contact resistance.

2. **Electromigration** It was previously shown that the electromigration flux ($J_e$) is given by

$$J_e = \frac{CDZ^* q\rho j_e}{RT}. \tag{9.6}$$

Just as on the chip level at metal–semiconductor interfaces, electromigration at contacts often represents a severe mass–flux divergence. Electromigration mass-transport effects are also expected to be narrowly confined to the contact interface where the current density ($j_e$) is large and the temperature high. Further away the conductor widens lowering $j_e$ as well as the temperature. An important term in Eqn (9.6) is the effective valence $Z^*$ whose value may range widely in magnitude, i.e., from $\sim -5$ to $-50$. Measurement has shown that $Z^*$ decreases slowly with increasing temperature.

3. **Thermomigration** Mass transport driven by temperature gradients is known as thermomigration. The corresponding flux $J_{th}$ is given by

$$J_{th} = -\frac{CQ^* D}{RT^2}\frac{dT}{dx}, \tag{9.7}$$

where $Q^*$ is the heat of transport for the element in question. A positive value for $Q^*$ implies atom migration to colder regions.

The ratio of the electro- to thermo-migration fluxes is given by:

$$\frac{J_e}{J_{th}} = \frac{Z^* q\rho j_e}{-(Q^*/T)dT/dx} \tag{9.8}$$

this simply represents a ratio of the two generalized forces.

## Example 9.1

Assuming representative values of $Z^* = -10$, $Q^* = 10$ kJ/mol, $j_e = 1 \times 10^{10}$ A/m$^2$, $\rho = 1 \times 10^{-7}$ $\Omega$-m, $T = 1000$ K, and $dT/dx = 10^7$ K/m, does electromigration or thermomigration dominate?

**Answer**

Substituting in Eqn (9.8) we have:

$$\frac{J_e}{J_{th}} = \frac{-10 \times 1.6 \times 10^{-19} \times 1 \times 10^{-7} \times 1 \times 10^{10} \times 6.02 \times 10^{23}}{-(10{,}000/1000) \times 10^7} = 9.6.$$

(Note that to make units match, Avogadro's number must be included in the numerator.) This calculation suggests that electromigration dominates thermotransport.

Without knowing the temperature distributions and material constants in a contact it is sometimes easier to make measurements. This was done by Runde et al. [5], who deposited minute quantities of radioactive zinc on one of two aluminum contact electrodes. A buildup of Zn observed on the counter electrode, consistent with the current polarity, demonstrated that electromigration had occurred. The transfer of metal at contacts changes the alloy composition and generally raises $R_C$.

### 9.2.5 Contact Force

Applied force or pressure is required to ensure that low-resistance electrical contact is made and then sustained. We have already seen that contact resistance depends on the contact area. Since contact heating can potentially alter the area through local mass transfer and/or melting, the relationships among contact resistance, force, and temperature are critical to the understanding of contact performance; what is required is a combination of a few simple definitions and formulas.

Let us first consider hardness $H$, a property measured in stress units of N/m$^2$, or commonly kg/mm$^2$. Hardness is essentially a measure of a material's ability to withstand penetration by a harder indenter that is pressed into its surface with normal load $P$. For a spherical indenter, the hardness is often defined as the load per projected area of diameter $\bar{d}$, or $H = 4P/\pi\bar{d}^2$. We know, however, that the true mechanical bearing area is but a small fraction of the nominal contact area. As a result, the asperity contacts are subjected to large pressures even with small loads. To account for normal contacts at multiple asperities, the hardness formula is modified to:

$$H = \frac{4P}{n\pi\bar{d}^2}, \tag{9.9}$$

where $n$ is the number of such contact points or spots, and $\bar{d}$ is the average spot diameter. Reconsidering the constriction resistance, it is apparent that

$$R_c = \frac{\rho}{n\bar{d}}. \tag{9.10}$$

This formula assumes that the resistor spots are effectively connected in parallel. Eliminating $\bar{d}$ between the last two equations yields:

$$R_c = \rho\left(\frac{\pi H}{4nP}\right)^{1/2}, \tag{9.11}$$

with the expected result that the contact resistance decreases as the load increases. This equation holds when $P$ is relatively large and the contact metal deforms plastically. However, when $P$ is small such that the average pressure is roughly less than $H/3$, the contact deforms elastically; in this case theory shows that $R_c \approx P^{-1/3}$. Many experiments have been performed establishing the validity of both the $P^{-1/2}$ and $P^{-1/3}$ dependencies for $R_c$. Contact integrity necessitates the maintenance of appropriate contact pressures.

## 9.3 DEGRADATION OF CONTACTS AND CONNECTORS

### 9.3.1 Introduction

Reliable separable electrical connectors, like the kind shown in Figure 9.2, require the force pressing male and female halves together to make contact, be constant, and not relax with time. Since lifetimes ranging from 10 to as much as 40 years are often desired for such connectors, the contacts must suffer little change in $R_c$. In this section we address practical situations that lead to an increase in $R_c$ because of a change in one or more of the terms in Eqn (9.11). After treating the mechanics of spring contacts, the first case we consider involves creep-prone contacts. These pose a reliability issue because the contact pressure relaxes with time, raising the contact resistance. Most contacts in relative motion suffer tribological or rubbing-type damage through wear processes that potentially alter both the geometry and the composition of the contacting surfaces. In this way $\rho$, $H$, $n$, and even $P$ change in ways that are not easily distinguishable, but which effectively raise $R_c$ and limit contact life.

### 9.3.2 Mechanics of Spring Contacts

Metal prongs that make electrical contact in separable connectors have been approximated by dual cantilevers; an example is the card edge connector. Therefore, let us consider the simple cantilever beam [7]

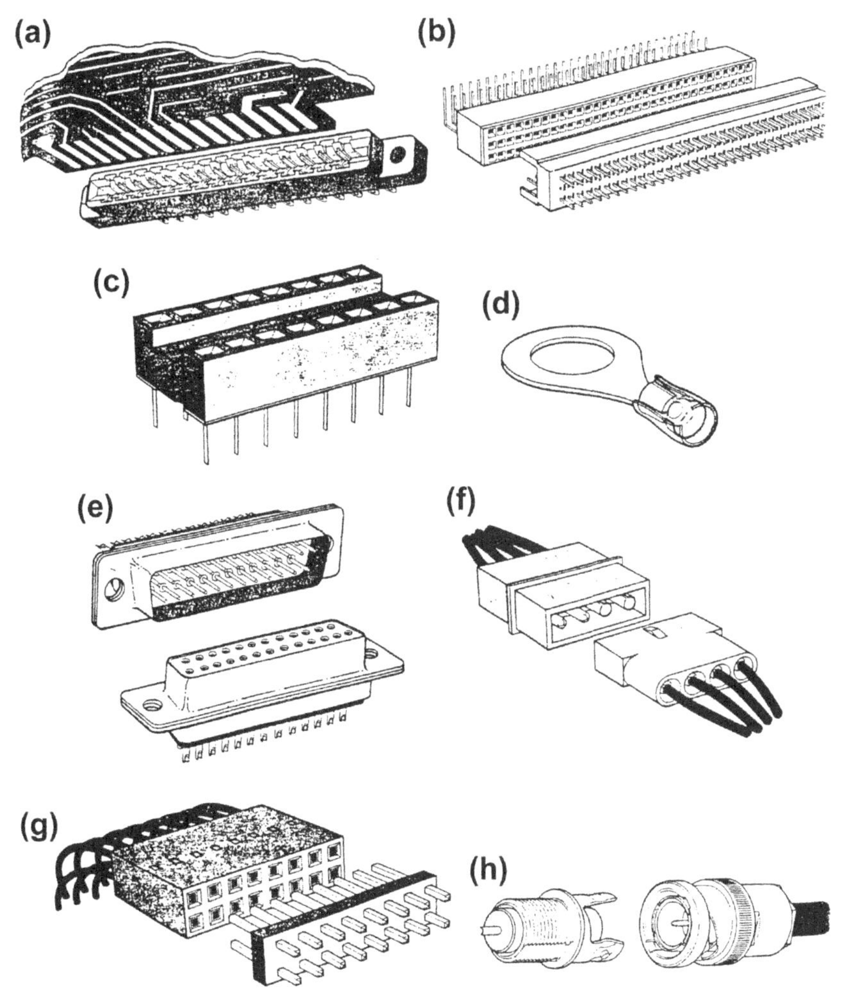


**Figure 9.2** An assortment of separable electrical connectors. (a) Edge card/PBB fingers; (b) two-part/euro or DIN; (c) IC socket; (d) crimp; (e) pin and socket/track and panel; (f) power/snap together; (g) post and box; (h) Coaxial (BNC). *From Ref. [6].*

shown in Figure 9.3. With force $P$ acting on a such a beam of width ($w$) and thickness ($T$), the deflection ($\delta$) is given by:

$$\delta = \frac{4PL^3}{wT^3E} \tag{9.12}$$

in elementary texts on strength of materials, where $E$ is Young's modulus. For spring contacts deflecting a fixed amount, a large $E$ is desirable because $P$ is then large, thus reducing $R_c$.

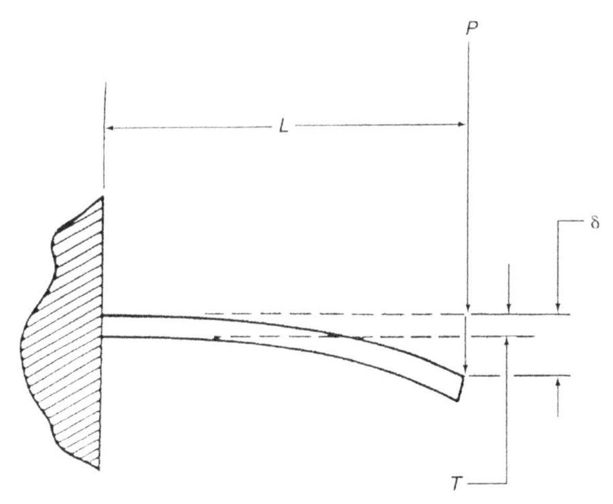


Figure 9.3 Cantilever model of a contact spring.

It is also useful to know the maximum force ($P_{max}$) realized by an elastic spring; clearly, the flow or yield stress ($\sigma_o$) of the metal limits $P_{max}$. The maximum stress in the outer fibers of a beam is 6 $PL/(wT^2)$ and is limited by $\sigma_o$. Therefore,

$$P_{max} = \frac{\sigma_o w T^2}{6L}. \tag{9.13}$$

For those copper alloys commonly used in connector prongs, Young's modulus varies only between 110 and 140 GPa. Therefore, one practical way to ensure large normal spring forces is to select high strength alloys; this accounts for the widespread use of Cu–Be for this purpose. In addition, there is an added reliability bonus because such alloys suffer less permanent set. Because it reduces the beam deflection, permanent set can raise $R_c$ by relaxing the contact force. Permanent set behavior in actual contact springs is displayed in Figure 9.4(a) and (b), where it is seen that stiffer, stronger alloys generally behave better in this regard.

### 9.3.3 Normal Force Reduction in Contact Springs

In order to quantitatively assess permanent set as manifested by service life, it is instructive to consider a study [8,9] of the heat-age testing performed on gold-plated, phosphor-bronze contact springs. The latter were subjected to a matrix of accelerated aging tests spanning a temperature range of 100 –200 °C, for time periods up to 3 months. Measurements of the

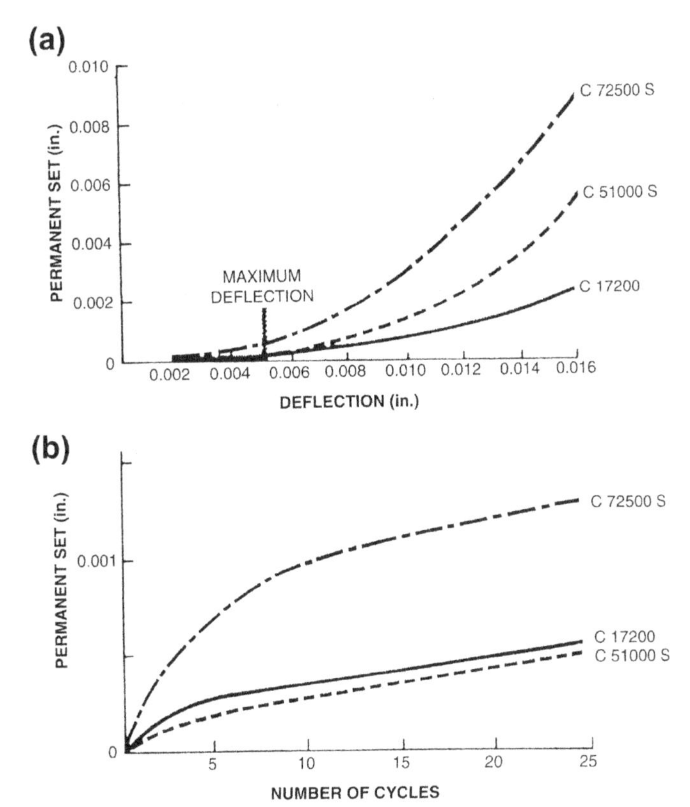


**Figure 9.4** (a) Permanent set in three different copper base alloy spring connectors as a function of deflection. Note: C17200 is beryllium copper ($E$ = 126 GPa, $\sigma_o$ = 721 MPa), C51000 is phosphor bronze ($E$ = 118 GPa, $\sigma_o$ = 658 MPa), C72500 is Cu–Ni–Sn ($E$ = 134 GPa, $\sigma_o$ = 630 MPa). (b) Permanent set in spring connectors as a function of the number of contact mating cycles. *From Ref. [7].*

spring deformation were made and converted to the equivalent normal force ($P_N$) applied.

In these two publications, it is interesting that essentially the same data were plotted in two different ways; the results are displayed in Figure 9.5. A log–log plot was used in Figure 9.5(b) to track the reduction in $P_N$ as a function of time ($t$). The straight line suggests the empirical relationship, In ($P_N$) = $-n$ ln ($t$) + $b$, so that $n$, a constant, is the slope, and $b$, another constant, is the intercept on the ordinate. An equivalent explicit formula expresses the force-relaxation kinetics as:

$$P_N = Ct^{-n}, \tag{9.14}$$

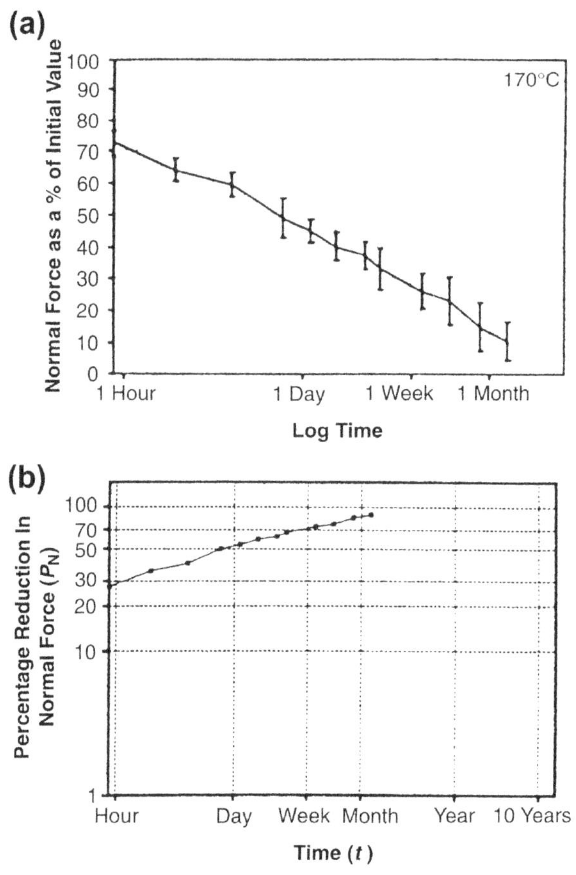


**Figure 9.5** (a) Force relaxation of a dual-cantilever electrical spring contact plotted as a function of log time. *(From Ref. [9].)* (b) Essentially the same data as in Figure 9.5(a) plotted as log percent force reduction versus log time. *From Ref. [8].*

where $C = \exp(b)$. Values of $n$ ranged between 0.1 and 0.2 and increased with temperature. The neglect of the explicit role of temperature is a shortcoming of this analysis.

In Figure 9.5(a) the data were plotted as $P_N$ versus log $t$, and the resulting straight line implies the linear equation, $P_N = r + s \ln t$ ($r$ and $s$ are constants). To account for the effect of temperature it was assumed that the rate of force recovery is thermally activated, and therefore.

$$\frac{dP_N}{dt} = -K \exp\left[-\frac{E - gP_N}{RT}\right]. \tag{9.15}$$

In this expression $K$ and $g$ are constants, and $RT$ has the usual meaning. The activation energy ($E$) in the Boltzmann factor is linearly reduced by the instantaneous recovering force and the strain energy associated with it. A precedent for modifying $E$ so that it is variable was established in Section 1.3.3 (Eqns (9.1)–(9.7)). The integration of this differential equation yields the logarithmic recovery law:

$$P_N = A \ln(Bt + C), \tag{9.16}$$

where $A = -RT/g$, $B = \frac{gK\left(\exp - \frac{E}{RT}\right)}{RT}$, and $C$ is a constant that can be evaluated from the initial condition (at $t = 0$, $P_N = P_N(0)$). When $Bt$ is large compared with $C$, this solution is in accord with experiment and can be extrapolated to long times in order to predict connector lifetime. Further insight into the relaxation mechanism can be obtained by evaluating $E$ and interpreting its magnitude.

This example illustrates the distinction between failure and reliability modeling. In the former, attempts are made to determine lifetime in terms of physical mechanisms that cause time-dependent changes in the material. Information provided by the failure modeling leading to Eqn (9.16) often yields ideas for corrective action that may eventually eliminate the problem. On the other hand, no such information or hope stems from the empirical statistical modeling of reliability implicit in Eqn (9.14).

### 9.3.4 Tribology

The subject of tribology encompasses phenomena that occur at material surfaces in contact and in relative motion. These include wear, friction, erosion, and lubrication. Of these subjects, wear is clearly most important for our needs. In electronic packaging, wear scars develop in cable connectors and on the tab portions of card edge assemblies and pluggable modules. Specific mechanisms for damage have been reviewed by Engel [9] and include:

1. **Adhesive wear** Surface smearing, galling, and seizure are the manifestations of the adhesive wear that results from the microwelding of asperities. Such welds fracture in shear at the original interface or below it depending on the relative magnitude of the interfacial and cohesive bond strengths. When the adhesive strength exceeds the cohesive

bond strength, wear particles may form after several cycles of contact. The wear volume ($V$) removed in this process is given by:

$$V = \frac{K_W PL}{3H}. \tag{9.17}$$

In this well-known expression due to Archard [10], $P$ is the normal operative force, $L$ is the length of relative surface displacement, $H$ is the indentation hardness of the softer body, and $K_W$ is a dimensionless constant whose magnitude depends on the materials in contact. Typical values for $K_W$ are $32 \times 10^{-3}$ for Cu–Cu pairs, $12 \times 10^{-3}$ for Ag–Ag pairs, and 1.5 for Cu on steel.

2. **Abrasive wear** In abrasive wear, hard surface asperities move against a softer material plowing it. Often particles enter the interface, and through cutting or grinding action the softer contact material is plowed or removed. The Archard expression describes the abrasive wear volume as well, but $K_W$ may assume different values depending on whether two-body (contacting surfaces) or three-body (contacting surfaces plus abrasive particles) are involved. Wear in this case depends strongly on the ratio $R_W = H_{abraded}/H_{abrasive}$, where $H$ is hardness of the indicated material. For $R_W < 1$, extensive wear occurs, but when $R_W$ is slightly greater than unity, wear effectively ceases.
3. **Delamination wear** Due to the presence of cyclic loading, cracks can develop and propagate parallel to the surface. For sufficiently high tangential frictional stresses, thin plate-like particles delaminate from the wearing surface.
4. **Corrosive wear** Continuous removal of surface films and chemical reaction between these wear products and the environment results in oxidation wear. Oxide films are generally brittle and protective when less than 10 μm thick. Under cyclic stressing the oxide may fracture, causing modest damage if the oxide particles are soft. Hard oxide particles, however, are abrasive and accelerate wear.
5. **Fretting wear** This important manifestation of contact wear is defined as accelerated surface damage occurring at the interface between contacting materials subjected to small oscillatory movements. Fretting wear is often associated with adhesive, abrasive, and corrosive wear mechanisms. Because we are often concerned with the number of displacement cycles sustained prior to contact failure, low amplitude fatigue is one way to describe fretting.

### 9.3.5 Fretting Wear Phenomena

The micromotions that cause fretting in electronic connectors range from a few micrometer to as much as 100 μm in length. They stem from mechanical vibrations, differential thermal expansion, and load relaxation of the contacting metals, as well as junction heating effects due to switching the power on and off. Manifestations of fretting damage include metal debris particles, wear scars, corrosion products, cracked surface oxide or sulfide films, and polymer formation from atmospheric organic precursors.

All metals (i.e., gold, palladium, tin, aluminum, nickel, silver, and copper) used for contacts and connectors suffer differing degrees of fretting damage depending on the particular metals involved, electrical power level, contact forces, and frequency of loading, surface roughness, presence of lubricants, temperature, and environment. As Figure 9.6(a) shows, fretting damage begins with cracking of surface films as a result of connector micromotions. Oxide regrows on the exposed metal but is in turn cracked with continued rubbing. Due to the accumulated damage during fretting, the contact resistance rises, as shown schematically in Figure 9.6(b). Following initial film cracking, a relatively long period of stable contact resistance ensues. Erratic resistance fluctuations are a precursor to the eventual failure, which is invariably manifested by a discontinuous resistance rise and open contact.

Tests to evaluate fretting damage usually involve the measurement of contact resistance changes under the wear track as a function of the number of loading (or displacement) cycles applied by mechanical, electromagnetic, or thermal means. Accelerated current cycling tests have also been performed. The results of extensive testing have yielded a reasonably coherent qualitative picture of the material behavior and wear phenomena observed. Some of this information [12–14] is summarized below:

1. Base metal (Sn, Sn–Pb solder) contacts oxidize rapidly and produce debris that accumulates in the contact zone.
2. Composition changes occur during fretting with dissimilar metals due to mass transfer, wear, and film formation. In general, the direction of metal transfer is from the softer to harder surface. Thus a Au–Pd contact maintains low resistance because the harder palladium becomes coated with gold. Conversely, Au–Sn contacts degrade as the softer tin coats the gold, effectively leaving a base metal contact behind. Intermetallic compound (IMC) formation (e.g., Al–Cu) has been observed to occur at low temperatures.

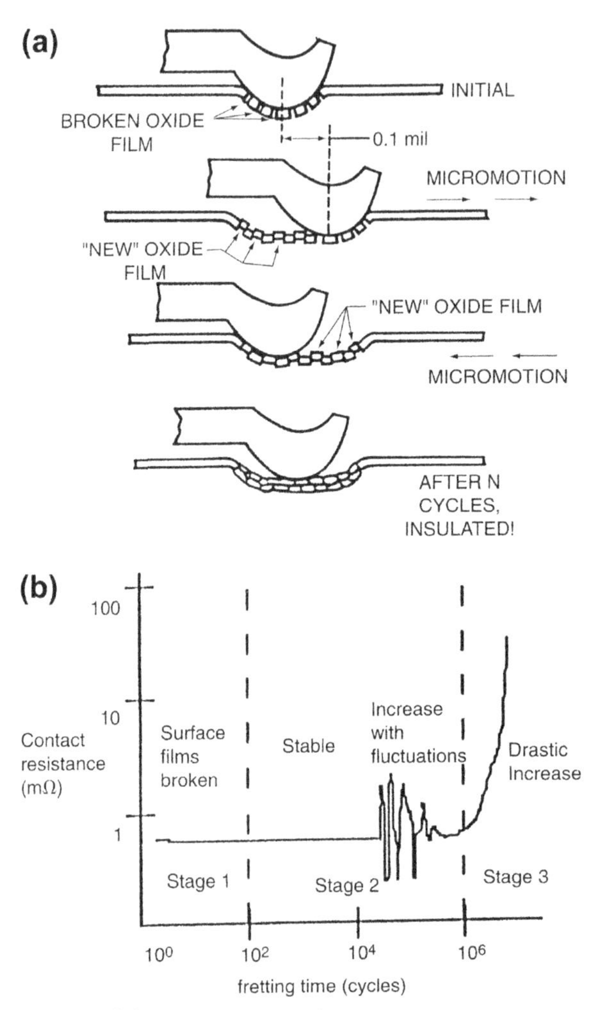


**Figure 9.6** (a) Model of fretting corrosion damage in a contact undergoing micromotions. *(From Ref. [7].)* (b) Stages in the rise of contact resistance during fretting. *From Ref. [11].*

3. Simultaneous corrosion virtually always accelerates fretting damage to contacts. The reduction in contact area causes either electrical or mechanical failure or a combination of both effects. Through the use of noble metals, corrosion effects are minimized. However, the corrosion of the underlying base metal through pores in gold-plated contacts has been observed.
4. Metals such as Pd, Rh, and Ru catalyze the formation of frictional polymers from adsorbed organic air pollutants. Unlike fretting corrosion, the surfaces experience little or no wear.

5. Aluminum connectors in contact with Cu–Ni and Cu–Sn alloys are prone to wear by abrasion, adhesion, and delamination during fretting. Material transfer in both directions occurs.
6. Metals like Au, Ag, and Au–Pd–Ag alloys generally suffer little fretting damage and form little polymer.
7. Lubricants (contact aid compounds) can retard the rate of fretting and stabilize the contact resistance. Commercial lubricants are available to alleviate fretting problems by dispersing polymer buildup, reducing wear, and shielding surfaces from oxidation. Paradoxically, lubricants do not have to be conducting; their function is to protect the contact zone from the environment, not conduct electricity.

### 9.3.6 Modeling Fretting Corrosion Damage

In this section a portion of a recent quantitative model for fretting damage developed by Bryant [11] is presented. The model assumes that the contact surface consists of a large number of hills and valleys. Initially clean asperities are exposed to corrosive attack, and films are formed. After fretting begins, valleys of the lower contact begin to fill with wear debris. Upon reversal of motion, these corrosion products are sheared off and fall into the valleys. When there is enough of an accumulation, the metal surfaces are separated by this nonconductive debris, and the contact resistance rises until connector failure occurs.

To calculate a failure time, we assume that during the $n$th fretting cycle, a corrosive layer of thickness $d_c$ forms over the exposed valley area $A_{ex}$, producing a volume of corrosion product $V_c$ equal to $d_c A_{ex}$. The volume of reactant metal that generated $V_c$ is equal to $1/\frac{V_c}{\gamma}$, where $\gamma$ is the ratio of the volume of the corrosion compound to volume of metal. For copper, aluminum, and chromium, $\gamma > 1$. Therefore, the asperity volume lost per cycle is $V_{asp} = V_c - \frac{V_c}{\gamma} = \left[\frac{\gamma-1}{\gamma}\right] V_c$. When the total asperity volume decrease is equal to a critical valley void volume $V^*$, failure occurs, or:

$$V^* = \sum_{n=1}^{N} \left[\frac{\gamma - 1}{\gamma}\right] V_c(n) = \sum_{n=1}^{N} \frac{\gamma - 1}{\gamma} \mathrm{d}_c\, A_{ex}(n). \tag{9.18}$$

For average values over fretting cycles, this equation can be rearranged to yield the number of cycles ($N_f$) to failure.

$$N_f = \frac{V^*}{\left(\frac{\gamma-1}{\gamma}\right) \mathrm{d}_c A_{ex}}. \tag{9.19}$$

In typical connectors, $N_f$ is predicted to range between $10^3$ and $10^6$ cycles. The normalized contact-resistance increase, as a function of corrosion debris volume, was theoretically shown to be constant until it reaches a value of 0.22 times the volume of the plastically deformed layer. A rapid resistance rise then signals failure in a behavior that parallels that shown in Figure 9.6(b).

## 9.4 CREEP AND FATIGUE OF SOLDER

### 9.4.1 Introduction

Rather than have contacts that open and close, it is often necessary to have permanent interconnections. As we have seen in Chapter 8, solders, perhaps the most ubiquitous of packaging materials, are almost universally employed for this purpose. While our previous concern was primarily with solder processing defects, we now focus on the reliability implications of soldered connections. For this purpose it is worthwhile reviewing Section 8.3, dealing with the metallurgical change in phase microstructure and composition as well as intermetallic-compound growth in solders exposed to elevated temperatures. Troublesome thermal stresses, treated in Section 8.5, also arise to promote a variety of microscopic structural changes within the context of much broader macroscopic plastic-flow effects.

In practice, thermal cycling of electrical hardware generates corresponding stresses that couple both creep and fatigue effects to produce dynamic strains that often have important consequences for packaging reliability. The resulting deformation will generally vary in a complex way depending on the frequency and amplitude of the thermal stress cycle. Once creep and fatigue are individually well understood, thermal-cycling phenomena, which combine differing proportions of creep and fatigue deformation, will be more intelligible. That is why this section is primarily reserved for separate descriptions of creep and fatigue phenomena in solders.

In discussing stress relaxation and fretting damage in contacts, we have already treated manifestations of creep and fatigue previously in this chapter.

### 9.4.2 Creep—An Overview

When loads are applied to materials at elevated temperatures, they creep or plastically strain, provided that both the stresses and temperatures are sufficiently high. In crystalline solids, creep produces a number of irreversible deformation effects, e.g., dislocation migration, and annealing, viscous grain-boundary sliding, a reduction in residual stress, void formation, softening, and mechanical relaxation. Creep in metals and ceramics generally means stress levels near the yield point and temperatures of approximately 0.5 $T_M$ and above, where $T_M$ is the melting point in degrees K. Electrical solder (60Sn–40Pb) melts at ~458 K (185 °C), and it is easy to appreciate its potential for creep even at room temperature, because 298/458 K = 0.65. Metal lead wires and tungsten filaments of incandescent bulbs operate at very high temperatures and are therefore strengthened to resist creep. Ceramic substrates that typically melt above 2000 °C will obviously not creep when used in electronic packaging. At the other extreme, polymers commonly creep in a viscous fashion at low temperatures, i.e., between the glass transition and melting points. Even the tiny Si seed crystal that supports the hot, heavy single crystal being pulled during Czochralski growth does not appear to creep.

Creep is usefully viewed as an admixture of two simultaneous effects; one is strain hardening, and the other is softening due to recrystallization and aging effects. Three regimes of creep deformation can be distinguished, depending on the magnitudes of the hardening and softening rates. When a tensile or shear load is applied under creep conditions, the material instantaneously stretches or strains the predictable elastic amount, after which the strain increases with time at an ever decreasing rate. In the *primary creep* stage the hardening rate exceeds that for softening. This primary-deformation stage then gives way imperceptibly to a regime of *steady-state*, or *secondary*, *creep*, where the hardening rate is balanced by the softening rate. As a result, the overall creep strain rate ($d\varepsilon/dt$) is constant. The design of operating conditions for creep-prone components is usually performed in this regime because time-dependent deformation can be confidently projected into the future. Steady-state creep phenomena also pose reliability concerns because potentially large plastic strains can develop. Finally, in the *tertiary creep* stage the strain increases rapidly with time, leading to failures characterized by much stretching and thinning, void cavitation at grain boundaries, and intergranular cracking. In our subsequent discussion of solder creep we will be exclusively concerned with these last two deformation regimes.

### 9.4.3 Constitutive Equations of Creep

In the broadest sense, the creep strain ($\varepsilon$) that develops is a function of time ($t$) as well as applied stress ($\sigma$) and temperature ($T$). Often, shear stresses ($\tau$)

and strains ($\gamma$) are involved, and we shall simply represent these by the generic $\sigma$ and $\varepsilon$ for simplicity. Furthermore, $\varepsilon$ is assumed to be a product of three separate factors, one of which is the ubiquitous Boltzmann factor; therefore, the constitutive equations for the creep strain and strain rates assume the respective forms

$$\varepsilon = f(\sigma)g(t)\exp\left[\frac{-E_c}{RT}\right] \quad (9.20a)$$

and

$$\frac{d\varepsilon}{dt} = \dot{\varepsilon} = f(\sigma)\left[\frac{dg(t)}{dt}\right]\exp\left[\frac{-E_c}{RT}\right]. \quad (9.20b)$$

Explicit expressions for the $f(\sigma)$ and $g(t)$ factors, and the researchers associated with them, are [15]

| | | | |
|---|---|---|---|
| $f(\sigma) = A\,\sigma^n$ | Norton | $g(t) = \left(1 + et^{1/3}\right)$ | Andrade (primary creep) |
| $= B\sinh(b\,\sigma)$ | Prandtl | $= t$ | Steady-state (secondary creep) |
| $= C\exp(c\,\sigma)$ | Dorn | $= t^m$ | Bailey (tertiary creep) |
| $= D\sinh(f\sigma)^n$ | Garafalo | | |

where $A$, $B$, $C$, $D$, $b$, $c$, $e$, $f$, $n$, and $m$ are constants, $E_c$ is the creep activation energy, and $RT$ has the usual meaning. For crystalline solids, $E_c$ is close in magnitude to a diffusional activation energy because creep cannot occur without simultaneous atomic motion. Depending on whether deformation primarily occurs within grains or at grain boundaries, activation energies applicable to these mechanisms would be appropriate. As we shall see, the stress exponent $n$ typically ranges from 1 to 8, depending on the material, stress/temperature range, and creep mechanism.

In each creep regime an equation for both $\varepsilon$ and $d\varepsilon/dt$ can now be constructed by selecting the appropriate forms of ($\sigma$) and $g(t)$. For short deformation times where primary creep occurs, the strain rate decreases with time. Therefore, such behavior can be described by the Andrade form of $g(t)$, where $d\varepsilon/dt \approx t^{-2/3}$. Shortly prior to failure, in the tertiary creep regime, the reverse behavior is operative, and the strain rate increases with time. The use of the Bailey time dependence is consistent with deformation in this range because $d\varepsilon/dt \approx t^{m-1}$, and $m > 1$.

Between these extremes in behavior, the extended steady-state creep regime for many materials is very commonly described by the phenomenological equation:

$$\frac{d\varepsilon}{dt} = A\,\sigma^{n}\exp\left[-\frac{E_c}{RT}\right]. \tag{9.21}$$

This expression generally holds up to intermediate stress levels ($\sigma/G < 10^{-3}$, where $G$ is the shear modulus) at temperatures above 0.5 $T_M$. Among the more significant terms absorbed within constant $A$ is $(1/d)^p$, where $d$ is the grain size and $p$ is a number dependent on the type of creep. So-called diffusional creep prevails at high temperatures and low stress levels ($\sigma/G < 10^{-4}$) such that $n = 1$. In such a case creep can be understood in terms of the Nernst–Einstein model (Eqns (9.1)–(9.6)), with stress the driving force and velocity the strain response. Two categories of diffusional creep have been distinguished. Herring–Nabarro diffusional creep, with $p = 2$, occurs at the higher temperatures, where lattice diffusion occurs. On the other hand, Coble creep, with $p = 3$, occurs at lower temperatures, where grain-boundary diffusion dominates.

At higher stress levels, power-law creep due to diffusion-aided dislocation motion is operative, and values of $n = 3$ to 7 have been predicted. However, as we shall see in the next section, Eqn (9.21) does not fully cover steady-state creep phenomena in solders at all stress levels and temperatures.

### 9.4.4 Solder Creep

In response to the thermal cycling of electronic equipment, stresses develop that cause creep deformation in solder interconnections. Two books edited by J. H. Lau [16,17] are devoted to this subject; they are highly recommended for their broad review of all aspects of thermal stress and creep in solders, as well as the reliability implications in electronic packaging. In addition, much creep test data for solder alloys used in packaging has accumulated in the literature [18–20]. Creep behavior is most commonly plotted as creep strain versus time, and a sample curve is shown in Figure 9.7. Data points for the three creep regimes lie below (primary), above (tertiary), or on the straight line (secondary).

In an extensive study [18] of the behavior of five different solder compositions, creep testing was performed on an array of flip-chip solder ball connections in both shear (loading parallel to the chip–substrate plane) and tension (loading perpendicular to the chip–substrate plane). While the

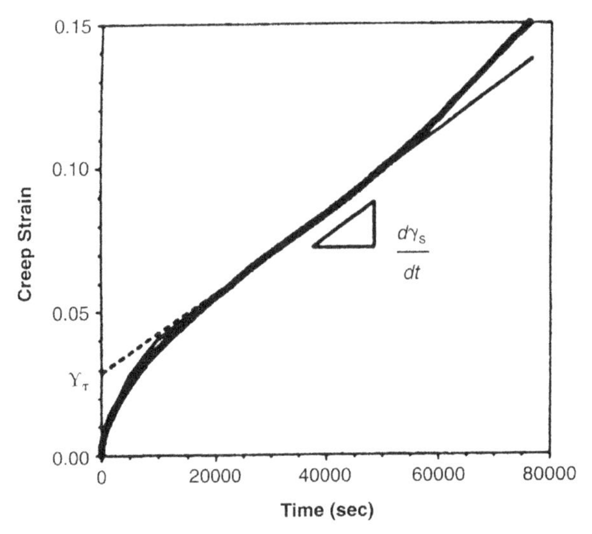


**Figure 9.7** Creep shear strain as a function of time in a constant load creep test. The solder containing 62Sn–36Pb–2Ag was tested at 28 °C with a shear stress of 15,751 bs/in$^2$. *From Ref. [18].*

power law fits data spanning some five orders of magnitude of strain deformation, it breaks down at stress levels of $\sigma/G$ higher than $10^{-3}$. Therefore, rather than Eqn (9.21), the experimental results in this intermediate stress range were analyzed by the Garafalo dependence in the form:

$$\frac{d\varepsilon}{dt} = C\left(\frac{G}{T}\right)\sinh\left(\frac{a\sigma}{G}\right)^n \exp\left[-\frac{E_c}{RT}\right], \tag{9.22}$$

where $C$, $a$, and $n$ are constants. An experimental test of this equation in dimensionless shear strain–rate–shear stress units is shown in Figure 9.8, demonstrating a good fit over the entire stress range for 60 Sn–40 Pb solder.

Accurately predicting the extent of solder deformation is a matter of precisely determining its creep characteristics. Experimentally measured values for the constants of a number of solder alloys are entered in Table 9.1. Theory has shown that stress exponents ($n$) of ~3 imply viscous dislocation glide, while $n$ values of ~5–7 are associated with vertical climb of dislocations past barriers. Later we will consider the physical implications of these exponents on stress relaxation of solders during thermal cycling. The activation-energy values for these alloys generally support

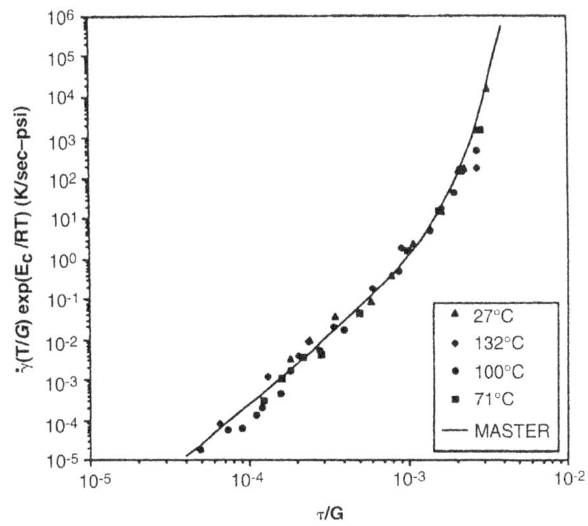


**Figure 9.8** Steady-state creep behavior of 60Sn–40Pb solder plotted as $\frac{d\gamma}{dt}\left(\frac{T}{G}\right)\exp\left[\frac{E_c}{RT}\right]$ versus $\tau/G$. When the data are fit to Eqn (9.22), i.e., $d\gamma/dt = C(T/G)\cdot[\sinh\cdot(a\tau/G)]^n\cdot\exp[-E_c/RT]$, values of constants derived from the plot are $C = 0.198$ K/psi-s, $a = 1300$, $n = 3.3$ and $E_c = 0.548$ eV. *From Ref. [18].*

**Table 9.1** Creep constants for solder alloys

| **Composition** | ***C* (K/psi-s)** | ***a*** | ***N*** | **$E_c$ (kcal/mol)** |
|---|---|---|---|---|
| 96.5Sn–3.5Ag | $4.89 \times 10^4$ | 1300 | 5.5 | 9.2 |
| 62Sn–36Pb–2Ag | $98.9 \times 10^{-3}$ | 1300 | 3.3 | 12.6 |
| 60Sn–40Pb | $198 \times 10^{-3}$ | 1300 | 3.3 | 12.6 |
| 97Pb–3Sn | $5.04 \times 10^7$ | 1200 | 7.0 | 27.7 |
| 95Pb–5Sn | $1.21 \times 10^7$ | 1000 | 7.0 | 27.7 |
| 96.5Sn–3.5Ag | $3.13 \times 10^{-3}$ | 1500 | 5.5 | 11.5 |

From Refs [17–19].

stress-enhanced diffusional or creeping motion of atoms through grain boundary paths.

In general, homologous-temperature arguments suggest that lower melting temperature solder alloys deform more readily at a given stress than those with higher melting points. Thus, if the electronic package environment demanded that solder joints accommodate large amounts of strain, the low melting temperature compositions should be chosen; under these conditions the higher melting point solders may stress the joints by not relaxing sufficiently.

## Example 9.2

In flip-chip interconnections using 95 Pb–5 Sn solder, suppose that stresses encountered in service fluctuate between $\sigma/G = 7 \times 10^{-4}$ and $1 \times 10^{-3}$ at temperatures between 25 and 35 °C. A measure of contact degradation is taken as the ratio of creep strains at the two stress levels. What is the difference in the predicted strain ratio using Eqn (9.21) as opposed to Eqn (9.22)?

**Answer** We refer to the high stress/temperature state as 2 and the low stress/temperature state as 1. In the case of Eqn (9.21), noting that $n = 7.0$ and $E_c = 27.7$ kcal/mol.

$$\frac{\dot{\varepsilon}_2}{\dot{\varepsilon}_1} = \frac{A\sigma_2^n \exp\left[-E_c/RT_2\right]}{A\sigma_1^n \exp\left[-E_c/RT_1\right]} = \frac{(1 \times 10^{-3})^7 \exp\left[-27,700/(1.99(308))\right]}{(7 \times 10^{-4})^7 \exp\left[-27,700/(1.99(298))\right]} = 55.3.$$

In contrast, the use of Eqn (9.22) where $a = 1000$ and $n = 7.0$, yields:

$$\frac{\dot{\varepsilon}_2}{\dot{\varepsilon}_1} = \frac{(298)(\sinh 1)^7 \exp\left[-27,700/(1.99(308))\right]}{(308)(\sinh 0.7)^7 \exp\left[-27,700/(1.99(298))\right]} = 9.4.$$

Therefore, depending on which constitutive equation is chosen, a difference by a factor of 5.9 can result in estimating strain acceleration.

### 9.4.5 Fatigue—An Overview

Failures due to fatigue are the most common and insidious that metals suffer from. After propagating for some time, unnoticed cracks suddenly and without warning cause fracture of components. Although fatigue is popularly associated with the failure of highly stressed rotating machinery, it also occurs in electrical equipment in the form of damage to moving contacts as well as vibrating and thermally cycled components. The cyclic loading can be sinusoidal about a mean zero stress level, but any other high- or low-frequency quasiperiodic stress distribution can also cause fatigue failure. Some materials, like steel, display an endurance limit, a critical stress level below which failure does not occur irrespective of the number of cycles; aluminum and polymers show no such endurance limit. These as well as other fatigue phenomena are reviewed in standard texts on the mechanical properties of materials [21].

A microscopic model for fatigue damage in crystalline materials starts with cyclic dislocation displacements on slip planes in concert with the applied load. As stacks of slip planes emerge on the surface, they form intrusions and extrusions of material that become nuclei for incipient cracks. One such favored surface crack begins to open as it extends into the interior, inexorably driven by the external cyclic stresses. In round sections, the smoothly parted crack surfaces grow outward in concentric circles amid the continuous back-and-forth stressing and rubbing action. Ultimately, very little of the original material remains intact to support the load, and rapid fracture ensues due to overloading. Surface microcracks, scratches, hard phases, notches, and other stress concentrators facilitate fatigue-crack formation, especially in corrosive environments.

In examining cyclic loading effects more carefully, it has been observed that metals are metastable and may either harden or soften with time [22]. Whether metals cyclically harden or soften is apparently related to whether they respectively exhibit large or low strain hardening effects in a simple tensile test. Another way to predict what a metal will do cyclically is to take the ratio of the ultimate tensile strength ($\sigma_{UTS}$) to the 0.2% offset yield stress ($\sigma_o$). If $\sigma_{UTS}/\sigma_o > 1.4$, hardening usually occurs, and if $\sigma_{UTS}/\sigma_o < 1.2$, softening is expected; in between, either or mixed behavior occurs. Solders cyclically soften, and depending on composition and cycling conditions the magnitude of the effect can be quite pronounced.

The case of a cyclically softening metal is shown in Figure 9.9, where the plastic strain amplitude is held fixed during periodic tensile and compressive straining cycles (Figure 9.9(a)). In concert, the stress follows the triangular strain cycle, but with a decreasing amplitude (Figure 9.9(b)). When time is eliminated from both cycling responses, the resulting pairs of stress–strain variables exhibit hysteresis behavior upon plotting one against

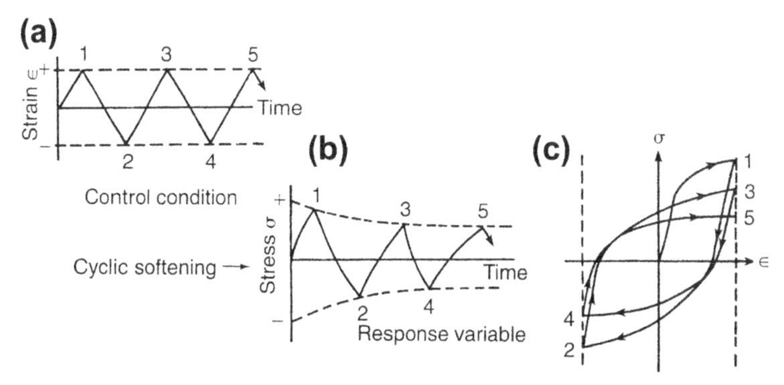


**Figure 9.9** The response of cyclic softening material to cyclic strain cycles. (a) Imposed triangular strain-time cycling. (b) Corresponding stress amplitude as a function of time illustrating cyclic softening. (c) Resultant stress–strain hysteresis loops shown exaggeratedly. *From Ref. [21].*

the other. The hysteresis loops obtained are shown exaggeratedly in Figure 9.9(c), where interestingly, their shapes vary with the number of cycles. But after the first few percent of the cycles-to-failure, the hysteresis tends to stabilize, and the loop retraces itself. Loading materials between fixed stress-amplitude limits also generates hysteresis curves that vary in shape with time. This time the width of the loop generally widens. Much fatigue testing and service loading is of this kind. Irrespective of the conditions of cyclic stressing, fatigue failures occur when enough cycles accumulate. Some of the stress and strain quantities used in the discussion below are defined within the context of the hysteresis loop displayed in Figure 9.10. To create this loop, the metal is first loaded elastically, deformed to a total tensile strain of $\Delta\varepsilon/2$ (or $\Delta\varepsilon_e/2 + \Delta\varepsilon_p/2$), and then elastically unloaded. Next, elastic and plastic compression occurs until a total strain of $-\Delta\varepsilon/2$ is attained. Finally, elastic unloading followed by tensile stretching to strain $+\Delta\varepsilon/2$ completes the cycle.

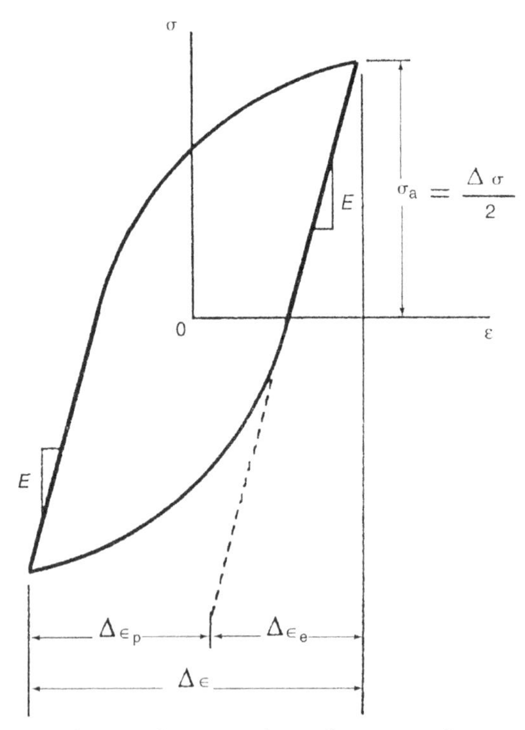


**Figure 9.10** Stress and strain hysteresis loop illustrating elastic and plastic strain contributions.

### 9.4.5.1 Quantifying Fatigue Failure

Engineers have traditionally presented fatigue data in the form of S–N curves, where the applied stress amplitude ($\sigma_a$) is plotted versus the log of the number of cycles to failure (log $N$). Large stress amplitudes mean failure in a relatively small number of loading cycles, while fatigue life is extended to many cycles when stress levels are low. Such curves are generally not linear. But semilog plots can often be straightened when the data are plotted in log–log fashion. Thus Basquin demonstrated that the ln $\sigma_a$–log $N$ plot was linear and suggested the exponential law of fatigue. This is normally written in the form:

$$\sigma_a = \sigma_f (2N_f)^b, \tag{9.23}$$

where $N_f$ is the number of stress reversals (1 reversal $= \frac{1}{2}$ cycle), $\sigma_f$ is the fatigue strength coefficient, and $b$ is Basquin's exponent. Typically, $\sigma_f$ is the true fracture stress, and $b$ varies between $-0.05$ and $-0.12$ for engineering materials.

More than a half century later, in 1955, Coffin and Manson suggested that a log–log plot of plastic strain ($\Delta\varepsilon_p$) extension in fatigue versus $2N_f$ was also linear. Again, plastic strain-life data are related by a power law, i.e.,

$$\frac{\Delta\varepsilon_p}{2} = \varepsilon_f (2N_f)^c, \tag{9.24}$$

where $\varepsilon_f$ is the fatigue ductility coefficient, a constant, and $c$ is the fatigue ductility exponent. Typically, $\varepsilon_f$ is the true strain at fracture, and $c$ varies between $-0.5$ and $-0.7$. Considerable reference to the Coffin–Manson equation will be made subsequently.

During fatigue the total strain ($\Delta\varepsilon$) that develops is the sum of elastic ($\Delta\varepsilon_e$) and plastic ($\Delta\varepsilon_p$) contributions, so that $\Delta\varepsilon = \Delta\varepsilon_e + \Delta\varepsilon_p$. By assuming that $\varepsilon_e = \sigma_a/E$, where $E$ is Young's modulus, we obtain:

$$\frac{\Delta\varepsilon}{2} = \frac{\sigma_f}{2E}(2N_f)^b + \varepsilon_f (2N_f)^c. \tag{9.25}$$

The total strain and its component elastic and plastic contributions are schematically depicted in Figure 9.11 as a function of the number of stress reversals; data for solder fatigue that obey Eqns (9.21) and (9.22) are plotted in Figure 9.12. Note that wide hysteresis loops are generated when plastic effects predominate, while narrow loops occur when elastic behavior is the rule. Because $\varepsilon_p$ dominates at short failure times, ductility or toughness

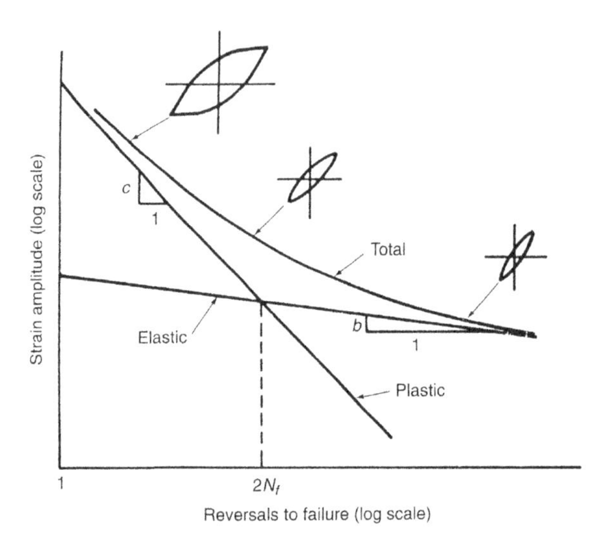


**Figure 9.11** Total strain and individual elastic and plastic contributions depicted as a function of the number of stress reversals. Changes in loop shape and size are schematically indicated. Note that the slopes of linear portions yield *b* and *c*.

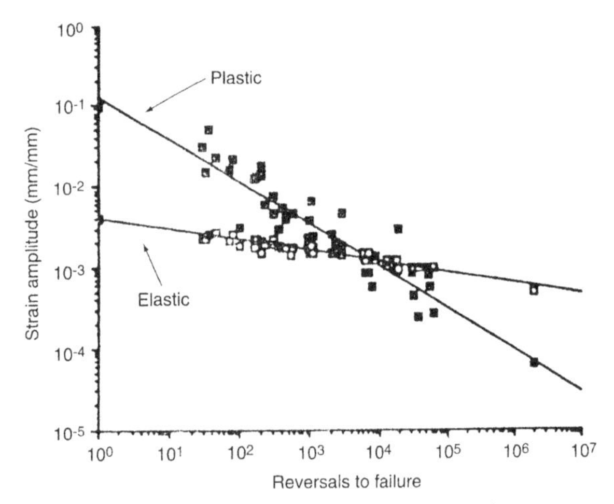


**Figure 9.12** Elastic and plastic strain contributions to solder fatigue versus number of reversals to failure. *From Ref. [23].*

controls performance in low-cycle fatigue. However, for long-lived behavior during high-cycle fatigue, performance is controlled by strength.

By recasting Eqn (9.24), we obtain

$$N_f = \frac{1}{2}\left(\frac{\Delta\varepsilon_p}{2\varepsilon_f}\right)^{1/c}, \tag{9.26}$$

an equation of the form that has been frequently used to predict solder fatigue lifetimes. With no loss in generality, elastic and plastic shear strains ($\gamma$) could be substituted for the tensile strains. This formula expresses the important connection between fatigue failure and the plastic strain that accumulates. Strictly speaking, Eqn (9.25) might be a more suitable criterion for failure, but since elastic strains are small, total and plastic strains are interchangeable for all practical purposes.

Fatigue does not usually occur at a single stress amplitude but is the result of cumulative damage occurring at a set of different stress or strain levels. For such cases the Palmgren–Miner linear cumulative damage rule is often invoked. This rule states that the fractional amount ($R$) of failure damage produced in $2N_\mathrm{i}$ reversals is expressed by $R = 2N_\mathrm{i}/2N_\mathrm{f}$, where $2N_\mathrm{f}$ represents the number of reversals to failure at the particular applied stress or strain. Since failure is the cumulative sum of $n$ fractional amounts of damage, we write the Palmgren–Miner rule as

$$1 = \Sigma R = \sum_{i=1}^{n} \frac{2N_\mathrm{i}}{2N_\mathrm{f}} \tag{9.27}$$

---

### Example 9.3

The solder of Figure 9.12 undergoes $10^3$ reversals at a strain of $1.3 \times 10^{-3}$. How many more cycles can the solder sustain before it fails at the larger strain of $10^{-2}$?

**Answer** Noting that $2N_\mathrm{f}$ at these respective strain levels is $\sim 10^4$ and $\sim 10^2$, Eqn (9.27) yields $1 = 10^3/10^4 + 2N_\mathrm{i}/10^2$. Solving, $2N_\mathrm{i} = 90$ reversals, or 45 cycles.

---

## 9.4.6 Isothermal Fatigue of Solder

Mechanical fatigue at constant temperature is important not only as a mode of solder-joint failure due to vibration, but it is helpful in describing and predicting the more complex problem of thermal fatigue. Solomon [24] has reviewed a number of issues related to fatigue at constant temperature, including cyclic softening and the effects of frequency and temperature; these phenomena are presented in turn.

### *9.4.6.1 Cyclic Softening*

Typical hysteresis loops for eutectic solder are shown in Figure 9.13 under testing conditions that maintain constant plastic strains or loop widths. As cycling continues, the load ($P$) required decreases. From such

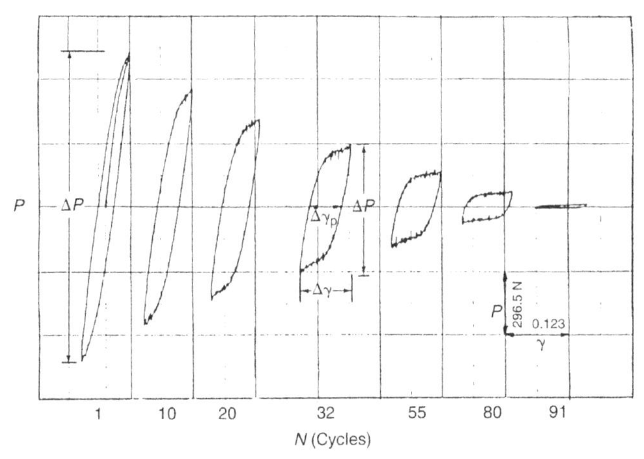


**Figure 9.13** Cyclic load hysteresis behavior of eutectic solder under total strain control. *From Ref. [25].*

data a load drop parameter is defined by $\phi = 1 - (\Delta P/\Delta P_{max})$, where $\Delta P$ is the load range in a particular cycle, and $\Delta P_{max}$ is the maximum value attained in the first few cycles. Significantly, a given level of $\phi$, just as a critical value of $\Delta\varepsilon_p$, may be taken as a criterion for failure. But what value of $\phi$ should be chosen? The problem is quite complicated because both $\phi$ and $\Delta\varepsilon_p$ are related, so that what constitutes fatigue life is a strong function of what we mean by failure. Nevertheless, fatigue–life parameters for a number of solders are entered in Table 9.2 for a Coffin–Manson relation of the form $N_f^{-c}\ \Delta\gamma = C$, where the values of the constants $c$ and $C$ depend on $\phi$.

**Table 9.2** Cofffin–Manson fatigue parameters for solders

| Composition | $-c$ | $C$ | Comments |
|---|---|---|---|
| 10Sn–90Pb | 0.42 | 0.93 | Air or vacuum, 300 K 95% load drop [26] |
| 63Sn–37Pb | $0.24 + 0.10\phi$ | $0.23 + 0.78\phi$ | 300 K [27] |
| 42Sn–58Bi | $0.32 + 0.15\phi$ | $0.10 + 0.80\phi$ | 300 K [27] |
| 43Sn–43Pb–14Bi | $0.19 + 0.08\phi$ | $0.23 + 0.27\phi$ | 300 K [27] |
| 55Sn–35Pb–10In | $0.32 + 0.14\phi$ | $0.14 + 2.76\phi$ | 300 K [27] |
| 59.3Sn–39.6Pb–1.1Sb | $0.23 + 0.13\phi$ | $0.38 + 1.04\phi$ | 300 K [27] |
| 96Sn–4Ag | $0.72 + 0.08\phi$ | $0.48 + 8.06\phi$ | 300 K [27] |

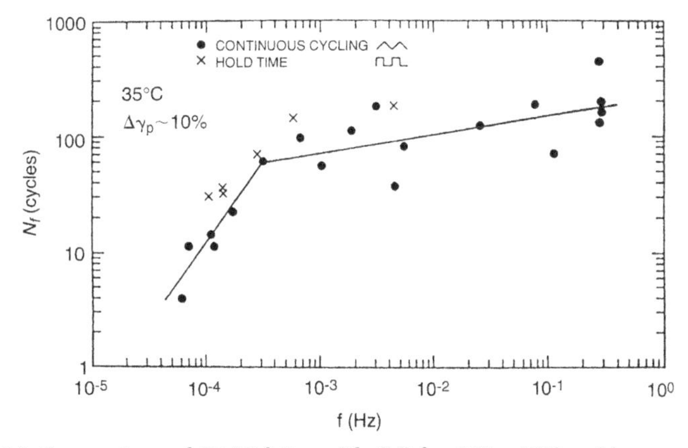


**Figure 9.14** Comparison of 35 °C fatigue life ($N_f$) for 60Sn–40Pb solder versus cycling frequency. *From Ref. [24].*

### *9.4.6.2 Frequency Dependence*

The effect of cycling frequency ($f$) on fatigue life is depicted in Figure 9.14 for 60Sn–40Pb solder. Amid considerable scatter two trends are discernable. At high frequency there is little or no effect; at lower frequencies, fatigue life is strongly dependent on frequency. A similar effect has been observed in the fatigue of superalloys. Such findings prompted the Coffin [28] frequency-modified version of the Coffin–Manson equation, i.e.,

$$N_f = \frac{1}{2} f^{1-k} \left( \frac{\Delta\varepsilon_p}{2\varepsilon_f} \right)^{1/c}, \tag{9.28}$$

where $K$, a constant, is a measure of the strength of the effect. Thus at 35 °C, $K \approx 1$ for frequencies ranging from $\sim 3 \times 10^{-4}$ to 1 Hz, while for $f$ below approximately $3 \times 10^{-4}$ Hz (or ~1 cycle an hour), $K \approx 0$. Furthermore, the frequency where the transition from $K = 1$ to $K = 0$ occurs rises with increasing temperature. It has been common to use a value of $K$ extrapolated from accelerated high-frequency testing to apply under use conditions where much lower frequencies are operative. In such cases fatigue lifetimes are dangerously overestimated.

### *9.4.6.3 Temperature Dependence*

It is abundantly documented that for solders, ultimate tensile strengths fall and creep (plastic) strain rates rise with increasing temperatures. In view of this, one expects fatigue life, or the number of cycles to failure, to be strongly temperature dependent. Confounding intuition, while there is a

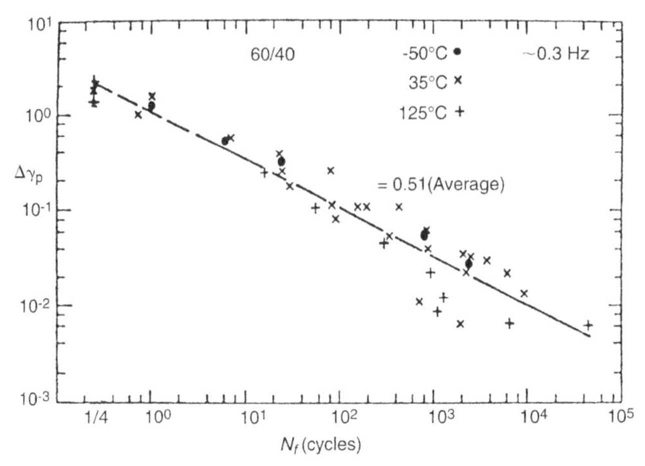


**Figure 9.15** Temperature dependence of low-cycle fatigue life for 60Sn–40Pb solder. A 50% load drop was taken as the failure criterion. *From Ref. [24].*

temperature dependence, it is not a large one nor, considering the statistical uncertainties in the data for 60Sn–40Pb solder shown in Figure 9.15, a particularly significant one [24]. The same is true of the high-lead solders. Solomon has suggested that claims for larger reported temperature dependencies in high-lead-tin solders may be exaggerated due to scatter in the data.

## 9.5 RELIABILITY AND FAILURE OF SOLDER JOINTS

### 9.5.1 Introduction to Thermal Cycling Effects

The previous sections dealing with creep and fatigue assumed that (1) isolated (unattached) solder materials were involved, and (2) deformation occurred under isothermal conditions. Now we address the more complicated issues associated with thermally cycled solder joints attached to IC chips, packages, and circuit boards. The objective is to understand the processes that lead to mechanical failure. We recall that the thermal stresses in solders (adhesives) that we calculated at great length in Chapter 8 were elastic shear stresses. However, in this chapter we have already noted that accumulated plastic strain is what causes failure of solder during creep or fatigue. How are elastic shear stresses, which after all produce small elastic strains, related to the much larger plastic strains? In addressing this question and others related to thermal cycling of solder joints, a review of thermal expansion mismatch effects is a good place to begin.

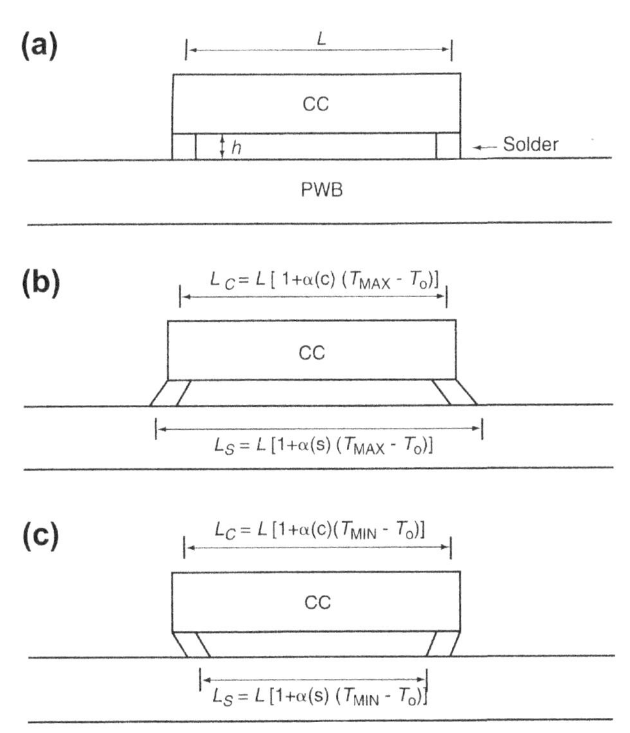


**Figure 9.16** Assorted thermal effects in soldered joint connecting a chip carrier (CC) to a substrate (PWB). Assembly at temperature $T_o$ with thermal expansion coefficients $\alpha(c)$ and $\alpha(s)$, such that $\alpha(s) > \alpha(c)$. (a) Initially unstressed solder joint. (b) Solder joint is heated to $T_{MAX}$, expanding PWB relative to CC. (c) Solder joint is cooled to $T_{MIN}$, contracting PWB relative to CC. *(From Ref. [24].)* Figure 9.16(d) idealized thermal strain for a through-board temperature gradient where $T_c > T_{ST} > T_{SB}$. (e) Idealized thermal strain for an in-plane temperature gradient where $T_s > T_O$. (f) Deformation of assembly due to thermal shock caused by temperature reversal. *From Ref. [29].*

### *9.5.1.1 Thermal Cycling Distortion of Solder Joints*

Following the study by Solomon [24], consider a chip or component (c) attached to printed wiring board (PWB) substrate (s) by means of a solder post of height $h$, all at temperature $T_o$. The thermal expansion coefficients $\alpha(c)$ and $\alpha(s)$ are such that $\alpha(s) > \alpha(c)$ for the geometry shown in Figure 9.16(a). We note that heating and cooling essentially fatigues the solder joints through the shear strain reversals depicted in Figure 9.16(b) and (c), respectively. It is not hard to calculate the strain that develops. First, taking the difference in the expanded lengths on heating to $T_{MAX}$ yields the net displacement of both joints as $L\,[\alpha(s) - \alpha(c)](T_{MAX} - T_o)$. Similarly, during cooling to $T_{MIN}$, the net displacement of both joints is $L\,[\alpha(s) - \alpha(c)]\,(T_{MIN} - T_o)$.

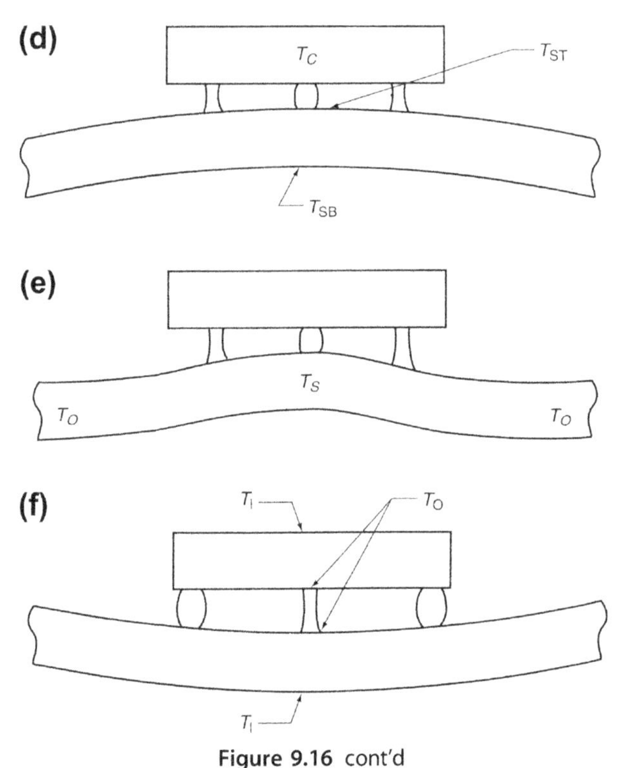


Figure 9.16 cont'd

The net total thermal displacement (Δ) after heating followed by cooling is the difference in these, or $\Delta = L\,[\alpha(s) - \alpha(c)](T_{MAX} - T_{MIN})$. By definition, the shear strain (γ) per joint is $\gamma = \Delta/(2h)$, so that:

$$\gamma = \left(\frac{L}{2h}\right)\left[\alpha(s) - \alpha(c)\right]\left(T_{MAX} - T_{MIN}\right). \tag{9.29}$$

### 9.5.1.2 Conversion of Elastic Strain to Plastic Strain

It might be imagined that thermal flexing of the assembly in Figure 9.16 would yield the same bend angles (positive and negative) for identical temperature excursions above and below the ambient (stress-free) temperature datum. This, after all, is how bimetallic strips respond to temperature. But as seen in Figure 9.17, this does not happen. Instead of reversible elastic bending, an asymmetrically shaped hysteresis loop develops, indicating that the thermal behavior of the solder is influencing the overall response of the assembly.

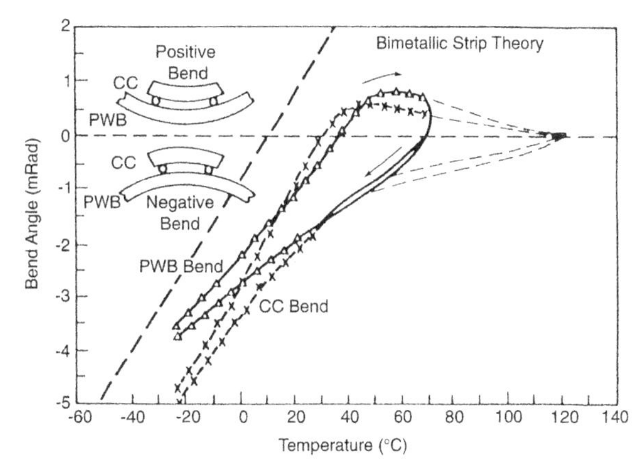


**Figure 9.17** Bend angles as a function of temperature in a soldered plate assembly subjected to reversed bending. Deviation from bimetallic strip theory is indicated. Hysteresis in bend angles (positive and negative) occurs due to temperature excursions above and below the ambient temperature (stress-free) datum. *From Ref. [30].*

This behavior is related to the question raised at the beginning of this section, which is now addressed. First we note that solder joints are elastically strained as well as stressed during thermal cycling. With time at elevated temperature, however, the elastic strain is essentially converted into plastic strain. This effectively limits the bend angle at elevated temperatures. To see why, we model the viscoelastic solder by a spring (of constant $k_s$) displaying elastic behavior, and by a dashpot (of viscosity $\eta$) that embodies its viscous nature. Solder behaves viscously during creep when dislocations and vacancies move or grain boundaries slide. In the classic Maxwell model of a viscoelastic solid [28] consisting of a spring and dashpot in series, consider what happens when this combination is stretched to a total strain ($\varepsilon_o$) that remains fixed thereafter. This is like the problem of a bolt initially tightened by force $P_o$ to a high stress level ($\sigma_o$) that then relaxes with time through viscous flow. The subsequent force (stress) relaxation behavior as a function of time ($t$) is described by:

$$P(t) = P_o \exp\left[\frac{-k_s t}{\eta}\right], \quad \text{or} \quad \sigma(t) = \sigma_o \exp\left[-\frac{t}{t_o}\right], \tag{9.30}$$

where the decay time constant $t_o$ is related to the ratio of Young's modulus to the effective viscosity. Again, with no loss in generality, shear stresses ($\tau$) can be substituted for tensile stresses in Eqn (9.30). Since the elastic strain ($\varepsilon_e$) is $\sigma(t)/E$, and $\varepsilon_o = \varepsilon_e + \varepsilon_p$, the plastic strain ($\varepsilon_p$) is given by

$$\varepsilon_p = \varepsilon_o - \left(\frac{\sigma_o}{E}\right)\exp\left(-\frac{t}{t_o}\right). \quad (9.31)$$

Thus, with time the elastic strain is irreversibly diminished and replaced by plastic strain. It is this accumulated plastic strain that eventually causes fatigue failure when it reaches a critical level.

#### *9.5.1.3 Thermal Distortion of Nonuniformly Heated Solder Joints*

In the cases of solder-joint distortion just treated, the chip, solder, and substrate were all assumed to be at the same temperature whether it was $T_{MAX}$, $T_{MIN}$, or $T_o$. But other scenarios for circuit-board distortion occur in practice. For example, in Figure 9.16(d) the chip carrier dissipates heat, so that there is a thermal gradient in the vertical direction. As a result of the hot spot, the top of the PWB is hotter ($T_{ST}$) than the bottom ($T_{SB}$), causing it to bow downward as shown. In addition to through-board gradients there may be in-plane thermal gradients generated by neighboring components. This causes similar convex bowing that buckles the PWB in the manner shown in Figure 9.16(e). In both of these cases it is assumed that the solder is totally relaxed, so that the outer solder posts are extended while the inner one is compressed.

Another environment of interest in packaging is thermal shock. Here the temperature is changed so rapidly that the inner parts ($T_o$) thermally lag the response of the outer parts ($T_1$). Therefore, by inverting the temperature distribution, the resulting distortion in Figure 9.16(f) is reversed.

### 9.5.2 Thermal-Stress Cycling of Surface-Mount Solder Joints

#### *9.5.2.1 Scope*

We now treat the case of thermal fatigue that occurs during thermal cycling of surface-mounted packages. Merely turning electrical equipment on and off is the simplest way to generate such periodic thermal stressing. The isothermal loading cycle that was considered in Section 9.4.6 is clearly a simpler process than the case now being addressed. Engelmaier and coworkers [31–33] have extensively studied the problem of thermal fatigue in surface-mount (SM) solder joints and the related issue of design for long-term reliability. Two types of SM joints were considered—leadless and leaded (see Figure 8.18). Leadless joints are stiff and sustain high stress levels, but occupy little space on the board. On the other hand, leaded joints are compliant, enabling them to reduce stress levels below the solder yield stress, thus minimizing cyclic fatigue damage as a result. The lower stress levels, in turn, slow and may even arrest harmful viscoplastic stress-relaxation and creep processes. As a result, leaded joints suffer lead fracture, while solder fatigue is the usual failure mechanism in leadless joints. The difference in these two joints is reflected in their response to thermal cycling, as indicated in the hysteresis loops of Figure 9.18(a) and (b). Simulated hysteresis loops for a number of different thermally cycled solders are plotted in Figure 9.19 and reveal similar behavior.

#### 9.5.2.2 Leadless Joints

In the case of the leadless SM solder joint at high temperature, elastic–plastic, shear stress–strain deformation occurs. Beyond the initial elastic response in Figure 9.16(a), the solder essentially yields at a constant stress

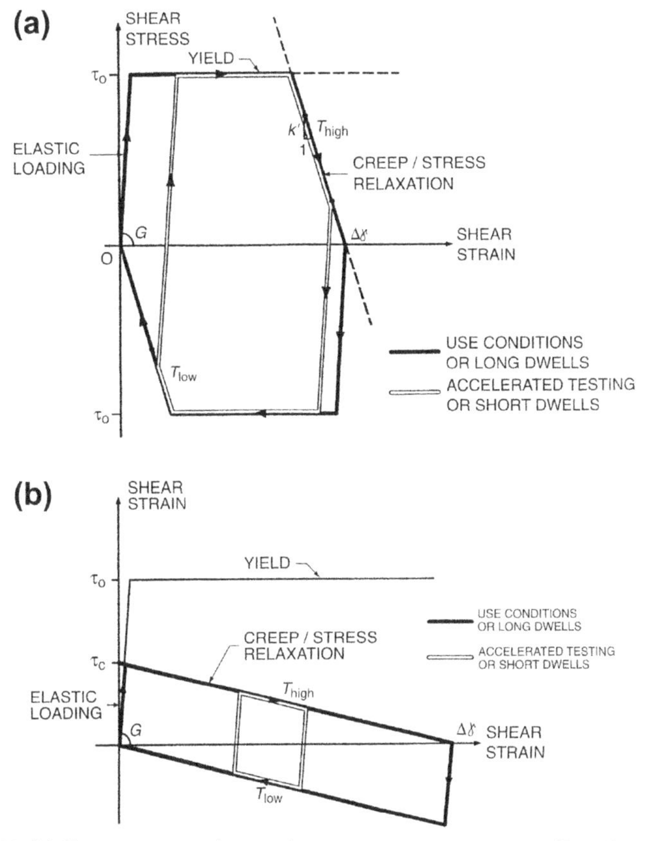


**Figure 9.18** (a) Shear stress and strain hysteresis response in a stiff, surface-mounted leadless joint. Use (long dwell) as well as accelerated (short dwell) cycles are indicated. (b) Shear stress and strain hysteresis response in a compliant, surface-mounted leaded joint. *From Ref. [31].*

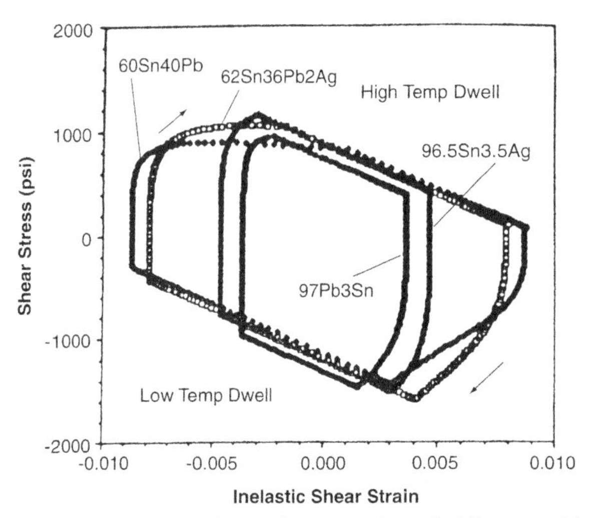


**Figure 9.19** Simulated hysteresis loops for a number of different solders, thermally cycled once each day, plotted as shear stress versus shear strain. The high and low temperature dwells are at 25 °C and −75 °C, respectively. *From Ref. [18].*

level ($\tau_o$), and depending on the ductility, large strains can be induced. With continued exposure to elevated temperature ($T_{high}$), the solder relaxes along a line dependent on the assembly stiffness. During this time, stress relaxation and creep processes are activated to extend the plastic strain. Given enough time the stress can even relax to zero. These stress and strain effects are reversed in the low-temperature ($T_{low}$) portion of the thermal cycle. When the initial temperature is again reached, the stresses and strains are ideally restored to zero; but the damage has been done in the form of plastic strain accumulation in the solder. The fact that the hysteresis loop appears symmetric does not mean that equal times were spent at high and low temperatures. In fact, longer times are required at $T_{low}$ than at $T_{high}$ to produce equivalent plastic strains. It should be noted that during accelerated thermal cycling, dwell times are shorter and stress relaxation may be incomplete. This results in the reduced hysteresis curves.

Due to the large thermal stresses in leadless joints, rapid and extensive stress relaxation causes plastic-strain accumulation. In such a case the relation for isothermal cyclic life (Eqn (9.24)) is applicable, but it must be modified by the operative temperature–time variables. The equation

$$N_f = \frac{1}{2}\left(\frac{\Delta\gamma}{2\varepsilon_f}\right)^{1/c} \tag{9.32}$$

has been used to model thermal fatigue, where for 63Sn–37Pb solder, $2\varepsilon_f = 0.65$ and $c = -0.442 - 6 \times 10^{-4} T_m + 1.74 \times 10^{-2} \ln(1 + 360/t_D)$. This complicated form of the fatigue ductility exponent accounts for both the mean cyclic solder-joint temperature ($T_m$) and half cycle dwell time ($t_D$ in minutes). To use Eqn (9.32), the plastic shear range ($\Delta\gamma$) must be known, and Eqn (9.29) can be used for this purpose. In rectangular packages $L$ is taken as the diagonal length of the chip because solder joints beneath the chip corners suffer the maximum stresses and are most susceptible to fatigue.

Finally, the mean cyclic life of leadless joints can be expressed by:

$$N_f = \frac{1}{2}\left\{L\frac{[\alpha(s) - \alpha(c)](T_{MAX} - T_{MIN})}{2h(2\varepsilon_f)}\right\}^{1/c}. \tag{9.33}$$

Quantitative agreement between this formula and the fatigue life of ceramic chip carriers of varying size on assorted substrates has been achieved [31].

#### 9.5.2.3 Leaded Joints

In an entirely parallel manner we now consider the corresponding issues for leaded SM solder joints, starting with the hysteresis behavior (Figure 9.18(b)). Initially the stress–strain response is elastic at elevated temperature. But if the lead is sufficiently compliant, it deflects easily and keeps stress levels below those required to cause yielding in the solder. The stress rises only to $\tau_c$ and then relaxes as the solder now creeps along isothermal lines of lesser slope than for the stiffer leadless joints. Given sufficient time, the stress relaxes to zero. Again reversal of stress and strain at lower temperature completes the loop.

Less plastic strain energy, as measured by the loop area, accumulates in leaded joints because the lower stress levels prevent extensive plastic straining. Therefore, Eqn (9.32) is an inappropriate criterion for joint lifetime. Instead, the viscoplastic strain energy ($\Delta W$), proportional to the area of the hysteresis loop, has been proposed as a measure of fatigue damage. Since $\Delta W$ effectively substitutes for $\Delta\varepsilon_p$, the formula:

$$N_f = \tfrac{1}{2}\left(\frac{\Delta W}{W_j}\right)^{1/c}, \tag{9.34}$$

with $W_j$ a constant, has been used to predict fatigue lifetimes. Basically, the loop area ($\Delta W$) is a product of the largest stress ($\Delta\tau_{MAX}$) and $\Delta\gamma$. We have already developed a formula for $\Delta\gamma$ in Eqn (9.29) and will assume that $\Delta\tau_{MAX}$ is proportional to $\Delta\gamma$ according to Hooke's law. Thus, $\Delta\tau_{MAX} = K_F (L/A) [\alpha(s) - \alpha(c)] (T_{MAX} - T_{MIN})$, where $K_F$ is the diagonal flexural stiffness (spring constant) of the package and $A$ is the effective solder area. Combining these terms,

$$\Delta W \approx \Delta\tau_{MAX}\Delta\gamma \approx \frac{K_F}{Ah}\{L[\alpha(s) - \alpha(c)](T_{MAX} - T_{MIN})\}^2. \tag{9.35}$$

Through substitution in Eqn (9.34), we finally express the mean cyclic life of leaded joints by:

$$N_f = 1\Big/2\left[\frac{K_F\{L[\alpha(s) - \alpha(c)](T_{MAX} - T_{MIN})\}^2}{AhW_j}\right]^{1/c}. \tag{9.36}$$

Agreement between this formula and cyclic tests for various SM component assemblies has been realized [34].

In summary, strategies used to extend SM solder-joint life essentially follow two directions, namely.

1. tailor the coefficients of thermal expansion (CTE), and
2. increase attachment compliancy.

Tailoring to reduce expansion mismatch shrinks the cyclic hysteresis loop primarily by reducing the strain range. On the other hand, attachment compliancy works by reducing the cyclic stress level.

### 9.5.3 Acceleration Factor for Solder Fatigue

The simplest way to transform the results of accelerated stress-cycling testing (s) to predict fatigue behavior under use conditions (u) is through a direct ratio of the involved failure times. Therefore, we may write for the acceleration factors.

$$\mathrm{AF} = \frac{N_f(u)}{N_f(s)} = \frac{[\Delta\gamma(u)]^{1/c(u)}}{[\Delta\gamma(s)]^{1/c(s)}} \tag{9.37a}$$

and

$$\mathrm{AF} = \frac{[\Delta W(u)]^{1/c(u)}}{[\Delta W(s)]^{1/c(s)}} \tag{9.37b}$$

for leadless and leaded joints, respectively. These expressions implicitly account for temperature and cycle duration in the values selected for $c$.

An early and still popular method for determining AF in solder-joint fatigue processes is due to Norris and Landzberg [35], who proposed the formula:

$$\mathrm{AF} = \frac{N(\mathrm{u})}{N(\mathrm{s})} = \left[\frac{f(\mathrm{u})}{f(\mathrm{s})}\right]^{1/3} \left(\frac{\Delta T(\mathrm{s})}{\Delta T(\mathrm{u})}\right)^{2} \left(\frac{\mathrm{MTTF(u)}}{\mathrm{MTTF(s)}}\right). \tag{9.38}$$

The symbols $N$, $f$, $\Delta T$, and MTTF refer to the number of cycles to failure, cycling frequency, temperature range, and isothermal fatigue-test lifetime, respectively. Often $\Delta T$ is taken as the maximum temperature in °C. With respect to MTTF(s) the isothermal test is performed at the peak temperature used in accelerated thermal cycling. As suggested by Eqn (9.28), a single value for $K = 0.67$ has been selected here; in view of considerations raised in Section 9.4.6.2, this assumption requires scrutiny.

To end our discussion of acceleration factors it is instructive to present the following example of Lau et al. [36].

---

### Example 9.4

Accelerated temperature cycling of solder joints on a 256 pin quad flat package (QFP) was carried out at a frequency of 1 cycle/45 min over the range −40 to 125 °C. The Weibull cumulative failure distribution was found to be of the form $F_S(t) = 1 - \exp[-(t/42{,}900)^{1.27}]$, where $t$ is the time in minutes. This chip will be used in a computer, where it undergoes 1 thermal cycle per day and is exposed to $\Delta T(\mathrm{u}) = 85$ °C. If it is assumed that MTTF(u)(85 °C)/MTTF(s)(125 °C) = 1.7.

1. What is the acceleration factor?
2. Predict the failure rate under use conditions.
3. Derive an expression for $F_U(t)$.

**Answer**

1. Substituting in Eqn (9.38), noting that 1 cycle/45 min = 32 cycles/day, AF = $(1/32)^{1/3}\,(165/85)^2\,(1.7) = 2.02$.
2. From Eqn 4.42 and the definition of Weibull failure rate (Eqn 4.15), $\lambda_U(t) = (1/2.02)^{1.27} \times (1.27/42{,}900)(t/42{,}900)^{0.27}$ or $0.0{,}000{,}121(t/42{,}900)^{0.27}$.
3. From Eqn 4.39, $F_U(t) = F_S(t/\mathrm{AF})$. Therefore, $F_U(t) = 1 - \exp[-(t/2.02 \times 42{,}900)^{1.27}] = 1 - \exp[-(t/86{,}700)^{1.27}]$.

---

## 9.5.4 Fracture of Solder Joints

### 9.5.4.1 Fracture Mechanics—An Overview

Mechanical fracture of semiconductor chips, plastic packages, and ceramic substrates are no doubt the most severe processing defects and reliability concerns we encounter. The subject of fracture mechanics is a very fruitful way to address these failures as well as those of solder joints. By assuming that defects are an unavoidable consequence of manufacturing, fracture mechanics addresses their implications with respect to mechanical failure of the surrounding matrix. The traditional analysis of the energetics at play when a defect or incipient crack of critical size ($a$) rapidly opens under applied stress $\sigma$, yields the quintessential formula of fracture mechanics:

$$K_{1C} = \sigma(\pi a)^{1/2}. \tag{9.39}$$

It concisely relates processing defects ($a$), material properties ($K_{1C}$), and design stresses ($\sigma$) in a neat way. In the application of this equation, the material property known as the fracture toughness ($K_{1C}$) plays a key role. Essentially, when $K_{1C}$ is greater than $\sigma\,(\pi a)^{1/2}$ the material is tough enough to withstand the effect of stress concentration about the crack. When both sides of Eqn (9.39) are equal, a precarious mechanical balance is maintained and the mechanically sensitive crack, though on the verge of propagating, is still stable. But when $\sigma(\pi a)^{1/2}$ exceeds $K_{1C}$, mechanical instability causes the crack to open rapidly.

The critical difference in toughness between metals and other classes of materials stems from the presence of nondirectional bonds. Dislocation motion enables atoms in metals to slide readily by one another, facilitating the plastic deformation that blunts the advance of cracks. No such mechanism exists in the case of other materials. Therefore, metals are tough and have large $K_{1C}$ values. Conversely, ceramics, semiconductors, and polymers are not tough, with $K_{1C}$ values an order of magnitude or more lower than for metals; even small defect cracks are critical and readily lead to brittle fracture.

In solder-contact applications we are concerned not only about the fracture toughness of solders, leads, and substrates, but also with the intermetallic phases that form at interfaces between them. Values of $K_{1C}$ for some of these materials are listed in Table 9.3. Although they are not easy to measure and subject to wide variations, the $K_{1C}$ values for bulk solders are probably larger than those for the Cu–Sn IMCs. Values for the latter, in turn, appear to be larger than for other intermetallics, which are barely tougher than typical ceramics. Importantly, it is clear that thicker interfacial compound layers are more fracture-prone than thinner layers.

**Table 9.3** $K_{1C}$ (MPa-m$^{1/2}$) values for solder and related materials

| | Layer thickness | | | | |
|---|---|---|---|---|---|
| Solder | 1 μm | 4 μm | 8 μm | 10 μm | >10 μm |
| 60Sn−40Pb | 8 [37] | 6 [37] | 6 [37] | 6 [37] | 71 [38] |
| ***Intermetallics*** | | | | | |
| $Cu_6Sn_5$ [39,40] | 12, 27 | 7, 22 | 6, 19 | | 1.4 |
| $Cu_3Sn$ [39] | 12 | 6 | 6 | | 1.7 |
| $Ni_3Sn_4$ [40] | 21 | — | — | | |
| Sn−Ag [37] | | | | | 1 |
| Sn−Bi [37] | | 5 | | | |
| Sn−Ag−Bi [37] | 5 | | | 2 | 3 |
| ***Substrates*** | | | | | |
| Cu | | | | | >100 est. |
| Ni | | | | | >100 est. |
| $Al_2O_3$ | | | | | 3.7 |

An interesting way to think about fracture was already presented in Section 1.3.5 dealing with load-strength interference effects. The fracture strength, or toughness, of materials and structures in solder joints is statistically distributed because of the assorted phases and processing defects present, and their respective size variations. Applied loads are also subject to statistical spread because of the their irregular static and dynamic character, and the variety of joint geometries and associated stress concentrations. Initially there is no load-strength interference because the respective Gaussian distributions do not overlap. However, through cyclic thermal stressing, the strength Gaussian distribution of the solder falls and probably broadens while a similar, but perhaps lesser change in loading occurs. The centroids of the two distributions approach each other, raising the probability of overlap and joint fracture.

### *9.5.4.2 Crack Propagation Rates*

Cracks in solders have been reported to nucleate at solder–base metal interfaces and interfacial IMCs, as well as at highly stressed surface points [38]. Once nucleated, crack propagation occurs through the bulk of the solder as well as along the solder-base metal interface. Some examples of solder fractures in through-hole and SM joints are reproduced in Figures 9.20 and 9.21, respectively. These are taken from a larger collection of beautiful solder structure and failure photographs by de Kluizenaar [41a,41b].

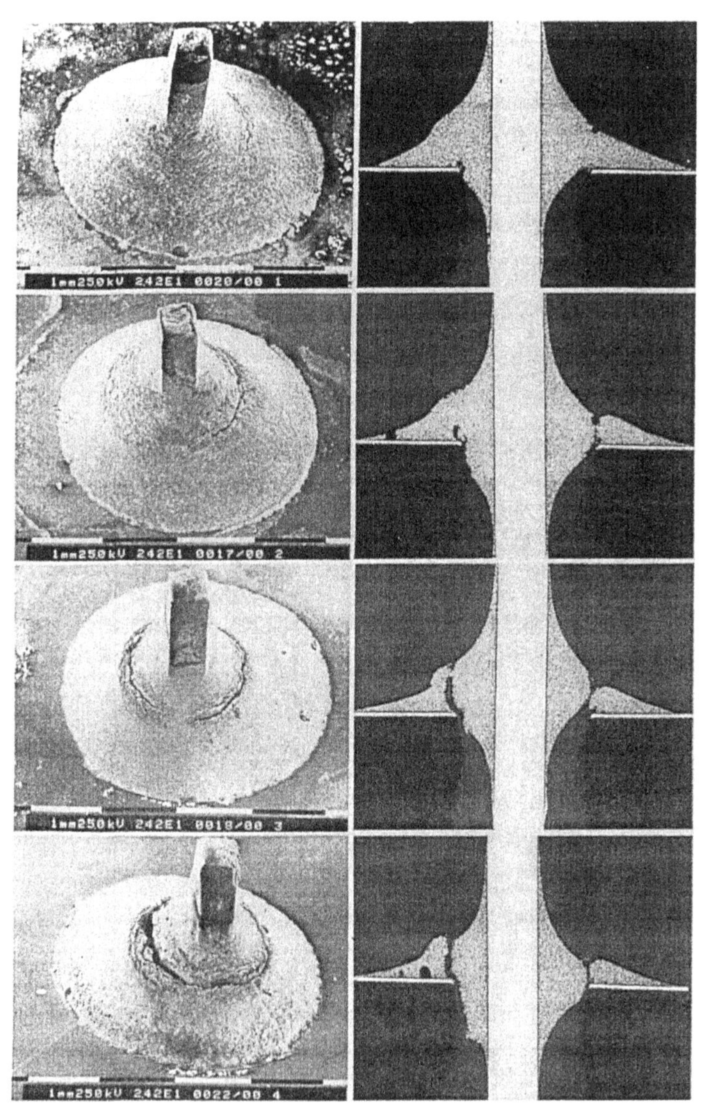


**Figure 9.20** Various stages of fatigue fracture in a through-hole solder joint connecting a lead to a single-sided circuit board. *From Ref. [41b].*

Irrespective of where it occurs, fatigue-fracture propagation has been modeled to yield the number of cycles to failure. We start by noting that at arbitrary, noncritical stress levels the stress intensity ($K$) is given by $K = (\pi a)^{1/2}\ \sigma$ (see Eqn (9.39)). The assumption is then made that crack growth depends on some power of cyclic stress intensity range to yield the Paris equation.

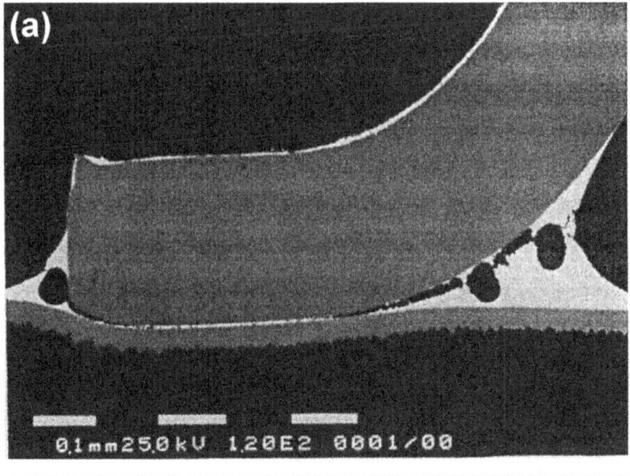

**Figure 9.21** A collection of surface-mount solder-joint fractures. (a) Porosity at solder–lead interface. (b) Creep rupture of reflowed solder joint with lead containing a large elastic deformation. Figure 9.21(c) General view of fatigue fracture of a soldered joint. (d) Cross-sectional view of fatigue fracture shown in Figure 9.21(c). *From Ref. [41b].*

$$\frac{da}{dN} = B(\Delta K)^m = B(\pi a)^{m/2}(\Delta\sigma)^m, \tag{9.40}$$

where $B$ and $m$ are constants derived directly from experiment. Physically, the crack growth rate is necessarily proportional to the amplitude of the applied peak-to-peak stress, $\Delta\sigma$, or difference between the maximum and minimum applied stress. Distinctions in this formula may arise depending on whether mode I (tension) or II (shear) cracks are involved. In addition, corrections for cracks of different geometries alter the constants slightly. This differential equation then directly links crack growth to applied stress, material constants, and instantaneous crack size. Below the

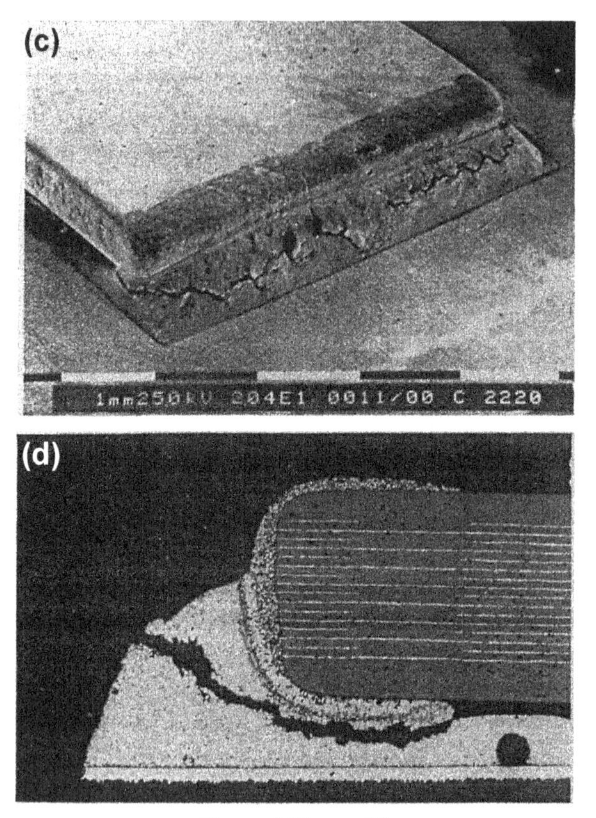


Figure 9.21 cont'd

indicated stress levels for applicability of Eqn (9.40), cracks do not grow appreciably, while above them crack growth is nonlinear.

If the crack (defect) size is initially $a_o$ the number of cycles $N_f$ required to extend it to a final length $a_f$ is easily obtained by first separating variables $a$ and $N$, and integrating, i.e.,

$$\int_0^{N_f} dN = \int_{a_o}^{a_f} \frac{da}{B\left[\Delta\sigma(\pi a)^{1/2}\right]^m}$$

Substitution of limits yields

$$N_f = \frac{2\left\{a_o^{-(m-2)/2} - a_f^{-(m-2)/2}\right\}}{(m-2)B\pi^{m/2}(\Delta\sigma)^m}, \quad m \neq 2. \tag{9.41}$$

The sparse amount of data on $B$ and $m$ values makes it difficult to apply this formula for particular solder-joint failure applications. Crack growth-rate data for eutectic Pb–Sn solder are shown in Figure 9.22, where amid much scatter Eqn (9.40) appears to be followed. From such curves both $B$ and $m$ can be extracted. But the question of what constitutes failure means the use of $\phi$. These same authors [42] also claim that eutectic solder is more resistant to crack growth than the 96Sn–4Ag and 42Sn–58Bi compositions. In using Eqn (9.41) to solve practical problems, $a_o$ and $a_f$ must be selected. The latter may be taken as the joint length, while a measure of $a_o$ is the interdendritic spacing within the microstructure, e.g., 0.01–0.1 μm.

A second approach to the problem of crack propagation rates in solder joints was proposed by Satoh et al. [43]. Instead of Eqn (9.40) they suggested the simple equation:

$$\frac{da}{dN} = Aa + B, \tag{9.42}$$

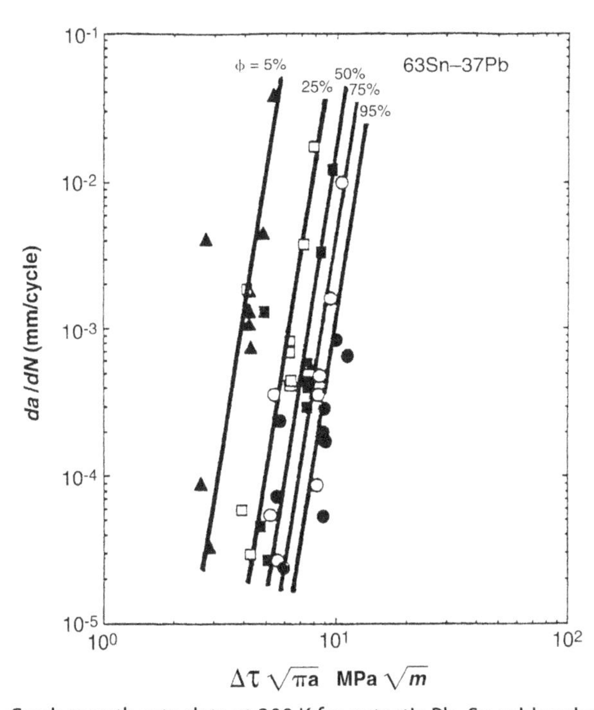


**Figure 9.22** Crack growth-rate data at 300 K for eutectic Pb–Sn solder plotted as $da/dN$ versus shear fracture toughness range. Triangular stress-time cycling was performed at a frequency of 0.1 Hz, and data are plotted for different values of $\phi$. *From Ref. [42].*

where constants $A$ and $B$ are related to crack propagation resistance and crack nucleation, respectively. The integration of Eqn (9.42) gives the fatigue life $N_f$ as

$$N_f = \int_0^{N_f} dN = \int_{a_o}^{a_f} \frac{da}{Aa + B} = A^{-1} \ln\left(\frac{Aa_f + B}{Aa_o + B}\right). \tag{9.43}$$

A fit of this relation to observed fatigue striations in QFP solder joints shown in Figure 9.23 reveals that $A = 1.48 \times 10^{-3}$ and $B = 0.18$. In addition, three-dimensional finite-element modeling of the plastic strain versus number of fatigue cycles obeyed the Coffin–Manson relationship. The authors suggest a synthesis of all the pertinent parameters, i.e., frequency, temperature, plastic strain, and crack dimensions that affect thermal fatigue to yield:

$$N_f = C\left\{\ln\left(\frac{Aa_f + B}{Aa_o + B}\right)\right\}(\Delta\varepsilon)^{-c} f^m \exp\left[-\frac{E_f}{kT}\right], \tag{9.44}$$

with $C$ a constant and $E_f$ an appropriate activation energy for fatigue. By now we should all realize the appropriateness and influence of the individual factors in this equation.

A parallel, finite-element analysis of crack growth in solder joints tested by Satoh et al. [43] has been performed by Lau [44]. This time the thermal

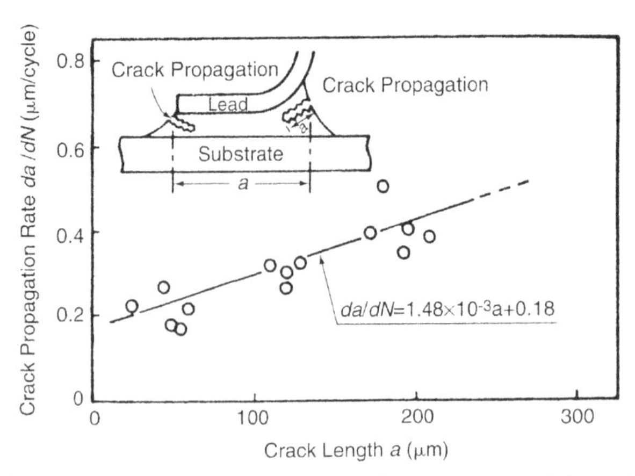


**Figure 9.23** Solder crack propagation rate as a function of crack length in a surface mounted quad flat package (QFP) joint. *From Ref. [43].*

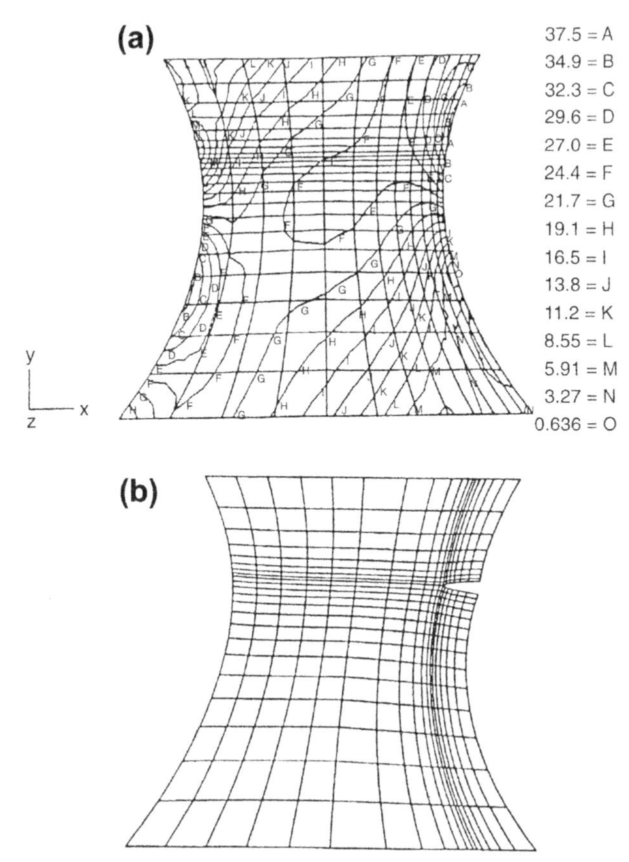


**Figure 9.24** (a) Two-dimensional stress distribution in the corner flip-chip solder joint (no crack). Stresses given in MPa are 11% less than three-dimensional results. (b) Deformation cracking of the flip-chip solder joint. *From Ref. [44].*

fatigue life of self-stretching flip-chip joints connecting a silicon die to an $Al_2O_3$ substrate was modeled. Assuming a temperature excursion from −50 to 150 °C and one cycle per hour, the approximate three-dimensional stress distribution of the uncracked corner joint is shown in Figure 9.24(a). Employing a fracture mechanics approach, failure of the cracked joint (Figure 9.24(b)) was predicted to occur in 3000 cycles. This estimate compared well with the experimentally observed 2700 cycle failure life. Significantly, but perhaps fortuitously, thermal fatigue life prediction by both fracture mechanics and Coffin–Manson approaches were in good agreement.

## 9.6 DYNAMIC LOADING EFFECTS IN ELECTRONIC EQUIPMENT

### 9.6.1 Introduction

We now address some other reliability concerns that stem from periodic stressing of structures and components. Rather than cyclic stresses arising from thermal expansion mismatch, external dynamic loading and the resultant displacements they produce in electronic hardware are of interest now. Electronic equipment is subject to many different forms of vibration over wide frequency ranges and acceleration levels [45]. These include dynamic-loading effects such as impact, shock, and acoustic noise by the action of cooling fans, repeated impact of keys during typing, accidental misuse, and vibration surrounding military activities, etc. The kinds of failure that result are due to flexure of circuit boards, fatigue, loosening, and even opening of contacts, and the shorting of leads, attached components, and devices. Complete loosening of screw and nut fasteners due to vibration can eventually cause them to rattle around inside equipment with the potential for producing serious damage. Loosening is an interesting phenomenon that is not entirely understood. Apparently, the dynamic loading stretches the bolt, momentarily reducing the friction between the nut and threads; after it unwinds a bit, the bolt relaxes, retightening the nut. This cyclic action is the driving force that causes fasteners to loosen; when they do, electrical contact (e.g., grounding) may be lost.

Characteristics that distinguish dynamically generated from thermal-mismatch-induced damage include:

1. Thermal-mismatch damage and failure is more common and a greater reliability concern. Vibrational stresses tend to be small by comparison.
2. Dynamic-loading effects influence the mechanical reliability of second (e.g., circuit boards, leads) and third (e.g., board contacts, ground planes) levels of packaging more than chip-level structures and components.
3. External rather than internal stresses are the sources of damage.
4. Dynamic effects occur and are modeled at constant temperature, e.g., room temperature.

A useful review of vibration analysis as it applies to electronic packaging is due to Suhir and Lee [46], and their treatment of the subject is presented here.

### 9.6.2 Vibration of Electronic Equipment

#### 9.6.2.1 Vibrating Mass

The behavior of a particle of mass $m$ attached to a spring whose spring constant is $k_s$ is well known and provides a convenient base from which to launch the discussion. In such a one degree-of-freedom system the vibrational displacement ($x$) is described by the equation:

$$m\frac{d^2x}{dt^2} + k_s x = 0, \tag{9.45}$$

where the first and second terms are the inertial and restoring forces, and $t$ is the time. The solution is of the form $x = A\cdot\sin\cdot\omega_o\cdot t$, where $A$ is a constant and $\omega_o = (k_s/m)^{1/2}$ is related to the natural frequency ($f_o$) of the free vibrations by $\omega_o = 2\pi f_o$.

It often happens, however, that the vibrational motion of structural elements within electronic equipment is driven by periodic forces ($F$) of the form $F = F_o\cdot\sin\omega t$, where $F_o$ is the force amplitude and $\omega$ is the excitation frequency. If it is further assumed that vibrations are damped by frictional or viscous-deformation effects that are proportional to the velocity, then for damping constant $\eta$:

$$\frac{d^2x}{dt^2} + \frac{\eta}{m}\frac{dx}{dt} + \frac{k_s x}{m} = \frac{F_o}{m}\sin\omega t. \tag{9.46}$$

This second-order, linear differential equation is identical to that which describes an R-L-C circuit powered by ac voltage. Under the most general conditions the solution to Eqn (9.46) contains a combination of terms of the form $B\cdot\sin\cdot(\omega t + \alpha)$ and $C\cdot\sin\cdot(\omega t + \beta)\cdot\exp\cdot[-(\eta/2m)t]$, where $\alpha$ and $\beta$ are phase angles and $B$ and $C$ are constants. Several physical cases of interest can be distinguished.

1. Free vibrations. If for the moment the driving force is neglected (i.e., $F_o = 0$), the displacement relative to the equilibrium position generally executes a damped oscillatory motion. Under the special condition that $\omega_o^2 = \eta/m$, the motion is not oscillatory but is critically damped and decays exponentially. Further, when $\eta/m > \omega_o^2$, the nonoscillatory motion is overdamped.
2. Steady-state vibrations. When $C$ is zero, the amplitude of vibration is

$$B = \frac{F_o}{m\left\{\left(\omega_o^2 - \omega^2\right)^2 + \left(\eta/m\omega\right)^2\right\}^{1/2}}. \tag{9.47}$$

In practical applications, larger amplitudes generally imply less reliable function of the vibrating body. Because the dynamic displacement is inversely proportional to the frequency squared, shifting $\omega_o$ upward is an obvious strategy to limit vibration. It is instructive to compare $B$ with the static displacement of the mass, $x_s = F_o/k_s$ under constant load $F_o$.

The ratio $B/x_s$, which is known as the amplification factor $A$, is then given by:

$$A = \frac{k_s}{m\left\{\left(\omega_o^2 - \omega^2\right)^2 + \left(\eta/m\omega\right)^2\right\}^{1/2}},$$

$$\text{or} \quad \frac{1}{\left\{\left(1 - \omega^2/\omega_o^2\right)^2 + \left[\eta/m(\omega/\omega_o)^2\right]\right\}^{1/2}}.$$

Among the possible behaviors are the following:

1. When $\omega_o > \omega$, $A \approx 1$. In this case, the amplitude of forced vibrations is equal to that of the static displacement.
2. If $\omega \approx \omega_o$, $A$ reaches its largest value of $[\eta/m]^{-1/2}$. This is the resonant condition, a state that should be avoided because dynamic loads and stresses are magnified.
3. When $\omega >> \omega_o$, $A >> (\omega_o/\omega)^2$. Thus at very high applied frequencies, $A$ approaches zero, and the system is relatively immune to vibrational damage.

#### 9.6.2.2 Vibration of Beams

Instead of the point masses above, we now consider the vibration of components that can be idealized as beams. The resulting vibrations cause transverse displacements $u(x, t)$ normal to the beam, or $x$-axis, that obey the partial differential equation [47]

$$\frac{m\partial^2 u(x,t)}{\partial t^2} = \frac{EI\partial^4 u(x,t)}{\partial x^4}. \tag{9.48}$$

Here $u = u(x, t)$ is the displacement function, $E$ is Young's modulus of the material, $I$ is the moment of inertia of the beam cross-section ($EI$ is the flexural rigidity), and $m$ is the mass per unit length. General solutions for the natural frequencies of vibration are

$$f_n = \frac{k_n}{2\pi}\left(\frac{EI}{ml^4}\right)^{1/2}. \tag{9.49}$$

where the value of $K_n$ depends on the particular boundary conditions. Several specific beam geometries are shown in Figure 9.25, where corresponding $K_n$ values are given. For example, if the beam is freely supported or hinged at opposing ends (Figure 9.23(a)), the displacements are those of

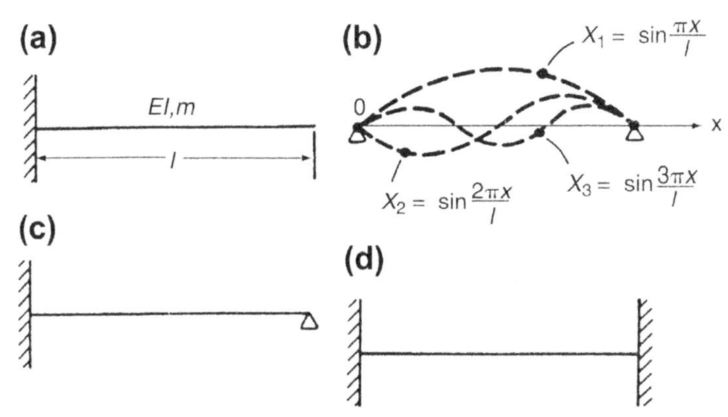


**Figure 9.25** Vibrational frequencies ($2\pi f_n = K_n[EI/ml^4]^{1/2}$) of various beams. (a) Cantilever; $K_1 = 3.516$, $K_2 = 22.04$, $K_n = \{(2n + 1)\pi/2\}^2$ for $n = 3, 4, \ldots$; (b) freely supported (hinged at both ends); $K_1 = 9.870$, $K_2 = 39.48$, $K_n = (n\pi)^2$ for $n = 3, 4, \ldots$; (c) clamped-hinged; $K_1 = 15.42$, $K_2 = 49.96$, $K_n = \{(4n + 1)\pi/2\}^2$ for $n = 3, 4, \ldots$; and (d) clamped at both ends; $K_1 = 22.37$, $K_2 = 61.67$, $K_n = \{(2n + 1)\pi/2\}^2$ for $n = 3, 4, \ldots$. *From Ref. [46].*

the simple vibrating string. In this case $K_n = (n\pi)^2$, where $n$ is an integer 1, 2, 3, … that denotes the first, second, third, … mode of vibration. Normally we are interested only in the first mode, because its vibrational amplitude exceeds that of the other modes.

From the standpoint of reliability, low-frequency oscillations should be avoided because they lead to large vibrational amplitudes. Therefore, other things being equal, stiffening the beam or board, reducing its mass and sectional area, and particularly, shortening its length, will all help to bring about the desideratum of higher-frequency vibrational modes. Clamping beams at either one or both ends also raises the vibrational frequencies. Electronic components whose geometry and structural behavior can be idealized as a cantilever are least desirable from the standpoint of mechanical reliability. This situation occurs in circuit boards plugged in or clamped at only one end. A simple comparison of the structures depicted in Figure 9.25 shows that cantilevers vibrate at the lowest frequencies.

### *9.6.2.3 Vibration of Packaging Components*

The formulas just presented apply equally well to elongated plates such as long, narrow printed circuit boards. But in order to use them, $EI$ in Eqn (9.49) must be replaced by the flexural rigidity of the plate, i.e., $Eh^3/[12(1-\nu^2)]$, where $\delta$ is the plate thickness. For wide plates of dimension $l \times W$, the formula that applies is.

$$f_n = \frac{K_n}{2\pi}\left\{\frac{Eh^3}{12(1-\nu^2)m(l^4+W^4)}\right\}^{1/2}, \tag{9.50}$$

where the mass is now on a per unit area basis. If the plate is simply supported, $K_1 = 9.87$.

Actual components employed in packaging have complex shapes that do not lend themselves to simple vibration analyses that yield closed-form solutions. As an example let us reconsider the gull-wing joint that was introduced earlier, in Section 8.5.3.3. Lau [48] has again done the finite-element analysis, which consists first in identifying the geometry and interfaces of interest and the boundary conditions that apply. Then the natural frequencies were determined in a free-vibration analysis. A more complete study would require estimation of the loads on the structure and their effect on the resultant forced vibrations.

Calculated fundamental frequencies of free vibrations are shown in Figure 9.26 where results for soldered (or good) joints are contrasted with unsoldered (or bad) joints. The fundamental frequencies of reliably soldered gull-wing leads are approximately six times those of the leads with

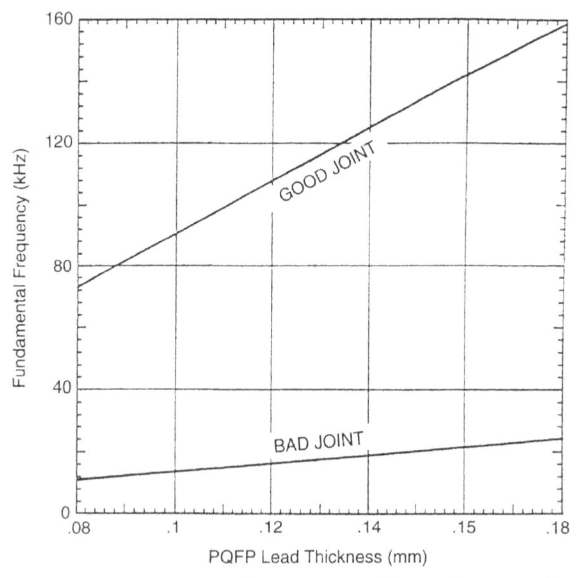


**Figure 9.26** Calculated fundamental frequencies of free vibrations for various quad flat packages having different lead thicknesses. Results for soldered (or good) joints are contrasted with unsoldered (or bad) joints. *From Ref. [48].*

cracked solder joints. In effect, beams fixed either at both ends or at only one end are being roughly compared; beneficial low-amplitude–high-frequency vibrations characterize the former relative to the unconstrained lead with the cracked solder end. A semiquantitative calculation of the frequency ratio of the fully clamped beam (cb) to the cantilever (c) is, from Eqn (9.49), $K(\text{cb})/K(\text{c}) = 22.37/3.516$, or 6.37. The good agreement with the exact solution indicates the value of using simple beam geometries for crude estimates. An additional finding is the small change in fundamental frequency with lead thickness. Similar results apply to gull-wing leads in plastic small-outline packages, and J leads in plastic leaded chip-carriers.

The traditional method for raising the natural frequency of PCBs (Printed Circuit Boards) has been to add stiffening ribs or provide point supports. Rib stiffeners are common in computer memory, logic, and processor cards, while point supports are used to stiffen power-supply boards. A design approach to locate such support points has been advanced by Pitarresi and Di Edwardo [49]. If the PCBs vibrate both in the $x$ and $y$ directions, the frequencies will generally differ because of the board geometry and the nonuniform placement of electronic components and devices. Optimizing arrangement of components through finite-element modeling has been explored by Chang and Magreb [50] as a way to maximize $f_n$.

## EXERCISES

**9.1** Consider 1 cm lengths of metal wire with diameter $d$ making contact.

**a.** Derive an expression for the ratio of the constriction resistance to total resistance if contact is made through a circular area of radius $a$.

**b.** For a 1 mm diameter Cu wire, what value of $a$ is required for the constriction resistance to reach a value of 10 mΩ?

**9.2** The contact voltage that produces a temperature equal to the melting point of the metal is known as the melting voltage. Correspondingly, the contact voltage that produces a temperature equal to the boiling point of the metal is known as the boiling voltage. What are the melting and boiling voltages for gold and platinum contacts?

**9.3** An a-spot in a contact shrinks in radius as a ring of oxide grows around it consuming metal in the process. If the oxide grows by a diffusional reaction, derive an expression for the time-dependent change in contact resistance.

**9.4** Accelerated steady-state creep testing of solder yielded the following results:

| Test | Temperature (°C) | Stress (MPa) | Strain rate ($s^{-1}$) |
|---|---|---|---|
| 1 | 22.5 | 6.99 | $2.50 \times 10^{-5}$ |
| 2 | 22.5 | 9.07 | $6.92 \times 10^{-5}$ |
| 3 | 46 | 6.99 | $2.74 \times 10^{-4}$ |

**a.** From these data determine the creep parameters $n$ and $E_c$.

**b.** What is the maximum stress that can be applied to a 0.5 mm high solder joint of this material to keep the creep extension less than 0.05 mm after 5 years of exposure to a 30 °C ambient?

**9.5** The tensile stress–strain behavior for solder pulled at a rate of $3.26 \times 10^{-5}$ gives a maximum tensile stress of 31 MPa. At a strain rate of $1.65 \times 10^{-3}$/s the maximum stress is 47 MPa. How do you account for this difference in strength?

**9.6** In a study of the thickness dependence of Cu–Sn IMCs on the thermal fatigue of SM solder joints, it was found that failures were described by Weibull statistics. In particular, the data points were fitted to the function $F(x) = 1 - \exp[-(x/\alpha)^{\beta}]$, where $F(x)$ is the cumulative distribution of failures, and $x$ is the number of cycles ($N$) to failure. Analysis of the results yielded the following:

| Compound thickness (μm) | $\alpha$ | $\beta$ | $N_{50\%}$ cycles |
|---|---|---|---|
| 0.95 | 1727 | 3.31 | 1475 |
| 1.16 | 1304 | 3.5 | 1123 |
| 1.43 | 922 | 2.5 | 749 |
| 1.7 | 767 | 2.87 | 639 |
| 1.97 | 723 | 2.91 | 605 |

Data from Ref. [51].

**a.** Plot $F(x)$ versus $x$ at the different IMC thickness values.

**b.** Replot the data in any manner that yields straight lines.

**c.** Provide a possible reason for the observed dependence of failure on IMC thickness.

**9.7** An electrical-probe assembly consisting of springy metal contactors undergoes compressive deformation during operation. The length of the contactor is $L_i$ initially and then shortens to $L_o$ when constant contacting load $P$ is applied to it. When the load is removed, the contactor length is $L_f$ instead of the desired $L_i$ ($L_i > L_f > L_o$). The observed permanent set or contact deformation appears to worsen when contact load is maintained for longer time, or if the contact tips get hot. Provide an explanation for these effects and mention the material properties you would recommend for set-free contact behavior.

**9.8** A leadless chip carrier soldered to a circuit board is thermally fatigued by turning the board on and off three times a day. Each cycle requires 30 min to reach the steady-state operating temperature of 50 °C. Predict how many cycles the eutectic solder can sustain prior to failure if the strain in each cycle is 0.005.

**9.9** Cracks in copper-plated through-holes were observed to grow during thermal cycling. The crack length ($2a$) varied with the number of cycles ($N$) as $2a = 1.16 \times 10^{-12} \cdot N^{3.77}$. Derive expressions for

**a.** $da/dN$ as a function of $N$.

**b.** $da/dN$ as a function of $a$.

**c.** $da/dN$ as a function of $\Delta K$.

**9.10** A high-power resistor is soldered at each end close to a PCB by means of solder posts. During operation, heat is emitted by the resistor. Discuss how the PCB warps in response.

**9.11** A leadless ceramic chip carrier is soldered to an FR4 printed circuit board and thermally cycled between 0 and 100 °C, with 20 min dwells at the maximum temperature. The chip carrier measures 0.635 × 0.635 cm and the joints are 0.013 cm high. If the CTE for the ceramic is $6 \times 10^{-6}$ and that for FR4 is $12 \times 10^{-6}$ °C$^{-1}$, predict how many cycles will be sustained prior to failure?

**9.12** It is desired that an electronic assembly, consisting of a chip carrier bonded through a fully compliant solder to a substrate, operate at a chip temperature $T_c$ and a substrate temperature $T_s$, where $T_c > T_s$. To minimize the shear strain produced in heating the chip from $T_o$ from $T_c$, and the substrate from $T_o$ from $T_s$, what value of thermal expansion coefficient ($\alpha_s$) would you recommend? The chip has a thermal expansion coefficient of $\alpha_c$.

**9.13** Eutectic solder fatigued at 25 °C and 10 Hz displayed a crack-growth dependence on the number of cycles given by $da/dN = 2.77 \times$

$10^{-7} \cdot \Delta K^{3.26}$, where $\Delta K$ has units of MPa-m$^{1/2}$. A ball-bond joint composed of this solder was subjected to fatigue loading, and finite-element analysis revealed that $\Delta K$ depends linearly on crack length. Specifically, $\Delta K = 0.151$ MPa-m$^{1/2}$ at the initial crack length of $5.65 \times 10^{-6}$ m, and $\Delta K = 0.261$ MPa-m$^{1/2}$ at a crack length of $1.02 \times 10^{-4}$ m corresponding to joint fracture.

**a.** How many cycles did the solder sustain prior to fatigue failure?

**b.** If the joint is cycled once per hour and reaches a maximum temperature of 150 °C, estimate how long it will take for the solder to fail.

**9-14** At the 1996 Surface Mount International Conference it was reported that plastic ball-grid array solder joints whose solder balls contained void defects (up to 24% of the pad area) had no adverse effect on board-level reliability; "In fact, PGBA solder joints with voids had 16% better reliability than those without voids"!

**a.** What are the likely sources of voids in freshly soldered joints?

**b.** Provide scientific explanations for these paradoxical findings.

**9.15** Suppose incandescent-bulb life is governed by low-cycle thermal fatigue. The thermal strain is induced by a 2000 °C rise or drop in the tungsten-filament temperature each time the bulb is turned on or off, respectively. If the equation governing lamp life is $N_f^{0.5} \Delta \varepsilon_p = 0.2$, predict the number of cycles to failure.

**9.16** Calculate the natural frequency of a vibrating cantilever spring that is loaded at the free end.

**9.17** A simply supported beam of length $L$ with a concentrated load $P$ at the center exhibits a deflection given by $\delta = PL^3/(48EI)$. If such a beam can be simulated as a spring-mass system, calculate the natural frequency of vibration.

**9.18** Experience shows that actual resonant frequencies of vibrating beams are lower than those theoretically calculated. Suggest a reason why this is so.

**9.19** Rapid variations in the temperature of packages can theoretically induce stress because of the acceleration forces associated with dimensional changes. Consider a 30 mm long package of specific gravity 2 that is 10 mm wide and 1 mm thick. If the package undergoes a temperature rise of 100 °C in 0.1 s, what is the stress from this source? How does this value compare with typical thermal stresses?

**9.20** A vibration test was conducted on circuit cards installed within a computer mainframe. The cards lay in the horizontal plane cantilevered off the motherboard. In ramping the frequency up to 60 Hz one of the circuit cards began to vibrate excessively at 50 Hz. The involved card made of fiberglass ($E = 38$ GPa), weighs 80 g and is 30 cm long, 3 cm wide and 0.1 cm thick. How thick should the card be to displace the first mode of vibration to 120 Hz? Mention another way to increase the vibrational frequency of the card.

**9.21** Determine the resonant frequency of a rectangular plug-in circuit card assembly that is simply supported on all four sides. The epoxy–glass printed circuit board weighs 0.545 kg, has dimensions of 22.9 × 17.8 × 0.2 cm and elastic properties defined by $E = 14$ GPa and Poisson's ratio of 0.12.

## REFERENCES


[1] R. Holm, Electrical Contacts-Theory and Applications, fourth ed., Springer Verlag, New York, 1967.
[2] F. Llewellyn Jones, The Physics of Electrical Contacts, Clarendon Press, Oxford, 1957.
[3] M. Braunovic', in: Proc. 40th IEEE Holm Conf. On Electrical Contacts, 1994, p. 1.
[4] M. Runde, IEEE Trans. Compon. Hybrids Manuf. Tech. CHMT-10 (1987) 89.
[5] M. Runde, H. Kongjorden, J. Kulsetas, B. Todtal, IEEE Trans. Compon. Hybrids Manuf. Tech. CHMT-9 (1986) 77.
[6] T.A. Yeger, K. Nixon, IEEE Trans. Compon. Hybrids Manuf. Tech. 7 (1984) 370.
[7] R.S. Mroczkowski, in: Proc. First Electronic Materials and Processing Congress, 1988, p. 159. Chicago, Sept. 24–30.
[8] N.A. Stennett, T.P. Ireland, D.S. Campbell, IEEE Trans. Compon. Hybrids Manuf. Tech. 14 (1991) 45.
[9] N.A. Stennett, D.S. Campbell, IEEE Trans. Compon. Packag. Manuf. Tech. 17A (1994) 128.
[10] J.F. Archard, J. Appl. Phys. 24 (1953) 981.
[11] M.D. Bryant, IEEE Trans. Compon. Packag. Manuf. Tech. 17 (1994) 86.
[12] M. Antler, IEEE Trans. Compon. Hybrids Manuf. Tech. 8 (1985) 87.
[13] M. Antler, IEEE Circuits Devices 3 (March 1987) 8.
[14] M. Braunovic', IEEE Trans. Compon. Hybrids Manuf. Tech. 15 (1992) 205.
[15] J.H. Lau, in: J.H. Lau (Ed.), Thermal Stress and Strain in Microelectronics Packaging, Van Nostrand Reinhold, New York, 1993.
[16] J.H. Lau (Ed.), Solder Joint Reliability, Van Nostrand Reinhold, New York, 1991.
[17] J.H. Lau (Ed.), Thermal Stress and Strain in Microelectronics Packaging, Van Nostrand Reinhold, New York, 1993.
[18] R. Darveaux, K. Banerji, IEEE Trans. Compon. Hybrids Manuf. Tech. 15 (1992) 1013.
[19] J.H. Lau, Solder. Surf. Mount. Tech. 13 (October 1993) 45.
[20] Z. Guo, Y.-H. Pao, H. Conrad, J. Electron. Packag. Trans. AIME 117 (June 1995) 100.
[21] G.F. Dieter, Mechanical Metallurgy, third ed., McGraw-Hill, New York, 1986.

[22] D.F. Socie, M.R. Mitchell, E.M. Caulfield, Fundamentals of Modern Fatigue Analysis, University of Illinois, 1978. FCP Report No. 26, U of Illinois.
[23] N.F. Enke, T.J. Kilinski, S.A. Schroder, J.R. Lesniak, IEEE Trans. Compon. Hybrids Manuf. Tech. 12 (4) (1989) 459.
[24] H.D. Solomon, in: J.H. Lau (Ed.), Solder Joint Reliability, Van Nostrand Reinhold, New York, 1991.
[25] J.S. Hwang, Modern Solder Technology for Competitive Electronics Manufacturing, McGraw Hill, New York, 1996.
[26] Z. Guo, A.F. Sprecher, D.Y. Jung, H. Conrad, IEEE Trans. Compon. Hybrids Manuf. Tech. 14 (4) (1991) 833.
[27] Z. Guo, A.F. Sprecher, H. Conrad, M. Kim, in: Materials Developments in Electronic Packaging Conf. Proc. Montreal, 1991.
[28] L.F. Coffin, Astm-Stp 520 (1973) 5.
[29] P.M. Hall, in: J.H. Lau (Ed.), Solder Joint Reliability, Van Nostrand Reinhold, New York, 1991.
[30] P.M. Hall, IEEE Comp. Hybrids Manuf. Tech. 7 (4) (1984) 314.
[31] W. Engelmaier, in: J.H. Lau (Ed.), Solder Joint Reliability, Van Nostrand Reinhold, New York, 1991.
[32] W. Engelmaier, Soldering Surf. Mount. Tech. 1 (February 1989) 14.
[33] J.-P.M. Clech, J.A. Augis, in: J.H. Lau (Ed.), Solder Joint Reliability, Van Nostrand Reinhold, New York, 1991.
[34] J.W. Balde, IEEE Trans. Compon. Hybrids Manuf. Tech. 10 (1987) 463.
[35] K.C. Norris, A.H. Landzberg, IBM J. Res. Develop. 13 (1969) 266.
[36] J.H. Lau, Y.-H. Pao, C. Larner, R. Govila, S. Twerefour, D. Gilbert, E. Erasmus, S. Dolot, Solder. Surf. Mount. Tech. 16 (February 1994) 42.
[37] D.R. Frear, J. Mater. 49 (May 1996).
[38] R.B. Clough, A.J. Shapiro, A.J. Bayba, G.K. Lucey, J. Electron. Packag. Trans. ASME 117 (1995) 270.
[39] R.E. Pratt, E.I. Stromswold, D.J. Quesnel, IEEE Trans. Compon. Packag. Manuf. Tech. A19 (1996) 134.
[40] D.R. Frear, F.M. Hosking, P.T. Vianco, in: Proc. Conf. Materials Developments in Microelectronic Packaging, Montreal, August 19, 1991, p. 229.
[41] [a] E.E. de Kluizenaar, Solder. Surf. Mount. Tech. 4 (February 1990) 27;
[b] E.E. de Kluizenaar, Solder. Surf. Mount. Tech. 5 (June 1990) 56.
[42] Z. Guo, A.F. Sprecher, H. Conrad, M. Kim, in: Proc. Conf. Materials Developments in Microelectronic Packaging, Montreal, August 1991, p. 155.
[43] R. Satoh, K. Arakawa, M. Harada, K. Matsui, IEEE Trans. Compon. Hybrids Manuf. Tech. 14 (1) (1991) 224.
[44] J.H. Lau (Ed.), Flip Chip Technologies, McGraw Hill, New York, 1996.
[45] D.S. Steinberg, Vibration Analysis for Electronic Equipment, second ed., Wiley, New York, 1988.
[46] E. Suhir, Y.C. Lee, in: Electronic Materials Handbook, Volume 1, Packaging, ASM International, Materials Park, OH, 1989, p. 45.
[47] M. Paz, Structural Dynamics, Van Nostrand Reinhold, New York, 1982.
[48] J.H. Lau, in: J.H. Lau (Ed.), Solder Joint Reliability, Van Nostrand Reinhold, New York, 1993.
[49] J.M. Pitarresi, A.V. Di Edwardo, J. Electron. Packag. Trans. ASME 115 (1993) 119.
[50] T.-S. Chang, E.B. Magreb, J. Electron. Packag. Trans. ASME 115 (1993) 312.
[51] P.L. Tu, Y.C. Chan, J.K.L. Lai, IEEE Trans. Compon. Packag. Manuf. Tech. 20 (1997) 87.

CHAPTER 10

# Degradation and Failure of Electro-Optical Materials and Devices

## 10.1 INTRODUCTION

Together with microelectronics, optoelectronics (or electro-optics) is transforming our lives and helping to shape a future based on high technology. This is evident from the dominant role of optical materials and devices in some of the most important arenas of human activity, e.g., fibers for optical communications; lasers and imaging systems for medical applications; optical sensors for a host of industrial processing, control and testing applications; lenses, filters; and optically active materials for observing, detecting, displaying, and recording images in microscopy, photography, media, and information systems.

Optical materials can be broadly divided into two categories—active and passive. The active category includes those materials that exhibit special optical properties in response to electrical, magnetic, mechanical, optical, and thermal signals and interactions. Examples based on semiconductors include light sources such as lasers and light-emitting diodes, and photodiode detectors. In addition, there are magneto-optical recording media, luminescent materials, polymer photoresists, and liquid crystal displays that are not based on semiconductors. Passive optical materials transmit and reflect the light in components such as lenses, waveguides, and optical coatings. The same material can be used in active and passive applications. For example, silicon is active in a solar cell when generating electric power, but passive in optical coatings that enhance or decrease reflectivity. Efficient operation of electro-optical systems often means the inseparable linkage of active devices and passive components.

Some of the reliability issues raised earlier in compound-semiconductor electronic devices (e.g., MESFETs, HEMTs) apply to electro-optical devices as well; similar p–n junction degradation, electrostatic discharge (ESD) effects, contact and packaging failure mechanisms, plague both types of

*Reliability and Failure of Electronic Materials and Devices*
ISBN 978-0-12-088574-9
http://dx.doi.org/10.1016/B978-0-12-088574-9.00010-0

devices. In addition, another driving force for degradation of these devices and components is the high-intensity light itself. Absorption of this light in the laser cavity and optical coating materials can cause destructive thermal damage. In this chapter both active and passive optical devices and components will be discussed with a focus on optical fibers in later sections.

## 10.2 FAILURE AND RELIABILITY OF LASERS AND LIGHT-EMITTING DIODES

### 10.2.1 A Bit of History

In the more than three decades of research and development of semiconductor optical sources 2 years are notable [1]: 1962 and 1970. It was in 1962 that high-electroluminescence (EL) radiation from GaAs p–n junctions was demonstrated and lasing action reported in homojunction GaAs and GaAsP diodes at cryogenic temperatures. Semiconductor lasers composed of AlGaAs/GaAs heterostructures capable of emitting light continuously (CW-continuous wave) at room temperature were announced in 1970; this achievement led to the steady improvement of both lasers and LEDs to the point of commercial application. Wavelengths for these AlGaAs lasers ranged from 0.8 to 0.9 μm, with telecommunications applications confined mostly between 0.82 and 0.85 μm at that time.

Despite the success, these early devices suffered from reliability problems soon after initial operation. Through improved crystal-growth techniques, substrates containing low-dislocation densities emerged, and with them rapid degradation was suppressed. But now long-term aging failures associated with bulk dislocations occurred. Such failures were attributed to mechanical stresses induced by bonding lasers to substrates or heat sinks, as well as to dielectric films deposited to confine the light. Facet degradation was identified as another failure mechanism.

In the late 1970s, InGaAsP/InP lasers appeared and were found to be relatively immune to the presence of crystal defects. These quaternary devices, as well as quantum-well lasers, which appeared in the late 1980s, continue to undergo development and improvement. Wavelengths range from 1.0 to 1.6 μm for these InGaAsP lasers, with telecommunications applications limited either to 1.3 or to 1.55 μm.

The reliability of lasers and LEDs has greatly improved in the past two decades. Nowadays LED technology produces devices whose optical output declines by only 5–20% after $\sim 10^3$ h at 55 °C, and half brightness is not reached in $\sim 10^6$ h at ambient temperatures. In most applications involving a human interface, a factor of two in degradation is not noticeable [2]. Most of these LEDs are based on GaAs or InP technology, which generally means long-wavelength visible (red) and invisible infrared output. However, in the short wavelength, or blue region of the spectrum, there had been a long-standing need for reliable LEDs and lasers. Finally, in 1994, Nichia Chemical Industries introduced blue LEDs based on GaN that were 100 times brighter than SiC LEDs. In one of the first studies on the reliability of these devices it was found that despite a high imperfection density that would have proved fatal to GaAs and InP LEDs, defects in GaN diodes are immobile [3].

Reliability issues in these electro-optical devices are not limited only to semiconductors. For example, telecommunications lasers are usually packaged within modules that may contain a monitor photodiode (to control light output), an optical isolator (to prevent back reflection of light), a thermoelectric cooler, a temperature sensor, circuitry for operating the laser, lenses, and optical fiber. Overall reliability of the laser module implies trouble-free functioning of these components as well. And just as in the case in IC chips, the reliability of lasers is influenced by the quality of the packaging materials.

### 10.2.2 Reliability Testing of Lasers and LEDs

Accelerated life tests are the most popular way to assess the reliability of lasers and LEDs. For this purpose constant-current (input) and constant-light (output) accelerated aging tests are performed. For lasers, degradation is usually tracked by measuring the increased threshold current with time or by monitoring the current required for a specified light output. On the other hand, the decrease in output light power (at fixed current) with time is usually taken as a measure of degradation in LEDs. The distinction between these two kinds of accelerated test data is shown in Figure 10.1(a) and (b).

#### *10.2.2.1 Constant Light Output-Power Testing*

Laser operating characteristics shown in Figure 2.27 (and rotated 90° in (Figure 10.1(a))) are the basis of both pass/fail qualification as well as degradation testing. Nash [4], in an extensive failure testing program on lasers employed in undersea optical communications systems, has addressed the following thorny issues:

1. What constitutes good or bad lasers? More broadly, what do we mean by failure?
2. Which equation(s) best describes laser degradation kinetics?
3. How can laser lifetimes be predicted with confidence?

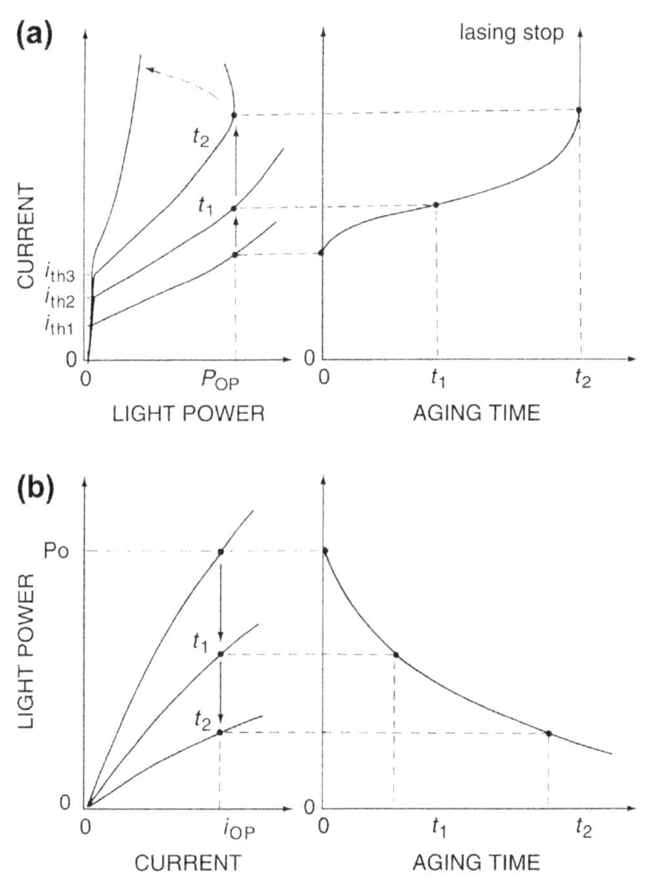


**Figure 10.1** (a) Left. Current versus light power output (e.g., in lasers). Right. Current increase with time required to maintain constant light power. (b) Left. Light power output as a function of operating current (e.g., in LEDs). Right. Light power decay with time as a result of constant-current testing. *From Ref. [1].*

When faced with the unenviable task of selecting reliable, high-performance lasers from a manufacturing lot of outwardly good lasers, a reasonable first step is to perform a short-time pass/fail screening, or "purge" test. One procedure consists of operating at higher than threshold currents ($i_{th}$), for a fixed time (e.g., 18 h), at elevated temperature (e.g., 100 °C). If, in order to maintain constant light output, the current level rises beyond a preset "hurdle" current, the laser is deemed to have failed and is rejected; if the current rise is less than this amount, the laser passes and is packaged. To qualify lasers for extended use, the degradation rate is obtained through similar testing, in which the value of current is recorded as a function of time. In this "burn-in" test, the constant-output light intensity is continuously or intermittently monitored, and variations are compensated for using feedback circuits to adjust the injected current ($i_L$). When plotted versus time ($t$), the fractional increase in current $\Delta i_L(t)/i_L(0)$ for an early vintage 1.3 μm laser is displayed in Figure 10.2. The rise in $i_L$ ($i_L > i_{th}$) clearly resembles the creep strain increase in mechanically loaded materials (see Figure 9.7). And like the short-time (primary) creep response, $i_L$ typically varies as a fractional power of $t$. For the case shown, with $t$ in hours,

$$\Delta i_L(t)/i_L(0) = 0.01t^{0.25}, \tag{10.1}$$

so that $d \cdot \Delta \cdot i_L/dt$ decreases with time. This behavior gradually gives way to a long steady-state linear rise in current where $d \cdot \Delta \cdot i_L/dt = \text{constant}$; specifically,

$$\Delta i_L(t)/i_L(0) = 8.5 \times 10^{-5}t + 0.05. \tag{10.2}$$

Given sufficient degradation, lasing eventually stops altogether as the current runs away in an uncontrollable way $\frac{d^2 i_L}{dt^2} > 0$.

Laser requirements and an operating definition of failure must be specified to enable laser lifetimes to be predicted based on the testing information. Suppose a 10 year life at 25 °C is required. Further, let us assume that failure occurs when $i_L$ rises to some prescribed level, e.g., $\Delta i_L(t)/i_L(0) = 0.50$. Therefore, an end-of-life (EOL) normalized degradation rate (NDR) can be defined as $NDR_{(EOL)} = \Delta i_L(t)/(i_L(0)\Delta t) = 0.50/10y = 0.57\%/kh$. Thus, a little over a half percent increase in current in 1000 h constitutes failure. It is, of course, impractical to test for 10 years (87,600 h) and thus a shorter time of 3000 h was selected. Even so,

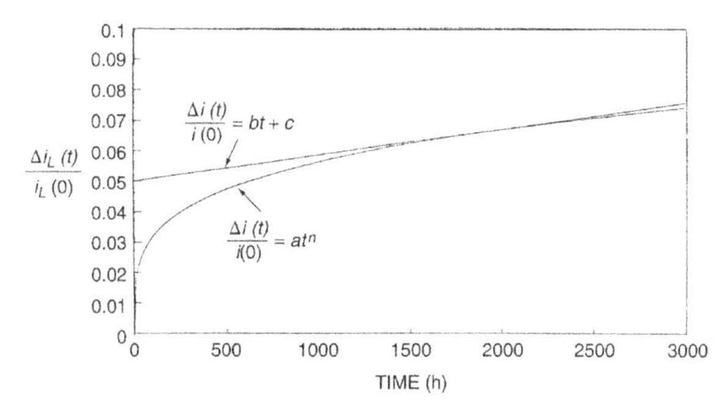


**Figure 10.2** Fractional increase in current $\Delta i_L(t)/i_L(0)$ required to sustain constant laser light output plotted versus time. *From Ref. [4].*

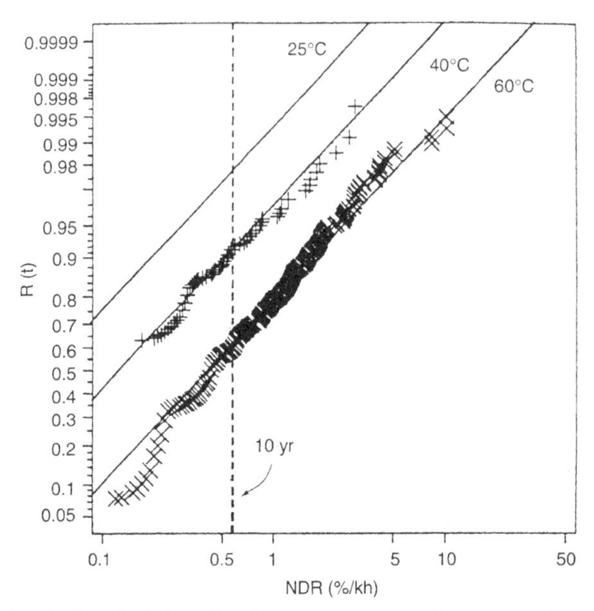


**Figure 10.3** Survival probability of early vintage 1.3 μm lasers tested at 40 and 60 °C plotted log normally as a function of the normalized degradation rate. *From Ref. [4].*

accelerated testing at elevated temperatures is necessary to induce a sufficient number of failures. Values of $\Delta i_L\ (t)/i_L\ (0)$ and NDR are then obtained, where $i_L\ (0)$ is taken as the straight line projected current (using Eqn (10.2)), not the initial current. Finally, the NDR failure test data are plotted in lognormal fashion in Figure 10.3. Note, however, that the ordinate is the survival rate $R(t)$, i.e., $R(t) = 1 - F(t)$. Since each data point ($x$) represents the end-of-life for a single laser, the degradation rate is quite variable, spanning some two orders of magnitude in value.

A single failure mechanism seems to be operative because plots obtained at 40 and 60 °C are parallel. It is left to the reader to show that the activation energy for laser failure ($E_L$) is 0.53 eV. Expanding these results to include the probable effect of current density ($j_L$), an empirical relation between MTTF and the relevant variables has the same form as Eqn (5.31), namely,

$$\text{MTTF}^{-1} = Aj_L^n \exp\left(-\frac{E_L}{kT}\right), \tag{10.3}$$

where $A$ and $n$ are constants. This equation is expected to model gradual laser degradation or wearout due to "drift" or "parametric shift."

Over the course of ~20 years experience in wafer growth, processing, manufacture, testing, and aging a variety of semiconductor lasers at AT&T Bell Laboratories, many failure modes, mechanisms, and device weaknesses with reliability implications have been uncovered. These are summarized in Table 10.1 together with testing screens that have been instituted to identify and reject the damaged lasers.

#### 10.2.2.2 Constant-Current Testing

For LEDs the light output power is monitored continuously or intermittently with a photodetector while maintaining a constant injected current. In this case, Figure 10.1(b) applies and indicates the decline of light intensity with time. The phenomenon resembles stress relaxation in mechanically loaded solids. Loss of optical power $P$, therefore, has the form of Eqn (9.28) [5], or.

$$P(t) = P(0)\exp - \left[\beta_o t \exp - \left(\frac{E}{kT}\right)\right], \qquad (10.4)$$

where ($t$) and (0) refer to time $t$ and $t = 0$, $\beta_o$ is a constant dependent on active region material, and $E$ refers to the activation energy for damage in lasers ($E_L$) or light-emitting diodes ($E_{LED}$). For GaAs, $\beta_o = 93$/h and $E = 0.57$ eV, while for InGaAsP, $\beta_o = 1.84 \times 10^7$/h and $E = 1.0$ eV [6]. Based on the Arrhenius plot of Figure 10.4 derived from these numbers, it is clear that InGaAsP devices are much longer lived primarily because of lower oxidation rates.

Again, an empirical relation between MTTF and $P$ has the familiar form

$$\mathrm{MTTF}^{-1} = BP^m \exp\left[-\frac{E_{LED}}{kT}\right], \qquad (10.5)$$

where $B$ and $m$ are constants. Values for $E_{LED}$ range from 0.35 to 0.8 eV [7] and reflect damage mechanisms that similarly afflict lasers. It is to these mechanisms that we now turn.

### 10.2.3 Microscopic Mechanisms of Laser Damage

Three main semiconductor laser failure modes have been distinguished: rapid, gradual, and catastrophic degradation. After lasing commenced, early lasers underwent very rapid degradation and failed too soon to be commercially viable. A primary cause of the rapid degradation is the *dark-line* defect, which was shown to originate from dislocation networks at p–n

**Table 10.1** Degradation and failure modes and mechanisms in lasers

| Failure Modes and Mechanism | Screens and Prevention |
|---|---|
| 1. Dark-spot and dark-line defects | Low dislocation-density substrates, improve epitaxial growth, eliminate sources of internal stress |
| 2. Partially attached or missing wire bond | Nondestructive and destructive pull tests, visual inspection |
| 3. Partially bonded laser chip (increases the thermal impedance and causes laser to operate at higher temperature and current) | Visual inspection, thermal impedance and cycling measurements, threshold measurements |
| 4. Strained laser chip | *L-i-V* testing, temperature cycling |
| 5. Debris—e.g., metal particles from metallization, cleaving, and packaging (causes shorting and reduces light output) | Visual inspection, thinner contact metallizations, dielectric coatings on vulnerable areas |
| 6. Solder whisker growth (causes shorting and reduces light output) | If Pb–Sn is used, there must be sufficient Pb to prevent Sn whiskers |
| 7. Copper diffusion to the laser's active region (causes nonradiative regions and accelerates degradation) | Use nickel diffusion barrier, bond the laser junction up or on electrically insulating substrate |
| 8. Assorted laser abnormalities and instabilities, kinks in *L-i* curve, light jumps or bistability in *L-i* curve above threshold (Figure 2.27), sustained, uncontrolled high frequency, e.g., 1–2 GHz oscillations in the optical output, shifts of lasing wavelength (dispersion) | Examination of $\frac{dL}{di} - i$, RF noise spectrum, study occurrence as a function of aging, high-frequency noise measurements |
| 9. Catastrophic (sudden) partial or total damage occurring predominantly at temperatures $\approx 40\,^{\circ}C$ | Current/optical power overstress |
| 10. Rapidly occurring degradation or premature failure from some unidentified grown or processed defect | High-temperature, high-current, short duration overstress to identify vulnerable lasers |
| 11. ESD, or equipment-induced overstressing | Care in handling, e.g., use of grounding wrist straps |
| 12. Gradual long-term degradation or wearout, reflected in increase in operating threshold voltage and current | Long-term elevated temperature tests to determine activation energy for making lifetime projections |

$L$ = Laser power.
Adapted from F. Nash, *Astrotec Laser Reliability Handbook*, AT&T Bell Labs.

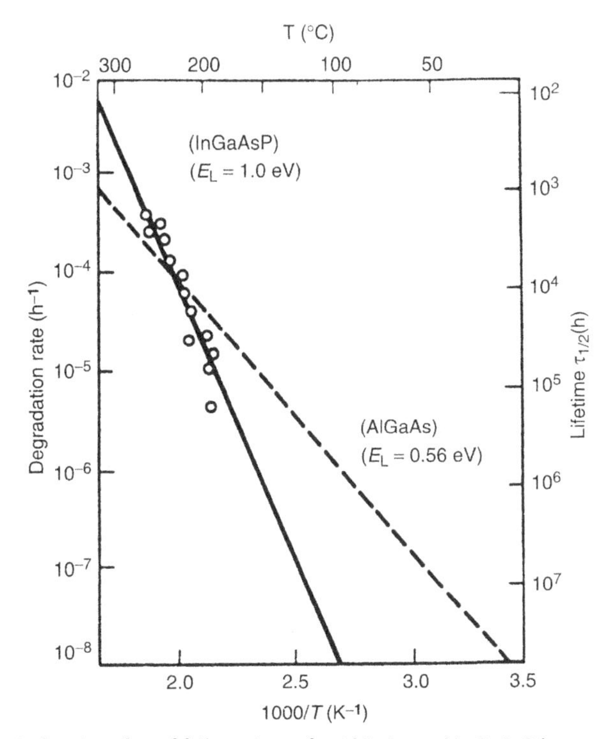


**Figure 10.4** Arrhenius plot of failure times for AlGaAs and InGaAsP lasers. *From Ref. [4].*

junctions. Improvement in materials and processes has extended laser life to the point where wearout failures are now the primary reliability concern. Indeed, gradual wearout was the failure mode tracked by constant-light testing in Section 10.2.2.1. The catastrophic degradation mode stems from damage to the mirror facets.

It is much easier to monitor light or current changes in these failure modes than it is to unravel the microscopic damage processes and expose the culprits responsible for them. Identifying specific failure mechanisms and the defects involved has only been achieved through the use of the transmission electron microscope (TEM).

### 10.2.3.1 Dark-Spot and Dark-Line Defects

A model for the formation of dark-spot defects (DSDs) and dark-line defects (DLDs) in AlGaAs/GaAs lasers begins with dislocations that originate from the substrate [8]. These dislocations nucleate at stacking faults and extend into the epitaxially grown compound-semiconductor films that comprise the active laser region. Known as threading dislocations, these

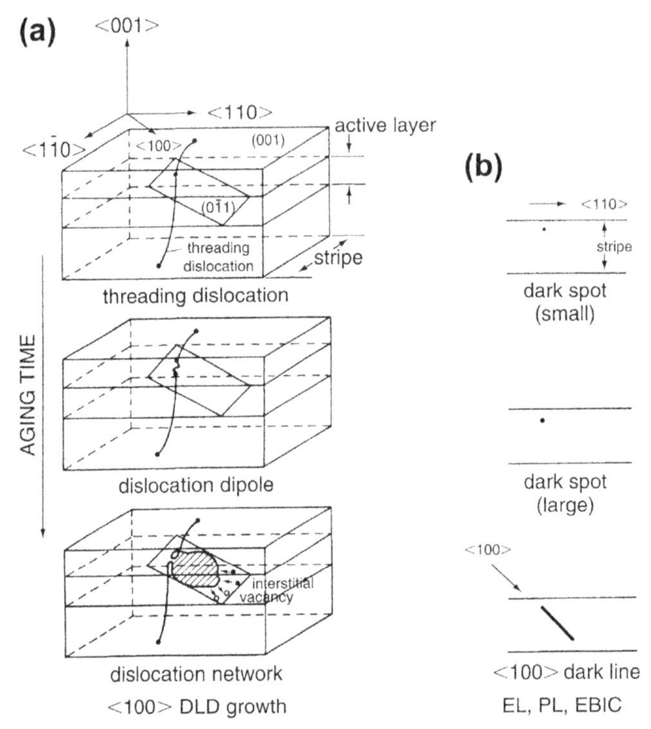


**Figure 10.5** (a) Mechanisms for dislocation elongation and generation of vacancies and interstitial atoms. (b) Dark-spot and dark-line defects generated during aging are revealed by electroluminesence (EL), photoluminesence (PL) and electron beam induced current (EBIC) (see Section 11.5.3.2). *After Ref. [8].*

crystallographic defects assume the geometry shown in Figure 10.5 relative to the active (001) plane. Recombination of carriers without desired light emission (nonradiative recombination) as well as photon absorption processes occurs preferentially at dislocations where extra electron states or levels exist. In such cases the photon energy is essentially converted into excitation of lattice vibrations (phonons) and manifested as heat. This enhances the motion of vacancy and interstitial point defects to and from associated dislocations, causing the latter to climb and increase in length, even at room temperature. During further aging the dislocation network grows and elongates in a complicated fashion.

Employing EL and photoluminescence techniques, DSDs and DLDs are made visible through promotion (or suppression) of electron level transitions associated with these defects. Such a DLD is shown in Figure 10.6. At first DSDs emerge and grow where threading dislocations (Figure 10.5 (a) and (b)) intersect the $(0\bar{1}1)$ plane. By viewing the device normal to the

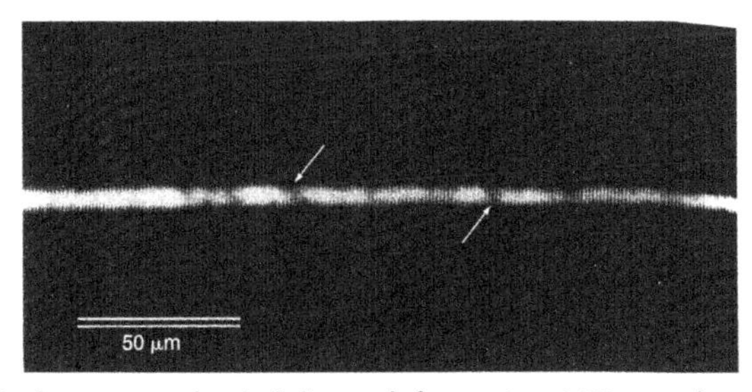


**Figure 10.6** Dark-spot and dark-line defects in 1.30 μm laser revealed by photoluminescence. *Courtesy S. N. G. Chu, Lucent Technologies, Bell Laboratories Innovations.*

page or along the $[1\bar{1}0]$ direction they can be seen. With aging, DLDs appear as schematically suggested by Figure 10.5(c). Lying in the $(0\bar{1}1)$ plane they grow by climbing along the [100] direction and are known as <100> DLDs. These as well as <110> DLDs were the cause of rapid degradation in early AlGaAs/GaAs lasers and LEDs. The <110> DLDs in these structures originate from dislocation glide on {111} slip planes during plastic deformation. When projected onto {100} planes they appear as a <110> DLD.

Both types of DLD appear in light-emitting diodes as well as lasers composed of InGaAsP/InP materials. Fortunately, DLD damage is not as serious a reliability concern in 1.3 and 1.55 μm InGaAsP/InP devices compared to GaAs-based devices. Reasons for this include low DLD dislocation-climb growth rates and a band structure that is less tolerant of nonradiative recombination processes.

Strategies to eliminate DLDs focus on both the growth stages and postgrowth fabrication steps involved in device manufacture. Employing low-dislocation GaAs (or InP) substrates (e.g., less than $10^3/\text{cm}^2$), improved epitaxy, reduction of internal stress, and elimination of oxygen through gettering action by small amounts of Al and Mg have all helped to increase both yield and reliability. Use of the more compliant soft solders for bonding and attachment to thermally matched substrates are postgrowth methods for minimizing DLDs; lower stresses mean fewer dislocations.

### *10.2.3.2 Facet Degradation*

Degradation of the mirror facets is the cause of an important failure mode in semiconductor lasers known as catastrophic optical damage (COD). Initially, the facets are roughly 100 °C higher than the cavity temperature, but when catastrophic damage occurs, accelerated heating causes the facet to melt. The basic damage mechanism involves preferential nonradiative recombination of carriers at facets because of the greater presence of surface states and defects there. Local heating reduces both the semiconductor bandgap energy and thermal conductivity, which in turn promotes optical excitation of carriers and further laser light absorption. Ultimately, positive feedback occurs at a critical power-output level and leads to thermal runaway, a process modeled in Section 10.3.4.3.

Facet damage is also believed to involve photo-enhanced oxidation of the III–V elements resulting in the formation of stable oxides. Factors like thermodynamic stability and solubilities of the mixed oxides, and the rate of ion transport through them, influence the rate of oxide growth. The mechanism of photoassisted oxidation apparently involves the breaking of bonds due to electron–hole generation as a result of irradiation. This facilitates point defect motion through the lattice. Light with energy greater than the semiconductor bandgap energy ($E_g$) is required for such a mechanism. The more intense the light source, the greater the oxidation rate. Under the same light intensity GaAs/AlGaAs oxidizes at a more rapid rate than GaP, InP, InAs, and $In_{0.47}$ $Ga_{0.53}As$ and $In_{0.72}Ga_{0.28}$ $As_{0.59}P_{0.41}$ lattice-matched to InP. Additionally, higher nonradiative recombination velocities are responsible for greater heating and facet degradation suffered in GaAs/AlGaAs relative to quaternary lasers. The enhanced aluminum oxidation at facets explains the trend to Al-free IR laser diodes.

Suppression and elimination of facet damage has been achieved by reducing light absorption and protecting the facet from atmospheric attack. Doping the active region with Zn lowers $E_g$ slightly. Thus the energy of the light emitted is reduced, and the facet does not absorb it as readily. To effect the latter, coatings of $Al_2O_3$, $SiO_2$, and $SiN_x$ have been used to limit facet oxidation. In this regard, mirror passivation of GaAs/AlGaAs quantum-well lasers through plasma oxidation [9] considerably improves resistance to COD relative to unprotected facets, as seen in Figure 10.7(a). Interestingly, an Arrhenius-like plot describes the relation between the log of laser life (MTTF) and the reciprocal of optical power density (PD). This suggests a linear proportionality between the facet temperature ($T$) and PD, i.e., $T = c\text{x}\text{PD}$. Therefore,

$$\text{MTTF}^{-1} = \nu \exp\left[-\frac{E_{\text{COD}}}{kT}\right], \tag{10.6}$$

where $\nu$ is a frequency factor and $E_{\text{COD}}$ is the activation energy for COD. Assuming that $c = 130$ (K-μm/mW), values of $E_{\text{COD}} = 1.38$ eV and

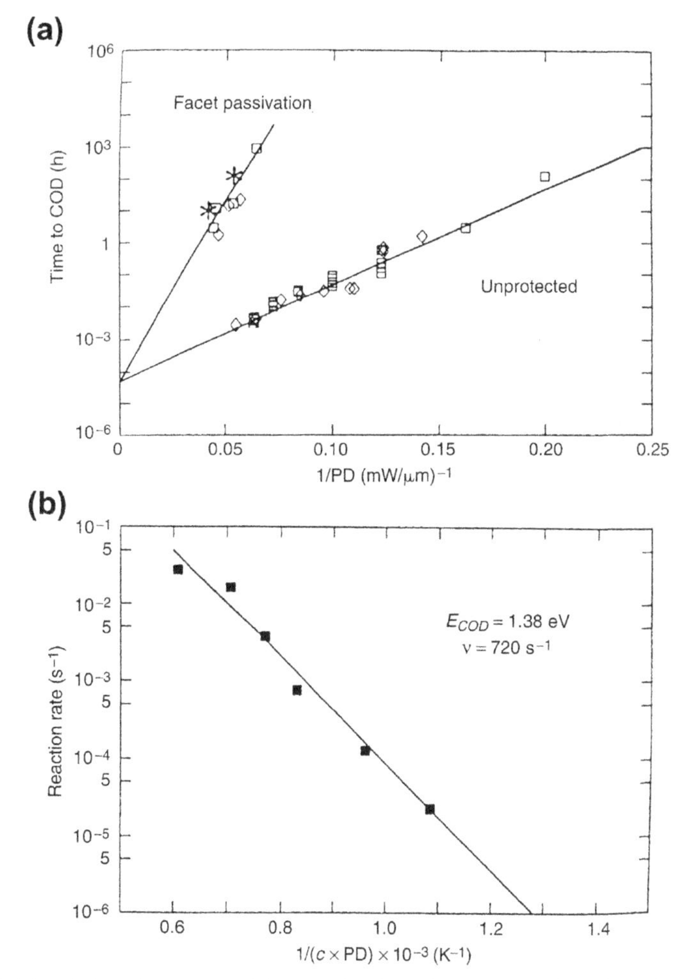


**Figure 10.7** (a) Time to catastrophic optical mirror damage as a function of the inverse optical power density transmitted through the mirror. (b) Arrhenius plot of mirror damage rate versus 1/(*cxPD*), i.e., ($\sim 1/T$). *From Ref. [9].*

v = 720/s were extracted. Note that at fixed laser currents, Eqns (10.3) and (10.6) have the same form.

#### *10.2.3.3 Degradation in 1.3 μm Channeled-Substrate Buried Heterostructure Lasers*

The comprehensive TEM study by Chu et al. [10] of Channeled-Substrate Buried Heterostructure lasers is an excellent summary of the interplay among various defect mechanisms and enables a better appreciation of degradation in these devices. Due to enhanced optical confinement in the buried crescent-shaped active stripe (Figure 10.8, also Figure 2.29), these InGaAsP/InP lasers have low room-temperature threshold currents, making them suitable for high bit-rate optical communication applications. To obtain a full picture of the distribution of damage at the crescent tips and sidewalls, it was necessary to obtain sections perpendicular and parallel to the stripe. The resultant TEM images are displayed in the montage of

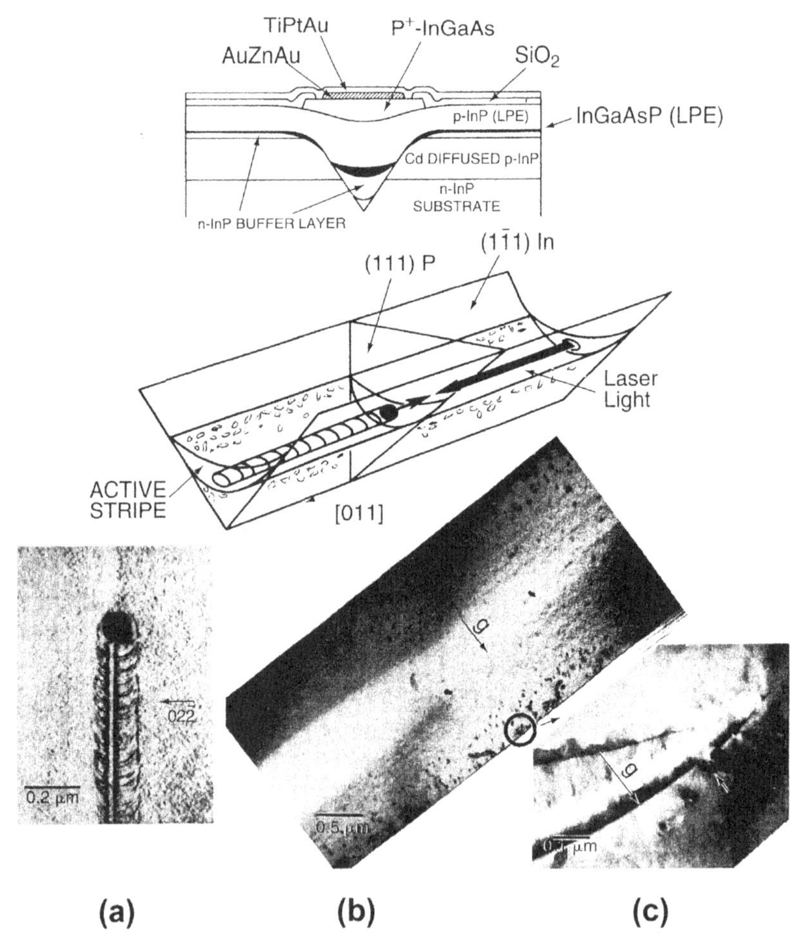


**Figure 10.8** Transmission electron microscope images of defects in the active region of a degraded 1.3 μm wavelength channeled-substrate buried heterostructure laser. Surrounding the schematic views of the active region are (a) wormlike defect, (b) dislocation loop growing into the {111} interfaces, (c) a magnified image of the crescent showing a dislocation loop growing into the *n*–InP buffer layer. *From reference [10]. Courtesy of S. N. G. Chu, Lucent Technologies, Bell Laboratories Innovations.*

Figure 10.8. Basically, three types of dislocations are generated inside the active region during gradual degradation.

1. 1/2 <100> Sessile loops—These dislocations are usually generated at the sidewall interface of the active stripe. Loops grow into the active region by continuous condensation of interstitials on the extra dislocation plane.
2. 1/2 <110> Slip Dislocations—Generated by stress in the strained layers of zinc blende structures, these dislocations lie in (111) slip planes. Once nucleated, internal plastic and thermal stresses cause them to propagate and multiply, forming complex dislocation structures.
3. 1/2 <110> Misfit Dislocations—Imaging under perpendicular diffraction conditions reveals that these dislocations are pure edge-type and stem from the bottom heterointerface of the active stripe.

The most intriguing manifestation of damage occurs during the rapid-degradation mode of laser failure. A wormlike defect shown in Figure 10.8 nucleates at the mirror facet and propagates into the center of the laser cavity. The tip of the worm is believed to be a metal-rich droplet. It probably forms by localized melting of the quaternary alloy due to nonradiative recombination of carriers at defect centers in the mirror facet. Melting continues on the side of the melt/solid interface facing the impinging laser light, while rapid solidification occurs on the backside. Epitaxial regrowth on the backside results in a growing tunnel of defective material.

## 10.2.4 Electrostatic Discharge Damage to Lasers

Electro-optical devices are no less vulnerable to ESD damage than electronic semiconductor devices. Since only limited work has been reported on ESD effects in semiconductor lasers, it is worthwhile to briefly present several results of a systematic study [11] of InP-based long wavelength (1.3, 1.48, and 1.55 μm) lasers. Light output current (L–i) as well as both forward and reverse bias i–V characteristics were used to determine the ESD sensitivity of these devices. Human body model ESD testing using voltage pulses up to 15 kV altered the i–V characteristics (Figure 10.9(a)), and yielded the indicated asymmetry in failure distributions upon forward (Figure 10.9(b)) and reverse biasing (Figure 10.9(c)). ESD failure voltages are seen to be much lower for reverse, relative to forward, polarity, a result that should not be too surprising in view of the arguments advanced in Section 6.5.4.1 for failure in p–n junctions.

Microscopically, the top view of ESD-generated defects in 1.3 μm lasers is shown in the TEM image of Figure 10.10. The active stripe bounded by the parallel dislocation lines contains small dislocation loops filled with yet

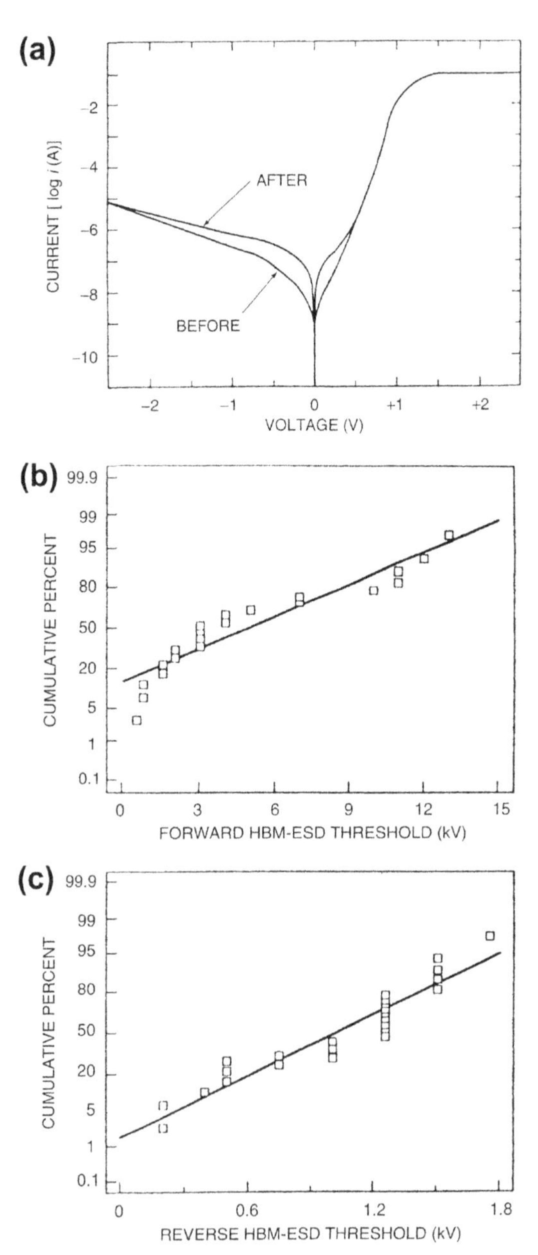


**Figure 10.9** Human body model (HBM) ESD testing using voltage pulses up to 15 kV alter the i–V characteristics (a), and yield the indicated asymmetry in failure distributions upon forward (b), and reverse biasing (c).

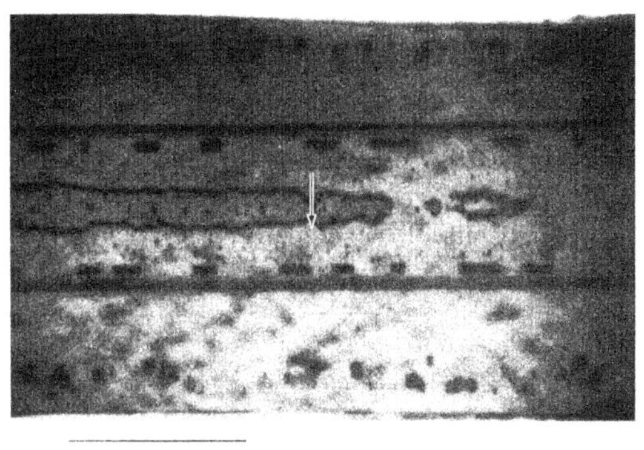


**Figure 10.10** Transmission electron microscope images of ESD-generated defects in a 1.3 μm laser. *Courtesy of S. N. G. Chu, Lucent Technologies, Bell Laboratories Innovations.*

smaller loops. In general, the greater damage within than outside the active region is indicative of current concentration in the active region. Similar, but less severe, defect damage occurs during electrical overstress (EOS). Rather than kilovolt pulses, in EOS a few tens to hundreds of volts are involved. In response, the active stripe contains long, narrow rectangular dislocation loops populated by tiny loops within; but overall, the defect density is reduced.

## 10.2.5 Contact and Bonding Reliability

Contact-related failures in electro-optical devices are qualitatively similar to those in other compound-semiconductor devices (e.g., MESFETs, HEMTs) considered earlier (Section 5.4.4). Up to five different metals plus As and P merge at metal contact–semiconductor interfaces. In addition, other metals are involved in nearby solder bonds and heat sinks. At elevated temperatures, particularly at high current densities, contact reactions, electromigration damage, intermetallic compound formation, precipitation, oxidation, etc., may threaten the integrity of the compound semiconductors and contacts. Superimposed on these effects are metal-interdiffusion reactions and thermal expansion mismatch stresses that cause solder degradation. Some of these issues are briefly addressed in turn.

1. *Contacts.* Both ohmic and Schottky barrier contacts are used in compound-semiconductor electronic devices. However, only the former are used in optical devices, where in addition to ohmic character, contacts must have low electrical and thermal resistivities to suppress the Joule heating that occurs during forward biasing. Common contact

metals like Au, Ti, and Pt exhibit large barrier heights (0.8–1 eV) and Schottky behavior when in contact with GaAs, InP, and InGaAsP; by alloying, however, the desired ohmic behavior develops. Gold-based contact metallurgies for LEDs have included Au–Zn, Au–Cr, and Ti–Pd–Au, as well as pure Au for *p*-side contacts. While the Zn-containing contacts degrade at current densities below 12 kA/cm$^2$ with an activation energy of 0.83 eV, the Ti–Pd–Au contacts are relatively immune to damage [7]. Accelerated testing causes DLD formation, decreased carrier lifetimes, increased leakage currents, and series resistance, effects that stem in part from interdiffusion of contact metal into the p–n junction. In another study [12] the contact resistance of Ti–Zn–Au on GaAs-based surface LEDs was lower than that of Ti–Pd–Au, accounting for an increase in reliability.

Laser contact metallurgies are also based on Au and display similar damage effects. Reactions between Au and both elements of the underlying InP semiconductor resemble electrode sinking (see Section 5.4.4.4) and take the form of $Au_3In$ compound and $Au_2P_3$ cluster formation [13]. High current densities also promote electromigration-induced spiking of shallow junctions by filament-like projections of conductive $Au_3In$.

2. *Bonding*. Beyond the contacts, metal leads must be connected to the device, which, in turn, is usually attached to a heat sink such as copper or diamond. This raises a host of potential problems associated with solders, attachment materials, and substrates, as discussed by Fukuda [1]. Troublesome compound formation sometimes occurs in the attachment between devices and the heat sink. In the case of lasers and LEDs, gold often serves as the cover metal for the heat sink that is bonded to one of the metal electrodes (sometimes also gold) via tin or indium-based solder layers. Resulting mass-transport mechanisms of the Au–Sn (or Au–In) reactions are schematically depicted in Figure 10.11. [1] Voids, always a serious reliability concern, are all the more threatening when generated at the electrode–solder interface. In this case Kirkendall voids form due to the unequal rates of rapid grain boundary diffusion of Sn into Au, and the slower bulk counterdiffusion of Au into Sn. Consequences of interdiffusion are brittle intermetallic compound formation, lack of adhesion, and loss in heat conduction. As a result, devices operate at higher temperatures, and the effective heat sink area shrinks due to void formation and solder melting. In bootstrap fashion, the hotter laser causes greater loss in thermal conductivity and heat extraction until laser destruction by thermal runaway ensues.

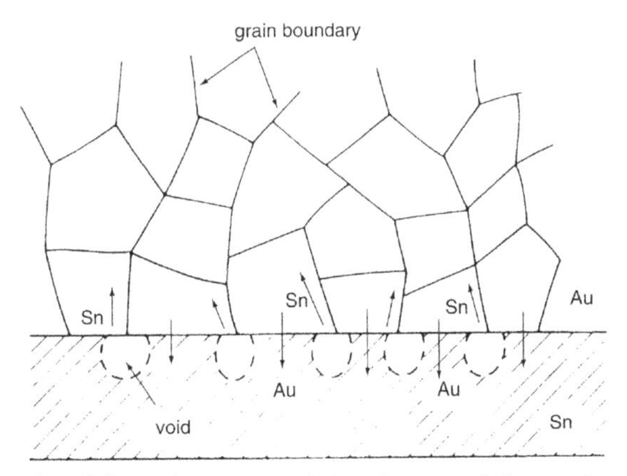


**Figure 10.11** Kirkendall voids generated by the interdiffusion between Au and Sn. *From Ref. [1].*

From what we have learned of die attachment in Chapter 8, there is a choice between employing the Au-rich, Au–Sn, Au–Si, or Au–Ge hard solders, or the softer Pb–Sn, Sn, and In solders. While stabilizing the interface against interdiffusion effects, the former less compliant and stronger solders pose the danger of mechanically stressing the LED or laser, as schematically shown in Figure 10.12(a). In contrast, the more compliant soft solders transfer less stress to the device (Figure 10.12(b)). Unfortunately, unlike hard solders, Sn and In are prone to sprouting whiskers (Figure 10.12(c)), which can short the device to its enclosure. Electromigration is a main driving force for whisker growth (see Figure 3.16), but the extent of damage is enhanced by internal strain, temperature, and humidity. Degradation of In- and Sn-based solders occurs at lower current densities (e.g., $10^4 A/cm^2$) than in aluminum metallizations.

Packaging failures include moisture leaks that promote corrosion of metal contacts and damage to optical components. The optical films considered next are prone to penetration of moisture through columnar grains by capillary action, a process that degrades optical properties such as index of refraction.

## 10.3 THERMAL DEGRADATION OF LASERS AND OPTICAL COMPONENTS

### 10.3.1 Introduction

In this section we address the damage caused by the absorption of laser light. As specific examples we shall consider optical coatings and lasers. The

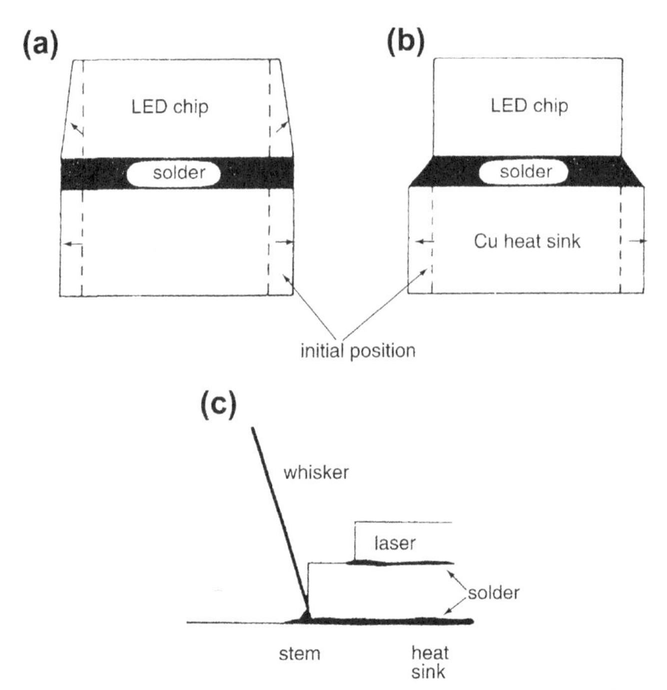


**Figure 10.12** (a) Stronger, less compliant solders pose the danger of mechanically stressing LEDs or lasers. (b) More compliant soft solders transfer less stress to optical devices. (c) Sn and In solders are prone to sprouting whiskers. *From Ref. [1].*

former consist of thin-film multilayers of oxides, fluorides, sulfides, etc., that are used as antireflection coatings, dielectric mirrors, and polarizers in optical components. In short-wavelength lasers used for inertial confinement fusion applications, damage to the optics is a serious problem, especially for extreme light pulse shapes.

Radiation incident on a surface with intensity $I_o$ (in units of PD, e.g., $W/cm^2$), is attenuated with distance $x$ beneath it, such that at any depth the intensity is

$$I(x) = I_o \exp[-\alpha x]. \quad (10.7)$$

The absorption coefficient $\alpha$, in reciprocal distance units, is one of the primary properties of concern. Absorption can give rise to beneficial as well as harmful effects; for example, electron–hole pairs are generated during photon absorption in semiconductors, making electro-optical effects possible. However, absorption also necessarily means conversion of radiant energy into potentially damaging heat. Actually, $\alpha$ is derived from the complex index of refraction ($N = n - ik$), which, in turn, has real and imaginary components. The former is the real index of refraction ($n$), while the latter is the index of absorption ($k$), a quantity related directly to $\alpha$, i.e., $\alpha = 4\pi k/\lambda$, where $\lambda$ is the photon wavelength.

Absorption is usually accompanied by some sort of electronic excitation process. In optical components (coatings, optical fibers) impurity atoms, ions, and defects are effectively photon absorbers, and every effort is made to remove these because $k$ (or $\alpha$) is reduced this way. Absorption coefficients in all materials are highly dependent on wavelength. Semiconductors, for example, strongly absorb radiation having energies greater than that of the bandgap ($E_g$), creating carriers in the process; but they are transparent to photons with energies less than $E_g$ (see Eqn (2.1)). In addition, $\alpha$ is a function of temperature, a fact that may promote catastrophic damage to optical devices when heating is uncontrolled (see Section 10.3.4.3).

Although it certainly makes a difference what the absorption mechanism is, under certain circumstances there are broad similarities in the heating effects produced irrespective of material. It is therefore worthwhile to treat first the generic problem of radiation absorption, and later particularize the phenomenon to specific materials and devices.

## 10.3.2 Thermal Analysis of Heating

In the simplest case we assume a homogeneous material heated by exposure to radiation of intensity $I$. The fundamental equation for one-dimensional heat conduction that must be solved to obtain the temperature $T(x, t)$, has the form

$$c\rho \frac{\partial^2 T}{\partial x^2}(x, t) = \kappa \frac{\partial^2 T}{\partial x^2}(x, t) + A(x, t), \tag{10.8}$$

where $A(x, t)$, in units of W/cm$^3$, accounts for the energy absorbed. A physically appropriate expression for this term is $A = (1 - R)\ \delta\ (x - x_o) I_o \exp[-\alpha x]$, where $R$ is the reflectance from the surface, and the delta function indicates absorption at depth $x_o$ and nowhere else. In this form the equation is generally difficult to solve; when real thin-film coating structures are considered in the next section, one can appreciate just how difficult.

It is instructive to consider absorption of a *constant* radiant-power density, which, in turn, produces a *constant* heat flux at the surface, where the latter is related to the value of $A$ at $x_o = 0$, i.e., $A = (1 - R)\ I_o$. Under these conditions the thermal history of the irradiated sample can be equivalently described by the heat conduction theory previously introduced in Section 6.5.3. The boundary value problem that models laser heating in the semi-infinite medium consists of Eqn (6.37) plus the following conditions. Initially,

$$T(x > 0, t = 0) = T_o, \tag{10.9a}$$

while the boundary conditions are

$$T(x = \infty, t > 0) = T_o (T_o \text{ is the ambient temperature}) \tag{10.9b}$$

$$-\kappa \left( \frac{\partial T}{\partial x} \right)_{x=0} = (1 - R) I_o \quad \text{for } t > 0. \tag{10.9c}$$

The last condition accounts for the concentration of radiant power in a surface layer, which gives rise to the flux of heat entering the body. When the problem is couched in these terms, the solution is identical to the form of Eqn (6.39), i.e.,

$$T = T_o + \frac{2(1 - R)I_o}{\kappa} \left\{ \left( \frac{Kt}{\pi} \right)^{1/2} \exp\left( -\frac{x^2}{4Kt} \right) - \frac{x}{2} \mathrm{Erfc}\left[ \frac{x}{2(Kt)^{1/2}} \right] \right\} \tag{10.10}$$

At $x = 0$ the surface temperature is simply

$$T = T_o + \frac{2(1 - R)I_o}{\kappa} \left( \frac{Kt}{\pi} \right)^{1/2} = T_o + 2(1 - R)I_o \left( \frac{t}{\pi c \rho \kappa} \right)^{1/2} \tag{10.11}$$

which clearly has the Wunsch–Bell form. As before, damage occurs when some critical temperature level, characteristic of the material, is reached.

Although these equations have a useful pedagogical simplicity, they only begin to address the complexity of actual thermal damage to either thin-film optical coatings or semiconductor lasers. More exact analyses consider heterogeneous media and material constants that are temperature dependent. Other complications arise from pulsed, as opposed to continuous, laser operation.

### 10.3.3 Laser-Induced Damage to Optical Coatings

Optical thin films are anything but homogeneous; unlike intrinsic bulk materials, films exhibit nonuniform stoichiometry and property anisotropies

[14]. These stem from the columnar, polycrystalline grain structure, in which grains that nucleate at the substrate are tiny, but then expand outward like inverted cones with further deposition and film growth. Hard-to-remove impurities distributed throughout the structure are sources of absorption, particularly when associated with imperfections and interfaces. More troublesome still are macroscopic inclusions or nodules stemming from the bulk-material sources, and the peculiarities of the vapor-deposition method employed. Suspected sources of these nodules include particles "spit" from sputtering targets or electron beam-heated materials that are intolerant of thermal shock.

When nodular heterogeneities such as that shown in Figure 10.13(a) are present, it may be assumed that both the absorption coefficients and thermal properties will differ from those of the surrounding matrix. Absorbed laser energy in the nodule vicinity causes a thermal impulse that rapidly heats and locally expands the material. The response has been mathematically modeled by coupling the time-dependent temperature profile evolution with the state of thermal stresses and displacements produced in the matrix [16]. As a result of the analysis, the following conclusions were reached:

1. An axisymmetric nodular defect causes significant laser-radiation enhancement along its centerline.
2. Both the temperature rise and maximum tensile stress in the nodule depend on the laser pulse length $t$; the maximum temperature rise varies as $t^{-0.2}$, and the maximum tensile stress is proportional to $t^{-0.4}$.
3. Nodules that are shallow and have a seed size of about 1 μm will develop the highest temperatures and stresses. Seeds less than 0.25 μm in size are significantly less harmful.

Permanent damage to thin-film optical coatings includes localized pitting from stress-induced nodular ejection (Figure 10.13(b)), plasma scalds (Figure 10.13(c)), delamination (Figure 10.13(d)), and film discoloration. The optical thin films discussed here are an excellent example of how processing defects are the source of later reliability problems. A common measure of the latter is the laser-induced damage-threshold level, a quantity easier to measure than predict. Even a small thermal distortion in a mirror can be fatal in some applications, and therefore efforts should be made to limit the presence of nodular defects of any size.

## 10.3.4 Thermal Damage to Lasers

In this section we return to laser damage in an attempt to quantitatively model some of the thermally induced failure modes and mechanisms

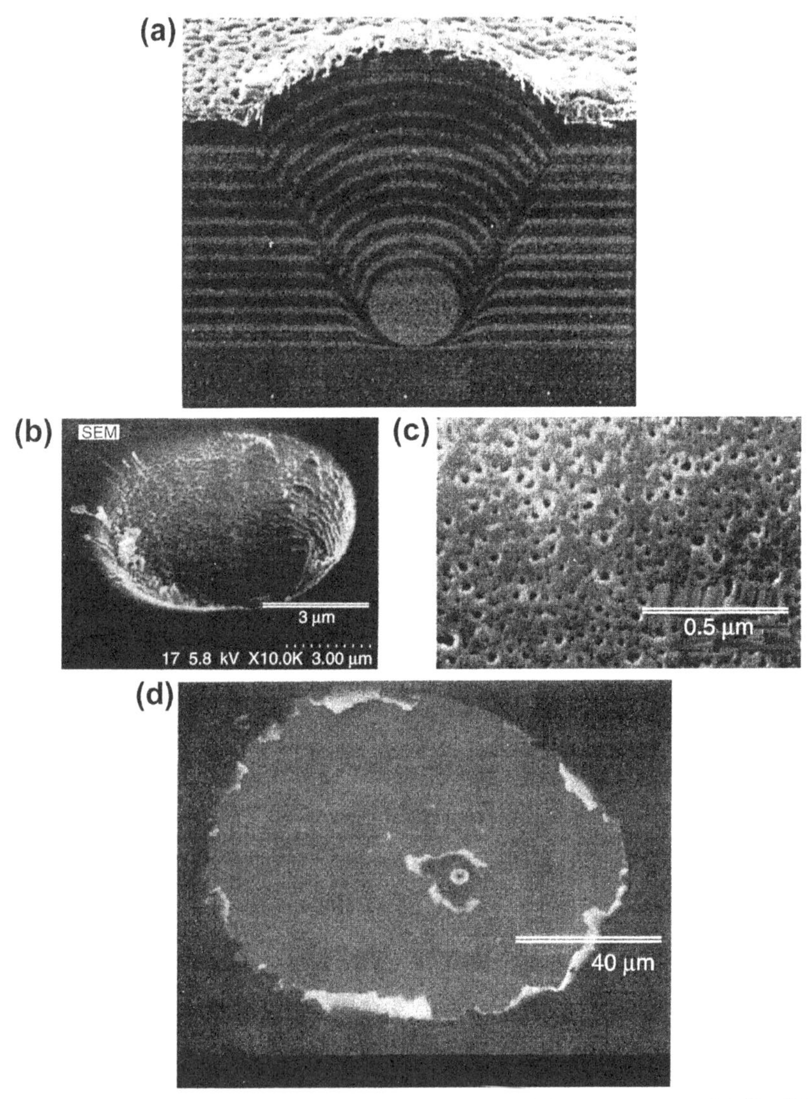


**Figure 10.13** (a) Cross-section of a nodule in a multilayer coating of $HfO_2/SiO_2$ deposited by electron beam evaporation. (b) A pit created on a mirror by ejection of a nodule after exposure to multiple shots of 47.2 J/cm$^2$ energy density. Illumination is with 3 ns, 1.064 μm laser pulses. Figure 10.13 (c) Plasma scalds on a mirror produced by a single shot at 41.6 J/cm$^2$. (d) Film delamination on a polarizer after one 21.5 J/cm$^2$ pulse at 8 ns. *From Ref. [15]. Courtesy M. R. Kozlowski.*

described earlier. Since the classic analysis by Henry et al. [17] has served as a standard reference for many later attempts to quantify laser damage, it is worth reviewing some of the issues raised in this chapter. These include local heating of nonradiative regions, dark-line damage, and thermal runaway. Even though AlGaAs double-heterostucture lasers were experimentally tested in this study, the results are broadly applicable to other semiconductor laser diodes. In thermal modeling, a key factor is the very strong dependence of the absorption coefficient; thus $\alpha = 140/\text{cm}$ at 0 °C, $10^3/\text{cm}$ at 50 °C, and $10^4/\text{cm}$ at 200 °C.

#### 10.3.4.1 Energy Flux Required to Heat a Cleaved Surface to the Melting Point

We have already addressed this issue in Eqn (10.11). The heat flux $((1 - R)\, I_o)$ or power density ($P$) at the surface generates the temperature rise $\Delta T$ (or $T_M - T_o$), given by

$$\Delta T = \frac{2P}{\kappa}\left(\frac{Kt}{\pi}\right)^{1/2} \tag{10.12}$$

A meaningful estimate of $P$ entails selecting the right values for $\kappa$ and $K$, both of which differ in the active and surrounding cladding regions of the laser and are strongly temperature-dependent as well. Taking values at the average temperature (900K) between the melting point $T_M$ and the ambient $T_o$, it is assumed that $\kappa = 0.12$ W/cm–K and $K = 0.058$ cm$^2$/s. Laser pulses of $t = 18$ ns were used to generate melting damage and therefore, by Eqn (10.12), $P$ is calculated to be approximately 4 MW/cm$^2$. When corrected for reflectivity of light at the surface, the optical power flux is still a factor of 2–3 lower than that measured to produce damage.

#### 10.3.4.2 Energy Flux Required for Dark-Line Damage

In the experimental portion of the Henry et al. study, dark lines approximately 0.2 μm in diameter were observed to grow with a velocity of 300 cm/s while absorbing a power of ~10 MW/cm$^2$. To reconcile these facts, DLD of dislocation tangles were assumed to behave like molten spheres. This assumption spawned the following thumbnail calculations and estimates:

1. The power to heat the GaAs spheres of radius $a$ (=0.1 μm) to the melting point in 18 ns is calculated to be $1.35 \times 10^{-2}$ W.
2. An additional $7.6 \times 10^{-4}$ W is needed for melting, assuming the latent heat of GaAs = 3250 J/cm$^3$.

3. At an average velocity of 300 cm/s the sphere travels a length of 0.054 μm in 18 ns. Additional melting of an equivalent cylinder of this volume requires $3.1 \times 10^{-4}$. Summing hives a total power expenditure of $1.46 \times 10^{-2}$ W.
4. Strong absorption of light occurs inside a waveguide that passes within a thermal diffusion distance ($\sim (Kt)^{1/2}$) beyond the sphere. Therefore, the absorbing cross-section in the active layer is about $4a\,[a + (Kt)^{1/2}] = 1.6 \times 10^{-9}$ cm$^2$.
5. The average absorbed flux in the moving sphere is thus calculated to be $1.46 \times 10^{-2}$ W/$1.6 \times 10^{-9}$ cm$^2 = 9.1$ MW/cm$^2$.

When accounting for reflection of incident light, the estimated power required for damage is in reasonable quantitative agreement with experiment.

### *10.3.4.3 Thermal Runaway*

Under conditions of rapid temperature change due to a time-dependent power flux incident at the surface, Eqn (10.12) is too simplistic. Instead, heat-conduction theory [18] shows that the temperature rise is given by

$$\Delta T = \frac{1}{\kappa}\left(\frac{K}{\pi}\right)^{1/2}\int_0^{\tau}\frac{P(t-\tau)\mathrm{d}\tau}{\tau^{1/2}}. \qquad (10.13)$$

We now recall that $P$ is proportional to $(1 - R)\ I_o \exp[-\alpha x]$. However, as already noted, $\alpha$ is temperature (and time) dependent; an exponential relation of the form $\alpha\ (T) = 140 \exp[(T - 300)21.4]$/cm has been suggested by Henry et al. [17] based on experimental data. Folding $\alpha\ (T)$ into Eqn (10.13) and employing semiconductor carrier dynamics to calculate the effective depth of absorption, the resulting integral equation was solved numerically. The solution is plotted in Figure 10.14 and indicates that temperature rises parabolically with time at low flux levels. But a two-orders-of-magnitude temperature increase occurs at higher power levels. Importantly, the exact solution follows the "low" curve only to switch to the "high" curve at 18 ns for a power flux of roughly 10 MW/cm$^2$. Thermal runaway is characterized by the substantial temperature increase that occurs within a nanosecond. To complicate matters further, the energy bandgap narrows with increasing temperatures, an effect that is greater in GaAs-based than in quaternary lasers. This further facilitates absorption and the damage feedback loop that accelerates failure.

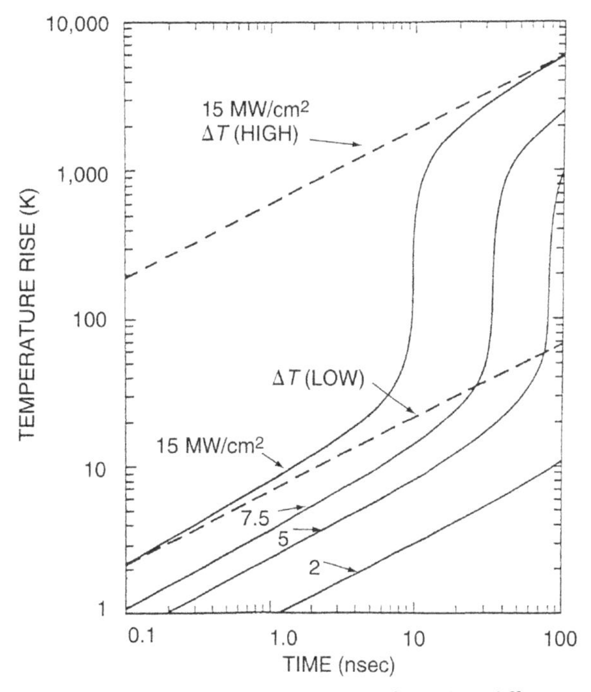


**Figure 10.14** Temperature rise at a cleaved surface for different incident flux densities. The dashed lines show the temperature rise for constant heating rates of 15 MW/cm$^2$. *From Ref. [17].*

Analogous thermal runaway phenomena occur during dielectric breakdown and electrostatic discharge failures. An instructive way to view thermal runaway has been presented by Schatz and Bethea [19], who applied the analysis to facet heating damage in lasers. The approach taken is graphically illustrated in Figure 10.15. Power dissipated at facets $P_D$ rises steeply with increasing temperature ($\Delta T$), as shown by the solid lines; by raising output power levels, greater amounts of power are dissipated, i.e., $P_D(4) > P_D(3) > P_D(2) > P_D(1)$. Simultaneously, heat flow in the device demands a thermal power $P_T$ to produce a temperature increase ($\Delta T$); $P_T$ depends on matrix thermal conductivity, dimensions of the heated zone, heat-sink geometry, etc., and varies with $\Delta T$ according to the dotted line. Steady-state conditions require that the power dissipated be sufficient to generate the heat needed to sustain the temperature. Operating temperatures, or solutions, arise at the intersection of the two curves. The filled circles represent stable solutions because any incremental increase in $P_D$ is less than the $P_T$ required to maintain the temperature increase; thus,

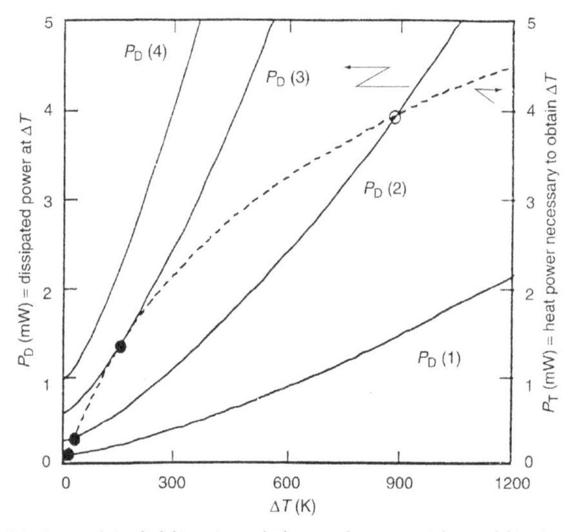


**Figure 10.15** Dissipated (solid lines) and thermal power (dotted line) as a function of the temperature increase for various levels of output power. The solid points correspond to stable steady-state solutions, while the open point reflects the thermal runaway of the unstable solution. *From Ref. [19].*

stability occurs when $d\ P_T/d\ \Delta T > d\ P_D/d\ \Delta T$. On the other hand, the open circle solution is unstable; i.e., $d\ P_D/d\ \Delta T > d\ P_T/d\ \Delta T$. Thermal runaway begins at a power level infinitesimally above $P_D(3)$, where there is no mathematical solution. Physically, the power dissipated is too great ever to be conducted away, and $\Delta T$ spirals upward in an unbounded manner.

# 10.4 RELIABILITY OF OPTICAL FIBERS

## 10.4.1 Introduction

In relatively recent years optical fibers have made extraordinarily rapid strides from a laboratory curiosity to a multitude of commercial applications in communications, signal processing, and sensors. With information capacities, or bandwidths, thousands of times larger than what is possible in copper wires and circuits, fiber optics have the capability of meeting the burgeoning demands of transmitting voice, data, and video for telephones, computers, and television. Over 10 million kilometers of fiber have already been installed in the United States, 95% of it underground, where the cables are well protected. In the future, however, fiber-in-the-loop leading to businesses and the home will be exposed to harsher environments and increased handling. The danger then exists that the high reliability required of optical fibers will be compromised.

Typical optical communication systems consist of a multilayer fiber with a light source (e.g., laser) at one end that converts the electric signal message into a series of light pulses. These are transmitted down the fiber, which may be tens of miles long, before the light has to be reamplified at repeater stations. At the other end of the fiber a photodiode detector converts the light back to the original electrical signal. In high-performance systems these devices operate at 1.55 μm, where maximum light transmission through the fiber occurs. How efficiently transmission occurs depends on light attenuation in the fiber. The key formula in this regard has already been given in this chapter as Eqn (10.7). Light launched with intensity $I_o$ is attenuated to magnitude $I$ with distance $L$ depending on the absorption coefficient ($\alpha$) of the fiber; small values of $\alpha$, or low loss, means fewer costly optical repeaters. It should be appreciated that the reliability of the system depends on the individual integrity of the electro-optical devices discussed earlier and the fiber we now consider. Our concern with reliability of fibers will be limited to a decrease in mechanical strength or an increase in optical loss ($L$). The latter is typically expressed in $dB$/km units and defined as $L = -\left(\frac{1}{L}\right)10\log\left(\frac{I}{I_o}\right)$ with L in km. A loss of ~0.5 dB/km is a typical value for undersea fiber.

Optical fiber used for communications consists of an active core of amorphous $SiO_2$ doped with $GeO_2$ in order to raise its index of refraction by about 1% above that of the surrounding $SiO_2$ cladding (Figure 10.16(a)). This enables light introduced into the core at certain angles to be confined to it without escaping into the cladding (Figure 10.16(b)). The core and cladding are typically 8.5 and 125 μm in diameter, respectively; although

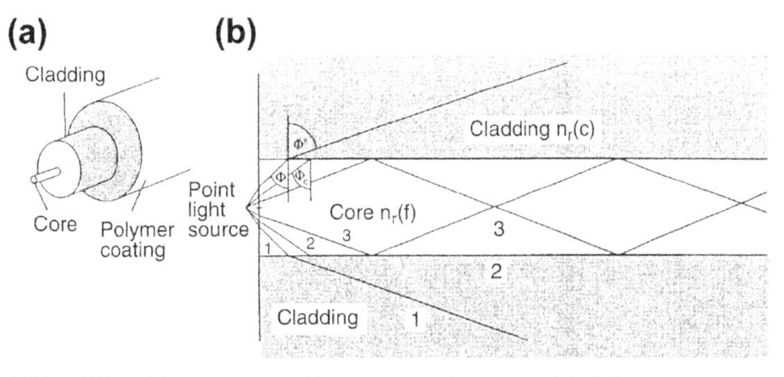


**Figure 10.16** (a) Optical fiber used for communications. (b) Refraction and propagation of light in an optical fiber. Rays 2 and 3 are confined to the fiber. Ray 1, incident on the fiber at too large an angle from the light source, escapes to the cladding. $n_r$(f) and $n_r$(c) are refractive indices of the fiber and core.

both are very brittle, they are, paradoxically, required also to be very strong. As we shall see, strength is dramatically reduced by surface damage produced by microcracks and exposure to corrosive environments. Therefore, as soon as the fiber, which is drawn from molten glass, cools sufficiently, it is coated by one or more layers of protective polymer. Some structural and compositional problems and defects that arise in fibers to adversely affect light guide performance are shown in Figure 10.17(a). Over the last two decades, fiber quality has been systematically improved to the point where defects and impurities have now been largely eliminated. Resulting optical losses due to impurity absorption have been virtually reduced to theoretical limits. In optical communications systems, fibers are encased in protective cable jackets. Nevertheless, cable-jacket defects and shrinkage may cause the fiber to suffer optical losses due to microbending (Figure 10.17(b) and (c)).

### 10.4.2 Optical Fiber Strength

All of the issues we shall consider surrounding the strength of optical fibers have been widely addressed in the literature [21–24]. The theoretical strength of silica glass based on the cohesive energy of the Si–O bond is estimated to be about 20 GPa, or 3 million psi. Short lengths of currently produced fibers have a strength of about 5.5 GPa (800 ksi) at ambient temperatures and moisture levels; this more than doubles to 13 GPa at liquid nitrogen temperatures (77K). Only steel piano wire, among structural engineering materials, has strengths approaching these magnitudes. Lower strength levels are, however, observed in longer optical fibers. Variability in glass strength has been universally attributed to the presence of microscopic surface cracks and modeled in terms of the classic Griffith crack theory for brittle fracture. This model, which was reviewed and applied to the problem of solder fracture in Section 9.5.4, is central to our understanding of the mechanical strength of optical fiber. Appropriately enough, tests on fine glass filaments demonstrated the validity of the theory in the early 1920s. Since Young's modulus ($E$) for $SiO_2$ is 70 GPa and it suffers little plastic deformation, exceptionally large elastic strains ($\sigma/E$) of 0.08–0.17 are possible at the stress ($\sigma$) levels quoted above.

According to the Griffith analysis, the strength or critical stress $\sigma_c$ required to cause rapid fracture in a fiber containing surface cracks of depth $a$ is given by

$$\sigma_c = \frac{K_{1c}}{Ya^{1/2}}. \tag{10.14}$$

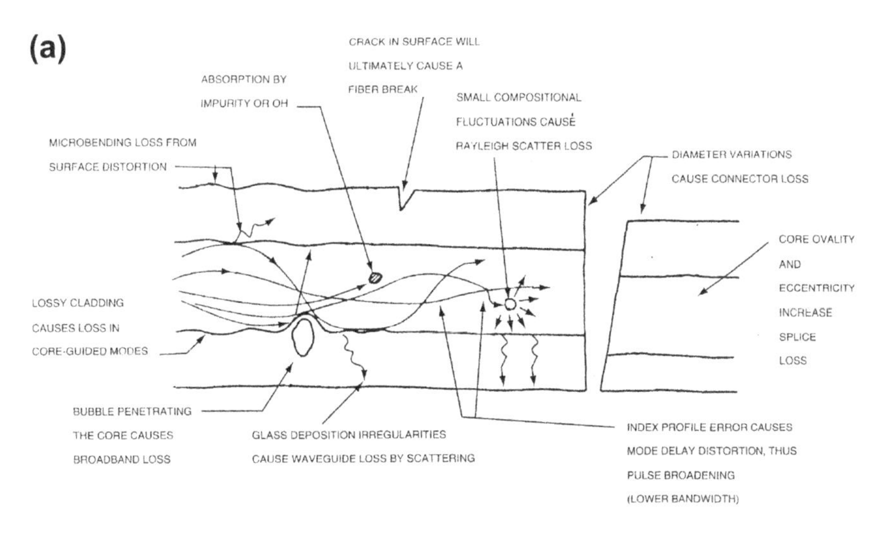


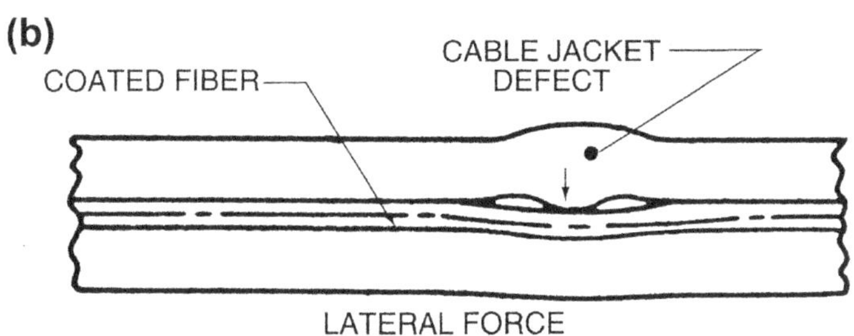


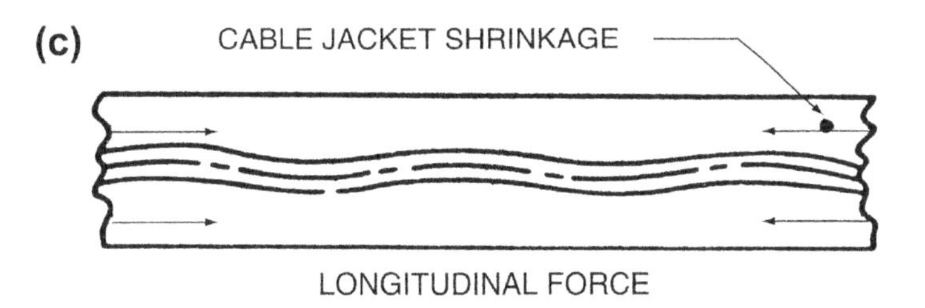


**Figure 10.17** (a) Structural and compositional problems within a fiber that can affect light guide performance. *(From Ref. [20].)* Optical loss in fiber due to microbending caused by the lateral force induced by a cable-jacket defect (b), and shrinkage of the cable jacket (c).

In this formula $K_{1C}$ is the critical stress-intensity factor, and for $SiO_2$ it has a value of 0.789 MPa-$m^{1/2}$. The critical stress for a given crack size is also known as the inert strength, a quantity usually defined at room temperature in the absence of time-dependent degradation. For smaller applied stresses ($\sigma < \sigma_c$) a subcritical stress-intensity factor $K(K < K_{1C})$ applies, where

$$K = \sigma Y a^{1/2}. \tag{10.15}$$

In this case no fracture occurs. The constant $Y$ depends on the crack geometry, and if the latter is semicircular, $Y = 1.24$. In an alternative view of failure, fast fracture occurs when a flaw reaches a critical size ($a_c$) under a given applied stress $\sigma$. These connections between tensile strength and crack size are plotted in Figure 10.18.

Values of $\sigma_c$ for optical fiber depend on the surface crack size dimensions and the test temperature and strain rate, as well as on environmental conditions. Anyone who has tested brittle materials like glass knows that experimental strength values have a wide statistical spread due in part to grip-induced breakage. This is not the case when ductile metals are similarly tested, where ultimate tensile strength values congregate with little variation about the mean. For this reason it is common to account for fiber strengths in terms of Weibull statistics, where the cumulative failure distribution function ($F$) as a function of fracture stress ($\sigma_f$) and length ($L$) is given by Eqn (4.30) or,

$$F = 1 - \exp\left[-\frac{L}{L_o}\left(\frac{\sigma_f}{\sigma_o}\right)^m\right]. \tag{10.16}$$

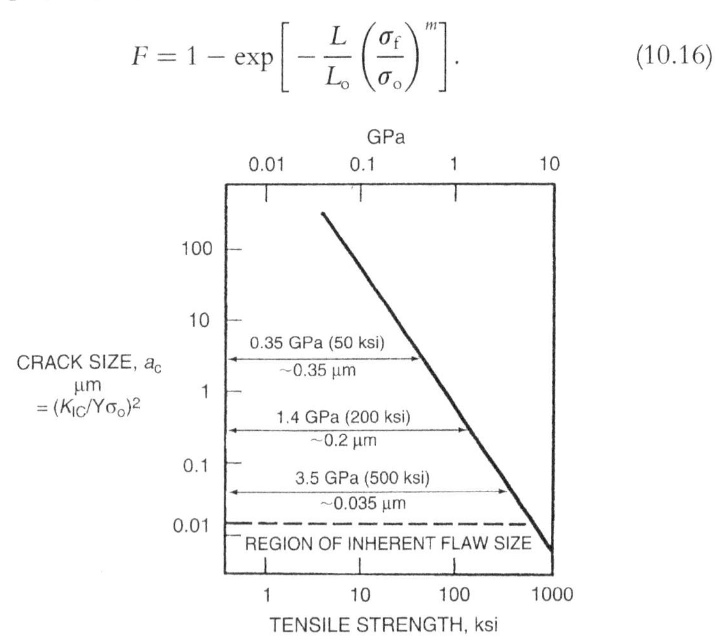


**Figure 10.18** Relationship between tensile strength and critical crack size of optical fiber.

In this formula $\sigma_o$ is the strength corresponding to a survival probability of $1/e$ or 36.8% measured at a gauge length of $L_o$. The Weibull scaling parameter $m$ is related to the variance, or width, of the distribution with respect to strength and determines the variation of strength with fiber length. We have previously noted that this equation is based on the notion of weak links (Section 4.5.2). Weibull statistics make physical sense for brittle materials because a singular weak-link defect, i.e., a large flaw, is fatal. The situation is entirely analogous to dielectric breakdown, as a comparison between Figure 6.9 and Figure 10.19 reveals. Electric fields that destroy dielectrics are like mechanical stresses that promote fracture of glass; plots of $F$ (or better yet ln [ln $(1 - F)^{-1}$]) versus stress accentuates the transition from defect-related failures to intrinsic failures above approximately 800 ksi.

With longer fibers there is a greater chance of incorporating a weak link, so that $F$ increases with fiber length. The dependence of fiber strength on length is easily derived. For the median strength, where $F = 0.5$, $L/L_o$ $(\sigma_f/\sigma_o)^m = 0.693$. Therefore, strengths ($\sigma_f$ (1), $\sigma_f$ (2)) of fibers with lengths $L_1$ and $L_2$ are related by

$$\left[\frac{\sigma_f(1)}{\sigma_f(2)}\right] = \left(\frac{L_2}{L_1}\right)^{1/m}. \tag{10.17}$$

Typical values of $m = 20$ signify a strong dependence of strength on fiber length.

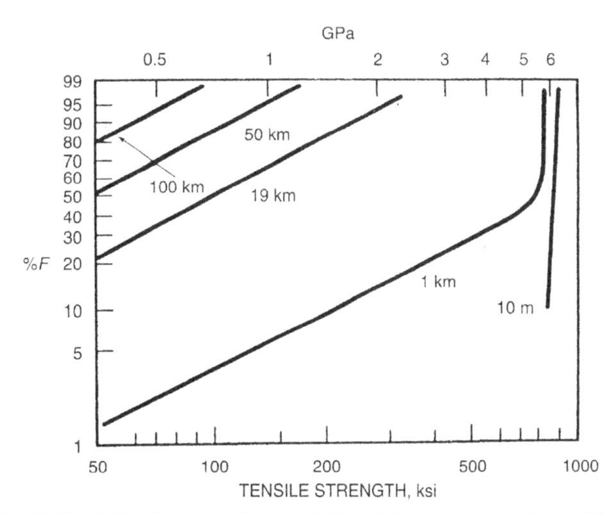


**Figure 10.19** Weibull distribution of optical fiber failure as a function of fracture stress and fiber length. *Courtesy of Gregg Bubel, Lucent Technologies, Bell Laboratories Innovations.*

The fact that Weibull distributions have been traditionally employed to describe fiber reliability does not mean that lognormal analysis is inappropriate or inapplicable. Since so much is at stake in optical communications, lognormal failure behavior should be critically evaluated to assess its utility in predicting long-term fiber life.

## 10.4.3 Static Fatigue—Crack Growth and Fracture

### *10.4.3.1 Power Law*

Glass optical fibers exhibit delayed fracture when sufficiently stressed in a moist environment. Under these conditions, time-dependent growth of cracks continues until a critical size is reached and the Griffith criterion for fast fracture becomes operative. *Static fatigue* is the phenomenon that describes subcritical flaw growth to such critical dimensions. Understanding the microscopic mechanisms of fatigue and formulating macroscopic equations to predict failure times are then the crucial tasks of reliability modeling. Over the years several models and equations for slow crack growth of glass fibers have emerged. The most widely used formula is based on Eqn (9.38), which is slightly modified here as

$$\frac{\mathrm{d}a}{\mathrm{d}t} = AK^n = A\left[\sigma Y a^{1/2}\right]^n. \tag{10.18}$$

This power law equation empirically fits assorted fatigue phenomena over a wide range of stress levels in different classes of solids. It applies to experimental data for silica exposed to water and ammonia (Figure 10.20)

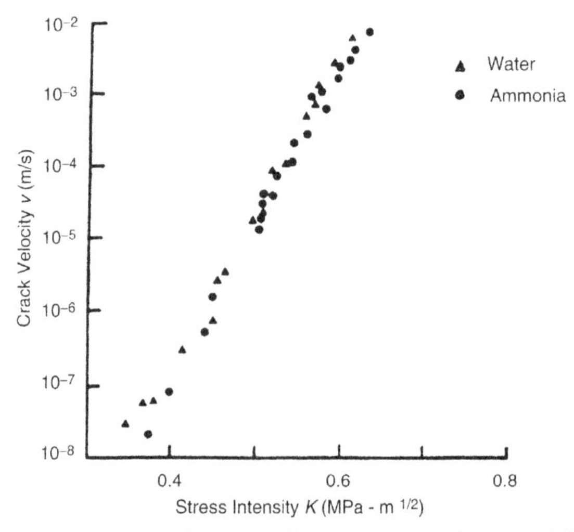


**Figure 10.20** Crack velocity as a function of stress intensity for optical fiber exposed to water and ammonia. *From Ref. [25].*

from which both constants $A$ and $n$ can be extracted. In our applications it is assumed that a constant, rather than alternating or dynamic stress, is applied to the fiber. Based on Eqn (9.39), an expression for the failure time ($t_f$) can be obtained by first separating variables $a$ and $t$ and then integrating from the initial crack size $a_i$ to the final crack size $a_f$; concurrently, time extends from $t = 0$ to $t_f$. The result is

$$t_f \approx \frac{2a_i^{-\left(\frac{n-2}{2}\right)}}{(n-2)AY^n\sigma^n}, \tag{10.19}$$

where it is assumed that the initial flaw length is much smaller than the crack depth prior to fast fracture ($a_f > a_i$). and that $n$ is large ($n > 15$). Thus the $a_f$ term can be safely neglected. It is common to relate the initial crack length to an equivalent inert strength $\sigma_c$ (i) via Eqn (10.14):

$$t_f = 2\frac{\left\{\left[\frac{Y\sigma_c(i)}{K_{IC}}\right]^{(n-2)}\right\}}{[(n-2)AY^n\sigma^n]} \tag{10.20}$$

This means that the failure time for a fiber of strength $\sigma_c$ (i) is predicted to be inversely proportional to the applied stress raised to power $n$; i.e., $t_f \approx \sigma^{-n}$. Comparison with Eqn (10.20) is made in the plot of log $\sigma$ versus log $t_f$ (Figure 10.21(a)), where the temperature dependence of static fatigue is also shown. Two different slopes (equal to $-1/n$) are evident; higher temperatures cause the break in slope (knee) to occur at shorter times. The phenomenon of the knee is treated again in Section 10.4.4.1.

#### *10.4.3.2 Exponential Law*

If crack growth is viewed on a molecular level, it is the chemical reactions that occur between strained bonds in vitreous silica and water that explain environmentally enhanced crack growth. Thus, through analogy to metals, static fatigue is like stress-corrosion cracking. Crack extension ensues when chemical bonds break at a faster rate than they are reformed. The basic degradation reaction is believed to open the silica network so that Si–OH bonds replace Si–O bonds, i.e.,

$$\mathrm{Si{-}O{-}Si + H_2O \rightarrow Si{-}OH + HO{-}Si.} \tag{10.21}$$

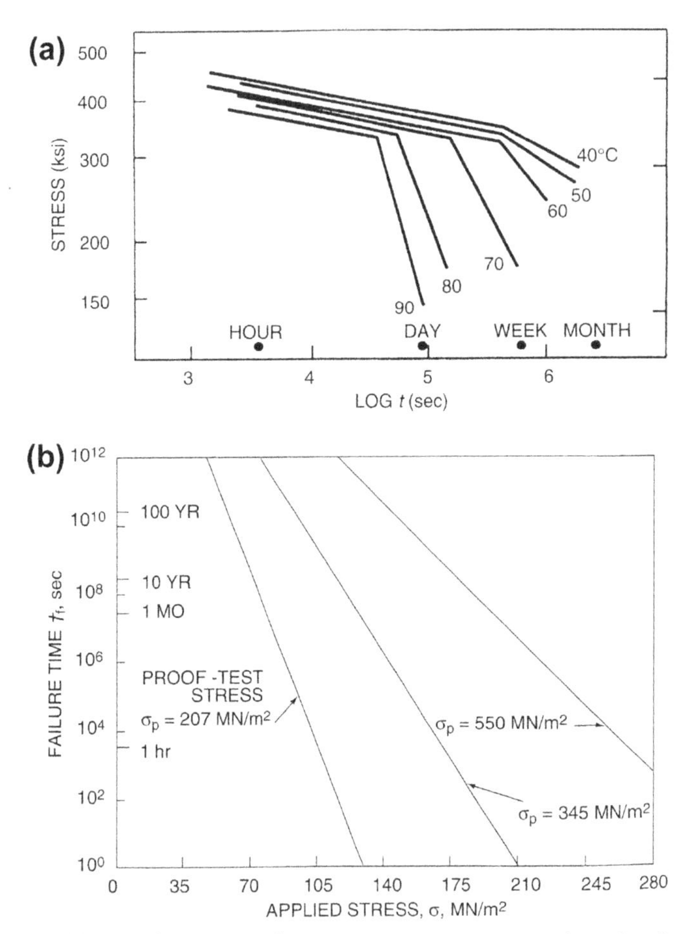


**Figure 10.21** (a) Static fatigue test data at various temperatures plotted as Eqn (10.20) (power law model). *(Adapted from Ref. [26].)* (b) Static fatigue test data plotted as Eqn (10.23) (exponential model). *Courtesy of Gregg Bubel, Lucent Technologies, Bell Laboratories Innovations.*

Employing concepts from chemical kinetics, the crack-growth velocity ($v$) can simply be written as

$$v = \frac{da}{dt} = v_o \exp\left[-\frac{E_\sigma}{kT}\right], \tag{10.22}$$

where $E_\sigma$ is a stress-reduced activation energy for Si–O bond dissociation and $v_o$ is a constant. Specifically, it is assumed that $E_\sigma = E_o\ (1 - K/K_{1C})$, where $E_o$ is the activation energy in the absence of stress. Due to the application of a static stress, $E_\sigma$ is reduced by a factor $K/K_{1C}$ [27]. Physically, when $K$ reaches the critical value $K_{1C}$, fracture proceeds with zero activation energy and is not thermally activated. The form of Eqn (10.22) suggests that the failure time will vary as

$$t_f = t_o \exp\left[\frac{E_o - \sigma\Omega}{kT}\right]. \tag{10.23}$$

This means that the functional relationship between the failure time and applied stress will show ln $t_f$ linearly proportional to $\sigma$, as observed in the static fatigue data of Figure 10.21(b).

A more rigorous approach to fiber fracture parallels that of power law kinetics of crack growth (Eqn (10.18)). Upon substituting the expression for $E_\sigma$, the differential equation describing crack extension that must be solved is of the form

$$\frac{da}{dt} = A_2 \exp\left[n_2\left(\frac{K}{K_{1C}}\right)\right], \tag{10.24}$$

where $A_2 = v_o \exp[-E_o/kT]$, $n_2 = E_o/kT$, and where it must be recalled that $K$ is proportional to $a^{1/2}$. Integration between crack lengths $a_i$ and $a_f$ yields the failure time

$$t_f = \frac{2}{Z^2 A_2}\left\{\left[\left(Za_i^{1/2} + 1\right)\exp Za_i^{1/2}\right] - \left[\left(Za_f^{1/2} + 1\right)\exp Za_f^{1/2}\right]\right\}, \tag{10.25}$$

with $Z = (E_o/kT)(Y\sigma/K_{1C})$. An alternative expression derived from Eqn (10.24) that is perhaps more usable has the form

$$t_f = \frac{2K_{1C}^2}{A_2 Y^2 n_2 \sigma_a^2}\left[\frac{\sigma}{\sigma_i} + \frac{1}{n_2}\right]\exp\left(-\frac{\sigma n_2}{\sigma_i}\right). \tag{10.26}$$

In this equation $\sigma_i$ is the intrinsic glass strength in the absence of fatigue and $\sigma_a$ is interpreted as the in-service stress. Alternatively, if $t_f$ is the design life, $\sigma_a$ can be interpreted as the maximum allowed service stress [28].

Now that expressions for failure times have been developed, one must not be lulled into the comfortable complacency that mechanical reliability of optical fibers is predictable and not a problem. Stress values must be known, and weak points in splices and extra cable twists contribute to them in ways not easily accounted for. In a more dramatic vein, what does one do about shark bites that have damaged undersea cables?

## 10.4.4 Environmental Degradation of Optical Fiber

### *10.4.4.1 Degradation due to Water*

Although we have already suggested the harmful implications of moisture with respect to fiber strength, the mechanism of water-induced degradation is more closely examined in this section. Among the effects of water is the "knee" in the fatigue behavior of optical fiber. As indicated in Figure 10.22, there is an abrupt change in slope above $\sim 10^5$ s that has important reliability implications. The knee occurs in coated as well as uncoated fiber and under stressed as well as unstressed, i.e., zero-stress, aging conditions. An important correlation between surface roughening of the fiber and its decrease in strength during aging was made by Yuce and coworkers [29] using scanning tunneling and atomic force microscopes. In unaged fiber the average roughness, or height, of peaks relative to valleys was about 0.4 nm, but this difference increased by 1 nm during aging. This suggests that the fatigue knee is caused by surface etching. Accordingly, preknee behavior is characterized by propagation of existing defects, while surface roughening, which creates new defects, stems from postknee aging. Formation of these surface pits occurs by a dissolution, or etching, process in which silica is removed, probably in the form of $Si(OH)_4$.

Dissolution is quite distinct from the crack-growth mechanism considered earlier, which only requires bond rupture along the crack plane. This etching process differs because material must be removed, and

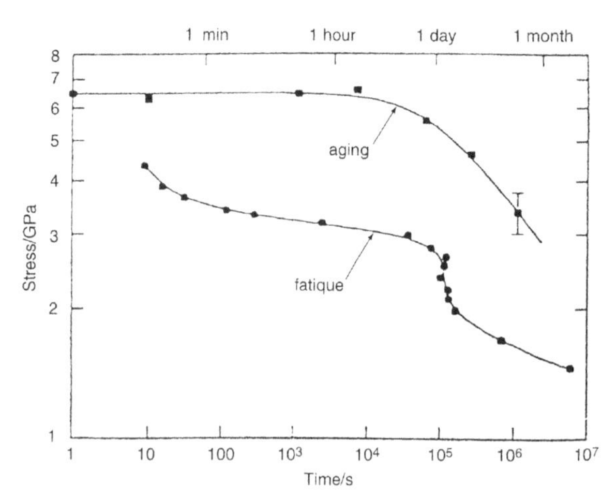


**Figure 10.22** Static fatigue (applied stress versus time to failure) ●, and zero-stress aging (residual strength versus. aging time) ■ for epoxy acrylate-coated fiber in 100 °C water. *From Ref. [33].*

because the surface-roughening reaction is free of the steric hindrance at a crack tip.

#### 10.4.4.2 Effect of Hydrogen

Fracture of optical fiber is very much a stress-corrosion phenomenon because cracks extend through chemical reactions that promote a zipperlike severing of Si–O bonds. In addition to the harmful influence of water, hydrogen is also known to degrade optical fiber. Generated by galvanic corrosion cells present in submerged cables, hydrogen can enter fiber through polymeric coatings. Laboratory simulation studies [30] have shown that corroding galvanized steel and hydrogen-producing bacteria can substantially raise optical losses. It was concluded that such bacteria found at water crossings near fiber cables, containing metal shields constructed of jute- and tar-covered galvanized-steel armor wires, could cause harmful light attenuation. Interestingly, sulfur-reducing bacteria tend to suppress the effect of the hydrogen-producing bacteria.

Lemaire [31] has extensively investigated the optical degradation due to hydrogen and found that it induces loss absorption peaks in the range 1.07–1.24 μm and a rising loss "edge" beyond 1.55 μm. Two useful equations were suggested for estimating time (*t*)-dependent fiber loss at a hydrogen pressure of 1 atm. The increase in loss at 1.0 μm is

$$\frac{dL}{dt} = 5.61 \times 10^8 T \exp\left[-\left(\frac{98.9\ \text{kJ/mol}}{RT}\right)\right] \text{dB/km} - \text{h}, \qquad (10.27)$$

with corresponding losses at 1.31 and 1.55 μm that are 0.7 and 0.55 times the value at 1.0 μm, respectively. Similarly, there is a comparable loss due to OH centered at 1.39 μm that is given by

$$\frac{dL}{dt} = 1.2 \times 10^8 T \exp\left[-\left(\frac{97.3\ \text{kJ/mol}}{RT}\right)\right] \text{dB/km} - \text{h}. \qquad (10.28)$$

This translates into losses at 1.31 and 1.55 μm that are, respectively, only 0.02 and 0.004 times the loss value at 1.39 μm. In addition, loss increases are also directly proportional to the hydrogen pressure ($P_{H_2}$). To produce losses of the magnitudes indicated, molecular hydrogen must diffuse through the coating and cladding and finally enter the light-guiding core region. A preferred value for the diffusion coefficient (*D*) of $H_2$ in silica fibers is given by

$$D(H_2) = 2.83 \times 10^{-4} \exp(-40.2\ \text{kJ/mol}/RT)\text{cm}^2/\text{s}. \qquad (10.29)$$

Taking all of these factors into account, it is concluded that after 25 years of service at $P_{H_2} = 1$ atm, the calculated losses would be intolerable. To prevent hydrogen permeation and mechanical degradation it is common to coat fibers. In addition to polymers, hermetic coatings of carbon and metal films markedly block ingress of hydrogen and minimize optical degradation.

### 10.4.4.3 Radiation Damage

The use of optical fibers in military communications and in radiation environments such as found in reactors, hospitals, nuclear waste sites, and outer space have raised questions about their long-term reliability. What is of particular concern is the increase in optical loss upon exposure to radioactive isotopes and radiation. The same collection of atomic and nuclear particles as well as ionizing radiation (see Section 7.6) that afflicts semiconductor devices is involved here, and the damage produced is broadly similar. However, in contrast to highly perfect semiconductor lattices, the amorphous glass network contains Schottky and Frenkel defects plus strained bonds and a wide variety of substitutional, interstitial, and multivalent cations as well as nonstoichiometry-induced defects. Electrons and holes created by radiation damage migrate through such a matrix and are either trapped by preexisting or newly created defects. New energy levels are created at these electron-defect centers, enabling absorption of light and a corresponding increase in loss, manifested by a darkening of the fiber.

The extent of radiation-induced fiber degradation depends on a host of variables, including the nature and energy of the radiation, the dose and dose rate, temperature, composition and uniformity of the exposed fiber, and wavelength of operation. Certain of these variables can have potent effects. For example, phosphorous-doped fiber exhibits a three-orders-of-magnitude greater loss than silica fiber after exposure to $10^5$ rad of $Co^{60}$ radiation for an hour [21]. Time-dependent loss increases for 1.3 μm fiber are shown in Figure 10.23 for both high- and low-dose rate environments.

Optical loss due to radiation has both permanent and transient components. Fortunately, the latter can be diminished by heating, which causes defect annealing, and by exposure to light, which promotes photobleaching or recovery due to electron–hole recombination. Overall, the effects are such that higher losses are suffered at lower temperatures, assuming equivalent doses. For 1.3 μm fiber irradiated with $Co^{60}$ at a dose rate of 213 rad/s, a loss of about 0.4 mdB/km-rad can be expected at 20 °C; at −70 °C the figure is about 2 mdB/km-rad [21].

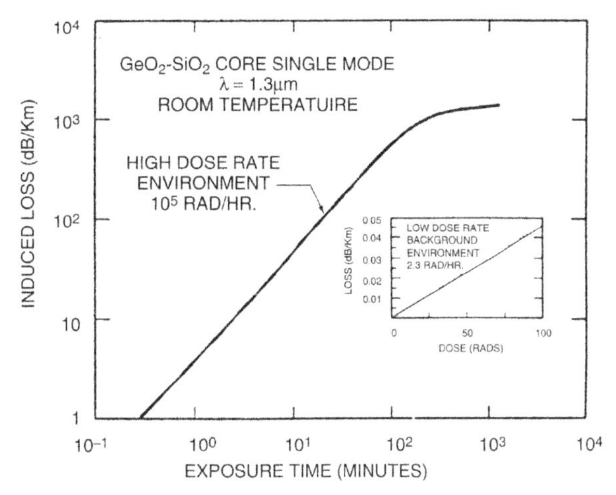


**Figure 10.23** Time-dependent loss increases for 1.3 μm fiber is shown for both high- and low-radiation dose-rate environments. *From Ref. [21].*

In conclusion, a simple mathematical relation that describes the time-dependent change in concentration ($C$) of the species that promotes loss in irradiated fibers has the form.

$$\frac{dC}{dt} = aR - b(T)C^2. \tag{10.30}$$

In this chemical kinetics-like equation, $R$ is the radiation dose rate, and $a$ and $b$ are temperature independent and dependent constants, respectively. Physically, at relatively low temperatures the loss builds linearly but then saturates. At elevated temperatures the second term on the right dominates due to defect annihilation, and a second-order kinetics drop in loss occurs. It is the admixture of these two effects that approximates the actual behavior of the fiber.

#### *10.4.4.4 Field-Aged Cable*

Rather than extrapolating the results of accelerated laboratory testing, it is always better to rely on field experience in evaluating strength degradation of optical fibers. In the few reported cases [32] where fibers were removed from cables that had aged naturally for up to 9 years, the results are decidedly encouraging. Handleability of fibers as measured by coating strippability, cleavage, splicing, and mechanical strength was comparable to freshly manufactured fiber. Based on atomic force microscopy, the narrow roughness range from 0.26 to 0.49 nm within both cable-aged and new fiber reveals no discernible difference in surface topography.

An encouraging finding was that models based on room-temperature aging overestimate the actual extent of strength degradation.

We end by noting that a 1993 study by Bellcore revealed that about one-third of telecommunications outages were due to cable failures. Of these, 59% were caused by accidental dig-ups. There are no reported cases where fiber fatigue was responsible for a network failure.

## EXERCISES

**10.1.** Short-time current aging of lasers is found to obey an equation of the form $i - i(0) = i(0)\ t^n$, where $i(0)$ is the initial current. Given the following data:

| $t$ (in 10,000 h) | $\frac{i-i(0)}{i(0)}$ |
|---|---|
| 0.5 | 8 |
| 1 | 10 |
| 1.5 | 11 |
| 2.0 | 12 |
| 3.0 | 13 |
| 4.0 | 14 |
| 5.0 | 15 |

Calculate the valve of $n$.

**10.2 a.** If the EOL criterion for the lasers depicted in Figure 10.3 were $\Delta i_L(t)/i_L(0) = 0.30$, how many lasers will survive 10 years at 25 °C?

**b.** Suppose these lasers are used in submarine telecommunications applications where they must operate at 20 °C. Furthermore, it is desired to have a laser survival rate of 99% with a 15 year life. In a laser-testing program what value of $\Delta i_L(t)/i_L(0)$ should be taken as a measure of EOL?

**10.3** Transform the failure data for passivated and unprotected facets in Figure 10.7 into corresponding Arrhenius plots. Calculate values for $E_{COD}$ and ( for each case.

**10.4** A suggestion has been made by Sieber et al. in Materials Science and Engineering B20, 29 (1993) that a photoplastic effect in $Al_xGa_{1-x}As/GaAs$ may be the first step in producing laser damage. The authors assert that compositional variations generate stresses that in concert with incident photons induce dislocation motion. To assess this idea calculate the stress level generated if the electroluminescent photon energy detected in neighboring regions differed by 0.078 eV. Assume that the energy gaps ($E_g$) and lattice parameters ($a_o$) for the involved materials are:

$$E_g(\mathrm{Al}_x\mathrm{Ga}_{1-x}\,\mathrm{As}) = E_g(\mathrm{GaAs}) + 1.247x\mathrm{eV}, \quad E_g(\mathrm{GaAs}) = 1.54\ \mathrm{eV}$$
$$a_o(\mathrm{Al}_x\mathrm{Ga}_{1-x}\,\mathrm{As}) = a_o(\mathrm{GaAs}) + 0.008x, \quad a_o(\mathrm{GaAs}) = 0.5653\ \mathrm{nm}$$

and that the modulus of elasticity for GaAs is 124 GPa.
How does this value compare with the yield stress?

**10.5** Mention reasons why GaAs-based lasers are more prone to degradation than quaternary lasers.

**10.6** Verify the rough estimates presented in Section 10.3.4.2 for the energy flux required to generate dark-line damage.

**10.7** In the theory of thermal runaway presented by Bethea and Schatz, suppose power absorption at GaAs laser facets varies with temperature rise according to $P_D$ (1) = $1 \times 10^{-4}$ $(\Delta T)$ mW, $P_D$ (2) = $2 \times 10^{-4}$ $(\Delta T)^{1.3}$ mW, $P_D$ (3) = $3 \times 10^{-4}$ $(\Delta T)^{1.6}$ mW, and $P_D(4) = 4 \times 10^{-4}$ $(\Delta T)^{1.9}$ mW. Furthermore, assume that $P_T = 0.3$ $(\Delta T)^{1/2}$ mW.

Calculate a set of $P_D$, $\Delta T$ values that leads to stable laser operation. Similarly, calculate a set of $P_D$, $\Delta T$ values that leads to an operating instability.

**10.8** The rise in temperature of a laser-irradiated material is given by Eqn (10.10). When the laser pulse is turned off at time $t = \tau_p$, the material cools.

**a.** What initial and boundary conditions hold for heat flow under these conditions?

**b.** Show that $T = T(\tau_p) - \frac{2(1-R)I_o}{\kappa}\left\{\left[\frac{K(t-\tau_p)}{\pi}\right]^{1/2} \exp - \left[\frac{x^2}{4K(t-\tau_p)}\right] - \frac{x}{2}\mathrm{Erfc}\left[\frac{x}{[2(K(t-\tau_p)^{1/2}}\right]\right\}$ is a solution to Eqn (10.8) during cooling if T(τp) is the temperature reached at time τp.

**c.** Derive an equation for the time-dependent surface temperature after the laser pulse is turned off.

**10.9** Pulsed laser irradiation of thin, absorbing ceramic films produces a kind of fatigue where thermal stresses are produced both during the heating and cooling portions of the cycle. Describe the nature of the stresses produced during exposure to pulsed radiation. Qualitatively sketch a theory that would predict the number of pulses the film could sustain before fracture. The theory should correlate the failure criterion to the laser intensity and the mechanical, thermal, and optical properties of the film material.

**10.10** Relative to silicon-integrated circuits, the manufacturing of semiconductor lasers tends to be less automated and involves more art. There are vastly fewer processing and lithography steps, a simple metallization and contact scheme, and greater variability in the dimensions of device features. What are the implications of these factors on laser reliability?

**10.11** In Griffith crack theory the static energy surrounding a plate containing a flat crack of size $a$ is $W = -\pi\ \sigma^2\ a^2/E + 4\ \gamma\ a$. Stress $\sigma$ acts to open the crack, while surface tension $\gamma$ acts to seal it shut.

**a.** What is the stress required to just cause the crack to open?

**b.** To determine the fracture velocity $v$, the kinetic energy, $\frac{1}{2}$ $k\rho v^2\ a^2\ (\sigma/E)^2$, must be accounted for, where $\rho$ is the density, $E$ is the elastic modulus, and $k$ is a constant. Explain why the kinetic energy has this form.

**c.** The rate of release of kinetic (strain) energy limits the crack velocity achieved. Show that the value for v that minimizes both the static and kinetic strain energies as well as the surface energy is equal to $\left\{\frac{2\pi E}{k\rho}\left[1 - \frac{a_c}{a}\right]\right\}^{1/2}$, where ac is the critical crack size.

**d.** Comment on the difference between v here and $da/dt$ in the text.

**10.12** Failure-time data for unstressed optical fiber as a function of temperature gave an Arrhenius plot with an activation energy value of 18 kcal/mol. When the fiber was stressed to 3.44 GPa, the activation energy was found to be 10.7 kcal/mol.

**a.** What is the value for the activation volume, and is it reasonable?

**b.** Suggest a way to obtain a value for $K_{1C}$ from such data.

**10.13** From Figure 10.19 what is the value of $m$? What is the median strength for 10 km lengths of this fiber?

**10.14** Optical fiber exposed to deuterium oxide ($D_2O$) is longer lived relative to similar fiber exposed to $H_2O$ at the same strain level. Suggest a possible reason for this. Plot this result as the cumulative failure probability versus log failure time if the ratio of the times to failure for these differently exposed fibers is roughly 2.

**10.15** In dynamic fatigue of optical fiber, a phenomenon that parallels that of static fatigue, the fiber strength is dependent on the stress rate, i.e., $d\sigma/dt$.

**a.** If $\sigma = t \times d\sigma/dt$, show that the fiber strength varies as $\left(\frac{d\sigma}{dt}\right)^{\frac{1}{n+1}}$ through application of the fatigue power law.

**b.** What fiber strength can be expected at a stress rate of 1 ksi/s if the strength is 800 ksi at a stress rate of 1000 ksi/s? Assume $n = 20$.

**10.16** Assume that a fiber withstands a stress of 700 ksi for 1 s. How long would the fiber survive at a stress of 300 ksi if $n = 24$?

**10.17** A bent fiber has a maximum tensile strength, $\sigma_{max}$, given by $\sigma = Er/R$, where $R$ is the bend radius of curvature and $r$ is the fiber radius.

**a.** What is the maximum tensile stress for a fiber bent to a radius of 0.75 cm?

**b.** A fiber stressed to this level failed after 1 month. For a 25-year life what strength fiber would be required? Assume $E = 70$ GPa and $r = 62.5$ μm.

**10.18** Both amorphous polymers and glasses degrade optically due to radiation damage. But only polymers suffer appreciable mechanical deterioration as well. Why? Explain possible reasons why silica fibers do not suffer mechanical damage upon exposure to ionizing radiation.

**10.19** Reliability of optical fiber after 25 years implies a loss increase of no more than 0.1 dB/km. What is the maximum hydrogen pressure that can be tolerated such that this level of loss at 1.31 and 1.55 μm is not exceeded during this time at 30 °C?

**10.20** Suggest a strategy for altering protective coatings in order to inhibit surface roughening of optical fibers due to etching.

**10.21** With little loss in generality, Eqn (10.30) expresses the radiation-induced time-dependent loss ($L$) in optical fiber as $dL/dt = d'R - b'(T)L^2$, where $d'$ and $b'$ are constants. Obtain a solution for loss as a function of time through separation of variables and direct integration. Assume that initially $L = 0.12$ dB/km. Does a log $L$ – log $t$ plot of the solution resemble Figure 10.22?

## REFERENCES

[1] M. Fukuda, Reliability and Degradation of Semiconductor Lasers and LEDs, Artech House, Boston, 1991.

[2] R.S. Mann, D.K. McElfresh, 33rd Annual Proceedings of the IEEE Reliability Physics Symposium, 1995, p. 177.

[3] D.L. Barton, J. Zeller, B.S. Phillips, P.-C. Chiu, S. Askar, D.-S. Lee, M. Osinski, K.J. Malloy, 33rd Annual Proceedings of the IEEE Reliability Physics Symposium, 1995, p. 191.

[4] F.R. Nash, Estimating Device Reliability: Assessment of Credibility, Kluwer, Boston, 1993.
[5] B. Mroziewicz, M. Bugajski, W. Nakwaski, Physics of Semiconductor Lasers, North Holland, Amsterdam, 1991.
[6] S. Yamakoshi, M. Abe, O. Wada, S. Komiya, T. Sakurai, IEEE J. Quantum Electronics, QE 17 (1981) 167.
[7] M. Fukuda, IEEE J. Lightwave Tech. 6 (10) (1988) 1488.
[8] P.W. Hutchinson, P.S. Dobson, Phil. Mag. 32 (1975) 745.
[9] A. Moser, E.E. Latta, J. Appl. Phys. 71 (10) (1992) 4848.
[10] S.N.G. Chu, S. Nakahara, M.E. Twigg, L.A. Koszi, E.J. Flynn, A.K. Chin, B.P. Segner, W.D. Johnston, J. Appl. Phys. 63 (3) (1988) 611.
[11] Y. Twu, L.S. Cheng, S.N.G. Chu, F.R. Nash, K.W. Wang, P. Parayanthal, J. Appl. Phys. 74 (3) (1993) 1510.
[12] C.E. Lindsay, Qual. Reliability Eng. Int. 9 (1993) 143.
[13] A. Piotrowska, P. Auvray, A. Guivarc'h, G. Pelous, P. Henoc, J. Appl. Phys. 52 (1981) 5112.
[14] M. Ohring, The Materials Science of Thin Films, Academic Press, Boston, 1992.
[15] F.Y. Genin, C. Stolz, Proceedings of 3rd International Workshop on Laser Beam and Optics Characterization, July, 1996. Quebec City.
[16] R.H. Sawicki, C.C. Shang, T.L. Swatlowski, Laser-induced Damage Opt. Mater. SPIE 2428 (1994) 333.
[17] C.H. Henry, P.M. Petroff, R.A. Logan, F.R. Merritt, J. Appl. Phys. 50 (5) (1979) 3721.
[18] H.S. Carslaw, J.C. Jaeger, Conduction of Heat in Solids, second. ed., Oxford, London, 1959.
[19] R. Schatz, C.G. Bethea, J. Appl. Phys. 76 (1994) 2509.
[20] J.A. Jefferies, R.J. Klaiber, The Western Electric Engineer XXIV (1) (1980) 13.
[21] S.R. Nagel, SPIE 717 (1986) 8.
[22] C.R. Kurkjian, D. Inniss, Opt. Eng. 30 (6) (1991) 681.
[23] G.M. Bubel, M.J. Matthewson, Opt. Eng. 30 (6) (1991) 737.
[24] F.P. Kapron, H.Y. Yuce, Opt. Eng. 30 (6) (1991) 700.
[25] T.A. Michalske, B.C. Bunker, J. Appl. Phys. 56 (1984) 2286.
[26] H.C. Chandan, D. Kalish, J. American Ceramic Soc. 65 (3) (1982) 171.
[27] K. Abe, G.S. Glaesemann, S.T. Gulati, T.A. Hanson, Opt. Eng. 30 (6) (1991) 728.
[28] M.J. Matthewson, V.V. Rondinella, C.R. Kurkjian, SPIE 1791 (1992) 52.
[29] H.H. Yuce, J.P. Varachi, J.P. Kilmer, C.R. Kurkjian, M.J. Matthewson, OFC' 92 Tech. Digest, Post deadline paper PD91, San Diego, CA, 1992.
[30] G. Schick, K.A. Tellefsen, A.J. Johnson, C.J. Wieczorek, Opt. Eng. 30 (6) (1991) 790.
[31] P.J. Lemaire, Opt. Eng. 30 (6) (1991) 780.
[32] J.L. Smith, A. Dwivedi, P.T. Garvey, Int. Wire Cable Symp. Proc. 848 (1995).
[33] M.J. Matthewson, C.R. Kurkjian, J. Am. Ceram. Soc. 71 (3) (1988) 177.

CHAPTER 11

# Characterization and Failure Analysis of Materials and Devices

## 11.1 OVERVIEW OF TESTING AND FAILURE ANALYSIS

### 11.1.1 Scope

When there are compelling reasons to find the physical causes for low yields, poor performance, degradation, or failure of components, devices, and systems, an experimental investigation of some sort is usually undertaken. Depending on the information sought and the complexity involved, the study may involve little more than a tedious testing routine, have all the earmarks of unraveling a detective mystery, or resemble a research project. In addressing the totality of issues raised in this book, the experimental studies conducted generally have one or more of the following objectives:

#### *11.1.1.1 Eliminating Defects and Improving Yield*

These universal desiderata are achieved through experimentation to optimize operating variables that control processes. Numerous structural, chemical, and electrical measurements and tests are made at convenient stages during manufacturing in order to assess process variation, detect contamination, and discover reasons for low yield. In many instances structural analysis instruments like the scanning electron microscope (SEM) are converted into diagnostic manufacturing tools that are directly integrated into processing lines to identify defects and maintain feature dimensions. The intent is to eliminate defects at the source.

#### *11.1.1.2 Quality Control*

After manufacturing is completed, the product normally undergoes a battery of tests to ensure that it meets design goals. Most important are the electrical measurements of voltage, current, light output, etc., specifications that are used to qualify the product.

#### *11.1.1.3 Reliability Testing*

Assessing the future performance of the product is the purpose of reliability testing. At first there is burn in, a procedure where elevated stress levels of

*Reliability and Failure of Electronic Materials and Devices*
ISBN 978-0-12-088574-9
http://dx.doi.org/10.1016/B978-0-12-088574-9.00011-2

voltage, current, temperature, mechanical loads, etc., are applied to completed manufactured items. The rationale is that if products can withstand the ordeal, they are fit for shipment to customers. In this way infant mortality failures are weeded out. Some of the wafer-level tests are also performed, e.g., Single wafer electrical accelerated test (SWEAT), to rapidly assess susceptibility to electromigration damage.

Extensive longer term reliability testing is conducted to gather data needed to predict failure lifetimes; the time dependent dielectric breakdown (TDDB) test for dielectrics, discussed in Section 6.3.5, is an example. Irrespective of the component, device, or system involved, the intent is to simulate and accelerate failures such that sufficient data are generated for statistical analysis. Most of the published research in the field and virtually all of the data presented in the book stem from such studies. Routine time-to-failure data as a function of the acceleration factors, though essential, offer little insight into failure mechanisms. To model microscopic failure mechanisms and make more accurate lifetime predictions, it is often necessary to determine material constants (e.g., diffusion coefficient, thermal conductivity) and parameters (e.g., activation energies) along the way. New research is involved, usually necessitating an array of impressive analytical instruments and careful measurement.

#### *11.1.1.4 Failure Analysis*

When failure occurs in the field, an investigation is normally carried out to determine the cause. The laboratory procedure is known as failure analysis, sometimes termed failure mode analysis (FMA), and generally involves electrical testing as well as structural and chemical characterization. In a four-pronged investigative thrust, the *failure mode* (e.g., open circuit, leakage current) is first identified, and then the *failure defect* (e.g., micro-cracks, growths) is exposed. Next, the *failure mechanism* (e.g., electromigration, corrosion) is suggested, and finally the *failure cause* (e.g., poor design, excessive current density) is proposed so that corrective action can be taken. A collection of failure modes, defects, and mechanisms is presented in Table 11.1.

Note that these four reasons for testing and experimentation are interdependent. Information gained during failure analysis and reliability testing is combined, and in turn fed back so that process changes can be made to eliminate product defects in the first place. The tighter the feedback loop, the greater is the probability that failure analysis will be unnecessary.

**Table 11.1** Failure modes, defects, mechanisms, and causes

| Failure mode | Failure defect | Failure mechanism and cause |
|---|---|---|
| **Reported fault** | **Analysis** | **Interpretation** |
| **OC** | Missing interconnects | Mask errors |
| | Corrosion | **1.** Moisture (internal gas; poor seal)<br>**2.** Contaminants + moisture |
| | Mechanical damage, scratching | Poor processing/handling |
| | Open bond | **1.** Pad contaminant<br>**2.** Bonding overpressure<br>**3.** Package stress<br>**4.** Al/Au intermetallic formation<br>**5.** Fatigue failure |
| | Open metallization | **1.** Electromigration<br>**2.** Processing defects, e.g., step coverage, lithography<br>**3.** Corrosion<br>**4.** Wrong passivation composition |
| | Metallization microcracks | Stress, migration in as-deposited metallization |
| | Local disruption of interconnects | **1.** Electrical overstress<br>**2.** Electrostatic discharge |
| | Die cracking | Thermal/mechanical stress |
| **SC** | Surface contamination | **1.** Contaminant between tracks<br>**2.** Metal migration/dendrites |
| | Isolation layer pinholes | Direct SC |
| | Hillocks | **1.** Electromigration SC<br>**2.** Whisker growth in double-layer metal |
| | Local disruption of interconnects | **1.** Electrical overstress<br>**2.** Electrostatic discharge |
| | Metal spikes | **1.** Enhanced diffusion along dislocations<br>**2.** Poor contact alloying<br>**3.** Overstress giving spiking<br>**4.** Electrothermal overstress |
| | Metal shorts | **1.** Poor bond placement<br>**2.** Metallization definition defects<br>**3.** Bond wire to substrate short<br>**4.** Electromigration<br>**5.** Stress-induced whisker growth |

*Continued*

**Table 11.1** Failure modes, defects, mechanisms, and causes—cont'd

| Failure mode | Failure defect | Failure mechanism and cause |
|---|---|---|
| | Package lead short | 1. Whisker growth—tin plating<br>2. Whisker growth due to corrosion<br>3. Whisker growth—frit seal |
| **Leakage** | Surface contamination | 1. Surface leakage<br>2. Ionic contamination at oxide/silicon interface |
| | Poor junction delineation | Processing defects |
| | Oxide breakdown | 1. Voltage overstress<br>2. Oxide defects |
| | Hot spots | 1. Power dissipation at metallization defects<br>2. Current concentration at crystallographic defects |
| | Metal spikes | Metal diffusion along crystal defects |
| **High-temperature failure** | Defective die-attach | Thermal overstress |

OC, open circuit; SC, short circuit.
From Refs [1,2].

This chapter is concerned with describing the interrelated testing for quality and reliability, as well as for failure analysis purposes. An important focus is failure analysis. To the extent that one starts with failure, and the harmful stressing variables and acceleration factors are determined, failure analysis may be regarded as the inverse of reliability testing. At first, the investigation is carried out in as nondestructive a manner as possible, because the clues to failure may vanish if the evidence is destroyed. However, it often happens that the failed package must be peeled like an onion to reveal the damage site. Once located, the focus of attention shifts to defect characterization in structural and chemical terms. The order of steps typically followed when conducting an actual failure analysis is itemized in Table 11.2, which also serves as a partial road map to the chapter. We shall address these issues in roughly the same sequence as that suggested in the table. In addition, this table catalogs a great many of the techniques that will be discussed.

To illustrate the methodology, case histories of defects or failures will be interspersed throughout.

**Table 11.2** Sequence of failure analysis investigation

| Technique | Information/purpose/comments |
|---|---|
| **1. Nondestructive examination** | |
| Visual/optical inspection | Gross defects in packages |
| X-Radiography | Imaging of internal package geometry and defects |
| Electrical tests | Adherence to specifications |
| Package tests | Hermeticity |
| SAM | Imaging of internal package features and defects |
| **2. Decapsulation of package** | |
| **3. Imaging and structural characterization** | |
| Optical/interference microscopy | Nondestructive examination to ~1 μm |
| SEM | Nondestructive examination to ~0.01 μm |
| SEM-VC; SEM-CL | Imaging of electrical functioning (N-D) |
| Metallography sectioning | Destructive |
| X-ray diffraction | Identification of compounds or structure (N-D) |
| **4. Chemical characterization** | |
| SEM-EDX | Local chemical analysis (N-D) |
| AES | Surface composition, depth profiling |
| XPS | Surface composition (usually N-D) |
| Secondary ion mass spectroscopy | Surface composition (depth profiling, D) |
| RBS | Composition of compound films, layers (N-D) |
| **5. Additional analytical techniques** | |
| Thermal imaging (IR microscopy) | Location of hot spots (N-D) |
| Gas analysis | Moisture content |
| Trace chemical analysis | Chemical composition (D) |
| **6. Comparison of failure mechanisms with published literature** | |
| **7. Failure simulation tests** | |

D, destructive; N-D, nondestructive.
After Ref. [1].

## 11.1.2 Characterization Tools and Methods

Although defects and failure in electronic materials and devices is our concern and these are usually exposed after electrical testing or probing of circuits, there will, paradoxically, be virtually no discussion of electrical testing. This subject is simply well beyond the scope of this book. We may, however, distinguish electrical characterization on the chip and individual device levels. For example, the magnitude and time dependence of power supply current (IDDq-supply current (Idd) in the quiescent state) drawn by a complementary metal oxide semiconductor (CMOS) circuit can identify devices with excessive leakage and indirectly detect defects [3]. In particular, such IDDq testing is well suited to revealing electro static discharge (ESD), electrical over stress (EOS), and latchup failures and detecting gate oxide and interlayer shorts. Further identification of faulty devices involves generating a datalog profile with automated test equipment. Often these indirect measurements reveal latent electrical open defects, such as conductor line narrowing, notches, and cracks. Electrical testing at the bench then tracks the subsequent destructive unpeeling and characterization of the chip. This means curve tracer analysis relative to a bus pin (usually ground) and determination of DC and AC characteristics. Comparison with the behavior of good devices helps to identify the failure mode.

In keeping with the materials slant of the book, the major focus of attention will be on imaging individual device surface and subsurface features as well as determining local chemical compositions. Analysis on the chip level will be emphasized, but many applications to packaging will be included.

Many of the characterization tools and methods used to improve the quality of electronic products and expose their failure modes and mechanisms are the same as those employed in varied scientific investigations spanning all disciplines. These instruments all function by probing the surface of the specimen with an input photon, electron, or ion beam, and detecting a signal composed of some combination of output photons, electrons, or ions. Interpreting the intensity and spectral response of this output signal is the key to understanding the phenomenon under investigation. Even though there is a broad and accessible literature describing the underlying physics of many of these techniques and the information they convey [4–7], a brief review of these issues will prove instructive. However, space constraints will mean an emphasis primarily on the special features that make these tools and techniques unique in analyzing failure of electronic products.

## 11.2 NONDESTRUCTIVE EXAMINATION AND DECAPSULATION

### 11.2.1 Radiography

If the human body is any guide, a noninvasive approach is the preferred way to initiate the diagnosis of an integrated circuit (IC) failure. This means that it is sensible to start by examining the package first. The reason is that many of the failure mode symptoms can potentially be attributable to defects in packaging components and structures. Interestingly, just as is the case for humans, similar diagnostic tools are used in the first stages of analysis, where it is desirable that the internal contents of packages be seen without opening them up.

X-radiography is the first nondestructive technique we will consider. The operating principle involved parallels that of nondestructive examination of other manufactured products, e.g., castings, welds, and medical applications. A sufficiently penetrating X-ray beam of intensity $I_o$ impinges on the sample and the attenuated transmitted radiation of intensity $I(x)$ is detected as an electronic or photographic image. This already familiar relation between the two intensities (see Eqn (10.7)) is

$$I(x) = I_o \exp[-\mu x], \quad (11.1)$$

where $x$ is the penetration distance and $\mu$ is the absorption coefficient of the involved material. When there are large differences in $\mu$ between and among the package components, the technique works best, for then the image contrast is greatest. Generally, $\mu$ scales in a complex way with both the radiation wavelength and atomic number of the absorber. Thus, X-rays readily penetrate the light elements of the plastic and Si wafer, while gold lead wires, lead frames, and solder are quite absorbing. To optimize image contrast, the X-ray energy must be carefully chosen because of its influence on the magnitude of $\mu$. For example, with 20-keV X-rays, $\mu = 10.2$/cm in Si and 1510/cm for Au; on the other hand, with 100-keV X-rays, $\mu = 0.429$/cm in Si and 98.8/cm for Au. Note that mass absorption values ($\mu_m$ in units of cm$^2$/g), not $\mu$, are normally quoted in the literature. To obtain $\mu$ we must multiply $\mu_m$ by the density $\rho$ (Example 11.1).

---

## Example 11.1

What is the difference in intensity of X-rays penetrating a pure Si wafer 0.05 cm thick relative to this same wafer that is patterned with a 1-μm-thick Au bond pad? Assume that the operating X-ray voltage is 20 kV. Is the contrast better at 100 kV?

**Answer** Using Eqn (11.1), for the Si wafer alone $I/I_o = \exp[-10.2\,(0.05)] = 0.60$. For the Si/Au combination, $I/I_o = \exp(-[10.2\,(0.05) + (1510)10^{-4}] = 0.52$. This intensity difference of over 13% is detectable on film.

---

At 100 kV the corresponding intensity ratios are $I/I_0 = \exp(-0.429(0.05)) = 0.979$ and $\exp(-(0.429(0.05) + (98.8)10^{-4})) = 0.969$. The more penetrating, high-energy X-rays are thus seen to considerably reduce the contrast.

Unfortunately, a disadvantage of the technique is the limited spatial resolution, which is on the order of 10 μm. Short wavelengths normally mean low dispersion and good spatial resolution. However, magnification imaging is limited for X-rays by the inability to form lenses because the refractive index of all materials at these wavelengths is approximately the same, i.e., typically less than unity by 1 part in $10^5$. Nevertheless, X-ray microradiography is sufficiently developed to reveal misplaced or broken Au wire bonds or detect gross defects in the die attachment. Low X-ray absorption makes it impractical to locate defects in aluminum bonds by this method.

The ability of X-rays to reveal soldering defects is dramatically demonstrated in the radiographs of ball grid arrays shown in Figure 11.1. What is significant about this application is that X-rays enable missing solder balls as well as bridging and other defects to be seen in bonded structures that are hidden from direct view. The high absorption coefficient of solder enables high-contrast images, while the use of real-time digital imaging methods makes this technique ideal in monitoring quality control.

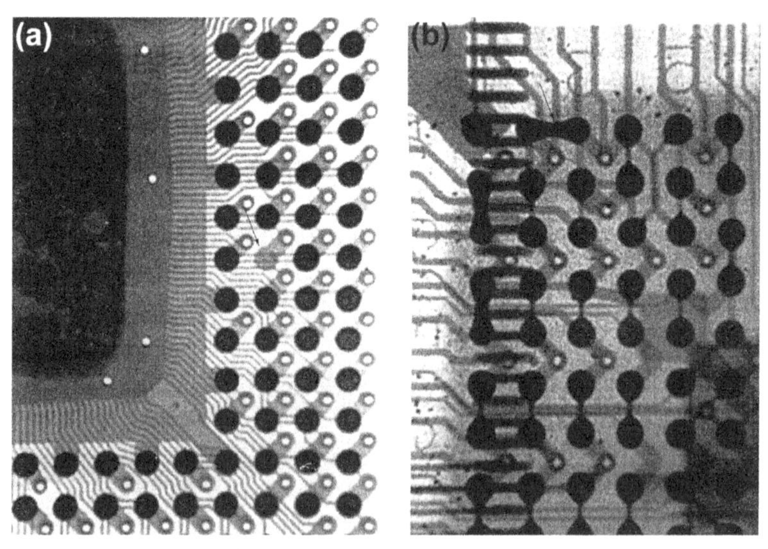


**Figure 11.1** X-ray images of a ball grid array package. The solder balls are ~0.5 mm in diameter. (a) Missing solder ball. (b) Solder bridging. *Courtesy of Gilly Zweig, Glenbrook Technologies Inc.*

### 11.2.2 Scanning Acoustic Microscopy

This important analytical technique has much in common with ultrasound imaging methods employed in medical applications, and for flaw detection in metals. In a number of reviews [2,8–10] scanning acoustic microscopy (SAM) has been shown to be very helpful in studying failures of plastic packages. There are several variants of the technique but, as schematically shown in Figure 11.2, they all rely on generating acoustic waves with a piezoelectric transducer and then launching them into the sample (IC package) under study via a lens (e.g., quartz) and coupling fluid medium (e.g., water). Acoustic energy reflected from sample features is collected by the lens and converted by the transducer back into an electrical signal. In this pulse-echo, radar-like technique, images result from raster scanning the transducer in a plane parallel to the surface of the specimen. Subsurface information is obtained by moving the lens vertically in the $z$ direction, causing the point of focus to vary in concert. By coupling the transducer–acoustic lens assembly to both the input and the output surfaces, a transmission mode of operation is possible.

There are three types of acoustic waves that contribute to observed images. The first two are the familiar bulk longitudinal and shear waves, and the third is due to the propagation of surface or Rayleigh waves. These

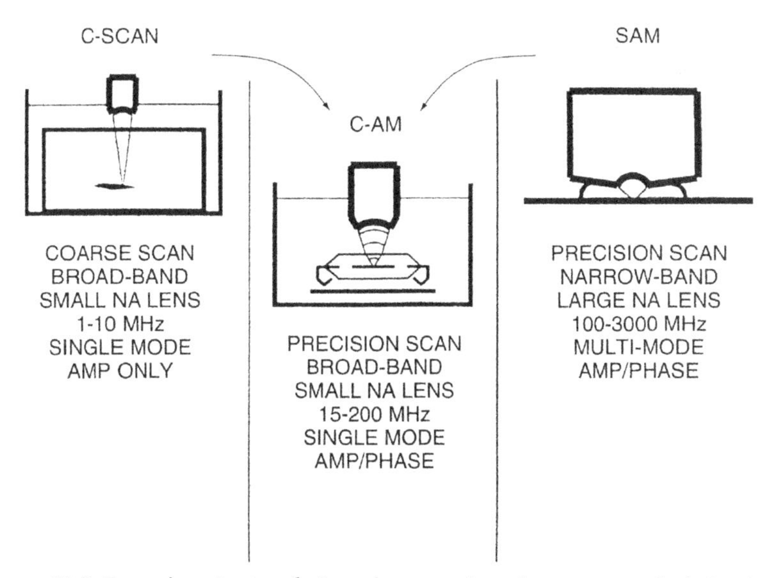


**Figure 11.2** Several variants of C-mode acoustic microscopy and their characteristics. *From Ref. [8].*

acoustic waves interact in complex ways with the elastic properties of the solid as modified by density, adhesion, viscosity, and structural variations. Imaging modes can be distinguished based on the interaction between waves with subsurface, interfacial, and surface features.

Both penetration and feature resolution are determined by the operating frequency and lens characteristics. At one extreme, low frequencies (1–10 MHz) and lenses of low numerical aperture (NA) enable coarse scans of low resolution, but great depths of imaging (several mm). This is the regime of medical ultrasound and nondestructive testing. In this so-called C-SCAN (low frequency acoustic scanning) mode of operation, detection of cracks in encapsulated ICs and ceramics is possible. At the other extreme, of high-frequency (100–2000 MHz) and large-NA lenses, the spatial resolution improves to about 1 μm with a focal depth of a few micrometers. Imaging depends on the interaction between Rayleigh waves and surface features. In this SAM mode, defects in silicon wafers, film adhesion, and delamination problems, and metallization structures in ICs can be revealed. Electrostatic discharge damage to ICs is often more sensitively detected with SAM methods than with optical microscopy. Nevertheless, interpretation of high-frequency images is complicated by the fact that contrast depends on both material and topographical differences that are not easy to separate.

A hybrid of the two established reflection acoustic techniques known as C-AM (lMegaHertz frequency acoustic microscopy) operates in the range of 15–200 MHz and strikes a good compromise between lateral and depth resolution. Among the package defects that can be analyzed with C-AM are popcorn and radiating cracks, voids in the molding compound, die attach separation and voiding, wire bond pull-out, and leadframe delamination. This technique is particularly suited to distinguishing between bonded and delaminated interfaces. For example, Figure 11.3 shows image differences between delaminated areas and regions of good adhesion.

In interpreting acoustic imaging a few basic physical principles should be borne in mind. Reflection of sound from layered media of differing densities is similar to the reflection of light from layered optical films of differing refractive indices. The quantity analogous to the real index of refraction is the acoustic impedance ($Z$), which for longitudinal waves is defined by the product of the speed of sound (m/s) in the medium and its density (kg/m$^3$). Acoustic impedance units are kg/m$^2$-s or Rayl, and for silicon $Z = 20 \times 10^6$ Rayl. Now consider an acoustic wave incident normal to the interface between two media 1 and 2 as shown in Figure 11.4. Of the

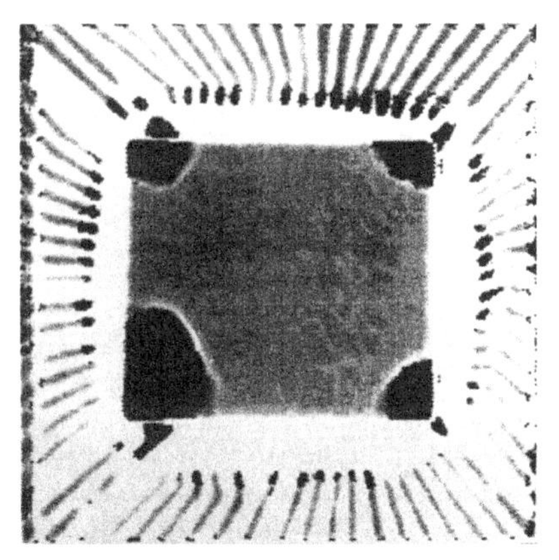

**Figure 11.3** Scanning acoustic microscope image of a die measuring 3.6 mm by 3.1 mm. The reflected intensity image appears as a grayscale image on a white background. Delaminated regions in the corners appear black. *From Ref. [8].*

acoustic energy incident ($E_o$), the fractions reflected ($E_R$) and transmitted ($E_T$) are given by

$$E_R = \left(\frac{Z_2 - Z_1}{Z_2 + Z_1}\right)^2 \tag{11.2}$$

and

$$E_T = \frac{4Z_2Z_1}{(Z_2 + Z_1)^2}. \tag{11.3}$$

Note that $E_R + E_T = 1$. Identical formulas hold for nonabsorbing optical thin films on a transparent substrate. Acoustic wave amplitudes also depend on $Z$. For the geometry indicated, the reflected wave amplitude ($A_R$) is

$$A_R = \frac{(Z_2 - Z_1)}{(Z_2 + Z_1)} \tag{11.4}$$

while the amplitude of the wave transmitted into region 2 is

$$A_T = \frac{2Z_2}{(Z_2 + Z_1)}. \tag{11.5}$$

As with optical films, there is the additional phenomenon of a possible phase change of the reflected wave; nothing in Eqns (11.2) and (11.3)

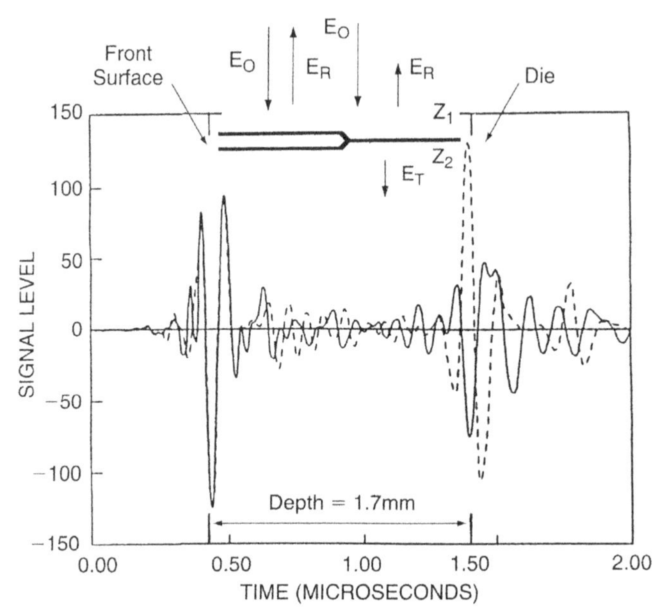


**Figure 11.4** Acoustic echo signals (15 MHz) from region of good adhesion (solid line) and a delaminated area (dashed line). Acoustic wave interactions at delaminated and bonded interfaces are indicated. *From Ref. [8].*

prepare us for such an effect. Nevertheless, there is a signal inversion in the reflection from the region of delamination. Phase inversions occur when sound travels from regions of high to low acoustic impedance, but there is no inversion when the reverse is true. Thus for sound passing from the package mold material ($Z_M$) into an air crevice ($Z_A$), there is a decrease in $Z$ (i.e., $Z_M > Z_A$). The fully reflected wave (echo) exhibits a phase inversion. But when passing from the polymer mold material into silicon, the acoustic impedance increases (i.e., $Z_M < Z_{Si}$) and there is no phase inversion in the reflected wave of diminished amplitude.

## 11.2.3 Analysis of Particle Impact Noise Detection Particles

Even prior to performing X-radiography and acoustic microscopy, and certainly before decapsulating a package (or any electrical product), it is worth shaking it. This simple nondestructive test senses the presence of loose conductive or corrosive particles, which may pose a reliability risk because of the danger of shorting bond wires to contact pads or to the package case. Particle Impact Noise Detection (PIND) test equipment exists to determine whether such particles are present in brazed-lid, sealed hermetic packages. In the PIND test the device is mechanically vibrated. A transducer detects particle impact and the signal is monitored either visually on an oscilloscope or audibly via a speaker. Loose particles can often be captured for further study by meticulously grinding open a corner of the package, placing tape over the hole, and shaking until the noise stops.

### 11.2.4 Decapsulation

If preliminary nondestructive inspection suggests that analysis at the chip level is required, then decapsulation of the package is required. Prior X-ray or acoustic imaging may be necessary to ascertain the orientation and dimensions of the package contents. This will set guidelines for how much mechanical grinding can be tolerated. Decapsulation, or depackaging, poses different challenges depending on whether the packages have a cavity, e.g., ceramic dual-in-line package, or are molded, e.g., plastic dual-in-line package. For the former, tools like small vice grips, miniature diamond cutting and grinding wheels, saws, and polishing equipment are employed.

In the molded thermoplastic or thermosetting (epoxies) packages moisture is first removed by baking at low temperature. Now the plastic package is ready for decapsulation. For both thermoplastic and thermosetting molding compounds, a solvent is needed to soften and dissolve the polymer. In the process, plastic particles sometimes break into little pieces that may damage delicate semiconductor component features. Decapsulation is far easier in thermoplastics because many organic solvents exist. For these materials the dissolution process involves a solvation step in which surface absorption of solvent molecules causes the polymer to swell, followed by diffusion of decomposed thermoplastic molecules through the solvent.

The more common thermosetting polymer packaging materials require more powerful inorganic reagents such as concentrated and fuming sulfuric and nitric acids, as well as chromic acid and proprietary commercial formulations [11]. It is usually necessary to protect the easily corrodible metal leads, solders, lead frames, and even chip contact pads and metallizations during exposure to such acids. The typical chemical decapsulation station is outfitted with the appropriate handling fixtures and tools (tweezers), solvent containers (acid percolator, beakers, eyedroppers), and methods to heat the solutions controllably. Plasmas have also been used to decapsulate chips, but the advantage of etching selectivity is generally offset by the long times required. Aggressive decapsulation methods sometimes raise doubts as to the interpretation of failure. For example, did cracking of the oxide or nitride coating cause the failure, or did the coating crack during decapsulation?

Following depackaging, the remaining chip and supports are ultrasonically cleaned, rinsed, and examined optically. If chip-level features must be revealed, the various layers must be sequentially exposed in a manner that resembles an archeological dig. Different film materials are selectively removed by etching in such a way that exposure of one level does not compromise the integrity of deeper levels. Both aqueous and plasma, or dry, etching is employed. Factors that influence the selection of an etchant for a particular material include the feature size involved, the etch rate, equipment required, the effect it has on the surrounding materials, and the undercutting action of the etch. Table 11.3 lists a number

**Table 11.3** Solutions and etchants for selective removal of semiconductor and interconnect layers

| Material | Chemical composition | Comments |
|---|---|---|
| **Silicon dioxide glass films and layers** | 1 (wt) $NH_4Cl$:1 (wt) glacial HOAc | |
| | 1 (wt) ammonium bifluoride:<br>3 (wt) glacial HOAc | |
| | 1 HF (49%):<br>10 $NH_4F$ solution (1lb $NH_4F$/680 ml $H_2O$) | Will remove $SiO_2$ at the rate of 7–9 Å/s |
| | $NH_4F$ (50%)<br>HOAc (50%) | Will remove $SiO_2$ at the rate of 700–900 Å/min |
| | Plasma etch | |
| **Silicon nitride films** | Plasma etch | |
| | Boiling orthophosphoric acid | Slight attack of Al at 100–120 °C |
| **Aluminum metallization** | 80 ml orthophosphoric acid, 1 ml HOAc<br>1 ml nitric acid<br>10 ml water | At 150 °C, dissolution is very rapid (a few seconds) |

**Table 11.3** Solutions and etchants for selective removal of semiconductor and interconnect layers—cont'd

| Material | Chemical composition | Comments |
|---|---|---|
| | HCl at 50 °C | Useful for two-layer metallization |
| | HCl at 250 °C | |
| | Bromine–methanol solution | Dissolves all metals |
| **Polysilicon** | Concentrated $HNO_3$: 80 ml<br>Concentrated HOAc: 60 ml<br>Concentrated HF (49%): 5 ml | Estimated etch time 15 s |
| **Silicon** | KOH (20%) at 40 °C | Used for complete die removal to observe the die bond and backing metal |
| | 30 wt% HF (49%)<br>50 wt% $HNO_3$ (69.5%)<br>20 wt% $H_2O$ | Removes Si at the rate of 76 μm/min |
| **Silicon etchants** | | |
| White etch | 1 HF:<br>3 $HNO_3$ | Chemical polish; can be used as a p-n junction etch<br>Etch time: 5 s |
| Dash etch | 1 HF:<br>3 $HNO_3$:<br>8–12 HOAc | Etch for p-n junction<br>Etch time: 10–20 s; 1 min and greater for dislocations |
| CP-4A | 3 HF<br>5 $HNO_3$<br>3 HOAc | Slow chemical polish; good for p-n junctions but all polishing damage must be removed to prevent die cracking<br>Etch time: 10 s–3 min |
| SD 1 | 25 HF: 18 $HNO_3$<br>5 HOAc/<br>$0.1Br_2{:}10H_2O$<br>1 g $Cu(NO_3)_2$ | Reveals edge and mixed dislocations on all planes<br>Etch time: 2–4 min |
| Sirtl | 1 HF<br>1 to 4 parts<br>40% Chromic acid | Good p-n junction etch for sections; excellent dislocation and stacking fault etch<br>Etch time: 5–20 s |
| **Assorted metals and alloys** | | |
| **Soft solder** (used for die bonding) | 1 part (vol) HOAc (glacial)<br>1 part (vol) $H_2O_2$ | Selectively removes soft solder without attacking backing metallization or heat sink |

*Continued*

**Table 11.3** Solutions and etchants for selective removal of semiconductor and interconnect layers—cont'd

| Material | Chemical composition | Comments |
|---|---|---|
| **Copper, copper base** | $NH_4OH$: 10 ml | |
| **Alloys** | $H_2O_2$ (3%): 10 ml | |
| **Steel (carbon)** | Ethanol: 98 ml | Grain boundary etchant |
| | Nitric acid: 2 ml | |
| **Steel (stainless)** | Glycerol: 30 ml | |
| | HCl: 30 ml | |
| | $HNO_3$: 10 ml | |
| **Nickel base alloys** | Ethanol: 60 ml | |
| **Kovar** | HCl: 15 ml | |
| | $FeCl_3$: 5 g | |
| **Nichrome** | 2.25–3.75 M $FeCl_3$ | |
| **Chromium** | Bromine–methanol solution | Dissolves all metals |
| **Molybdenum** | Potassium ferricyanide: 30 g | |
| **Tungsten** | NaOH: 10 g, $H_2O$: 100 ml | |
| **Gold** | 1 Potassium ferricyanide (10%) | |
| | 1 Ammonium persulfate (10%) | |
| | 4.6 g KI, 1.3 g $I_2$, 100 ml $H_2O$ | Will dissolve Au from a die but will not attack unalloyed aluminum |
| | 1 $HNO_3$—3 HCl (aqua regia) | At 25 °C |
| **W-Ti** | 3% $H_2O_2$ | At 50–60 °C |
| | Bromine–methanol (+$FeBr_3$) | |

Note: Neither the author nor the publisher is responsible for any consequences incurred in the use of information contained within this table. HOAc, acetic acid, $CH_3COOH$.
Recipes taken from Ref. [12] and other sources.

of reagents and etchants used to selectively remove semiconductor and interconnect materials.

The results of progressive delayering of an IC chip are shown in Figure 11.5. After removing glass, aluminum, oxide, and polysilicon films, the origins of electrostatic discharge damage are ultimately revealed.

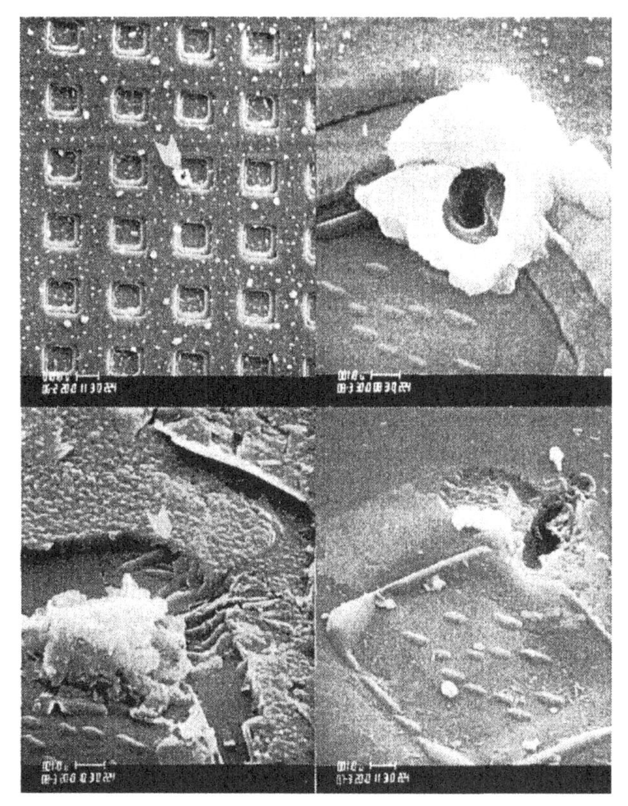

**Figure 11.5** Failure site in an IC chip revealed by successive delayering. At the top left, passivation glass and aluminum have been removed. Detail of the site appears at the top right. At the lower left TEOS oxide has been removed. Finally, at the lower right the polysilicon has been stripped, exposing ESD damage. *Courtesy of G.P. Thome, Motorola Corp.*

## 11.3 STRUCTURAL CHARACTERIZATION

### 11.3.1 Optical Microscopy

Because seeing is believing, virtually every failure analysis investigation starts with some sort of structural examination procedure. Of the potential tools that can be used for this purpose, none are more popular than low-power stereo and standard, higher power, optical microscopes. Various levels of the decapsulated package structure are routinely examined by optical methods to qualify the procedure and uncover the sought-after evidence of failure. Optical transmission and metallurgical microscopes are the most widely employed scientific instruments for these purposes. The former is limited to thin-film devices and structures on transparent substrates, while the latter are useful in imaging opaque objects like circuits on silicon wafers. Since the principles of the common transmission and reflection modes of operation are well known to readers, the discussion will focus on descriptions of less common imaging methods.

Additional optical microscopy techniques applicable to powered devices will be discussed in Section 11.5.

#### *11.3.1.1 Nomarski Contrast*

By illuminating specimens with plane polarized light, a significant enhancement in feature contrast is possible because the image can be made to assume a three-dimensional quality. In the Nomarski differential interference contrast method, optical microscopes are outfitted with a Wollaston prism that causes the incident plane polarized light to be separated into two beams. One beam passes through the feature of interest, while the other beam is incident nearby. Upon reflection, the two beams are recombined by a second prism above the objective. As a result, a phase change introduced into the object beam by the specimen is converted into an amplitude of color difference. The Nomarski technique is very sensitive to small surface asperities such as etch pits. Such shallow surface depressions in epitaxial films cannot be discerned in the SEM but are readily visible employing Nomarski contrast. Another application involves resolution of slight thickness variations in thin oxides grown on differently doped silicon regions. After removing the oxide, the step is easily imaged.

#### *11.3.1.2 Scanning Light Microscopy*

Bridging the gap between conventional optical and scanning electron microscopy are imaging methods employing scanned optical beams. Actually, either the beam is scanned past a fixed specimen, or the specimen is scanned past a fixed beam in these microscopes. In scanning light microscopy, images are built up one point at a time, and the whole field is covered by scanning. Lasers are employed as beam sources, prompting the descriptor laser scanning microscope. In the reflected light mode, advantages relative to the SEM include no sample preparation, no charging or damage of insulating surfaces, imaging through the depth of transparent layers, and the lack of necessity for a vacuum environment [2].

An important difference between the scanning and ordinary optical microscopes is the decoupling of magnification (determined by scanning) and resolution (determined by the optics). This allows modifications such as confocality, in which a point detector, consisting of a pinhole in front of a photodiode or photomultiplier, is placed at the precise focal point of the image. The detector then receives light from a point source in the specimen, resulting in images that have both a reduced depth of focus and a slightly improved resolution. Points that are not in focus are not imaged, and as a result, the out-of-focus blur is reduced, dramatically improving image quality.

#### *11.3.1.3 Fluorescent Tracers*

In this technique fluorescent tracer dyes are applied to device surfaces in order to demarcate or decorate defective sites for further microscopic observation. After application, the tracer dyes penetrate cracks, corrosion products, and regions of delamination. Under ultraviolet (UV) illumination the dyes fluoresce and are made visible with the use of appropriate filters. Crevices and pinholes in glass layers and dielectrics have been exposed this way, while in corrosion studies, pH-sensitive fluorescent dyes have been used [13]. Employing fluorescent microthermographic imaging systems, defects in powered devices have been identified. This technique capitalizes on the temperature-dependent fluorescence of certain rare-earth-metal-chelated compounds such as europium thenoyltrifluoroacetonate (EuTTA). When heated, EuTTA fluoresces bright orange (612 nm) when excited by a UV source.

Fluorescent dye penetrants have long been used to reveal microcracks on machined and ground metal surfaces.

### 11.3.2 Electron Microscopy

The SEM and the transmission electron microscope (TEM) are indispensable for high-resolution imaging of small structural features. Shrinkage of device dimensions has fostered the increasing use of electron microscopy relative to optical techniques for chip-level examination. In turn, with the inexorable advances in device reduction we will witness the greater use of the scanning point probe microscopies. (The scanning tunneling microscopes and atomic force microscopes (AFM) make use of atomically sharp probes brought exceedingly close to surfaces as their terrain is scanned. The AFM already plays a role in defining the metrics of features on IC chips and has been used to probe the topography of planarized surfaces. Destined to play a prominent future role in the characterization of nanosized features and defects, there are presently an insufficient number of applications to issues of concern in this book. Therefore, these techniques will not be discussed further.)

The TEM, like its optical counterpart, is a true microscope because the entire image in both appears at once; and like the optical microscope, diffraction-limited optics govern the resolution, i.e., the wavelength divided by the NA. This is not true of the SEM, whose resolution is not diffraction limited, and where the image is built bit by bit through scanning a probe beam.

#### *11.3.2.1 Scanning Electron Microscopy*

There is little doubt that the most versatile analytical instrument used in failure analysis studies is the SEM. The reason is that the SEM simultaneously allows for microstructural imaging, chemical analysis, and electrical characterization of materials and devices on both the chip and packaging levels. While it may neither possess the highest spatial resolution nor be the most sensitive with regard to identifying atoms, the SEM represents an excellent compromise between these two important analytical capabilities within a single instrument that is relatively easy to use and modest in cost. As an imaging tool it exceeds the resolution of the optical microscope by a factor of almost 100, so that features approaching 1 nm in size can be resolved under optimum conditions. Furthermore, the images possess a three-dimensional beauty conferred by the great depth of focus of the electron beam.

The electron optical column of a typical SEM is shown in Figure 11.6. Electrons are thermionically emitted from tungsten filaments under a potential difference that ranges from ~1 to 50 kV. If higher intensity or greater resolution is required, $LaB_6$ or field emission cathodes, respectively, are available. After passing through the condenser and objective lenses, the beam is demagnified to an ~1-nm spot and then raster scanned over the specimen. Beneath the point of beam impingement a teardrop-shaped region ~1 μm in depth fills the specimen with the products of electron excitation and deexcitation. These include secondary electrons (SEs), Auger electrons, and elastic backscattered electrons with successively higher energies of ~50 eV, ~1000 eV, and ~10–30 keV (depending on the primary beam energy), respectively.

Each of these kinds of electrons, as well as atomic fluorescent X-rays and other photons, form the basis of important imaging and analytical

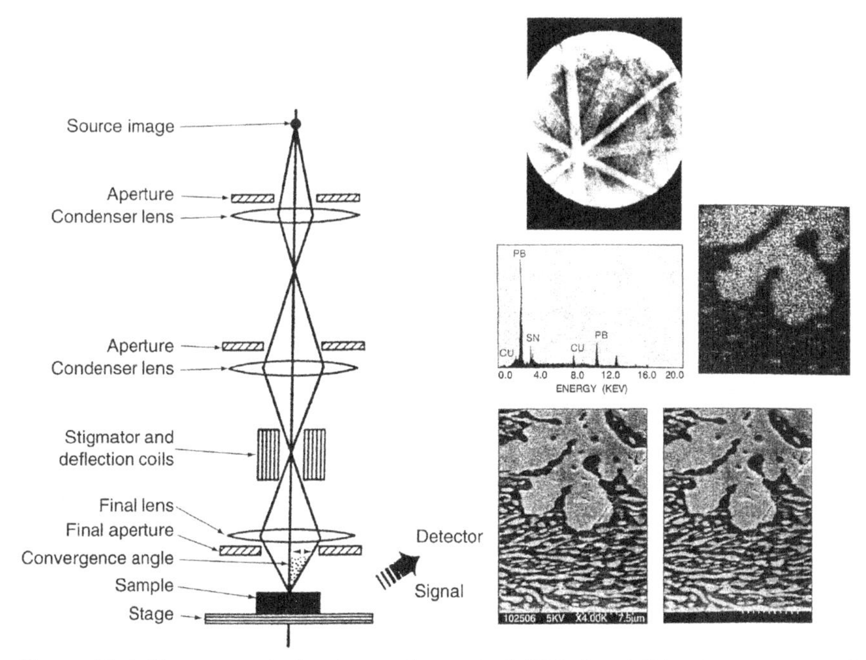


**Figure 11.6** Electron optical column of a typical SEM. Capabilities in imaging and spectroscopy are shown. Starting at lower left and going counterclockwise are 1, scanning electron image of solder; 2, backscattered electron image; 3, X-ray map of lead; 4, electron backscatter diffraction pattern; 5, X-ray fluorescence spectrum. *After Ref. [14].*

techniques. Virtually all the SEM images in this book are the result of collecting the low-energy SEs. These emanate from surface layers in intensities that vary with topography. More SEs are emitted at sharp corners, edges, and sloped regions than planar surfaces. The magnification is the ratio of the image area on a television screen to that of the rastered area on the specimen. Unlike secondaries but like fluorescent X-rays, Auger electrons are atom specific; by measuring their energies it is possible to uniquely fingerprint the atoms from which they arise. This property forms the basis of a separate instrument, the Auger electron spectrometer (Section 11.4.2), which is an indispensable tool in surface analysis. Practically all SEMs are equipped with fluorescent X-ray detectors that make simultaneous elemental microanalysis possible. This analytical technique, known as energy-dispersive X-ray (EDX) analysis, will be further discussed in Section 11.4.2.

The subsequent coverage of SEM techniques will be divided into two categories. Additional SEM capabilities and modes of operation that are not as common as the primary ones noted above will be treated next. These are very helpful in performing failure analyses but do not normally require powering of the devices being viewed. In contrast, powerful SEM techniques have been developed to view devices under electrical test, and they will be deferred to Section 11.5.

#### 11.3.2.1.1 Detection of Backscattered Electrons

The backscattered electrons have the same energy as incident electrons and display intensity or contrast variations that depend on atomic number ($Z$). Since the detected signal increases directly with $Z$, elastic (without energy loss) scattering from nuclei is probably the principal electron–atom interaction. Because they possess higher energy than secondaries, backscattered electrons emerge from deeper subsurface layers of the specimen. For this reason the spatial resolution is poorer than for SEs. Nevertheless, the resolution is good enough to expose electromigration voids beneath the passivation layer in the image of Figure 11.7(a).

#### 11.3.2.1.2 Diffraction

When an incident SEM electron beam penetrates a crystalline specimen, electrons are inelastically scattered, and they lose angular correlation with the primary beam. In the process, a point source of electrons effectively forms and Bragg diffracts from the sample lattice. Electrons scattered at large angles in the shape of conelike beams intercept a flat detector and produce a pattern consisting of intersecting hyperbolic sections [16]. In a mode mentioned in Section 5.6.3, these electron backscatter diffraction patterns enable the crystallographic orientation of individual grains to be determined. One variant of the technique, known as orientation imaging microscopy (OIM), is demonstrated in Figure 11.7(b). With the exception of two grains in the computerized image of Al–Cu metallization, the vectors normal to all other grains are parallel to within 5° of the [111] direction, indicating the strong (111) texture in these films. This raises the question of whether such rogue grains are the preferred sites for film degradation during electromigration. Since such information is readily obtained in a nondestructive way, OIM promises to be an important new research tool in the study of metallization damage.

#### 11.3.2.1.3 Cathodoluminescence (CL)

Certain materials like semiconductors, ceramics, and polymers emit photons when high-energy electrons impinge on them. As a result of this electron

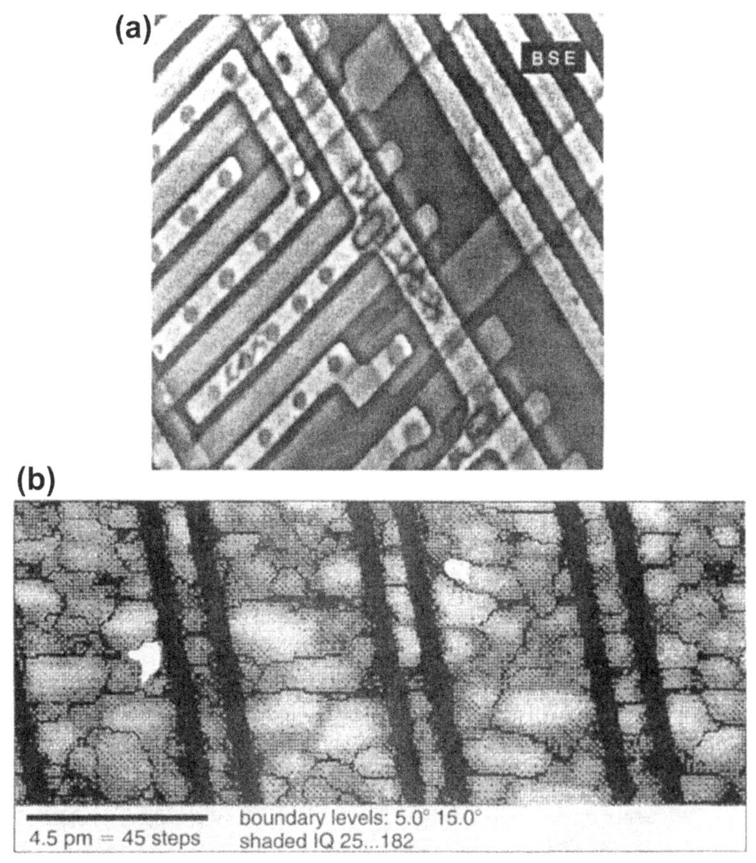


**Figure 11.7** Some less common SEM capabilities. (a) Backscattered electron image of metallizations revealing underlying voids. *(From Ref. [15].)* (b) OIM image from polycrystalline metallizations. Thick parallel black lines between metal lines are spaces, while thin curved lines are grain boundaries. The two misoriented grains are highlighted in white. *(From Ref. [16].)* (c) Interfacial bonding defects in high-density interconnect of a multichip module revealed in an E-SEM. Polyimide metallization delamination is due to temperature–humidity testing. *From Ref. [17].*

bombardment, electron–hole pairs are generated, and when the excess carriers recombine, light is emitted in a process known as CL. Depending on the material, the photons may appear within the UV, visible, and infrared (IR) portions of the spectrum. This phenomenon has been proved to be useful in the characterization, failure analysis, and quality assurance of optoelectronic materials and devices. For high-spatial-resolution applications, CL is carried out in the SEM, where provision is made to detect the emitted photons with scintillation counters or silicon photodiodes.

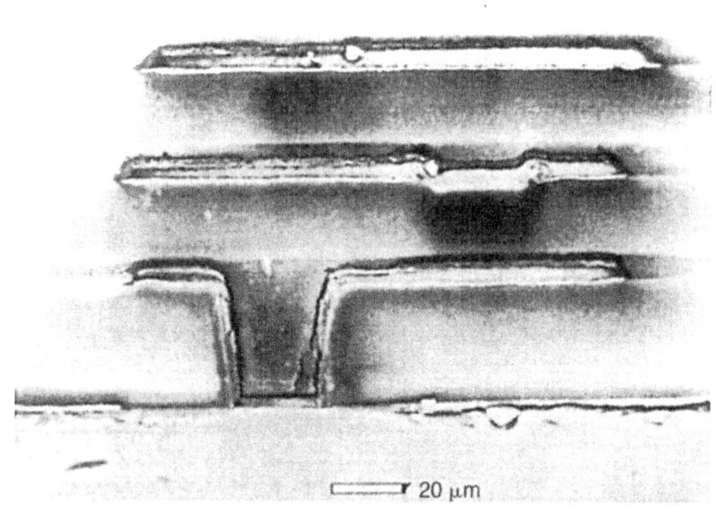


**Figure 11.7** Cont'd.

In a study by Pey et al. [18], CL was used to investigate nonradiative defects that diminished the light output of $GaAs_{0.72}P_{0.28}$ light-emitting diodes (LEDs). Black spots roughly 10 μm in size due to dislocations and stacking faults were observed in both p and n regions, and these appeared to be the cause of degradation. In this work GaP-based LEDs were also investigated by CL techniques. Anomalous, multijunction p-n-p-n regions that caused a flickering light output and larger reverse breakdown voltages than good LEDs were observed. In many ways the information gained with CL is similar to that obtained with electron-beam-induced current (EBIC) (described in Section 11.5.3.2). However, CL does not require the presence of a junction or biasing, but EBIC does. A CL image is reproduced later, in Figure 11.7(c).

Thermal Wave Imaging

As in all SEM-based techniques, an electron beam is first scanned across the specimen surface. Energy absorbed at or near the surface results in a periodic heating, which becomes the source of thermal waves that propagate from the heated region [19]. The thermal waves interact with, and scatter from, subsurface features that have differing thermal conduction characteristics from those of the surroundings. In this way thermoacoustic waves are generated at precipitates, doped regions, grain boundaries, dislocation arrays, and cracks. These defects are detected with a piezoelectric transducer and made visible by using the signal they induce to modulate the intensity at a frequency of 10 kHz–10 MHz.

### Environmental SEM (E-SEM)

One of the illusive goals of electron microscopy has been high-resolution structural imaging of specimens not in vacuum, but in the ambient atmosphere. The barriers to achieving this goal are numerous and include corrosion and oxidation of filaments, discharges in the electron optical column, surface charge buildup, and scattering of electrons by the gas molecules present in air and humid ambients. To overcome these problems the E-SEM has been developed. In operation, a gas pressure of $\sim 20$ torr, rather than $\sim 10^{-5}$ torr in a conventional SEM, is maintained in the chamber. The latter is separated from the electron optical column, which operates at high vacuum. Interactions between the probe beam electrons, signal electrons, and the specimen with the chamber gas intimately depend on the pressure, local electric fields, and the operating voltage. Basically, the positively ionized gas molecules bombard regions of negative charge buildup on the specimen surface, neutralizing them. SEs generated through gas ionization are then collected by a positively biased gaseous detector to form the image.

Li and Pecht [20] have demonstrated the effectiveness of the E-SEM in studying thermally induced failures in monolithic microwave integrated circuit (MMIC) devices, water-induced adhesion loss at metal–polyimide interfaces, and temperature–humidity cycling of printed circuit boards. An E-SEM image is shown in Figure 11.7(c).

## 11.3.3 Transmission Electron Microscopy

Normally a research tool, the TEM is indispensable for the structural imaging of processing and crystallographic defects that are nanometer sized. In comparison, the resolution of the SEM is an order of magnitude poorer. For simplicity, the TEM may be compared to a slide projector with the slide (specimen) illuminated by light (electron beam) that first passes through the condenser lens (electromagnetic condenser lens). The transmitted light forms an image that is magnified by the projector lens (electromagnetic objective and projector lenses) and viewed on a screen (or photographed).

There are several imaging and analytical modes that are noteworthy.

1. Bright-field imaging. The directly transmitted beam is used to provide images of the microstructure and morphology of features.
2. Dark-field imaging. Diffracted beam images are used to obtain crystallographic information about structural features.

3. Lattice imaging. A combination of diffracted and direct beams is used to yield images with atomic resolution.
4. Diffraction. Diffraction yields crystallographic and orientation effect information on defects and phases. Special techniques include microdiffraction and convergent-beam diffraction.
5. X-ray spectroscopy. This analytical capability is basically the same as EDX analysis and allows elemental identification.
6. Electron energy loss spectroscopy (EELS). Through energy analysis of the transmitted electron beam, composition analysis is possible. EELS is useful for detecting low-Z elements.

Further details of the electron optics and operational modes of these and other TEM modes of operation have been discussed in many books; the recent one by Williams and Carter [21] is a recommended source of this information.

A primary disadvantage of TEM methods is that specimens must be very thin, e.g., 100–1000 nm thick, to enable electrons to penetrate them. For failure analysis purposes this frequently presents a significant time-consuming experimental challenge, which additionally destroys devices and components in the process of sample preparation. This, combined with its poor selectivity of location, relegates the TEM to identifying specific microscopic failure mechanisms, particularly those involving the roles of dislocations, precipitates, and crystalline perfection. In this way TEM methods have played a crucial role in exposing subtle defects that limit yield and lead to reliability problems. Due to the limitations noted above, the SEM is now the preferred tool for structural evaluation in failure analysis.

If, however, one has the luxury of time, which is normally absent in the FMA laboratory environment, it is now possible to prepare thin specimens having almost any orientation with respect to an IC chip or device. Examples were shown previously in the laser defects depicted in Figures. 10.8 and 10.10. The extra effort made in obtaining these images gives truth to the adage that "one picture is worth a thousand words."

Perhaps the most dramatic images of devices are those taken in cross-section. In this important technique, specimens that are already thin in one dimension are now thinned in the transverse direction; it is like imaging this page in edge view rather than in the plane the words appear on. What is involved in the case of ICs is cleaving a number of wafer specimens transversely, bonding these slivers in an epoxy button, and thinning them by grinding and polishing. Finally, the resulting disk is ion milled until a hole appears. By preparing many specimens simultaneously, the probability of capturing images from the desired circuit features is enhanced. In this way the striking images shown in Figures. 3-8a and 10-8c were obtained.

### 11.3.4 Focused Ion Beams

The use of focused ion beams (FIBs) provides two additional capabilities to the failure analyst. First, the FIB can be operated as a high-resolution microscope (focused ion microscope) that has some advantages (and disadvantages) relative to the SEM and TEM. Briefly, the FIM resembles an SEM, but instead of electrons, gallium ions generated in a liquid-metal ion source are employed as the imaging vehicle. In the optical column these ions are electrostatically focused into a fine beam that raster scans the specimen surface, releasing secondary electrons (SEs) for structural imaging. The FIB image of the interconnect shown in Figure. 5–18b reveals grain orientation contrast unlike SEM imaging; also unlike the TEM, no sample preparation is required in FIB microscopy.

At the same time as the Ga ions help form images, they contaminate specimens, introduce artifacts, and sputter away the material being observed. The last of these undesirable features for imaging is turned to good advantage in preparing TEM cross-sections, where it has eased the drudgery and eliminated the element of chance in locating features for examination [22]. Operating like a milling machine, the 10- to 25-keV Ga beam precisely ablates material from submicron regions of specimens. Suitably thinned features can then be readily examined in the TEM. This sample preparation technique, illustrated in Figure 11.8 has greatly advanced failure analysis methodology.

In concluding our discussion of electron microscopy, the comparative images of dislocation arrays in epitaxial InGaAs films reproduced in Figure 11.9 are worthy of consideration. The TEM micrograph (Figure 11.9(a)) resolves these subsurface defects to the greatest extent. While these same defects are imaged by the SEM EBIC (discussed in Section 11.5.3.2) and CL (Section 11.3.2.1) techniques in Figures 11.9(b) and 11.9(c), respectively, the resolution is considerably poorer. On the other hand, these SEM methods reveal their overall geometry and require little in the way of sample preparation.

## 11.4 CHEMICAL CHARACTERIZATION

### 11.4.1 Introduction

In the course of investigations to determine causes for yield loss during fabrication or reasons for failure of products in service, it is often an

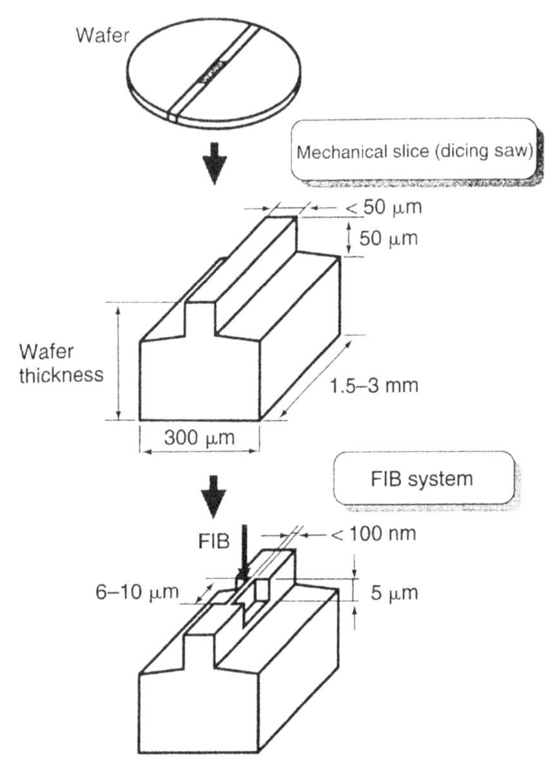


**Figure 11.8** Cross-section sample preparation for TEM employing rough mechanical slicing followed by FIB milling techniques to remove submicron layers of material [23].

objective to identify the composition of contaminants, defects, phases, compounds, etc. These entities appear during the microstructural portion of the examination and are an invitation to the chemical analyst to determine what elements are present and in what proportions. Just as was the case for structural characterization, there are a variety of techniques available for chemical characterization, and these are listed in Table 11.4 together with their acronyms and particular attributes. The general references [3–6] provide a comparative overview of these analytical techniques. Further specific details on each can be found in the additional references cited.

### 11.4.2 Making Use of Core-Electron Transitions

The first three methods (EDX, Auger electron spectroscopy (AES), and X-ray photoelectron spectroscopy (XPS)) specifically capitalize on the core-electron level structure of atoms to be analyzed [24]. By means of

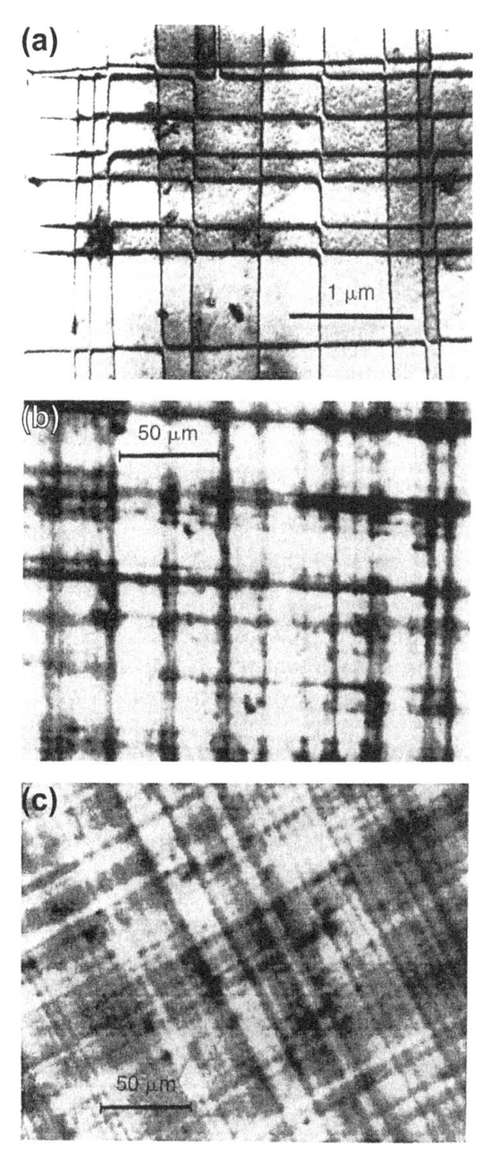


**Figure 11.9** Buried dislocation array in a GaInAs epitaxial film imagined by: (a) transmission electron microscopy, (b) scanning electron microscopy (EBIC), and (c) total light CL. *Images recorded by Dr. M. Al-Jassim.*

either impinging energetic electrons or photons, core electrons of these unknown atoms are ejected, creating vacant electron levels, or holes. This destabilizes the atom and initiates mechanisms for the remaining atomic electrons to reduce their energy by falling into these vacant levels. As a

**Table 11.4** Summary of major chemical characterization techniques

| Method | Elemental sensitivity | Detection limit (at%) | Lateral resolution | Effective probe depth |
|---|---|---|---|---|
| Scanning electron microscope—energy dispersive X-ray (SEM/EDX) | Na–U | ~0.I | ~1 μm | ~1 μm |
| Auger electron spectroscopy (AES) | Li–U | ~0.1–1 | 50 nm | ~1.5 nm |
| X-ray photoelectron spectroscopy (XPS) | Li–U | ~0.1–1 | ~100 μm | ~1.5 nm |
| Rutherford backscattering (RBS) | He–U | ~1 | 1 mm | ~20.0 nm |
| Secondary ion mass spectrometry (SIMS) | H–U | ~$10^{-4}$% | ~I μm | 1.5 nm |
| Laser ionization mass analysis (LIMA) | H–U | ~$10^{-2}$% | ~I μm | |

result of these transitions, either X-ray photons (as in EDX) or electrons (as in AES and XPS) are emitted with energies characteristic of the atom in question. All that is needed is a detector with the capability of measuring the energy of the expelled photons or electrons. Qualitative elemental identification is then made by comparing detected peak positions with the unique spectral signatures of characteristic X-ray or Auger energy peaks that are stored in computerized data bases. Quantitative elemental amounts are proportional to the strength of the signal detected.

To see how atoms are fingerprinted through their electronic structure, consider the core-electron levels for titanium atoms displayed in Figure 11.10(a). In this scheme the most energetic electrons are at the top with 0 representing the vacuum level; electrons having this energy are essentially free of the material. Levels at the bottom are populated by electrons tightly bound to the nucleus; the K level, having a binding energy of 4966.4 eV, is associated with s electrons.

#### *11.4.2.1 Energy-Dispersive X-Ray*

In the common configuration of this technique, the impinging SEM electron beam knocks out electrons from all levels simultaneously. Electrons in more energetic levels drop into the vacated state levels. Conservation of energy demands emission of photons with energies corresponding to the difference between those of the two involved levels. The

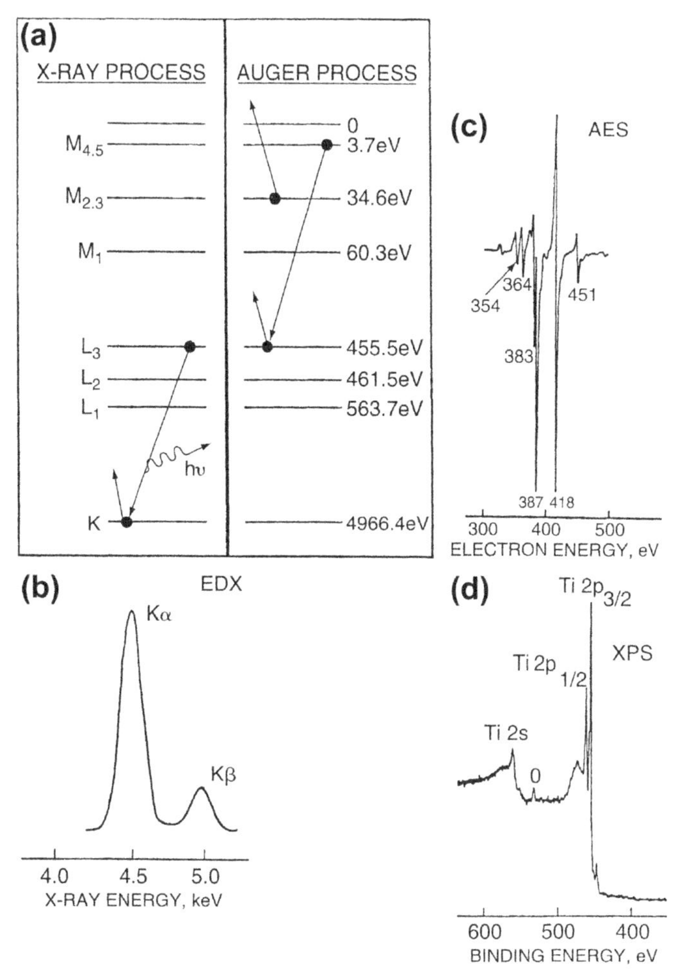


**Figure 11.10** Electron excitation processes in titanium. Shown are the energy level scheme and EDX, AES, and XPS spectral lines for Ti.

high beam energy means absorption and excitation processes between deep core levels that are manifested in emitted X-rays. These are captured by a cryogenically cooled Si (Li) diode detector attached to the SEM column and converted through standard pulse height analysis electronics into the spectrum of peaks schematically indicated in Figure 11.10(b). As an example, the Kα( peak for Ti reflects the transition of electrons from the L3 to K levels. The well-known quantum formula

$$\frac{hc}{\lambda} = E_2 - E_1 \tag{11.6}$$

with $h$ and $c$ being the Planck's constant and the speed of light, respectively, enables the wavelength ($\lambda$) between energy levels $E_2$ ($L_3$) and $E_1$(K) to be calculated. For this purpose a useful equation is

$$\lambda(\mu\text{m}) = \frac{1.24}{[E_2 - E_1](\text{eV})}. \tag{11.7}$$

Noting that all binding energies are negative in sign, substitution yields $[-455.5 - (-4966.4)] = 4510.9$ eV. This corresponds to a wavelength of $K_\alpha = 0.275$ nm.

The roughly 1 μm wide and deep volume excited by the impinging electron beam means that the spatial resolution is limited by these dimensions. Furthermore, atomic identification is normally possible only for elements with $Z$ greater than 11, implying that the EDX technique is ideally suited to analyzing metallic contaminant particles, corrosion products, and bulk reaction products. However, surface contaminant detection is obscured by the substrate background signal, and that is why Auger analysis was developed.

#### *11.4.2.2 Auger Electron Spectroscopy*

Particular sensitivity to the composition of surfaces and the ability to detect elements of lower atomic number are advantages of AES. The same set of electron energy levels is involved, but instead of focusing on only two, we now consider transitions involving three levels. As before, a high-energy core electron falls to fill a vacant energy state. The photon emitted does not exit, however, but ejects a so-called Auger electron via a kind of internal photoelectric effect. For example, the LMM spectral line has an energy magnitude given by $E_{LMM} = E_{L3} - E_{M2,3} - E_{M4,5} = 455.5 - 34.6 - 3.7$, or 417 eV. The signal detected as a function of energy is shown in Figure 11.10(c), from which it is possible to identify the element.

If we recall that in the TEM, 100-keV electrons penetrate ~100 nm of material, it is not hard to imagine that low-energy Auger electrons would emerge from surface layers only a few nanometers deep at most. This is why AES is regarded as a surface analysis tool. Through the use of a finely focused electron beam that is rastered, we essentially have the surface imaging capability of the SEM, but with poorer spatial resolution. By controllably sputtering away the surface, AES can be expanded to include (destructive) compositional depth profiling.

In yield monitoring as well as in failure analysis, the primary use of EDX and AES is to detect and identify defects and contaminants. An example of

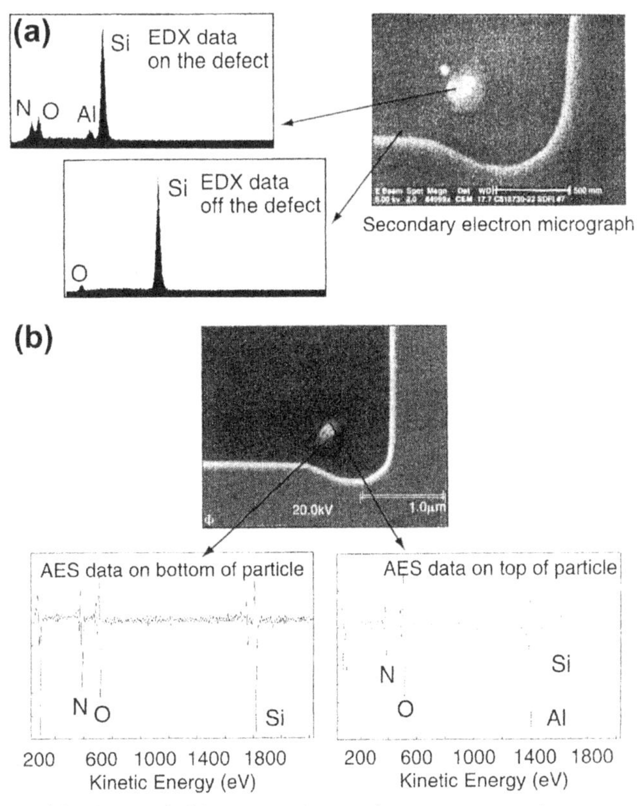


**Figure 11.11** (a) EDX and (b) AES analyses of a 0.25-μm multicomponent defect particle. *Courtesy Physical Electronics.*

how this is accomplished on a water containing a 0.25-μm multicomponent defect is outlined in Figure 11.11. Elemental EDX analysis revealed the presence of aluminum contamination in a defect particle embedded within a silicon oxynitride matrix. To further probe the composition distribution through the top and bottom of the particle, AES spectra were recorded at both surfaces. Silicon oxynitride was detected at the bottom of the particle, while a strong aluminum signal was found at the top surface of the particle. The aluminum that was not present at this process step contaminated the wafer after the blanket deposition of silicon oxynitride film but before this film was plasma etched. Of note is the detailed elemental information from an extremely small area. As a result, contamination from process tooling was suggested as the likely source of the aluminum.

#### 11.4.2.3 X-Ray Photoelectron Spectroscopy

Unlike the two prior methods, XPS essentially involves only one core energy level. An incident beam of energetic photons, usually long-wavelength X-rays from an aluminum or magnesium target, ejects photoelectrons from the specimen, and their subsequent kinetic energies are measured to yield spectra of the kind shown in Figure 11.10(d). Reliance is made on the Einstein equation for the photoelectric effect, which simply relates the electron kinetic energies to the difference between the incident photon and material binding or core-level energies. Since the latter are element specific, atomic identification is possible. Moreover, information on atomic binding and valence is provided, so that chemical substances of similar composition can be distinguished, e.g., $Cu_2O$ and CuO. This, plus the fact that photons are less destructive than electrons upon impingement of surfaces, makes XPS the preferred technique to detect organic contaminants. Because they are surface techniques, both AES and XPS require very high vacuum (e.g., $\sim 10^{-10}$ torr) during operation.

### 11.4.3 Chemical Analysis by Means of Ions

Rather than exploiting the electronic structure of atoms with the help of quantum theory to explain the measured spectra, a second approach to chemical characterization views the atom in classical terms as little more than a charged spherical mass. Thus by essentially measuring the atom or ion mass and charge, it can be unambiguously identified. Three popular ion beam techniques known as secondary ion mass spectrometry (SIMS), Rutherford backscattering (RBS), and laser ionization mass analysis (LIMA) fingerprint atoms through some variant of mass spectroscopy analysis. They will be briefly considered in turn.

#### 11.4.3.1 SIMS

The mass spectrometer, long used in the chemistry laboratory for the analysis of gases, has been dramatically transformed to create an SIMS apparatus capable of analyzing the chemical composition of solid surface layers [25]. In SIMS, a 2- to 15-keV energy beam of ions, e.g., $Ar^+$, $O^{2-}$, and $Cs^+$, bombards the surface and sputters neutral atoms, for the most part, and also positive and negative ions from the outermost surface layers. Once in the gas phase, the ions and molecular fragments are mass analyzed in order to identify the species present as well as determine their abundance. The charged-sputtered particles are first extracted, and they then enter an energy analyzer consisting of a magnetic-sector mass filter whose function is to select a particular mass for detection. The desired ion of mass $m$, charge $q$, and velocity v traces a specific area of radius $r$ in the magnetic field ($B$) of an electromagnet, given by $r = mv/(qB)$. In this way the ions are separated, and a spectrum of peaks representing individual masses is obtained. The advantage of high-detection sensitivity is offset by an extremely complex spectrum of peaks corresponding to the masses of ions and ion fragments. Standards, composed of the specific elements and matrices in question, are thus necessary for quantitative determinations of composition.

Depth profiling of ion-implanted dopants in various portions of devices is the primary application of SIMS in semiconductor technology. Better than part-per-million atom fraction detection sensitivities are achieved. In the area of failure analysis SIMS has been used [26] to detect

1. carbon and oxygen at critical interfaces of multilayer metallizations;
2. mobile ion contamination of good and leaky transistors and diodes;
3. transport of mobile ions through passivation layers; and
4. relative levels of hydrogen, chlorine, and fluorine in thick oxides and nitrides.

#### *11.4.3.2 RBS*

This technique relies on the use of very high energy (e.g., 2 MeV) beams of low mass ions that are incident on a film–substrate combination [27]. The ions, typically $^4He^+$, lose energy through

1. transiting the specimen during penetration,
2. binary elastic collisions with matrix atoms that result in ion backscatter
3. retraversing the material to exit it.

The magnitudes of all these energy losses are well known; the first and third are due to both electronic excitation and nuclear collisions. However, the largest contribution to energy loss of the probe ion stems from the momentum transfer in the elastic collision between ion and atom. By measuring the number and energy distribution of the backscattered ions, it is possible to work backward and obtain information on the nature of the elements present and their concentration and depth distribution, all without appreciably damaging the specimen. In this way, compound compositions, thicknesses, and distributions can be accurately ascertained without the use of standards. Within its realm of applicability, RBS is the preferred choice relative to EDX, AES, XPS, and SIMS because it is intrinsically quantitative and requires no calibration.

Poor spatial resolution limits RBS to scientific investigation of broad-area (blanket) multicomponent thin-film layers rather than to similar film structures in actual devices. Thus it is the perfect tool to quantitatively study the reactions between aluminum and gold films, for example, but purple plague formation at an actual contact is more profitably studied with SEM/EDX.

#### *11.4.3.3 LIMA*

Rather than remove surface atoms for analysis through ion impact as in SIMS, they are blasted off the surface by miniexplosions using a finely focused laser beam. The ions so created move through a time-of-flight drift tube, where their mass-to-charge ratio is determined by measuring their velocity in a fixed electric field. This information enables identification of the atoms.

## 11.5 EXAMINING DEVICES UNDER ELECTRICAL STRESS

### 11.5.1 Introduction

All the analytical and diagnostic methods presented so far have one major shortcoming—the materials and devices are not viewed under use conditions. In certain cases such as failure, postmortem examination of the remains may be the only practical course. Even during quality or yield testing, structural and chemical analyses are always easier to perform when devices are not electrically powered, heated, or otherwise exposed to operating variables. But this momentary advantage of ease in measurement may be offset by difficulty in interpreting and applying the results of analysis. This section is devoted to exploring those techniques that have been developed to provide information on local charge transport and heating effects in powered devices. As we shall see, the structural resolution possible in these in situ methods generally falls short of electron microscope capabilities; but greater sensitivity to local damage and ability to monitor time-dependent change are important advantages.

### 11.5.2 Emission Microscopy

This exciting new advance in probing and imaging damage sites in ICs is built around an optical microscope [28,29]. A schematic of the light emission microscopy (LEM) technique is shown in Figure 11.12. This method employs light of variable wavelength, typically in the range 400–1000 nm, which impinges on an IC chip electrically powered by a circuit tester. In response to optical stimulation, weak luminescence occurs locally at defects. The emitted photon signal is captured and amplified

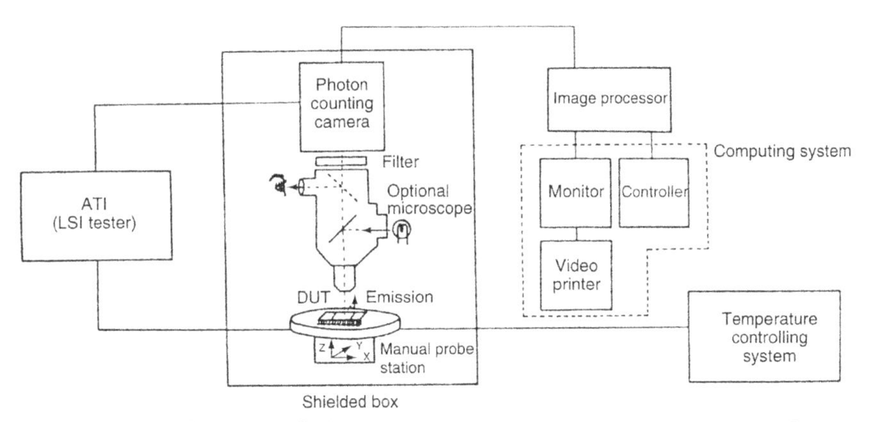


**Figure 11.12** Schematic of the optical emission microscope. *From Ref. [28]. ATI - Automatic Test Instrument, DUT - Device Under Test, LSI - Large Scale Integrated Circuit*

with a sensitive photomultiplier–detector, enabling the emission site to be superimposed on the reflected optical image. In this way regions of enhanced emission approaching 1 μm in size can be pinpointed. Rather than look for the proverbial "needle in the haystack" typical of standard optical or electron microscopy methods, the luminescent signal conspicuously broadcasts its location. This results in an impressive signal-to-noise "search" ration and rapid identification of defect sites and device structures vulnerable to damage. Light emission from n-channel transistors within a row of flip-flop devices are responsible for the image captured in Figure 11.13.

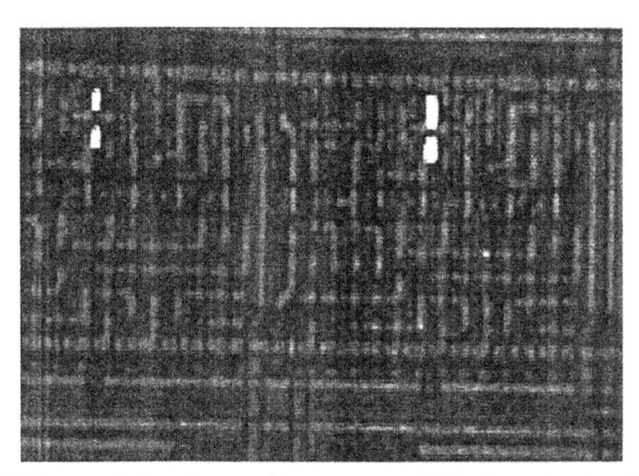

**Figure 11.13** Light emission image from a row of flip-flop circuits. The n-channel transistors are the light emission sources. *From Ref. [30].*

Light emission phenomena can be basically classified into three categories, each characterized by different spectral distributions [31].

#### 11.5.2.1 Recombination Radiation

In this case the emission occurs due to minority carrier recombination at a forward-biased p-n junction. For Si devices this means emission at 1.1 μm. Very sensitive IR detectors, developed for night vision purposes, have been used in these systems. Forward-biased emitter–base junctions in bipolar transistors, thyristors (which have p-n-p-n structures), as well as parasitic bipolar effects such as latchup in CMOS devices, are all amenable to LEM.

#### 11.5.2.2 Field Accelerated Carriers

These energetic charge carriers are generated by high electric fields and allow the investigation of reverse-biased p-n junctions, transistor punch-through (Section 2.3.3.3), and metal oxide semiconductor (MOS) transistor saturation behavior. Corresponding phenomena that can be probed are hot-electron degradation, electrostatic discharge events at input–output protection circuits, and microplasma emission during dielectric breakdown.

#### 11.5.2.3 Oxide Currents

Here the application is to MOS capacitor structures. When biased by applying high positive voltages to the polysilicon gate, electrons tunnel from Si into the $SiO_2$, a process accompanied by photon emission at 1.8 eV. This makes the technique useful in characterizing gate oxide damage prior to breakdown.

Spectra obtained from these three categories of emission phenomena are depicted in Figure 11.14. Depending on particular application, the spectral response of the photocathode must be optimized.

There are, however, a few caveats worth noting about emission microscopy. Some optical emission also occurs in unpowered and normally functioning powered devices. This means that a fault mechanism cannot be inferred in all cases; in a positive vein, however, artifacts can be singled out this way. The emission behavior is not materials specific and has been observed in silicon oxide, nitride, and oxynitride. In general, direct assignment of the defect type is not always immediately obvious, so that a physical model is needed to decipher the phenomenon leading to the observed emission.

A few additional applications will serve to highlight the great potential for emission microscopy in locating processing defects. Uraoka et al. [29,32]

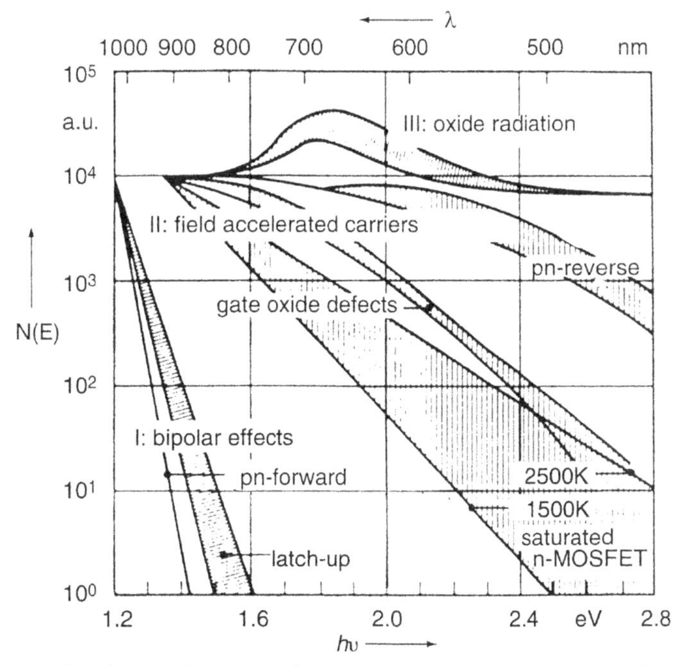


**Figure 11.14** Fundamental spectra of important emission phenomena classified into three groups. *From Ref. [31].*

observed enhanced emission from some of the 45,000 via-hole chain contacts that connected two levels of metallization. The via-hole contact resistance was subsequently measured to be higher at these "defect" locations relative to the lower normal values recorded across the wafer. Open as well as short circuits were detected by Murase [28]. Enhanced photon emission was observed at shorts due to stress-induced voiding in a metal line, while Joule heating at opens caused emission of IR radiation.

### *11.5.2.1 Optical-Beam-Induced Current*

One of the important modes of operation is known as optical-beam-induced current (OBIC). Image contrast arises from the current due to laser-induced carrier collection at p-n junctions. This technique is clearly analogous to the more popular EBIC effect, which will be discussed in greater detail in the following section. An interesting application of OBIC methods to failure analysis in multilevel dynamic random access memory (DRAMs) was reported by Mitsuhashi et al. [33]. In these high-density chips, the levels of aluminum metallization obscure the underlying devices because they occupy such a large fraction of the viewing area.

Therefore, observation from the polished wafer backside using an IR laser enabled defects to be revealed in the powered chips. With a lateral resolution of 0.8 μm, which was sufficient to separate the storage capacitor from the transfer transistor, devices prone to latchup could be precisely located.

## 11.5.3 Voltage Contrast Techniques

### *11.5.3.1 Voltage Contrast and Current Collection in the SEM*

There are two categories of SEM techniques that generate images in powered devices simply based on the surface voltage reached or on the current collected at p-n junctions. Both methods are useful because they expose electrical defects that cause malfunction. The virtue of voltage contrast (VC) techniques is the ability to turn electrical signals into dark and light images. All that is necessary in the simple static mode is to place the device or IC into the SEM, power it, and observe the resulting image contrast. VC arises because the potential developed on the surfaces of devices alters the SE emission. To see why, we first note that the SE detector is biased positively. If a conducting surface is raised to a negative potential, SE emission is encouraged, and the resultant image appears brighter. Conversely, relative darkness prevails in regions raised to a positive potential. Physically, for low beam energies more secondaries leave the surface than are injected, but at higher energies the reverse is true. With beam energies of less than 1 keV the surface acquires a positive charge that prevents low-energy SE emission; an equilibrium potential of 0.5 V develops such that incident and escaping electron fluxes are equal. Device bias alters the SE image contrast depending on the difference between the equilibrium and applied potentials. An important application of VC techniques has been to expose CMOS devices sensitive to latchup [34].

VC methods have been extended to enable dynamic imaging of devices under specific clock cycles. In such cases stroboscopic techniques are required to separate individual logic states. Another variant, known as capacitive coupling voltage contrast (CCVC), permits imaging and dynamic voltage measurement beneath passivation layers of buried conductors. In essence, the passivation is the dielectric of a capacitor having bound charge on the surface and a subsurface source of bias potential. An example of CCVC contrast is shown in Figure 11.15.

A major outgrowth of VC methods has been the development of electron beam testing (EBT) methods for locating faults in microprocessors and memory chips [35]. Faults are traced back in these complex circuits through successive comparisons of VC images in functional and failing

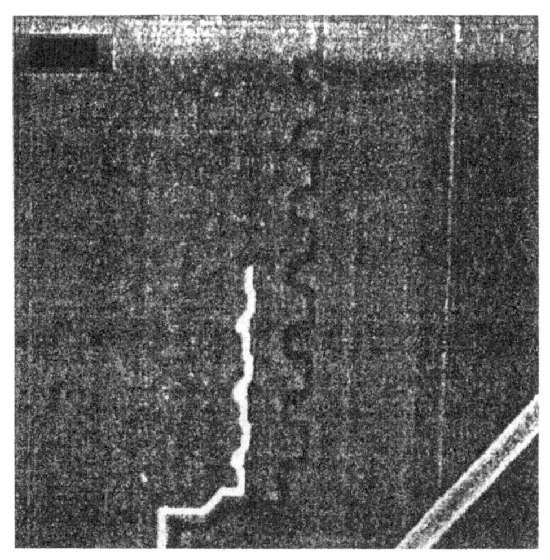

**Figure 11.15** CCVC image of a failing bit line (bright–negative transition) and a functional bit-not line (dark–positive transition). *From Ref. [15].*

devices. The resulting time sequences of states are differentiated and computer processed to highlight divergences in what resembles a movie loop of the nucleation and growth of faults. Rather than conventional VC, which is a local technique, EBT is global in nature. It uses powerful computers to analyze chips that are themselves computers.

### *11.5.3.2 Electron-Beam-Induced Current*

This mode relies on the SEM electron beam probe injecting minority carriers into a semiconductor, which then diffuse away from the point of generation and recombine at rates that depend on carrier lifetimes and diffusion lengths [2]. What is required is an electric field to separate and collect carriers, and this is present internally at a p-n junction or Schottky barrier. In the EBIC mode, shown schematically in Figure 11.16, the electron beam is parallel to the specimen p-n junction and scanned across it. The detected EBIC current signal is largest at the junction because the carriers are collected before they have a chance to diffuse away or recombine. Shorter diffusion lengths in GaAs relative to Si mean that junctions are more highly resolved in the former. The signal falls with distance as the probability of carrier collection diminishes; encountering defects is a cause of signal loss. Defects act as nonradiative recombination centers that shorten carrier lifetimes and reduce the EBIC signal.

Generation of electron–hole pairs and their separation in EBIC parallels these processes in solar cells (see Section 2.8.2). The analogy can be

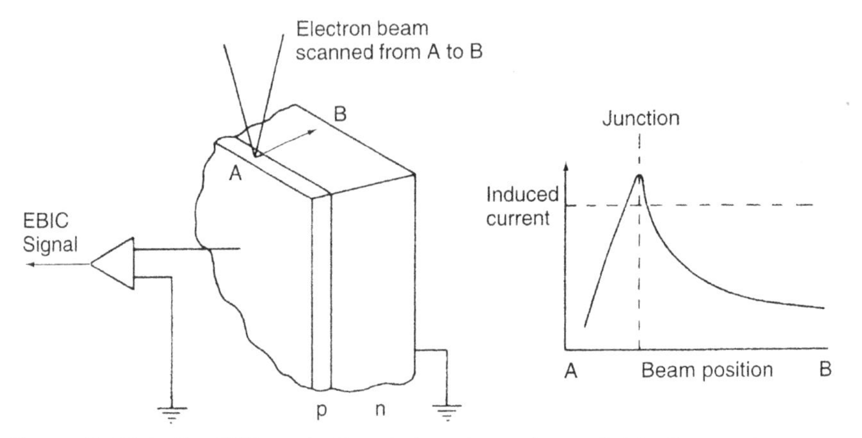


**Figure 11.16** In the EBIC mode the electron beam is parallel to the specimen p-n junction and scanned across it as schematically shown. *From Ref. [36].*

taken a step further by considering the external circuit resistance *R*. If *R* is very small a short circuit current flows, while an open circuit voltage develops when *R* is infinitely large. Using changes in short circuit current during scanning to modulate the SEM intensity is a popular way to perform EBIC.

To illustrate EBIC use in assessing the quality of InGaAs/InP photodiodes consider Figure 11.17. The diode structure shown in Figure 11.17(a) has a cylindrical symmetry and curved p-n junction that can be seen in the SEM EBIC images of Figures 11.17(b) and (c). Of interest is the contrast in the vicinity around the junction or outer circumference. In Figure 11.17(b) the narrow light band is indicative of good devices; on the other hand, the diffuse light region of Figure 11.17(c) extends deeply into the n-type InGaAs and is indicative of poor device behavior. The spatially distributed intensity (not collected current) is quantitatively displayed in Figure 11.17(d) where brighter regions are associated with defects, charge recombination centers, and inhomogeneous doping.

### *11.5.3.3 VC by FIBs*

This probeless VC technique, the subject of a recent publication [37], uses an FIB to isolate defects in ultralarge-scale ICs, e.g., 64- and 256-Mb DRAMs with line widths as small as 0.25 μm. In device mazes of such small dimensions, common defect exposure methods such as LEM, OBIC, and liquid crystal (LC) probing are severely challenged. The positive FIB in the VC technique charges floating structures on the chip,

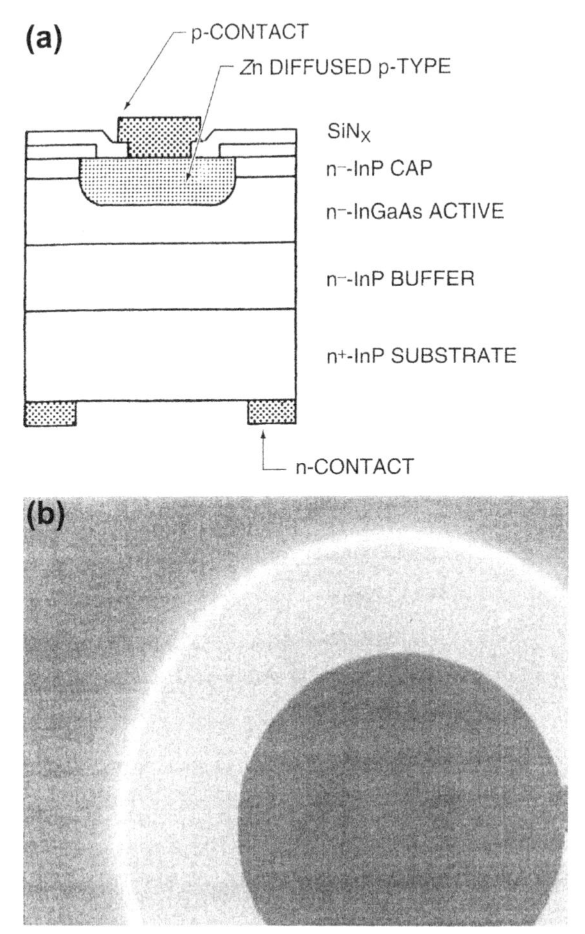


**Figure 11.17** Use of EBIC in assessing the quality of InGaAs/InP photodiodes. (a) Device structure. (b) EBIC image of good diode. (c) EBIC image of bad diode. (d) Intensity–distance variation for mid- and low-noise diodes. *Courtesy of R. Farrow, Lucent Technologies, Bell Laboratories Innovations.*

establishing a potential difference between them and the ground. As a result, the SE image tends to be attenuated at such structures, making them visible in the electron imaging mode. In this sense the technique parallels that of SEM VC (Section 11.5.3.1). However, the FIB technique can be leveraged to ion mill defective regions in order to delayer them and expose cross-sections for further in situ examination; therein lies its great advantage. In practice, LEM or OBIC first locates the approximate defect position, after which it is precisely marked at high resolution by

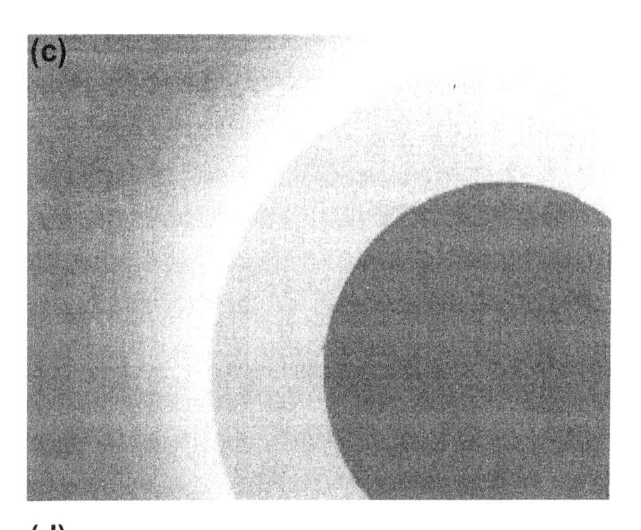


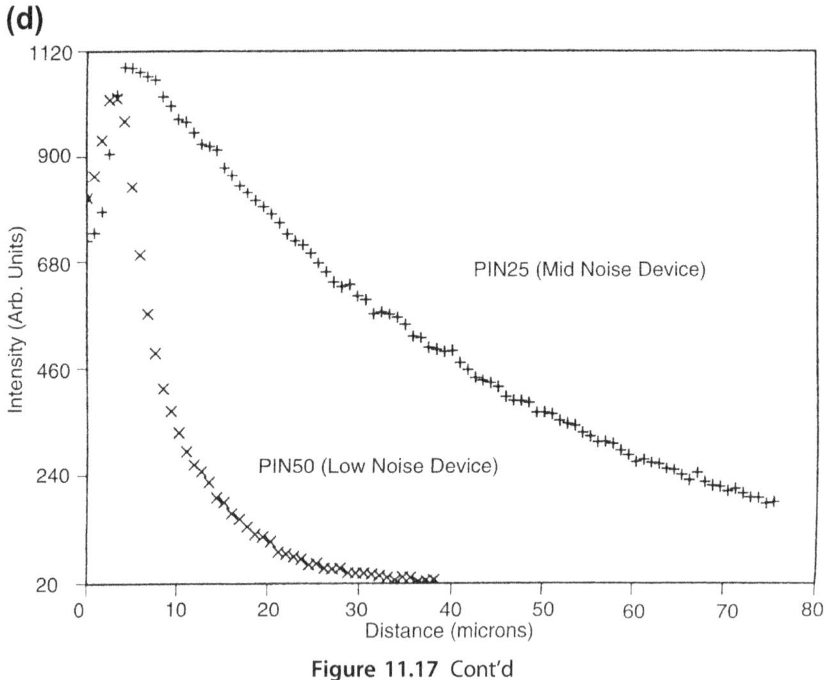


**Figure 11.17** Cont'd

FIB. Failure modes for which the technique is applicable include deep trench capacitor polysilicon to substrate leakage, gate conductor to self-aligned contact shorts, metal line opens, via chain opens, and metal line to line shorts.

### 11.5.4 Thermography

Joule heating is often a consequence of circuit function. However, when circuits contain components or devices that are in the process of degrading or have failed, the local temperature rises above the "normal" Joule heating level. Thermography broadly encompasses a class of experimental techniques having the capability of imaging the heat distribution in objects. In particular, several methods that have been developed to pinpoint the location of heat generation in ICs and electrical equipment are considered next.

#### *11.5.4.1 LC Imaging*

In this technique LCs are the temperature-sensing element [38]. For this purpose nematic LCs, in which rod-shaped molecules align along a unique axis, are used. Below a critical first-order phase transition temperature known as the clearing point ($T_C$), the crystals are aligned, and the liquid is optically anisotropic. This means that if the LC is imaged between a pair of crossed polarizers (analyzer and polarizer at right angles), the image will appear bright. In this case the ordered LCs rotate the incident plane of light polarization, and the analyzer transmits some light as a result. Above $T_C$ the LC is disordered and optically isotropic, so that the light is extinguished; now the unrotated light has a plane of polarization exactly perpendicular to that of the analyzer. The practical implication of this is that the image appears dark at a hot spot.

These simple ideas are capitalized upon in the optical-microscope-based, liquid crystal imaging system of Figure 11.18. A few drops of saturated solution composed of LC in acetone is first spread over the wafer or packaged IC. Examples of LCs that give good results include substituted phenyl benzoates with $T_C = 49\,°C$ (available as ROCE 1510 from Hoffman-LaRoche) and cyano-alkyl-biphenyls with $T_C = 29\,°C$ (available as K18, BDH Chemical Ltd) [39]. Observations with typical resolutions of $\sim 1\ \mu m$ can be expected for devices under test. Typical hot spots located by LC imaging include those due to electrostatic discharge, interlayer shorts, oxide shorts, and latchup.

#### *11.5.4.2 IR Thermography*

IR temperature measurement and imaging techniques are finding increasing use in the design, production, and failure analysis of printed circuit boards, hybrid as well as ICs, and other electronic components and systems. The reasons are not hard to see because contactless temperature

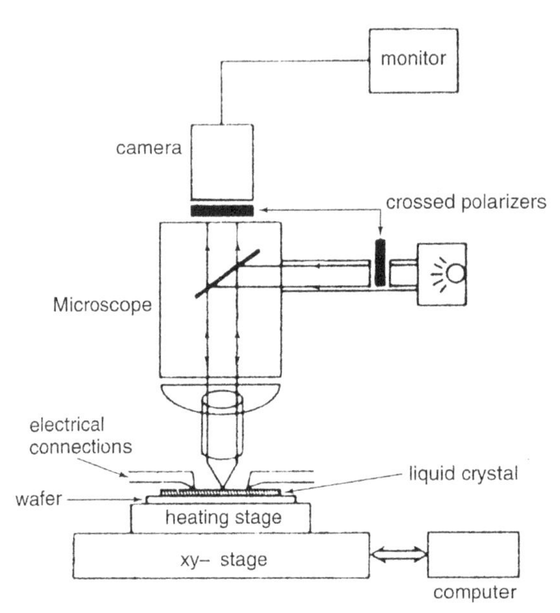


**Figure 11.18** LC microscopy system for locating sites of heat dissipation in ICs. *From Ref. [37].*

measurements of small objects can be recorded in intervals as short as one-thirtieth of a second [40]. Temperature differences as small as 0.1 °C have been detected between unpowered and powered circuits. In addition to locating surface hot spots, subsurface delamination or lack of adhesion can be exposed in thermal images when the involved materials are transparent to IR radiation.

Two broad categories of IR temperature-sensing techniques can be distinguished depending on whether thermal measurement or thermal imaging is the goal. In the former, the objective is to measure the spot temperature at an observable target point. Heated objects emit a radiated spectrum that changes with temperature; the IR spectrum peaks at shorter wavelengths for hotter objects and longer wavelengths for cooler objects. Pyrometers essentially take advantage of this phenomenon in the IR spectral range, ~1 to ~20 μm. Typically employing photo-sensitive HgCdTe or InSb detectors, temperatures ranging from slightly above the ambient to hundreds and even thousands of degrees Celsius are measurable. Knowledge of the emissivity, or the ratio of the actual to theoretical emission of a true blackbody at the same temperature, limits the accuracy of the temperature reading. Simple point-and-shoot convenience does not usually yield accurate temperatures because surface emissivity varies widely with material, e.g., 0.1 for Al to 0.9 for Si.

Thermal imaging enables a spatial distribution of temperature to be mapped. This is a very useful capability for locating design flaws in powered devices and circuits, as shown in Figure 11.19. Commercial thermal imaging systems display the output in colors that correspond to temperature, as evident in the image of laser facet heating reproduced in Figure 11.20. Most systems operate in the middle- and far-IR spectral regions, with 3–5 μm and 8–12 μm being popular choices. Areacoverage is accomplished either mechanically, by moving a beam with rotating mirrors, or electronically, through the use of detector arrays. In the case of the laser facet heating of Figure 11.20, the temperature distribution was detected by a 256 (256 InSb focal plane array camera attached to a diffraction-limited IR microscope. Because the 3- to –5- μm camera band pass is well below the InP bandgap, it can image InP-based lasers employed in optical communications systems [41].

## 11.5.5 Trends in Failure Analysis

To conclude this chapter and establish a bridge to the next chapter, it is worth reconsidering several techniques deemed by Nakajima and Takeda [42] to play a significant role in exposing defects in the coming

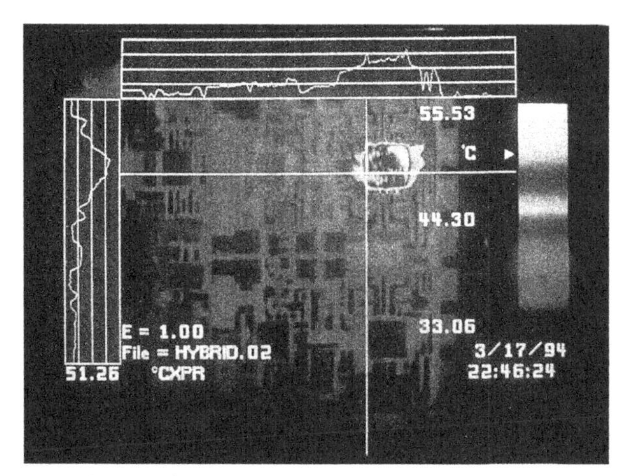


**Figure 11.19** Hybrid circuit failure analysis employing IR thermal imaging. Hot spot is pinpointed at intersection of horizontal and vertical temperature cursors. *Courtesy of FLIR Systems Inc.*

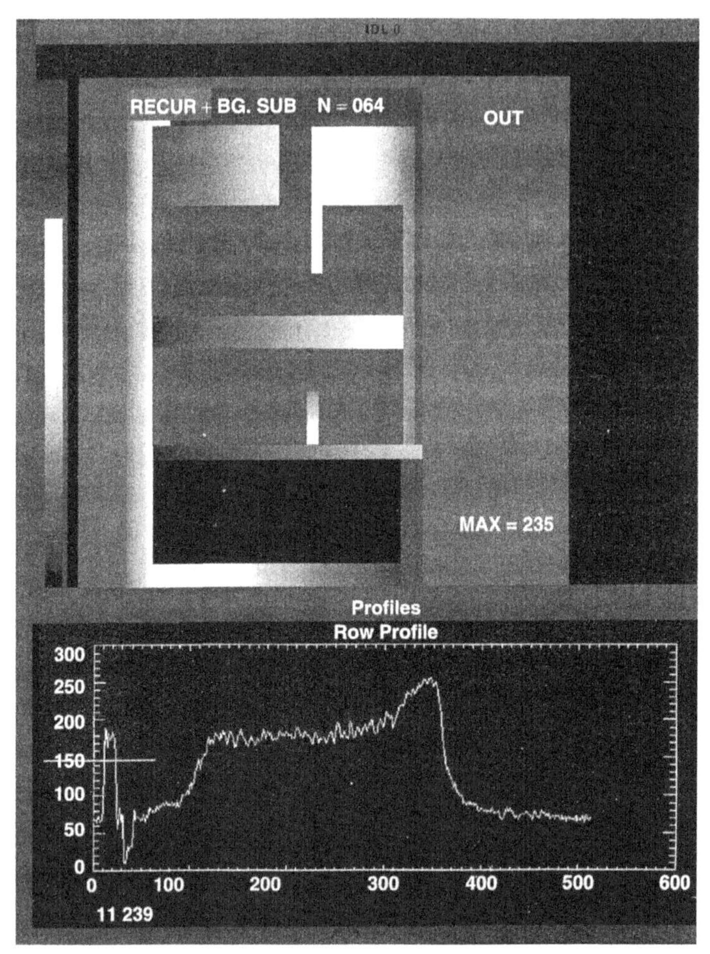


**Figure 11.20** Thermal image of facet heating in a 0.98-μm laser. The maximum facet temperature recorded is 235 °C and the laser dimensions are approximately 500 μm × 500 μm. *Courtesy of C. Bethea, Lucent Technologies, Bell Laboratories Innovations.*

0.25-μm (MOSFET channel length) era. In Table 11.5, reproduced from their recent paper, the current state of the art, the impediments to their use in the 0.25-μm regime and countermeasures required for their implementation are summarized. All the techniques have been mentioned earlier in the chapter, and there are apparently no serious barriers to enhancing their performance to meet challenges of the immediate future.

**Table 11.5** Trends in failure techniques

| Technique | Present characteristics | Difficulty in application to 0.25-μm ICs | Counter-measures |
|---|---|---|---|
| EBT | $S = <0.25$ μm V ~400 mV (experimental <100 mV) | Weak EB images due to planarized surface; small S/N ratio; low signal levels from multilevel metallizations | Improve stage positioning, increase probe current, remove dielectrics |
| LCI | $S = 1$ μm, $C = 1$ ma; $T$, binary response | Small temperature rise at hot spots; poor heat conduction due to multimetal layers | Use LCI together with FMI |
| FMI | (Experimental) $S = 0.3$ μm, $T = 0.01$ °C | Calibration each time; temperature–wavelength reproducibility problems | Improve sample preparation |
| LEM | $S = 0.5$ to 0.25 μm, $C = 1–10$ μA (3–5 V) | Low light emission due to low $V_D$ (2–5 V) and multireflection of light by metal layers | More sensitive photodetector; use optical filters |
| OBIC | $S = 0.25$ μm, $C = 1–10$ nA | Low current detection; low $V_D$ produces small OBIC signal | Improve sensitivity of amplifier |

$S$, spatial resolution; $V$, voltage resolution; $C$, current sensitivity; $T$, temperature resolution, $V_D$, supply voltage; LCI, liquid crystal imaging; FMI, fluorescent microthermographic imaging.
From Ref. [41].

## EXERCISES

1. Distinguish among X-ray radiography, diffraction, and fluorescence by writing the key equation that defines each. Provide a specific example that illustrates the use of each technique in the failure analysis of a die attach fracture.
2. The wafer of the illustrative problem (Example 1) is contained within molded plastic that is 0.4 cm thick, and the entire package is irradiated with 20-keV X-rays. If the absorption coefficient of the polymer is 0.6/cm, what is the relative intensity difference in regions where the Au pad is present or absent?
3. A 0.3-cm-thick ceramic package contains a void pocket that extends 0.015 cm deep. X-rays having an energy of 20 keV are used to expose

the void defect. If the absorption coefficient is 10/cm what is the intensity difference between the dense and void-containing ceramic?

4. The speed of sound in solids is theoretically given by $(E/\rho)^{1/2}$, where $E$ is the modulus of elasticity and $\rho$ is the density. Acoustic waves were launched at an epoxy package containing a silicon chip lying 1 mm below the surface.
   a. How long will it take to detect the echo reflected from the Si surface?
   b. What are the values of ER, ET, AR, and AT?
5. Ultrasonic detection of cracks and seams in metal castings, welds, and forgings is normally carried out at frequencies ranging from 200 kHz to 10 MHz. In electronic packaging applications the frequency used is much higher, e.g., 10–100 MHz. What accounts for this difference in operating frequency?
6. Sound waves impinge on the bottom surface of an alumina substrate. If a silicon die is bonded to the top surface of the substrate, calculate the value of the reflected acoustic energy ($E_R$) for the case of a good bond between these components. During service, debonding occurred at the interface, leaving an air space. What is the resulting value of $E_R$?
7. "Seeing is believing, but atomic fingerprints do not lie." Give two examples of failures in electronic materials where it is more important to reveal structural morphology than chemical composition. Mention two other examples where the reverse is true.
8. When binary metals react, intermetallic compounds may form, e.g., $Cu_6Sn_5$ or $Cu_3Sn$, $AuAl_2$ or $Au_2Al$.
   a. What is the simplest (cheapest) experimental technique you would employ to distinguish between compounds that contain the same elements but different compositions? Would your answer change if the compound: (1) Had dimensions of 50 μm as opposed to 0.5 μm? (2) Consisted of discrete particles instead of a layered film? (3) Was embedded within the interior of a metallic matrix or located on the surface?
   b. Suppose analysis had to be performed nondestructively. What techniques would you employ?
9. a. Structural imaging is often done while devices are powered. Chemical characterization, however, is not normally carried out on powered devices. Why?

b. Would typical operating voltages and currents alter the chemical information obtained from devices using EDX, AES, and XPS methods? Explain.

10. In Figure 11.11, AES analysis was sufficiently sensitive to detect Al in the SiON defect. If 2% Al was detected in a layer 1 nm thick, how many atoms were actually detected?

11. For tungsten the core-electron energy levels in electronvolts are given by:

| $K$ | $L_1$ | $L_2$ | $L_3$ | $M_1$ |
|---|---|---|---|---|
| 69,525 | 12,100 | 11,544 | 10,207 | 2820 |

a. Create an energy level diagram for W.

b. Calculate the energy and wavelength of the photon emitted when an electron in a W atom falls from an $n = 3$ to the $n = 1$ state.

c. A typical EDX detector system displays X-ray energies over a range of 1–30 keV. In performing an analysis on a Ti–W contact diffusion barrier, what are the energies of the X-rays that will be displayed?

12. Suggest ways to:

1. nondestructively examine the quality of the solder interconnections between a flip chip and ceramic substrate.
2. identify the nature of the discoloration on the surface of a passivated IC chip.
3. nondestructively locate the exact position of a break in the grounding strap within a 20-foot section of a buried telephone cable after it was dug up.

13. Consider a bipolar transistor which contains both $^{121}Sb$ and $^{123}Sb$ in the emitter but only $^{121}Sb$ in the collector. What information does this provide on the doping processing methods employed?

14. Using LCs on a 1.5-μm-thick $SiO_2$ layer, a light spot 1 μm in diameter was observed to darken within 1 ms as the temperature rose 1 °C. Approximately how much power was dissipated during the transition?

15. A coiled tungsten filament within an incandescent lamp is attached to copper alloy lead wires. Failure occurred at the attachment interface. What analytical instruments would you use in order to determine the cause of failure?

16. Uncited
    a. Explain why the energy of oxide emission is greater than that for forward-biased bipolar light emission in Figure 11.14.
    b. In a p-n diode more light emission occurred under forward bias than negative bias. Why?
17. Thin-film, foil, and ceramic capacitors undergo failures with often serious consequences. From what you know about them, suggest analysis techniques to study the following types of failures:
    1. pinholes in thin tantalum oxide film capacitors.
    2. opens and shorts in wound tantalum and plastic foils used to make capacitors.
    3. voids in metal plates used to contact the dielectrics in ceramic capacitors.
    4. delamination between alternating conductor and dielectric plates of ceramic capacitors.
18. Carbon, wire-wound, and thin-film resistors undergo degradation that causes circuits to fail. From what you know about resistors suggest analysis techniques to study the following types of failures:
    1. open wire-wound resistors due to fracture of the wires and broken wire-end cap welds.
    2. water absorption and electrostatic discharge of thin-film resistors.
19. Unstable transistor behavior and failure was observed in almost all the chips of an experimental 1-Gb DRAM processed wafer. It is believed that contamination of gate oxides, which are 4.5 nm thick with 0.25 μm lateral dimensions, occurred. In particular, transition metals at the Si–$SiO_2$ interface are suspected.

Suggest an experimental course of action in order to verify whether contamination was the root cause of failure, and recommend techniques and instruments necessary for such a study.

## REFERENCES

[1] B.P. Richards, P.K. Footner, Microelectron. J. 15 (1) (1984) 5.
[2] B.P. Richards, P.K. Footner, The Role of Microscopy in Semiconductor Failure Analysis, Oxford University Press, Oxford, 1992.
[3] G. Scheissler, C. Spivak, S. Davidson, IEEE Proc. Custom Integr. Circuits Conf. 4 (1) (1993) 26.
[4] C.R. Brundle, C.A. Evans, S. Wilson (Eds.), Encyclopedia of Materials Characterization, Butterworth–Heinemann, Boston, 1992.
[5] J.B. Bindell, in: S.M. Sze (Ed.), VLSI Technology, McGraw Hill, New York, 1988.

[6] H.W. Werner, in: R.A. Levy (Ed.), Microelectronic Materials and Processes, Kluwer, Dordrecht, 1989.
[7] E.S. Meieran, P.A. Flinn, J.R. Carruthers, Proc. IEEE 75 (7) (1987) 908.
[8] T.M. Moore, Tutorial Notes, IEEE International Reliability Physics Symposium, 2c.1, 1992.
[9] P. Yalamanchili, A. Christou, S. Martel, C. Rust, IEEE Circuits Devices 10 (July 1994) 36.
[10] T.M. Moore, R. McKenna, S.J. Kelsall, 29th Annual Proceedings of the IEEE Reliability Physics Symposium, p. 160, 1991.
[11] N. Cartharge, V. Hauser, AT&T Technol. Tech. Dig. 68 (1982) 7.
[12] M. Jacques, IEEE 17th Reliability Physics Proceedings vol. 197 (1979).
[13] W. Kern, R.B. Comizolli, G.L. Schnable, RCA Rev. 43 (1982) 310.
[14] J.B. Bindell, Adv. Mater. Processes 143 (3) (1993) 20.
[15] E.I. Cole, J.M. Soden, in: T.W. Lee, S.V. Pabbisetty (Eds.), Microelectronic Failure Analysis, Desk Reference, third ed., ASM International, Materials Park, OH, 1993.
[16] D.P. Field, D.J. Dingley, Solid State Technol. 38 (11) (1995) 91.
[17] M.J. Li, X. Wu, M. Pecht, K. Paik, E. Bernard, Int. J. Microelectron. Packag. 1 (13) (1995).
[18] K.L. Pey, W.K. Chim, J.C.H. Phang, D.S.H. Chan, Microelectron. Reliab. 34 (1994) 1193.
[19] A. Rosencwaig, Ann. Rev. Mater. Sci. 15 (1985) 103.
[20] M.J. Li, M. Pecht, J. Electron. Packag. Trans. AIME 117 (1995) 225.
[21] D.B. Williams, C.B. Carter, Transmission Electron Microscopy—a Textbook for Materials Science, Plenum, New York, 1996.
[22] S. Morris, E. Tatti, E. Black, N. Dickson, H. Mendez, B. Schwiesow, R. Pyle, International Symposium on Testing and Failure Analysis, p. 417, November 1991.
[23] T. Ishitani, T. Yaguchi, H. Koike, Hitachi Rev. 45 (1) (1996) 19.
[24] D. Briggs, M.P. Seah (Eds.), Practical Surface Analysis by Auger and Photoelectron Spectroscopy, Wiley, New York, 1984.
[25] A.W. Benninghoven, F.G. Rudenauer, H.W. Werner, Secondary Mass Ion Spectrometry—Basic Concepts, Instrumental Aspects, Applications and Trends, Wiley, New York, 1987.
[26] K. Evans, International Symposium on Testing and Failure Analysis, p. 45, 1989.
[27] J.R. Bird, J.S. Williams (Eds.), Ion Beams for Materials Analysis, Academic Press, Sydney, 1989.
[28] M. Murase, NEC Res. Develop. 35 (1) (1994) 52.
[29] Y. Uraoka, I. Miyanaga, K. Tsuji, S. Akiyama, IEEE Trans. Semicond. Manuf. 6 (4) (1993) 324.
[30] G. Shade, K.S. Wills, in: T.W. Lee, S.V. Pabbisetty (Eds.), Microelectronic Failure Analysis, Desk Reference, third ed., ASM International, Materials Park, OH, 1993.
[31] J. Kolzer, A. Dallmann, G. Deboy, J. Otto, D. Weinmann, Qual. Reliab. Eng. Int. 8 (1992) 225.
[32] Y. Uraoka, N. Tsutsu, Y. Nakata, S. Akiyama, IEEE Trans. Semicond. Manuf. 4 (3) (1991) 183.
[33] J. Mitsuhashi, S. Komori, N. Tsubouchi, Qual. Reliab. Eng. Int. 8 (1992) 239.
[34] K.S. Wills, C.J. Pilch, A. Hyslop, 24th Annual Proceedings of the IEEE Reliability Physics Symposium, p. 115, 1986.
[35] T.C. May, G.L. Scott, E.S. Meieran, P. Winer, V.R. Rao, 22nd Annual Proceedings of the IEEE Reliability Physics Symposium, 1984, p. 95.
[36] B.P. Richards, A.D. Trigg, GEC J. Res. 3 (3) (1985) 167.
[37] K.J. Giewont, D.B. Hunt, K.M. Hummler, J. Vac. Sci. Technol. B 15 (4) (1997) 916.

[38] C.G.C. de Kort, Philips J. Res. 44 (2/3) (1989) 295.
[39] F. Beck, Qual. Reliab. Eng. Int. 2 (1986) 143.
[40] B. Linnander, IEEE Circuits Devices 9 (July 1993) 35.
[41] W.-C.W. Fang, C.G. Bethea, Y.K. Chen, S.L. Chuang, IEEE J. Sel. Top. Quantum Electron. 1 (1995) 117.
[42] S. Nakajima, T. Takeda, Microelectron. Reliab. 37 (1997) 39.

## CHAPTER 12

# Future Directions and Reliability Issues

## 12.1 INTRODUCTION

In this final chapter, trends in semiconductor technology are projected into the future in order to assess the reliability issues that may arise in coming years. Projections are always a risky business, so we may expect more speculative issues and subject matter to be raised here than elsewhere in the book. George Santayana's famous admonition that "those who cannot remember the past are condemned to repeat it" has approximately applied to the reliability of microelectronics as well. Will the mind-boggling advances in semiconductor technology that have changed the world in so many ways continue unabated, with no surprises in store? If so, then we can profitably heed Santayana by projecting our current models of failure mechanisms to smaller feature sizes. However, what if history does not exactly repeat? Are there technological limits to how far device features can be shrunk? More fundamentally, what are the physical limits to device behavior and feature sizes? Will the smaller devices and features spawn new failure mechanisms and reliability concerns? And finally, what strategies will have to be adopted to meet these future reliability challenges?

This chapter will attempt to address some of these difficult questions. Our road map into the future began in Chapter 1, where the kinds of electronic products that are and will be in demand were considered. Here, feature-scaling trends in device processing will be projected into the future. Both considerations should provide a basis for assessing the growth in applications and capabilities of future electronic products. At some point, however, the apparently unbridled growth of microelectronics will eventually slow as some of the fundamental as well as practical size limitations are approached. The physical basis for these limits will, therefore, be discussed. We must however, temper these projections by what we have learned

*Reliability and Failure of Electronic Materials and Devices*
ISBN 978-0-12-088574-9
http://dx.doi.org/10.1016/B978-0-12-088574-9.00012-4

about failure and reliability in all of the intervening chapters. Lastly, the final section of the book addresses the subject of improving reliability and how to build it in.

## 12.2 INTEGRATED CIRCUIT TECHNOLOGY TRENDS

### 12.2.1 Introduction

To intelligently assess future reliability prospects we must first be sensitive to the implications of shrinking solid-state device dimensions. Our initial reaction, based on the content of previous chapters in this book, is probably to be skeptical or even pessimistic. After all, with smaller structures, atoms or electrons have shorter distances to traverse between "reliable" and "failed" states. This means that these states, the precursors to degradation and failure, will be accessed in shorter times. Happily, our intuition has proven wrong, and paradoxically, successive generations of integrated circuits have become increasingly more reliable. The reasons for it are many and include improved electric circuit designs, better materials, and cleaner, more controlled processing. Predictions of slowdowns in the dizzying pace of IC technology have not materialized; instead, market forces, competition, and human ingenuity have fostered the tenacious adherence to historical growth trends. Another aspect of the growth in density of devices in Integrated Circuits (ICs) has been the relentless pursuit of the fundamental causes of failure. When the true root cause of a failure is discovered it always leads to ways around the problem. Sometimes the way around the problem presents more problems, which have to be solved. The mobile Sodium problem that plagued n channel devices was finally solved by eliminating sodium and providing barriers to its movement in the gate dielectric. A lack of attention to sodium, even today, could lead to disaster in a manufacturing environment. Similarly, any number of root causes previously eliminated can rear their ugly head due to a lack of attention to the important details necessary to prevent premature failure. As another example consider the water used in manufacturing ICs. It has to be pure without contaminants (to parts per billion ($10^9$) or more depending upon the contaminant) and in truth it is cleaner than normal distilled water or saline solution used in the medical industry. As discussed earlier, particulate matter must be kept below a critical density in the air within a clean room. These particulates can introduce minute defects, which cause premature or incipient failure. The IC industry has had to continually reduce the particulates and their size in clean rooms. Without this relentless improvement in water and air in the manufacturing environment, so that they are commensurate with the ever-decreasing dimensions, the industry could not continue.

## 12.2.2 IC Chip Trends

Can feature reduction continue without limit and without incurring reliability penalties in devices? To address this question let us first examine Tables 12.1–12.3, where near-term expected trends in IC technology and metal oxide Semiconductor (MOS) devices are projected.

Accomplishing some of these projected goals will no doubt necessitate new technologies and materials. In Dynamic random-access memory (DRAM) capacitors, for example, oxide thickness limitations necessitate the use of higher dielectric constant materials such as $SrTiO_3$, lead zirconate titanate (PZT), or other ferroelectrics in order to maintain equivalent charge storage capabilities.

Several critical dimensions in these tables can be related to the failure mechanisms discussed in earlier chapters. For example, the minimum feature size, i.e., the channel length figures prominently in hot-electron effect failures. Other feature dimensions associated with interconnects, e.g., width, and gate oxide, thickness, strongly influence electromigration and dielectric breakdown failures, respectively. It is instructive to plot the dimensional changes in these features, as shown in Figures 12.1 and 12.2. In view of Moore's law (Section 1.1.3.2) it is not surprising that all dimensions, including other metallization parameters, such as metal-interconnect pitch and cross-section, are shrinking at similar semi-logarithmic rates. More problematical is the metallization aspect ratio (height to width), a quantity that is projected to increase. In 1995, this ratio was 1.5:1, but it is ever increasing to 4:1 and greater. Such changes are certain to change the film microstructure, state of stress, and prospects for electromigration. As we shall see next, the tables and figures do not, however, tell the complete story.

### *12.2.2.1 Metallization*

In the CMOS technology of 1995, there were typically a few hundred meters of 0.35-μm-wide interconnects per chip distributed among 4–5 metallization levels. By 2020, it is projected that there will be 10–20 metallization levels containing 100,000 m of interconnects that are only 0.05-μm wide. Provided that the metallization consists of copper with a typical blanket-film grain size of ~0.2 μm, the patterned interconnects will

**Table 12.1** IC technology trends

| Year of first DRAM shipment | 1995 | 1998 | 2001 | 2004 | 2007 | 2010 | Driver |
|---|---|---|---|---|---|---|---|
| Minimum feature (μm) | 0.35 | 0.25 | 0.18 | 0.13 | 0.10 | 0.07 | |
| **Memory** | | | | | | | |
| Bits/chip (DRAM) | 64M | 256M | IG | 4G | 16G | 64G | D |
| Bits/chip (SRAM) | 16M | 64M | 256M | 1G | 4G | | |
| Cost/bit at volume (microcents) | 17,000 | 7000 | 3000 | 1000 | 500 | 200 | |
| **Logic (high volume: microprocessor)** | | | | | | | |
| Logic transistors/$cm^2$ | 4M | 7M | 13M | 25M | 50M | 90M | L (μP) |
| Bits/$cm^2$ (cache SRAM) | 2M | 6M | 20M | 50M | 100M | 300M | |
| Cost/transistor at volume (millicents) | 1 | 0.5 | 0.2 | 0.1 | 0.05 | 0.02 | |
| **Logic (low volume: ASIC)** | | | | | | | |
| Transistors/$cm^2$ (auto layout) | 2M | 4M | 7M | 12M | 25M | 40M | L (A) |
| Nonrecurring engineering cost/transistor (millicents) | 0.3 | 0.1 | 0.05 | 0.03 | 0.02 | 0.01 | |
| **Number of chip I/Os** | | | | | | | |
| Chip to package (pads) high performance | 900 | 1350 | 2000 | 2600 | 3600 | 4800 | L (A) |
| **Number of package pins/balls** | | | | | | | |
| Microprocessor/controller | 512 | 512 | 512 | 512 | 800 | 1024 | A |
| ASIC (high performance) | 750 | 1100 | 1700 | 2200 | 3000 | 4000 | |
| Package cost (cents/pin) | 1.4 | 1.3 | 1.1 | 1.0 | 0.9 | 0.8 | |
| **Chip frequency (MHz)** | | | | | | | |
| On-chip clock, cost performance | 50 | 200 | 300 | 400 | 500 | 625 | μP |
| On-chip clock (high performance) | 300 | 450 | 600 | 800 | 1000 | 1100 | |
| Chip-to-board speed (high performance) | 150 | 200 | 250 | 300 | 375 | 475 | L |

| Chip size ($mm^2$) | | | | | | | |
|---|---|---|---|---|---|---|---|
| DRAM | 190 | 280 | 420 | 640 | 960 | 1400 | |
| Microprocessor | 250 | 300 | 360 | 430 | 520 | 620 | |
| ASIC | 450 | 660 | 750 | 900 | 1100 | 1400 | |
| **Maximum number wiring levels (logic)** | | | | | | | |
| On-chip | 4–5 | 5 | 5–6 | 6 | 6–7 | 7–8 | μP |
| Electrical defect density ($d/m^2$) | 240 | 160 | 140 | 120 | 100 | 25 | A |
| Minimum mask count | 18 | 20 | 20 | 22 | 22 | 24 | L |
| Cycle time days (theoretical) | 9 | 10 | 10 | 11 | 11 | 12 | L |
| **Maximum substrate diameter (mm)** | | | | | | | |
| Bulk or epitaxial or SOI wafer | 200 | 200 | 300 | 300 | 400 | 400 | D |
| **Power supply voltage (V)** | | | | | | | |
| Desktop | 3.3 | 2.5 | 1.8 | 1.5 | 1.2 | 0.9 | μP |
| Battery | 2.5 | 1.8–2.5 | 0.9–1.8 | 0.9 | 0.9 | 0.9 | A |
| **Maximum power** | | | | | | | |
| High performance with heat sink (W) | 80 | 100 | 120 | 140 | 160 | 180 | μP |
| Logic without heat sink ($W/cm^2$) | 5 | 7 | 10 | 10 | 10 | 10 | A |
| Battery (W) | 2.5 | 2.5 | 3.0 | 3.5 | 4.0 | 4.5 | L |
| **Design and test** | | | | | | | |
| Volume tester cost/pin ($K) | 3.3 | 1.7 | 1.3 | 0.7 | 0.5 | 0.4 | L |
| Number of test vectors (μP/M) | 16–32 | 16–32 | 16–32 | 8–16 | 4–8 | 4 | L |
| % IC function with BIST/DFT | 25 | 40 | 50 | 70 | 90 | 90+ | L |

SOI = silicon-on-insulator, A = ASIC (application specific IC), D = DRAM, L = logic, μP = microprocessor, BIST = built-in self-test, DFT = design for testability.
From Ref. [29].

**Table 12.2** ORTC1 Summary 2013 ORTC technology trend targets

| **Year of production** | **2013** | **2015** | **2017** | **2019** | **2021** | **2023** | **2025** | **2028** |
|---|---|---|---|---|---|---|---|---|
| Logic industry "Node Name" label | "16/14" | "10" | "7" | "5" | "3.5" | "2.5" | "1.8" | |
| Logic $^1/_2$ pitch (nm) | 40 | 32 | 25 | 20 | 16 | 13 | 10 | 7 |
| Flash $^1/_2$ pitch (2D) (nm) | 18 | 15 | 13 | 11 | 9 | 8 | 8 | 8 |
| DRAM 1/2 pitch (nm) | 28 | 24 | 20 | 17 | 14 | 12 | 10 | 7.7 |
| FinFET fin Half-pitch (new) (nm) | 30 | 24 | 19 | 15 | 12 | 9.5 | 7.5 | 5.3 |
| FinFET fin width (new) (nm) | 7.6 | 7.2 | 6.8 | 6.4 | 6.1 | 5.7 | 5.4 | 5.0 |
| 6-t SRAM cell Size ($\mu m^2$) (@60f2) | 0.096 | 0.061 | 0.038 | 0.024 | 0.015 | 0.010 | 0.0060 | 0.0030 |
| MPU/ASIC High Performance 4t NAND Gate Size ($\mu m^2$) | 0.248 | 0.157 | 0.099 | 0.062 | 0.039 | 0.025 | 0.018 | 0.009 |
| 4-input NAND Gate density ($\kappa$-gates/mm) (@155f2) | 4.03E + 03 | 6.37E + 03 | 1.01 E + 04 | 1.61 E + 04 | 2.55E + 04 | 4.05E + 04 | 6.42E + 04 | 1.28E + 05 |
| Flash generations label (bits per chip) (SLC/MLC) | 64G/128G | 128G/256G | 256G/512G | 512G/1T | 512G/1T | 1T/2T | 2T/4T | 4T/8T |
| Flash 3D number of layer targets (at relaxed poly half pitch) | 16–32 | 16–32 | 16–32 | 32–64 | 48–96 | 64–128 | 96–192 | 192–384 |
| Flash 3D layer half-pitch targets (nm) | 64 nm | 54 nm | 45 nm | 30 nm | 28 nm | 27 nm | 25 nm | 22 nm |

| DRAM generations label (bits per chip) | 4G | 8G | 8G | 16G | 32G | 32G | 32G | 32G |
|---|---|---|---|---|---|---|---|---|
| 450 mm Production high-volume manufacturing begins (100 Kwspm) | | | | 2018 | | | | |
| $V_{dd}$ (high performance, high-$V_{dd}$ transistors) | 0.86 | 0.83 | 0.80 | 0.77 | 0.74 | 0.71 | 0.68 | 0.64 |
| 1/(CV/I) (1/psec) | 1.13 | 1.53 | 1.75 | 1.97 | 2.10 | 2.29 | 2.52 | 3.17 |
| On-chip local clock MPU HP (at 4% CAGR) | 5.50 | 5.95 | 6.44 | 6.96 | 7.53 | 8.14 | 8.8 | 9.9 |
| Maximum number wiring levels (unchanged) | 13 | 13 | 14 | 14 | 15 | 15 | 16 | 17 |
| MPU high-performance (HP) Prinied Gate length (GLpr) (nm) | 28 | 22 | 18 | 14 | 11 | 9 | 7 | 5 |
| MPU high-performance physical Gate length (GLph) (nm) | 20 | 17 | 14 | 12 | 10 | 8 | 7 | 5 |
| ASIC/Low Standby power (LP) physical Gate length (nm) (GLph) | 23 | 19 | 16 | 13 | 11 | 9 | 8 | 6 |

From Ref. [29].

**Table 12.3** MOSFET device trends (historical and future)

**Historical**

| Year | 1977 | 1980 | 1983 | 1986 | 1989 | 1992 |
|---|---|---|---|---|---|---|
| Minimum feature (μm) | 3 | 2 | 1.5 | 1 | 0.7 | 0.5 |
| Gate-oxide thickness (nm) | 70 | 40 | 25 | 25 | 20 | 15 |
| Junction depth (μm) | 0.6 | 0.4 | 0.3 | 0.25 | 0.2 | 0.15 |
| Channel length (μm) | 2 | 1.5 | 1.2 | 0.9 | 0.6 | 0.5 |
| NMOS ($i_{Dsat}$ mA/μm width) @ $V_G = 5$ V | 0.1 | 0.14 | 0.23 | 0.27 | 0.36 | 0.56 |
| PMOS ($i_{Dsat}$ mA/μm width) @ $V_G = 5$ V | | 0.06 | 0.11 | 0.14 | 0.19 | 0.27 |

**Future (high-speed scenario)**

| Year | 1992 | 1995 | 1998 | 2001 | 2004 | 2007 |
|---|---|---|---|---|---|---|
| Minimum feature (μm) | 0.5 | 0.35 | 0.25 | 0.18 | 0.13 | 0.10 |
| Gate-oxide thickness (nm) | 15 | 9 | 8 | 7 | 4.5 | 4 |
| Junction depth (μm) | 0.15 | 0.15 | 0.1 | 0.08 | 0.08 | 0.07 |
| Channel length (μm) | 0.5 | 0.3 | 0.2 | 0.1 | 0.07 | |
| NMOS ($i_{Dsat}$, mA/μm) | 0.56 | 0.48 | 0.55 | 0.65 | 0.51 | 0.62 |
| PMOS ($i_{Dsat}$, mA/μm) | 0.27 | 0.22 | 0.26 | 0.32 | 0.24 | 0.32 |

**Future (low-power scenario)**

| Gate-oxide thickness (nm) | 12 | 7 | 6 | 4.5 | 4 | 4 |
|---|---|---|---|---|---|---|
| NMOS ($i_{Dsat}$, mA/μm) | 0.35 | 0.27 | 0.31 | 0.21 | 0.29 | 0.33 |
| PMOS ($i_{Dsat}$, mA/μm) | 0.16 | 0.11 | 0.14 | 0.09 | 0.13 | 0.16 |

From Ref. [29].

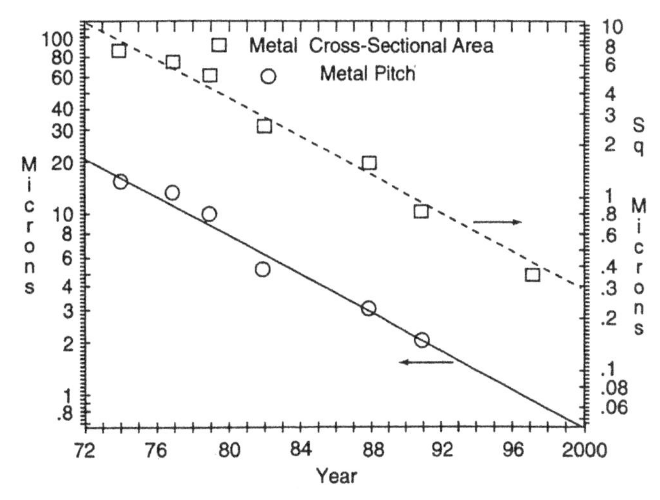


**Figure 12.1** Metallization pitch and minimum line cross-section trends. *From Ref. [30].*

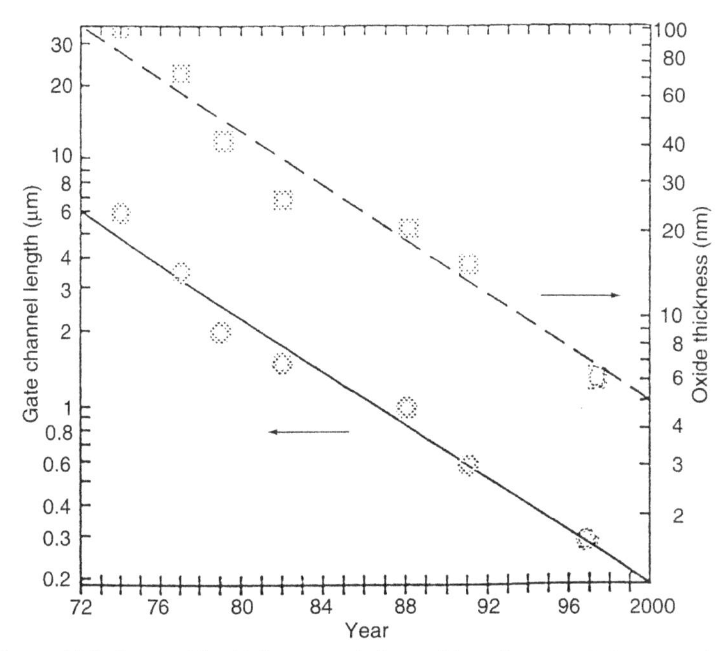


**Figure 12.2** Gate-oxide thickness and channel length. *From Refs [30] and [1].*

likely exhibit a strong bamboo morphology with grain boundaries approximately perpendicular to the stripe axis. We may then imagine a 100,000- meter-long ladder containing some $20 \times 10^{10}$ grain-boundary steps. While we know that oriented bamboo grains are electromigration resistant, grain boundaries are also the sites of degradation and stress voiding. Even with redundancy to compensate for defects, will all steps be reliable?

In addition, there is the issue of the interlevel vias and the local mass transport effects that threaten them. To deal with mass transport in general, suppose the diffusion length ($x$) for failure is the same as the interconnect width. Since $x$ is related to $(Dt)^{1/2}$, a reduction in $x$ by a factor of 5 means that an equivalent factor of 25 decrease in $Dt$ must occur to achieve the same functionality. Maintaining the same lifetime ($t$) means lowering the diffusion coefficient ($D$) by a factor of 25, a feat that necessitates a significant drop in operating temperature. Despite lower operating voltages, microprocessor power levels are projected to increase significantly, meaning higher chip temperatures and added cooling burdens.

To meet these challenges, copper was introduced as a replacement for aluminum in metallizations; a higher activation energy for diffusion is the reason Cu will electromigrate less than Al. The higher electrical conductivity of Cu is an added bonus, however, because it will reduce the RC time constant that delays transmission of electrical signals. As a result, Cu metallization is projected to increase circuit speed. Nevertheless, copper is difficult to plasma etch and there are concerns that it can be a killer contaminant and poison silicon as well. There is an additional parallel strategy to increase chip speed and that is to reduce circuit capacitance. For this reason there is an intense search for low-dielectric constant ($\varepsilon$) insulators to replace the $SiO_2$ fill ($\varepsilon = 3.9$) that now surrounds the various levels of metallization. Assorted fluoropolymers ($\varepsilon = \sim 2.2 - \sim 3$) as well as amorphous carbon ($\varepsilon = 2.6$) have been touted for such use.

Metallization issues are only one example of the trade-offs involved in implementing successive generations of devices and projecting their reliability. We have not even addressed processing complexities that an increase in the number of lithography steps implies. Lithography will have to cope with killer defects whose critical size decreases proportionately with feature geometry. For example, in 256 Mb DRAMs the killer-defect size is $\sim 0.02$ μm, but by 2020, it is projected to be well below 5 nm.

#### 12.2.2.2 Gate Oxide

Hu [1] has pointed out two goals and two constraints associated with MOS scaling and, in particular, the influence of gate-oxide dimensions. The first goal is increased transistor current needed for quicker charging and discharging of parasitic capacitances. This requires short channels and high gate-oxide fields. To achieve the second goal, reduced sizes are required to increase device-packing density and minimize capacitance as well as power needs.

These two constraints demand suppression of leakage currents when the transistor is off and a typical reliability of 20 years. The so-called "short-channel" effects that lower the transistor threshold voltage and enhance bulk punch-through are causes of enhanced leakage currents. As with former generations of oxides, hot-carrier and contact-degradation effects as well as dielectric breakdown will be the primary reliability concerns. Another source of leakage is stress-induced leakage current (SILC), which is an increase in the gate leakage current of a metal-oxide semiconductor field-effect transistor (MOSFET), due to defects in the gate oxide as a result of electrical stressing. SILC is a large attendant problem arising as a result of device miniaturization. Oxide defects causing leakage current are generated by charge trapping in the oxide. Nitridization of the gate oxide with nitrous oxide ($N_2O$) has been used to reduce the rate of charge trapping. However, this can lead to NBTI and PBTI, as discussed in Chapter 6.6.

The use of high dielectric constant ($\kappa$) $\kappa$ gate dielectrics further facilitates reduction in size of devices. This term high-$\kappa$ dielectric refers to materials which have a larger dielectric constant than silicon dioxide the historical gate dielectric. Further, tunneling occurs, at gate oxides of 30 Å, which further increases leakage current and power consumption. High-$\kappa$ gate dielectrics accomodate storing more charge in a smaller volume, thus enhancing miniaturization of devices.

Some of the materials used are hafnium silicate, zirconium silicate, hafnium dioxide, and zirconium dioxide. These materials are often deposited by atomic layer deposition.

Defects in the high-$k$ dielectric change its electrical properties. Defects can be measured by using zero bias thermally stimulated current spectroscopy.

#### 12.2.2.3 AC versus DC Reliability Effects

Inasmuch as Table 12.1 projects a substantial increase in circuit operational frequencies, we may ask what effect this will have on the reliability of metallizations and gate oxides. A comparison of mean times to failure obtained during both ac and dc testing has fortunately revealed that the ratio mean time to failure (MTTF)(ac)/MTTF(dc) is substantially greater than one [2]. This can be seen in Figure 12.3(a) and (b), where failure times are plotted versus frequency in the cases of electromigration and dielectric

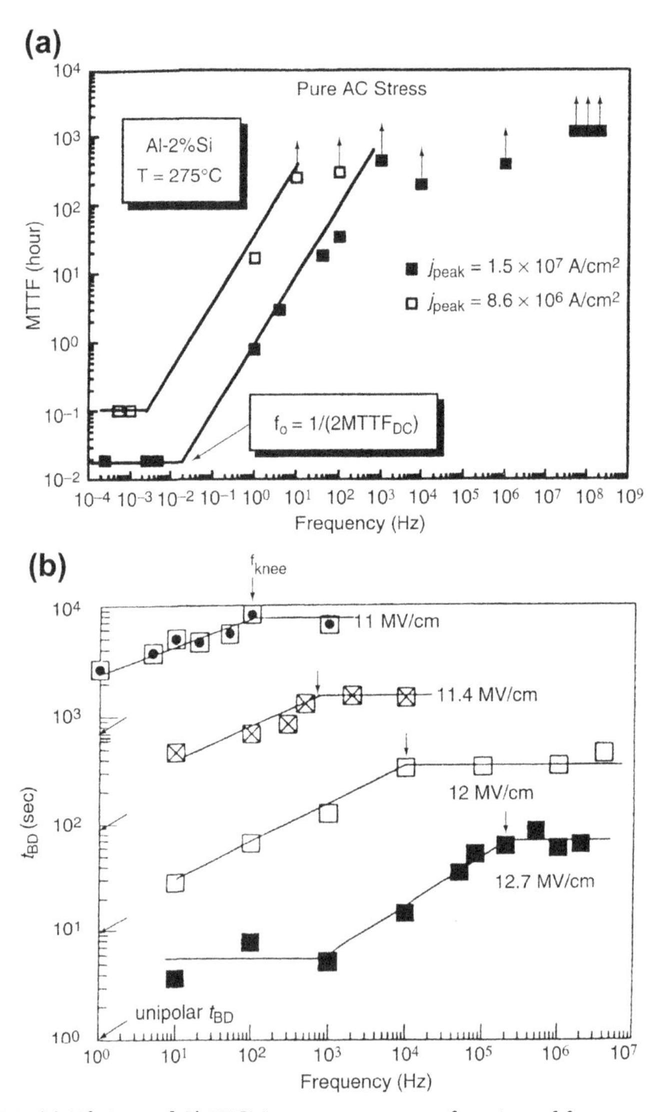


**Figure 12.3** (a) Lifetime of Al-2%Si interconnects as a function of frequency. Beyond frequency $f_o$, the conductor life is greatly extended. (b) Time to breakdown ($t_{BD}$) as a function of frequency. Note that the saturation (or "knee") frequency $f_{knee}$ decreases with decreasing electric field. *From Ref. [2].*

breakdown, respectively. The several orders of magnitude increase in interconnect lifetime relative to the dc value apparently stems from the inability to reach critical levels of defects; before vacancies can accumulate to generate void damage, the conductor heals because current reversal disperses them. Thus, the ac signals that drive transistors will not cause damage; it is the larger bus lines that carry dc current that will pose the electromigration problems.

For analogous reasons, a 10-fold increase in gate-oxide life can be expected during ac relative to dc stressing. In this case holes do not have time to drift from the anode to the oxide interior because of field reversal. Instead of generating bulk electron traps that lead to breakdown, holes are limited to producing more benign interface traps.

Hu has also shown that the ac-to-dc hot-electron lifetime ratio for damage in n- as well as p-MOSFETs exceeds unity; however, the ratio decreases inversely with frequency increase.

### 12.2.3 Contamination Trends

Two major trends appear to define future contamination problems during chip processing. The first is the dominance of equipment and processes as sources of contamination relative to people and clean rooms. In the second, airborne molecular contamination may be to the future what particle contamination was to the past. Table 12.4 lists some projected wafer surface airborne molecular contamination limits that will be tolerated.

Furthermore, contaminants are classified according to whether they are acids, bases, condensates, or dopants. Allowable molar concentrations range

**Table 12.4** Wafer surface contamination budget

| Year | 1995 | 1998 | 2001 | 2004 | 2007 | 2010 |
|---|---|---|---|---|---|---|
| Minimum feature (μm) | 0.35 | 0.25 | 0.18 | 0.13 | 0.10 | 0.07 |
| Front end processes organics (atoms of $C/cm^2 \times 10^{14}$) | 1 | 0.5 | 0.3 | 0.1 | 0.05 | 0.03 |
| Back end processes anions $F^-$, $Cl^-$, $SO_4^{2-}$, $PO_4^{3-}$ ($atoms/cm^2 \times 10^{12}$) | 1 | 1 | 1 | <1 | <1 | <1 |
| Organics (atoms of $C/cm^2 \times 10^{15}$) | 1 | 1 | 1 | <1 | <1 | <1 |
| Metals ($atoms/cm^2 \times 10^{10}$) | 5 | 2 | 1 | <1 | <1 | <1 |
| Al, Ca | 10 | 5 | 3 | 2 | <2 | <2 |

Adapted from Ref. [3].

from 1 to 10,000 parts per trillion (ppt) depending on the process step involved. Thus, prior to gate oxidation the contamination limits on the wafer are $\sim 3 \times 10^{10}$ atoms/cm$^2$ for metals, 50–13,000 ppt for acids and bases, and only 0.1–90 ppt for dopant-bearing contaminants [3].

## 12.2.4 Lithography

As can be readily appreciated, lithography is the critical enabling technology that permits feature shrinkage. Currently, optical lithography is being pushed to its limits by using shorter wavelength sources (ultraviolet light) and special techniques (phase-shift methods, spatial filtering). By such means the resolution limit for optical lithography will be extended to about 0.15 μm, corresponding to feature sizes in 1 Gb DRAMs. Beyond this, electron beam and X-ray lithographies will be necessary. Electron, or E-beam, lithography is already used in making masks and for direct writing on wafers. Comparison of the three lithography techniques is made in Figure 12.4, where a figure of merit, defined as the wafer throughput (number of 6 inch wafers per hour) divided by the product of the E-beam resolution and overlay accuracy, is plotted as a function of the design rule (feature size). Maintaining the same throughput as feature sizes shrink means order of magnitude increases in the figure of merit. Among the future lithography challenges is the extremely precise overlay positioning accuracy

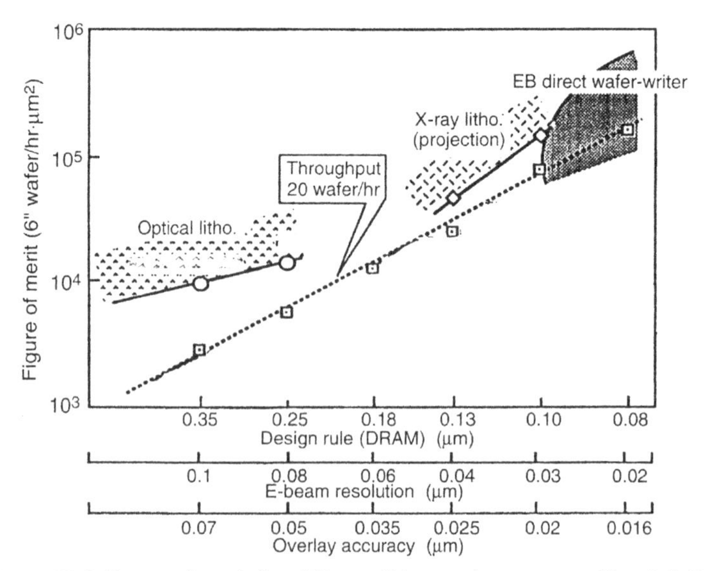


**Figure 12.4** Figure of merit for different lithography systems. *After Ref. [31].*

required to prevent feature errors. In addition, pattern edge roughness, stemming from the large dimensions of photoresist molecules, is a potential source of defects. Therefore, resists with smaller molecules will have to be formulated without sacrificing resolution.

### 12.2.5 Packaging

#### *12.2.5.1 Rent's Rule*

Before addressing packaging trends it is helpful to know first how many terminals are required to communicate between a component and the rest of the system. As the complexity of a component increases, so do the number of inputs, outputs, controls, and test points necessary for function. When the component is the IC chip, it has been found that the number of input–output terminals ($N_{I/O}$) increases as a power of the number of circuits ($C$), or alternately, the number of gates. The relationship is given by

$$N_{I/O} = bC^{p} \tag{12.1}$$

and is known as Rent's rule after the discoverer [4]. In this formula $b$ and $p$ are constants that had respective values of 2 and 2/3 in the original work. Rent's rule has been since applied to microprocessors and gate arrays and has semiquantitatively predicted the input–output terminal count required in various generations of these chips. Fortunately, fewer pins than suggested by Rent's rule have actually been required in contemporary ICs; in microprocessors values of $p$ equal to 0.3–0.4 apply, while in memory chips $N_{I/O}$ is roughly independent of $C$. Extensions of Rent's rule have been applied to terminal counts in various packaging levels such as multichip modules, cards, and boards where the number of circuits on a chip, number of chips on a card or module, and number of cards must be respectively accounted for [5].

#### *12.2.5.2 Packaging Trends*

Chip-level technology and the marketplace are the driving forces for change in electronic packaging. For example, in the growing high-volume, low-price commodity office products of personal computers, printers, and communications equipment, plastic rather than ceramic packaging appears to be favored. Despite the high reliability of laminated ceramic packages, it turns out that lower cost, high density on the circuit board, compatibility with surface-mount technology, and lower dielectric constants (for high clock speeds) provide advantages for plastic packaging. However, there will still be demand for high-end, or electrical and thermal performance-driven ceramic and multichip-module packages. McCullen et al. [6] have recently reviewed the attributes of present packages and suggested the following future trends and potential reliability issues:

1. *Package Outline.* Plastic packages will be thinner and wider and will contain higher pin counts and finer pitch, or distance between pins and leads. These packaging trends are shown in Figure 12.5 and are borne out in the projected packaging production numbers graphed in Figure 12.6. Thinner and wider packages may mean higher mechanical stresses and finer pitch leads will increase the probability of pin–pin shorts and leakage currents. Packages thinner than 1 mm have been introduced in Japan but are rare in the United States as of this writing. These ultrathin packages present a host of potential yield problems associated with wafer back-grinding and cracking, wire bonding, and handling; moisture retention and chemical trapping may also compromise long-term reliability.

2. *Lead Frames.* Multiple metal planes for improved electrical and thermal performance can be expected.

3. *Interconnection Technology.* Traditional wire bonding is limited in pitch because of ball size and placement control, as well as tool interference with adjacent bonds. Increasing lead count means finer pitch bond pads on the

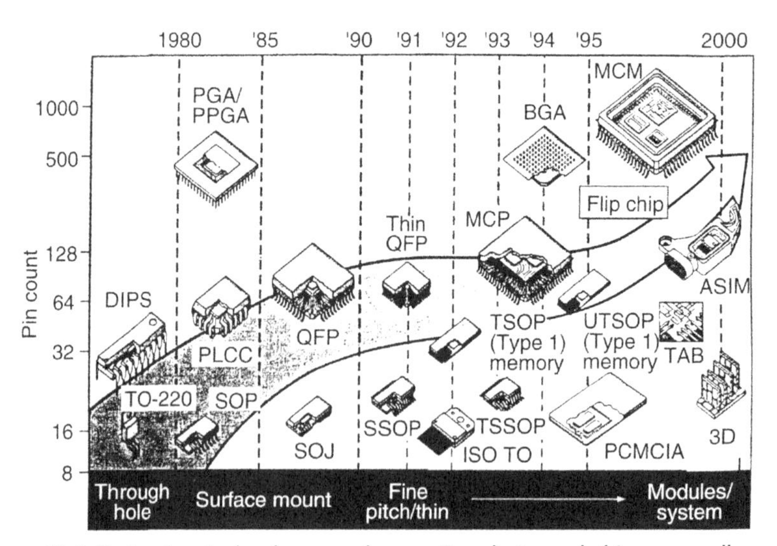


**Figure 12.5** Packaging technology road map. Trends toward thinner, smaller, and higher pin count packages are illustrated. SOP = small outline package, SOJ = small outline J (leaded package), TO = thin outline, TSOP = thin small outline package. *From Ref. [32].*

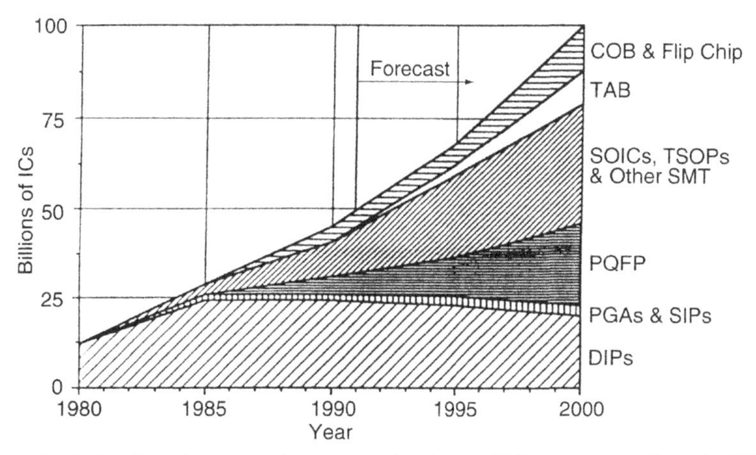


**Figure 12.6** Semiconductor-packaging projections. COB = chip on board, SOIC = small outline-integrated circuit. *From.Ref. [33] Reproduced in Ref. [34]*

die and longer wires; as a result, wire deformation and damage may be enhanced. Tape-automated bonding will be required for the very fine pitches required, but the wafer-bumping processes and bump-passivation overlap may create reliability problems.

Much greater use will be made of flip-chip technology to alleviate the limitations of wire bonding.

4. *Passivation.* Plastic packages will require more hermetic and robust chip passivations because the polymer mold compounds, which directly contact the die, are not moisture barriers. The use of silicon nitride as the moisture barrier and polyimide for mechanical protection is one strategy; another approach employs reflowable glass/nitrides.

5. *Die Attach.* As die sizes increase there will probably be greater use made of the silver-filled epoxies. These evolve minimal amounts of volatile products on curing, thus inhibiting void formation. Voids are particularly troublesome during thermal cycling and are a cause of die fracture.

6. *Circuit Board Mounting.* Surface-mount technology will grow at the expense of through-hole mounting. In 1990, each of these solder mounting technologies was employed equally, but by the end of the twentieth century, surface mounting is projected to account for 95% of package attachments. This means that packages must be capable of resisting the solder reflow temperatures (~220 °C). Fewer hygroscopic molding components will also be required. Projecting into a future driven by increased numbers of I/O contacts and shrinking board space in which to accommodate them, flip-chip and ball-grid array mounting will become correspondingly favored. Micro

ball-grid (μ-BGA) arrays with very small package footprints will no doubt gain wider acceptance.

In summary, the general trend is to more complex plastic/organic packages with larger dies using thinner and potentially more fragile passivations. These packages will have to withstand the thermomechanical shock of reflow soldering and the danger of popcorn cracking. They will also have to contend with higher temperatures as a result of increased microprocessor operating speeds. For example, newer versions processors operating at frequencies greater than 500 MHz will generate twice the heat of chips they replace. Heat sinks or fans may be required, raising questions about battery life for portable applications. Package failure mechanisms and reliability problems that were raised in earlier chapters will remain concerns and probably worsen because of the added heat burden and increasing numbers of soldered electrical connections.

## 12.3 SCALING

### 12.3.1 Introduction

Shrinking, or scaling down dimensions, is the usual route to improving device speed performance and approaching the reduced power-delay limits discussed in Section 12.4. The original device-scaling concept involved a coordinated change in dimensions, voltage, and doping while preserving a constant electric field distribution and power density [7,8]. Reducing physical dimensions generally means changes in resistance, capacitance, the voltages applied to them, the currents that flow as a result, and even the temperatures generated. After introducing scaling concepts here, they will be modified slightly and included in failure models discussed in Section 12.3.2, in order to expose their reliability implications.

Following Dennard et al. [8], M.H. Woods [9], and H.J.M. Veendrick [10], we start by reducing device dimensions ($L$) listed in Table 12.5 by a factor of $1/K$ with $K > 1$. In addition, the capacitance ($C$), voltage ($V$), and current ($i$) are reduced by $1/K$, but the doping level is increased by $K$. As a result, the area is reduced by $1/K \times 1/K$, or $K^{-2}$, and the delay time constant ($t_d = RC = VC/i$) is reduced by $(1/K \times 1/K)/(1/K) = 1/K$. Proceeding in this manner, the power-delay product ($V \times i \times t_d$) is improved because it is reduced by a factor of $1/K \times 1/K \times 1/K$, or $K^{-3}$. The power density ($V \times i/L^2$) remains constant with scaling because $1/K \times 1/K/K^{-2} = 1$. Scaling factors for interconnections and other electrical parameters are also listed in Table 12.5.

**Table 12.5** Impact of scaling on device and electronic parameters (shrink factor $K > 1$)

| Electrical parameter | Scaled voltage | Unscaled voltage |
|---|---|---|
| Device dimensions ($L$, oxide thickness, channel length, junction depth, contact dimension, etc) | $1/K$ | $1/K$ |
| Device area ($A$) | $1/K^2$ | $1/K^2$ |
| Supply voltage ($V$) | $1/K$ | 1 |
| Device current ($i$) | $1/K$ | $K \rightarrow 1$ |
| Capacitance ($C = \varepsilon A/L$) | $1/K$ | $1/K$ |
| Doping concentration | $K$ | |
| Metal current density | $K$–$K^{1.5}$ | $K^3$ |
| Contact area current density | $K$–$K^{1.5}$ | $K^3$ |
| Stored oxide charge | $1/K^2$ | $1/K$ |
| Oxide field ($E_{ox}$) | 1 | $K$ |
| Gate delay ($t_d$) | $1/K$–$1/K^{1.5}$ | $1/K^2$ |
| Power dissipation per gate ($P$) | $1/K^{1.5}$–$1/K^2$ | $K$ |
| Power delay ($P \times t_d$) | $1/K^3$ | $1/K$ |
| MOS Transconductance | | $K$ |
| Threshold voltage | | 1 |
| Drain-source current (in both linear and saturation regions) | | $K$ |

After Dennard et al. [8], M.H. Woods [9], and H.J.M. Veendrick [10].

In new generations of IC chips, the operating voltage is often not reduced, so that electric fields and power densities essentially increase. Assuming constant operating voltages, a different set of scaling factors is applicable, and these are entered in Table12.5. Some scaling factors for MOSFETs under these conditions are also included. In what follows, the impact of scaling on several IC failure mechanisms treated earlier in the book is considered.

## 12.3.2 Implications of Device Scaling

### *12.3.2.1 Electromigration*

The impact of scaling on electromigration reliability can be assessed by starting with the Black equation for the $MTTF_i$ (Eqn (5.31)).

$$MTTF = \frac{B^{-1} exp \frac{E_e}{kT}}{j_e^n}. \tag{12.2}$$

It is assumed that $B$ and $E_e$ depend only on interconnect microstructure and not on scaling. For a worst-case, constant-voltage assumption, the current increases linearly, while cross-sectional dimensions decrease as the square of the scale factor; therefore, the current density scales as $K^3$. But $n = 2$, so that an overall reduction in MTTF by a factor of $K^6$ is predicted [11]. Such a disastrous lifetime dependence on shrinking conductors is avoided by keeping $j_e$ constant.

For contacts subjected to electromigration damage at thin sidewalls, the time to failure is given by a similar expression,

$$\text{MTTF} = \frac{C^{-1}L^2}{j_e^n} exp \frac{E_e}{kT}, \tag{12.3}$$

where $C$ is a constant independent of scaling and $L^2$ is proportional to the contact area. Because this area shrinks as $K^2$, an overall decrease in lifetime by a factor of $K^8$ may be anticipated. In both cases even higher scaling factors are possible if increased temperature operation is considered.

#### 12.3.2.2 Oxide Breakdown

Earlier, in Chapter 6, we considered both the reciprocal field and linear electric field models for dielectric breakdown. With constant electric field scaling there is no predicted adverse effect on reliability, because the expressions for MTTF under scaled (s) and unscaled (u) conditions are identical. However, for the worst-case scenario of constant oxide voltage $V_o$, the reciprocal field theory (Eqn (6.10)) yields

$$\frac{\text{MTTF(s)}}{\text{MTTF(u)}} = exp\left[\frac{G_R X_o(\frac{1}{K} - 1)}{V_o}\right], \tag{12.4}$$

where $X_o$ is the oxide thickness and $G_R$ was defined earlier. In contrast, the linear field model suggests that dielectric breakdown times are given by

$$\frac{\text{MTTF(s)}}{\text{MTTF(u)}} = exp\left[\frac{G_L V_o(K - 1)}{X_o}\right], \tag{12.5}$$

where $G_L$ was defined earlier, in (Eqn (6.12)). In either case, shrinkage of the gate-oxide thickness results in markedly shorter lifetimes. One consequence of scaling arises in screening. Weak oxides are normally screened through high-voltage stress for short time. As scaling has reduced oxide thicknesses, operating electric fields come closer to those for intrinsic wear-out. This reduces field acceleration factors, making the screening of latent defects more difficult.

### 12.3.2.3 Hot-Carrier Effects

We noted in Section 6.4.4 that the probability of injecting hot carriers from Si into $SiO_2$ is exponentially enhanced with increasing lateral electric field in the MOS channel. For short-channel lengths ($L_c$) the electric field varies inversely with $L_c$. Hot-electron degradation thus scales in a manner similar to Eqn (12.4), or

$$\frac{\mathrm{MTTF(s)}}{\mathrm{MTTF(u)}} = exp\left[\frac{aL_c(\frac{1}{K} - 1)}{V}\right], \tag{12.6}$$

where $a$ is a constant. Scaling down power-supply voltages is an obvious way to reduce the threat to reliability.

### 12.3.2.4 Soft Errors

In Section 7.6.6.4 the critical charge ($Q_c$) required to produce upsets was shown to decrease as device feature dimensions shrank (Eqn (7.46)), thus enhancing the soft error rate (SER). Woods [11] has suggested that SER varies either exponentially or directly with $Q_c$ for large or small charge levels, respectively. This would mean larger numbers of soft errors with respective scaling according to SER = exp ($K^2$) or SER = $K^2$.

Over the past 30 years a many-orders-of-magnitude decrease in reliability margins has been predicted (Figure 12.7) based on the effects of

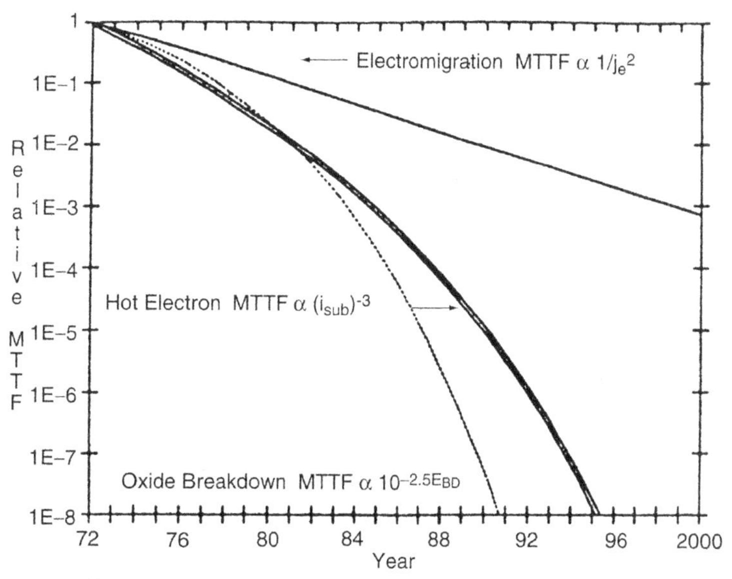


**Figure 12.7** Effects of device and metallization scaling on mean failure times. *From Ref. [30].*

scaling devices and metallizations. Astonishingly, in apparent suspension of physical laws, the infant mortality and long-term failure rates have actually declined. By the end of the twentieth century, a 10 FIT goal for ICs is projected. Perhaps when we take notice of the advances in the enabling processing technologies it is not so astonishing.

## 12.4 FUNDAMENTAL LIMITS

### 12.4.1 Introduction

Moore's law, introduced in Section 1.1.3.2 has an important corollary [12,13], which asserts "future opportunities to achieve multibillion transistor chips or gigascale integration in the twenty-first century will be governed by a hierarchy of limits." The levels of this hierarchy can be categorized as (1) physical, (2) material, (3) device, (4) circuit, (5) packaging [14], and (6) system. Each level, in turn, is defined by two sets of limitations, one theoretical and the second practical. The fundamental laws of physics govern theoretical limits, while manufacturing constraints and the marketplace serve to define practical limits [7,15–17].

We shall deal only with the first three of these limits to acquire a pedagogical flavor of the arguments and calculations involved. The problem of limits is usefully approached by considering the fundamental binary switching step. Electrical power, $P$, is transferred in the switching transition process and it occurs in a transit or propagation delay time of $t_d$. Actual device behavior can be displayed on a plot of $P$ versus $t_d$. Later, these characteristics will be compared to the limits imposed by the physics, which we now attempt to quantify.

### 12.4.2 Physical Limits

The three most fundamental limits are those derived from thermodynamics, quantum mechanics, and electromagnetic theory. In the first of these let us assume that due to thermal fluctuations, semiconductor devices may switch from logic 1 to logic 0 or vice versa. In the process, thermal energy $kT$ can be equated to equivalent electrical energy $qV$, where $q$ is the electronic charge and $V$ is the voltage. Because the thermal limit of switching energy $E_s$ is $kT$, we require that the operating energy be several times $kT$; otherwise thermally induced switching will probably create errors. In fact, the probability of thermal noise ($P_n$) varies as the Boltzmann factor,

$$P_n \approx exp\left[-\frac{E_s}{kT}\right], \tag{12.7}$$

and decreases exponentially as $E_s/kT$ increases. Typically, we require that $E_s \geq 4\ kT$. Therefore, at 300 K, $E_s \geq 1.66 \times 10^{-20}$ J, or 0.104 eV, which is equivalent to one electron moving through a potential of 0.104 V. This energy expenditure is some $10^7$ times smaller than current practice, where higher voltages are applied and many more electrons are involved in switching. We may, therefore, view the energy involved in one electron switching as a *thermodynamic* limit for this transition.

The *quantum-mechanical* limit stems from the Heisenberg uncertainty principle. Common expressions for this principle state that uncertainties ($\Delta$) in the momentum ($p$) and position ($x$), or equivalently, the energy ($E$) and time ($t$) of subatomic particles and the events they participate in are related by

$$(\Delta p)(\Delta x) > \frac{h}{2\pi} \quad \text{and} \quad (\Delta E)(\Delta t) > \frac{h}{2\pi}. \tag{12.8}$$

For our purposes the last inequality can be expressed by $E_s = h/(2\pi t_d)$, where $h$ is Planck's constant. The average power transfer during a switching transition involving a single electron wave packet is thus limited in time by the fundamental expression

$$P = \frac{h}{2\pi (t_d)^2}, \tag{12.9}$$

which is alternatively written in the form, $\log P = \log h/2\pi - 2\log t_d$.

*Electromagnetic* limits are based on the fact that the velocity ($v$) of pulse propagation in an interconnect of length $L$ can never exceed the speed of light ($c$), or

$$v = \frac{L}{t_d} < c, \tag{12.10}$$

where $t_d$ is the conductor transit time.

### 12.4.3 Material Limits

At the second level of hierarchy there are material limits that are independent of size and geometry of device structures. For semiconductors the key material limits are governed by carrier mobility, carrier saturation velocity ($v_s$), self-ionizing electric field (breakdown) strength ($E_b$), and thermal conductivity. Of these properties, only $v_s$ and $E_b$ will play a role in our calculation of material limits. Let us now consider a cube of semiconductor of dimensions $\Delta a$, acted on by voltage $V$ to produce $E_b$, where $V = E_b \times \Delta a$. At this critical electric field, the transit time $t_d = \Delta a/v_s$. If we assume that the minimum voltage applied corresponds to the equivalent thermal energy, i.e., $V = kT/q$, then

$$t_d = \frac{kT}{qE_b v_s}. \tag{12.11}$$

Substituting values for silicon, $E_b = 5 \times 10^5$ V/cm and $v_s = 1 \times 10^7$ cm/s, the minimum transit time is $t_d \approx 5 \times 10^{-15}$ s at room temperature. Since values of $E_b$ and $v_s$ in gallium arsenide (GaAs) are comparable to those in Si, we may expect similar transit times.

### 12.4.4 Device Limits

At the next hierarchical level are devices whose limits are independent of particular circuit configuration. The most important device in microelectronics is the MOSFET and it is limited most critically by the minimum effective channel length ($L_{min}$). At the outset of a switching transition, the relevant energy ($E$) is effectively stored on the gate of an MOS capacitor, whose capacitance per unit area corresponding to $L_{min}$ is $C_o$. Making use of the expression for the energy stored in a capacitor, the switching-energy limit is then given by

$$E = Pt_d = \frac{1}{2} C_o (L_{min})^2 V^2, \tag{12.12}$$

where $V$ is the supply voltage. The minimum value of $t_d$ is the channel transit time, which in turn is given by

$$t_d(min) = \frac{L_{min}}{v_s}, \tag{12.13}$$

where $v_s$ is the electron saturation velocity. Furthermore, it is assumed that $C_o = \varepsilon_{ox}/d_{ox}$, where $\varepsilon_{ox}$ and $d_{ox}$ are the dielectric constant and thickness of the oxide, respectively, and $d_{ox}$ is some fraction $f$ of the channel length; i.e., $d_{ox} = fL_{min}$. Combining Eqns (12.12) and (12.13) plus accompanying definitions, the limiting power-delay time characteristics of MOSFETs are given by

$$P = 1/2\varepsilon_{ox} f V^2 v_s^3 t_d^2. \tag{12.14}$$

We now have sufficient information to sketch the boundary limits for power dissipation versus propagation delay time of silicon devices.

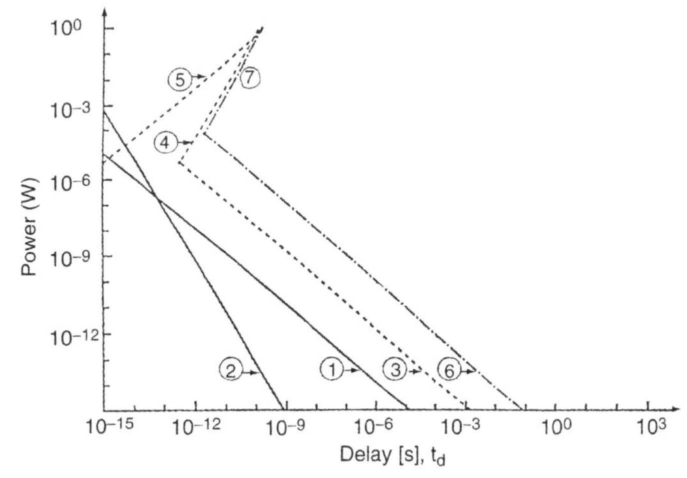


**Figure 12.8** Average power ($P$) per binary switching operation versus transition interval ($t_d$) for fundamental limits. 1. Thermodynamic limit, 2. Quantum limit, 3, 4, 5. Material limits, 6, 7. Silicon device limits. *From Refs. [12], [15].*

This important representation is shown in Figure 12.8, where the thermodynamic, quantum, material, and device limits suggested above by Meindl are plotted. It should be noted that the product of $P$ and $t_d$ represents the energy dissipated. Various magnitudes of energy are therefore indicated by lines of slopes equal to $-1$. In principle, devices that dissipate high power in short times, or low power for long times, lie at the extremes of one of these isoenergy lines.

Power-delay time characteristics for actual devices are plotted in Figure 12.9 and lie to the right and above the lines bounding the regions that represent behavior forbidden by intrinsic physical limits. In general, optical as well as superconducting Josephson devices dissipate less energy and exhibit faster switching speeds than silicon devices [18]. In the range from $10^{-11}$ to $10^{-5}$ s, semiconductor electronics dominates. Within semiconductor categories, MOS logic exhibits lower power-delay products than the assorted bipolar logic families. For times less than $10^{-11}$ s, optical devices have no competition.

There are still gaps between actual device behavior and physical limits, and these represent potential room for improvement. Much of the difference is due to the practical limitations of manufacturing and economics. Limits pose challenges and launch exciting quests to reach them; the question is how much improvement is possible. New devices, materials, and processing will probably be necessary, raising new reliability challenges.

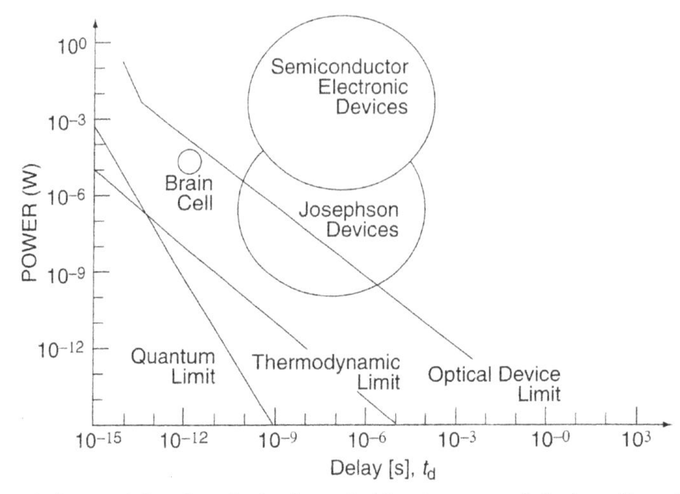


**Figure 12.9** Power-delay time limits for switching in assorted devices. *From Refs. [15], [18].*

## 12.4.5 Circuit Limits

At the next hierarchical levels are circuits whose limits are dependent on particular circuit configuration. The most important circuit limitation in microelectronics is the RC time constant. RC ultimately has a limitation, which depends on two material properties *k* (dielectric constant) and rho (resistivity) or sigma (conductivity). How fast a response we expect depends upon this time constant:

R = rho L/A and C = *k* e0 A/t and for an idealized RC element we have finally that RC = k e0 rho = k e0/sigma.

This simplistic exercise leads to the expected result that dielectric constant and conductivity or resistivity as embodied in the materials used also limit the circuit performance. Although expected and usually assumed, we see at the most fundamental level the materials used ultimately limit circuit performance.

## 12.5 IMPROVING RELIABILITY

Rather than viewing reliability through the lens of mathematics and the statistical base that flows from Chapter 4, we now have the perspective of failure physics (Chapters 5–11) and the constraints imposed by device-processing trends and limits. From this admixture of viewpoints, practical strategies to improve reliability have emerged. Two such approaches are addressed in this section. One stresses failure prediction from the standpoint of the design process and the other attempts to build in reliability (BIR). Considering both is an appropriate way to end this book. But first let us consider the subject of reliability growth, the goal that these two strategies hope to achieve.

### 12.5.1 Reliability Growth

Experience shows that during development and production of electronic products there is a growth of reliability [19]. This is normally attributed to the construction and testing of product prototypes. Information on failure rates and mechanisms in these prototypes are then used to modify the design of the next version, which in turn is tested in the next iteration. From one prototype cycle to the next, the reliability, as measured by the instantaneous $MTTF_i$, increases or grows as bugs are worked out. In time, as the learning curve is traversed, values for $MTTF_i$ eventually reach the saturation value of the predicted life. In Section 4.5.3 it was suggested that the learning curve follows an Avrami-like dependence, which like Weibull behavior is S-shaped when the CDF is plotted versus log time. Thus, at first reliability growth is slow, then it accelerates, and, finally, it saturates as there is little room for improvement.

It is common to model reliability growth of prototypes through the use of a Duane plot [20]. This consists of a log–log plot of $N(t)/t$ versus time ($t$), where $N(t)$ is the number of failures from the initiation of testing through time $t$. Not too surprisingly, a straight line usually results because making log–log plots is a good way to linearize sets of data points. Therefore,

$$ln\left[\frac{N(t)}{t}\right] = a - blnt, \tag{12.15}$$

where $a$ and $b$ are constants. Rearranging terms,

$$N(t) = \left[\exp^{a}\right]t^{1-b}. \tag{12.16}$$

By definition, $d[N(t)]/dt = \lambda(t)$, where $\lambda$ is the failure rate and $MTTF_i = 1/\lambda(t)$. Therefore, differentiation yields

$$\lambda(t) = (1 - b)e^{a}t^{-b} \tag{12.17}$$

and

$$MTTF_i = \frac{t^{b}}{(1 - b)e^{a}}. \tag{12.18}$$

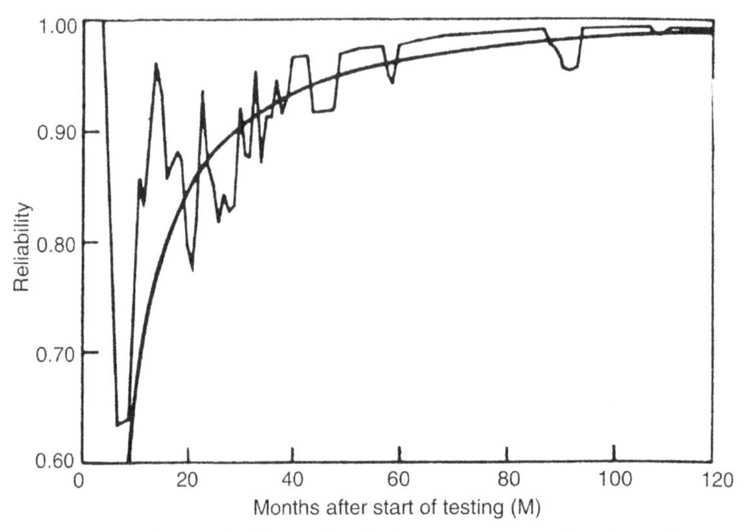


**Figure 12.10** Growth in reliability of a flight system as a function of time. *From Ref. [20].*

Quite often, $b$ is about 0.5, so that product life is predicted to increase at a parabolic rate. Amid much scatter, the data for a flight system shown in Figure 12.10 reveal the jagged growth in reliability [21] that appears to resemble Avrami-like behavior.

## 12.5.2 Failure Prediction: Stressor–Susceptibility Interactions

The topic we now address outwardly resembles load–strength interference, a subject introduced in Section 1.3.5. Brombacher and his coworkers [22,23] have pioneered a semiquantitative physical approach to predicting failure in devices based on their susceptibility to applied stressors. The latter are what we have previously called stresses. Examples of stressors include current density, temperature, dissipated power, electric field, moisture, voltage, stored charge, speed of current charging or discharging, and internal or applied mechanical stress. Failure susceptibility to one or more stressors can be catastrophically abrupt or more forgiving and gradual. Probability density distributions, for both stressor and susceptibilities are then generated as a function of the stressor magnitude. Depending on the extent of overlap of these distributions, there is a corresponding probability of failure.

To illustrate the method, consider the problem of metal burnout in a fuse. Hot spot temperatures generated as a function of current and ambient

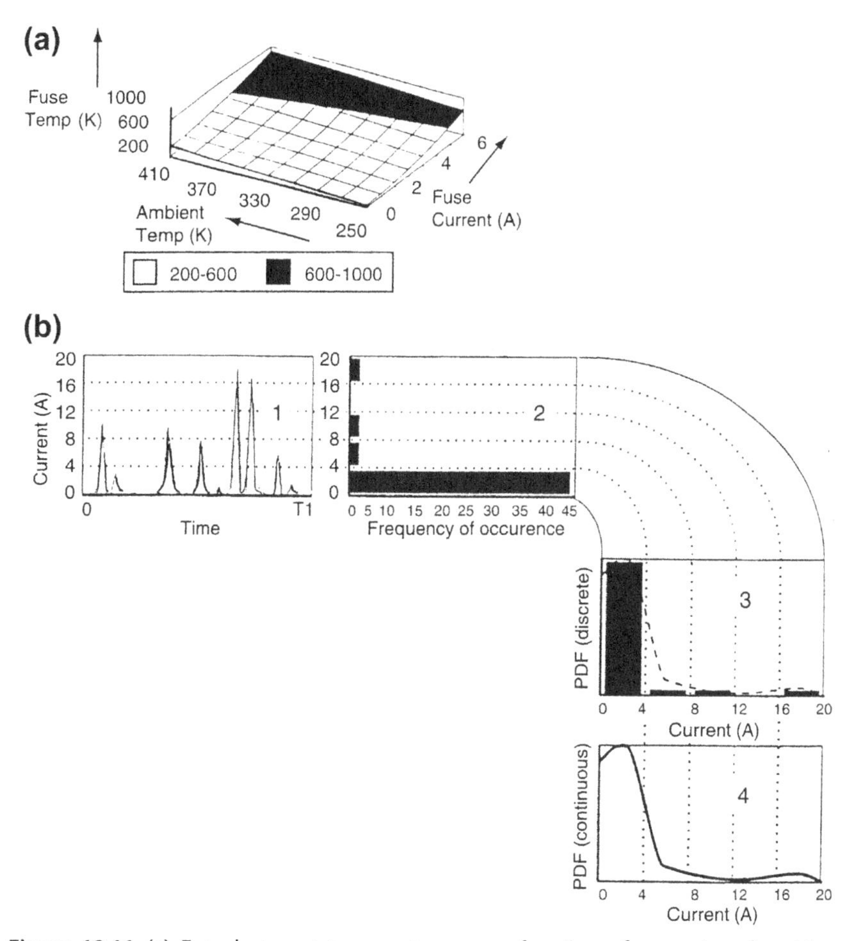


**Figure 12.11** (a) Fuse hot spot temperatures as a function of current and ambient temperature stressors. (b) Translating time signals to stressor probability density functions. *From Ref. [23].*

temperature stressors are depicted in Figure 12.11(a), where the melting temperature shown shaded is assumed to be 600 K. The current-time signal that is derived from simulation or actual hardware measurements is then converted into a stressor probability density function through the self-explanatory steps shown in Figure 12.11(b). Similarly, a susceptibility density is generated, perhaps by the testing and statistical methods used in Section 4.2.1. Both probability density functions are plotted in Figure 12.12. The circuit designer would then avoid those values of current where the stressor and susceptibility density functions overlap.

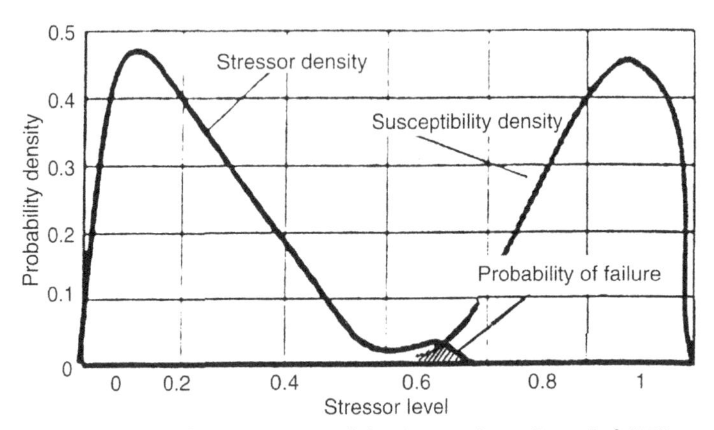


**Figure 12.12** Stressor susceptibility interactions. *From Ref. [23].*

While this computer-aided integrated reliability optimization approach is still experimental, it shows promise of aiding the design process by analyzing potential quality and reliability problems at an early stage of development.

## 12.5.3 Building-In Reliability

This last subject of the book has been raised for well over one-fourth of a century by all sectors of the manufacturing, consumer, and government communities concerned with reliability [24–26]. Like motherhood and apple pie, everyone likes the idea of building greater reliability into products, but it is not always clear how to achieve the goal. Incremental improvement in the reliability of products has always occurred, as evidenced by Figures 12.1–12.7.

Designing in reliability should certainly be the first step in the process. This means, for example, that devices and circuits should be designed to eliminate the possibility of latchup and provide protection against electrostatic discharge events. One of the useful tools in circuit design is software that locates hot spots on the IC chip [27]. In the case of GaAs, which has a low thermal conductivity, the improper layout of devices can result in excessive local heating and attendant reliability problems. To address problems of this kind, the software first calculates device self-heating and then the temperature rise in neighboring regions. If the latter is excessive, devices are repositioned to minimize the temperature.

During manufacture the issue faced is that of which philosophical approach should be adopted to ensure continued increased product reliability. Should we follow the traditional path of screening and accelerated stress testing of finished products? Or is the route to reliability improvement a reduction in defects and process variation through the use of in-line monitors? It is worthwhile describing the attributes of each approach so that the preferred path to future reliability can be intelligently chosen.

The traditional approach exemplified by the 1970s through the early 1990s has focused on time to failure. Embodied in this philosophy is measuring failure rates, and this is accomplished through infant mortality screening, accelerated testing, burn-in, statistical modeling, and use of acceleration factors. The mind-set is to measure a product failure rate after the fact and compare it to some failure rate goal. If the goal is met, the product is shipped. There is no incentive to improve product processes unless the failure rate exceeds the set goals. Overall, the approach is to test in reliability and it is reactive in nature; success means predicting failure rather than preventing it. As with all traditions, there is an ingrained acceptance of this approach on the part of the manufacturer and vendor because of past successes.

In contrast, the newer, more proactive approach attempts to BIR by anticipating, identifying, monitoring, and controlling the critical parameters of manufacturing processes. Rather than time to failure, the focus is on why the failure, and instead of end-of-line tests, the emphasis is on continuing in-line monitoring to reduce processing variation and defects. Sputtering of thin-film metallizations has been selected as a model process for executing the BIR approach [26], as suggested by Figure 12.13. This self-explanatory road map to improved electromigration reliability, not unexpectedly, stresses careful monitoring and control of all aspects of the deposition process. Examples of monitoring are the use of residual gas analyzers, particulate detectors, resistivity measurements, wafer-level electromigration tests (e.g., Standard water level electromigration accelerated test), and microscopy on production lines. In fact, any nondestructive test or measurement method capable of distinguishing chemical and structural differences at unprecedented limits of resolution is potentially useful as a defect or process monitor. Very practical concerns are driving this proactive approach. With the very reliable products now being manufactured, the number of units that must be sampled to measure the low failure rates has risen astronomically. Assuring a 10 FIT rather than 1000 FIT failure rate goal may mean a 100-fold increase in samples under test, a procedure that is normally prohibitively expensive. Furthermore, failure analysis on high-density chips is increasingly difficult.

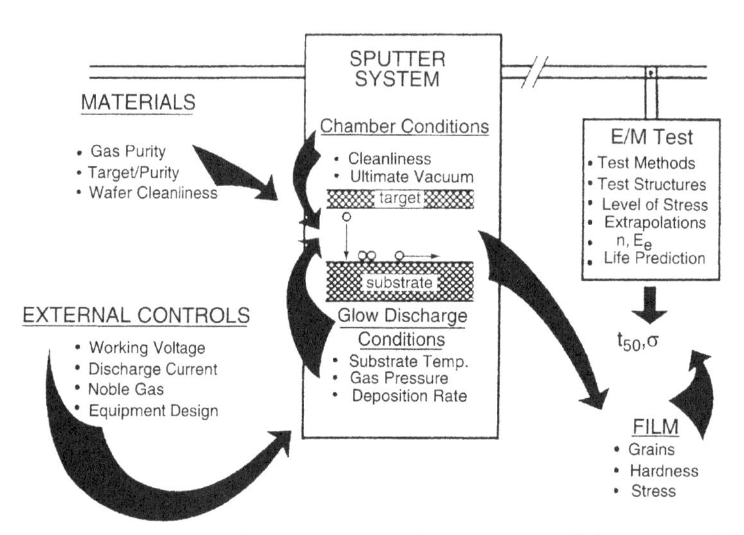


**Figure 12.13** Building-in reliability to counter electromigration failures. *From Ref. [26].*

The economic implications of the two philosophies are illustrated in Figure 12.14. Reliability improvement by the use of screens and end-of-line testing is relatively cheap if the failure rate is high, but expensive when failure rates are low. Just the reverse is true if the route to improving reliability involves reducing process variations and defects. While it is hard to argue with the proactive approach, screening is probably the only feasible route to achieving reliability in certain devices like lasers. In all cases field

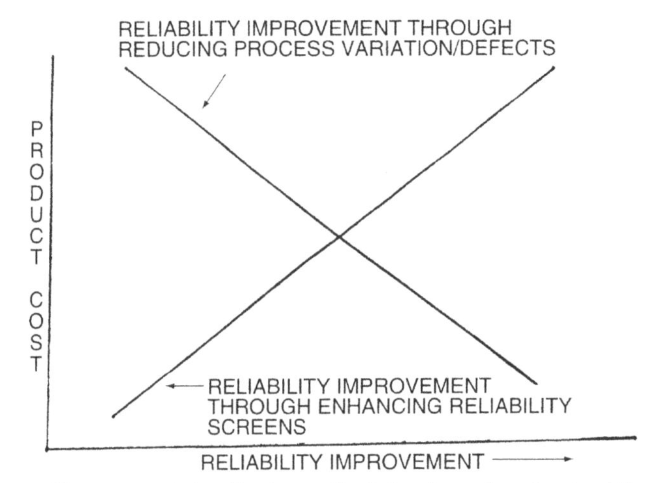


**Figure 12.14** The economic implications of building-in and testing-in philosophies for improving reliability. *From Ref. [30].*

performance information should be folded into the process of building-in reliability.

As R.A. Evans [28] has sagely observed, "Characterization and control of product and process is what reliability and quality are all about." Even so, making reliable products is one thing; proving that they have the desired reliability is another issue.

Finally, it must be remembered that the earlier in the R&D project that incipient failures are discovered the more time there is available to remedy the problem, which begs the questions: How much (initial characterization, early reliability testing to failure, determination of true root cause of failure (not assumed), confirmation that problems do not occur at application conditions, process control, proof that root causes of failure have been mitigated in the final process for the product) is necessary? The answering and mastery of these questions is the key to successful and reliable products. Solving these issues at the fundamental level defines success. The challenge is, elegantly using the fundamental laws of the universe to purpose. In truth the IC industry has done this since the 1950s. Rather than accept the limitations presented by nature, the ingenuity of man has usually found another, often, better way.

## EXERCISES

**12.1.** What average rate of annual increase in transistor count has occurred over the 40 years from 1957 to 1997?

**12.2.** The Semiconductor Industry Association road map for silicon wafers includes the following projections:

| Year | 1995 | 1998 | 2001 | 2004 | 2007 | 2010 |
|---|---|---|---|---|---|---|
| Minimum feature (μm) | 0.35 | 0.25 | 0.18 | 0.13 | 0.10 | 0.07 |
| Wafer diameter (mm) | 200 | 200 | 300 | 300 | 400 | 400 |
| Surface Fe, Ni, Cu, Zn, Na ($10^{10}$ at/cm$^2$) | 5 | 2.5 | 1 | 0.5 | 0.25 | TBD |
| Total bulk Fe ($10^{11}$ at/cm$^2$) | 1 | 0.5 | 0.2 | 0.1 | 0.05 | TBD |
| Oxygen (ppma) | 24 | 23 | 23 | 22 | 22 | TBD |
| Bulk microdefects (def/cm$^2$) | 5000 | 1000 | 500 | 100 | 100 | 100 |
| Oxygen-induced stacking faults (def/cm$^2$) | 20 | 20 | 5 | 1 | 1 | 1 |
| Particles per unit area (#/m$^2$) | 1700 | 1300 | 750 | 550 | 275 | 150 |

**a.** What accounts for the roughly constant projected oxygen concentration?

**b.** Why is it important to minimize the presence of the transition metals?

**c.** If the oxygen concentration remains constant, what factors will serve to reduce the bulk microdefect and oxygen-induced stacking fault densities?

**d.** Suppose microprocessor chips have the same killer-defect densities as noted above. Predict whether yields for these chips will increase or decrease with time from 1995 to 2010.

**12.3.** The memory cell area of DRAMs has declined exponentially with time; e.g, 34 $\mu m^2$, 1 Mb in 1985; 1.6 $\mu m^2$, 64 Mb in 1994; and 0.6 $\mu m^2$, 256 Mb in 1997. Concurrently, the DRAM die area has increased at a greater than linear rate; e.g., 60 $mm^2$ in 1985; 200 $mm^2$ in 1994; and 300 $mm^2$ in 1997. Plot these trends as a function of time. Explain the variation in die size with time.

**12.4.** How do you explain the paradox of decreasing reliability margins with scaling and a decreasing failure rate of ICs?

**12.5.** Show by means of sketches that as device densities and number of pads increase and package dimensions shrink, the length of bond wires must increase.

**12.6.** The total number of pins on the Pentium Pro (P6) microprocessor is 387 instead of the 900 predicted by an overly conservative version of Rent's rule. If each pin is 99.99% reliable, what is the relative difference in reliability between these two pin counts?

**12.7.** Suppose the failure rate of the fight system in Figure 12.10 is given by $\lambda(t) = 0.5\ (\exp a)t^{-0.5}$.

**a.** Derive an expression for the reliability dependence on time.

**b.** From the reliability function and Figure 12.10 calculate the value of $a$.

**12.8.** Develop a formula for the dependence of Al-2%Cu interconnects life on ac frequency and current density using data plotted in Figure 12.3(a). Does the low-frequency MTTF data obey the Black-equation dependence on current density?

**12.9.** Consider the dependence of oxide-breakdown times on frequency shown in Figure 12.3(b).

**a.** Which model—the reciprocal electric field or linear electric field—better describes breakdown times at a frequency of 1 Hz?

**b.** Which model—the reciprocal electric field or linear electric field—better describes breakdown times at a frequency of $10^6$ Hz?

**c.** Can you account for the decrease in saturation (knee) frequency with decreasing electric field?

**12.10.** Sketch the cross-sectional view of a field-effect transistor, e.g., Figure 2.16(a), with source and drain contact metals in place. Resketch the transistor dimensions under constant-field scaling conditions where features are reduced by a constant factor.

**12.11.** The MOS transistor gate-oxide thickness and voltage are projected to be 4 nm and 0.9 V, respectively, in 2007. Under these conditions the electron tunneling current density through the oxide has been measured to be roughly $10^{-9}$ A/cm$^2$. Speculate on the implications of the above with respect to the reliability of such gate oxides.

**12.12.** Suppose a CMOS device whose features are 0.1 μm in size switches with energies of $5 \times 10^{-17}$ J. Locate the position of this device on Figure 12.9.

**12.13.** There are important needs for new dielectric materials for use in DRAM storage capacitors and as inter-(metallization) layer filling.

**a.** In each case explain whether a higher or lower dielectric constant material, relative to that currently used, is desired.

**b.** Corrosion of Al–Cu metallization in contact with dielectrics is a potential reliability problem. How would you predict whether particular metallization–dielectric pairs are reactive? What can be done to prevent reaction?

## REFERENCES


[1] C. Hu, Semicond. Int. (June 1994) 105.
[2] C. Hu, Microelectron. Reliab. 16 (1996) 1611.
[3] D. Kinkead, Semicond. Int. (June 1996) 231.
[4] C. Radke, IEEE Proc. Des. Automation Conf. (1969) 257.
[5] D.P. Seraphim, A.L. Jones, in: D.P. Seraphim, R. Lasky, C.-Y. Li (Eds.), Principles of Electronic Packaging, McGraw-Hill, New York, 1989.
[6] J. McCullen, T.M. Moore, S.V. Golwalker, Tutorial Notes, IEEE Int. Reliab. Phys. Symp. (1996), 7.1.
[7] V.L. Rideout, in: N. Einspruch (Ed.), VLSI Electronics Microstructure Science, Vol 7, Academic Press, Orlando, 1983.
[8] R.H. Dennard, F.H. Gaensslen, H.N. Yu, V.L. Rideout, E. Bassous, A. LeBlanc, IEEE J. Solid State Circuits SC-9 (1974) 256.
[9] M.H. Woods, Proc. IEEE 74 (12) (1986) 1715.
[10] H.J.M. Veendrick, MOS ICs-from Basics to ASICs, VCH Publishers, Weinheim, 1992.
[11] M.H. Woods, IEEE Tutorial Notes, IEEE Int. Reliab. Phys. Symp. (1985), 6.1.

[12] J.D. Meindl, Proc. IEEE 83 (4) (1995) 619.
[13] J.D. Meindl, J. Vac. Sci. Tech. 14 (1996) 192.
[14] W.E. Pence, J.P. Krusius, IEEE Trans. Comp. Hybrids Manuf. Tech. 10 (2) (1987) 176.
[15] J.D. Meindl, Circuits Devices 12 (November 1996) 19.
[16] R.W. Keyes, Adv. Electron. Phys. 70 (1988) 159.
[17] S.J. Wind, D.J. Frank, H.-S. Hong, Microelectron. Eng. 32 (1996) 271.
[18] P.W. Smith, IEEE Circuits Devices 3 (May 1987) 9.
[19] H.S. Blanks, Microelectron. Reliab. 20 (1980) 219.
[20] J.J. Duane, IEEE Trans. Aerospace 2 (1964) 563.
[21] D.J. Lloyd, Qual. Reliab. Eng. Int. 2 (1986) 19.
[22] A.C. Brombacher, Reliability by Design, CAE Techniques for Electronic Components and Systems, Wiley, Chichester, 1992.
[23] A.C. Brombacher, E.V. Geest, R. Arendsen, A.V. Steenwuk, O. Herrmann, Qual. Reliab. Eng. Int. 9 (1993) 239.
[24] W.H. Schroen, J.G. Aiken, G.A. Brown, 10th Annual Proceedings of the IEEE Reliability Physics Symposium, 1972, p. 42.
[25] H.A. Schafft, D.A. Baglee, P.E. Kennedy, 29th Annual Proceedings of the IEEE Reliability Physics Symposium, 1991, p. 1.
[26] D.L. Erhart, H.A. Schafft, W.K. Gladden, IEEE International Integrated Reliability Workshop, Lake Tahoe, October 1995.
[27] T. Volden, J. Hootman, Circuits Devices 11 (September 1995) 23.
[28] R.A. Evans, IEEE Trans. Reliability 42 (4) (1993) 541.
[29] Semiconductor Industry Association National Technology Road map for Semiconductors, Solid State Technol. 38 (2) (1995) 42.
[30] D.L. Crook, 29th Annual Proceedings of the IEEE Reliability Physics Symposium, 1990, p. 2.
[31] E. Takeda, Microelectron. Reliab. 37 (1997) 985.
[32] R. Iscoff, Semicond. Int. (May 1994) 48.
[33] R. Heitmann, Electron. Packag. Prod. (December 1993) 34.
[34] R. Frank, Understanding Smart Sensors, Artech House, Boston, 1996.

# APPENDIX

## VALUES OF SELECTED PHYSICAL CONSTANTS

| Constant | Symbol | Value |
|---|---|---|
| Absolute zero of temperature | | −273.2 °C |
| Avogadro's number | $N_A$ | $6.023 \times 10^{23}$/mol |
| Boltzmann's constant | k | $1.381 \times 10^{-23}$ J/K |
| | | $8.620 \times 10^{-5}$ eV/K |
| Electronic charge | q | $1.602 \times 10^{-19}$ C |
| Electron mass | $m_e$ | $9.110 \times 10^{-31}$ kg |
| Faraday's constant | F | $9.649 \times 10^{4}$ C/mol |
| Gas constant | R | 1.987 cal/mol-K |
| | | 8.314 J/mol-K |
| Gravitational acceleration | g | 9.806 m/s |
| Permittivity of vacuum | $\varepsilon_o$ | $8.854 \times 10^{-12}$ F/m |
| Planck's constant | h | $6.626 \times 10^{-34}$ J-s |
| Velocity of light in vacuum | c | $2.998 \times 10^{8}$ m/s |

# ACRONYMS

A: Amps, Amperes
AlAs: Aluminum arsenide
AlGaAs: Aluminum gallium arsenide
ASIC: Application-specific *integrated circuit*
ATI: Automatic test instrument
B: Boron
BEOL: Back end of line
CCD: Charge-coupled device
CD: Compact disc
CDF: Cumulative distribution function
CMOS: Complementary metal oxide semiconductor
CMP: Chemical mechanical polishing
CVD: Chemical vapor deposition
DRAM: Dynamic random access memory
DUT: Device under test
DVD: Digital video disc
EEPROM: Electrically erasable programmable read only memory
ELSI: Extreme-large-scale integration
EM: Electromigration
EOS: Electrical overstress
ESD: Electrostatic discharge
FEOL: Front end of line
FIT: Failure in time, failure in 10th to the 9th device hours
GaAs: Gallium arsenide
GaInAsP: Gallium indium arsenic phosphide
GB: Grain boundary
IC: Integrated circuit
IDDq: Supply current (Idd) in the quiescent state
InGaAs: Indium gallium arsenide
LED: Light-emitting diode
LPCVD: Low-pressure chemical vapor deposition
LSI: Large-scale integration
MEMS: Microelectromechanical system
MESFET: Metal semiconductor field effect transistor
MLC: Multilayer ceramic
MMIC: Monolithic microwave integrated circuit

MOS: Metal oxide semiconductor
MOSFET: Metal oxide semiconductor field effect transistor
MPU: Multiprocessor unit, Microprocessor unit, Multicore processor unit
MSI: Medium-scale integration
MTTF: Mean time to failure
NBTI: Negative-bias temperature instability
NMOS: N-type metal oxide semiconductor
P: Phosphorous
PBTI: Positive-bias temperature instability
PCB: Printed circuit board
PECVD: Plasma-enhanced chemical vapor deposition
PMOS: P-type metal oxide semiconductor
RF: Radio frequency
SLC: Single-level cell (flash memory technology)
SSI: Small-scale integration
SWEAT: Single wafer electrical accelerated test
TBD: To be determined
TDDB: Time-dependent dielectric breakdown
TEOS: Tetraethyl orthosilicate
TiPtAu: Titanium platinum gold
ULSI: Ultra-large-scale integration
VLSI: Very-large-scale integration

# INDEX

*Note*: Page numbers followed by "f" and "t" indicate figures and tables respectively

## A

## D

## F

## J



## K

## M

## P

## Q



## R

## T

## X



## Y



## Z

CPSIA information can be obtained at www.ICGtesting.com
Printed in the USA
BVOW10*1835141214

379387BV00001B/1/P

9 780120 885749